Proceedings of the International Conference on Complex Geometry and Related Fields

AMS/IP
Studies in
Advanced
Mathematics

Volume 39

Proceedings of the International Conference on Complex Geometry and Related Fields

Zhijie Chen, Sheng-Li Tan, Jianpan Wang, Stephen S.-T. Yau, Editors

American Mathematical Society · International Press

Shing-Tung Yau, General Editor

2000 *Mathematics Subject Classification.* Primary 32–XX, 14–XX, 53–XX, 17–XX, 20–XX, 11-XX, 05Cxx, 93–XX, 30Fxx.

Library of Congress Cataloging-in-Publication Data

International Conference on Complex Geometry and Related Fields (2004 : East China Normal University)

Proceedings of the International Conference on Complex Geometry and Related Fields / Zhijie Chen... [et al.], editors.

p. cm. — (AMS/IP studies in advanced mathematics, ISSN 1089-3288 ; v. 39)

Includes bibliographical references.

ISBN-13 978-0-8218-3949-2

1. Geometry, Differential—Congresses. 2. Geometry, Algebraic—Congresses. I. Chen, Zhijie, 1941– .

QZ641 .I572 2004
516.3′6—dc22 2007060748

∞ The paper used in this book is acid-free and falls within the guidelines
established to ensure permanence and durability.
Visit the AMS home page at http://www.ams.org/
Visit the International Press home page at URL: http://www.intlpress.com/

10 9 8 7 6 5 4 3 2 1 12 11 10 09 08 07

Contents

PREFACE

The Institute of Mathematics at East China Normal University was established in 1994. For the celebration of its 10th anniversary, an International Conference on Complex Geometry and Its Related Fields took place in the main campus of East China Normal University at Shanghai. The conference was a very successful one. Participants, in addition to those from Mainland of China, include Roger Brockett, Yum-Tong Siu and Shing-Tung Yau from Harvard University, Fabrizio Catanese from University of Bayreuth, Germany, Sidney Webster from University of Chicago, Ngaiming Mok from Hong Kong University etc. Our speakers consist of important mathematicians from USA, Australia, Canada, Singapore, and many major Universities and Institutes in China (including Taiwan).

Zhijie Chen
Sheng-Li Tan
Jianpan Wang
Stephen S.-T. Yau

AMS/IP Studies in Advanced Mathematics
Volume **39**, 2007

Representations of two-parameter quantum orthogonal and symplectic groups

Nantel Bergeron, Yun Gao, and Naihong Hu*

ABSTRACT. We investigate the finite-dimensional representation theory of two-parameter quantum orthogonal and symplectic groups that we found in [BGH] under the assumption that rs^{-1} is not a root of unity and extend some results [BW1, BW2] obtained for type A to types B, C and D. We construct the corresponding R-matrices and the quantum Casimir operators, by which we prove that the complete reducibility Theorem also holds for the categories of finite-dimensional weight modules for types B, C, D.

1. Preliminaries: Two-parameter Quantum Groups for Classical Types

Let $\mathbb{K} \supset \mathbb{Q}(r, s)$ denote an algebraically closed field, where the two-parameters r, s are nonzero complex numbers satisfying $r^2 \neq s^2$.

In this section, we recall the definitions of the two-parameter quantum groups $U_{r,s}(\mathfrak{g})$ for $\mathfrak{g} = \mathfrak{sl}_{n+1}$ from [BW1], and for $\mathfrak{g} = \mathfrak{so}_{2n+1}$, $\mathfrak{sp}_{2n}$ and $\mathfrak{so}_{2n}$ from [BGH]. Let Ψ be a finite root system of a simple Lie algebra $\mathfrak{g}$ of rank n with Π a base of simple roots. Regard Ψ as a subset of a Euclidean space $E = \mathbb{R}^n$ with an inner product $(\,,)$. Let $\epsilon_1, \cdots, \epsilon_n$ denote an orthonormal basis of E. We need the following data on (prime) root systems.

Type A:

$$\Pi = \{\alpha_i = \epsilon_i - \epsilon_{i+1} \mid 1 \leq i \leq n\},$$
$$\Psi = \{\pm(\epsilon_i - \epsilon_j) \mid 1 \leq i < j \leq n+1\}.$$

Type B:

$$\Pi = \{\alpha_i = \epsilon_i - \epsilon_{i+1} \mid 1 \leq i < n\} \cup \{\alpha_n = \epsilon_n\},$$
$$\Psi = \{\pm\epsilon_i \pm \epsilon_j \mid 1 \leq i \neq j \leq n\} \cup \{\pm\epsilon_i \mid 1 \leq i \leq n\}.$$

1991 *Mathematics Subject Classification.* Primary 17B37, 81R50; Secondary 17B35.

Key words and phrases. Two-parameter quantum group, R-matrix, quantum Casimir operator, complete reducibility.

N.B. supported by the NSERC and the CRC of Canada.

Y.G. supported by the NSERC of Canada and the Chinese Academy of Science.

*N.H., Corresponding Author, supported in part by the NNSF (Grant 10431040), the TRAPOYT and the FUDP from the MOE of China, the SRSTP from the STCSM, the Shanghai Priority Academic Discipline from the SMEC.

Type C:

$$\Pi = \{\alpha_i = \epsilon_i - \epsilon_{i+1} \mid 1 \le i < n\} \cup \{\alpha_n = 2\epsilon_n\},$$
$$\Psi = \{\pm\epsilon_i \pm \epsilon_j \mid 1 \le i \ne j \le n\} \cup \{2\epsilon_i \mid 1 \le i \le n\}.$$

Type D:

$$\Pi = \{\alpha_i = \epsilon_i - \epsilon_{i+1} \mid 1 \le i < n\} \cup \{\alpha_n = \epsilon_{n-1} + \epsilon_n\},$$
$$\Psi = \{\pm\epsilon_i \pm \epsilon_j \mid 1 \le i \ne j \le n\}.$$

In the cases of type A, C and D, we set $r_i = r^{\frac{(\alpha_i,\alpha_i)}{2}}$, $s_i = s^{\frac{(\alpha_i,\alpha_i)}{2}}$; while for type B, we set $r_i = r^{(\alpha_i,\alpha_i)}$, $s_i = s^{(\alpha_i,\alpha_i)}$.

Assigned to Π, there are two sets of mutually-commutative symbols $W = \{\omega_i^{\pm 1} \mid 1 \le i \le n\}$ and $W' = \{\omega_i'^{\pm 1} \mid 1 \le i \le n\}$. Define a pairing $\langle\, ,\rangle : W' \times W \longrightarrow \mathbb{K}$ as follows:

$$(1_A) \qquad \langle \omega_i', \omega_j\rangle = r^{(\epsilon_j,\alpha_i)} s^{(\epsilon_{j+1},\alpha_i)}, \qquad i \le n+1,\ j \le n, \qquad \textit{for } \mathfrak{sl}_{n+1},$$

$$(1_B) \qquad \langle \omega_i', \omega_j\rangle = \begin{cases} r^{2(\epsilon_j,\alpha_i)} s^{2(\epsilon_{j+1},\alpha_i)}, & i \le n,\ j < n, \\ r^{2(\epsilon_n,\alpha_i)}, & i < n,\ j = n, \\ r^{(\epsilon_n,\alpha_n)} s^{-(\epsilon_n,\alpha_n)}, & i = j = n. \end{cases} \qquad \textit{for } \mathfrak{so}_{2n+1},$$

$$(1_C) \qquad \langle \omega_i', \omega_j\rangle = \begin{cases} r^{(\epsilon_j,\alpha_i)} s^{(\epsilon_{j+1},\alpha_i)}, & i \le n,\ j < n, \\ r^{2(\epsilon_n,\alpha_i)}, & i < n,\ j = n, \\ r^{(\epsilon_n,\alpha_n)} s^{-(\epsilon_n,\alpha_n)}, & i = j = n. \end{cases} \qquad \textit{for } \mathfrak{sp}_{2n},$$

$$(1_D) \qquad \langle \omega_i', \omega_j\rangle = \begin{cases} r^{(\epsilon_j,\alpha_i)} s^{(\epsilon_{j+1},\alpha_i)}, & i \le n,\ j < n, \\ r^{(\epsilon_{n-1},\alpha_i)} s^{-(\epsilon_n,\alpha_i)}, & i \ne n-1,\ j = n, \\ r^{(\epsilon_n,\alpha_{n-1})} s^{-(\epsilon_{n-1},\alpha_{n-1})}, & i = n-1,\ j = n. \end{cases} \qquad \textit{for } \mathfrak{so}_{2n},$$

$$(2) \qquad \langle \omega_i'^{\pm 1}, \omega_j^{-1}\rangle = \langle \omega_i'^{\pm 1}, \omega_j\rangle^{-1} = \langle \omega_i', \omega_j\rangle^{\mp 1}, \qquad \textit{for any } \mathfrak{g}.$$

LEMMA 1.1. *For the prime root systems of the Lie algebras $\mathfrak{g} = \mathfrak{sl}_n$, $\mathfrak{so}_{2n+1}$, $\mathfrak{so}_{2n}$, and $\mathfrak{sp}_{2n}$, there hold the identities:*

$$(\epsilon_{j+1}, \alpha_i) = -(\epsilon_i, \alpha_j), \qquad (i, j < n), \qquad \textit{for any } \mathfrak{g},$$

$$(\epsilon_{j+1}, \alpha_n) = \begin{cases} -(\epsilon_n, \alpha_j), & (j < n), \quad \textit{for } \mathfrak{g} = \mathfrak{so}_{2n+1}, \\ -2(\epsilon_n, \alpha_j), & (j < n), \quad \textit{for } \mathfrak{g} = \mathfrak{sp}_{2n}, \end{cases}$$

$$(\epsilon_j, \alpha_n) = \begin{cases} -(\epsilon_n, \alpha_{j-1}), & (j \le n,\ j \ne n-1), \\ (\epsilon_{n-1}, \alpha_{n-1}), & (j = n-1) \end{cases} \qquad \textit{for } \mathfrak{g} = \mathfrak{so}_{2n}.$$

□

Observe that Lemma 1.1 ensures the compatibility of the defining relations of the two-parameter quantum groups defined below.

Let $U_{r,s}(\mathfrak{g})$ be the unital associative algebra over $\mathbb{K}$ generated by symbols $e_i, f_i, \omega_i^{\pm 1}, \omega_i'^{\pm 1}$ $(1 \le i \le n)$, subject to the following relations $(X1)$—$(X4)$:

$(X1)$ $\omega_i^{\pm 1}\omega_j'^{\pm 1} = \omega_j'^{\pm 1}\omega_i^{\pm 1}, \qquad \omega_i^{\pm 1}\omega_i^{\mp 1} = 1 = \omega_i'^{\pm 1}\omega_i'^{\mp 1}.$

$(X2)$ For $1 \le i,\, j \le n$, we have

$$\omega_j\, e_i\, \omega_j^{-1} = \langle \omega_i', \omega_j\rangle\, e_i, \qquad \omega_j\, f_i\, \omega_j^{-1} = \langle \omega_i', \omega_j\rangle^{-1} f_i,$$
$$\omega_j'\, e_i\, \omega_j'^{-1} = \langle \omega_j', \omega_i\rangle^{-1}\, e_i, \qquad \omega_j'\, f_i\, \omega_j'^{-1} = \langle \omega_j', \omega_i\rangle\, f_i.$$

$(X3)$ For $1 \le i,\, j \le n$, we have

$$[\, e_i, f_j\,] = \delta_{ij}\frac{\omega_i - \omega_i'}{r_i - s_i}.$$

$(X4)$ For any $i \ne j$, we have the (r, s)-Serre relations:

$$\left(\mathrm{ad}_l\, e_i\right)^{1-a_{ij}}(e_j) = 0,$$
$$\left(\mathrm{ad}_r\, f_i\right)^{1-a_{ij}}(f_j) = 0,$$

where the definitions of the left-adjoint action $\mathrm{ad}_l\, e_i$ and the right-adjoint action $\mathrm{ad}_r\, f_i$ are given in the following sense:

$$\mathrm{ad}_l\, a\,(b) = \sum_{(a)} a_{(1)}\, b\, S(a_{(2)}), \quad \mathrm{ad}_r\, a\,(b) = \sum_{(a)} S(a_{(1)})\, b\, a_{(2)}, \quad \forall\, a, b \in U_{r,s}(\mathfrak{g}),$$

where $\Delta(a) = \sum_{(a)} a_{(1)} \otimes a_{(2)}$ is given by Proposition 1.2 below.

The following fact is straightforward.

PROPOSITION 1.2. *The algebra $U_{r,s}(\mathfrak{g})$ ($\mathfrak{g} = \mathfrak{sl}_{n+1},\ \mathfrak{so}_{2n+1},\ \mathfrak{sp}_{2n}$, or $\mathfrak{so}_{2n}$) is a Hopf algebra under the comultiplication, the counit and the antipode defined below:*

$$\Delta(\omega_i^{\pm 1}) = \omega_i^{\pm 1} \otimes \omega_i^{\pm 1}, \qquad \Delta(\omega_i'^{\pm 1}) = \omega_i'^{\pm 1} \otimes \omega_i'^{\pm 1},$$
$$\Delta(e_i) = e_i \otimes 1 + \omega_i \otimes e_i, \qquad \Delta(f_i) = 1 \otimes f_i + f_i \otimes \omega_i',$$
$$\varepsilon(\omega_i^{\pm}) = \varepsilon(\omega_i'^{\pm 1}) = 1, \qquad \varepsilon(e_i) = \varepsilon(f_i) = 0,$$
$$S(\omega_i^{\pm 1}) = \omega_i^{\mp 1}, \qquad S(\omega_i'^{\pm 1}) = \omega_i'^{\mp 1},$$
$$S(e_i) = -\omega_i^{-1} e_i, \qquad S(f_i) = -f_i\, \omega_i'^{-1}.$$

□

REMARK 1.3. When $r = s^{-1} = q$, Hopf algebra $U_{r,s}(\mathfrak{g})$ modulo the Hopf ideal generated by the elements $\omega_i' - \omega_i^{-1}$ $(1 \le i \le n)$, is just the quantum groups $U_q(\mathfrak{g})$ of Drinfel'd-Jimbo type.

DEFINITION 1.4. *A skew-dual pairing of two Hopf algebras $\mathcal{A}$ and $\mathcal{U}$ is a bilinear form $\langle\, ,\rangle : \mathcal{U} \times \mathcal{A} \longrightarrow \mathbb{K}$ such that*

$$\langle f, 1_{\mathcal{A}}\rangle = \varepsilon_{\mathcal{U}}(f), \qquad \langle 1_{\mathcal{U}}, a\rangle = \varepsilon_{\mathcal{A}}(a),$$
$$\langle f, a_1 a_2\rangle = \langle \Delta_{\mathcal{U}}^{op}(f), a_1 \otimes a_2\rangle, \qquad \langle f_1 f_2, a\rangle = \langle f_1 \otimes f_2, \Delta_{\mathcal{A}}(a)\rangle,$$

for all $f,\ f_1,\ f_2 \in \mathcal{U}$, and $a,\ a_1,\ a_2 \in \mathcal{A}$, where $\varepsilon_{\mathcal{U}}$ and $\varepsilon_{\mathcal{A}}$ denote the counits of $\mathcal{U}$ and $\mathcal{A}$, respectively, and $\Delta_{\mathcal{U}}$ and $\Delta_{\mathcal{A}}$ are their respective comultiplications.

Let $\mathcal{B} = B(\mathfrak{g})$ (resp. $\mathcal{B}' = B'(\mathfrak{g})$) denote the Hopf subalgebra of $U = U_{r,s}(\mathfrak{g})$ generated by $e_j,\ \omega_j^{\pm 1}$ (resp. $f_j,\ \omega_j'^{\pm 1}$) with $1 \le j \le n$ for $\mathfrak{g} = \mathfrak{sl}_{n+1}$, and with $1 \le j \le n$ for $\mathfrak{g} = \mathfrak{so}_{2n+1},\ \mathfrak{so}_{2n}$, and $\mathfrak{sp}_{2n}$, respectively. The following result was obtained for the type A case by [BW1], and for the types B, C and D cases by [BGH].

PROPOSITION 1.5. *There exists a unique skew-dual pairing* $\langle\,,\rangle : \mathcal{B}'\times\mathcal{B}\longrightarrow\mathbb{K}$ *of the Hopf subalgebras* $\mathcal{B}$ *and* $\mathcal{B}'$ *in* $U_{r,s}(\mathfrak{g})$, *for* $\mathfrak{g}=\mathfrak{sl}_{n+1}$, $\mathfrak{so}_{2n+1}$, $\mathfrak{so}_{2n}$, *or* $\mathfrak{sp}_{2n}$ *such that* $\langle f_i, e_j\rangle = \frac{\delta_{ij}}{s_i-r_i}$, *and the conditions* (1_X) *(where* $X = A$, B, C, *or* D*) and* (2) *are satisfied, and all other pairs of generators are* 0. *Moreover, we have* $\langle S(a), S(b)\rangle = \langle a, b\rangle$ *for* $a\in\mathcal{B}'$, $b\in\mathcal{B}$. □

DEFINITION 1.6. *For any two skew-paired Hopf algebras* $\mathcal{A}$ *and* $\mathcal{U}$ *by a skew-dual pairing* $\langle\,,\rangle$, *one may form the Drinfel'd double* $\mathcal{D}(\mathcal{A},\mathcal{U})$ *as in* [KS, 8.2], *which is a Hopf algebra whose underlying coalgebra is* $\mathcal{A}\otimes\mathcal{U}$ *with the tensor product coalgebra structure, and whose algebra structure is defined by*

$$(a\otimes f)(a'\otimes f') = \sum\langle S_{\mathcal{U}}(f_{(1)}), a'_{(1)}\rangle\langle f_{(3)}, a'_{(3)}\rangle\, aa'_{(2)}\otimes f_{(2)}f', \tag{3}$$

for $a, a'\in\mathcal{A}$ *and* $f, f'\in\mathcal{U}$. *The antipode* S *is given by*

$$S(a\otimes f) = (1\otimes S_{\mathcal{U}}(f))(S_{\mathcal{A}}(a)\otimes 1).$$

Clearly, both mappings $\mathcal{A}\ni a\mapsto a\otimes 1\in\mathcal{D}(\mathcal{A},\mathcal{U})$ and $\mathcal{U}\ni f\mapsto 1\otimes f\in\mathcal{D}(\mathcal{A},\mathcal{U})$ are injective Hopf algebra homomorphisms. Let us denote the image $a\otimes 1$ (resp. $1\otimes f$) of a (resp. f) in $\mathcal{D}(\mathcal{A},\mathcal{U})$ by $\hat{a}$ (resp. $\hat{f}$). By (3), we have the following cross commutation relations between elements $\hat{a}$ (for $a\in\mathcal{A}$) and $\hat{f}$ (for $f\in\mathcal{U}$) in the algebra $\mathcal{D}(\mathcal{A},\mathcal{U})$:

$$\hat{f}\,\hat{a} = \sum\,\langle S_{\mathcal{U}}(f_{(1)}), a_{(1)}\rangle\,\langle f_{(3)}, a_{(3)}\rangle\,\hat{a}_{(2)}\hat{f}_{(2)}, \tag{4}$$

$$\sum\langle f_{(1)}, a_{(1)}\rangle\,\hat{f}_{(2)}\,\hat{a}_{(2)} = \sum\,\hat{a}_{(1)}\,\hat{f}_{(1)}\,\langle f_{(2)}, a_{(2)}\rangle. \tag{5}$$

In fact, as an algebra the double $\mathcal{D}(\mathcal{A},\mathcal{U})$ is the universal algebra generated by the algebras $\mathcal{A}$ and $\mathcal{U}$ with cross relations (4) or, equivalently, (5).

THEOREM 1.7 ([BW1, BGH]). *The two-parameter quantum group* $U = U_{r,s}(\mathfrak{g})$ *is isomorphic to the Drinfel'd quantum double* $\mathcal{D}(\mathcal{B},\mathcal{B}')$, *for* $\mathfrak{g}=\mathfrak{sl}_{n+1}$, $\mathfrak{so}_{2n+1}$, $\mathfrak{so}_{2n}$, *or* $\mathfrak{sp}_{2n}$. □

Let us denote $U_{r,s}(\mathfrak{n})$ (resp. $U_{r,s}(\mathfrak{n}^-)$) the subalgebra of $\mathcal{B}$ (resp. $\mathcal{B}'$) generated by e_i (resp. f_i) for all $i\le n$. Let

$$U^0 = \mathbb{K}\,[\omega_1^{\pm1},\cdots,\omega_n^{\pm1},\omega_1'^{\,\pm1},\cdots,\omega_n'^{\,\pm1}],$$

$$U_0 = \mathbb{K}\,[\omega_1^{\pm1},\cdots,\omega_n^{\pm1}],\qquad U_0' = \mathbb{K}\,[\omega_1'^{\,\pm1},\cdots,\omega_n'^{\,\pm1}]$$

denote the respective Laurent polynomial subalgebras of $U_{r,s}(\mathfrak{g})$, $\mathcal{B}$, and $\mathcal{B}'$. Clearly, $U^0 = U_0U_0' = U_0'U_0$. Thus, by definition, we have $\mathcal{B} = U_{r,s}(\mathfrak{n})\rtimes U_0$, and $\mathcal{B}' = U_0'\ltimes U_{r,s}(\mathfrak{n}^-)$, such that the double $\mathcal{D}(\mathcal{B},\mathcal{B}')\cong U_{r,s}(\mathfrak{n})\otimes U^0\otimes U_{r,s}(\mathfrak{n}^-)$, as vector spaces.

Let $\langle\,|\,\rangle_0 : \mathcal{B}\times\mathcal{B}'\longrightarrow\mathbb{K}$ denote the skew-dual pairing given by $\langle b\,|\,b'\rangle_0 = \langle S(b'), b\rangle$. Then, via a variation of its Drinfel'd double structure, we obtain the standard triangular decomposition of $U_{r,s}(\mathfrak{g})$ in [BGH, Corollary 2.6] as follows.

COROLLARY 1.8. $U_{r,s}(\mathfrak{g})\cong U_{r,s}(\mathfrak{n}^-)\otimes U^0\otimes U_{r,s}(\mathfrak{n})$, *as vector spaces. In particular, it induces* $U_q(\mathfrak{g})\cong U_q(\mathfrak{n}^-)\otimes U_0\otimes U_q(\mathfrak{n})$, *as vector spaces.* □

Let $Q=\mathbb{Z}\Psi$ denote the root lattice and set $Q^+ = \sum_{i=1}^n\mathbb{Z}_{\ge0}\alpha_i$. Then for any $\zeta = \sum_{i=1}^n\zeta_i\alpha_i\in Q$, we denote

$$\omega_\zeta = \omega_1^{\zeta_1}\cdots\omega_n^{\zeta_n},\qquad \omega_\zeta' = (\omega_1')^{\zeta_1}\cdots(\omega_n')^{\zeta_n}. \tag{6}$$

The following Q-graded structure on $U_{r,s}(\mathfrak{g})$ is necessary to develop to its weight representation theory discussed in the sequel.

COROLLARY 1.9 ([BGH, Corollary 2.7]). *For any* $\zeta = \sum_{i=1}^{n} \zeta_i \alpha_i \in Q$, *the defining relations* $(X2)$ *in* $U_{r,s}(\mathfrak{g})$ *take the form below:*

$$\omega_\zeta \, e_i \, \omega_\zeta^{-1} = \langle \omega_i', \omega_\zeta \rangle \, e_i, \qquad \omega_\zeta \, f_i \, \omega_\zeta^{-1} = \langle \omega_i', \omega_\zeta \rangle^{-1} f_i,$$

$$\omega_\zeta' \, e_i \, \omega_\zeta'^{-1} = \langle \omega_\zeta', \omega_i \rangle^{-1} e_i, \qquad \omega_\zeta' \, f_i \, \omega_\zeta'^{-1} = \langle \omega_\zeta', \omega_i \rangle \, f_i.$$

Then $U_{r,s}(\mathfrak{n}^\pm) = \bigoplus_{\eta \in Q^+} U_{r,s}^{\pm\eta}(\mathfrak{n}^\pm)$ *is* $Q^\pm$*-graded, where*

$$U_{r,s}^{\eta}(\mathfrak{n}^\pm) = \left\{ a \in U_{r,s}(\mathfrak{n}^\pm) \;\middle|\; \omega_\zeta \, a \, \omega_\zeta^{-1} = \langle \omega_\eta', \omega_\zeta \rangle \, a, \; \omega_\zeta' \, a \, \omega_\zeta'^{-1} = \langle \omega_\zeta', \omega_\eta \rangle^{-1} a \right\},$$

for $\eta \in Q^+ \cup Q^-$.

Moreover, $U = \bigoplus_{\eta \in Q} U_{r,s}^{\eta}(\mathfrak{g})$ *is* Q*-graded such that*

$$U_{r,s}^{\eta}(\mathfrak{g}) = \Big\{ \sum F_\alpha \omega_\mu' \omega_\nu E_\beta \in U \;\Big|\; \omega_\zeta \left(F_\alpha \omega_\mu' \omega_\nu E_\beta\right) \omega_\zeta^{-1} = \langle \omega_{\beta-\alpha}', \omega_\zeta \rangle \, F_\alpha \omega_\mu' \omega_\nu E_\beta,$$
$$\omega_\zeta' \left(F_\alpha \omega_\mu' \omega_\nu E_\beta\right) \omega_\zeta'^{-1} = \langle \omega_\zeta', \omega_{\beta-\alpha} \rangle^{-1} \, F_\alpha \omega_\mu' \omega_\nu E_\beta, \text{ with } \beta - \alpha = \eta \Big\},$$

where F_α *(resp.* E_β*) is a certain monomial* $f_{i_1} \cdots f_{i_l}$ *(resp.* $e_{j_1} \cdots e_{j_m}$*) such that* $\alpha_{i_1} + \cdots + \alpha_{i_l} = \alpha$ *(resp.* $\alpha_{j_1} + \cdots + \alpha_{j_m} = \beta$*).* □

2. Finite-Dimensional Weight Representation Theory and Category $\mathcal{O}$

As we know, the standard triangular decomposition of $U_{r,s}(\mathfrak{g})$ suggests that $U_{r,s}(\mathfrak{g})$ possesses highest weight representation theory. Indeed, this has been developed by Benkart and Witherspoon in [BW2] for $\mathfrak{g} = \mathfrak{gl}_n$ or $\mathfrak{sl}_n$. In principle, one can expect the same theory to be valid as well for $\mathfrak{g} = \mathfrak{so}_{2n+1}$, $\mathfrak{so}_{2n}$ and $\mathfrak{sp}_{2n}$. To establish this, we will follow Benkart and Witherspoon's main ideas. However, to treat these cases in a unified fashion, we need to have better insights here and there to generalize the techniques used in the type A case. Throughout the article, we assume that $\mathbb{K}$ is an algebraically closed field containing $\mathbb{Q}(r,s)$ as a subfield and rs^{-1} is not a root of unity.

Let Λ be the weight lattice of $\mathfrak{g}$ for $\mathfrak{g} = \mathfrak{so}_{2n+1}$, $\mathfrak{so}_{2n}$, or $\mathfrak{sp}_{2n}$, respectively. We adopt similar notions and notations in [BW1]. Associated to any $\lambda \in \Lambda$ is an algebra homomorphism $\hat{\lambda}$ from the subalgebra U^0 over $\mathbb{K}$ generated by the elements ω_i, ω_i' $(1 \le i \le n)$ to $\mathbb{K}$ given by

$$\hat{\lambda}(\omega_i) = \langle \omega_\lambda', \omega_i \rangle, \qquad \hat{\lambda}(\omega_i') = \langle \omega_i', \omega_\lambda \rangle^{-1}, \tag{1}$$

here we extend the definition of $\langle \, , \rangle$ from $\lambda \in Q$ to $\lambda \in \Lambda$ via taking appropriate half-integer powers when necessary, observing that $\Lambda \subseteq \bigoplus_{i=1}^{n} \frac{1}{2}\mathbb{Z}\alpha_i \subseteq \bigoplus_{i=1}^{n} \frac{1}{2}\mathbb{Z}\epsilon_i$.

Let M be a U-module of dimension $d < \infty$ where $U = U_{r,s}(\mathfrak{g})$. As $\mathbb{K}$ is algebraically closed, by linear algebra, we have

$$M = \bigoplus_{\chi} M_\chi,$$

where each $\chi : U^0 \longrightarrow \mathbb{K}$ is an algebra homomorphism, and M_χ is the generalized eigenspace given by

$$M_\chi = \left\{ m \in M \mid (\omega_i - \chi(\omega_i)1)^d m = 0 = (\omega_i' - \chi(\omega_i')1)^d m, \; \forall \, i \right\}. \tag{2}$$

When $M_\chi \neq 0$ we say that χ is a weight and M_χ is the corresponding weight space. In the case when M decomposes into genuine eigenspaces relative to U^0, we say that U^0 acts semisimply on M.

Relations in $(X2)$ imply

$$e_j M_\chi \subseteq M_{\chi\cdot\widehat{\alpha_j}}, \qquad f_j M_\chi \subseteq M_{\chi\cdot(\widehat{-\alpha_j})}, \tag{3}$$

where $\widehat{\alpha_j}$ is as in (1), and $\chi\cdot\psi$ is the homomorphism with values $(\chi\cdot\psi)(\omega_i) = \chi(\omega_i)\psi(\omega_i)$ and $(\chi\cdot\psi)(\omega_i') = \chi(\omega_i')\psi(\omega_i')$. In fact, if $(\omega_i - \chi(\omega_i)1)^k m = 0$, then $(\omega_i - \chi(\omega_i)\langle\omega_j',\omega_i\rangle 1)^k e_j m = 0$, and similarly for ω_i' and for f_j. On the one hand, (3) means that the sum of the eigenspaces is a submodule of M, and so if M is simple, the sum must be M itself, meanwhile we may replace the power d in (2) by 1, that is, U^0 acts semisimply on each simple M. On the other hand, a direct consequence of (3) is that for each simple M there is a homomorphism χ so that all the weights of M are of the form $\chi\cdot\hat{\zeta}$, where $\zeta\in Q$.

When all the weights of a module M are of the form $\hat{\lambda}$, where $\lambda\in\Lambda$, we say that M has weights in Λ. Any simple U-module having one weight in Λ has all its weights in Λ.

The observation below, which arises from Benkart and Witherspoon [BW2, Proposition 3.5] in the case when $\mathfrak{g} = \mathfrak{gl}_n$, or $\mathfrak{sl}_n$, also holds in our cases when $\mathfrak{g} = \mathfrak{so}_{2n+1}$, $\mathfrak{so}_{2n}$ and $\mathfrak{sp}_{2n}$.

LEMMA 2.1. *For $\mathfrak{g} = \mathfrak{sl}_n$, $\mathfrak{so}_{2n+1}$, $\mathfrak{so}_{2n}$ and $\mathfrak{sp}_{2n}$, suppose that $\hat{\zeta} = \hat{\eta}$, where $\zeta,\eta\in\Lambda$. Assume that rs^{-1} is not a root of unity, then $\zeta = \eta$.*

PROOF. The proof for $\mathfrak{g} = \mathfrak{sl}_n$ was given in [BW1, Proposition 3.5]. We now give the proof case by case for $\mathfrak{g} = \mathfrak{so}_{2n+1}$, $\mathfrak{sp}_{2n}$ and $\mathfrak{so}_{2n}$, respectively.

For $\zeta = \sum_{i=1}^n \zeta_i\alpha_i \in \Lambda$, by definition, we have

$$\hat{\zeta}(\omega_i) = \langle\omega_\zeta',\omega_i\rangle = \begin{cases} r^{2(\epsilon_i,\zeta)}s^{2(\epsilon_{i+1},\zeta)}, & i<n,\\ r^{2(\epsilon_n,\zeta)}(rs)^{-\zeta_n}. & i=n;\end{cases} \tag{B}$$

$$\hat{\zeta}(\omega_i') = \langle\omega_i',\omega_\zeta\rangle^{-1} = \begin{cases} r^{2(\epsilon_{i+1},\zeta)}s^{2(\epsilon_i,\zeta)}, & i<n,\\ s^{2(\epsilon_n,\zeta)}(rs)^{-\zeta_n}. & i=n.\end{cases}$$

$$\hat{\zeta}(\omega_i) = \langle\omega_\zeta',\omega_i\rangle = \begin{cases} r^{(\epsilon_i,\zeta)}s^{(\epsilon_{i+1},\zeta)}, & i<n,\\ r^{2(\epsilon_n,\zeta)}(rs)^{-2\zeta_n}. & i=n;\end{cases} \tag{C}$$

$$\hat{\zeta}(\omega_i') = \langle\omega_i',\omega_\zeta\rangle^{-1} = \begin{cases} r^{(\epsilon_{i+1},\zeta)}s^{(\epsilon_i,\zeta)}, & i<n,\\ s^{2(\epsilon_n,\zeta)}(rs)^{-2\zeta_n}. & i=n.\end{cases}$$

$$\hat{\zeta}(\omega_i) = \langle\omega_\zeta',\omega_i\rangle = \begin{cases} r^{(\epsilon_i,\zeta)}s^{(\epsilon_{i+1},\zeta)}, & i<n,\\ r^{(\epsilon_{n-1},\zeta)}s^{-(\epsilon_n,\zeta)}(rs)^{-2\zeta_{n-1}}. & i=n;\end{cases} \tag{D}$$

$$\hat{\zeta}(\omega_i') = \langle\omega_i',\omega_\zeta\rangle^{-1} = \begin{cases} r^{(\epsilon_{i+1},\zeta)}s^{(\epsilon_i,\zeta)}, & i<n,\\ r^{-(\epsilon_n,\zeta)}s^{(\epsilon_{n-1},\zeta)}(rs)^{-2\zeta_{n-1}}. & i=n.\end{cases}$$

Denote $\mu = \zeta-\eta$, from $\hat{\zeta}(\omega_n) = \hat{\eta}(\omega_n)$ and $\hat{\zeta}(\omega_n') = \hat{\eta}(\omega_n')$, in the type B or C case, we get $r^{2(\epsilon_n,\mu)}(rs)^{-\mu_n} = 1$, $s^{2(\epsilon_n,\mu)}(rs)^{-\mu_n} = 1$; or $r^{2(\epsilon_n,\mu)}(rs)^{-2\mu_n} = 1$, $s^{2(\epsilon_n,\mu)}(rs)^{-2\mu_n} = 1$. So $(rs^{-1})^{2(\epsilon_n,\mu)} = 1$ which, together with the assumption, means the integer $2(\epsilon_n,\mu) = 0$, that is,

$$\mu_{n-1} = \mu_n, \quad \textit{(for type B)}, \quad \textit{or} \quad \mu_{n-1} = 2\mu_n, \quad \textit{(for type C)}. \tag{4}$$

Again from $\hat{\zeta}(\omega_{n-1}) = \hat{\eta}(\omega_{n-1})$ and $\hat{\zeta}(\omega'_{n-1}) = \hat{\eta}(\omega'_{n-1})$, in the type B or C case, we get $(\alpha_{n-1}, \mu) = 0$, that is,

$$\mu_{n-2} = \mu_n, \quad (\textit{for type } \mathrm{B}), \quad \textit{or} \quad \mu_{n-2} = 2\mu_n, \quad (\textit{for type } \mathrm{C}). \tag{5}$$

But similar to the deduction in the case of type A (see [BW1]), noting $\mu_0 = 0$, we have

$$\mu_{i+2} - \mu_{i+1} - \mu_i + \mu_{i-1} = 0, \quad (i = 1, 2, \cdots, n-2), \tag{6}$$

$$\mu_{2k} = k\mu_2, \qquad \mu_{2k+1} = k\mu_2 + \mu_1. \tag{7}$$

Thus, by (4), (5) & (6), we get $\mu_n = \mu_{n-1} = \cdots = \mu_1 = \mu_0 = 0$ in the type B case. For the type C case, if $n = 2m$, by (5) & (7), we get $\mu_{n-2} = (m-1)\mu_2 = 2\mu_n = 2m\mu_2$, i.e., $\mu_2 = 0$, so $\mu_n = 0$; if $n - 1 = 2m$, then by (4), (5), & (7), we get $m\mu_2 = \mu_{n-1} = \mu_{n-2} = (m-1)\mu_2 + \mu_1$, i.e., $\mu_2 = \mu_1$, again by (4) & (7), we get $\mu_2 = 0$, so $\mu_n = 0$, which is reduced to the precondition of the proof in the type A case. Hence, using the same argument as in the case of type A ([BW1]), we have $\mu = 0$. Therefore, $\zeta = \eta$ in both cases B and C.

For the type D case, from $\hat{\zeta}(\omega_i) = \hat{\eta}(\omega_i)$ and $\hat{\zeta}(\omega'_i) = \hat{\eta}(\omega'_i)$ for $i = n-1$, n, we have $(rs^{-1})^{(\alpha_{n-1},\mu)} = 1$ and $(rs^{-1})^{(\alpha_n,\mu)} = 1$, that means, together with the assumption, the integers $(\alpha_{n-1}, \mu) = 0$ and $(\alpha_n, \mu) = 0$. So we get $\mu_{n-2} = 2\mu_{n-1} = 2\mu_n$. If $n = 2m$, then $(m-1)\mu_2 = \mu_{n-2} = 2m\mu_2$, i.e., $\mu_2 = 0$. If $n - 1 = 2m$, applying (7) to $\mu_{n-1} = \mu_n$, we get $\mu_1 = 0$; applying (7) to $\mu_{n-2} = 2\mu_{n-1}$, we get $\mu_2 = 0$. So we have $\mu_n = 0$ for any n. Using the same proof as in the case of type A, we obtain $\mu = 0$, i.e., $\zeta = \eta$. □

REMARK 2.2. Lemma 2.1 indicates that under the assumption that rs^{-1} is not a root of unity, we may simplify the notation by writing M_λ for the weight space rather than writing $M_{\hat{\lambda}}$ for $\lambda \in \Lambda$. So it makes sense to let (3) take the classical form: $e_j M_\lambda \subseteq M_{\lambda+\alpha_j}$ and $f_j M_\lambda \subseteq M_{\lambda-\alpha_j}$.

Similar to the proof of [BW2, Corollary 3.14], we have

COROLLARY 2.3. *Let M be a finite-dimensional $U_{r,s}(\mathfrak{g})$-module for $\mathfrak{g} = \mathfrak{sl}_{n+1}$, $\mathfrak{so}_{2n+1}$, $\mathfrak{so}_{2n}$ or $\mathfrak{sp}_{2n}$. Assume that rs^{-1} is not a root of unity, then the elements e_i, f_i $(1 \le i \le n)$ act nilpotently on M.* □

Obviously, when rs^{-1} is not a root of unity, a finite-dimensional simple U-module is a highest weight module by Corollary 2.3 and (3).

We state the definition of the category $\mathcal{O}$ of weight U-modules as in [BW1, Section 4].

DEFINITION 2.4. *Let $\mathcal{O}$ denote the category of modules M for $U_{r,s}(\mathfrak{g})$ (where $\mathfrak{g} = \mathfrak{so}_{2n+1}$, $\mathfrak{so}_{2n}$, or $\mathfrak{sp}_{2n}$) which satisfy the following conditions:*

($\mathcal{O}1$) *U^0 acts semisimply on M, and the set $\mathrm{wt}(M)$ of weights of M belongs to Λ : $M = \bigoplus_{\lambda\in \mathrm{wt}(M)} M_\lambda$, where $M_\lambda = \{\, m \in M \mid \omega_i.m = \langle \omega'_\lambda, \omega_i\rangle\, m,\ \omega'_i.m = \langle \omega'_i, \omega_\lambda\rangle^{-1} m,\ \forall\, i\,\}$;*

($\mathcal{O}2$) *$\dim_{\mathbb{K}} M_\lambda < \infty$ for all $\lambda \in \mathrm{wt}(M)$;*

($\mathcal{O}3$) *$\mathrm{wt}(M) \subseteq \cup_{\mu\in F}(\mu - Q^+)$ for some finite set $F \subset \Lambda$.*

The morphisms in $\mathcal{O}$ are U-module homomorphisms.

Actually, the category $\mathcal{O}$ just focuses on the class of the so-called *type* 1 U-modules like in the case of Drinfel'd-Jimbo quantum groups (see [J], [Jo], [KS]), which is closed under taking sub-object or sub-quotient object, making finite direct sum and taking tensor product.

Let V^ψ be the one-dimensional $\mathcal{B}$-module on which e_i acts as multiplication by 0 ($1 \le i \le n$), and U^0 acts via ψ, an algebra homomorphism from U^0 to $\mathbb{K}$. As usual, we can define the Verma module $M(\psi)$ with highest weight ψ to be the U-module induced from V^ψ, that is,

$$M(\psi) = U \otimes_{\mathcal{B}} V^\psi.$$

Set $v_\psi = 1 \otimes v \in M(\psi)$, where $v(\neq 0) \in V^\psi$. Then $e_i.v_\psi = 0$ ($1 \le i \le n$) and $a.v_\psi = \psi(a)\, v_\psi$ for any $a \in U^0$ by construction. By Corollary 1.8, $M(\psi) \cong U_{r,s}(\mathfrak{n}^-) \otimes v_\psi$. Corollary 1.9 indicates that each Verma module $M(\psi) \in \mathrm{Ob}(\mathcal{O})$ if and only if $\psi \in \hat{\Lambda}$.

Let N' be a proper submodule of $M(\psi)$, then (3) implies that

$$N' \subset \sum_{\mu \in Q^+ - \{0\}} M(\psi)_{\psi\cdot(\widehat{-\mu})},$$

as $M(\psi)_\psi = \mathbb{K}v_\psi$ generates $M(\psi)$. Hence, $M(\psi)$ has a unique maximal submodule N, namely the sum of all proper submodules, and a unique simple quotient, $L(\psi)$. Actually, all finite-dimensional simple U-modules are of this form, as the Theorem below indicates (which was proved by Benkart and Witherspoon [BW2, Theorem 2.1] in the case when $\mathfrak{g} = \mathfrak{gl}_n$, $\mathfrak{sl}_n$, but still holds with the same proof for our cases of $\mathfrak{g}$).

THEOREM 2.5. *For $\mathfrak{g} = \mathfrak{sl}_{n+1}$, $\mathfrak{so}_{2n+1}$, $\mathfrak{so}_{2n}$ or $\mathfrak{sp}_{2n}$, let M be a $U_{r,s}(\mathfrak{g})$-module, on which U^0 acts semisimply and which contains an element $m \in M_\psi$ ($\psi \in \mathrm{Hom}_{\mathrm{Alg}}(U^0, \mathbb{K})$) such that $e_i.m = 0$ for all i. Then there is a unique homomorphism of $U_{r,s}(\mathfrak{g})$-modules $F : M(\psi) \longrightarrow M$ with $F(v_\psi) = m$. In particular, if rs^{-1} is not a root of unity and M is a finite-dimensional simple $U_{r,s}(\mathfrak{g})$-module, then $M \cong L(\psi)$ for some weight ψ.* □

As in [BW2, Lemma 2.3], it is easy to verify the commutation relations below.

LEMMA 2.6. *For $m \ge 1$, set $[m]_i = \frac{r_i^m - s_i^m}{r_i - s_i}$. Then for $1 \le i \le n$, we have*

$$e_i f_i^m = f_i^m e_i + [m]_i\, f_i^{m-1} \frac{r_i^{1-m}\omega_i - s_i^{1-m}\omega_i'}{r_i - s_i},$$

$$e_i^m f_i = f_i e_i^m + [m]_i\, e_i^{m-1} \frac{s_i^{1-m}\omega_i - r_i^{1-m}\omega_i'}{r_i - s_i}.$$

□

Set $\alpha^\vee = \frac{2\alpha}{(\alpha,\alpha)}$, for any simple root $\alpha \in \Pi$, then for any $\lambda \in \Lambda$, $(\lambda, \alpha^\vee) \in \mathbb{Z}$ by definition. Let $\Lambda^+ \subset \Lambda$ be the subset of dominant weights, that is, $\Lambda^+ = \{\, \lambda \in \Lambda \mid (\lambda, \alpha_i^\vee) \ge 0, \text{ for } 1 \le i \le n \,\}$.

Similar to [BW2, Lemma 2.4] in the type A case, we have

LEMMA 2.7. *For $\mathfrak{g} = \mathfrak{so}_{2n+1}$, $\mathfrak{so}_{2n}$ and $\mathfrak{sp}_{2n}$, assume that rs^{-1} is not a root of unity. Let M be a nonzero finite-dimensional $U_{r,s}(\mathfrak{g})$-module on which U^0 acts semisimply. Suppose there is some nonzero vector $v \in M_\lambda$ with $\lambda \in \Lambda$ such that $e_i.v = 0$ for all i ($1 \le i \le n$). Then $\lambda \in \Lambda^+$.*

PROOF. It suffices to prove that $(\lambda, \alpha_n^\vee) \geq 0$, as the proof of $(\lambda, \alpha_i^\vee) \geq 0$ $(1 \leq i < n)$ is the same as that of [BW2, Lemma 2.4].

Since f_n acts nilpotently on M by Corollary 2.3, there is some integer $m \geq 0$ such that $f_n^{m+1}.v = 0$ but $f_n^m.v \neq 0$. Applying e_n to $f_n^{m+1}.v = 0$, using Lemma 2.6 and the fact that $e_n.v = 0$, we get $r_n^{-m}\hat{\lambda}(\omega_n) = s_n^{-m}\hat{\lambda}(\omega_n')$. Equivalently,

$$r_n^{-m} r^{2(\epsilon_n,\lambda)}(rs)^{-\lambda_n} = s_n^{-m} s^{2(\epsilon_n,\lambda)}(rs)^{-\lambda_n}, \qquad (\textit{for type } \mathrm{B})$$

$$r_n^{-m} r^{2(\epsilon_n,\lambda)}(rs)^{-2\lambda_n} = s_n^{-m} s^{2(\epsilon_n,\lambda)}(rs)^{-2\lambda_n}, \qquad (\textit{for type } \mathrm{C})$$

$$r^{-m} r^{(\epsilon_{n-1},\lambda)} s^{-(\epsilon_n,\lambda)}(rs)^{-2\lambda_{n-1}} = s^{-m} r^{-(\epsilon_n,\lambda)} s^{(\epsilon_{n-1},\lambda)}(rs)^{-2\lambda_{n-1}}, \qquad (\textit{for type } \mathrm{D})$$

or equivalently,

$$(r_n s_n^{-1})^{-m+(\lambda,\alpha_n^\vee)} = 1, \qquad (\textit{for types } \mathrm{B, C, D}).$$

The assumption of rs^{-1} forces $(\lambda, \alpha_n^\vee) = m \geq 0$. Therefore, $\lambda \in \Lambda^+$. □

COROLLARY 2.8. *For $\mathfrak{g} = \mathfrak{so}_{2n+1}$, $\mathfrak{so}_{2n}$ and $\mathfrak{sp}_{2n}$, assume that rs^{-1} is not a root of unity, then any finite-dimensional simple $U_{r,s}(\mathfrak{g})$-module with weights in Λ is isomorphic to $L(\lambda)$ for some $\lambda \in \Lambda^+$.* □

The representation theory of $U_{r,s}(\mathfrak{sl}_2)$, developed by Benkart and Witherspoon in [BW2], plays a crucial role in the classification of finite-dimensional simple modules for $U_{r,s}(\mathfrak{sl}_n)$ (see [BW2, Section 2]) like in the classical case of the simple Lie algebras or in the quantized case of the Drinfel'd-Jimbo quantum groups. Note the observation arising from the structure constants of $U_{r,s}(\mathfrak{g})$ for $\mathfrak{g} = \mathfrak{so}_{2n+1}$, $\mathfrak{so}_{2n}$ and $\mathfrak{sp}_{2n}$: for any vertex i from the corresponding Dynkin diagram of type B, C, or D, respectively, $\langle \omega_i', \omega_i \rangle = r_i s_i^{-1}$ always holds. This fact guarantees that even in the two-parameter quantum orthogonal or symplectic groups $U_{r,s}(\mathfrak{g})$, there exist isomorphic copies of $U_{r,s}(\mathfrak{sl}_2)$ as well. This suggests that these quantum groups possess a familiar finite-dimensional (weight) representation theory provided that rs^{-1} is not a root of unity.

Now let us recall the representation theory for $U_{r,s}(\mathfrak{sl}_2)$. The first two assertions of the following Proposition comes from [BW2, Proposition 2.8 (i)], the last one may be regarded as an intrinsic generalization of [BW2, Proposition 2.8 (ii)] with a deep insight.

PROPOSITION 2.9. *Assume that rs^{-1} is not a root of unity. For $U = U_{r,s}(\mathfrak{sl}_2)$ generated by e, f, ω and ω', for a given $\phi \in \mathrm{Hom}_{\mathrm{Alg}}(U^0, \mathbb{K})$, set $\phi = \phi(\omega)$, $\phi' = \phi(\omega')$, and in the Verma module $M(\phi)$, put $v_j = f^j/[j]! \otimes v_\phi$ for $j \geq 0$. Then*

(i) *$M(\phi)$ is a simple U-module if and only if $\phi \cdot r^{-j} - \phi' \cdot s^{-j} \neq 0$ for any $j \geq 0$.*

(ii) *If $\phi(\omega') = \phi(\omega)(rs^{-1})^{-m}$ for some integer $m \geq 0$, then $\mathrm{Span}_{\mathbb{K}}\{ v_j \mid j \geq m+1 \} \cong M(\phi - (m+1)\alpha)$ is the unique maximal submodule of $M(\phi)$. The quotient is the $(m+1)$-dimensional simple module $L(\phi)$ spanned by vectors $v_0, v_1, \cdots, v_m$ and having U-action given by*

$$\begin{aligned} \omega.v_j &= \phi \cdot (rs^{-1})^{-j} v_j, \qquad \omega'.v_j = \phi \cdot (rs^{-1})^{-(m-j)} v_j, \\ e.v_j &= \phi \cdot r^{-m}[m+1-j]\, v_{j-1}, \quad (v_{-1} = 0) \\ f.v_j &= [j+1]\, v_{j+1}. \quad (v_{m+1} = 0) \end{aligned} \tag{8}$$

Any $(m+1)$-dimensional simple U-module is isomorphic to $L(\phi)$ for some such ϕ.

(iii) *If* $\nu = \nu_1\lambda_1 + \cdots + \nu_n\lambda_n \in \Lambda^+$, *where* λ_i *is the* i*-th fundamental weight for* $\mathfrak{g}$, *then* $\hat{\nu}(\omega_i') = \hat{\nu}(\omega_i)(r_i s_i^{-1})^{-\nu_i}$, *and the* U_i*-module* $L(\nu_i\lambda_i)$ *is* (ν_i+1)*-dimensional and has* U_i*-action given by* (8) *with* $\phi_i = \hat{\nu}(\omega_i)$, *where* U_i *is the copy of* $U_{r,s}(\mathfrak{sl}_2)$ *in* $U_{r,s}(\mathfrak{g})$ *corresponding to the* i*-th vertex of the Dynkin diagram.*

PROOF. For the proof of the last assertion, it suffices to show that there hold

$$\frac{\hat{\nu}(\omega_i')}{\hat{\nu}(\omega_i)} = (r_i s_i^{-1})^{-(\alpha_i^\vee,\nu)} = \frac{\widehat{\nu_i\lambda_i}(\omega_i')}{\widehat{\nu_i\lambda_i}(\omega_i)}, \qquad (\textit{for any } i) \tag{9}$$

for $\mathfrak{g} = \mathfrak{sl}_n$, $\mathfrak{so}_{2n+1}$, $\mathfrak{so}_{2n}$ and $\mathfrak{sp}_{2n}$.

In the type A case, we have $\lambda_i = \epsilon_1 + \cdots + \epsilon_i$ for $1 \le i \le n$ and $\lambda_n = 0$. By definition,

$$\begin{aligned}\frac{\hat{\nu}(\omega_i')}{\hat{\nu}(\omega_i)} &= \frac{s^{(\epsilon_i,\nu)} r^{(\epsilon_{i+1},\nu)}}{r^{(\epsilon_i,\nu)} s^{(\epsilon_{i+1},\nu)}} = (rs^{-1})^{-(\alpha_i,\nu)} = (rs^{-1})^{-\nu_i} \\ &= \frac{s^{(\epsilon_i,\nu_i\lambda_i)} r^{(\epsilon_{i+1},\nu_i\lambda_i)}}{r^{(\epsilon_i,\nu_i\lambda_i)} s^{(\epsilon_{i+1},\nu_i\lambda_i)}} = \frac{\widehat{\nu_i\lambda_i}(\omega_i')}{\widehat{\nu_i\lambda_i}(\omega_i)}.\end{aligned}$$

For types B, C and D, it suffices to consider types B_2, C_2 and D_4, respectively.

In the type B_2 case, we have $\lambda_1 = \epsilon_1$, $\lambda_2 = \frac{1}{2}(\epsilon_1 + \epsilon_2)$. By the defining formula (B) in Lemma 2.1, for $i = 1$, it follows directly from the argument in the type A case; while for $i = 2$, we get

$$\frac{\hat{\nu}(\omega_2')}{\hat{\nu}(\omega_2)} = \frac{s^{2(\epsilon_2,\nu)}}{r^{2(\epsilon_2,\nu)}} = (rs^{-1})^{-(\alpha_2^\vee,\nu)} = \frac{s^{2(\epsilon_2,\nu_2\lambda_2)}}{r^{2(\epsilon_2,\nu_2\lambda_2)}} = \frac{\widehat{\nu_2\lambda_2}(\omega_2')}{\widehat{\nu_2\lambda_2}(\omega_2)}.$$

In the type C_2 case, we have $\lambda_1 = \epsilon_1$, $\lambda_2 = \epsilon_1 + \epsilon_2$. It suffices to consider the case $i = 2$. Similarly, we have

$$\frac{\hat{\nu}(\omega_2')}{\hat{\nu}(\omega_2)} = \frac{s^{2(\epsilon_2,\nu)}}{r^{2(\epsilon_2,\nu)}} = (r_2 s_2^{-1})^{-(\alpha_2^\vee,\nu)} = \frac{s^{2(\epsilon_2,\nu_2\lambda_2)}}{r^{2(\epsilon_2,\nu_2\lambda_2)}} = \frac{\widehat{\nu_2\lambda_2}(\omega_2')}{\widehat{\nu_2\lambda_2}(\omega_2)}.$$

In the type D_4 case, we have $\lambda_1 = \epsilon_1$, $\lambda_2 = \epsilon_1 + \epsilon_2$, $\lambda_3 = \frac{1}{2}(\epsilon_1 + \epsilon_2 + \epsilon_3 - \epsilon_4)$, $\lambda_4 = \frac{1}{2}(\epsilon_1 + \epsilon_2 + \epsilon_3 + \epsilon_4)$. It suffices to consider the cases $i = 3$, 4. By the formula (D) in Lemma 2.1, we have

$$\frac{\hat{\nu}(\omega_i')}{\hat{\nu}(\omega_i)} = (rs^{-1})^{(\alpha_i,\nu)} = \frac{\widehat{\nu_i\lambda_i}(\omega_i')}{\widehat{\nu_i\lambda_i}(\omega_i)},$$

for $i = 3$, 4.

The proof is completed. □

Proposition 2.9 (iii) and its proof imply the following result.

COROLLARY 2.10. *Assume that* rs^{-1} *is not a root of unity and* $\lambda \in \Lambda^+$, *set* $\nu_i = (\lambda, \alpha_i^\vee)$, *then each vector* $f_i^{\nu_i+1}.v_\lambda$ *in the Verma* U*-module* $M(\lambda)$ *generates the Verma submodule* $M(\lambda - (\nu_i+1)\alpha_i)$ *for all* i, *where* $\mathfrak{g} = \mathfrak{sl}_{n+1}$, $\mathfrak{so}_{2n+1}$, $\mathfrak{so}_{2n}$ *or* $\mathfrak{sp}_{2n}$.

PROOF. It follows from a direct calculation of $e_i f_i^{\nu_i+1}.v_\lambda = 0$ by Lemma 2.6 and (9). □

More generally, we have

PROPOSITION 2.11. *Let $M(\lambda)$ be a Verma module with $\lambda \in \Lambda^+$. Then for every element ω of the Weyl group $\mathcal{W}$ of $\mathfrak{g}$, there exists a Verma submodule in $M(\lambda)$ with highest weight*

$$\lambda_\omega = \omega(\lambda + \rho) - \rho, \tag{10}$$

where ρ is the half-sum of all positive roots of $\mathfrak{g}$. Every simple U-module as a composition factor of $M(\lambda)$ determines a highest weight module in $\mathcal{O}$. These highest weights are of the form (10).

PROOF. The proof of this proposition is analogous to that of the corresponding assertion in the classical theory (see Dixmier [D]). □

LEMMA 2.12. *For any simple U-module $L(\lambda)$ with $\lambda \in \Lambda^+$, take any $\beta = \sum_{i=1}^n m_i\alpha_i \in Q^+$ such that $m_i \leq (\lambda, \alpha_i^\vee)$, $\forall\, i$, then the linear mapping $U_{r,s}^{-\beta}(\mathfrak{n}^-) \ni x \mapsto x.v_\lambda$ is injective.*

PROOF. By the definition of the Verma module, it is enough to show that $\lambda - \beta$ is not a weight of the maximal U-submodule N. This follows from Proposition 2.11, because no set of weights $\{\, \lambda_\omega - \sum_{i=1}^n n_i\alpha_i \mid n_i \in \mathbb{Z}^+ \,\}$, $\omega \in \mathcal{W} - \{1\}$ contains $\lambda - \beta$. □

LEMMA 2.13. *If an element $a \in U_{r,s}^{-\beta}(\mathfrak{n}^-)$ satisfies the relations $e_ia = ae_i$ for $i = 1, 2, \cdots, n$, then we have $a = 0$. If $f_ib = bf_i$, $i = 1, 2, \cdots, n$, for some $b \in U_{r,s}^{\beta}(\mathfrak{n})$, then $b = 0$.*

PROOF. Write $\beta = \sum_{i=1}^n m_i\alpha_i \in Q^+$, and take a dominant weight $\lambda \in \Lambda^+$ such that $(\lambda, \alpha_i^\vee) \geq m_i$ for all i. Consider the simple U-module $L(\lambda)$ with highest weight vector v_λ. Since $(e_ia).v_\lambda = (ae_i).v_\lambda = 0$ for all i, the vector $a.v_\lambda$ generates a proper submodule of $L(\lambda)$. Thus $a.v_\lambda = 0$, as $L(\lambda)$ is simple. Hence $a = 0$ by Lemma 2.12.

In order to prove the second assertion, we introduce a $\mathbb{Q}$-algebra isomorphism $\theta : U_{r,s}(\mathfrak{g}) \longrightarrow U_{r,s}(\mathfrak{g})$ defined by

$$\begin{aligned} \theta(r) &= s^{-1}, & \theta(s) &= r^{-1}, \\ \theta(\omega_i) &= \omega_i', & \theta(\omega_i') &= \omega_i, \\ \theta(e_i) &= f_i, & \theta(f_i) &= (r_is_i)e_i. \end{aligned} \tag{11}$$

In fact, we can find that the image of θ is $\mathbb{Q}$-algebraically isomorphic to the associated quantum group $U_{s^{-1},r^{-1}}(\mathfrak{g})$, i.e., $\mathrm{Im}(\theta) \cong (U_{s^{-1},r^{-1}}(\mathfrak{g}), \langle\,|\,\rangle)$, where the pairing $\langle \omega_i' | \,\omega_j\rangle$ is defined via substituting (r, s) by (s^{-1}, r^{-1}) in the defining formula for $\langle \omega_i', \omega_j\rangle$ (see formulae (1_X) and (2) in Section 1).

Now applying the $\mathbb{Q}$-algebra isomorphism θ to the equation $f_ib = bf_i$, we get $\theta(b) = 0$, by the first assertion. Hence, $b = 0$. □

Returning to the pairing $\langle\, ,\rangle : \mathcal{B}' \times \mathcal{B} \longrightarrow \mathbb{K}$ in Proposition 1.5, and combining with the Q-gradation on U (see Corollary 1.9), we have

PROPOSITION 2.14. *For any $\beta \in Q^+$, the restriction of the pairing $\langle\, ,\rangle$ in Proposition 1.5 to ${\mathcal{B}'}^{-\beta} \times \mathcal{B}^\beta$ is nondegenerate.*

PROOF. We have to show that for any $a \in {\mathcal{B}'}^{-\beta}$ such that $\langle a, b\rangle = 0$ for some $b \in \mathcal{B}^\beta$, implies that $b = 0$. This will be proved by induction with respect to the usual ordering of Q_+. If β is a simple root, then it is true by formula (2) in Section

1. Let $\beta > 0$ with $\mathrm{ht}(\beta) > 1$ and suppose that it holds for all $\gamma \in Q^+$ such that $\beta - \gamma \in Q^+$.

Note that using the defining properties of skew-dual pairing and the comultiplication in U (see Proposition 1.2), we may check by induction:

$$\langle c\,\omega'_\nu, \omega_\mu\, d\rangle = \langle \omega'_\nu, \omega_\mu\rangle\,\langle c, d\rangle, \qquad \forall\, c \in U_{r,s}(\mathfrak{n}^-),\ d \in U_{r,s}(\mathfrak{n}), \tag{12}$$

$$\langle c, d\rangle = 0, \qquad c \in U_{r,s}^{-\sigma}(\mathfrak{n}^-),\ d \in U_{r,s}^{\delta}(\mathfrak{n}),\ \sigma,\, \delta \in Q^+,\ \sigma \neq \delta. \tag{13}$$

It suffices to assume that $b \in U_{r,s}^{\beta}(\mathfrak{n})$. By Proposition 1.2, we can write

$$\Delta(b) = \sum_{0 \leq \gamma \leq \beta} (\omega_\gamma \otimes 1)\, b_\gamma, \qquad b_\gamma \in U_{r,s}^{\gamma}(\mathfrak{n}) \otimes U_{r,s}^{\beta-\gamma}(\mathfrak{n}), \tag{14}$$

where $b_0 = b \otimes 1$ and $b_\beta = 1 \otimes b$. Let $\gamma \in Q^+$, $0 < \gamma < \beta$, $x \in \mathcal{B}'^{-\gamma}$ and $y \in \mathcal{B}'^{-(\beta-\gamma)}$.

By (2), (12) & (13), we have

$$0 = \langle xy, b\rangle = \langle x \otimes y, \Delta(b)\rangle = \langle x \otimes y, (\omega_\gamma \otimes 1)\, b_\gamma\rangle = \langle x \otimes y, b_\gamma\rangle. \tag{15}$$

By assumption, for any $\gamma' < \beta$ the restriction of $\langle\,,\rangle$ to $\mathcal{B}'^{-\gamma'} \times \mathcal{B}^{\gamma'}$ is nondegenerate, so is its extension to a bilinear form on $[\,\mathcal{B}'^{-\gamma} \otimes \mathcal{B}'^{-(\beta-\gamma)}\,] \times [\,\mathcal{B}^{\gamma} \otimes \mathcal{B}^{\beta-\gamma}\,]$. Hence it follows from (15) that $b_\gamma = 0$. Because of (14) this means that $\Delta(b) = b \otimes 1 + \omega_\beta \otimes b$. By (13), together with $\Delta(f_i) = 1 \otimes f_i + f_i \otimes \omega'_i$, we get $\hat{f}_i\, \hat{b} = \hat{b}\, \hat{f}_i$, and then $f_i\, b = b\, f_i$ for any i, after using φ (see the proof of [BGH, Theorem 2.5]). Thus, by Lemma 2.12, $b = 0$.

Similar reasoning indicates that for any $b \in \mathcal{B}^{\beta}$ such that $\langle a, b\rangle = 0$ for some $a \in \mathcal{B}'^{-\beta}$ implies that $a = 0$. □

In what follows, we consider the finite-dimensionality question of the simple $U_{r,s}(\mathfrak{g})$-modules $L(\lambda)$ with $\lambda \in \Lambda^+$. This problem has been solved by Benkart and Witherspoon in [BW2, Section 2] in the case when $\mathfrak{g} = \mathfrak{gl}_n$, or $\mathfrak{sl}_n$. The same idea can be used to prove that $M(\lambda)$ has a $U_{r,s}(\mathfrak{g})$-submodule $M'(\lambda)$ of finite codimension, as $L(\lambda)$ is the quotient of $M(\lambda)$ by its unique maximal submodule, where $M'(\lambda)$ is defined by

$$M'(\lambda) = \sum_{i=1}^{n} U_{r,s}(\mathfrak{g}) f_i^{k_i+1}.v_\lambda \cong \sum_{i=1}^{n} M(\lambda - (k_i+1)\alpha_i), \tag{16}$$

where $k_i = (\lambda, \alpha_i^\vee)$ for all i. That is, to prove the module $L'(\lambda) = M(\lambda)/M'(\lambda)$ is nonzero and finite-dimensional. $L'(\lambda) \neq 0$ is clear, since any weight in $M'(\lambda)$ is less than or equal to $\lambda - (k_i+1)\alpha_i$ for some i, $v_\lambda \notin M'(\lambda)$.

LEMMA 2.15. (i) ([BW2, Lemma 2.10]) *The elements e_j, f_j $(1 \leq j \leq n)$ act locally nilpotently on $U_{r,s}(\mathfrak{g})$-module $L'(\lambda)$.*

(ii) ([BW2, Lemma 2.11]) *Assume that rs^{-1} is not a root of unity, $V = \bigoplus_{j \in \mathbb{Z}^+} V_{\lambda - j\alpha} \in \mathrm{Ob}(\mathcal{O})$ is a $U_{r,s}(\mathfrak{sl}_2)$-module for some weight $\lambda \in \Lambda$. If e, f act locally nilpotently on V, then $\dim_{\mathbb{K}} V < \infty$, and the weights of V are preserved under the simple reflection taking α to $-\alpha$.*

PROOF. The proof of (i) is parallel to the type A case; the second part assertion is direct from [BW2]. □

PROPOSITION 2.16. *Assume that rs^{-1} is not a root of unity. Then for the $U_{r,s}(\mathfrak{g})$-module $L'(\lambda) \in \mathrm{Ob}(\mathcal{O})$ with $\lambda \in \Lambda^+$, we have* $\dim_{\mathbb{K}} L'(\lambda) < \infty$, *so* $\dim_{\mathbb{K}} L(\lambda) < \infty$.

PROOF. Consider $L'(\lambda)$ as a U_i-module, where U_i is the copy generated by e_i, f_i, ω_i, ω_i'. For μ a weight of $L'(\lambda)$, applying Lemma 2.15 to the U_i-module

$$L_i'(\mu) = U_i.L'(\lambda)_\mu = \bigoplus_{j\in\mathbb{Z}^+} L_i'(\mu)_{\lambda'-j\alpha_i}$$

for some weight $\lambda' \leq \lambda$, we get that the simple reflection w_i preserves the weights of $L_i'(\mu)$, so $w_i(\mu)$ is a weight of $L'(\lambda)$. That is, the Weyl group $\mathcal{W}$ of $\mathfrak{g}$ preserves the set of weights of $L'(\lambda)$. From Lie theory, we know that each $\mathcal{W}$-orbit only contains one dominant weight. But there are only finitely many dominant weights $\leq \lambda$, and as each weight space of $L'(\lambda)$ is of finite-dimension, we have $\dim_{\mathbb{K}} L'(\lambda) < \infty$. Thereby, $\dim_{\mathbb{K}} L(\lambda) < \infty$. □

For $\mathfrak{g} = \mathfrak{sl}_{n+1}$, $\mathfrak{so}_{2n+1}$, $\mathfrak{so}_{2n}$ or $\mathfrak{sp}_{2n}$, Corollary 2.8 and Proposition 2.16 imply the following

COROLLARY 2.17. *A finite-dimensional simple object in the category $\mathcal{O}$ is precisely a $U_{r,s}(\mathfrak{g})$-module $L(\lambda)$ for some $\lambda \in \Lambda^+$, and $L(\lambda) \cong L(\mu)$ if and only if $\lambda = \mu$.* □

Finite-dimensional simple (weight) modules of generic type. As noted in [BW2, Section 2], for $\mathfrak{g} = \mathfrak{gl}_n$, $\mathfrak{sl}_n$, Benkart and Witherspoon gave a description of a classification of finite-dimensional simple $U_{r,s}(\mathfrak{g})$-modules. We find that a similar structural feature for finite-dimensional simple $U_{r,s}(\mathfrak{g})$-modules also holds when $\mathfrak{g} = \mathfrak{so}_{2n+1}$, $\mathfrak{so}_{2n}$, or $\mathfrak{sp}_{2n}$, after modifying some of the treatments.

Given a one-dimensional $U_{r,s}(\mathfrak{g})$-module L, Theorem 2.5 indicates that $L = L(\chi)$ for some $\chi \in \mathrm{Hom}_{\mathrm{Alg}}(U^0, \mathbb{K})$ with the elements e_i, f_i $(1 \leq i \leq n)$ trivially acting on $L(\chi)$. Relation $(X3)$ $(X = B, C, D)$ gives

$$\chi(\omega_i) = \chi(\omega_i'), \qquad (1 \leq i \leq n). \tag{17}$$

Conversely, if $\chi \in \mathrm{Hom}_{\mathrm{Alg}}(U^0, \mathbb{K})$ satisfies the equation (17), then Proposition 2.9 (ii) guarantees $\dim_{\mathbb{K}} L(\chi) = 1$. We denote by L_χ the one-dimensional $U_{r,s}(\mathfrak{g})$-module $L(\chi)$.

The following Lemma was proved by Benkart and Witherspoon in the case of type A. We will give a unified proof for the classical types of $\mathfrak{g}$ based on an intrinsic observation in Proposition 2.9 (ii) & (iii).

LEMMA 2.18. *Assume rs^{-1} is not a root of unity. Given a finite-dimensional simple $U_{r,s}(\mathfrak{g})$-module $L(\psi)$ with highest weight ψ, there exists a pair (χ, λ), where $\chi \in \mathrm{Hom}_{\mathrm{Alg}}(U^0, \mathbb{K})$ such that (17) holds, and $\lambda \in \Lambda^+$, so that $\psi = \chi \cdot \hat{\lambda}$, and $\mathrm{wt}(L(\psi)) \subseteq \chi \cdot \hat{\Lambda}$.*

PROOF. As $L(\psi)$ is finite-dimensional and simple, for each pair of eigenvalues $(\psi(\omega_i), \psi(\omega_i'))$ when considering $L(\psi)$ as a U_i-module (where U_i is a $U_{r,s}(\mathfrak{sl}_2)$-copy of $U_{r,s}(\mathfrak{g})$), Proposition 2.9 (ii) tells us that there exists a nonnegative integer ν_i for each index i such that $\psi(\omega_i') = \psi(\omega_i)(r_i s_i^{-1})^{-\nu_i}$. Set $\lambda = \sum_{i=1}^n \nu_i \lambda_i$ where λ_i is the ith fundamental weight of $\mathfrak{g}$, then $\lambda \in \Lambda^+$.

Now we take $\chi(\omega_i) = \psi(\omega_i)\hat{\lambda}_i(\omega_i)^{-1}$ and $\chi(\omega_i') = \psi(\omega_i')\hat{\lambda}_i(\omega_i')^{-1}$, that is, $\chi = \psi \cdot \hat{\lambda}_i^{-1} \in \mathrm{Hom}_{\mathrm{Alg}}(U^0, \mathbb{K})$ and satisfies

$$\begin{aligned}\chi(\omega_i') &= \psi(\omega_i')\hat{\lambda}_i^{-1}(\omega_i') = \psi(\omega_i)(r_i s_i^{-1})^{-\nu_i}\hat{\lambda}_i^{-1}(\omega_i') \\ &= \psi(\omega_i)\hat{\lambda}^{-1}(\omega_i) \qquad (\text{ by } (9)\ ) \\ &= \chi(\omega_i),\end{aligned}$$

as required. The last assertion that $\mathrm{wt}(L(\psi)) \subseteq \chi \cdot \hat{\Lambda}$ is quite clear. □

Similar to [BW2, Theorem 2.19], for $\mathfrak{g} = \mathfrak{so}_{2n+1}, \mathfrak{so}_{2n}$, or $\mathfrak{sp}_{2n}$, we have the classification Theorem for finite-dimensional simple $U_{r,s}(\mathfrak{g})$-modules as follows.

THEOREM 2.19. *Let rs^{-1} be a non-root of unity. Each finite-dimensional simple $U_{r,s}(\mathfrak{g})$-module $L(\psi)$ with $\psi \in \mathrm{Hom}_{\mathrm{Alg}}(U^0, \mathbb{K})$ is isomorphic to $L_\chi \otimes L(\lambda)$, where $\chi \in \mathrm{Hom}_{\mathrm{Alg}}(U^0, \mathbb{K})$ with $\chi(\omega_i) = \chi(\omega_i')$ $(1 \le i \le n)$ and $\lambda \in \Lambda^+$.* □

3. R-matrices, Quantum Casimir Operators, Complete Reducibility

For any two objects M, $M' \in \mathrm{Ob}(\mathcal{O})$, Benkart and Witherspoon in [BW1, Section 4] constructed a $U_{r,s}(\mathfrak{sl}_n)$-module isomorphism

$$R_{M',M} : M' \otimes M \longrightarrow M \otimes M'$$

by a remarkable method due to Jantzen [J, Chap. 7] for the quantum groups $U_q(\mathfrak{g})$ of Drinfel'd-Jimbo type.

The aim of this section is to generalize this result to the setting of $\mathfrak{g} = \mathfrak{so}_{2n+1}, \mathfrak{so}_{2n}, \mathfrak{sp}_{2n}$.

Noting that the weight lattice

$$\Lambda \subseteq \bigoplus_{i=1}^{n} \frac{1}{2}\mathbb{Z}\alpha_i \subseteq \bigoplus_{i=1}^{n} \frac{1}{2}\mathbb{Z}\epsilon_i,$$

as it was done in formula (1) of Section 2, for $\lambda \in \Lambda$, we have an algebra homomorphism $\hat{\lambda} \in \mathrm{Hom}_{\mathrm{Alg}}(U^0, \mathbb{K})$. Furthermore, we extend the pairing $\langle\, , \rangle$ to $\Lambda \times \Lambda$, such that for any $\lambda = \sum_{i=1}^n p_i\alpha_i$, $\mu = \sum_{i=1}^n q_i\alpha_i \in \Lambda$ with p_i, $q_i \in \frac{1}{2}\mathbb{Z}$, we define

$$\langle \omega_\lambda', \omega_\mu \rangle = \Pi_{i=1}^n \hat{\lambda}(\omega_i)^{q_i}, \tag{1}$$

which is well-defined in the algebraically closed field $\mathbb{K}$.

Now we define the map $f : \Lambda \times \Lambda \longrightarrow \mathbb{K}^*$ by

$$f(\lambda, \mu) = \langle \omega_\mu', \omega_\lambda \rangle^{-1}, \tag{2}$$

which satisfies

$$\begin{gathered} f(\lambda + \mu, \nu) = f(\lambda, \nu)\, f(\mu, \nu), \\ f(\lambda, \mu + \nu) = f(\lambda, \mu)\, f(\lambda, \nu), \\ f(\alpha_i, \mu) = \langle \omega_\mu', \omega_i \rangle^{-1}, \qquad f(\lambda, \alpha_i) = \langle \omega_i', \omega_\lambda \rangle^{-1}. \end{gathered} \tag{3}$$

And we define the linear transformation $\tilde{f} = \tilde{f}_{M,M'} : M \otimes M' \longrightarrow M \otimes M'$ by $\tilde{f}(m \otimes m') = f(\lambda, \mu)\,(m \otimes m')$ for $m \in M_\lambda$ and $m' \in M_\mu'$.

Owing to $\Delta(e_i) = e_i\otimes 1+\omega_i\otimes e_i$, we have $\Delta(x) \in \sum_{0\le\nu\le\zeta} U_{r,s}^{\zeta-\nu}(\mathfrak{n})\omega_\nu\otimes U_{r,s}^{\nu}(\mathfrak{n})$, for all $x\in U_{r,s}^{\zeta}(\mathfrak{n})$, by induction. For each i, the expression of $\Delta(x)$ defines two skew-derivations $\partial_i,\ {}_i\partial : U_{r,s}^{\zeta}(\mathfrak{n}) \longrightarrow U_{r,s}^{\zeta-\alpha_i}(\mathfrak{n})$ such that

$$(4)\qquad \begin{aligned} \Delta(x) &= x\otimes 1 + \sum_{i=1}^{n} \partial_i(x)\,\omega_i\otimes e_i + \textit{the rest},\\ \Delta(x) &= \omega_\zeta\otimes x + \sum_{i=1}^{n} e_i\,\omega_{\zeta-\alpha_i}\otimes {}_i\partial(x) + \textit{ the rest}, \end{aligned}$$

where in each case "*the rest*" refers to terms involving products of more than one e_j in the second (resp. first) factor. More precisely, parallel to [BW1, Lemma 4.6] or comparing with [KS, Lemmas 6.14, 6.17], we have

LEMMA 3.1. *For all $x\in U_{r,s}^{\zeta}(\mathfrak{n})$, $x'\in U_{r,s}^{\zeta'}(\mathfrak{n})$, and $y\in U_{r,s}(\mathfrak{n}^-)$, the following hold:*

(i) $\partial_i(xx') = \langle\omega'_{\zeta'},\omega_i\rangle\,\partial_i(x)\,x' + x\,\partial_i(x')$.
(ii) ${}_i\partial(xx') = {}_i\partial(x)\,x' + \langle\omega'_i,\omega_\zeta\rangle\,x\,{}_i\partial(x')$.
(iii) $\langle f_iy,\,x\rangle = \langle f_i,e_i\rangle\,\langle y,\,{}_i\partial(x)\rangle = (s_i-r_i)^{-1}\langle y,\,{}_i\partial(x)\rangle$.
(iv) $\langle yf_i,\,x\rangle = \langle f_i,e_i\rangle\,\langle y,\,\partial_i(x)\rangle = (s_i-r_i)^{-1}\langle y,\,\partial_i(x)\rangle$.
(v) $f_ix - xf_i = (s_i-r_i)^{-1}\big(\partial_i(x)\,\omega_i - \omega'_i\,{}_i\partial(x)\big)$. □

Also, for each i, the expression of $\Delta(y)$ for $y\in U_{r,s}^{-\zeta}(\mathfrak{n}^-)$ defines two skew-derivations $\partial_i,\ {}_i\partial : U_{r,s}^{-\zeta}(\mathfrak{n}^-) \longrightarrow U_{r,s}^{-\zeta+\alpha_i}(\mathfrak{n}^-)$ such that

$$(5)\qquad \begin{aligned} \Delta(y) &= y\otimes\omega'_\zeta + \sum_{i=1}^{n} \partial_i(y)\otimes f_i\,\omega'_{\zeta-\alpha_i} + \textit{the rest},\\ \Delta(y) &= 1\otimes y + \sum_{i=1}^{n} f_i\otimes {}_i\partial(y)\,\omega'_i + \textit{the rest}. \end{aligned}$$

Parallel to [BW1, Lemma 4.8], we have

LEMMA 3.2. *For all $y\in U_{r,s}^{-\zeta}(\mathfrak{n}^-)$, $y'\in U_{r,s}^{-\zeta'}(\mathfrak{n}^-)$, and $x\in U_{r,s}(\mathfrak{n})$, the following hold:*

(i) $\partial_i(yy') = \partial_i(y)\,y' + \langle\omega'_\zeta,\omega_i\rangle\,y\,\partial_i(y')$.
(ii) ${}_i\partial(yy') = \langle\omega'_i,\omega_{\zeta'}\rangle\,{}_i\partial(y)\,y' + y\,{}_i\partial(y')$.
(iii) $\langle y,\,e_ix\rangle = \langle f_i,e_i\rangle\,\langle\partial_i(y),\,x\rangle = (s_i-r_i)^{-1}\langle\partial_i(y),\,x\rangle$.
(iv) $\langle y,\,xe_i\rangle = \langle f_i,e_i\rangle\,\langle\,{}_i\partial(y),\,x\rangle = (s_i-r_i)^{-1}\langle\,{}_i\partial(y),\,x\rangle$.
(v) $e_iy - ye_i = (r_i-s_i)^{-1}\big(\omega_i\,\partial_i(y) - {}_i\partial(y)\,\omega'_i\big)$. □

By Proposition 2.14, the spaces $U_{r,s}^{\zeta}(\mathfrak{n})$ and $U_{r,s}^{-\zeta}(\mathfrak{n}^-)$ are non-degenerately paired. We may select a basis $\{u_k^\zeta\}_{k=1}^{d_\zeta}$, ($d_\zeta = \dim U_{r,s}^{\zeta}(\mathfrak{n})$), for $U_{r,s}^{\zeta}(\mathfrak{n})$ and a dual basis $\{v_k^\zeta\}_{k=1}^{d_\zeta}$ for $U_{r,s}^{-\zeta}(\mathfrak{n}^-)$. Then for each $x\in U_{r,s}^{\zeta}(\mathfrak{n})$ and $y\in U_{r,s}^{-\zeta}(\mathfrak{n}^-)$, we have

$$(6)\qquad x = \sum_{k=1}^{d_\zeta}\langle v_k^\zeta,x\rangle\,u_k^\zeta,\qquad y = \sum_{k=1}^{d_\zeta}\langle y,u_k^\zeta\rangle\,v_k^\zeta.$$

For $\zeta\in Q^+ = \bigoplus_{i=1}^{n}\mathbb{Z}^+\alpha_i$, we define

$$(7)\qquad \Theta_\zeta = \sum_{k=1}^{d_\zeta} v_k^\zeta\otimes u_k^\zeta.$$

Set $\Theta_\zeta = 0$ if $\zeta \notin Q^+$. Similar to [BW1, Lemma 4.10], for the cases when $\mathfrak{g} = \mathfrak{so}_{2n+1}$, $\mathfrak{so}_{2n}$ and $\mathfrak{sp}_{2n}$, we also have

LEMMA 3.3. *For $1 \leq i \leq n$, the following relations hold*
(i) $(\omega_i \otimes \omega_i)\,\Theta_\zeta = \Theta_\zeta\,(\omega_i \otimes \omega_i)$, $\qquad (\omega_i' \otimes \omega_i')\,\Theta_\zeta = \Theta_\zeta\,(\omega_i' \otimes \omega_i')$;
(ii) $(e_i \otimes 1)\,\Theta_\zeta + (\omega_i \otimes e_i)\,\Theta_{\zeta-\alpha_i} = \Theta_\zeta\,(e_i \otimes 1) + \Theta_{\zeta-\alpha_i}\,(\omega_i' \otimes e_i)$;
(iii) $(1 \otimes f_i)\,\Theta_\zeta + (f_i \otimes \omega_i')\,\Theta_{\zeta-\alpha_i} = \Theta_\zeta\,(1 \otimes f_i) + \Theta_{\zeta-\alpha_i}\,(f_i \otimes \omega_i)$. □

Now we define

$$\Theta = \sum_{\zeta \in Q^+} \Theta_\zeta. \tag{8}$$

Given $U_{r,s}(\mathfrak{g})$-module M and M' in $\mathcal{O}$, we apply Θ to their tensor product:

$$\Theta = \Theta_{M,M'} : M \otimes M' \longrightarrow M \otimes M'.$$

Note that $\Theta_\zeta : M_\lambda \otimes M'_\mu \longrightarrow M_{\lambda-\zeta} \otimes M'_{\mu+\zeta}$ for all λ, $\mu \in \Lambda$, and there are only finitely many $\zeta \in Q^+$ such that $M'_{\mu+\zeta} \neq 0$, thanks to condition ($\mathcal{O}3$). So Θ is a well-defined linear transformation on $M \otimes M'$. After appropriately ordering the chosen countable bases of weight vectors for both M and M', we see that each Θ_ζ with $\zeta > 0$ has a strictly triangular matrix, while $\Theta_0 = 1 \otimes 1$ acts as the identity transformation on $M \otimes M'$, hence $\Theta_{M,M'}$ is an invertible transformation.

THEOREM 3.4. *Let M and M' be $U_{r,s}(\mathfrak{g})$-modules in $\mathcal{O}$ where $\mathfrak{g} = \mathfrak{so}_{2n+1}$, $\mathfrak{so}_{2n}$ or $\mathfrak{sp}_{2n}$. Then the map*

$$R_{M',M} = \Theta \circ \tilde{f} \circ P : M' \otimes M \longrightarrow M \otimes M'$$

is an isomorphism of $U_{r,s}(\mathfrak{g})$-modules, where $P : M' \otimes M \longrightarrow M \otimes M'$ is the flip map such that $P(m' \otimes m) = m \otimes m'$ for any $m \in M$, $m' \in M'$.

PROOF. Obviously, $R_{M',M}$ is invertible. It remains to show that $R_{M',M}$ is a $U_{r,s}(\mathfrak{g})$-module homomorphism, that is, to check that

$$\Delta(a) R_{M',M}(m' \otimes m) = R_{M',M}\Delta(a)(m' \otimes m) \tag{9}$$

holds for all $a \in U_{r,s}(\mathfrak{g})$, $m \in M_\lambda$ and $m' \in M'_\mu$. It suffices to verify (9) for the generators e_n, f_n, ω_n, ω_n', because the subalgebra generated by the first $4(n-1)$ generators e_i, f_i, ω_i, ω_i' $(1 \leq i < n)$ is isomorphic to $U_{r,s}(\mathfrak{sl}_n)$, and this can be reduced to the proof of the type A case (see [BW1, Theorem 4.11]). We will present the computation just for $a = f_n$. Using Lemma 3.3 (iii), we get

$$\begin{aligned}
\text{LHS of (9)} &= f(\lambda,\mu)\Delta(f_n)\Theta(m \otimes m') \\
&= f(\lambda,\mu)(1 \otimes f_n)\big(\textstyle\sum \Theta_\zeta\big)(m \otimes m') \\
&\quad + f(\lambda,\mu)(f_n \otimes \omega_n')\big(\textstyle\sum \Theta_{\zeta-\alpha_n}\big)(m \otimes m') \\
&= f(\lambda,\mu)\big(\textstyle\sum \Theta_\zeta\big)(1 \otimes f_n)(m \otimes m') \\
&\quad + f(\lambda,\mu)\big(\textstyle\sum \Theta_{\zeta-\alpha_n}\big)(f_n \otimes \omega_n)(m \otimes m') \\
&= f(\lambda,\mu)\langle \omega_n', \omega_\lambda\rangle\big(\textstyle\sum \Theta_\zeta\big)(\omega_n' m \otimes f_n m') \\
&\quad + f(\lambda,\mu)\langle \omega_\mu', \omega_n\rangle\big(\textstyle\sum \Theta_{\zeta-\alpha_n}\big)(f_n m \otimes m').
\end{aligned}$$

On the other hand, we have

$$\begin{aligned}\text{RHS of (9)} &= R_{M',M}(m'\otimes f_n m + f_n m'\otimes \omega_n' m)\\ &= (\Theta\circ \tilde{f})(f_n m\otimes m' + \omega_n' m\otimes f_n m')\\ &= f(\lambda-\alpha_n,\mu)\Theta(f_n m\otimes m') + f(\lambda,\mu-\alpha_n)\Theta(\omega_n' m\otimes f_n m')\\ &= f(\lambda-\alpha_n,\mu)\big(\sum \Theta_\zeta\big)(f_n\otimes 1)(m\otimes m')\\ &\quad + f(\lambda,\mu-\alpha_n)\big(\sum\Theta_\zeta\big)(\omega_n'\otimes f_n)(m\otimes m').\end{aligned}$$

Thus (3) indicates that (9) holds. □

REMARK 3.5. Similar to the treatment in [BW1, Section 5] for the type A case, we can prove the maps $R_{M',M}$ satisfy the quantum Yang-Baxter equation for our cases. That is, given three $U_{r,s}(\mathfrak{g})$-modules M, M', M'' in $\mathcal{O}$, we have $R_{12}\circ R_{23}\circ R_{12} = R_{23}\circ R_{12}\circ R_{23}$ as maps from $M\otimes M'\otimes M''$ to $M''\otimes M'\otimes M$ (see [BW1, Theorem 5.4]). On the other hand, we also can prove the hexagon identities (see [BW1, Theorem 5.7]) for the maps $R_{M',M}$ by the same approach. Consequently, $\mathcal{O}$ is a braided monoidal category with braiding $R = R_{M',M}$ for each pair of modules M', M in $\mathcal{O}$.

Quantum Casimir operators and complete reducibility. The $U_{r,s}(\mathfrak{g})$-module isomorphisms $R_{M',M}$ constructed in Theorem 3.4, which are called the R-matrices, are mainly determined by Θ. For the expression (7) of Θ_ζ, we set

(10) $$\Omega_\zeta = \sum_k S(v_k^\zeta)u_k^\zeta, \qquad \Omega_\zeta' = \theta(\Omega_\zeta),$$

(11) $$\Omega = \sum_{\zeta\in Q^+}\Omega_\zeta, \qquad \Omega' = \sum_{\zeta\in Q^+}\Omega_\zeta',$$

where θ is the $\mathbb{Q}$-algebra isomorphism of $U_{r,s}(\mathfrak{g})$ into its associated quantum group $U_{s^{-1},r^{-1}}(\mathfrak{g})$ (for definition, see [BGH]) introduced in the formula (11) in Section 2. Obviously, Θ_ζ, Ω_ζ, Ω and Ω' are independent of the choice of bases $\{u_k^\zeta\}$ and $\{v_k^\zeta\}$. Ω preserves the weight spaces of any $M\in\mathcal{O}$.

DEFINITION 3.6. *The element Ω is called a quantum Casimir element for the two-parameter quantum group $U_{r,s}(\mathfrak{g})$.*

PROPOSITION 3.7. *Let ψ and φ be the algebra automorphisms of $U_{r,s}(\mathfrak{g})$ such that $\psi(\omega_i)=\omega_i$, $\psi(\omega_i')=\omega_i'$, $\psi(e_i)=\omega_i'\omega_i^{-1}e_i$, $\psi(f_i)=f_i\omega_i'^{-1}\omega_i$ and $\varphi(\omega_i)=\omega_i$, $\varphi(\omega_i')=\omega_i'$, $\varphi(e_i)=e_i\omega_i^{-1}\omega_i'$, $\varphi(f_i)=\omega_i\omega_i'^{-1}f_i$. Then*

(12) $$\psi(a)\,\Omega = \Omega\, a, \qquad \varphi(a)\,\Omega' = \Omega'\, a, \quad \textit{for } a\in U_{r,s}(\mathfrak{g}).$$

PROOF. Since ψ is an algebra automorphism, it is enough to prove the first assertion for the generators $a=\omega_i$, ω_i', e_i, f_i. For $a=\omega_i$ or ω_i', it is obviously true. Applying the mapping $\mathfrak{m}\circ(S\otimes 1)$ to both sides of Lemma 3.3 (ii) & (iii) (where $\mathfrak{m}$ is the product of $U_{r,s}(\mathfrak{g})$ and S is its antipode) and summing over $\zeta\in Q^+$ we obtain $\Omega\, e_i = \omega_i'\omega_i^{-1}e_i\Omega$ and $\Omega\, f_i = f_i\omega_i'^{-1}\omega_i\Omega$. This means that $\Omega\, e_i = \psi(e_i)\,\Omega$ and $\Omega\, f_i = \psi(f_i)\,\Omega$. Applying the automorphism θ we get the assertion for Ω'. □

COROLLARY 3.8. *For $M\in \mathrm{Ob}(\mathcal{O})$, assume that $m\in M_\lambda$. Then*
(i) $\Omega\, e_i.m = (r_is_i^{-1})^{-(\lambda+\alpha_i,\alpha_i^\vee)}e_i\,\Omega.m$,
(ii) $\Omega\, f_i.m = (r_is_i^{-1})^{(\lambda,\alpha_i^\vee)}f_i\,\Omega.m$.

PROOF. By Proposition 3.7, for $m \in M_\lambda$, we have

$$\psi(e_i)\,\Omega.m = \langle \omega'_{\lambda+\alpha_i}, \omega_i\rangle^{-1}\langle \omega'_i, \omega_{\lambda+\alpha_i}\rangle^{-1} e_i\,\Omega.m,$$
$$\psi(f_i)\,\Omega.m = \langle \omega'_\lambda, \omega_i\rangle\,\langle \omega'_i, \omega_\lambda\rangle\, f_i\,\Omega.m.$$

Using formulas (B), (C), & (D) in Lemma 2.1, we can conclude the required result. □

REMARK 3.9. According to Section 1, we have made a convention: we have $r_i = r^{(\alpha_i,\alpha_i)}$, $s_i = s^{(\alpha_i,\alpha_i)}$ only in the type B case,so $(r_i s_i^{-1})^{(\lambda,\alpha_i^\vee)} = (rs^{-1})^{2(\lambda,\alpha_i)}$ for any i. However, for any other case, we always have $(r_i s_i^{-1})^{(\lambda,\alpha_i^\vee)} = (rs^{-1})^{(\lambda,\alpha_i)}$ for any i since $r_i = r^{\frac{(\alpha_i,\alpha_i)}{2}}$, $s_i = s^{\frac{(\alpha_i,\alpha_i)}{2}}$. Based on this observation, we make the following definition.

DEFINITION 3.10. *For $M \in \mathrm{Ob}(\mathcal{O})$, define a linear operator $\omega : M \longrightarrow M$ by setting*

$$\omega.v_\mu = (rs^{-1})^{\frac{\Delta_{X,B}}{2}(\mu+\rho,\mu+\rho)} v_\mu, \quad \text{for } v_\mu \in M_\mu, \tag{13}$$

where ρ is the half-sum of all positive roots of $\mathfrak{g}$, and $\Delta_{X,B} = 2$ if $X = B$, otherwise, $\Delta_{X,B}$ will take value 1.

PROPOSITION 3.11. *Assume that the Verma module $M(\lambda) \in \mathrm{Ob}(\mathcal{O})$, then the operator $\Omega\omega$ is a multiple of the identity operator, that is,*

$$\Omega\omega = (rs^{-1})^{\frac{\Delta_{X,B}}{2}(\lambda+\rho,\lambda+\rho)} I. \tag{14}$$

PROOF. Let v_λ be a highest weight vector of the Verma module $M(\lambda)$. Then $M(\lambda) = U_{r,s}(\mathfrak{n}^-)v_\lambda = \sum_{\beta\in Q^+} U_{r,s}^{-\beta}(\mathfrak{n}^-)v_\lambda$. For $f_\beta \in U_{r,s}^{-\beta}(\mathfrak{n}^-)$, denote $v_{\lambda-\beta} := f_\beta.v_\lambda$, which is a weight vector of weight $\lambda - \beta$. We claim that

$$\Omega\omega.f_i.v_{\lambda-\beta} = f_i.\Omega\omega.v_{\lambda-\beta}, \tag{15}$$

for any $\beta \in Q^+$ and any i. Indeed, noting that

$$\frac{1}{2}\big[(\sigma-\alpha_i+\rho,\sigma-\alpha_i+\rho)-(\sigma+\rho,\sigma+\rho)\big]+(\sigma,\alpha_i) = \frac{1}{2}\big[(\alpha_i,\alpha_i)-2(\alpha_i,\rho)\big] = 0,$$

and setting $\lambda - \beta = \sigma$, we have

$$\begin{aligned}
\Omega\omega.f_i.v_{\lambda-\beta} &= (\Omega f_i)(rs^{-1})^{\Delta_{X,B}c}\,\omega.v_{\lambda-\beta}\\
&= (f_i\omega_i'^{-1}\omega_i\Omega)\,(rs^{-1})^{\Delta_{X,B}c}\,\omega.v_{\lambda-\beta}\\
&= f_i(rs^{-1})^{\Delta_{X,B}(\lambda-\beta,\alpha_i)}(rs^{-1})^{\Delta_{X,B}c}\,\Omega\omega.v_{\lambda-\beta}\\
&= f_i\,\Omega\omega.v_{\lambda-\beta},
\end{aligned}$$

where $c = \frac{1}{2}\big[(\lambda-\beta-\alpha_i+\rho,\lambda-\beta-\alpha_i+\rho)-(\lambda-\beta+\rho,\lambda-\beta+\rho)\big]$. (15) yields

$$\begin{aligned}
\Omega\omega.f_\beta.v_\lambda &= f_\beta.\Omega\omega.v_\lambda\\
&= (rs^{-1})^{\frac{\Delta_{X,B}}{2}(\lambda+\rho,\lambda+\rho)} f_\beta.\Omega.v_\lambda\\
&= (rs^{-1})^{\frac{\Delta_{X,B}}{2}(\lambda+\rho,\lambda+\rho)} f_\beta.\Omega_0.e_\lambda\\
&= (rs^{-1})^{\frac{\Delta_{X,B}}{2}(\lambda+\rho,\lambda+\rho)} f_\beta.v_\lambda, \quad (\Omega_0 = 1).
\end{aligned}$$

So the relation (14) follows. □

COROLLARY 3.12. (i) *For the simple $U_{r,s}(\mathfrak{g})$-module $L(\lambda) \in \mathrm{Ob}(\mathcal{O})$, there holds*

$$\Omega\omega = (rs^{-1})^{\frac{\Delta_{X,B}}{2}(\lambda+\rho,\lambda+\rho)}I.$$

(ii) *For each finite-dimensional $M \in \mathrm{Ob}(\mathcal{O})$, the eigenvalues of the operator $(\Omega\omega)|_M$ are integral powers of $(rs^{-1})^{\frac{1}{2}}$.*

PROOF. (i) is evident. For (ii), as $M \in \mathrm{Ob}(\mathcal{O})$ is finite-dimensional, it has a composition series whose factors are finite-dimensional simple $U_{r,s}(\mathfrak{g})$-modules in $\mathcal{O}$, on which $\Omega\omega$ acts as multiplication by $(rs^{-1})^{\frac{\Delta_{X,B}}{2}(\mu+\rho,\mu+\rho)}$ for some $\mu \in \Lambda^+$, as indicated by (i) and Corollary 2.8. After taking an appropriate basis of M compatible with a chosen composition series, the acting matrix of $(\Omega\omega)|_M$ has the required property. □

From Corollary 3.8 and Definition 3.10, we have a further result as follows.

THEOREM 3.13. *The operator $\Omega\omega : M \longrightarrow M$ commutes with the action of $U_{r,s}(\mathfrak{g})$ on any module $M \in \mathrm{Ob}(\mathcal{O})$, where $\mathfrak{g} = \mathfrak{sl}_{n+1}$, $\mathfrak{so}_{2n+1}$, $\mathfrak{so}_{2n}$, or $\mathfrak{sp}_{2n}$.*

PROOF. At first, it needs to show that $\Omega\omega$ commutes with e_i, f_i $(1 \leq i \leq n)$. For $m \in M_\mu$, by Corollary 3.8 and Definition 3.10, we get

$$\begin{aligned}
\Omega\omega.(e_i.m) &= (rs^{-1})^{\frac{\Delta_{X,B}}{2}(\mu+\alpha_i+\rho,\,\mu+\alpha_i+\rho)}\Omega e_i.m\\
&= (rs^{-1})^{\Delta_{X,B}[\,\frac{1}{2}(\mu+\alpha_i+\rho,\,\mu+\alpha_i+\rho)-(\mu+\alpha_i,\,\alpha_i)\,]}e_i\Omega.m\\
&= (rs^{-1})^{\frac{\Delta_{X,B}}{2}(\mu+\rho,\,\mu+\rho)}e_i\Omega.m\\
&= e_i.(\Omega\omega.m).\\
\Omega\omega.(f_i.m) &= (rs^{-1})^{\frac{\Delta_{X,B}}{2}(\mu-\alpha_i+\rho,\,\mu-\alpha_i+\rho)}\Omega f_i.m\\
&= (rs^{-1})^{\Delta_{X,B}[\,\frac{1}{2}(\mu-\alpha_i+\rho,\,\mu-\alpha_i+\rho)+(\mu,\,\alpha_i)\,]}f_i\Omega.m\\
&= (rs^{-1})^{\frac{\Delta_{X,B}}{2}(\mu+\rho,\,\mu+\rho)}f_i\Omega.m\\
&= f_i.(\Omega\omega.m).
\end{aligned}$$

Obviously, $\Omega\omega$ commutes with the action of ω_i, ω_i' $(1 \leq i \leq n)$, for it preserves the weight spaces of M. □

The following Lemma is due to [BW2, Lemma 3.7] for the case of $\mathfrak{g} = \mathfrak{gl}_{n+1}$, or $\mathfrak{sl}_{n+1}$, which is still valid in our cases.

LEMMA 3.14. *Assume that rs^{-1} is not a root of unity. Let M be a nonzero finite-dimensional quotient of the Verma $U_{r,s}(\mathfrak{g})$-module $M(\lambda) \in \mathrm{Ob}(\mathcal{O})$. Then M is simple. In particular, $L'(\lambda) = L(\lambda)$ for $\lambda \in \Lambda^+$.*

PROOF. Lemma 2.6 means $\lambda \in \Lambda^+$. The proof is based on the counter-evidence method and Proposition 3.11, which is the same as that of [BW2, Lemma 3.7], with slight differences: for the function $g(\lambda)$ used in the proof there we use $(rs^{-1})^{\frac{\Delta_{X,B}}{2}(\lambda+\rho,\lambda+\rho)}$ instead, noting the fact from Lie algebra theory (see [D], or [K]) that for any weight $\mu \leq \lambda$ where $\lambda \in \Lambda^+$, $(\lambda+\rho, \lambda+\rho) = (\mu+\rho, \mu+\rho)$ if and only if $\mu = \lambda$. □

Based on the above results, using a similar argument due to Kac [K] in the proof of complete reducibility of category $\mathcal{O}$ for affine Kac-Moody Lie algebras (or comparing with the proof of [BW2, Theorem 3.8] in the spirit of Lusztig [L1]), we have

THEOREM 3.15. *Assume that rs^{-1} is a non-root of unity. For $\mathfrak{g} = \mathfrak{sl}_{n+1}$, $\mathfrak{so}_{2n+1}$, $\mathfrak{so}_{2n}$ or $\mathfrak{sp}_{2n}$, let M be a nonzero finite-dimensional $U_{r,s}(\mathfrak{g})$-module on which U^0 acts semisimply. Then M is completely reducible.* □

Acknowledgments

Part of this work was done when N. Hu visited the Department of Mathematics and Statistics of York University in Canada from Sept. 2003 to Sept. 2004, visited the DMA, l'Ecole Normale Supérieure de Paris in France as an invited professor from Oct. 20 to Nov. 20, 2004, and visited the Fachbereich Mathematik der Universität Hamburg in Germany as a DFG-visiting professor from Nov. 21, 2004 to Feb. 28, 2005. He would like to express his deep thanks to M. Rosso and H. Strade for their invitation and extreme hospitality, as well as the supports from York University, l'Ecole Normale Supérieure de Paris and die Deutche Forschungsgemeinschaft (DFG).

References

[BW1] G. Benkart and S. Witherspoon, *Two-parameter quantum groups and Drinfel'd doubles*, Algebr. Represent. Theory, **7** (2004), 261–286.

[BW2] G. Benkart and S. Witherspoon, *Representations of two-parameter quantum groups and Schur-Weyl duality*, Hopf algebras, pp. 65–92, Lecture Notes in Pure and Appl. Math., 237, Dekker, New York, 2004.

[BW3] G. Benkart and S. Witherspoon, *Restricted two-parameter quantum groups*, Fields Institute Communications, "Representations of Finite Dimensional Algebras and Related Topics in Lie Theory and Geometry", vol. 40, Amer. Math. Soc., Providence, RI, 2004, pp. 293–318.

[BGH] N. Bergeron, Y. Gao and N. Hu, *Drinfel'd doubles and Lusztig's symmetries of two-parameter quantum groups*, (math.RT/0505614), J. of Algebra, **301** (2006), 378–405.

[D] J. Dixmier, *Enveloping Algebras*, North-Holland, Amsterdam, 1977.

[Dr] V.G. Drinfeld, *Quantum groups*, in "Proceedings ICM", Berkeley, Amer. Math. Soc. 1987, pp. 798–820.

[J] J.C. Jantzen, *Lectures on Quantum Groups*, vol. 6, Graduate Studies in Math., Amer. Math. Soc. Providence, RI, 1996.

[Jo] A. Joseph, *Quantum Groups and Their Primitive Ideals*, Ergebnisse der Mathematik und ihrer Grenzgebiete, Springer-Verlag, Berlin, 1995.

[K] V.G. Kac, *Infinite Dimensinal Lie Algebras*, Third edition, Cambridge University Press, 1990.

[KS] A. Klimyk and K. Schmüdgen, *Quantum Groups and Their Representations*, Springer-Verlag Berlin Heidelberg New York, 1997.

[L1] G. Lusztig, *Introduction to Quantum Groups*, Birkhäuser Boston, 1993.

[L2] G. Lusztig, *Quantum groups at roots of* 1, Geom. Dedicata **35** (1990), 89–114.

[RTF] N.Yu. Reshetikhin, L.A. Takhtajan and L.D. Faddeev, *Quantization of Lie groups and Lie algebras*, Algebra and Anal. **1**, 178–206 (1989) (Leningrad Math. J. 1 [Engl. transl. 193–225 (1990)]).

Department of Mathematics and Statistics, York University, North York, Toronto, ON Canada, M3J 1P3

E-mail address: bergeron@mathstat.yorku.ca

Department of Mathematics and Statistics, York University, North York, Toronto, ON Canada, M3J 1P3

E-mail address: ygao@yorku.ca

Department of Mathematics, East China Normal University, Shanghai 200062, PR China

E-mail address: nhhu@euler.math.ecnu.edu.cn

AMS/IP Studies in Advanced Mathematics
Volume **39**, 2007

Optimal control of the Liouville equation

R. W. Brockett

ABSTRACT. We explore the role of a special class of optimization problems involving controlled Liouville equations. We present some new results on the controllability of the Liouville equation and discuss the optimal control of its moments for some important special cases. The problems treated suggest a new domain of applicability for optimal control, one which can include constraints more appropriate for the synthesis of control systems for a large class of real world systems.

1. Introduction

In this paper we argue that some of the important limitations standing in the way of the wider use of optimal control can be circumvented by explicitly acknowledging that in most situations the apparatus implementing the control policy will judged on its ability to cope with a distribution of initial states, rather than a single state. This makes it appropriate to represent the system dynamics in terms of its Liouville equation and to formulate performance measures in that setting. That is, we argue in favor of replacing the usual control model $\dot{x} = f(x, u)$ by the first order partial differential equation

$$\frac{\partial \rho(t, x)}{\partial t} = -\left\langle \frac{\partial}{\partial x}, f(x, u)\rho(t, x) \right\rangle$$

and for the consideration of performance measures which include *non trajectory dependent terms* such as the second and third terms on the right-hand side of

$$\eta = \int_0^T \int_S \rho(t, x)L(x, u)dxdt + \int_S \left(\frac{\partial u}{\partial x}\right)^2 dx + \int_0^T \left(\frac{\partial u}{\partial t}\right)^2 dt$$

Later on we give a number of examples to support our contention but, in brief, we may say that by forcing the specification of a range of initial conditions this formulation allows one to include some important types of performance measures not expressible in standard deterministic optimal control theory. In specific cases this allows the designer to express, in mathematical terms, issues related to the role of feedback, robustness and cost of implementation. This approach has some

This work was supported in part by the US Army Research Office under Boston University prime GC169369 NGD, the National Science Foundation grant number ECS 0424342, and by DARPA under Stanford University Prime PY-1606.

aspects in common with stochastic control but occupies a position of intermediate complexity, lying between deterministic calculus of variations on finite dimensional manifolds and, the usually much less tractable, control of finite dimensional diffusion processes.

In almost all applications of automatic control there is a trade off between the gains to be had from trajectory optimization and the cost of the apparatus required to implement the optimal policy. A simpler apparatus that generates commands that closely approximate the optimal trajectories but costs less to build and maintain can be preferable to a more complex apparatus that precisely generates the optimal trajectories. Such considerations apply both to technological applications and to the study of motion control in animals where evolutionary pressure seeks to optimize the allocation of neural circuitry. In spite of these arguments, there is almost nothing in the control literature attempting to translate these ideas into mathematics. In this paper we formulate and solve some problems relating to such questions.

We begin with some background. Typically problems in deterministic optimal control are formulated in terms of a variable x, called the *state*, related to a variable u, called the *control*, through a differential equation, $\dot{x} = f(x, u)$ and an initial equation $x(0) = x_0$, together with a performance measure

$$\eta = \int_0^T L(x, u)dt$$

The problem is to find a control $u^*(x, t)$ that minimizes η. Assuming that an optimal control u^* exists, it might be expressed as a function of t and x_0 and otherwise independent of x or it might be expressed in terms of t and $x(t)$. Engineers express the matter by saying that when the optimal u is expressed as a function of t and x_0 it is in *open loop* form, and when it is expressed in terms of $x(t)$, it is expressed in *closed loop* form. The distinction between open loop and closed loop is of little significance for the mathematical problem stated because the trajectories are the same in either case, but the difference is quite significant when it comes to using such results. In the first place, some apparatus must be made available to generate the function u and the cost and complexity of this apparatus can be expected to depend on this distinction. Secondly, the mathematical model for the system and the characterization of the initial state is inevitably inexact and the effect of the error will depend on whether u is realized in closed loop form or open loop form.

The situation found in stochastic control is different. In this subject the distinction between open loop performance and closed loop performance is quite obvious. It is standard to formulate problems in terms of a probabilistic model for the observations and to restrict the control to be a *past measurable function* of the observables. In this way the distinction between open loop and closed loop control is woven into the problem formulation from the beginning. Typically problems are described in terms of a stochastic differential equation such as $dx = f(x, u)dt + gdw$ and a vector of observables $dy = h(x)dt + d\nu$. The objective is to minimize the expected value of some functional. In this case there is a difference between the optimal performance obtainable when u is constrained to be a function of time and the statistics of $x(0)$ only, versus the performance obtainable when it is allowed to depend on the value of x at the present time. Typically no open loop control can perform as well as the closed loop one. One concrete aspect of this distinction is is illustrated by the fact that in the Fokker-Planck equation for the probability density of x, namely

$$\frac{\partial \rho(t,x)}{\partial t} = -\left\langle \frac{\partial}{\partial x}, f(x,u)\rho(t,x) \right\rangle + \frac{1}{2}\left\langle \frac{\partial^2}{\partial x^2}, gg^T \rho(t,x) \right\rangle \quad ; \quad \rho(0,x) = \text{given}$$

u is acted upon by partial derivative with respect to x so that the type of functional dependence matters. However, the use of stochastic control theory requires considerable effort to gather the statistical data needed to define the problem, and even then is often quite difficult to solve.

In this paper we describe an alternative formulation of control problems that attempts to capture more fully the implications of the various ways of expressing u without encountering the full range of difficulties associated with stochastic control. As in our earlier work [1], this approach involves controlling a Liouville equation and leads to a class of variational problems that seem to have been largely ignored until now.

2. Controlling the Liouville Equation

Let X be an oriented differentiable manifold with a fixed, nondegenerate, volume form dv. Let $\phi : X \to X$ be a diffeomorphism. If ρdv is a nonnegative density on X and if ϕ is orientation preserving, then ϕ acts on densities according to

$$\rho(\cdot) \mapsto \rho(\phi^{-1}(\cdot))/\mathrm{det J}_\phi$$

where J_ϕ is the Jacobian of ϕ. In this sense $\mathrm{Diff}_O(X)$, the set of orientation preserving diffeomorphisms, generates an orbit through a given ρ. It is known, for example, that if the manifold is compact then this action acts transitively on the strictly positive densities. (See Moser [2].) If the densities are only assumed to be nonnegative this action is clearly not transitive.

Associated with a control equation of the form $\dot{x} = f(x,u)$ there is a first order partial differential equation containing a control term which governs the evolution of smooth densities,

$$\frac{\partial \rho(t,x)}{\partial t} = -\left\langle \frac{\partial}{\partial x}, f(x,u)\rho(t,x) \right\rangle$$

This is a control dependent version of the Liouville equation of mathematical physics. If the underlying manifold is $\mathbb{R}^n$ we can sketch its derivation as follows. Assume that ρ is a smooth function and that $\psi(x)$ is a smooth test function with compact support. On one hand, the time evolution of the average value of ψ is given by

$$\frac{\partial}{\partial t}\int_{\mathbb{R}^n} \psi(x)\rho(t,x)dx = \int_{\mathbb{R}^n} \psi(x)\frac{\partial}{\partial t}\rho(t,x)dx$$

on the other hand,

$$\frac{d}{dt}\psi(x) = \left\langle \frac{\partial \psi}{\partial x}, f(x,u) \right\rangle$$

and so

$$\frac{\partial}{\partial t}\int_{\mathbb{R}^n} \psi(x)\rho(t,x)dx = \int_{\mathbb{R}^n} \rho(t,x)\left\langle \frac{\partial \psi(t,x)}{\partial x}, f(x,u) \right\rangle dx$$

An integration-by-parts then shows that for all such ψ

$$\int_{\mathbb{R}^n} \psi(x)\frac{\partial}{\partial t}\rho(t,x)dx = -\int_{\mathbb{R}^n} \psi(x)\left\langle \frac{\partial \rho(t,x)}{\partial x}, f(x,u) \right\rangle dx$$

In view of the fact that ψ is arbitrary we see that a smooth ρ evolves as given.

If $\rho(0,x)$ is nonnegative with $\int \rho(0,x)dx = 1$, then the Liouville equation can be thought of as the evolution equation for an initial probability density of x under the flow defined by the deterministic equation $\dot{x} = f(x,u)$. As is well known, the solution of the Liouville equation can be expressed in terms of the general solution of $\dot{x} = f(x,u)$. If the solution of $\dot{x} = f(x,u(x))$ is such that the initial value x_0 goes to $\phi(t,x_0)$ at time t then the solution of the Liouville equation is

$$\rho(t,x) = \rho(0,\phi^{-1}(t,x))/\mathrm{det J}_\phi(\mathrm{x})$$

where ϕ^{-1} denotes the result of solving $x = \phi(t,x_0)$ for x_0 and J is the Jacobian of the map ϕ. For example, if $\dot{x} = x$ the map is $\phi(t,x) = e^t x$ and the initial density is $\pi^{-1}/(1+x^2)$ then the density at time t is $e^{-2t}\pi^{-1}/(1+e^{-2t}x^2)$.

Given $\dot{x} = f(x,u)$ together with a performance measure $\eta = \int_0^T L(x,u)dt$ such that an optimal control exists, we have an optimal value function

$$\eta^*(x(0)) = \min_u \int_0^T L(x,u)dt$$

If S is the support of the initial density We can evaluate the average performance of the optimal system using

$$\mathcal{E}\eta(x) = \int_S \rho(0,x)\eta^*(x)dx$$

If we place no further conditions on the way u depends on x and t then the Liouville point of view contributes nothing essentially new because

$$\Gamma^*(\rho(0,x)) = \min_u \int_0^T \int_S \rho(t,x)L(x,u)dxdt$$

That is, $\Gamma^*(\rho(0,x)) = \mathcal{E}\eta^*$. However, if we place conditions on u with the goal of making the system more robust with respect to errors or with the goal of simplifying the apparatus necessary to generate the control, such as requiring that u be affine in x or limiting the norm of $\partial u/\partial x$, etc. then the most we can say is that

$$\Gamma(\rho(0,x)) \geq \mathcal{E}\eta(x)$$

and we have a genuinely new class of problems.

To explore these issues we investigate the control of the Liouville equation with performance measures of the type suggested in the introduction. We consider three examples.

Example 1: Consider the scalar equation $\dot{x} = u$ with associated Liouville equation

$$\frac{\partial\rho(t,x)}{\partial t} = -\frac{\partial}{\partial x}u(t,x)\rho(t,x)$$

Suppose we constrain u to be affine function of x and want to minimize

$$\Gamma = \int_0^T \int_S \rho(t,x)x^2 dxdt + \int_0^T \left(\frac{\partial u}{\partial x}\right)^2 + \left(u - \frac{\partial u}{\partial x}x\right)^2 dt$$

The terms in Γ can be interpreted as follows. The ρx^2 term penalizes trajectories for deviating from $x = 0$. The $\partial u/\partial x$ term penalizes the design for using high gain feedback whereas the remaining term penalizes the control for using large open loop signal. According to standard linear quadratic control theory, the minimum value of

$$\eta = \int_0^T x^2 + v^2 dt$$

for a system $\dot{x} = a(t)x + v$ is given by $\eta^* = k(0)x^2(0)$ where

$$\dot{k} = 2ak + 1 - k^2 \ ; \ k(T) = 0$$

Thus our problem can be recast as that of minimizing

$$\Gamma = k(0) \int_{\mathbb{R}} \rho(0,x)x^2 dx + \int_0^T a^2 dt$$

Solving for a in terms of k and denoting the second moment of x at time 0 as m, the problem is to minimize

$$\Gamma = mk(0) + \int_0^T \left(\frac{\dot{k} - k^2 + 1}{2k} \right)^2 dt$$

Letting $r = \ln k$ the integrand becomes $(\dot{r}/2 - \sinh r)^2$. Applying Euler-Lagrange theory we see that $\ddot{r} - 4\cosh r \sinh r = 0$. Thus there exists a constant d such that

$$\dot{r} = \sqrt{d - 4\sinh^2 r}$$

so that $k = e^r$ and $a = (\dot{k} - k^2 + 1)/2k$.

Example 2: Consider the same system as above and again restrict the control law to be affine in x. Now we consider the least squares control of the mean and variance of the density. This class of problems will be discussed in more detail in section 5. The system is

$$\dot{x} = -kx + v$$

$$\dot{\sigma} = -2k\sigma$$

and we want to find k and v so as to steer the mean and variance to desired values while minimizing

$$\eta = \int_0^T k^2 + v^2 dt$$

In anticipation of using the Pontryagin maximum principle, we form the Hamiltonian

$$h = -pkx + pv - 2qk\sigma + v^2/2 + k^2$$

Thus $v_{\text{opt}} = -p$ and $k_{\text{opt}} = q\sigma + px/2$ and

$$\dot{p} = kp$$

$$\dot{q} = 2kq$$

¿From this we see that $q\sigma$ is a constant and thus v and k satisfy

$$\dot{k} = -v^2/2$$

$$\dot{v} = kv$$

This implies that $\ddot{k} = -2\dot{k}k$ and so $\dot{k} = -k^2 + c$ and

$$k(t) = \sqrt{c}\tanh(\sqrt{c}\, t + \alpha) \ ; \ v = \sqrt{2c - 2k^2}$$

Example 3: Again we consider the scalar control problem $\dot{x} = u$ with the distribution of initial conditions being given by a density $\rho_0(x)$. The Liouville equation is as above but now we consider the minimization of

$$\eta = \int_0^\infty \int_{-1}^1 \rho(t,x)ax^2 dxdt + \int_{\mathbb{R}} \left(\frac{\partial u}{\partial x}\right)^2 dx$$

If we constrain $u(x)$ to be a function of x alone and if $\int_0^\infty x^2 dt$ is finite then

$$\int_0^\infty x^2 dt = \int_0^{x(0)} \frac{x^2}{u(x)} dx \text{ and } \int_0^\infty \mathrm{u}^2 \mathrm{dt} = \int_0^{\mathrm{x}(0)} \mathrm{u(x)dx}$$

so that

$$\int_0^\infty \int_{\mathbb{R}} \rho(t,x)ax^2 dxdt = \int_{\mathbb{R}} \rho_0(x_0) \left(\int_0^{x(0)} a\frac{x^2 dx}{u(x)} + bu(x) \right) dx_0$$

Thus the functional to be minimized can be written as

$$\eta = \int_{\mathbb{R}} \rho_0(x) \left(\int_0^x a\frac{w^2}{u(w)} dw \right) dx + \int_{\mathbb{R}} \left(\frac{\partial u}{\partial x}\right)^2 dtdx$$

In the special case where ρ_0 is a delta function centered at x_0 the term involving ρ_0 can be simplified giving

$$\eta = \int_0^{(x(0)} \left(a\frac{x^2}{u(w)} \right) dx + \int_{\mathbb{R}} \left(\frac{\partial u}{\partial x}\right)^2 dtdx$$

The indicated partial derivative is actually a total derivative and so an application of the Euler-Lagrange operator gives

$$\frac{d^2u}{dx^2} + a\frac{x^2}{u^2} = 0$$

Because the solution x should remain at zero when reaching zero, it is necessary to impose the condition $u(0) = 0$ on solutions of the Euler Lagrange equation. It is easy to verify that the appropriate solution is

$$u = -(9a/4)^{1/3} x^{4/3}$$

In this case the integral squared error is proportional to $x^{5/2}(0)$.

3. Controllability of the Liouville Equation

Before considering optimal control problems involving the Liouville equation further we establish some results describing the extent to which one can influence the evolution of a density through the use of feedback control. With this in mind we now establish a controllability property of the Liouville equation associated with controllable linear systems.

Consider a system evolving in $\mathbb{R}^n$ described by

$$\dot{x} = Ax + Bu$$

The corresponding Liouville equation is

$$\frac{\partial \rho(t,x)}{\partial t} = -\left\langle \frac{\partial}{\partial x}, (Ax + Bu)\,\rho(t,x) \right\rangle$$

A pair (A, B) with A and n by n matrix and B an n by m matrix is said to be *controllable* if the columns of $B, AB, ..., A^{n-1}B$ contains a basis for $\mathbb{R}^n$. For such systems we introduce a one parameter family of symmetric, nonnegative definite matrices

$$W(0,t) = \int_0^t e^{-A\tau}BB^Te^{-A^T\tau}d\tau$$

The controllability property implies that $W(0,t)$ is nonsingular for positive t. From the variation of constants formula we see that the solution x corresponding to the control

$$u(t) = B^Te^{-A^Tt}W^{-1}(0,T)(e^{-At}x_f - x_0)$$

is

$$x(t) = e^{At}\left(x_0 + \int_0^t e^{-A\sigma}BB^Te^{-A^T\sigma}d\sigma W^{-1}(0,T)(e^{-AT}x_f - x_0)\right)$$

steers the system from $x(0) = x_0$ to $X(T) = x_f$

As noted above, it is not true that any initial density can be transformed into any other density at a later time. For example, open sets on which the initial density vanishes will necessarily map into open sets on which subsequent densities vanish. However, it is possible to reach from $\rho_0(\cdot)$ a large class of densities.

We begin with a useful lemma. We use the notation $||M||$ to denote the induced Euclidean norm of a matrix.

Lemma 1: Let $\dot{x} = Ax + Bu$ be a controllable linear system. Let $\psi : \mathbb{R}^n \to \mathbb{R}^n$ be a diffeomorphism with Jacobian $J_\psi(x)$. Suppose $T > 0$ is given and that for some $k < 1$ and all x, y the map ψ satisfies the Lipschitz inequality

$$||e^{-AT}\psi(x) - x - e^{-AT}\psi(y) + y|| \leq k||x - y||$$

Then there exists a control law $u(x,t)$ that steers the Liouville equation from the given initial condition $\rho(0,x) = \rho_0(x)$ to $\rho(T,x) = \rho_0(\psi(x)))/\mathrm{det}(\mathrm{J}_\psi(\mathrm{x}))$.

Proof: The solution of $\dot{x} = Ax + Bu$ with the control

$$u(t,x_0) = B^Te^{-A^Tt}W^{-1}(0,T)(e^{-At}\psi(x_0) - x_0)$$

is

$$x(t) = e^{At}\left(x_0 + W(0,t)W^{-1}(0,T)(e^{-AT}\psi(x_0) - x_0)\right)$$

and hence this control steers x_0 to $\psi(x_0)$ at time T. We will use the trajectories defined in this way to define an equivalent closed loop control. The first step involves solving this equation for x_0. Some manipulation of the equation for $x(t)$ gives the implicit equation

$$x_0 = e^{-At}x(t) - W(0,t)W^{-1}(0,T)(e^{-AT}\psi(x_0) - x_0)$$

Because $W(0,T) \geq W(0,t) \geq 0$ for $0 \leq t \leq T$ we see that $||W(0,T)W^{-1}(0,T)|| \leq 1$. Thus under the given hypothesis the contraction mapping theorem shows that for all t between 0 and T this equation can be solved to get $x_0 = \chi(t,x(t))$. The differentiability of χ follows from the assumed differentiability of ψ and the contraction condition. The control

$$u(t,x) = B^Te^{A^Tt}W^{-1}(0,T)(e^{-At}\psi(\chi(t,x(t)) - \chi(t,x(t)))$$

then steers an arbitrary x_0 to $\psi(x_0)$ at time T.

This proof is analogous to the Picard proof for the existence of solutions for ordinary differential equations in that it proceeds through a sequence of small steps involving a contraction to get a final result that does not involve a contraction.

This lemma has several implications. For example, if ψ is the identity then $e^{-AT}\psi(x) - x$ is a contraction for small values of T and so we can map ρ into itself after an arbitrarily small time. This means that it is possible to maintain ρ near any achievable value of ρ for an arbitrarily long period of time. Similarly, if a diffeomorphism ϕ can be factored as $\phi = \psi_1(\psi_2(...\psi_k(\cdot)...))$ with the ψ_i such that $\psi_i(x) - x$ defines a contraction then we can also reach the diffeomorphism ϕ using a piecewise smooth u. Putting these ideas to gether we have the following result.

Theorem 1: Given $T > 0$, a diffeomorphism $\phi : \mathbb{R}^n \to \mathbb{R}^n$ such that ϕ is the composition of differentiable contractions, and given a controllable linear system $\dot{x} = Ax + Bu$ on $\mathbb{R}^n$, there exists a piecewise smooth control $u(t,x)$ such that the solution of the Liouville equation

$$\frac{\partial \rho(t,x)}{\partial t} = -\langle \frac{\partial}{\partial x}, (Ax + Bu)\,\rho(t,x)\rangle \;\; ; \;\; \rho(0,x) = \rho_0(x)$$

takes on the value $\rho_0(\phi^{-1}(x))/\det \mathrm{J}_\phi(\mathrm{x})$ at time T.

Remark 1: Linearity of the evolution equation does not play a critical role here. What is important is that that the there should be a solution to the classical controllability problem that depends smoothly on the initial and final states and that the initial condition for the individual trajectories should be uniquely and smoothly determined by the current value of $x(t)$.

Remark 2: It would be of control theoretic interest to understand the controllability of the Liouville equation under the further restriction that u is not allowed to depend on x in its entirety but only allowed to depend on some noninvertible function of x. This situation will be touched on briefly in section 5. It corresponds to what is called *output feedback*, as opposed to *state feedback*.

4. Linear Systems: Controlling the Mean and Variance

In the previous section we showed that the Liouville equation associated with a controllable linear systems is, itself, controllable in a strong sense using feedback control. The control given, however, depends on an arbitrary diffeomorphism reflecting the complexity of the task. In some circumstances it might not be necessary to shape the entire density but only a few moments. We now show how one can get much more explicit results by restricting attention to the control of the first two moments of the solution of the Liouville equation. More specifically, we show that we can control the mean and variance of a linear system using control laws that are affine in x.

Again we consider a controllable system $\dot{x} = Ax + Bu$ together with its Liouville equation and observe that the first two moments can be characterized in terms of

$$\bar{x} = \int_S \rho(t,x)x\,dx$$

$$\Sigma = \int_S \rho(t,x)(x-\bar{x})(x-\bar{x})^T dx = \int_S \rho(t,x)xx^T dx - \bar{x}\bar{x}^T$$

Making use of the definitions and the equation of evolution with u replaced by $u(t) = K(t)x + v(t)$ we see that

$$\frac{d\bar{x}}{dt} = (A + BK)\,\bar{x} + Bv(t)$$

and

$$\dot{\Sigma} = (A + BK)\,\Sigma + \Sigma\,(A + BK)^T$$

We want to investigate circumstances under which these equations are controllable in the sense that $\bar{x}$ and Σ can be steered to any desired final state consistent with the conditions required by the definitions, $\Sigma(T) = \Sigma^T(T) > 0$. The controllability of the $\bar{x}$ equation follows from the assumption. Now regarding K as a time varying control, the controllability of

$$\dot{\Phi} = (A + BK)\Phi$$

on the space of nonsingular matrices with positive determinant would imply the controllability of the Σ equation on the space of positive definite matrices. This follows from the fact that if the solution of the equation $\dot{\Phi} = F\Phi$ with $\Phi(0) = I$ is $\Phi(T)$ then the solution of

$$\dot{\Sigma} = F\Sigma + \Sigma F^T$$

at time T is $\Phi(T)\Sigma(0)\Phi^T(T)$ and the fact that if Σ_1 and Σ_2 are both positive definite matrices it is possible to find a nonsingular matrix Φ with positive determinant such that $\Phi\Sigma_1\Phi^T = \Sigma_2$.

With this in mind we consider the controllability of $\dot{\Phi} = (A + BK)\Phi$.

Lemma 2: Suppose that (Ω, B) with Ω skew-symmetric is a controllable pair with $e^{\Omega t}$ periodic of period T. Then if $||M|| < 1$ and $\det M > 0$ there exists $K(\cdot)$, a m by n matrix defined on $[0, T]$ such that the solution of the matrix equation

$$\dot{X}(t) = (\Omega + BK(t))X(t) \ ; \ X(0) = I$$

satisfies $X(T) = M$.

Proof: Recall from the previous section the definition of $W(0, t)$ and the inequality valid for $0 < t \leq T$, $W(0, T) \geq W(0, t) > 0$. The linear matrix equation

$$\dot{X}(t) = \Omega X(t) + BB^T e^{\Omega t} W^{-1}(0, T)(M - I) \ ; \ X(0) = I$$

has the solution

$$X(t) = e^{\Omega t}\left(I + W(0, t)W^{-1}(0, T)(M - I)\right)$$

and so $X(T) = M$. Consider $\det X(t) = \det(I + W(0, t)W^{-1}(0, T)(M - I))$. On the interval $[0, T]$ we have $W(0, t) \leq W(0, T)$ and so if $||M - I|| < 1$ we have $||W(0, t)W^{-1}(0, T)(M - I)|| < 1$ and on this interval $X(t)$ is nonsingular. Take K to be

$$K(t) = B^T e^{\Omega t} X^{-1}(t)$$

The following lemma asserts that we can get essentially the same conclusion with significantly weaker assumptions.

Lemma 3: Let T be positive and let (A, B) be a controllable pair. Then if $||M||$ has a positive determinant there exists $K(\cdot)$ defined on $[0, T]$ such that the solution of the matrix equation

$$\dot{X}(t) = (A + BK(t))X(t) \ ; \ X(0) = I$$

satisfies $X(T) = M$.

Proof: Rather than looking at this problem as a control problem involving right invariant vector fields on a Lie group as in [2], we continue along the lines of the previous lemma. Let $T > 0$ be given. It is well known in control theory that under the given hypothesis it is possible to find a real, constant matrix K_0 such that the eigenvalues of $A + BK_0$ equal any given subset of n complex numbers subject only to the constraint that the complex eigenvalues occur in conjugate pairs. We write K as $K_0 + K_1$ with K_0 such that the eigenvalues of $A + BK_0$ are purely imaginary, unrepeated and a multiple of $2\pi/T$. Under this assumption we can find P such that $P(A+K_0)P^{-1} = \Omega$ with (Ω, B) satisfying the assumptions of the lemma. Thus $Y = PXP^{-1}$ satisfies

$$\dot{Y}(t) = (\Omega + PBK(t)P^{-1})Y(t) \;\; ; \;\; Y(0) = I$$

This result, together with the previous discussion allows us to claim the following result on the controllability of the mean and variance associated with a linear system.

Theorem 2: Let T be positive and let (A, B) be a controllable pair. Then for any given m and $\Sigma_1 > 0$ there exists $K(\cdot)$ and v defined on $[0, T]$ such that the solution of

$$\frac{d}{dt}\bar{x}(t) = (A + BK(t))\bar{x}(t) + Bv \;\; ; \;\; x(0) = x_0$$

$$\dot{\Sigma}(t) = (A + BK(t))\Sigma + \Sigma(A + BK(t))^T \;\; ; \;\; \Sigma(0) > 0$$

satisfies $x(T) = m$, $\Sigma(T) = \Sigma_1$.

Remark 3: It is of interest to explore the extent to which the affine control law can be restricted to laws of the form $u = KCx$ with C constant but not invertible. As mentioned above, in the control theory literature this would be called the output feedback case to distinguish it from the state feedback case treated above. In this case we have

$$\frac{d\bar{x}}{dt} = (A + BKC)\,\bar{x} + Bv(t)$$

and

$$\dot{\Sigma} = (A + BKC)\,\Sigma + \Sigma\,(A + BKC)^T$$

and again the controllability of the $\bar{x}$ equation follows from the assumption. The controllability of

$$\dot{\Phi} = (A + BKC)\Phi$$

on the space of nonsingular matrices with positive determinant would imply the controllability of the Σ equation on the space of positive definite matrices but now the previous argument does not apply and it seems we must treat this as a controllability problem on Lie groups, as in [3], and be content with less complete results.

5. Optimal control

Having established the existence of controls that steer the mean and variance appropriately, we now turn to a question involving optimal control. In this section we consider the more general output feedback situation but limit ourselves to the so called normal problems of the maximum principle.

Theorem 3: Given $T > 0$, a controllable pair (A, B) and n by n matrices X_0 and X_f with $\det X_f X_0 > 0$, suppose that the control $K(t)$ steers the system

$$\dot{X} = (A + BKC)X \;\; ; \;\; X(0) = X_0$$

to X_f in T units of time while minimizing

$$\eta = \int_0^T \frac{1}{2}||K^T K||^2 dt$$

Then if the problem is normal

$$K(t) = -B^T M(t) C^T$$

with M satisfying the quadratic, isospectral equation

$$\dot{M} = [M, (A - BB^T MC^T C)^T]$$

Proof: Assume an optimal control exists and the problem is normal. According to the maximum principle there is a matrix P satisfying

$$\dot{P} = -(A + BKC)^T P$$

such that the optimal value of K minimizes $\mathrm{tr}(P^T(A + BKC)X + \frac{1}{2}K^T K)$. Thus the optimal K is given by $K = -B^T PX^T C^T$. A short calculation shows that $M = PX^T$ satisfies the given equation. Any matrix flow of the form $\dot{M} = [M, \psi(M)]$ evolves in such a way that the spectrum of M does not change.

In view of the fact that transpose inverse of X satisfies the same equation as P, there exists a constant matrix R such that $P = (X^{-1})^T R^T$ to get $M^T = XR^{-1}X^{-1}$ which implies the following

Corollary: Under the hypothesis of the theorem, there exists a constant matrix R such that the optimal trajectories satisfy

$$\dot{X} = (A - BB^T(XR^{-1}X^{-1})^T C^T C)X$$

Remark 4: This result can be recast as a result about the variance equation in the following way. Given $T > 0$, a controllable pair (A, B) and n by n symmetric, positive definite matrices Σ_0 and Σ_f, suppose that the control $K(t)$ steers the system

$$\dot{\Sigma} = (A + BKC)\Sigma + \Sigma(A + BKC)^T \ ; \ \Sigma(0) = \Sigma_0$$

to Σ_f in T units of time while minimizing

$$\eta = \int_0^T ||K^T K||^2 dt$$

Then for some value of $M(0)$

$$K(t) = -B^T M(t) C^T$$

with

$$\dot{M} = [M, (A - BB^T MC^T C)^T]$$

We point out that if one applies these ideas directly to the equations for the mean and variance the result changes only slightly. This is the promised generalization of Example 2 of Section 2.

Theorem 4: Given $T > 0$, a controllable linear system $\dot{x} = Ax + Bu$, two symmetric, positive definite matrices $\Sigma(0)$ and Σ_f and vectors $\bar{x}(0), \bar{x}_f$, if the control law (v, K) steers the system

$$\frac{d}{dt}\bar{x} = (A + BKC)\bar{x} + Bv \ ; \ \bar{x}(0) = \bar{x}_0$$

$$\dot{\Sigma} = (A + BKC)\Sigma + \Sigma(A + BKC)^T \ ; \ \Sigma(0) = \Sigma_0$$

to $(\bar{x}_f, \Sigma_f)$ in T units of time while minimizing

$$\eta = \int_0^T \frac{1}{2} v^T v + \mathrm{tr}(\mathrm{K^T K})\mathrm{dt}$$

then

$$K = -B^T(M + qx^T)C^T \ ; \ v = -B^T q(t)$$

with q and M satisfying

$$\dot{q} = -(A + BKC)^T q$$

$$\dot{M} = [M, (A - BB^TMC^TC)^T]$$

6. Generating Elements of Diff $^{(1)}(X)$ with $\dot{x} = f(x, u)$

Theorem one describes a strong controllability property of the Liouville equation associated with a controllable linear system. Its proof uses the fact that we have a convenient explicit expression for a control that executes a given transfer. The development of results applicable to more general controllable systems of the form

$$\dot{x} = f(x) + \sum_{i=1}^{m} g_i(x)u_i = f(x) + G(x)u$$

requires different techniques because no such explicit expression is generally available. However, in this more general setting it is useful to observe that the reachable set for solutions of the Liouville equation can sometimes be better investigated by observing that it is invariant under the transformation $(f(x), G(x)) \mapsto (f(x) + G(x)p(x), G(x)Q(x))$ with $p(x)$ a vector of appropriate smoothness and Q smooth and invertible. This is often described by saying that the reachable set is invariant under the action of the fedback group. (See [4] for elaborations on this theme.) With this in mind we recast the situation described in theorem 1 in terms of vector fields and Lie algebras.

Theorem 5: Let (A, B) be a controllable pair in dimension n. Given a $\mathcal{C}^1$ vector field on $\mathbb{R}^n$,

$$F = f_1(x)\frac{\partial}{\partial x_1} + f_2(x)\frac{\partial}{\partial x_1} + \cdots + f_n(x)\frac{\partial}{\partial x_n}$$

there exists a finite collection of $\mathcal{C}^1$ functions $\phi_1, \phi_2, ..., \phi_m$ such that F belongs to the Lie algebra generated by

$$S = \left\{ \sum_i a_{ij}x_j\frac{\partial}{\partial x_i}, \sum_i b_{ij}\phi_k\frac{\partial}{\partial x_i} \right\}_{k=1,2,\dots,m}$$

Proof: Let B have columns $b_1, b_2, ...b_m$. It is convenient to represent the vector fields as vectors in $\mathbb{R}^n$, i.e., as Ax, b_i etc. Consider the Lie bracket of Ax and $b_i\phi_j$

$$[Ax, b_i\phi_j] = Ab_i\phi_j - b_i\langle d\phi_j, Ax\rangle$$

and so if we include along with each ϕ_i the function $\langle d\phi_i, Ax\rangle$ we see that we have $Ab_i\phi_j$ in S. Likewise, the bracket

$$[Ax, Ab_i\phi_j] = A^2b_i\phi_j - Ab_i\langle d\phi, Ax\rangle$$

so $A^2b\phi_j$ belongs to the Lie algebra, etc. In this way we see that $A^kb\phi_j$ belongs to the Lie algebra for each k. In view of the Cayley-Hamiltonian theorem and the controllability assumption, as k ranges from 0 to $n-1$ the set $A^kb_1, A^kb_2, ...$ contains a basis for $\mathbb{R}^n$ so we have proven the theorem.

This provides another point of view on situation treated in theorem one pointing out that controllable linear systems have a very strong vector field generating property.

References

[1] R. W. Brockett, *Minimum Attention Control*, Proceedings of the 1997 Conference on Decision and Control, IEEE, Piscataway, NJ, 1997, pp. 2628–2632.
[2] Jurgen Moser, *On the Volume Elements of a Manifold.* Trans. American Math. Soc., **120** (1965), 286–294.
[3] R. W. Brockett, *System Theory on Group Manifolds and Coset Spaces.* SIAM Journal on Control, **10** (1972), 265–284.
[4] R. W. Brockett, *Feedback Invariants for Nonlinear Systems.* Proceedings of the 1978 IFAC Congress, Helsinki, Finland, Pergamon Press, Oxford, 1978.

ENGINEERING AND APPLIED SCIENCE, HARVARD UNIVERSITY
E-mail address: brockett@deas.harvard.edu

AMS/IP Studies in Advanced Mathematics
Volume **39**, 2007

Real structures on torus bundles and their deformations

Fabrizio Catanese and Paola Frediani

ABSTRACT. We describe the family of real structures σ on principal holomorphic torus bundles X over tori, and prove its connectedness when the complex dimension is at most three. Hence follows that the differentiable type (more precisely, the orbifold fundamental group) determines the deformation type of the pair (X, σ) provided we have complex dimension at most three, fibre dimension one, and a certain "reality" condition on the fundamental group is satisfied.

1. Introduction

Given an oriented compact differentiable manifold M, a quite general problem is to consider the space $\mathcal{C}(M)$ of compatible complex structures J on M. Its connected components are called "Deformation classes in the large (of complex structures on M)".

A complete answer is known classically for curves, and, by the work of Kodaira, extending Enriques' classification, for special algebraic surfaces.

Since $\mathcal{C}(M)$ is infinite dimensional, one can equivalently consider instead the finite dimensional complex analytic space $\mathfrak{T}(M)$, called Teichmüller space, corresponding to the quotient of $\mathcal{C}(M)$ by the group $Diff^0(M)$ of diffeomorphisms isotopic to the identity. Its local structure is described by the Kuranishi theory ([**Ku65**]).

Inside $\mathfrak{T}(M)$ one has the open set $\mathfrak{K}T(M)$ of complex structures admitting a Kähler metric.

Hironaka showed ([**Hir62**]) that $\mathfrak{K}T(M)$ fails to be a union of connected components of $\mathfrak{T}(M)$.

Later, Sommese showed ([**Somm75**]) that, even in the case of complex tori of dimension ≥ 3, $\mathfrak{T}(M)$ has more than one connected component. It was known to Kodaira and Spencer that in this case $\mathfrak{K}T(M)$ is connected and coincides with the subset of translation invariant complex structures.

1991 *Mathematics Subject Classification.* 14D15, 32G08, 32G13, 32L05, 32Q57, 53C30, 53C56.

Key words and phrases. Real structures, holomorphic torus bundles, deformations.

The present research took place in the framework of the Schwerpunkt "lobale Methoden in der komplexen Geometrie", and of the PRIN 2003: "pazi dei moduli e teoria di Lie" (MURST).

It seems to be more the rule than the exception that $\mathfrak{T}(M)$ has a lot of connected components. For instance, already for surfaces of general type it was shown by Manetti ([**Man01**]), and by Catanese and Wajnryb in the simply connected case ([**C-W04**]), that the number of connected components of $\mathfrak{T}(M)$ may be arbitrarily large.

The above work shows also that perhaps a more approachable version of the general problem is the following:

QUESTION : given a complex structure $X := (M, J)$, determine the connected component $\mathcal{C}^0(X)$ of $\mathcal{C}(M)$ containing it. I.e., given a complex manifold X, find its deformations in the large.

In [**Cat02**] (cf. also [**Cat04**]) it was shown that every deformation in the large of a complex torus is a complex torus. This was used to determine the stability by deformations in the large of holomorphic torus bundles over curves of genus at least two ([**Cat04**]), and to obtain the same result ([**C-F05**]) for threefolds which are holomorphic torus bundles over two dimensional tori, provided that the fundamental group satisfies a suitable 'reality' condition.

Consider now, as customary, a real manifold as a pair (X, σ), where X is a compact complex manifold and σ is a antiholomorphic involution.

DEFINITION 1.1. *(X, σ) is said to be* **simple** *if, once we fix the differentiable type of the pair (X, σ), then we have a unique deformation class.*

I.e., M being the oriented differentiable manifold underlying X, simplicity holds iff the space $\mathcal{R}(M, \sigma)$ of compatible complex structures (which make σ antiholomorphic) is connected.

(X, σ) is said to be **quasi- simple** *if the above holds when we restrict ourselves to complex structures which are deformations of X (cf. [**DK02**]). I.e., (X, σ) is quasi- simple iff $\mathcal{R}(M, \sigma) \cap \mathcal{C}^0(X)$ is connected.*

*More generally, as pointed out to us by Itenberg (cf. [**DIK04**]), one can pose the problem of quasi-simplicity for the more general case of a finite subgroup G of the Klein group of dianalytic (i.e., biholomorphic or antibiholomorphic) automorphisms of X ($G \cong \mathbb{Z}/2$ in the previous case).*

A basic question here is: which real manifolds are simple? Which ones are quasi simple?

The answer is positive for curves, by work of Seppäla and Silhol ([**S-S89**]), and also for many surfaces of special type, thanks to work of many authors, like Comessatti, Silhol, Nikulin, Degtyarev, Itenberg, Kharlamov, Catanese-Frediani, Welschinger. These works (cf. references) give evidence to the following

CONJECTURE SSS: Special surfaces are simple.

Note that, by contrast, for surfaces of general type, not only simplicity (compare [**Man01**], [**C-W04**]), but even quasi-simplicity fails to hold (see [**K-K05**]).

Motivated by this conjecture, in [**C-F03**] we introduced the notion of orbifold fundamental group in order to prove the simplicity of real hyperelliptic surfaces and in order to achieve a complete classification of them (there are 78 types).

The orbifold fundamental group of a real variety (X, σ) is an exact sequence

$$1 \to \pi_1(X) \to \Pi_\sigma := \pi_1^{orb}(X) \to \mathbb{Z}/2 \to 1,$$

where, if the action of σ is fixpoint free, Π_σ is the fundamental group of the quotient X/σ.

We are mainly interested here in the special case where X is a $K(\pi, 1)$ and more precisely where the universal cover $\tilde{X}$ of X is contractible. In this case we have

$$X = \tilde{X}/\pi_1(X), \ X/\sigma = \tilde{X}/\Pi_\sigma \ .$$

In the even more special case where $\tilde{X}$ has no moduli, the simplicity or quasi simplicity question is translated into the question whether the conjugacy class of the embedding $\Pi_\sigma \to Dian(\tilde{X})$ is parametrized by a connected variety.

Now, the cases which we treated already, real hyperelliptic surfaces and Kodaira surfaces ([**C-F03**] , [**F04**]) are both torus bundles over tori, whose study is reduced ([**Cat04**]) to the study of P.H.T.B.T. = principal holomorphic torus bundles over tori. The basic question which is of interest to us is to determine when a real P.H.T.B.T. is simple.

One could consider in greater generality the following situation: we have a nilpotent group Π, such that the factors of the central series are torsion free abelian of even rank.

Then the real Lie group $\Pi \otimes \mathbb{R}$ contains Π as a cocompact lattice (i.e., as a discrete subgroup with compact quotient $M := (\Pi \otimes \mathbb{R})/\Pi$) and we consider the differentiable manifold $M := (\Pi \otimes \mathbb{R})/\Pi$.

Already the case of tori has warned us that there could be too many exotic complex structures, so we restrict ourselves to consider only the right invariant complex structures on the homogeneous space M.

Obviously, the right invariant almost complex structures J correspond exactly to the complex structures on the tangent space at the origin, but there remains to write down explicitly the integrability conditions for a given almost complex structure J.

This is rather simple in the case of P.H.T.B.T. , treated in [**Cat04**] and in [**C-F05**]: we have an analogue of the Riemann bilinear relations (classically known for line bundles on complex tori) which are easy to explain as follows (cf. the next section for more details).

Consider the central extension of fundamental groups given by the homotopy exact sequence of the fibre bundle $f : X \to Y$:

$$1 \to \pi_1(T) := \Lambda \to \Pi := \pi_1(X) \to \pi_1(Y) := \Gamma \to 1.$$

The central extension is classified by an element

$$A \in H^2(\Gamma, \Lambda) \cong \Lambda^2(\Gamma^\vee) \otimes \Lambda.$$

As differentiable manifolds the base $Y = (\Gamma \otimes \mathbb{R})/\Gamma$, and the fibre $T = (\Lambda \otimes \mathbb{R})/\Lambda$, acquire a varying complex structure by writing a Hodge decomposition

$$\Gamma \otimes \mathbb{C} = V \oplus \bar{V}, \ \ \Lambda \otimes \mathbb{C} = U \oplus \bar{U}.$$

The Riemann bilinear relations (which are equivalent to the integrability of the given almost complex structure on $M := (\Pi \otimes \mathbb{R})/\Pi$) amount to saying that the component of A in $\Lambda^2(V^\vee) \otimes \bar{U}$ is zero. These equations characterize then the so-called Appell-Humbert family of such bundles, which parametrizes all the right invariant complex structures on $M := (\Pi \otimes \mathbb{R})/\Pi$.

The Riemann relations allow a decomposition

$$A = B + \overline{B}, \ \ B = B' + B'', \ B' \in \Lambda^2(V)^\vee \otimes U, \ B'' \in (V^\vee \otimes \overline{V})^\vee \otimes U.$$

B' is called the holomorphic part, B'' is called the Hermitian part, and this decomposition reduces to questions of multilinear algebra the description of the spaces of holomorphic forms on X, of the subspace of closed holomorphic forms, and the explanation of an important phenomenon discovered by Nakamura ([**Nak75**]), namely the fact that local deformations of a parallelizable manifold need not be parallelizable.

One has for instance the following

THEOREM 0, [**Cat04**]. *X is parallelizable if and only if $B'' = 0$.*

The Riemann bilinear relations define a parameter space which parametrizes the complex structures in the Appell-Humbert family, and which is denoted by $\mathcal{T}'\mathcal{B}_A$, since it parametrizes torus bundles.

When the dimension of X is small one can ask whether the Appell-Humbert family yields a connected component of $\mathcal{C}((\Pi \otimes \mathbb{R})/\Pi)$. A first result in this direction is given by the following

THEOREM A, [**Cat04**]. *Assume that $d = 1$ and $dim(ImA) = 2$. Then the Appell-Humbert family yields a closed subset of $\mathcal{C}((\Pi \otimes \mathbb{R})/\Pi)$.*

One sees moreover that the parameter space $\mathcal{T}'\mathcal{B}_A$ is smooth if $m = dim_{\mathbb{C}}(Y) \leq 2$, $d = dim_{\mathbb{C}}(T) = 1$ (and it is singular already for $d = 1, m \geq 3$, at the points where $B'' = 0$). The smoothness of $\mathcal{T}'\mathcal{B}_A$ makes it possible to analyse the versality of the (complete) Appell-Humbert family and to prove it ([**C-F05**]) under the following

"Reality" condition for A in the case $d = 1$: choose a basis e_1, e_2 of Λ and write $A = A_1 \otimes e_1 + A_2 \otimes e_2$. Then we want a real solution for the equation

$$Pfaffian(\lambda_1 A_1 + \lambda_2 A_2) = 0.$$

THEOREM B, [**C-F05**]. *Assume $d = 1, m = 2$ and that*

i) A is non degenerate

ii) $dim(ImA) = 2$

iii) A satisfies the "reality" condition.

Then the Appell-Humbert family yields an open subset of $\mathcal{C}((\Pi \otimes \mathbb{R})/\Pi)$.

Combining Theorem A with Theorem B we find that, under the hypotheses of Theorem B, the Appell-Humbert family yields a connected component of $\mathcal{C}((\Pi \otimes \mathbb{R})/\Pi)$.

In this paper we describe completely, via explicit equations, the family which parametrizes real structures on a torus bundle in the Appell-Humbert family and then we show the connectivity of the family in special cases.

The following Theorems C and D are the main results of the paper, the first one is formulated here in a less technical form

THEOREM C. *Assume again $d = 1, m = 2$, that A is nondegenerate, and fix the orbifold fundamental group Π_σ. Then the real structures for torus bundles in the Appell-Humbert family are parametrized by a connected family.*

We obtain therefore as a corollary

THEOREM D. *Same assumptions as in theorem B: $d = 1, m = 2$, and i), ii), iii) are satisfied.*

Then simplicity holds for real torus bundles in the Appell-Humbert family.

The paper is organized as follows: in section 2 we recall the theory developed in [**Cat04**] and [**C-F05**] concerning principal holomorphic torus bundles over tori, while section 3 is devoted to recalling the results already established for the Appell-Humbert family.

The bulk of the paper is section 4, which studies the real structures on torus bundles in the Appell-Humbert family. Strangely enough, the extra symmetries coming from the existence of a real structure makes calculations somewhat easier.

In fact, the real structure allows splittings

$$\Gamma \otimes \mathbb{R} = V^+ \oplus V^- \ , \ \Lambda \otimes \mathbb{R} = U^+ \oplus U^-,$$

and, after we show that the orbifold fundamental group Π_σ has a dianalytic affine representation in the complex vector space $(V \otimes \mathbb{C}) \oplus (U \otimes \mathbb{C})$, we can describe the complex structures in the Appell-Humbert family which are compatible with the real structure as a pair of linear maps

$$B_2 : V^- \to V^+ \ , \ B_1 : U^- \to U^+ \ ,$$

satisfying explicit second degree matrix equations.

To illustrate the power of this way of calculating, we give a short new proof of the second author's result that simplicity holds for Kodaira surfaces.

Then we proceed to the case of threefolds, obtaining our main results (Theorems C and D).

2. Principal holomorphic torus bundles over tori: generalities

Throughout the paper, our set up will be the following: we have a holomorphic submersion between compact complex manifolds

$$f : X \to Y,$$

such that the base Y is a complex torus, and one fibre F (whence all the fibres, by theorem 2.1 of [**Cat04**]) is also a complex torus.

We shall denote this general situation by saying that f is differentiably a torus bundle.

We let $n = dimX$, $m = dimY$, $d = dimF = n - m$.

In general (cf.[**FG65**])) f is a holomorphic bundle if and only if all the fibres are biholomorphic.

This holds necessarily in the special case $d = 1$, because the moduli space for 1-dimensional complex tori exists and is isomorphic to $\mathbb{C}$.

Assume now more specifically that we have a holomorphic torus fibre bundle, thus we have (cf. [**Cat04**], pages 271-273) the exact sequence of holomorphic vector bundles

$$0 \to \Omega^1_Y \to f_* \Omega^1_X \to f_* \Omega^1_{X|Y} \to 0.$$

We have a principal holomorphic bundle if moreover $f_* \Omega^1_{X|Y}$ is a trivial holomorphic bundle.

REMARK 2.1. *In general (cf. e.g. [**BPV84**]) if T is a complex torus, we have an exact sequence of complex Lie groups*

$$0 \to T \to Aut(T) \to M \to 1$$

where M is discrete. Taking sheaves of germs of holomorphic maps with source Y we get

$$0 \to \mathcal{H}(T)_Y \to \mathcal{H}(Aut(T))_Y \to M \to 1$$

hence the exact sequence

$$0 \to H^1(Y, \mathcal{H}(T)_Y) \to H^1(Y, \mathcal{H}(Aut(T)))_Y \to H^1(Y, M) \to H^2(Y, \mathcal{H}(T)_Y).$$

Since holomorphic bundles with base Y and fibre T are classified by the cohomology group $H^1(Y, \mathcal{H}(Aut(T))_Y)$, $H^1(Y, M)$ determines the discrete obstruction for a holomorphic bundle to be a principal holomorphic bundle.

In view of this, the study of holomorphic torus bundles is reduced to the study of principal holomorphic torus bundles.

In the case of a principal holomorphic bundle we write $\Lambda := \pi_1(T)$, $\Gamma := \pi_1(Y)$ and the exact sequence

$$\to H^0(\mathcal{H}(T)_Y) \to H^1(Y, \Lambda) \to H^1(Y, \mathcal{O}_Y^d) \to H^1(\mathcal{H}(T)_Y) \to^c \to H^2(Y, \Lambda)$$

determines a cohomology class $\epsilon \in H^2(Y, \Lambda)$ which classifies the central extension

$$1 \to \pi_1(T) = \Lambda \to \Pi := \pi_1(X) \to \pi_1(Y) = \Gamma \to 1$$

(it is central by the triviality of the monodromy automorphism).

PROPOSITION 2.2. [**C-F05**] *Let $f : X \to Y$ be a principal holomorphic torus bundle over a torus as above.*

Then the universal covering of X is isomorphic to $\mathbb{C}^{m+d}$ and X is biholomorphic to a quotient $X \cong \mathbb{C}^{m+d}/\Pi$.

Let us briefly recall the classical way to look at the family $\mathcal{T}_m$ of complex tori of complex dimension $= m$. We fix a lattice Γ of rank $2m$, and we look at the complex (m dimensional) subspaces $V \subset \Gamma \otimes \mathbb{C}$ such that $V \oplus \bar{V} = \Gamma \otimes \mathbb{C}$: to V corresponds the complex torus $Y_V := (\Gamma \otimes \mathbb{C})/(\Gamma \oplus \bar{V})$. We finally select one of the two resulting connected components by requiring that the complex orientation of V induces on $\Gamma \cong p_V(\Gamma)$ a fixed orientation.

We consider similarly the complex tori $T_U := (\Lambda \otimes \mathbb{C})/(\Lambda \oplus \bar{U})$ which can occur as fibres of f.

Consider now our principal holomorphic torus bundle $f : X \to Y$ over a complex torus Y_V of dimension m, and with fibre a complex torus T_U of dimension d and let $\epsilon \in H^2(Y, \Lambda) = H^2(\Gamma, \Lambda)$ be the cohomology class classifying the central extension

$$1 \to \Lambda \to \Pi \to \Gamma \to 1. \tag{2.1}$$

We reproduce an important result from [**C-F05**]

LEMMA 2.3. *It is possible to "tensor" the above exact sequence with $\mathbb{R}$, obtaining an exact sequence of Lie Groups*

$$1 \to \Lambda \otimes \mathbb{R} \to \Pi \otimes \mathbb{R} \to \Gamma \otimes \mathbb{R} \to 1$$

such that Π is a discrete subgroup of $\Pi \otimes \mathbb{R}$ and such that X is diffeomorphic to the quotient manifold

$$M := \Pi \otimes \mathbb{R}/\Pi.$$

PROOF. Consider, as usual, the map

$$A : \Gamma \times \Gamma \to \Lambda,$$

$$A(\gamma, \gamma') = [\hat{\gamma}, \hat{\gamma'}] = \hat{\gamma}\hat{\gamma'}(\hat{\gamma})^{-1}(\hat{\gamma'})^{-1},$$

where $\hat{\gamma}$ and $\hat{\gamma'}$ are respective liftings to Π of elements $\gamma, \gamma' \in \Gamma$. We observe that since the extension (2.1) is central, the definition of A does not depend on the choice of the liftings of γ, resp. γ' to Π.

As it is well known, A is bilinear and alternating, so A yields a cocycle in $H^2(\Gamma, \Lambda)$ which "classifies" the central extension (2.1). Let us review how does this more precisely hold.

Assume that $\{\gamma_1, ..., \gamma_{2m}\}$ is a basis of Γ and choose fixed liftings $\hat{\gamma}_i$ of γ_i in Π, for each $i = 1, \ldots, 2m$. Then automatically we have determined a canonical way to lift elements $\gamma \in \Gamma$ through:

$$\gamma = \gamma_1^{n_1} \ldots \gamma_{2m}^{n_{2m}} \mapsto \hat{\gamma} := \hat{\gamma}_1^{n_1} ... \hat{\gamma}_{2m}^{n_{2m}}.$$

Hence a canonical way to write the elements of Π as products $\lambda\hat{\gamma}$, where $\lambda \in \Lambda$ and $\hat{\gamma}$ is as above.

Since $\forall i, j$, one has

$$\hat{\gamma}_i \hat{\gamma}_j = A(\gamma_i, \gamma_j) \hat{\gamma}_j \hat{\gamma}_i,$$

we have a standard way of computing the products $(\lambda\hat{\gamma})(\lambda'\hat{\gamma}')$ as $\lambda''(\widehat{(\gamma\gamma')})$, where λ'' will be computed using A.

We can also view Π as a group of affine transformations of $(\Lambda \otimes \mathbb{R}) \oplus (\Gamma \otimes \mathbb{R})$. In fact, $(\Lambda \otimes \mathbb{R}) \oplus (\Gamma \otimes \mathbb{R})$ is a real vector space with basis $\{\lambda_1, ..., \lambda_{2d}, \gamma_1, ..., \gamma_{2m}\}$ where $\{\lambda_1, ..., \lambda_{2d}\}$ is a basis of Λ and the action of Π on $(\Lambda \otimes \mathbb{R}) \oplus (\Gamma \otimes \mathbb{R})$ is given as follows:

λ_i acts on $(\Lambda \otimes \mathbb{R}) \oplus (\Gamma \otimes \mathbb{R})$ sending (y, x) to $(y + \lambda_i, x)$, while the action of $\hat{\gamma}_j$ is defined using the multiplication $(\lambda\hat{\gamma}) \mapsto (\lambda\hat{\gamma})\hat{\gamma}_j$.

More precisely if $y \in \Lambda \otimes \mathbb{R}$, $x = \sum x_j \gamma_j \in \Gamma \otimes \mathbb{R}$, $\gamma' = \sum \nu_h \gamma_h \in \Gamma$, we have

$$(y, x)\hat{\gamma'} := (y + \phi_{\gamma'}(x), x + \gamma'),$$

where

$$\phi_{\gamma'}(x) = \sum_{j \geq h} x_j \nu_h A(\gamma_j, \gamma_h) = \sum_{j \geq h} x_j A_{jh} \nu_h = {}^t x T^- \gamma',$$

where T^- is the lower triangular part of the matrix A, so that we can write $A = T^- - {}^t T^-$.

Therefore we can endow $(\Lambda \otimes \mathbb{R}) \oplus (\Gamma \otimes \mathbb{R}) =: \Pi \otimes \mathbb{R}$ with a Lie group structure defined by

$$(y, x)(y', x') = (y + y' + T^-(x, x'), x + x'),$$

and the quotient $(\Pi \otimes \mathbb{R})/\Pi$ of this Lie group by the discrete subgroup Π is immediately seen to be diffeomorphic to X. □

REMARK 2.4. *We can change coordinates in* $(\Lambda \otimes \mathbb{R}) \oplus (\Gamma \otimes \mathbb{R})$ *in such a way that the action of the set* $\hat{\Gamma} \cong \Gamma \subset \Gamma \otimes \mathbb{R}$ *on* $\Pi \otimes \mathbb{R}$ *is given by*

$$(y, x)\hat{\gamma} = (y + A(x, \gamma) + 2S(\gamma, \gamma), x + \gamma),$$

where $S(\gamma, \gamma')$ *is a symmetric bilinear* $(\frac{1}{4}\Lambda)$*- valued form, and* $2S(\gamma, \gamma) \in \Lambda$.

PROOF. Let us define the symmetric form $S := -\frac{T^- + {}^t T^-}{4}$, so that $T^- + 2S = \frac{T^- - {}^t T^-}{2} = \frac{A}{2}$.

Consider the map $\Psi : (\Lambda \otimes \mathbb{R}) \oplus (\Gamma \otimes \mathbb{R}) \to (\Lambda \otimes \mathbb{R}) \oplus (\Gamma \otimes \mathbb{R})$ defined by $\Psi(y, x) = (2(y + S(x, x)), x) =: (\eta, x)$.

Then $\forall \hat{\gamma} \in \hat{\Gamma}$ we have an induced action

$$(\eta, x)\hat{\gamma} = \Psi((y, x)\hat{\gamma}) = \Psi(y + T^-(x, \gamma), x + \gamma) =$$

$$(2y + 2T^-(x, \gamma) + 2S(x + \gamma, x + \gamma), x + \gamma) = (\eta + A(x, \gamma) + 2S(\gamma, \gamma), x + \gamma),$$

and we conclude observing that $2S(\gamma,\gamma) = T^-(\gamma,\gamma) \in \Lambda$.

□

We recall from [**Cat04**] the First Riemann bilinear Relation: it is derived from the exact cohomology sequence

$$H^1(Y, \mathcal{O}_Y \otimes U) \cong H^1(Y, \mathcal{H}(U)_Y) \to H^1(\mathcal{H}(T)_Y) \to^c \to H^2(Y,\Lambda) \to H^2(Y, \mathcal{H}(U)_Y)$$

and says that the class ϵ maps to zero in $H^2(Y, \mathcal{H}(U)_Y)$. More concretely the

First Riemann Relation for principal holomorphic Torus Bundles

is expressed as follows:

Let $A : \Gamma \times \Gamma \to \Lambda$ be the alternating bilinear map representing the cohomology class ϵ : then

$$A \in \Lambda^2(\Gamma \otimes \mathbb{R})^\vee \otimes (\Lambda \otimes \mathbb{R}) \subset \Lambda^2(\Gamma \otimes \mathbb{C})^\vee \otimes (\Lambda \otimes \mathbb{C}) \subset \Lambda^2(V \oplus \bar{V})^\vee \otimes (U \oplus \bar{U}),$$

satisfies the property that its component in $\Lambda^2(\bar{V})^\vee \otimes (U)$ is zero.

It is important, for the forthcoming calculations, to understand in detail the bilinear algebra underlying the Riemann bilinear relation.

We observe preliminarly that one has a natural isomorphism $\Lambda^2(V \oplus \overline{V})^\vee \cong \Lambda^2(V)^\vee \oplus (V^\vee \otimes \overline{V})^\vee \oplus \Lambda^2(\overline{V})^\vee$, where the middle summand embeds by the wedge product : $w' \otimes \bar{w} \mapsto 2w' \wedge \bar{w} = w' \otimes \bar{w} - \bar{w} \otimes w'$.

Consider the alternating bilinear form

$$A \in \Lambda^2(\Gamma \otimes \mathbb{C})^\vee \otimes (\Lambda \otimes \mathbb{C}) = \Lambda^2(V \oplus \overline{V})^\vee \otimes (U \oplus \overline{U}),$$

satisfying the first bilinear relation and let us write

$$A = B + \overline{B},$$

where $B \in \Lambda^2(\Gamma \otimes \mathbb{C})^\vee \otimes U$, and $\overline{B} \in \Lambda^2(\Gamma \otimes \mathbb{C})^\vee \otimes \overline{U}$.

By the first bilinear relation $B = B' + B''$, with $B' \in \Lambda^2(V)^\vee \otimes U$, $B'' \in (V^\vee \otimes \overline{V})^\vee \otimes U$.

Concretely, $A = B' + B'' + \overline{B'} + \overline{B''}$, where B' is an alternating complex bilinear form. The fact that B'' is alternating reads out as:

$$B''(v', \bar{v}) = -B''(\bar{v}, v') \ \forall v, v' \in V$$

whereas conjugation of tensors reads out as:

$$\bar{B}(\bar{x}, \bar{y}) = \overline{B(x,y)} \ \forall x, y \Rightarrow \bar{B''}(\bar{v}, v') = \overline{B''(v, \bar{v'})}.$$

3. Appell-Humbert families

We shall recall in this section the definition of the family given by the pairs of subspaces satisfying the Riemann bilinear relations and some related results obtained in [**Cat04**], [**C-F05**].

DEFINITION 3.1. *Given A as above, we define $\mathcal{TB}_A$ as the subset of the product of Grassmann Manifolds $Gr(m, 2m) \times Gr(d, 2d)$ defined by*

$$\mathcal{TB}_A = \{(V, U) \in Gr(m, 2m) \times Gr(d, 2d) \mid V \cap \overline{V} = (0),$$
$$U \cap \overline{U} = (0), \mid \text{the component of } A \text{ in } \Lambda^2(\overline{V})^\vee \otimes U \text{ is } = 0\}.$$

This complex space is called the Appell-Humbert space of Torus Bundles.

Since this space is not connected, we restrict ourselves to its intersection with $\mathcal{T}_m \times \mathcal{T}_d$, i.e., we take two fixed orientations of Λ, resp. Γ, and consider pairs of complex structures which have the same orientation as the fixed ones.

One sees immediately that $\mathcal{TB}_A$ is a complex analytic variety of codimension at most $dm(m-1)/2$.

Note however that, for $d \geq 3, m >> 0$ we get a negative expected dimension. The structure of these complex spaces should be investigated in general, for our present purposes we recall from [**C-F05**] the following

LEMMA 3.2. *([**C-F05**]) If $d = 1$ then the open set $\mathcal{TB}_A \cap (\mathcal{T}_m \times \mathcal{T}_d)$ is connected.*

DEFINITION 3.3. *The standard (Appell-Humbert) family of torus bundles parametrized by $\mathcal{TB}_A$ is the family of principal holomorphic torus bundles $X_{V,U}$ over $Y := Y_V$ and with fibre $T := T_U$ determined by the cocycle in $H^1(Y, \mathcal{H}(T)_Y)$ obtained by taking $f_\gamma(v)$ which is the class modulo Λ of*

$$F_\gamma(v) := B'(v, p_V(\gamma)) + 2B''(v, \overline{p_V(\gamma)}) + B''(p_V(\gamma), \overline{p_V(\gamma)}), \forall v \in V.$$

In other words, $X_{V,U}$ is the quotient of $T_U \times V$ by the action of Γ such that

$$\gamma([u], v) = ([u + B'(v, p_V(\gamma)) + 2B''(v, \overline{p_V(\gamma)}) + B''(p_V(\gamma), \overline{p_V(\gamma)})], v + p_V(\gamma)).$$

REMARK 3.4. *As observed in [**C-F05**] the above formula was a correction of the formula given in Definition 6.4 of [**Cat04**], where an identification of $\Gamma \otimes \mathbb{R}$ with V was used, and thus $A(z, \gamma)$ was identified with $B(z, \gamma)$. In the latter formula one had thus*
$-B'(v, \gamma) - B''(v, \overline{p_V(\gamma)})$ *instead of*
$B'(v, \gamma) + 2B''(v, \overline{p_V(\gamma)}) + B''(p_V(\gamma), \overline{p_V(\gamma)})$.

We also recall from [**Cat04**] the definition of the complete Appell-Humbert space.

DEFINITION 3.5. *Given A as above we define*

$$\mathcal{T}'\mathcal{B}_A = \{(V, U, \phi) \mid (V, U) \in \mathcal{TB}_A, \ \phi \in H^1(Y_V, \mathcal{H}(U)_{Y_V}) \cong \overline{V}^\vee \otimes U\}.$$

The complete Appell-Humbert family of torus bundles parametrized by $\mathcal{T}'\mathcal{B}_A$ is the family of principal holomorphic torus bundles $X_{V,U,\phi}$ on $Y := Y_V$ and with fibre $T := T_U$ determined by the cocycle in $H^1(Y, \mathcal{H}(T)_Y)$ obtained by taking the sum of $f_\gamma(z)$ with the cocycle $\phi \in H^1(Y_V, \mathcal{H}(U)_{Y_V}) \cong H^1(Y, \mathcal{O}_Y^d)$.

Finally we have the following

THEOREM 3.6. [**Cat04**]
Any principal holomorphic torus bundle with extension class isomorphic to $\epsilon \in H^2(\Gamma, \Lambda)$ occurs in the complete Appell-Humbert family $\mathcal{T}'\mathcal{B}_A$.

We recall from [**C-F05**] also the following

PROPOSITION 3.7. [**C-F05**] *Let $A : \Gamma \times \Gamma \to \Lambda$ be non zero.*
If $m = 2, d = 1$, i.e., $\Gamma \cong \mathbb{Z}^4$, $\Lambda \cong \mathbb{Z}^2$, both Appell - Humbert spaces $\mathcal{TB}_A$ and $\mathcal{T}'\mathcal{B}_A$ are smooth.
If $d = 1$ and $m \geq 3$, $\mathcal{TB}_A$ is singular at the points where $B'' = 0$.

4. Symmetries of principal holomorphic torus bundles over tori

Let us now assume that $f : X \to Y$ is a real principal holomorphic torus bundle over a torus, and let $\sigma : X \to X$ be an antiholomorphic involution. We assume as above that $Y = Y_V$, $V \cong \mathbb{C}^m$, while the fibre of f is $T = T_U$, with $U \cong \mathbb{C}^d$. In 2.2 we have seen that the universal covering of X is isomorphic to $V \oplus U \cong \mathbb{C}^{m+d}$ and we can find a lifting $\tilde{\sigma}$ of σ to the universal covering $U \oplus V \cong \mathbb{C}^{m+d}$.

PROPOSITION 4.1. *Assume that the alternating form $A : \Gamma \times \Gamma \to \Lambda$ is nondegenerate, or equivalently that $\Lambda = Z(\Pi)$. Let $\tilde{\sigma}$ be a lifting of σ to $U \oplus V \cong \mathbb{C}^{m+d}$: then $\tilde{\sigma}$ is an affine transformation.*

PROOF. Since $\tilde{\sigma}$ is a lifting of σ to the universal covering, $\tilde{\sigma}$ acts by conjugation on the fundamental group Π of X, therefore it also acts on the centre Λ of Π, because it is characteristic.

Hence for every $\lambda \in \Lambda$, there exists a $\lambda' \in \Lambda$ such that

$$\tilde{\sigma}(u + p_U(\lambda), v) = \tilde{\sigma}(u, v) + p_U(\lambda'),$$

where $p_U : U \oplus \bar{U} \to U$ denotes as usual the projection on the first factor. Assume that $\tilde{\sigma} : U \oplus V \to U \oplus V$ is given by $\tilde{\sigma}(u, v) = (\sigma_1(u, v), \sigma_2(u, v))$. Then we must have

$$(\sigma_1(u + p_U(\lambda), v), \sigma_2(u + p_U(\lambda), v)) = (\sigma_1(u, v) + p_U(\lambda'), \sigma_2(u, v)),$$

therefore σ_2 is constant as a function of u, and we may write $\sigma_2(u, v) = o_2(v)$.

Looking at the first component we obtain $\sigma_1(u + p_U(\lambda), v) = \sigma_1(u, v) + p_U(\lambda')$, so σ_1 is affine antiholomorphic in u and we may write

$$\sigma_1(u, v) = A_1(v)\bar{u} + c_1(v) \tag{4.1}$$

where $A_1(v)$ is a linear map depending antiholomorphically on v (we may think of it as a $(d \times d)$ matrix once we fix a basis for U).

Recall that, for every $\gamma \in \Gamma$, and for any lift to $U \oplus V$ of the action of γ on $T \times V$, there exists $\lambda' \in \Lambda$ such that

$$\gamma(u, v) = (u + F_\gamma(v) + p_U(\lambda'), v + p_V(\gamma)),\ \forall (u, v) \in U \oplus V,$$

where $F_\gamma(v)$ is as in definition (3.3) and $p_V : V \oplus \bar{V} \to V$ is the first projection. For all $\gamma \in \Gamma$ there must therefore exist $\gamma' \in \Gamma$, and a $\lambda'' \in \Lambda$ such that $\tilde{\sigma} \circ \gamma = \lambda''\gamma' \circ \tilde{\sigma}$. Hence we have

$$\begin{aligned} \tilde{\sigma} \circ \gamma(u, v) &= \tilde{\sigma}(u + F_\gamma(v) + p_U(\lambda'), v + p_V(\gamma)) = \\ &= (\sigma_1(u + F_\gamma(v) + p_U(\lambda'), v + p_V(\gamma)), \sigma_2(v + p_V(\gamma))) = \\ &= \lambda''\gamma'(\tilde{\sigma}(u, v)) = (\sigma_1(u, v) + F_{\gamma'}(\sigma_2(v)) + p_U(\lambda'''), \sigma_2(v) + p_V(\gamma')), \end{aligned}$$

where $\lambda''' \in \Lambda$. Therefore we obtain that σ_2 is affine antiholomorphic,

$$\sigma_2(v) = A_2\bar{v} + d_2,$$

where A_2 is a linear map (a $(m \times m)$ matrix if we fix a basis of V) and we have

$$A_2\bar{v} + A_2\overline{p_V(\gamma)} = A_2\bar{v} + p_V(\gamma'),$$

so that

$$A_2\overline{p_V(\gamma)} = p_V(\gamma').$$

Looking at the first component we have

$$\sigma_1(u + F_\gamma(v) + p_U(\lambda'), v + p_V(\gamma)) = \sigma_1(u, v) + F_{\gamma'}(A_2\bar{v} + d_2) + p_U(\lambda''').$$

Now using (4.1) we have

$$(4.2)\quad A_1(v+p_V(\gamma))(\bar{u}+\overline{F_\gamma(v)}+\overline{p_U(\lambda')})+c_1(v+p_V(\gamma)) = \\ = A_1(v)\bar{u}+c_1(v)+F_{\gamma'}(A_2\bar{v})+F_{\gamma'}(d_2)+p_U(\lambda''').$$

So by derivation with respect to the variables $\bar{u}_i{}'s$ we obtain

$$A_1(v+p_V(\gamma)) = A_1(v),$$

for all $v \in V$, for all $\gamma \in \Gamma$, therefore A_1 is constant in v and we can write $A_1(v) = A_1$. Now (4.2) becomes

(4.3)

$$c_1(v+p_V(\gamma))-c_1(v) = F_{A_2\overline{p_V(\gamma)}}(A_2\bar{v})-A_1\overline{F_\gamma(v)}+F_{A_2\overline{p_V(\gamma)}}(d_2)+p_U(\lambda''')-A_1\overline{p_U(\lambda')}$$

and derivation with respect to the variables $\bar{v}_j's$ yields the vanishing of the derivatives of

$$c_1(v+p_V(\gamma))-c_1(v),\ \forall v \in V,\ \forall \gamma \in \Gamma,$$

so these derivatives are constant and

$$c_1(v) = q(v,v)+l(v)+d_1,$$

where $q(v,v)$ is quadratic in v, $l(v)$ is linear in v, and d_1 is a constant.

Now (4.3) gives

$$q(v+p_V(\gamma),v+p_V(\gamma))+l(v)+l(p_V(\gamma))+d_1-q(v,v)-l(v)-d_1 = \\ = F_{A_2\overline{p_V(\gamma)}}(A_2\bar{v})-A_1\overline{F_\gamma(v)}+F_{A_2\overline{p_V(\gamma)}}(d_2)+p_U(\lambda''')-A_1\overline{p_U(\lambda')},$$

so

$$q(p_V(\gamma),p_V(\gamma))+2q(v,p_V(\gamma))+l(p_V(\gamma)) = \\ = F_{A_2\overline{p_V(\gamma)}}(A_2\bar{v})-A_1\overline{F_\gamma(v)}+F_{A_2\overline{p_V(\gamma)}}(d_2)+p_U(\lambda''')-A_1\overline{p_U(\lambda')}.$$

Now, by looking in the above expression at the variable γ we immediately get $q(p_V(\gamma),p_V(\gamma)) = 0$, since it is the only quadratic term (substitute γ with $m\gamma$ and look at the asymptotic growth). Since $q(p_V(\gamma),p_V(\gamma)) = 0$ for all $\gamma \in \Gamma$ and $p_V(\Gamma)$ is a lattice in V, we must have $q = 0$ and $c_1(v) = l(v)+d_1 = L\bar{v}+d_1$, where L is a $(d \times m)$ matrix. So we finally get

$$\tilde{\sigma}(u,v) = (A_1\bar{u}+L\bar{v}+d_1, A_2\bar{v}+d_2)$$

and the proposition is proven. □

REMARK 4.2. *With the above notation we have*

$$\tilde{\sigma}(u,v) = \begin{pmatrix} A_1 & L \\ 0 & A_2 \end{pmatrix}\begin{pmatrix} \bar{u} \\ \bar{v} \end{pmatrix}+\begin{pmatrix} d_1 \\ d_2 \end{pmatrix}$$

and the following properties hold:

(1) $A_1\overline{p_U(\Lambda)} = p_U(\Lambda)$, $A_2\overline{p_V(\Gamma)} = p_V(\Gamma)$.

(2) $\forall \gamma \in \Gamma$ *we have*

$$A_1\ \overline{B''(p_V(\gamma),\overline{p_V(\gamma)})} = B''(A_2\overline{p_V(\gamma)},\overline{A_2}p_V(\gamma)).$$

(3) $\forall v \in V$, $\forall \gamma \in \Gamma$ *we have*

$$A_1\overline{B'(v,p_V(\gamma))}+2A_1\ \overline{B''(v,\overline{p_V(\gamma)})} = B'(A_2\bar{v},A_2\overline{p_V(\gamma)})+2B''(A_2\bar{v},\overline{A_2}p_V(\gamma)).$$

(4) $\forall \gamma \in \Gamma$

$$L\overline{p_V(\gamma)}-B'(d_2,A_2\overline{p_V(\gamma)})+2B''(d_2,\overline{A_2}p_V(\gamma))+B''(A_2\overline{p_V(\gamma)},\overline{A_2}p_V(\gamma)) \in p_U(\Lambda).$$

(5) $A_1\bar{A}_1 = I$, $A_2\bar{A}_2 = I$.
(6) *There exists a* $\gamma \in \Gamma$ *such that* $\forall v \in V$

$$A_1\overline{L}(v) + L\overline{A_2}v = F_\gamma(v).$$

(7) $\sigma_2^2 = Id \ mod \ p_V(\Gamma)$, *i.e.* σ_2 *induces an antiholomorphic involution on* Y.
(8) $A_1\overline{d_1} + L\overline{d_2} + d_1 \in p_U(\Lambda)$.

PROOF. Conditions (1), (2), (3) and (4) easily follow as in the proof of (4.1) by imposing that for all $\lambda \in \Lambda$ there exists a $\lambda' \in \Lambda$ such that $\tilde{\sigma} \circ \lambda = \lambda' \circ \tilde{\sigma}$, and for all $\gamma \in \Gamma$ there must exist $\gamma' \in \Gamma$, $\lambda'' \in \Lambda$ such that $\tilde{\sigma} \circ \gamma = \lambda''\gamma' \circ \tilde{\sigma}$. Here we also used the expression of $F_\gamma(v)$ given in definition (3.3):

$$F_\gamma(v) := B'(v, p_V(\gamma)) + 2B''(v, \overline{p_V(\gamma)}) + B''(p_V(\gamma), \overline{p_V(\gamma)}), \forall v \in V.$$

Conditions (5), (6), (7) and (8) immediately follow by imposing $\tilde{\sigma}^2 \in \Pi$. □

THEOREM 4.3. *Let* $f : X \to Y$ *be a principal holomorphic torus bundle over a torus such that* A *is non degenerate and assume that* $\sigma : X \to X$ *is an antiholomorphic involution on* X. *The differentiable type of the pair* (X, σ) *is completely determined by the orbifold fundamental group exact sequence. More precisely, the affine embedding of* Π_σ *is uniquely determined up to conjugation.*

PROOF. Let

$$1 \to \Pi \to \hat{\Pi} := \Pi_\sigma \to \mathbb{Z}/2 \to 1 \tag{4.4}$$

be the orbifold fundamental group exact sequence of the pair (X, σ). Every lifting $\tilde{\sigma}$ of σ to $\hat{\Pi}$ acts by conjugation on Π, so it acts by conjugation on the centre Λ of Π and thus it acts on the quotient $\Gamma = \Pi/\Lambda$. Therefore we have determined an extension

$$1 \to \Gamma \to \hat{\Gamma} \to \mathbb{Z}/2 \to 1$$

which is the orbifold fundamental group exact sequence of the real torus (Y, σ_2), where σ_2 denotes as above the second component of $\tilde{\sigma}$. Since for a real torus the orbifold fundamental group exact sequence determines the differentiable type (cf. [**Cat02**]), we have shown that we can fix the differentiable type of the pair (Y, σ_2).

By proposition 4.1 we know that any lifting $\tilde{\sigma}$ of σ to the universal covering $(\Lambda \otimes \mathbb{R}) \oplus (\Gamma \otimes \mathbb{R})$ is of the form

$$\tilde{\sigma}(y, x) = \begin{pmatrix} A_1 & L \\ 0 & A_2 \end{pmatrix} \begin{pmatrix} y \\ x \end{pmatrix} + \begin{pmatrix} d_1 \\ d_2 \end{pmatrix}$$

and we know the $(2m \times 2m)$-matrix A_2 and the translation vector d_2. We also have $A_1^2 = I$, $A_2^2 = I$, because $\tilde{\sigma}^2 \in \Pi$. Furthermore we know $\tilde{\sigma}^2 \in \Pi$, since we know the extension (4.4), therefore we know the vector

$$\begin{pmatrix} A_1 & L \\ 0 & A_2 \end{pmatrix} \begin{pmatrix} d_1 \\ d_2 \end{pmatrix} + \begin{pmatrix} d_1 \\ d_2 \end{pmatrix} = \begin{pmatrix} Ld_2 + A_1 d_1 + d_1 \\ A_2 d_2 + d_2 \end{pmatrix}$$

which is the translation part of $\tilde{\sigma}^2$.

For every $\lambda \in \Lambda$ we know $\tilde{\sigma}\lambda\tilde{\sigma}^{-1} \in \Lambda$ and $\tilde{\sigma}\lambda\tilde{\sigma}^{-1}(w) = w + M\lambda$, $\forall w \in (\Gamma \otimes \mathbb{R}) \oplus (\Lambda \otimes \mathbb{R})$, where

$$M = \begin{pmatrix} A_1 & L \\ 0 & A_2 \end{pmatrix}$$

Thus we know $A_1(\lambda) \in \Lambda$, $\forall \lambda \in \Lambda$ and since Λ generates $\Lambda \otimes \mathbb{R}$, we know A_1. So we know A_1, A_2, d_2, and $Ld_2 + A_1 d_1 + d_1$.

For all $\hat{\gamma} \in \Pi$ lifting a given $\gamma \in \Gamma$, $\gamma \neq 0$, we know $\tilde{\sigma}\hat{\gamma}\tilde{\sigma}^{-1}$. Let us set $\tilde{\sigma}(w) = Mw + b$, $\hat{\gamma}(w) = Dw + h$, where

$$D\begin{pmatrix} y \\ x \end{pmatrix} = \begin{pmatrix} I_{2d} & \phi_\gamma \\ 0 & I_{2m} \end{pmatrix}\begin{pmatrix} y \\ x \end{pmatrix}$$

where $\phi_\gamma(x)$ can be chosen to be equal to $A(x,\gamma)$ as it is proven in (2.4), and $h = \begin{pmatrix} l \\ \gamma \end{pmatrix}$, where $l \in \Lambda$.

We have $\tilde{\sigma}\hat{\gamma}\tilde{\sigma}^{-1}(w) = \tilde{\sigma}\hat{\gamma}(M^{-1}w - M^{-1}b) = \tilde{\sigma}(DM^{-1}w - DM^{-1}b + h) = MDM^{-1}w - MDM^{-1}b + Mh + b$. Thus we know MDM^{-1} and $-MDM^{-1}b + Mh + b$. One easily computes

$$-MDM^{-1}b + Mh + b = \begin{pmatrix} -A_1\phi_{\gamma'}A_2^{-1}d_2 + L\gamma + A_1 l \\ A_2\gamma \end{pmatrix}$$

and since we know both $-A_1\phi_{\gamma'}A_2^{-1}d_2$ and $A_1 l$, we also know $L\gamma$ for all $\gamma \in \Gamma$. Now Γ generates $\Gamma \otimes \mathbb{R}$, so we know $L : \Gamma \otimes \mathbb{R} \to \Lambda \otimes \mathbb{R}$.

We have already seen that we know $Ld_2 + A_1 d_1 + d_1$, so we also know $d_1' = A_1 d_1 + d_1$ and we know A_1. Now, $A_1^2 = I$, so we can decompose $\Lambda \otimes \mathbb{R} = Z \oplus S \oplus W^+ \oplus W^-$, where $W^\pm$ are the ± 1-eigenspaces of A_1 and where

$$A_1(z, s, w^+, w^-) = (s, z, w^+, -w^-).$$

Observe that since A_1 is antiholomorphic, W^+ and W^- have the same dimension (also $dim(Z) = dim(S)$). Thus we can write $d_1 = (z, s, d_1^+, d_1^-)$. $A_1(z, s, d_1^+, d_1^-) + (z, s, d_1^+, d_1^-) = (s + z, s + z, 2d_1^+, 0)$, so we know d_1^+ and $z + s$. We can change the origin by translating with $(y, x) \in (\Lambda \otimes \mathbb{R}) \oplus (\Gamma \otimes \mathbb{R}) \mapsto (y + t, x) \in (\Lambda \otimes \mathbb{R}) \oplus (\Gamma \otimes \mathbb{R})$ and we may assume that $d_1 = (z, s, d_1^+, d_1^-) + A_1(t) - t = (z, s, d_1^+, d_1^-) + (t_2 - t_1, t_1 - t_2, 0, -2t^-)$ $(t = (t_1, t_2, t^+, t^-))$. So we can choose $t^- = d_1^-/2$, $t_2 - t_1 = -z$, therefore the first and the last components of d_1 can be chosen equal to zero. Then the second component is $s + t_1 - t_2 = s + z$ and therefore we know it, finally the third component is d_1^+ and we already know it.

□

REMARK 4.4. *Let us fix the orbifold fundamental group exact sequence of a real principal holomorphic torus bundle over a torus such that the alternating bilinear form $A : \Gamma \times \Gamma \to \Lambda$ is non degenerate. Then the topological and differentiable action of Π_σ is fixed, and the action of any element $\tilde{\sigma}$ of the orbifold fundamental group $\hat{\Pi}$ on the universal covering $(\Lambda \otimes \mathbb{R}) \oplus (\Gamma \otimes \mathbb{R})$ is given by the affine transformation*

$$\tilde{\sigma}\begin{pmatrix} y \\ x \end{pmatrix} = \begin{pmatrix} A_1 & L \\ 0 & A_2 \end{pmatrix}\begin{pmatrix} y \\ x \end{pmatrix} + \begin{pmatrix} d_1 \\ d_2 \end{pmatrix}$$

If we fix such an element $\tilde{\sigma}$, the following holds:

(1) $A_1 : \Lambda \to \Lambda$, $A_1^2 = I$.
(2) $A_2 : \Gamma \to \Gamma$, $A_2^2 = I$.
(3) $\forall \gamma$, $\forall x \in \Gamma \otimes \mathbb{R}$, $A_1(A(x,\gamma)) = A(A_2(x), A_2(\gamma))$.
(4) *The* $\mathbb{R}$ *- linear map* $L' : \Gamma \otimes \mathbb{R} \to \Lambda \otimes \mathbb{R}$, $L'(x) := Lx - A(d_2, A_2(x))$ *satisfies* $L'(\Gamma) \subset \Lambda$.
(5) $A_2(d_2) + d_2 \in \Gamma$.
(6) $L(d_2) + A_1(d_1) + d_1 \in \Lambda$.
(7) $\exists \gamma \in \Gamma$ *such that* $L(A_2(x)) + A_1(L(x)) = -A(x,\gamma)$ $\forall x \in \Gamma \otimes \mathbb{R}$.

PROOF. For all $\lambda \in \Lambda$ we have $\tilde{\sigma}\lambda\tilde{\sigma}^{-1} \in \Lambda$ and this immediately implies that $A_1(\Lambda) \subset \Lambda$. $A_1^2 = I$, since $\tilde{\sigma}^2 \in \Pi$ and for all $g \in \Pi$ the action of g on $(\Lambda \otimes \mathbb{R}) \oplus (\Gamma \otimes \mathbb{R})$ is given by

$$g \begin{pmatrix} y \\ x \end{pmatrix} = \begin{pmatrix} I & \phi_\gamma \\ 0 & I \end{pmatrix} \begin{pmatrix} y \\ x \end{pmatrix} + \begin{pmatrix} l \\ \gamma \end{pmatrix}$$

where γ is the image of g in Γ, $l \in \Lambda$, and $\phi_\gamma(x) = A(x, \gamma)$ as in (2.4).

This also implies that $A_2^2 = Id$.

For every $g \in \Pi$, $g \notin \Lambda$, we know $\tilde{\sigma} g \tilde{\sigma}^{-1} = h \in \Pi - \Lambda$. If we set

$$g \begin{pmatrix} y \\ x \end{pmatrix} = \begin{pmatrix} I & \phi_\gamma \\ 0 & I \end{pmatrix} \begin{pmatrix} y \\ x \end{pmatrix} + \begin{pmatrix} l \\ \gamma \end{pmatrix}$$

where as above γ is the image of g in Γ and $l \in \Lambda$, and

$$h \begin{pmatrix} y \\ x \end{pmatrix} = \begin{pmatrix} I & \phi_\delta \\ 0 & I \end{pmatrix} \begin{pmatrix} y \\ x \end{pmatrix} + \begin{pmatrix} \lambda \\ \delta \end{pmatrix}$$

where δ is the image of h in Γ and $\lambda \in \Lambda$, we have

$$\tilde{\sigma} g \tilde{\sigma}^{-1} \begin{pmatrix} y \\ x \end{pmatrix} = \begin{pmatrix} I & A_1 \phi_\gamma A_2^{-1} \\ 0 & I \end{pmatrix} \begin{pmatrix} y \\ x \end{pmatrix} + \begin{pmatrix} -A_1 \phi_\gamma A_2^{-1} d_2 + L\gamma + A_1 l \\ A_2 \gamma \end{pmatrix}$$

$$= h \begin{pmatrix} y \\ x \end{pmatrix} = \begin{pmatrix} I & \phi_\delta \\ 0 & I \end{pmatrix} \begin{pmatrix} y \\ x \end{pmatrix} + \begin{pmatrix} \lambda \\ \delta \end{pmatrix}$$

This yields $A_1 \phi_\gamma A_2^{-1} = \phi_\delta$, $A_2 \gamma = \delta$, $-A_1 \phi_\gamma A_2^{-1} d_2 + L\gamma + A_1 l = \lambda \in \Lambda$. So $A_2(\Gamma) = \Gamma$ and for all $\gamma \in \Gamma$ we have $L'\gamma = L\gamma - \phi_{A_2\gamma}(d_2) = L(\gamma) - A(d_2, A_2(\gamma)) \in \Lambda$.

The condition $A_1 \phi_\gamma A_2^{-1} = \phi_{A_2(\gamma)}$, $\forall \gamma \in \Gamma$ can be written as

$$A_1(A(x, \gamma)) = A(A_2(x), A_2(\gamma)), \ \forall x \in \Gamma \otimes \mathbb{R}, \ \forall \gamma \in \Gamma.$$

Finally $\tilde{\sigma}^2 \in \Pi$ implies that $A_2(d_2) + d_2 \in \Gamma$, $L(d_2) + A_1(d_1) + d_1 \in \Lambda$, furthermore there must exist a $\gamma \in \Gamma$ such that $LA_2 + A_1 L = \phi_\gamma$. This implies that, for every $\delta \in \Gamma$, $L(A_2(\delta)) + A_1(L(\delta)) = -A(\delta, -\gamma) \in \Lambda$. □

We have three $\mathbb{R}$ - linear maps $A_1 : \Lambda \to \Lambda$, $A_2 : \Gamma \to \Gamma$, $L : \Gamma \otimes \mathbb{R} \to \Lambda \otimes \mathbb{R}$ with the above properties.

Let us now fix a complex structure on the bundle $f : X \to Y$, i.e. we fix a point (V, U) in $\mathcal{TB}_A = \{(V, U) \in Gr(m, 2m) \times Gr(d, 2d) \mid V \cap \overline{V} = (0),\ U \cap \overline{U} = (0), \mid \textit{the component of } A \textit{ in } \Lambda^2(\overline{V})^\vee \otimes U \textit{ is } = 0\}$.

We may now see A_1 as a real element in $(\Lambda \otimes \mathbb{C})^\vee \otimes (\Lambda \otimes \mathbb{C}) = (U \oplus \overline{U})^\vee \otimes (U \oplus \overline{U})$ and we want to impose that A_1 is antiholomorphic with respect to the chosen complex structure, so we want that the component of A_1 in $U^\vee \otimes U$ is zero. Analogously we see A_2 as a real element in $(\Gamma \otimes \mathbb{C})^\vee \otimes (\Gamma \otimes \mathbb{C}) = (V \oplus \overline{V})^\vee \otimes (V \oplus \overline{V})$ and we want that the component of A_2 in $V^\vee \otimes V$ is zero. Finally we also see L as a real element in $(\Gamma \otimes \mathbb{C})^\vee \otimes (\Lambda \otimes \mathbb{C}) = (V \oplus \overline{V})^\vee \otimes (U \oplus \overline{U})$ and we want that its component in $V^\vee \otimes U$ is zero.

Observe that since any other lifting $\tilde{\sigma}'$ of σ to $\hat{\Pi}$ is obtained from $\tilde{\sigma}$ by composition with an element in Π, which acts holomorphically with respect to the chosen complex structure, we may give the following definition.

DEFINITION 4.5. *We define spaces*

$$\mathcal{TB}_A^{\mathbb{R}}(A_1, A_2, L) =$$

$$= \Big\{(V, U) \in Gr(m, 2m) \times Gr(d, 2d) \;\Big|\; V \cap \overline{V} = (0), U \cap \overline{U} = (0),$$

$$\text{the component of } A \text{ in } \Lambda^2(\overline{V})^\vee \otimes U \text{ is } = 0,$$

$$\text{the component of } A_1 \text{ in } U^\vee \otimes U \text{ is } = 0,$$

$$\text{the component of } A_2 \text{ in } V^\vee \otimes V \text{ is } = 0,$$

$$\text{the component of } L \text{ in } V^\vee \otimes U \text{ is } = 0\Big\},$$

and

$$\mathcal{T}'\mathcal{B}_A^{\mathbb{R}}(A_1, A_2, L) =$$

$$= \{(V, U, \phi) \mid (V, U) \in \mathcal{TB}_A^{\mathbb{R}}(A_1, A_2, L)\; \phi \in H^1(Y_V, \mathcal{H}(U)_{Y_V}) \cong \overline{V}^\vee \otimes U\}.$$

REMARK 4.6. *Assume that we have fixed the orbifold fundamental group exact sequence of the pair (X, σ) and that the alternating bilinear form $A : \Gamma \times \Gamma \to \Lambda$ is non degenerate: then the above spaces parametrize families of real structures on principal holomorphic torus bundles, and in particular the latter family parametrizes all the possible real structures on a principal holomorphic torus bundle over a torus with a given real topological type.*

Let us now write

$$\Lambda \otimes \mathbb{R} = U^+ \oplus U^-,$$
$$\Gamma \otimes \mathbb{R} = V^+ \oplus V^-,$$

according to the eigenspace decomposition for A_1, respectively for A_2.

If we set $A = A^+ + A^-$, where A^+ is the component of A with values in U^+, A^- is the component of A with values in U^-, we can write condition (3) of (4.4) as follows: $\forall x \in \Gamma \otimes \mathbb{R}$, $\forall \gamma \in \Gamma$,

$$A_1(A(x, \gamma)) = A^+(x, \gamma) - A^-(x, \gamma) = A(A_2(x), A_2(\gamma)) = A(x^+ - x^-, \gamma^+ - \gamma^-),$$

where $x = x^+ + x^-$, $\gamma = \gamma^+ + \gamma^-$ according to the above decomposition.

Thus we have

$$A^+(x^+ + x^-, \gamma^+ + \gamma^-) = A^+(x^+ - x^-, \gamma^+ - \gamma^-),$$
$$A^-(x^+ + x^-, \gamma^+ + \gamma^-) = -A^-(x^+ - x^-, \gamma^+ - \gamma^-),$$

which in turn is equivalent to

$$A^+(x^+, \gamma^-) + A^+(x^-, \gamma^+) = 0, \;\; A^-(x^+, \gamma^+) + A^-(x^-, \gamma^-) = 0.$$

We conclude then that condition (3) is equivalent to

$$A^+|_{V^+ \times V^-} \equiv 0, \; A^-|_{V^+ \times V^+} \equiv 0, \; A^-|_{V^- \times V^-} \equiv 0, \; i.e., \tag{4.5}$$

$$A(x_1, x_2) = A^+(x_1^+, x_2^+) + A^+(x_1^-, x_2^-) + A^-(x_1^-, x_2^+) + A^-(x_1^+, x_2^-). \tag{4.6}$$

Condition (7) of (4.4) is equivalent to the existence of $\hat{\gamma} \in \Gamma$ such that

$$L^+(x^+ - x^-) + L^-(x^+ - x^-) + L^+(x^+ + x^-) - L^-(x^+ + x^-) =$$
$$-A^+(x^+ + x^-, \hat{\gamma}^+ + \hat{\gamma}^-) - A^-(x^+ + x^-, \hat{\gamma}^+ + \hat{\gamma}^-),$$

that is, to

$$2L^+(x^+) = -A^+(x^+,\hat\gamma^+) - A^+(x^-,\hat\gamma^-),$$
$$2L^-(x^-) = A^-(x^+,\hat\gamma^-) + A^-(x^-,\hat\gamma^+),$$

in particular we have

$$A^+(x^-,\hat\gamma^-) \equiv 0 \equiv A^-(x^+,\hat\gamma^-),$$

or equivalently,

$$A(-,\hat\gamma^-) \equiv 0.$$

Since we are assuming A nondegenerate, we must have $\hat\gamma^- = 0$, and we have

$$L^+(x^+) = -\frac{A^+}{2}(x^+,\hat\gamma^+)$$
$$L^-(x^-) = \frac{A^-}{2}(x^-,\hat\gamma^+) \tag{4.7}$$

We look now for respective complex structures J_1, J_2 on $\Lambda\otimes\mathbb{R}$, $\Gamma\otimes\mathbb{R}$ making L, A_1 and A_2 antiholomorphic.

The condition $J_2A_2 = -A_2J_2$ immediately implies that J_2 exchanges the two Eigenspaces for A_2, therefore we can write $J_2(x^+,x^-) = (Cx^-,Dx^+)$. The condition $J_2^2 = -Id$ is then equivalent to $D = -C^{-1}$, so, if we set $B_2 = -C$, we have

$$J_2(x^+,x^-) = (-B_2x^-, B_2^{-1}x^+).$$

Proceeding analogously for J_1 we obtain

$$J_1(y^+,y^-) = (-B_1y^-, B_1^{-1}y^+).$$

Finally $L\circ J_2 = -J_1\circ L$ is equivalent to

$$L(-B_2x^-, B_2^{-1}x^+) = (B_1L^-(x), -B_1^{-1}L^+(x)),$$

equivalently,

$$L^+(-B_2x^-, B_2^{-1}x^+) = B_1L^-(x^+,x^-),$$
$$L^-(-B_2x^-, B_2^{-1}x^+) = -B_1^{-1}L^+(x^+,x^-).$$

If we now write $L(x^+,x^-)$ as $L_+x^+ + L_-x^-$ we obtain

$$-L_+^+B_2(x^-) + L_-^+B_2^{-1}(x^+) = B_1L_+^-(x^+) + B_1L_-^-(x^-),$$
$$-L_+^-B_2(x^-) + L_-^-B_2^{-1}(x^+) = -B_1^{-1}L_+^+(x^+) - B_1^{-1}L_-^+(x^-).$$

We rewrite the first equation as

$$L_+^+B_2 + B_1L_-^- \equiv 0,$$
$$L_-^+B_2^{-1} - B_1L_+^- \equiv 0. \tag{4.8}$$

After rewriting the second equation as

$$L_+^-B_2 - B_1^{-1}L_-^+ \equiv 0,$$

$$L_-^-B_2^{-1} + B_1^{-1}L_+^+ \equiv 0,$$

we observe that these equations are clearly equivalent to (4.8).

Conditions (4.7) become

$$\begin{aligned} L_+^+(x^+) &= -\frac{A^+}{2}(x^+, \hat{\gamma}^+), \\ L_-^-(x^-) &= \frac{A^-}{2}(x^-, \hat{\gamma}^+). \end{aligned} \tag{4.9}$$

We shall now write the Riemann bilinear relations.

Recall that if $(V, U) \in \mathcal{TB}_A$, $V = \{x - iJ_2 x \mid x \in \Gamma \otimes \mathbb{R}\}$, $U = \{z - iJ_1 z \mid z \in \Lambda \otimes \mathbb{R}\}$ and we want

$$\begin{aligned} &A(x - iJ_2 x, y - iJ_2 y) = \\ &\qquad A(x, y) - A(J_2 x, J_2 y) - i(A(x, J_2 y) + A(J_2 x, y)) \in U, \ \forall x, y \in \Gamma \otimes \mathbb{R}. \end{aligned}$$

This means that

$$A(x, J_2 y) + A(J_2 x, y) = J_1 A(x, y) - J_1 A(J_2 x, J_2 y). \tag{4.10}$$

We shall now split (4.10) into U^+ and U^- components, so that using (4.5), (4.6) the first component is

(4.11)

$$\begin{aligned} &A_{++}^+(x^+, -B_2 y^-) + A_{--}^+(x^-, B_2^{-1} y^+) + \\ &A_{++}^+(-B_2 x^-, y^+) + A_{--}^+(B_2^{-1} x^+, y^-) = -B_1 A_{+-}^-(x^+, y^-) - B_1 A_{-+}^-(x^-, y^+) \\ &-B_1 A_{+-}^-(B_2 x^-, B_2^{-1} y^+) - B_1 A_{-+}^-(B_2^{-1} x^+, B_2 y^-) \end{aligned}$$

while the second component is

(4.12)

$$\begin{aligned} &A_{+-}^-(x^+, B_2^{-1} y^+) + A_{-+}^-(B_2^{-1} x^+, y^+) + \\ &A_{+-}^-(-B_2 x^-, y^-) + A_{-+}^-(x^-, -B_2 y^-) = B_1^{-1}[A_{++}^+(x^+, y^+) + A_{--}^+(x^-, y^-) \\ &-A_{++}^+(B_2 x^-, B_2 y^-) - A_{--}^+(B_2^{-1} x^+, B_2^{-1} y^+)]. \end{aligned}$$

From these equations we derive the following equations looking at the four possible bilinear types (x^+, y^+), (x^+, y^-), (x^-, y^+), (x^-, y^-).

$$\begin{aligned} &-A_{++}^+(x^+, B_2 y^-) + A_{--}^+(B_2^{-1} x^+, y^-) = \\ &= -B_1 A_{+-}^-(x^+, y^-) - B_1 A_{-+}^-(B_2^{-1} x^+, B_2 y^-) \end{aligned}$$

$$\begin{aligned} &A_{--}^+(x^-, B_2^{-1} y^+) - A_{++}^+(B_2 x^-, y^+) = \\ &= -B_1 A_{-+}^-(x^-, y^+) - B_1 A_{+-}^-(B_2 x^-, B_2^{-1} y^+) \end{aligned}$$

$$\begin{aligned} &A_{+-}^-(x^+, B_2^{-1} y^+) + A_{-+}^-(B_2^{-1} x^+, y^+) = \\ &= B_1^{-1}[A_{++}^+(x^+, y^+) - A_{--}^+(B_2^{-1} x^+, B_2^{-1} y^+)] \end{aligned}$$

$$\begin{aligned} &-A_{+-}^-(B_2 x^-, y^-) - A_{-+}^-(x^-, B_2 y^-) = \\ &= B_1^{-1}[A_{--}^+(x^-, y^-) - A_{++}^+(B_2 x^-, B_2 y^-)]. \end{aligned}$$

These four equations can be rewritten as tensor equations as follows (according to the standard notation for the transformation of bilinear forms)

$$-A^+_{++}B_2 +^t B_2^{-1}A^+_{--} = B_1(-A^-_{+-} -^t B_2^{-1}A^-_{-+}B_2) \tag{4.13}$$

$$A^+_{--}B_2^{-1} -^t B_2A^+_{++} = B_1(-A^-_{-+} -^t B_2A^-_{+-}B_2^{-1}) \tag{4.14}$$

$$A^-_{+-}B_2^{-1} +^t B_2^{-1}A^-_{-+} = B_1^{-1}(A^+_{++} -^t B_2^{-1}A^+_{--}B_2^{-1}) \tag{4.15}$$

$$-^tB_2A^-_{+-} - A^-_{-+}B_2 = B_1^{-1}(A^+_{--} -^t B_2A^+_{++}B_2) \tag{4.16}$$

We observe that the tensor equations above are all equal, in fact (4.14) yields (4.13) by composing with B_2 to the right and with ${}^tB_2^{-1}$ to the left. (4.15) yields (4.13) by composing with B_2 to the right. (4.16) yields (4.13) by composing with ${}^tB_2^{-1}$ to the left.

So we have shown that if we have a real structure we have only one equation for the Riemann bilinear relation:

$$A^+_{--} -^t B_2A^+_{++}B_2 = -B_1(A^-_{-+}B_2 -^t B_2 \, {}^tA^-_{-+}), \tag{4.17}$$

where we have used $A^-_{+-} = -{}^tA^-_{-+}$ since A^- is alternating.

Now, in order to simplify the notation, we set $A^+_{--} =: A_-$, $A^+_{++} =: A_+$, $D := A^-_{-+}$, so (4.17) becomes

$$A_- -^t B_2A_+B_2 = -B_1(DB_2 -^t (DB_2)), \tag{4.18}$$

We can now easily show that in the case $m = d = 1$ the space $\mathcal{T}'\mathcal{B}^{\mathbb{R}}_A(A_1, A_2, L) \cap (\mathcal{T}_1 \times \mathcal{T}_1)$ is connected. We observe, here and in the following, that it suffices to show that $\mathcal{T}\mathcal{B}^{\mathbb{R}}_A(A_1, A_2, L) \cap (\mathcal{T}_1 \times \mathcal{T}_1)$ is connected. Note that the case $m = d = 1$, since we assume A to be nondegenerate, corresponds to the case of Kodaira surfaces. The fact that the moduli space of real Kodaira surfaces of a given topological type is connected was already proved by the second author with different methods (cf. [**F04**]).

Proposition 4.7. *If $m = d = 1$, the space $\mathcal{T}\mathcal{B}^{\mathbb{R}}_A(A_1, A_2, L) \cap (\mathcal{T}_1 \times \mathcal{T}_1)$ is connected.*

Proof. We observe first of all that equation (4.18) does not appear since for $m = 1$, $A_- = A_+ = 0$, and B_1, B_2 are scalars.

So we only have to consider conditions (4.8). With an appropriate choice of orientation on each vector space we may assume $B_1 > 0, B_2 > 0$, so if $L^-_- \neq 0$, by (4.8) we obtain $B_1 = -\frac{B_2L^+_+}{L^-_-}$ and

$$L^+_- + \frac{B_2^2L^+_+L^-_+}{L^-_-} = 0,$$

or equivalently

$$L^+_-L^-_- + B_2^2L^+_+L^-_+ = 0,$$

which is solvable if either $L^+_-L^-_- = L^+_+L^-_+ = 0$, or both are non zero and have opposite sign and in this case we only have one positive solution.

If $L_-^- = 0$, then also $L_+^+ = 0$ and we have $B_1 B_2 = \frac{L_-^+}{L_+^-}$. Since we are assuming $B_1 > 0$, $B_2 > 0$, we must have $\frac{L_-^+}{L_+^-} > 0$ and the set is clearly connected. □

REMARK 4.8. *Observe that if $L = 0$ and $d = 1$, we set $B_2 =: B$, $B_1 =: b$, a scalar, and we only have the equation*

$$A_- -^t BA_+B = -b(DB -^t (DB)), \tag{4.19}$$

If we also assume $m = 2$, we can write $A_- = a_- \begin{pmatrix} 0 & 1 \\ -1 & 0 \end{pmatrix}$,

$A_+ = a_+ \begin{pmatrix} 0 & 1 \\ -1 & 0 \end{pmatrix}$, *and we obtain only one scalar equation*

$$a_- - det(B)a_+ = -b(d_{11}b_{12} + d_{12}b_{22} - d_{21}b_{11} - d_{22}b_{21})$$

and we must combine it with the condition

$$det(B) = b_{11}b_{22} - b_{12}b_{21} > 0.$$

THEOREM 4.9. *For $d = 1$, $m = 2$, the space $\mathcal{X} = \mathcal{TB}_A^{\mathbb{R}}(A_1, A_2, L) \cap (\mathcal{T}_2 \times \mathcal{T}_1)$ is connected.*

PROOF. By remark (4.8), if $L = 0$ we have the two conditions:

$$a_- - det(B)a_+ = -b(d_{11}b_{12} + d_{12}b_{22} - d_{21}b_{11} - d_{22}b_{21}) \tag{4.20}$$

$$det(B) = b_{11}b_{22} - b_{12}b_{21} > 0, \tag{4.21}$$

and we assume $b > 0$.

Assume that the matrix $D = 0$: then we have a product of the half-line $\{b > 0\}$ with the quadric $\{detB = a\}$, where a is a positive constant. In this case we are done since $\{detB = a\}$ is a central quadric in $\mathbb{R}^4$ with quadratic part of signature $(+, +, -, -,)$, and is therefore connected.

Assume then that $D \neq 0$, then the rigth hand side is equal to bb', where we define b' as the linear form $b' := d_{11}b_{12} + d_{12}b_{22} - d_{21}b_{11} - d_{22}b_{21}$.

We want to change coordinates in the $\mathbb{R}^4$ of the matrices B, completing the linear form b' to a basis (b', x, y, z) of the space of linear forms. The projective quadric $\{detB = 0\}$ and the hyperplane $\{b' = 0\}$ determine an affine quadric in $\mathbb{R}^3$ with the given signature, therefore there are only two cases, according to the property whether the hyperplane is transversal or tangent to the quadric.

In the first case, after dehomogenizing (i.e., setting $b' = 1$) we get a one sheeted (hyperbolic) hyperboloid:

$$\pm detB = b'^2 - z^2 + xy,$$

in the second case we get a hyperbolic paraboloid

$$detB = b'z + xy.$$

Let us observe that the space $\mathcal{X}$ maps to the open set Ω in $\mathbb{R}^4$ defined by the inequality $detB > 0$, and that if we define $\Omega' := \Omega \cap \{b' \neq 0\}$, then Ω' is homeomorphic to its inverse image in $\mathcal{X}$.

It is immediate to conclude, in view of the above normal forms, that the open sets $\Omega^+ := \Omega \cap \{b' > 0\}$, resp. $\Omega^- := \Omega \cap \{b' < 0\}$ are connected.

We also see by the way (this remark is not indispensable) that the inverse image $\mathcal{X}^0$ of $\{b' = 0\}$ has at most two connected components, since it is the product of the half-line $\{b > 0\}$ with a central quadric $\{detB = a > 0\} \cap \{b' = 0\}$ in $\mathbb{R}^3$ with possible signatures $(+,+,-)$,$(+,-,-)$,$(+,-)$.

In order to prove that $\mathcal{X}$ is connected, let us observe that the space $\mathcal{X}$ we are considering is the intersection of a real quadric Q in $\mathbb{R}^5$ with an open set. Moreover, the quadric Q is centred in the origin and the associated quadratic form has negativity index, respectively positivity index, at least two. Therefore Q is everywhere of pure dimension 4, thus the closed set $\mathcal{X}^0$ is in the closure of the complement $\mathcal{X} \setminus \mathcal{X}^0$, and it suffices to show that there is a point p of $\mathcal{X}^0$ and a neighbourhood of p meeting both $\mathcal{X}^+$ and $\mathcal{X}^-$.

But if this were not so, in the points of $\mathcal{X}^0$ the linear form b' restricted to $\mathcal{X}$ would vanish with its derivatives everywhere, which implies that $\mathcal{X}^0$ would be a linear subspace counted with multiplicity two, which is not the case.

We have thus shown that if $L = 0$, $\mathcal{X}$ is connected.

Assume now that $L \neq 0$. In this case we also have to consider equations (4.8):

$$L_+^+ B = -bL_-^-,$$
$$L_+^- B = b^{-1} L_-^+.$$

If L_+^+, L_+^- are linearly independent we can uniquely determine B

$$B = \begin{pmatrix} L_+^+ \\ L_+^- \end{pmatrix}^{-1} \begin{pmatrix} -bL_-^- \\ b^{-1}L_-^+ \end{pmatrix},$$

where $\begin{pmatrix} L_+^+ \\ L_+^- \end{pmatrix}$ denotes the (2×2) matrix whose first row is L_+^+ and whose second row is L_+^- and $\begin{pmatrix} -bL_-^- \\ b^{-1}L_-^+ \end{pmatrix}$ is the (2×2) matrix whose first row is $-bL_-^-$ and whose second row is $b^{-1}L_-^+$.

Then $det(B) = -\dfrac{det(\begin{pmatrix} L_-^- \\ L_-^+ \end{pmatrix})}{det(\begin{pmatrix} L_+^+ \\ L_+^- \end{pmatrix})} =: \alpha.$

So we must have $\alpha > 0$, and equation (4.20) is of the form

$$a_- - \alpha a_+ = -b(c_1 b + c_2 b^{-1}),$$

where c_1, c_2 are functions depending only on D and L.

So we have to solve an equation of the form

$$c_1 b^2 + c = 0,$$

and we have at most one positive solution, thus $\mathcal{X}$ is connected.

Suppose now $L_+^- = 0$. Then if there is a solution also $L_-^+ = 0$ and we may assume without loss of generality that $L_+^+ = (1,0)$ (since $L \neq 0$).

If we then set $B = (b_{ij})$ our equations (4.8) reduce to

$$(b_{11}, b_{12}) = -bL_-^-,$$

and equation (4.20) becomes

$$a_- - a_+ det(B) = -b(c_0 b_{21} + c_2 b_{22} + bc_1),$$

where the c_j's only depend on D and L.

Moreover, we can write $det(B) = b\rho$, where $\rho = l_0 b_{21} + l_2 b_{22}$ and the l_j's depend only on L. So we have an equation in b, b_{21}, b_{22}

$$a_- - a_+ b\rho = -b(c_0 b_{21} + c_2 b_{22} + bc_1),$$

which is an equation of the form

$$b^2 c_1 + b(a_0 b_{21} + a_2 b_{22}) + a_- = 0,$$

where the a_j's depend only on L and D.

The condition $det(B) > 0$ becomes $\rho > 0$, which is a linear inequality in b_{21}, b_{22}. If we fix $b \in \mathbb{R}_+$, our solution is the intersection of the line in $\mathbb{R}^2$ given by

$$a_0 b_{21} + a_2 b_{22} + \frac{a_- + b^2 c_1}{b} = 0$$

with the half plane $\rho > 0$.

To simplify things, we introduce a linear form τ in the $\mathbb{R}^2$ with coordinates b_{21}, b_{22}, namely

$$\tau := a_0 b_{21} + a_2 b_{22}.$$

There are three cases:

1) the linear forms ρ, τ are independent

2) $\tau = c\rho$ for a constant $c \neq 0$.

3) $\tau = 0$.

In case 1) we have $\tau = -\frac{a_- + b^2 c_1}{b}$, $\rho > 0, b > 0$, and $\mathcal{X}$ is homeomorphic to a quadrant in $\mathbb{R}^2$.

In case 2) we have that $\mathcal{X}$ is a product of $\mathbb{R}$ with the set $\{(\rho, b) | \rho > 0, b > 0, \rho = -\frac{a_- + b^2 c_1}{bc}\} \subset \mathbb{R}^2$, which is diffeomorphic to the interval $\{(b) | b > 0, 0 > c(a_- + b^2 c_1)\} \subset \mathbb{R}$.

In case 3) $\mathcal{X}$ is a product of $\mathbb{R}$ with the set

$$\{(\rho, b) | \rho > 0, b > 0, 0 = a_- + b^2 c_1\} \subset \mathbb{R}^2.$$

which is a half-line in $\mathbb{R}^2$,

Therefore $\mathcal{X}$ is connected in all three cases.

Assume now $L_+^- \neq 0$, $L_+^+ = \beta L_+^-$, $\beta \neq 0$. Then

$$L_+^+ B = -b L_-^-,$$

$$L_+^- B = b^{-1} L_-^+,$$

yield an equation of the form

$$b^2 L_-^- = -\beta L_-^+,$$

and if we solve for b we find at most one positive solution $\hat{b}$.

Without loss of generality we can assume $L_+^+ = (1, 0)$ and the equation $L_+^+ B = -\hat{b} L_-^-$ allows us to determine the first row of B.

Hence both $det(B)$ and equation (4.20) are polynomials of degree one in b_{21}, b_{22}, therefore the intersection of $\{det(B) > 0\}$ with the set where the equation (4.20) is satified is connected (a half-line) or empty.

Finally, if $L^+_+ = 0$, then also $L^-_- = 0$ and we may assume w.l.o.g. $L^-_+ = (1,0)$. Therefore we have

$$(b_{11}, b_{12}) = b^{-1}L^+_- =: b^{-1}(l_1, l_2),$$

and since then $a_- - det(B)a_+ = a_- - a_+b^{-1}(l_1b_{22} - l_2b_{21})$ our equation becomes

$$a_- - a_+b^{-1}(l_1b_{22} - l_2b_{21}) = -b(b^{-1}(d_{11}l_2 - d_{21}l_1) + d_{12}b_{22} - d_{22}b_{21}),$$

and the condition $det(B) > 0$ can be written as $l_1b_{22} - l_2b_{21} > 0$.

So we get

$$b_{22}(d_{12}b^2 - a_+l_1) + b_{21}(-b^2d_{22} + a_+l_2) = -b(a_- - d_{21}l_1 + d_{11}l_2),$$
$$l_1b_{22} - l_2b_{21} > 0.$$

We define functions

$$x(b_{22}, b_{21}) := l_1b_{22} - l_2b_{21}$$
$$y(b, b_{22}, b_{21}) := b_{22}(d_{12}b^2 - a_+l_1) + b_{21}(-b^2d_{22} + a_+l_2).$$

These are linear functions of (b_{22}, b_{21}) with determinant $b^2(d_{12}l_2 - d_{22}l_1)$.

Case 1) : $(d_{12}l_2 - d_{22}l_1) \neq 0$.

Then $(b, b_{22}, b_{21}) \to (b, x(b_{22}, b_{21}), y(b, b_{22}, b_{21}))$ is a self-diffeomorphism of $\mathbb{R}_+ \times \mathbb{R}^2$, which transforms $\mathcal{X}$ into the set $\mathcal{X}' \subset \mathbb{R}^3$, $\mathcal{X}' = \{(b,x,y)|b > 0, x > 0, y = -b(a_- - d_{21}l_1 + d_{11}l_2)\}$, which is diffeomorphic to the quadrant $\{(b,x)|b > 0, x > 0\}$, which is clearly connected.

Case 2) : $(d_{12}l_2 - d_{22}l_1) = 0$.

In this case

$$y(b, b_{22}, b_{21}) := b_{22}(d_{12}b^2 - a_+l_1) + b_{21}(-b^2d_{22} + a_+l_2) = (cb^2 - a_+)\,x(b_{22}, b_{21}).$$

There is a linear self-diffeomorphism of $\mathbb{R}^3$ such that

$$(b, b_{22}, b_{21}) \to (b, x(b_{22}, b_{21}), z(b_{22}, b_{21}))).$$

It carries $\mathcal{X}$ into the set
$\{(b,x,z)|b > 0, x > 0, (cb^2 - a_+)x = -b(a_- - d_{21}l_1 + d_{11}l_2)\}$, which is the product of $\mathbb{R}$ with the subset of $\mathbb{R}^2$

$$\mathcal{Y} := \{(b,x)|b > 0, x > 0, (cb^2 - a_+)x = -b(a_- - d_{21}l_1 + d_{11}l_2)\}.$$

Case 2.1) : $(a_- - d_{21}l_1 + d_{11}l_2) = 0$.

Then we get $(cb^2 - a_+) = 0$, which has at most one positive solution, and $\mathcal{Y}$ is diffeomorphic to the half-line $\{x > 0\}$.

Case 2.2) : $(a_- - d_{21}l_1 + d_{11}l_2) \neq 0$, $c = 0$.

In this subcase either $\mathcal{Y}$ is the empty set, or $\mathcal{Y}$ is diffeomorphic to $\{(b \in \mathbb{R})|b > 0\}$.

Case 2.3) : $(a_- - d_{21}l_1 + d_{11}l_2) \neq 0$, $c \neq 0$.

In this subcase we get that $\mathcal{Y}$ is diffeomorphic to the set $\{b|b > 0,\ cb^2 < a_+\}$ or to the set $\{b|b > 0,\ cb^2 > a_+\}$.

We are done, since in both cases either $\mathcal{Y}$ is empty or it is diffeomorphic to an interval. □

From the above theorem follows now easily

THEOREM C. *Assume again $d = 1, m = 2$, that A is nondegenerate and fix the orbifold fundamental group Π_σ. Then the real structures for torus bundles in the Appell-Humbert family are parametrized by a connected family.*

PROOF. It suffices to combine the connectedness result that we have just proven in Theorem 4.9 with theorem 4.3, with Remark 4.4, and with the fact that $\mathcal{T}'\mathcal{B}_A^{\mathbb{R}}(A_1, A_2, L)$ is connected if and only if $\mathcal{T}\mathcal{B}_A^{\mathbb{R}}(A_1, A_2, L)$ is connected. □

As an immediate corollary we have

THEOREM D. *Same assumptions as in theorem B: $d = 1, m = 2$, and i), ii), iii) are satisfied.*

Then simplicity holds for real torus bundles in the Appell-Humbert family.

References

[Apr98] M. Aprodu, *An Appell-Humbert theorem for hyperelliptic surfaces*, J. Math. Kyoto Univ. 38, **1** (1998), 101–121.

[BPV84] W. Barth, C. Peters and A. Van de Ven, *Compact Complex Surfaces*, Ergebnisse der Mathematik und ihrer Grenzgebiete, 3.F, B. 4, Springer-Verlag, 1984.

[Bla53] A. Blanchard, *Recherche de structures analytiques complexes sur certaines variétés*, C.R. Acad.Sci., Paris **238** (1953), 657–659.

[Bla56] A. Blanchard, *Sur les variétés analytiques complexes*, Ann.Sci. Ec. Norm. Super., III Ser., **73** (1956), 157–202.

[Cal58] E. Calabi, *Construction and properties of some 6-dimensional almost complex manifolds*, Trans. Amer. Math.Soc. **87** (1958), 407–438.

[Cat91] F. Catanese, *Moduli and classification of irregular Kähler manifolds (and algebraic varieties) with Albanese general type fibrations. Appendix by Arnaud Beauville.* Inv. Math. **104** (1991) 263–289; Appendix 289 .

[Cat95] F. Catanese, *Compact complex manifolds bimeromorphic to tori.* Proc. of the Conf. "Abelian Varieties", Egloffstein 1993, De Gruyter, (1995), 55-62.

[Cat02] F. Catanese, *Deformation types of real and complex manifolds*, Proc. of the Chen-Chow Memorial Conference, Contemporary trends in algebraic geometry and algebraic topology (Tianjin, 2000), Nankai Tracts Math., 5, World Sci. Publishing, River Edge, NJ, (2002), 195–238 .

[Cat04] F. Catanese, *Deformation in the large of some complex manifolds, I*, Ann. Mat. Pura Appl. (4) 183, **3**, Volume in Memory of Fabio Bardelli, (2004), 261–289.

[C-F03] F. Catanese, P. Frediani, *Real hyperelliptic surfaces and the orbifold fundamental group.* J. Inst. Math. Jussieu 2, **2**, (2003), 163–233.

[C-F05] F. Catanese, P. Frediani *Deformation in the large of some complex manifolds, II*, math.CV/0507508, to appear in the AMS Series 'Contemporary Mathematics', in the Proceedings of the 10th anniversary (2004) Conference for the Mathematics Institute at East China Normal University.

[C-W04] F. Catanese, B. Wajnryb, *Diffeomorphism of simply connected algebraic surfaces* 33 pages , math.AG/0405299.

[Che58] S.S. Chern, *Complex manifolds*, Publ. Mat. Univ. Recife (1958).

[CO12] A. Comessatti, *Fondamenti per la geometria sopra le superficie razionali dal punto di vista reale*, Math. Annalen **43** (1912), 1–72.

[CO14] A. Comessatti, *Sulla connessione delle superficie razionali reali*, Annali di Matemetica **23** (1914), 215–285.

[CO32] A. Comessatti *Reelle Fragen in der algebraischen Geometrie*, Jahresbericht d. Deut. Math. Vereinigung **41** (1932), 107–134.

[D-K96] A. Degtyarev, V. Kharlamov *Topological classification of real Enriques surfaces*, Topology **35** (1996), 711–729.

[DIK00] Degtyarev, A., Itenberg, I., Kharlamov, V., *Real Enriques surfaces.*, Lecture Notes in Mathematics. **1746**. Berlin: Springer, (2000)

[DK02] A. Degtyarev, V. Kharlamov, *Real rational surfaces are quasi-simple.* J. Reine Angew. Math. 551, 87-99 (2002).

[DIK04] A. Degtyarev, I. Itenberg, V. Kharlamov, *Finiteness and quasi-simplicity for symmetric K3-surfaces.* Duke Math. J. 122, No.1, 1-49 (2004).

[FG65] W. Fischer, H. Grauert, *Lokal-triviale Familien kompakter komplexer Mannigfaltigkeiten* , Nachr. Akad. Wiss. Göttingen, II. Math.-Phys. Kl. 1965, (1965), 89–94.

[F04] P. Frediani, *Real Kodaira surfaces*, Collect. Math. **55** (2004), no. 1, 61–96.

[Hir62] H. Hironaka, *An example of a non-Kählerian complex-analytic deformation of Kählerian complex structures*, Ann. Math. (2) 75, (1962),190–208.

[Hö93] T. Höfer, *Remarks on torus principal bundles*, J. Math. Kyoto Univ. 33 , no. 1 (1993), 227–259.

[K-K05] V. Kharlamov, V. Kulikov, *Surfaces with DIF $\neq$ DEF real structures*, 44 pages, math.AG/0507059.

[Ko64] K. Kodaira, *On the structure of complex analytic surfaces I*, Amer. J. Math. **86** (1964), 751–798.

[Ko68] K. Kodaira , *On the structure of complex analytic surfaces IV*, Amer. J. Math. **90** (1968), 1048–1066.

[K-M71] K. Kodaira, J. Morrow , *Complex manifolds*, Holt, Rinehart and Winston, New York-Montreal, Que.-London (1971).

[K-S58] K. Kodaira , D. Spencer *On deformations of complex analytic structures I-II*, Ann. of Math. **67** (1958), 328–466.

[Ku61] M. Kuranishi, *On a type of family of complex structures*, Ann. of Math. (2) 74 (1961), 262–328.

[Ku65] M. Kuranishi, *New proof for the existence of locally complete families of complex structures*, Proc. Conf. Complex Analysis (Minneapolis, 1964), Springer, Berlin, (1965), 142–154.

[Man01] M. Manetti, *On the Moduli Space of diffeomorphic algebraic surfaces*, Inv. Math. **143** (2001), 29–76.

[Man97] F. Mangolte, *Cycles algébriques sur les surfaces $K3$ réelles*, Math. Z. 225, 4, (1997), 559–576.

[ManvH98] F. Mangolte, J. van Hamel, *Algebraic cycles and topology of real Enriques surfaces*, Compositio Math. 110, 2, (1998), 215–237.

[Mos78] G. Mostow, *Strong rigidity of locally symmetric spaces*, Annals of Math.Stud. **78** Princeton Univ. Press (1978).

[Nak75] I. Nakamura, *Complex parallelisable manifolds and their small deformations*, J. Differ. Geom. **10**, (1975),85-112.

[Nak98] I. Nakamura, *Global deformations of $\mathbb{P}^2$-bundles over $\mathbb{P}^1$*, J. Math. Kyoto Univ. 38, No.1 (1998), 29–54.

[Ni79] V.V. Nikulin, *Integral symmetric bilinear forms and some of their applications*, Izv. Akad. Nauk SSSR **43** 1, 117–177. (1979) (Russian); English transl. in Math. USSR-Izv. **14** 103-167 (1980).

[S-S89] M. Seppälä, R. Silhol, *Moduli spaces for real algebraic curves and real abelian varieties*, Math. Z. **201**, (1989), 151-165.

[Si89] R. Silhol, *Real Algebraic Surfaces*, Lectures Notes in Mathematics 1392, Springer - Verlag 1989.

[Somm75] A. J. Sommese , *Quaternionic manifolds*, Math. Ann. **212**, (1975), 191-214.

[Su70] T. Suwa, *On hyperelliptic surfaces*, J. Fac. Sci. Univ. Tokyo, **16** (1970), 469-476.

[Su75] T. Suwa, *Compact quotient spaces of $\mathbb{C}^2$ by affine transformation groups*, J. Differ. Geom. **10**, (1975), 239-252.

[Su77-I] T. Suwa, *Compact quotients of $\mathbb{C}^3$ by affine transformation groups, I.*, Several complex Variables, Proc. Symp. Pure Math. 30, Part 1, Williamstown 1975, (1977), 293-295.

[Su77-II] T. Suwa, *Compact quotients of $\mathbb{C}^3$ by affine transformation groups, II.*, Complex Analysis and Algebraic Geometry, Iwanami Shoten (1977), 259-278.

[Ue75] K. Ueno, *Classification theory of algebraic varieties and compact complex spaces.*, Lecture Notes in Mathematics 439, Springer-Verlag, XIX, 278 p.(1975).

[Ue80] K. Ueno, *On three-dimensional compact complex manifolds with non positive Kodaira dimension*, Proc. Japan Acad., vol.56,S.A.n.10, (1980), 479-483.

[Ue82] K. Ueno, *Bimeromorphic Geometry of algebraic and analytic threefolds*, in 'C.I.M.E. Algebraic Threefolds, Varenna 1981', Lecture Notes in math. 947, (1982), 1-34.

[Ue87] K. Ueno, *On compact analytic Threefolds with non trivial Albanese Tori*, Math. Ann. 278 (1987), 41-70.

[Wel03] J. Y. Welschinger, *Real structures on minimal ruled surfaces.*, Comment. Math. Helv. 78, No.2, 418-446 (2003).

LEHRSTUHL MATHEMATIK VIII, UNIVERSITÄT BAYREUTH, D-95440, BAYREUTH, GERMANY
E-mail address: Fabrizio.Catanese@uni-bayreuth.de

DIPARTIMENTO DI MATEMATICA UNIVERSITÀ DI PAVIA, VIA FERRATA 1, 27100 PAVIA, ITALY
E-mail address: paola.frediani@unipv.it

AMS/IP Studies in Advanced Mathematics
Volume **39**, 2007

Cubic equations of rational triple points of dimension two

Zhijie Chen*, Rong Du†, Sheng-Li Tan*, and Fei Yu*

ABSTRACT. We will give a cubic defining equation for each type of rational triple points of dimension two. Namely, a rational triple point is isomorphic to the normalization of the local surface defined by a cubic equation.

1. Introduction

Rational double points of dimension two were studied first by Du Val ([**4**]) in 1934. There are 5 types of rational double points and each type has one standard quadratic defining equation. These equations are very useful in the classification of algebraic surfaces.

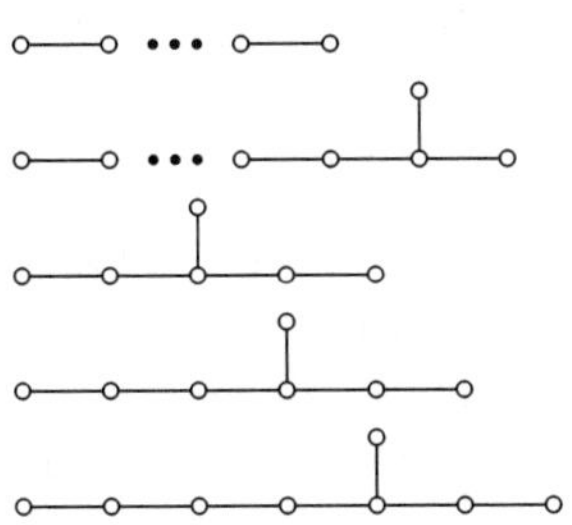

A_n : $z^2 + x^2 + y^{n+1} = 0, \quad (n \geq 1)$

D_n : $z^2 + y(x^2 + y^{n-2}) = 0, \quad (n \geq 4)$

E_6 : $z^2 + x^3 + y^4 = 0$

E_7 : $z^2 + x(x^2 + y^3) = 0$

E_8 : $z^2 + x^3 + y^5 = 0$

From the quadratic equations, we can resolve the surface singularity by using a canonical method for double covers (see [**2**], p.107).

A rational point of multiplicity higher than 2 is not a hypersurface singularity, so it is impossible to define the singularity itself by one equation (cf. [**1**]). On the other hand, a surface singularity is isomorphic to the normalization of a local hypersurface $f(x, y, z) = 0$ in $\mathbb{C}^3$. Sometimes, it is very convenient if we know f, especially when we know the processes of normalization and resolution directly

2000 *Mathematics Subject Classification.* 14B05, 14E15, 32S05.

This work is supported by the 973 Project Foundation, the NSF of China, the Doctoral Program Foundation of EMC and the Program of Shanghai Subject Chief Scientist. The authors are also supported by "DFG-Schwerpunktprogramm Globalc Methoden in der Komplexen Geometrie", and the DFG-NSFC Chinese-German project "Komplexen Geometrie".

from f. A typical example is the Hirzebruch-Jung singularity defined by the normalization of $z^n = xy^{n-q}$. We do not need to find the defining equations of the normalized singularity. In fact, the singularity is determined by n and q.

In 1966, M. Artin [**1**] classified the dual graphs of rational triple points of dimension 2 into 9 classes, and he proved that each rational triple point can be embedded into $\mathbb{C}^4$. In 1968, Tyurina [**6**] gave explicitly 3 defining equations for each singularity. Tyurina [**7**] proved also that a rational triple point is determined uniquely by its dual graph. So isomorphically, there are 9 rational triple points.

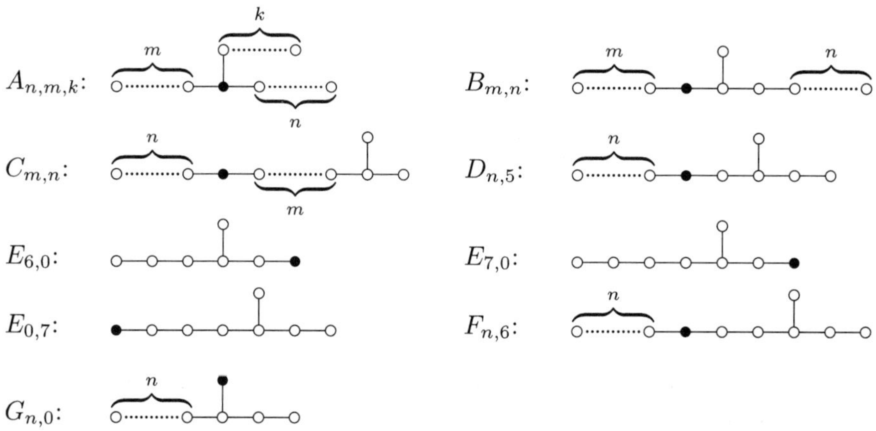

Where $\circ$ is a (-2)-curve, $\bullet$ is a (-3)-curve.

On the other hand, the third author found a simple method (the canonical resolution) to resolve the singularities coming from the normalization of a local surface defined by a cubic equation $z^3+s(x,y)z+t(x,y)=0$ (see §1). Theoretically, a rational triple point might be defined by one cubic equation (up to normalization). So it is interesting to find out the equations similar to rational double points. The main purpose of this paper is to give explicitly the cubic equations for rational triple points. In the last section, we will present the final steps of their canonical resolutions. The following is the list of the corresponding cubic equations (some of them may be not as simple as possible):

$A_{n,m,k}$: $(n \geq m \geq k)$

$$\begin{cases} z^3 + (x+y^{k+1})z^2 + x^2y^{n-k}z + x^2y^{m+2n-3k}(x+y^{k+1}) = 0, & (m>k)\,; \\ z^3 + x(x+y^{k+1})z + x(x+y^{k+1})^2y^p = 0, & (n=3p+k,\ m=k)\,; \\ z^3 + x(x+y^{k+1})z + (x+y^{k+1})^2y^{n+1} = 0, & (n \not\equiv k \pmod 3,\ m=k)\,. \end{cases}$$

$B_{m,n}$:

$$\begin{cases} z^3 + xz^2 - y^{m+3}(y^m+y^p)z - xy^{2m+3} = 0, & n=2p+1\,; \\ z^3 + (x+y^{p+2})z^2 - y^{2m+3}z - xy^{2m+3} = 0, & n=2p+2\,. \end{cases}$$

$C_{m,n}$:

$$\begin{cases} z^3 + x(y^2+x^{m+2})z + x(y^2+x^{m+2})^2y^p = 0, & n=3p+1\,; \\ z^3 + x(y^2+x^{m+2})z + (y^2+x^{m+2})^2y^{n+1} = 0, & n \not\equiv 1 \pmod 3). \end{cases}$$

$D_{n,5}$:

$$z^3 + xz^2 + y^{n+3}z + x^2y^{2n+2} = 0\,.$$

$E_{6,0}$:

$$z^3 + y^3z + x^2y^2 = 0\,.$$

$E_{7,0}$:

$$z^3 + x^2yz + y^4 = 0\,.$$

$E_{0,7}$:

$$z^3 + y^2(x^2 + y^3) = 0\,.$$

$F_{n,6}$:

$$\begin{cases} z^3 + x(x^2+y^3)z + x(x^2+y^3)^2y^p = 0, & n = 3p+2\,; \\ z^3 + x(x^2+y^3)z + (x^2+y^3)^2y^{n+1} = 0, & n \not\equiv 2 \pmod 3\,. \end{cases}$$

$G_{n,0}$:

$$\begin{cases} z^3 + x^{p+2}yz + x^2y^3 = 0, & n = 3p+1\,; \\ z^3 + xy^2(y + x^{p+1}) = 0, & n = 3p+2\,; \\ z^3 + x^{p+2}yz + xy^3 = 0, & n = 3p+3\,. \end{cases}$$

We would like to thank Prof. Stephen S.-T. Yau and Prof. Jin-Gen Yang for useful discussions and suggestions to find the cubic equations.

2. Canonical resolution of the singularities defined by cubic equations

2.1. Branch locus. In this section we will recall Tan's resolution of surface singularities given by a cubic equation [**5**]. Let

$$(V,0) = \{\, (x,y,z) \in \mathbb{C}^3 \;\big|\; z^3 + s(x,y)z + t(x,y) = 0 \,\}\,,$$

where s and t are polynomials in x and y. Denote by $\eta : Y_0 \to (V,0)$ the normalization of $(V,0)$. We will give a canonical resolution of the singularities of Y_0 over 0.

We note first that the projection $(x,y,z) \to (x,y)$ induces a triple cover

$$\pi_0 : Y_0 \to X_0 = \mathbb{C}^2.$$

Assume that we have the following factorizations in $\mathbb{C}[x,y]$ (or in $\mathbb{C}[[x,y]]$)

$$s = a_1a_2^2b_1a_0e^2, \quad t = a_1a_2^2b_1^2b_0e^3, \tag{2.1}$$

where $\gcd(a_i,b_j) = 1$, $\gcd(a_1,a_2) = 1$ and a_1, a_2, b_1 are square-free. The factors are unique up to constants (or units). e has no effect on the normalization Y_0.

The factorizations of s and t can be obtained in a different way. Let

$$a = \frac{4s^3}{\gcd(s^3,t^2)}, \qquad b = \frac{27t^2}{\gcd(s^3,t^2)}, \qquad c = \frac{4s^3+27t^2}{\gcd(s^3,t^2)}.$$

Then a, b and c are coprime polynomials such that $a + b = c$.

$$a = 4a_1a_2^2a_0^3, \qquad b = 27b_1b_0^2. \tag{2.2}$$

It is easy to see that a_i and b_j can also be obtained uniquely from the factorizations (2.2) of a and b. In fact, the triple cover π_0 is also determined by a and b.

Let $c = c_1c_0^2$, where c_1 is square-free. Denote the corresponding divisors by

$$A_i = \mathrm{Div}(a_i), \qquad B_i = \mathrm{Div}(b_i), \qquad C_i = \mathrm{Div}(c_i).$$

Let $D_1 = B_1 + C_1$, $D_2 = A_1 + A_2$. Then the branch locus of the triple cover π_0 is

$$R = 2D_2 + D_1 = 2A_1 + 2A_2 + B_1 + C_1.$$

π_0 is totally ramified over $D_2 = A_1 + A_2$, hence D_2 is called the *totally ramified branch locus*. D_1 is called the *simply ramified branch locus*. Isolated points over which π_0 is totally ramified are contained in D_1.

By ([**5**], Theorem 3.2), Y_0 is smooth if and only if D_2 is smooth, D_2 and D_1 have no common points, and all of the singular points of D_1 are double points of type $y^2 + x^{3n} = 0$ where π_0 is totally ramified.

2.2. Canonical resolution. In case Y_0 is not smooth over $p_0 = (0,0) \in X_0$, we have a canonical method to resolve the singularities.

$$\begin{array}{ccccccccccc}
\widetilde{Y} = Y_k & \xrightarrow{\tau_k} & Y_{k-1} & \xrightarrow{\tau_{k-1}} & \cdots & \longrightarrow & Y_2 & \xrightarrow{\tau_2} & Y_1 & \xrightarrow{\tau_1} & Y_0 \\
\widetilde{\pi} = \pi_k \big\downarrow & & \big\downarrow \pi_{k-1} & & & & \big\downarrow \pi_2 & & \big\downarrow \pi_1 & & \big\downarrow \pi_0 \\
\widetilde{X} = X_k & \xrightarrow{\sigma_k} & X_{k-1} & \xrightarrow{\sigma_{k-1}} & \cdots & \longrightarrow & X_2 & \xrightarrow{\sigma_2} & X_1 & \xrightarrow{\sigma_1} & X_0
\end{array}$$

(1) σ_{i+1} is the blow-up of X_i at a singular point p_i of the branch locus of π_i. Y_{i+1} is the normalization of $X_{i+1} \times_{X_i} Y_i$. $\widetilde{\pi} = \pi_k$ has a smooth branch locus. So $\widetilde{Y} = Y_k$ is smooth.

(2) The corresponding data $\left(a^{(i)},\ b^{(i)},\ c^{(i)}\right)$ of π_i is obtained from

$$\left(\sigma_i^* a^{(i-1)},\ \sigma_i^* b^{(i-1)},\ \sigma_i^* c^{(i-1)}\right)$$

by eliminating the common factors.

Here is the process of the canonical resolution.

Step 0: *Eliminate the common factors of s^3 and t^2 from the equation of discriminate $4s^3 + 27t^2 = \delta$, we get $a + b = c$.*

Step 1: *Blow up X at a singular point of the branch locus, $\sigma : X' \to X$, and eliminate the common factors of $\sigma^*(a)$ and $\sigma^*(b)$ (coming from the exceptional curve) from the equation $\sigma^*(a) + \sigma^*(b) = \sigma^*(c)$, we get the new data $a' + b' = c'$.*

Step 2: *If the new branch locus is not smooth, repeat Step 1.*

The process will terminate after a finite number of steps. Finally the branch locus becomes smooth.

2.3. Numerical data from the canonical resolution. Put

$$d_i = \min\left\{\mu_{p_i}(A^{(i)}),\ \mu_{p_i}(B^{(i)}),\ \mu_{p_i}(C^{(i)})\right\},$$

where $\mu_p(D)$ is the multiplicity of a divisor D at p. Let

$$m_i = \left[\frac{\mu_{p_i}(D_1^{(i)})}{2}\right], \tag{2.3}$$

$$n_i = \begin{cases} \mu_{p_i}(D_2^{(i)}), & \text{if } d_i \equiv \mu_{p_i}(A^{(i)}) \pmod 3; \\ \mu_{p_i}(D_2^{(i)}) - 1, & \text{otherwise.} \end{cases} \tag{2.4}$$

Let E_i be the exceptional curve of σ_i, $\mathcal{E}_i$ be the total transform of E_i in $\widetilde{X}$, and let $\sigma = \sigma_1 \cdots \sigma_k$. Then

$$\widetilde{D}_1 = \sigma^*(D_1) - 2\sum_{i=0}^{k-1} m_i \mathcal{E}_{i+1}, \tag{2.5}$$

$$\widetilde{D}_2 = \sigma^*(D_2) - \sum_{i=0}^{k-1} n_i \mathcal{E}_{i+1}. \tag{2.6}$$

Hence we get the following criterion:

LEMMA 2.1. *We use also E_i to denote the strict transform of E_i in $\widetilde{X}$.*

(i) $E_i \subset \widetilde{D}_1 \Longleftrightarrow \mu_{p_i}(D_1^{(i)})$ *is odd;*

(ii) $E_i \not\subset \widetilde{D}_1$ *and* $E_i \not\subset \widetilde{D}_2 \Longleftrightarrow \mu_{p_i}(D_1^{(i)})$ *is even and* $d_i \equiv \mu_{p_i}(A^{(i)}) \pmod 3$*;*

(iii) $E_i \subset \widetilde{D}_2 \Longleftrightarrow \mu_{p_i}(D_1^{(i)})$ *is even and* $d_i \not\equiv \mu_{p_i}(A^{(i)}) \pmod 3$. *Furthermore,*

(a) *if* $\mu_{p_i}(A^{(i)}) - d_i \equiv 1 \pmod 3$, *then E_i is a component of $\widetilde{A}_1$;*

(b) *if* $\mu_{p_i}(A^{(i)}) - d_i \equiv 2 \pmod 3$, *then E_i is a component of $\widetilde{A}_2$.*

LEMMA 2.2. ([**3**], Lemma 2.2) *The local intersection number $(D_1 D_2)_p$ of the divisors D_1 and D_2 at any point p is an even number.*

REMARK 2.3. This lemma is very useful in the canonical resolution when we do not know the defining equations.

3. An example of the canonical resolution

EXAMPLE 3.1. $z^3 + x^2yz + y^4 = 0, \quad p_0 = (0,0)$

(0) $s = x^2y,\ t = y^4$

(1) $a = 4x^6,\ b = 27y^5,\ c = 4x^6 + 27y^5$. (Step 0: eliminate $y^3 = \gcd(s^3, t^2)$)

(2) Multiplicities of (a,b,c) at p_0 are $(6,5,5)$.

(3) Pullback of $a + b = c$: $e^6\bar{a} + e^5\bar{b} = e^5\bar{c} \implies e\bar{a} + \bar{b} = \bar{c}$. (Eliminate e^5)

(4) New data: $a' = e\bar{a}$, $b' = \bar{b}$, $c' = \bar{c}$. So e is in a_1.

(5) Multiplicities of (a', b', c') at p_1 are $(1,5,1)$.

...

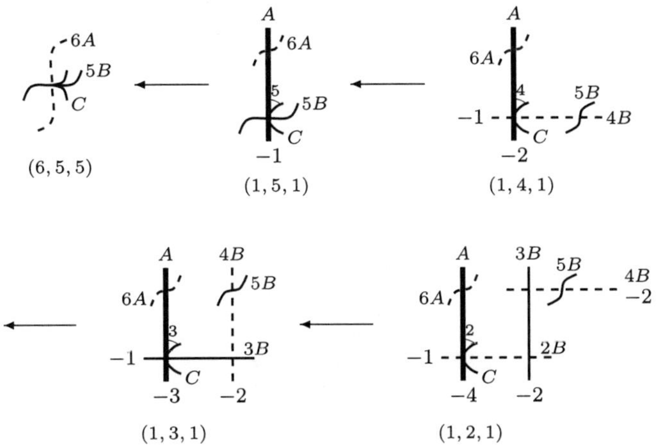

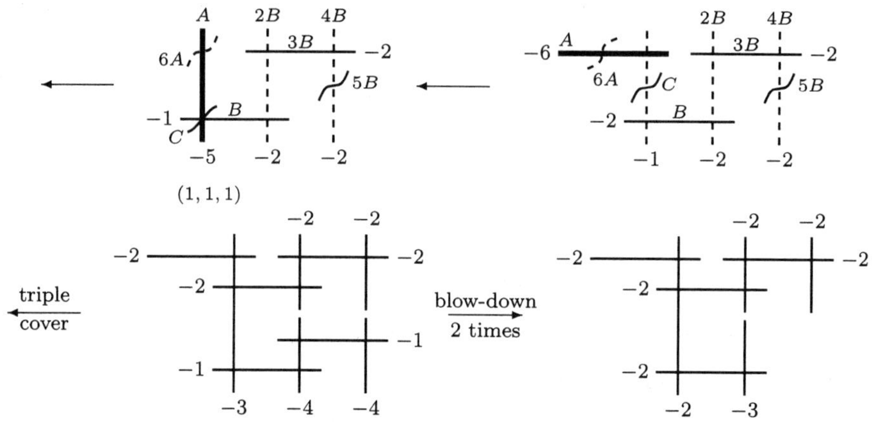

This is a rational triple point of type $E_{7,0}$.

4. Canonical resolution of rational triple points

In this section, we will present the graphs of the final steps of the canonical resolution of the singularities given by the cubic equations in §1. We use the following notations:

(thick line) -3 : a rational curve with self-intersection number -3 which is a component of the totally ramified branch locus D_2;

(thin line) -4 : a rational curve with self-intersection number -4 which is a component of the simply ramified branch locus D_1, note that the self-intersection number -2 will not be marked;

(dashed line) -1 : a rational curve with self-intersection number -1 which is not a component of the branch locus, note that the self-intersection number -2 will not be marked;

(thin mark on dashed line) : a simply ramified point on a rational curve;

(thick mark on dashed line) : a totally ramified point on a rational curve;

(dotted box) p : the objects within this box will be repeated p times.

$A_{n,m,k}$:
$n > k$, $m > k$:
$(n, m, k) = (2p + k, 2q + k, k)$:

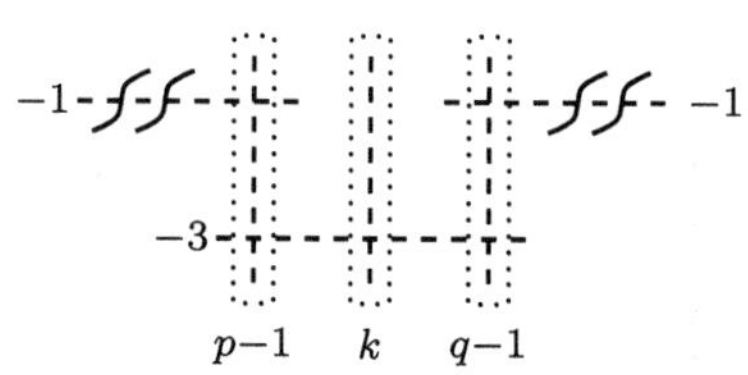

$(n, m, k) = (2p + k + 1, 2q + k, k)$:

$p \geq 1$

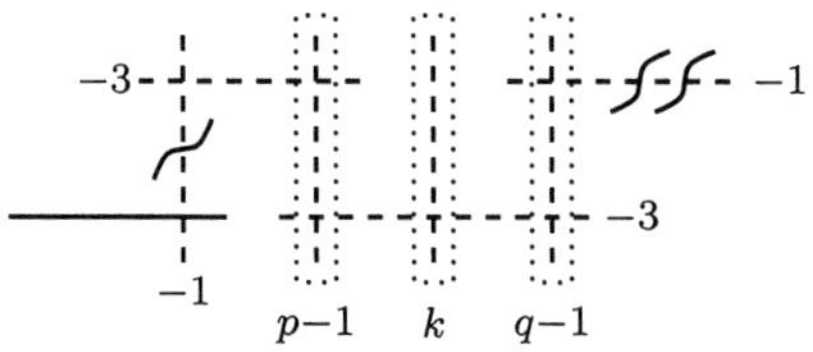

$p = 0$

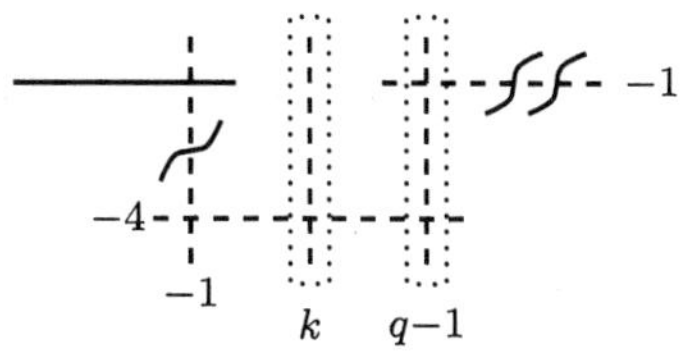

$(n, m, k) = (2p + k + 1, 2q + k + 1, k)$ $(p \geq q)$:

$q \geq 1$

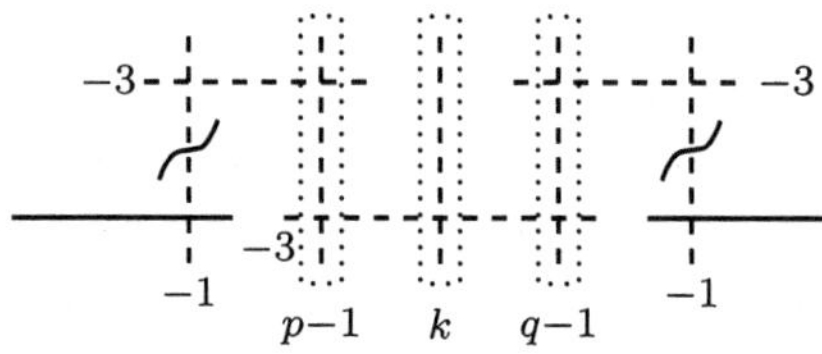

$q = 0$

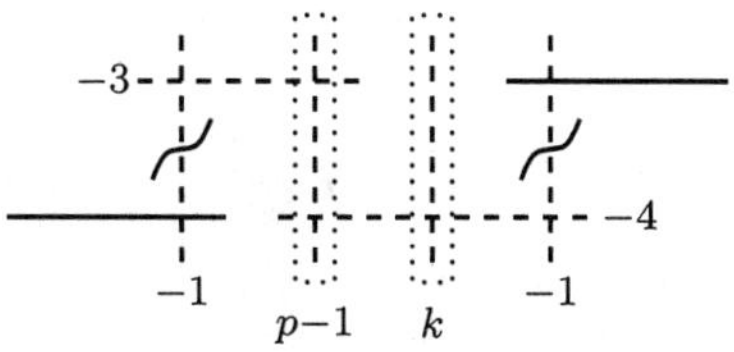

$p = q = 0$

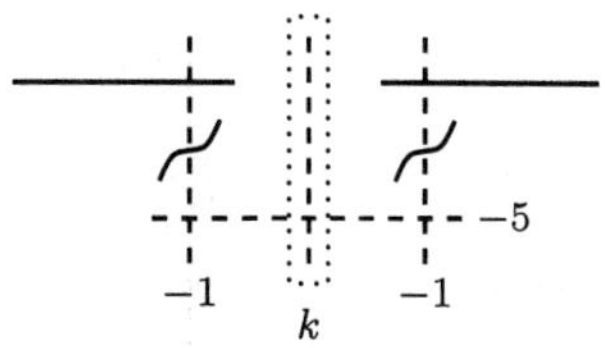

$n \geq m = k$:
$(n, m, k) = (3p + k, k, k)$:

$p \geq 1$

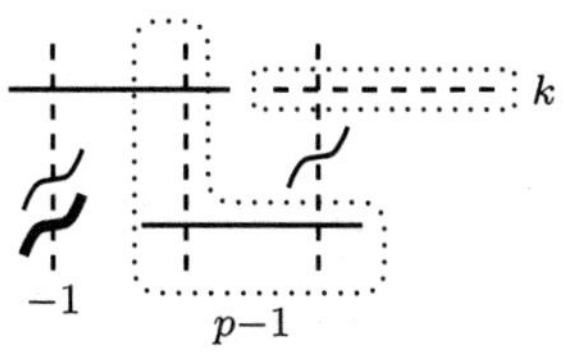

$p = 0$

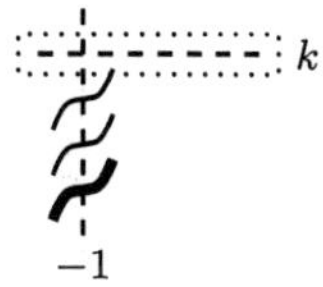

$(n, m, k) = (3p + k + 1, k, k)$:

$p \geq 1$

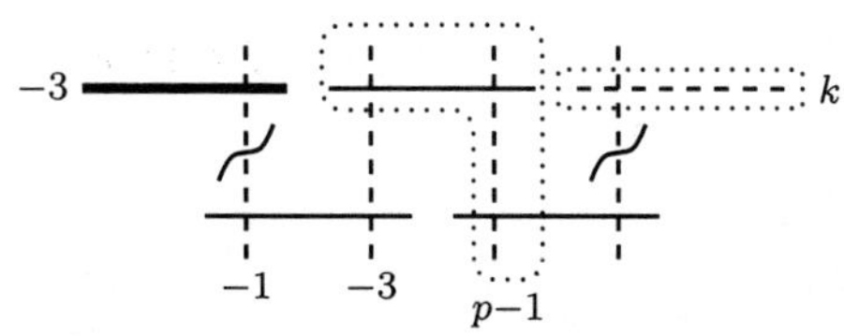

$p = 0$

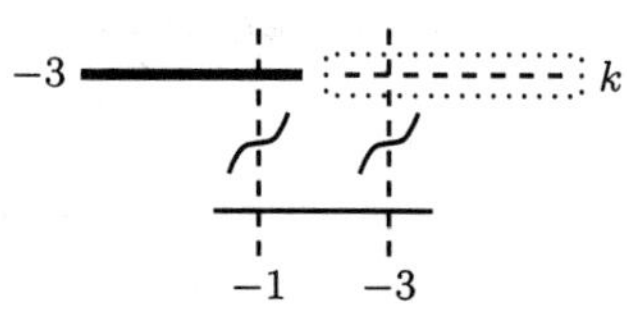

$(n, m, k) = (3p + k + 2, k, k)$:

$p \geq 1$ $p = 0$

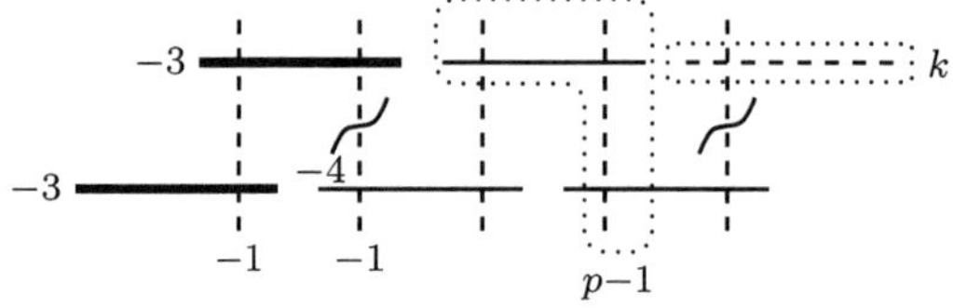

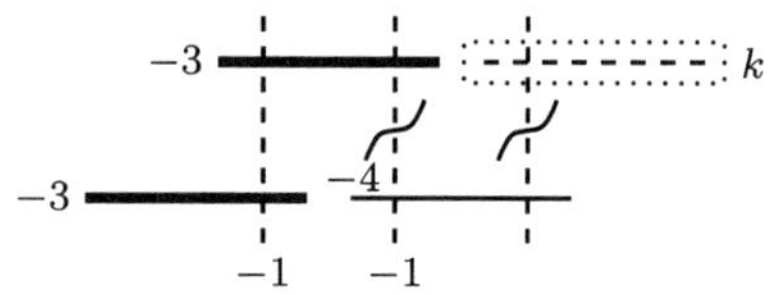

$B_{m,n}$:
$(m, n) = (2q + p, 2p + 1)$:

$q \geq 2$ $q = 1$

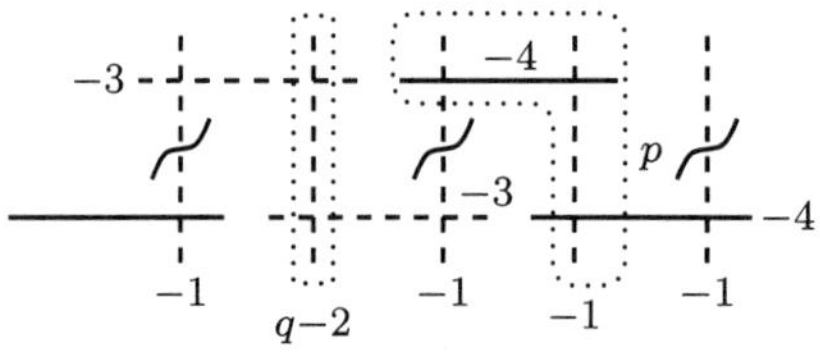

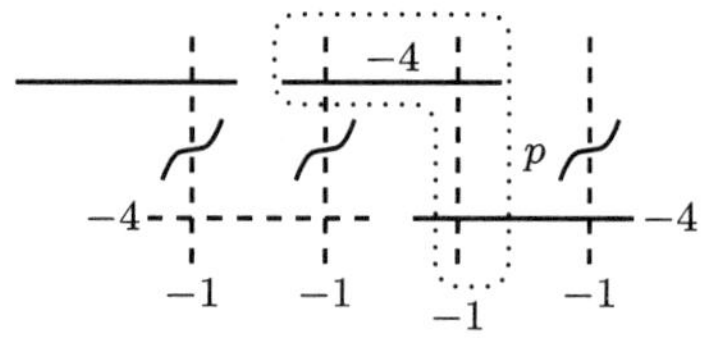

$(m, n) = (2q + p + 1, 2p + 1)$:

$q \geq 1$ $q = 0$

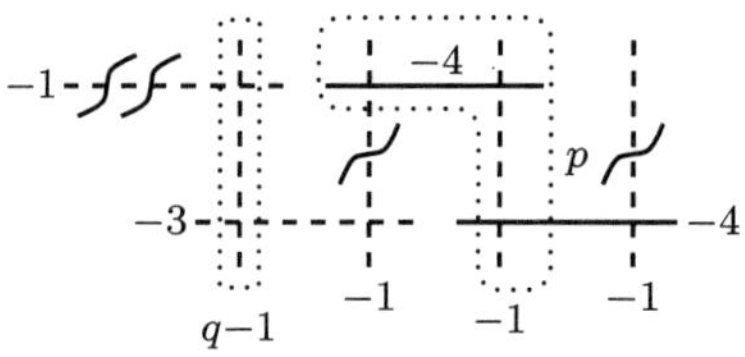

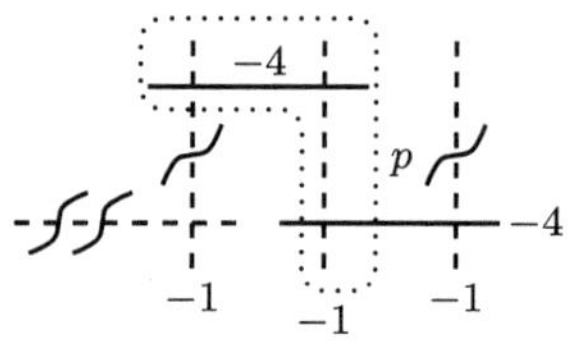

$(m, n) = (p, 2p + 1)$:

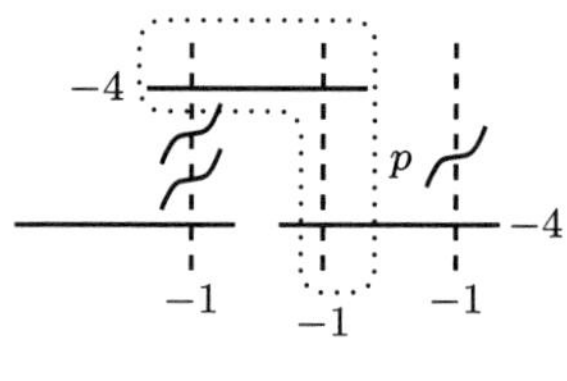

$(m, n) = (m, 2p + 1)$, $m < p$:

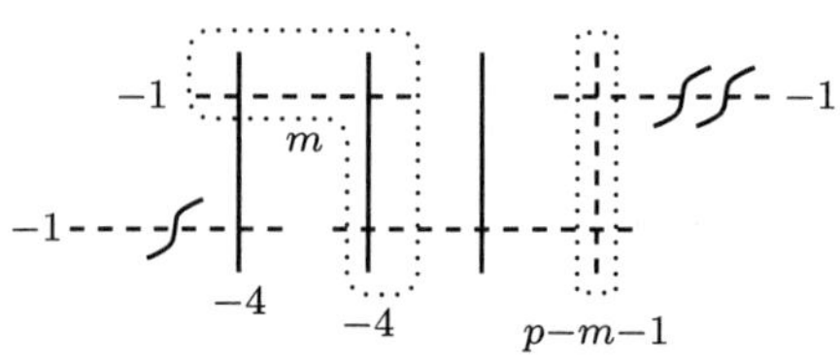

$(m, n) = (3q + p + 1, 2p + 2)$:

$q \geq 1$

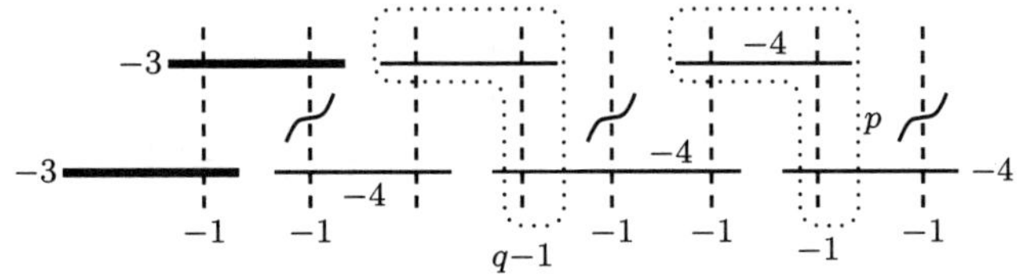

$q = 0$

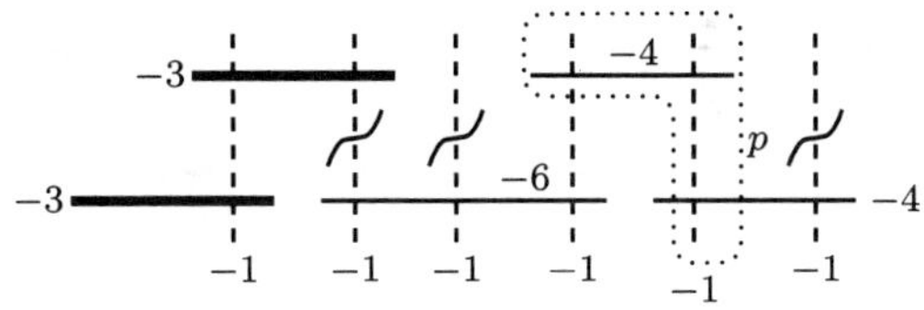

$(m, n) = (3q + p + 2, 2p + 2)$:

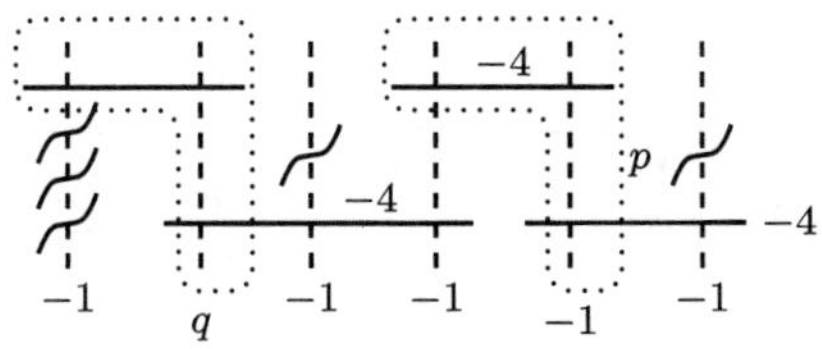

$(m, n) = (3q + p + 3, 2p + 2)$:

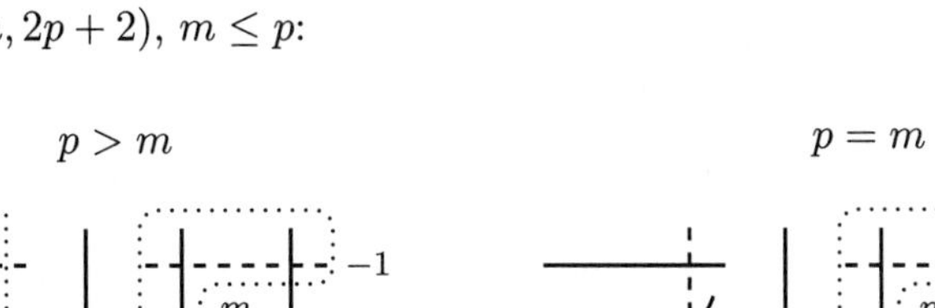

$(m, n) = (m, 2p + 2)$, $m \leq p$:

$p > m$

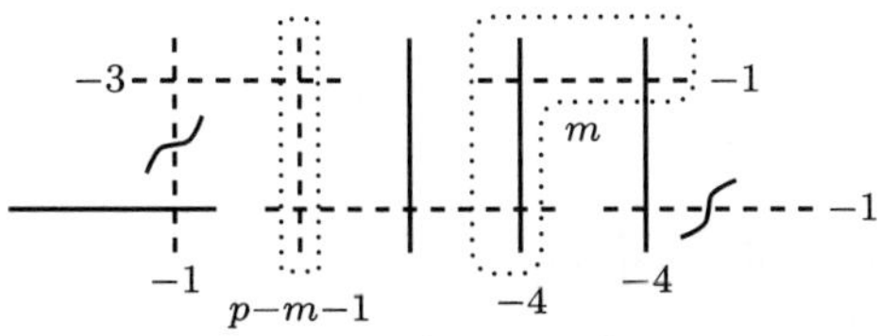

$p = m$

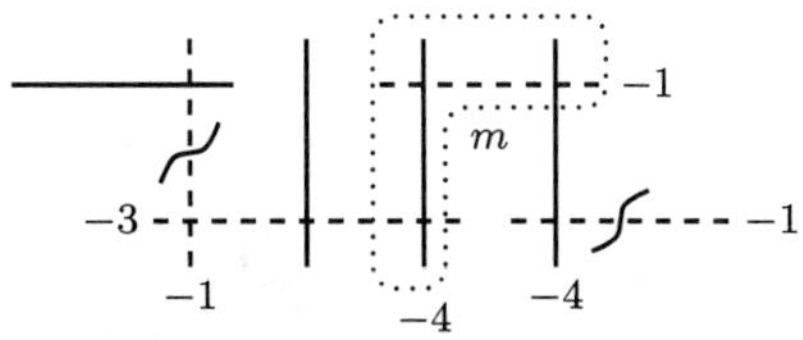

$C_{m,n}$:
$(m, n) = (2q, 3p)$:

$p \geq 1$

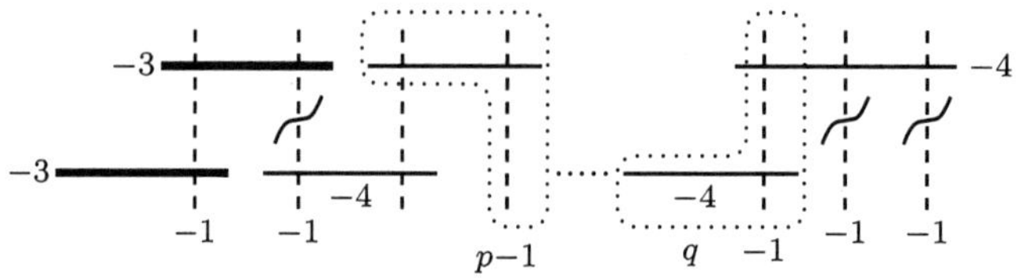

$p = 0,\ q \geq 1$ $p = q = 0$

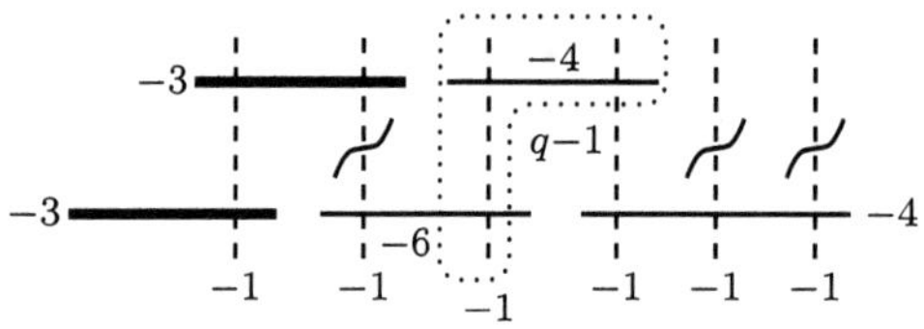

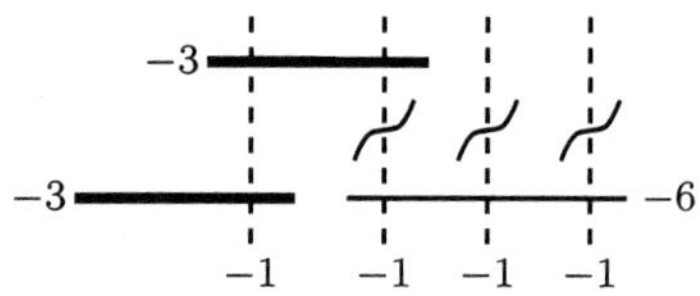

$(m, n) = (2q + 1, 3p)$:

$p \geq 1$

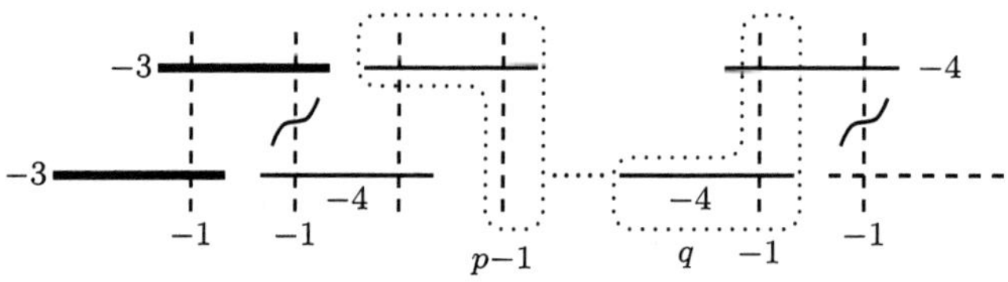

$p = 0$

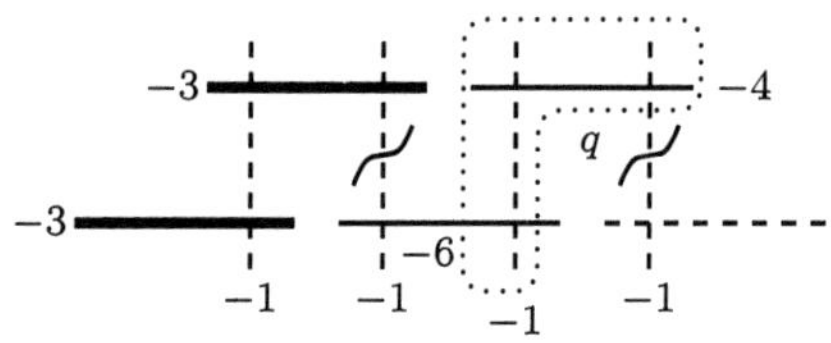

$(m, n) = (2q, 3p + 1)$:

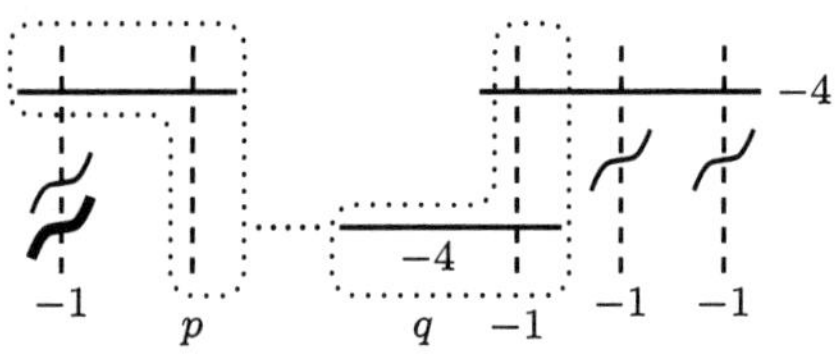

$(m, n) = (2q + 1, 3p + 1)$:

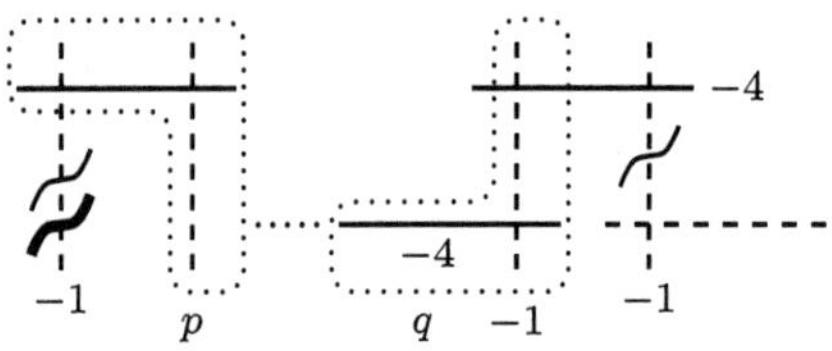

$(m, n) = (2q, 3p + 2)$:

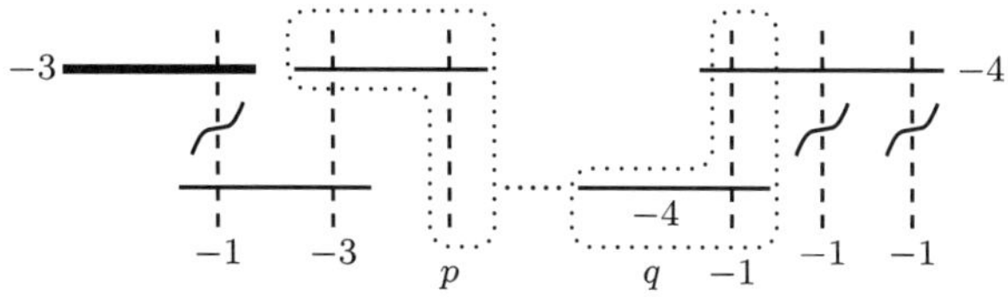

$(m, n) = (2q + 1, 3p + 2)$:

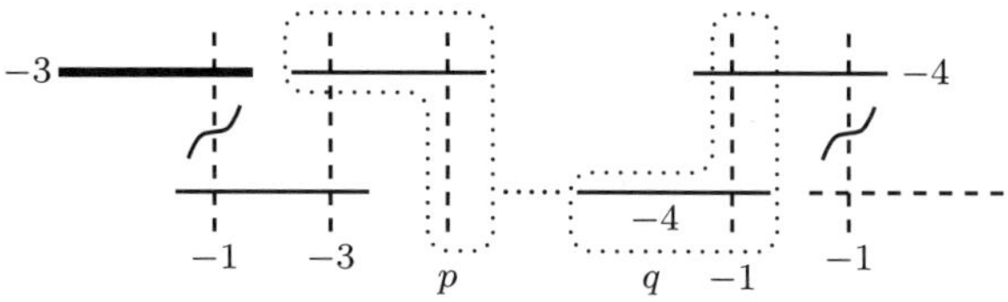

$D_{n,5}$:
$n = 2p$:

$p \geq 1$

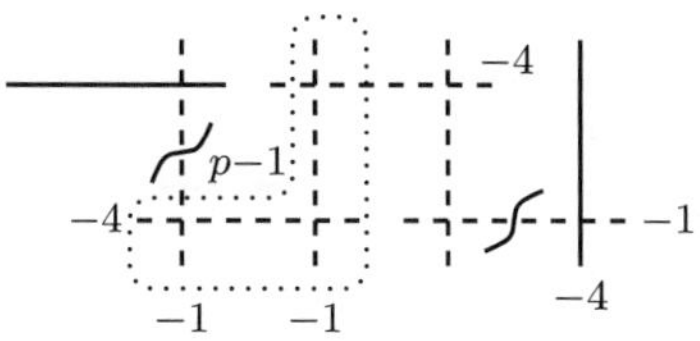

$p = 0$

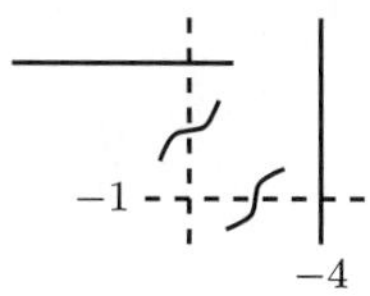

$n = 2p + 1$:

$p \geq 1$

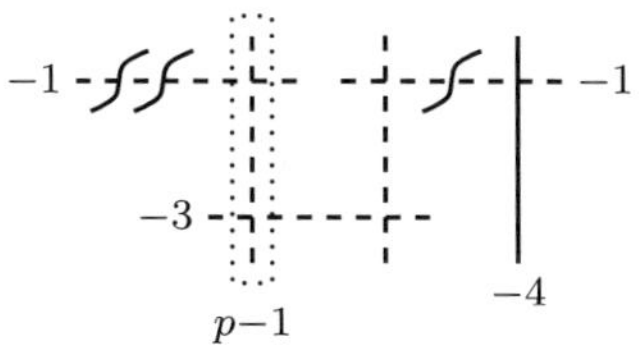

$p = 0$

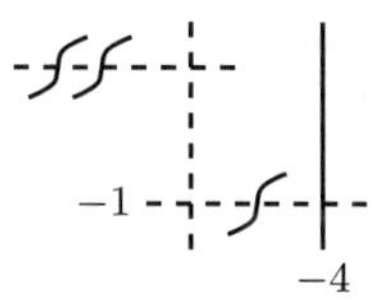

$E_{6,0}$:

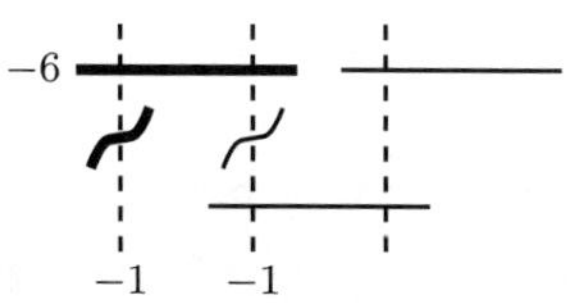

$E_{7,0}$:

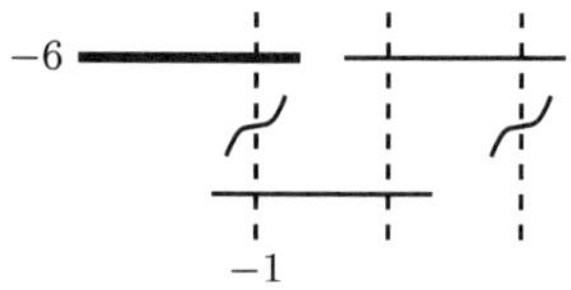

$E_{0,7}$:

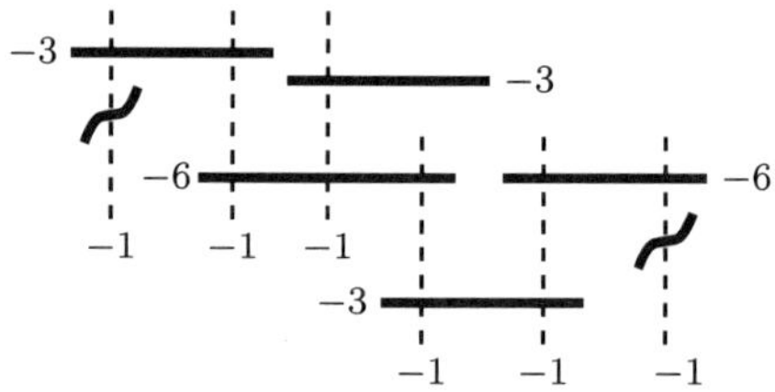

$F_{n,6}$:
$n = 3p$:

$p \geq 1$ $p = 0$

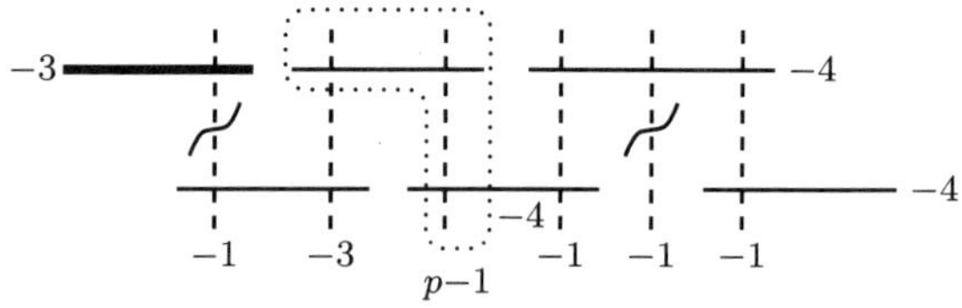

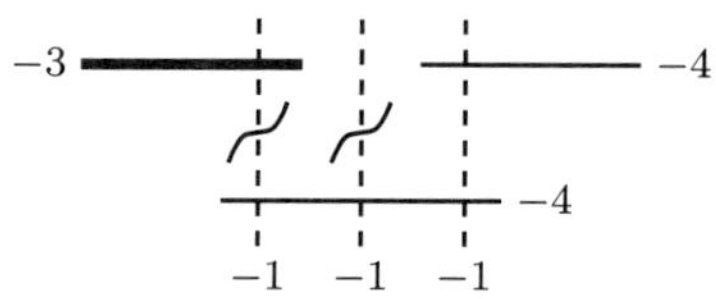

$n = 3p + 1$:

$p \geq 1$

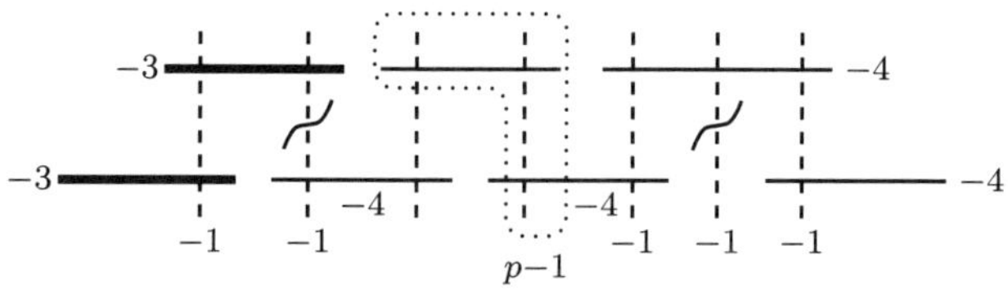

$p = 0$

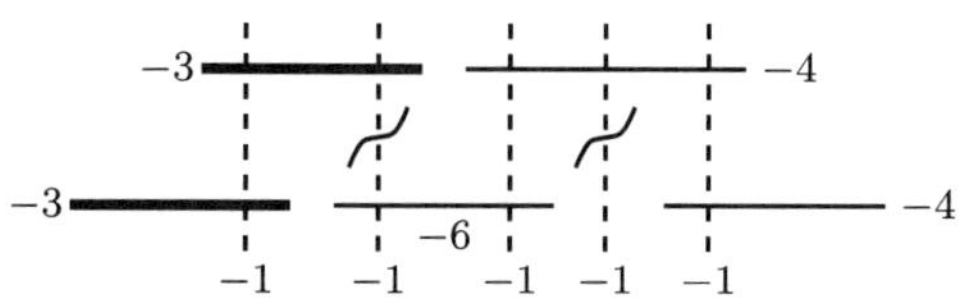

$n = 3p + 2$:

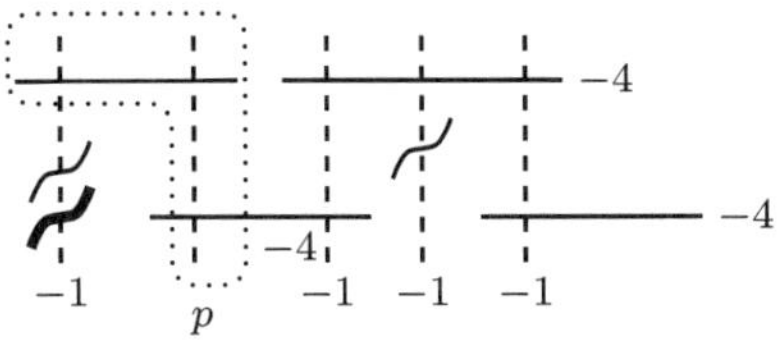

$G_{n,0}$:
$n = 3p + 1$:

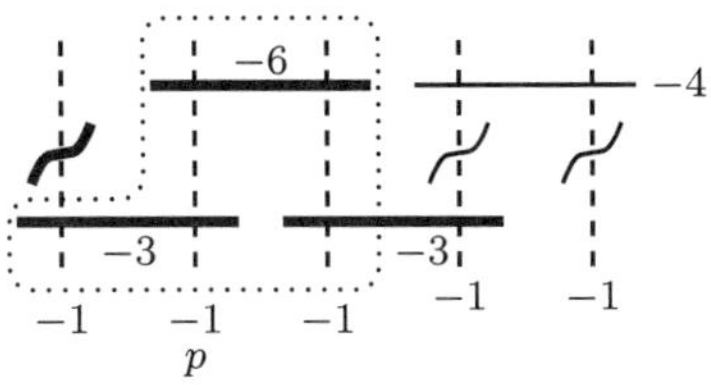

$n = 3p + 2$:

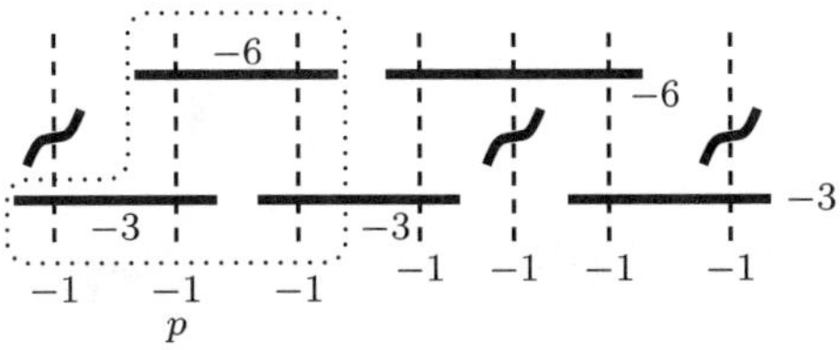

$n = 3p + 3$:

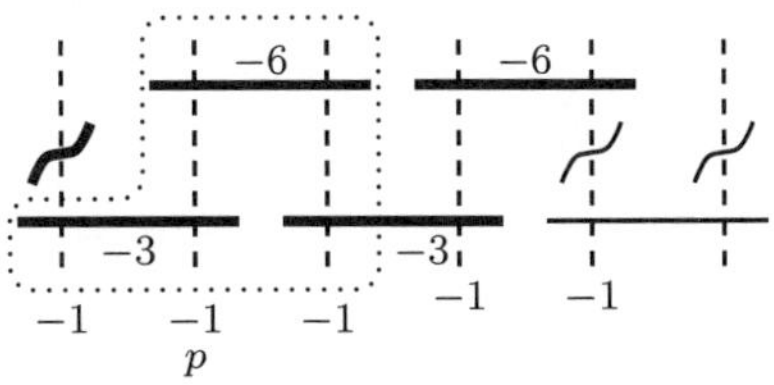

References

[1] M. Artin, *On isolated rational singularities of surface*, Amer. J. Math., **88**(1966), 129–136.

[2] W. Barth, K. Hulek, C. Peters, A. Van de Ven, *Compact Complex Surfaces,* Second enlarged edition, Ergebnisse der Mathematik, Springer-Verlag, Berlin 2004.

[3] Z. Chen, S.-L. Tan, *Upper bounds on the slope of a genus 3 fibration*, Comtemp. Math., Amer. Math. Soc., Providence, RI2006 (to appear).

[4] P. Du Val, *On isolated singularities of surfaces that do not affect the condition of adjunction I, II, III,* Proc. Cambridge Philos. Soc. 30 (1934), 483–491.

[5] S.-L. Tan, *Triple covers on smooth algebraic varieties*, Geometry and Nonlinear Partial Differential Equations, AMS/IP Studies in Adv. Math., vol. 29, Amer. Math. Soc. and Intern. Press, 2002, pp. 143–164.

[6] G. N. Tyurina, *Absolute isolatedness of rational singularities and triple rational points*, Func. Anal. Appl., **2** (1968), 324–332.

[7] G. N. Tyurina, *The rigidity of rationally contractible curves on a surface(Russian)*, Izy. Akad. Nauk SSSR, Ser. Mat **32** (1968), 943–970.

* Department of Mathematics, East China Normal University, Shanghai 200062, P. R. of China

E-mail address: zjchen@math.ecnu.edu.cn, sltan@math.ecnu.edu.cn, vvyufei@yahoo.com.cn

† Department of Mathematics, Statistics and Computer Sciences (M/C 249), University of Illinois at Chicago, 851 South Morgan Street, Chicago, IL 60607-7045, USA

E-mail address: rdu2@uic.edu

AMS/IP Studies in Advanced Mathematics
Volume **39**, 2007

Multi-parameter cells of finite Coxeter groups

Jie Du and Hebing Rui

Abstract. Cells of Coxeter groups are certain equivalence classes defined by the Kazhdan-Lusztig type basis of the associated Hecke algebra with a set of parameters. In this paper, we prove that, for finite Coxeter groups, cells arising from a multi-parameter Hecke algebra are determined by those arising from a Hecke algebra with parameters which are powers of a single parameter.

1. Introduction

Let (W, S) be a Coxeter system. For each $s \in S$, let u_s be an indeterminant such that $u_s = u_t$ if $s, t \in S$ are W-conjugate. Let $\mathcal{A} = \mathbb{Z}[\Gamma]$ be the group ring, where Γ is the abelian group generated by u_s for all $s \in S$. Let $\lhd$ be a total (or linear) order on Γ, which is compatible with the group structure of Γ. In [**Lu1**] (see [**KL**] for the equal parameter case) Lusztig introduced the canonical basis with respect to $\lhd$ for the Hecke algebra $\mathcal{H} = \mathcal{H}_{\mathcal{A}}$ over $\mathcal{A}$ associated to W (with parameters $\{u_s\}_{s\in S}$). Via such a basis, W can be divided into certain equivalence classes called generalized cells relative to $\lhd$ (or simply $\lhd$-cells), which play an important role in the representation theory of Hecke algebras. If Γ is not a cyclic group, these cells are called multi-parameter cells in this paper.

Consider a group homomorphism $\sigma : \Gamma \to \mathbb{Z}$. It induces a homomorphism $\tilde{\sigma} : \mathcal{A} \to \mathcal{Z} := \mathbb{Z}[q^{\frac{1}{2}}, q^{-\frac{1}{2}}]$ such that $\tilde{\sigma}(u_s) = q^{\sigma(u_s)/2}$ and let $\mathcal{H}^\sigma = \mathcal{H}^\sigma_{\mathcal{Z}} = \mathcal{H} \otimes \mathcal{Z}$. We shall assume that σ is admissible (see Definition 3.1 below), and call the generalized cells (or σ-cells) arising from the canonical basis for $\mathcal{H}^\sigma$ the Kazhdan-Lusztig cells if $\sigma(u_s) = \sigma(u_t)$ for all $s, t \in S$, or unequal parameter cells otherwise.

The unequal-parameter cells have been studied by many authors. For example, in [**Lu1**], Lusztig showed that all the left σ-cell representations are simple if the Hecke algebras of type B_n have unequal parameters q, $q^{3/2}$ or q and $q^{\frac{1}{2}}$. In [**DPS**], σ-cells were used to study certain endomorphism algebras. Also, beautiful relations between certain σ-cells and Kazhdan-Lusztig cells were discovered in [**B**]. Moreover, σ-cells are the main objects discussed in the recent book [**Lu3**].

2000 *Mathematics Subject Classification.* 20C08.

Key words and phrases. multi-parameter cells, Hecke algebras,

The work was completed while the second author was visiting the University of New South Wales in 1998 funded by ARC Grant A69530243.

However, it seems not clear for us to study the representations of Hecke algebras with multi-parameters via cells. This is because the multi-parameter cells are harder to be described. This motivates the authors to propose the following questions.

QUESTION 1.1. (a) For any total order $\lhd$ on Γ, is there a specialization σ such that the multi-parameter $\lhd$-cells coincide with σ-cells?
(b) For any specialization σ, is there a total ordering on Γ such that the resulting $\lhd$-cells agree with the σ-cells?

In this paper, we show that, for a finite Coxeter group, Question 1.1(a) is true. However, Question 1.1(b) is not true in general. We will provide some examples to illustrate it.

The contents of this paper are organized as follows. In §2, we recollect some definitions and results on multi-parameter cells. In §3, we prove that multi-parameter $\lhd$-cells of W coincide with σ-cells for some specialization σ. In §4, we describe all multi-parameter $\lhd$-cells for dihedral groups and multi-parameter $\lhd$-cells for the Weyl group of type B_3 in order to make a comparison with the unequal-parameter σ-cells described in [**Lu1**]. Our example shows that Question 1.1(b) is not true in general.

2. Multi-parameter cells

In this section, we recollect some definitions and results. We shall restrict our attention to finite Coxeter groups although some of definitions and results in this section are valid for infinite Coxeter groups.

Let (W,S) be a finite Coxeter system where $S=\{s_1,s_2,\ldots,s_n\}$ is the set of Coxeter generators. For any $s_i\in S$, let u_{s_i} be an indeterminate such that $u_{s_i}=u_{s_j}$ if s_i and s_j are W-conjugate. Write $u_i=u_{s_i}$ for brevity. Let $\Gamma=\langle u_1,u_2,\ldots,u_n\rangle$ be the free abelian group generated by $u_1,u_2,\ldots,u_n$, and let $A=\mathbb{Z}[\Gamma]$ be the group ring of Γ. Then the Hecke algebra $\mathcal{H}$ associated to W is an associative algebra with free A-basis $\{T_w\}_{w\in W}$ satisfying the following multiplicative rules:

$$\begin{cases}(T_{s_i}-u_i^2)(T_{s_i}+1)=0, & \text{if } s_i\in S,\\ T_yT_w=T_{yw}, & \text{if } l(yw)=l(y)+l(w).\end{cases}$$

Here l is the length function on (W,S).

Let $u_w=\prod_{j=1}^k u_{i_j}$ if $s_{i_1}s_{i_2}\cdots s_{i_k}$ is a reduced expression of w. It is known that u_w is independent of the choice of a reduced expression of w. Putting $\widetilde{T}_w=u_w^{-1}T_w$, we obtain a new basis $\{\widetilde{T}_w\}_{w\in W}$ for $\mathcal{H}$. Following [**Lu1**], let $\bar{\ }$ be the $\mathbb{Z}$-linear involution on the ring $\mathcal{A}$ satisfying $\bar{\gamma}=\gamma^{-1}$ for all $\gamma\in\Gamma$, and extend it to $\mathcal{H}$ by setting

$$\overline{\sum_w a_w\widetilde{T}_w}=\sum_w \bar{a}_w\widetilde{T}_{w^{-1}}^{-1},\ a_w\in\mathcal{A}.$$

For any $x,y\in W$, let $R_{x,y}\in\mathcal{A}$ be defined by $\widetilde{T}_{y^{-1}}^{-1}=\sum_{x\in W}R_{x,y}\widetilde{T}_x$. Clearly, $R_{x,y}\neq 0$ implies $x\leqslant y$ where $\leqslant$ is the Bruhat-Chevalley order on W, and for

$sy < y$, we have

$$\begin{aligned}\widetilde{T}_{y^{-1}}^{-1} = \widetilde{T}_s^{-1}\widetilde{T}_{(sy)^{-1}}^{-1} &= \sum_{z\leqslant sy} R_{z,sy}\widetilde{T}_s\widetilde{T}_z + (u_s^{-1} - u_s)\sum_{z\leqslant sy} R_{z,sy}\widetilde{T}_z \\ &= \sum_{z\leqslant sy, sz>z} R_{z,sy}\widetilde{T}_{sz} + (u_s^{-1} - u_s)\sum_{z\leqslant sy} R_{z,sy}\widetilde{T}_z \\ &+ \sum_{z\leqslant sy, sz<z} \left(R_{z,sy}\widetilde{T}_{sz} + (u_s - u_s^{-1})\widetilde{T}_z\right)\end{aligned}$$

Consequently,

$$R_{x,y} = \begin{cases} R_{sx,sy}, & \text{if } sx < x \text{ and } sy < y, \\ (u_s^{-1} - u_s)R_{x,sy} + R_{sx,sy}, & \text{if } sx > x \text{ and } sy < y. \end{cases} \tag{2.1}$$

Note that the polynomial $R_{x,y}$ is the polynomial $\bar{R}^*_{x,y}$ defined in [**Lu1**], and $\bar{R}_{x,y} = \epsilon_x\epsilon_y R_{x,y}$, where $\epsilon_w = (-1)^{l(w)}$.

LEMMA 2.1. *For any $x, y \in W$ with $x \leqslant y$, if $\prod_{i=1}^n u_i^{a_i}$ is a monomial appearing in $R_{x,y}$, then $\sum_{i=1}^n |a_i| \leqslant l(y) - l(x)$. Moreover, if the inequality becomes an equality, then $R_{x,y}$ has the highest term $\epsilon_y\epsilon_x u_y u_x^{-1}$ and the lowest term $u_x u_y^{-1}$.*

PROOF. By (2.1), $R_{x,x} = R_{e,e} = 1$. Hence the result follows. Suppose $x < y$. We prove the result by induction on $2l(y) - l(x)$.

Since $x < y$, we have $2l(y) - l(x) \geqslant l(y) + 1 \geqslant 2$. If $2l(y) - l(x) = 2$, then $y \in S$ and $x = e$. Thus, the results follow immediately from the equality $R_{e,s} = u_s^{-1} - u_s$ with highest term $\epsilon_s u_s$ and lowest term u_s^{-1}. Assume $2l(y) - l(x) > 2$. Then

(1) $2l(sy) - l(sx) < 2l(y) - l(x)$ if $sx < x$ and $sy < y$, and
(2) $2l(sy) - l(sx) < 2l(sy) - l(x) < 2l(y) - l(x)$, if $sx > x$ and $sy < y$.

In the first case, we have by (2.1) $R_{x,y} = R_{sx,sy}$. Since $l(y) - l(x) = l(sy) - l(sx)$, all results follow from induction. In the second case, since $l(sy) - l(sx) < l(sy) - l(x) < l(y) - l(x)$, we have $\sum_{i=1}^n |a_i| \leqslant l(y) - l(x)$ by (2.1) and induction assumption. If $\sum_{i=1}^n |a_i| = l(y) - l(x)$, then the highest and lowest terms must appear in $(u_s^{-1} - u_s)R_{x,sy}$. Thus, by induction, they are of the forms $-u_s\epsilon_x\epsilon_{sy}u_x^{-1}u_{sy} = \epsilon_y\epsilon_x u_y u_x^{-1}$ and $u_s^{-1}u_x u_{sy}^{-1} = u_x u_y^{-1}$, as desired. □

Consider a total order $\lhd$ on Γ which is *compatible* with the group structure of Γ in the sense that, if $a \lhd b$ and $c \lhd d$, then $ac \lhd bd$. Recall that $u_{s_i} = u_i$ for any $s_i \in S$. Such an order is said to be *admissible* if $u_i \rhd 1$ for all $s_i \in S$. Let $\lhd$ be a fixed admissible total ordering on Γ. Thus, if $1 \lhd \gamma$, then $\bar{\gamma} = \gamma^{-1} \lhd 1$. Put $\Gamma_+ = \{\gamma \in \Gamma \mid 1 \lhd \gamma\}$, $\Gamma_- = \Gamma_+^{-1}$ and $\Gamma_0 = \{1\}$. Let $\mathbb{Z}[\Gamma_-]$ (resp. $\mathbb{Z}[\Gamma_+]$) be the $\mathbb{Z}$-submodule of $\mathbb{Z}[\Gamma]$ spanned by Γ_- (resp. Γ_+).

Fix a linear ordering on W: $w_1, w_2, \ldots, w_\nu$ with $\nu = |W|$ such that $w_i \leqslant w_j$ implies $i \leqslant j$. Let $R = (R_{ij})$ be the $\nu \times \nu$ matrix with $R_{ij} = R_{w_i,w_j}$. It is known that $R_{i,i} = 1$ for all i, and $R_{i,j} = 0$ unless $w_i \leqslant w_j$. Therefore, R is a pomatrix in the sense of [**Du2**, 1.2].[1] Furthermore, $R\bar{R} = I$, the identity matrix. So the matrix R is a 1-pomatrix in the sense of [**Du2**, 1.2][2]. We now endow Γ a total

[1]A pomatrix is a matrix $A = (a_{ij})_{i,j\in\wp}$ over a poset $\wp$ satisfying $a_{ii} = 1$ and $a_{ij} = 0$ unless $i \leqslant j$.

[2]A 1-pomatrix R is a pomatrix R such that $R\bar{R} = I$. An ∞-pomatrix R (resp. 0-pomatrix) is a pomatrix such that $R_{ij} \in \mathbb{Z}[\Gamma_-]$ (resp. $R_{ij} \in \mathbb{Z}[\Gamma_+]$) for $i < j$ in $\wp$.

ordering so that R can be uniquely written as a product of an ∞-pomatrix P and the 0-pomatrix $\bar{P}^{-1}$.

LEMMA 2.2. *Maintain the notation introduced above. There exists a unique upper triangular unipotent matrix* $P^{\triangleleft} = (P^{\triangleleft}_{i,j})$ *with* $P^{\triangleleft}_{i,j} = P^{\triangleleft}_{w_i,w_j} \in \mathbb{Z}[\Gamma_-]$, $P^{\triangleleft}_{i,i} = P^{\triangleleft}_{w_i,w_i} = 1$ *such that* $R = P^{\triangleleft}\bar{P^{\triangleleft}}^{-1}$.

PROOF. We claim that $P^{\triangleleft}_{i,j}$ are the unique solution to the system

$$x_{i,j} = \sum_{\substack{k \\ w_i \leqslant w_k \leqslant w_j}} R_{i,k}\bar{x}_{k,j}$$

with $x_{i,i} = 1$ and $x_{i,j} \in \mathbb{Z}[\Gamma_-]$ for $w_i < w_j$. Thus, $R = P^{\triangleleft}\bar{P^{\triangleleft}}^{-1}$. The proof of the claim is entirely similar to [**Du1**, 1.2] (cf. [**Lu2**, 7.10]): write $i \leqslant j$ if $w_i \leqslant w_j$ and suppose that $P^{\triangleleft}_{k,j}$ are known for all k, $i < k \leqslant j$. Then

$$\begin{aligned}\sum_{\substack{k, \\ i<k\leqslant j}} R_{i,k}\bar{P}^{\triangleleft}_{k,j} &= \sum_{\substack{k, \\ i<k\leqslant j}} R_{i,k} \sum_{\substack{k', \\ k\leqslant k'\leqslant j}} \bar{R}_{kk'}P^{\triangleleft}_{k',j} = \sum_{\substack{k', \\ i<k'\leqslant j}} \sum_{\substack{k, \\ i<k\leqslant k'}} R_{i,k}\bar{R}_{kk'}P^{\triangleleft}_{k',j} \\ &= \sum_{\substack{k', \\ i<k'\leqslant j}} (-\bar{R}_{ik'})P^{\triangleleft}_{k'j} = -\sum_{\substack{k, \\ i<k\leqslant j}} \overline{R_{i,k}\bar{P}^{\triangleleft}_{k,j}}\end{aligned}$$

Now $P^{\triangleleft}_{i,j} \in \mathbb{Z}[\Gamma_-]$ is the unique solution satisfying

$$P^{\triangleleft}_{i,j} - \bar{P}^{\triangleleft}_{i,j} = \sum_{\substack{k, \\ i<k\leqslant j}} R_{i,k}\bar{P}^{\triangleleft}_{k,j}.$$

□

Note that $P^{\triangleleft}_{x,w}$ is a multi-variable polynomial and is the polynomial $P^{*}_{x,w}$ discussed in [**Lu1**]. They can be calculated recursively by the following formulas:

$$\begin{cases} P^{\triangleleft}_{w,w} = 1, \text{ for any } w \in W, \\ P^{\triangleleft}_{x,w} - \bar{P}^{\triangleleft}_{x,w} = \sum_{x<y\leqslant w} R_{x,y}P^{\triangleleft}_{y,w}. \end{cases} \tag{2.2}$$

Also, we have

$$\sum_{x\leqslant z\leqslant y} \epsilon_x\epsilon_z P^{\triangleleft}_{x,z}P^{\triangleleft}_{w_0y,w_0z} = \delta_{x,y} \text{ for all } x, y \in W,$$

where w_0 is the longest element in W.

The decomposition $R = P^{\triangleleft}\bar{P^{\triangleleft}}^{-1}$ given in Lemma 2.2 is called an *antipode* decomposition. Every admissible total ordering $\triangleleft$ gives rise to such a decomposition. However, the number of such decompositions is finite as we shall see below.

DEFINITION 2.3. Two admissible total orders $\triangleleft$ and $\triangleleft'$ on Γ are said to be equivalent if $P^{\triangleleft} = P^{\triangleleft'}$.

PROPOSITION 2.4. *The number of non-equivalent admissible total orders on* Γ *is finite.*

PROOF. First, we claim that, if a monomial $\prod_{i=1}^{n} u_i^{a_i}$ appears in $P^{\triangleleft}_{x,w}$ for $x < w$, then the inequality $\sum_{i=1}^{n} |a_i| \leqslant l(w) - l(x)$, independent of the total order $\triangleleft$. We prove it by induction on $l(w) - l(x)$. Indeed, if $l(w) - l(x) = 1$, then a reduced expression of x can be obtained from a reduced expression of w by dropping a factor $s \in S$. In this case, $P^{\triangleleft}_{x,y} = u_s^{-1}$, and our claim follows. Assume now $l(w) - l(x) > 1$. By Lemma 2.1, we have $\sum_{i=1}^n |b_i| \leqslant l(y) - l(x)$ if $\prod_{i=1}^n u_i^{b_i}$ is a monomial appearing in $R_{x,y}$ for any $x < y \leqslant w$. Since $y > x$, we have $l(w) - l(y) < l(w) - l(x)$. By inductive hypothesis, we have $\sum_{i=1}^n |c_i| \leqslant l(w) - l(y)$ if $\prod_{i=1}^n u_i^{c_i}$ appears in $P^{\triangleleft}_{y,w}$ for $y < w$. Therefore, by (2.2), any monomial $\prod_{i=1}^n u_i^{a_i}$ of $P^{\triangleleft}_{x,w} - \bar{P}^{\triangleleft}_{x,w}$ satisfies $\sum_{i=1}^n |a_i| \leqslant l(w) - l(x)$ (independent of $\triangleleft$). Since $P^{\triangleleft}_{x,w} \in \mathbb{Z}[\Gamma_-]$ and $\bar{P}^{\triangleleft}_{x,w} \in \mathbb{Z}[\Gamma_+]$, our claim follows immediately.

By the claim, for given $x \leqslant w$, the number of such polynomials $P^{\triangleleft}_{x,w}$ is finite. Since $P^{\triangleleft}_{x,w} = 0$ if $x \nleqslant w$ and $P^{\triangleleft}_{x,x} = 1$ for all $x \in W$, there are only finitely many antipode decompositions of R. Consequently, the number of non-equivalent total orderings on Γ is finite. □

With the matrix $P^{\triangleleft}$, we may define the *canonical basis* $\{C^{\triangleleft}_w \mid w \in W\}$ for Hecke algebra $\mathcal{H}$, where $C^{\triangleleft}_w = \sum_{y \leqslant w} P^{\triangleleft}_{y,w} \widetilde{T}_y$. Such a basis was first introduced by Lusztig in [**Lu1**][3]. He also gave the following formula:

$$C^{\triangleleft}_s C^{\triangleleft}_w = \begin{cases} (u_s + u_s^{-1}) C^{\triangleleft}_w, & \text{if } sw < w, \\ C^{\triangleleft}_{sw} + \sum_{z<w, sz<z} M^{s,\triangleleft}_{z,w} C^{\triangleleft}_z, & \text{if } sw > w, \end{cases} \tag{2.3}$$

where $M^{s,\triangleleft}_{z,w}$ is defined by the condition that $\sum_{z \leqslant y < w, sy < y} P^{\triangleleft}_{z,y} M^{s,\triangleleft}_{y,w} - u_s P^{\triangleleft}_{y,w}$ is a $\mathbb{Z}$-linear combinations of elements in Γ_-. Note that $\bar{M}^{s,\triangleleft}_{z,w} = M^{s,\triangleleft}_{z,w}$ since $\bar{C}^{\triangleleft}_w = C^{\triangleleft}_w$.

The canonical basis $\{C^{\triangleleft}_w\}$ is used to define cells *relative to* the total order $\triangleleft$, or simply $\triangleleft$-cells. As defined in [**KL**], let $\leqslant_{L,\triangleleft}$ be the pre-order relation on W generated by the relation " $x \leqslant_{L,\triangleleft} y$ if there is $s \in S$ such that $C^{\triangleleft}_x$ appears in the expression of $C^{\triangleleft}_s C^{\triangleleft}_y$ with non-zero coefficient." In other words, $x \leqslant_{L,\triangleleft} y$ if there is a sequence $x_1 = x, x_2, \cdots, x_k = y$ such that, for every $i = 1, 2, \cdots, k-1$, either $x_i = sx_{i+1} > x_{i+1}$ or $M^{s,\triangleleft}_{x_i,x_{i+1}} \neq 0$ for some $s \in S$. We denote by $x \leqslant_{R,\triangleleft} y$ if $x^{-1} \leqslant_{L,\triangleleft} y^{-1}$. Let $\leqslant_{LR,\triangleleft}$ be the pre-order on W generated by $\leqslant_{L,\triangleleft}$ and $\leqslant_{R,\triangleleft}$. For any $x \in \{L, R, LR\}$ and $w, w' \in W$, $w \sim_{x,\triangleleft} w'$ if $w \leqslant_{x,\triangleleft} w' \leqslant_{x,\triangleleft} w$. Thus $\sim_{L,\triangleleft}$, $\sim_{R,\triangleleft}$ and $\sim_{LR,\triangleleft}$ are equivalent relations on W. The corresponding equivalence classes are called left, right and two-sided cells of W relative to the given total order $\triangleleft$, or simply, left, right and two-sided $\triangleleft$-*cells*. If Γ is *not* cyclic, $\triangleleft$-*cells* are called *multi-parameter cells*; if Γ is cyclic and the total order $\triangleleft$ is the natural order, then $\triangleleft$-cells are called *Kazhdan-Lusztig cells* if all u_s are equal, or *unequal-parameter cells* otherwise.

Recall from [**Lu1**] that, if Γ is cyclic (thus, $\Gamma \cong \mathbb{Z}$), then $u_s = q^{i_s/2}$ for some integer i_s and indeterminate $q^{\frac{1}{2}}$. In this case, the Hecke algebra is a one-parameter Hecke algebra and the natural order on $\mathbb{Z}$ gives a unique decomposition $R = P\bar{P}^{-1}$, where $P = (P_{ij})$ is an ∞-pomatrix in the sense that P is an upper unitriangular matrix with $P_{ij} \in q^{-1/2}\mathbb{Z}[q^{-1/2}]$ for $i < j$. Using P, we may introduce a new basis $C_{w_j} = \sum_{i \leqslant j} P_{ij} \widetilde{T}_{w_i}$ (see [**Du2**, §2-3]), fixed under $^-$. Thus, the Coxeter group is divided into (single-parameter) cells, which were introduced in [**KL**] when $i_s = i_t$ for all $s, t \in S$, and in [**Lu1**] otherwise. These cells[4] are determined by the given

[3]The element $C^{\triangleleft}_w$ is denoted C^*_w there.

[4]The corresponding cell representations in this case have also been investigated in [**DPS**] in the context of stratifications of Hecke endomorphism algebras.

one-parameter Hecke algebra. Such a Hecke algebra is a specialization $\mathcal{H}^\sigma$ of $\mathcal{H}$ via a map $\sigma:\Gamma\to\mathbb{Z}$. It is natural to relate $\lhd$-cells with the cells arising from $\mathcal{H}^\sigma$.

3. A Comparison on multi-parameter and unequal-parameter cells

In order to compare two kinds of cells, we first have to look at the specialization of the canonical basis for $\mathcal{H}$ relative to $\lhd$. We hope that image of any canonical basis element in $\mathcal{H}$ is a canonical basis element in $\mathcal{H}^\sigma$. This requires that the specialization σ preserve the orderings. Recall from [**KL**] that the definition of cells in one-parameter case uses the natural ordering on the cyclic group. So we have the following definition.

DEFINITION 3.1. An abelian group homomorphism $\sigma:\Gamma\to\mathbb{Z}$, sending u_s to i_s, is called *admissible* if $i_s>0$ for all $s\in S$ and $\sum a_si_s<\sum a_s'i_s$ whenever $\prod u_s^{a_s}\lhd\prod u_s^{a_s'}$ in Γ.

Cells in a finite Coxeter group W are determined by the cells in the irreducible components of W. Without loss of generality, we may assume that W is irreducible (i.e., the corresponding Coxeter graph is connected [**H**]). We can assume that W is of type B_n, C_n, F_4, or $I(m)$ ($m>3$), since the Hecke algebras of types A_{n-1}, D_n, E_6,E_7,E_8, H_3 and H_4 have equal parameters. Consequently, the Hecke algebra $\mathcal{H}$ associated to W has two parameters, say, u and v, and the group $\mathbb{Z}\times\mathbb{Z}\cong\Gamma$ under the "exponential" isomorphism $\mathbf{e}:(a,b)\mapsto u^av^b$ for $a,b\in\mathbb{Z}$. If σ is an admissible specialization of Γ as given in Def. 3.1, then $i=\sigma(1,0)>0$ and $j=\sigma(0,1)>0$ and σ induces a map $\sigma:\mathbb{Z}\times\mathbb{Z}\to\mathbb{Z}$ with $\sigma(a,b)=a\sigma(1,0)+b\sigma(0,1)$, $a,b\in\mathbb{Z}$. Note that $\sigma(a,b)<\sigma(a',b')$ if $u^av^b\lhd u^{a'}v^{b'}$. For any $k\in\mathbb{Z}$, let

$$L_k=\sigma^{-1}(k)=\{(x,y)\in\mathbb{Z}\times\mathbb{Z}\mid\sigma(x,y)=k,\forall x,y\in\mathbb{Z}\}.$$

Define "half 0-lines" $L_0^+=\{(a,b)\in L_0|\ a>0\}$ and $L_0^-=\{(a,b)\in L_0|\ a<0\}$. Note that, if $(a,b)\in\Gamma_0$, then $ab<0$.

LEMMA 3.2. *Let σ be an admissible specialization of $\Gamma=\langle u,v\rangle$ given as in Def. 3.1.*

(1) *We have either $\mathbf{e}(L_0^-)\subset\Gamma_+$ and $\mathbf{e}(L_0^+)\subset\Gamma_-$, or $\mathbf{e}(L_0^-)\subset\Gamma_-$ and $\mathbf{e}(L_0^+)\subset\Gamma_+$.*
(2) *If $(a,b)\in L_k$, $(c,d)\in L_{k'}$ and $u^av^b\rhd u^cv^d$, then $k\geqslant k'$. When $k=k'$, we have*
    (a) *$a<c$ if and only if $\mathbf{e}(L_0^-)\subset\Gamma_+$;*
    (b) *$a>c$, if and only if $\mathbf{e}(L_0^+)\subset\Gamma_+$.*

PROOF. If (1) were false, then there would be $(a,b)\in L_0^-$ and $(a_1,b_1)\in L_0^+$ such that $u^av^b,u^{a_1}v^{b_1}\in\Gamma_+$. However, since $(u^av^b)^{b_1}=(u^{a_1}v^{b_1})^b$, $b_1<0$ and $b>0$, $(u^av^b)^{b_1}\in\Gamma_+\cap\Gamma_-=\emptyset$, a contradiction, forcing (1) to be true.

Since σ is admissible and $u^av^b\rhd u^cv^d$, $k=\sigma(a,b)\geqslant k'=\sigma(c,d)$ and $\mathbf{e}(a-c,b-d)\in\Gamma_+$. If $k=k'$, then $(a-c,b-d)\in L_0$. Therefore, $a<c$ (resp. $a>c$) if and only if $(a-c,b-d)\in L_0^-$ (resp. $\in L_0^+$). Now, (a) and (b) follow immediately from (1). □

This lemma suggests us to introduce some special admissible orders on Γ.

DEFINITION 3.3. An admissible total ordering $\lhd$ on Γ is called a *parallel-line ordering* and write $u^av^b\lhd u^{a_1}v^{b_1}$ for $(a,b)\in L_k$ and $(a_1,b_1)\in L_{k'}$ if one of the following conditions holds true.

(1) either $k < k'$, or $k = k'$ and $a < a_1$.
(2) either $k < k'$, or $k = k'$ and $a > a_1$.

The following result shows that the parallel-line orderings exhaust the representatives for the equivalence classes of admissible orderings (see Def. 2.3).

THEOREM 3.4. *Any admissible total ordering $\lhd$ on Γ is equivalent to a parallel-line ordering. Moreover, there are infinite number of parallel-line orders, which are equivalent to $\lhd$.*

PROOF. Let $R = P^{\lhd}\bar{P^{\lhd}}^{-1}$ be the unique antipode decomposition with respect to $\lhd$. Let

$$X = \{(-a, b) \in \mathbb{Z}^{>0} \times \mathbb{Z}^{>0} \mid u^a v^b \text{ appears in } P^{\lhd}_{x,y} \text{ for some } x, y \in W\},$$
$$Y = \{(a, -b) \in \mathbb{Z}^{>0} \times \mathbb{Z}^{>0} \mid u^a v^b \text{ appears in } P^{\lhd}_{x,y} \text{ for some } x, y \in W\}.$$

Since W is a finite Coxeter group, both X and Y are finite sets.

First, assume $X \neq \emptyset$ and $Y \neq \emptyset$. Write

$$k = \max\{\frac{b}{-a} \mid (-a, b) \in X\} \text{ and } l = \min\{\frac{-b}{a} \mid (a, -b) \in Y\}.$$

We claim $k < l$. Indeed, if this is not the case, then there are $(-a_0, b_0) \in X$ and $(a_0', -b_0') \in Y$ such that $k = \frac{b_0}{-a_0} \geqslant l = \frac{-b_0'}{a_0'}$. Since $P^{\lhd}_{x,y} \in \mathbb{Z}[\Gamma_-]$, $u^{a_0} \lhd v^{-b_0}$ and $u^{a_0'} v^{b_0'} \in \Gamma_-$. Consequently,

$$u^{-a_0 b_0'} = (u^{a_0})^{-b_0'} \lhd (v^{-b_0})^{-b_0'} = v^{b_0 b_0'}, \text{ and } u^{a_0 b_0'} \rhd v^{-b_0 b_0'} \rhd u^{a_0' b_0}.$$

We have $a_0 b_0' \geqslant a_0' b_0$. This together with $a_0' b_0 \geqslant a_0 b_0'$ implies $a_0' b_0 = a_0 b_0'$. Hence $(u^{a_0} v^{b_0})^{b_0'} = 1 \notin \Gamma_+$, a contradiction.

Let $\sigma : \mathbb{Z} \times \mathbb{Z} \to \mathbb{Z}$ be a specialization satisfying $k < \sigma(1, 0)/\sigma(0, 1) < l$ and $\sigma(1, 0), \sigma(0, 1) \in \mathbb{Z}^{>0}$. Let $\lhd'$ be the parallel-line order defined by σ as in Definition 3.3. To show $\lhd'$ and $\lhd$ are equivalent, we need to prove that $P^{\lhd} = P^{\lhd'}$. By the uniqueness of the decomposition of R-matrix, we only need to prove that any monomial $u^a v^b$ appearing in $P^{\lhd}_{x,y}, x, y \in W$ satisfies $u^a v^b \lhd' 1$.

Suppose $u^a v^b$ is a monomial in $P^{\lhd}_{x,y}$ for $x, y \in W$. Then we have $a, b < 0$ or $(-a, b) \in X$ or $(a, -b) \in Y$. Obviously, $u^a v^b \lhd' 1$ if $a, b < 0$. Assume $(-a, b) \in X$. Then $\frac{b}{-a} \leqslant k < \frac{\sigma(1,0)}{\sigma(0,1)}$, and hence, $\sigma(a, b) < 0$. Thus $u^a v^b \lhd' 1$. Similarly, $\sigma(a, b) < 0$ if $(a, -b) \in Y$, and hence $u^a v^b \lhd' 1$. So we conclude that any monomial of $P^{\lhd}_{x,y}$, $x, y \in W$ is in the set Γ'_- defined by $\lhd'$. Hence $P^{\lhd} = P^{\lhd'}$ and $\lhd$ and $\lhd'$ are equivalent.

We set $k = 0$ (resp. $l = +\infty$) if $X = \emptyset$ and $Y \neq \emptyset$ (resp. $X \neq \emptyset$ and $Y = \emptyset$). If $X = Y = \emptyset$, we set $k = 0$ and $l = +\infty$. This case occurs only when all total orders on Γ are equivalent. For each of these cases, one can prove the first assertion similarly.

Since there are infinitely many rational numbers between k and l, it follows that there are infinitely many parallel-line orders which are equivalent to $\lhd$. This proves the second assertion. $\square$

For any admissible $\sigma : \mathbb{Z} \times \mathbb{Z} \to \mathbb{Z}$, let $m_\sigma = \sigma(1, 0)/\sigma(0, 1)$. Let $\tilde{\sigma}$ be the induced homomorphism from $\mathbb{Z}[\Gamma]$ to $\mathbb{Z}[q^{\frac{1}{2}}, q^{-\frac{1}{2}}]$.

COROLLARY 3.5. Let k, l be integers determined by $\lhd$ as in the proof above. If the admissible $\sigma : \mathbb{Z} \times \mathbb{Z} \to \mathbb{Z}$ satisfies $k < m_\sigma < l$, then $x \leqslant_{L,\sigma} y$ implies $x \leqslant_{L,\lhd} y$.

PROOF. From the proof of Theorem 3.4, we have that the image of any monomial of $P^{\lhd}$ under $\tilde{\sigma}$ is of form $q^{i/2}$ with $i < 0$. By the uniqueness of antipode decomposition of the R-matrix, we have $\tilde{\sigma}(P^{\lhd}) = P^{\sigma}$, and hence $\tilde{\sigma}(B_w^{\lhd}) = B_w^{\sigma}$. Since $\tilde{\sigma}$ induces a ring homomorphism $\tilde{\sigma}$ from $\mathcal{H}$ to $\mathcal{H}^{\sigma}$, $\tilde{\sigma}(B_x^{\lhd}B_y^{\lhd}) = \tilde{\sigma}(B_x^{\lhd})\tilde{\sigma}(B_y^{\lhd}) = B_x^{\sigma}B_y^{\sigma}$. Thus $\tilde{\sigma}(M_{x,y}^{s,\lhd}) = M_{x,y}^{s,\sigma}$ for any $x, y \in W$, where $M_{x,y}^{s,\lhd}$ is defined in (2.3) and $M_{x,y}^{s,\sigma}$ is defined with $\lhd$ replaced by σ in (2.3) for the Hecke algebra $\mathcal{H}^{\sigma}$. Thus $M_{x,y}^{s,\lhd} \neq 0$ if $M_{x,y}^{s,\sigma} \neq 0$. By the definition of the relation $\leqslant_L$, we obtain that $x \leqslant_{L,\sigma} y$ implies $x \leqslant_{L,\lhd} y$. □

THEOREM 3.6. *For any $\lhd$ on Γ, there is a specialisation σ such that $x \sim_{L,\lhd} y$ if and only if $x \sim_{L,\sigma} y$.*

PROOF. It follows from Theorem 3.4 that there are infinite number of specializations σ with $k < m_{\sigma} < l$ satisfying $\tilde{\sigma}(P^{\lhd}) = P^{\sigma}$, and $\tilde{\sigma}(M_{x,y}^{s,\lhd}) = M_{x,y}^{s,\sigma}$ for any $x, y \in W$, and the number of such m_{σ} is also infinite. Since W is a finite Coxeter group, there are only finitely many non-zero $M_{x,y}^{s,\lhd}$. Thus, if $M_{x,y}^{s,\lhd} \neq 0$, then the number of m_{σ} for which $\tilde{\sigma}(M_{x,y}^{s,\lhd}) = M_{x,y}^{s,\sigma} = 0$ is finite. Therefore, there exists a specialization σ such that $M_{x,y}^{s,\sigma} \neq 0$ whenever $M_{x,y}^{s,\lhd} \neq 0$. So $x \leqslant_{L,\sigma} y$ if $x \leqslant_{L,\lhd} y$. Now, the result follows immediately from Corollary 3.5. □

REMARK 3.7. The above theorem answers the question given in 1.1(a). At the same time, the proof above shows that the 1.1(b) may not be true. For a specialization σ which vanishes some M-polynomials, there could be no $\lhd$ on Γ for which both $\lhd$-cells and σ-cells agree. Explicit examples will be given in next section.

4. Examples of Multi-parameter cells

In this section, we shall describe $\lhd$-cells and σ-cells for dihedral groups and the Weyl group of type B_3. Our results suggest that the problem 1.1(b) is not true in general.

First, we recall [**Lu1**, Proposition 4], which says that the polynomials $M_{x,y}^{s,\lhd}$'s satisfy the following recursive formula

$$M_{x,y}^{s,\lhd} = P_{sx,y}^{\lhd} - P_{x,sy}^{\lhd} + u_s P_{x,y}^{\lhd} - \sum_{x<z<y, sz<z} P_{x,z}^{\lhd} M_{z,y}^{s,\lhd}, \tag{4.1}$$

The following lemma can be found in [**Xi**, 1.14.13].

LEMMA 4.1. *Let W be a finite Coxeter group.*

(1) *For $x, y \in W$ with $x < sx \leqslant y$ and $sy < y$, we have $P_{x,y}^{\lhd} = u_s^{-1} P_{sx,y}^{\lhd}$.*
(2) *If $x < y$ and $l(y) = l(x)+1$, then a reduced expression of x can be obtained from a reduced expression of y by dropping a factor $s \in S$. Let $t \in S$ be with $tx < x < y < ty$. Then*

$$M_{x,y}^{t,\lhd} = \begin{cases} 0, & \text{if } u_t < u_s, \\ 1, & \text{if } u_t = u_s, \\ u_t u_s^{-1} + u_s u_t^{-1}, & \text{if } u_t > u_s. \end{cases}$$

(I) The dihedral groups. Let $W = I(m)$ be the dihedral group with $S = \{s, t\}$. Then the order of the product st is m. We denote by $u_s = u$ and $u_t = v$.

LEMMA 4.2. *Suppose that $\lhd$ is a total order on $\Gamma = \langle u, v \rangle$ such that $u \lhd v$. Then*

(1) *$u_s^{-1} M_{x,y}^{s,\lhd} \in \mathbb{Z}[\Gamma_-]$ for all $x, y \in W$ and $s \in S$.*

(2) *for any* n *with* $1 \leqslant n \leqslant [m/2]-1$, $P^{\lhd}_{(st)^{n-1},(st)^n} = u^{-1}v^{-1}$.

(3) *for any* n *with* $1 \leqslant n \leqslant [m/2]-2$, $M^{t,\lhd}_{(ts)^n,(st)^{n+1}} = 0$.

(4) *For any* n *with* $0 \leqslant n \leqslant [m/2]-3$, $M^{t,\lhd}_{t(st)^n,(st)^{n+2}} = 1$.

PROOF. (1) follows from Lemma 4.1(2) under the assumption $l(y)-l(x)=1$. In general, it follows from (4.1) and the inductive hypothesis. By Lemma 4.1(1), $P^{\lhd}_{e,st} = u^{-1}v^{-1}$. Thus (2) is true for $n=1$. Suppose $n>1$. We have

$$\begin{aligned} P^{\lhd}_{(st)^{n-1},(st)^n} &= uP^{\lhd}_{(st)^{n-1},t(st)^{n-1}} + P^{\lhd}_{t(st)^{n-2},t(st)^{n-1}} - M^{s,\lhd}_{(st)^{n-1},t(st)^{n-1}} \\ &= uv^{-1} + P^{\lhd}_{t(st)^{n-2},t(st)^{n-1}} \\ &= uv^{-1} + vP^{\lhd}_{t(st)^{n-2},(st)^{n-1}} + P^{\lhd}_{(st)^{n-2},(st)^{n-1}} - M^{t,\lhd}_{t(st)^{n-2},(st)^{n-1}}, \\ &= uv^{-1} + vu^{-1} + u^{-1}v^{-1} - (uv^{-1}+vu^{-1}) \\ &= u^{-1}v^{-1}. \end{aligned}$$

This proves (2). If $(ts)^n < z < (st)^{n+1}$ and $tz<z$, then $z=(ts)^n t$. By (4.1),

$$M^{t,\lhd}_{(ts)^n,(st)^{n+1}} - vP_{(ts)^n,(st)^{n+1}} + P_{(ts)^n,(ts)^n t}M^{t,\lhd}_{(ts)^n t,(st)^{n+1}} \in \mathbb{Z}[\Gamma_-].$$

Using (1) and Lemma 4.1, we have

$$\begin{aligned} &P_{(ts)^n,(ts)^n t}M^{t,\lhd}_{(ts)^n t,(st)^{n+1}} = v^{-1}M^{t,\lhd}_{(ts)^n t,(st)^{n+1}} \in \mathbb{Z}[\Gamma_-], \\ &vP_{(ts)^n,(st)^{n+1}} = u^{-1}. \end{aligned}$$

Thus, $M^{t,\lhd}_{(ts)^n,(st)^{n+1}} \in \mathbb{Z}[\Gamma_-]$. Since $\bar{M}^{s,\lhd}_{z,w} = M^{s,\lhd}_{z,w}$, we have $M^{t,\lhd}_{(ts)^n,(st)^{n+1}} = 0$, proving (3). By a direct computation, we have $M^{t,\lhd}_{(ts)^{n+1},(st)^{n+2}} = 0$, $P^{\lhd}_{t(st)^n,t(st)^{n+1}} = u^{-1}v^{-1}+v^{-1}u$ and $M^{t,\lhd}_{t(st)^{n+1},(st)^{n+2}} = uv^{-1}+vu^{-1}$. Using (4.1) and noting that $\bar{M}^{s,\lhd}_{z,w} = M^{s,\lhd}_{z,w}$, we obtain (4). □

PROPOSITION 4.3. *Let* W *be the dihedral group* $I(m)$ *with* $S=\{s,t\}$ *as its distinguished generator set. Let* $u=u_s$ *and* $v=u_t$.

(1) *If* $u=v$, *then* W *has 3 Kazhdan-Lusztig two-sided cells* $\{e\}\cup\{w_0\}\cup W\backslash\{e,w_0\}$. *The last two-sided cell consists of two Kazhdan-Lusztig left cells* $\{w\in W|\ ws<w<wt\}$ *and* $\{w\in W|\ wt<w<ws\}$.

(2) *For any total order* $\lhd$ *on* Γ *with* $u\lhd v$, W *has 5 two-sided* $\lhd$*-cells* $\{e\}\cup\{w_0\}\cup\{s\}\cup\{sw_0\}\cup W\backslash\{e,s,sw_0,w_0\}$. *The last two-sided* $\lhd$*-cell consists of two left* $\lhd$*-cells* $\{w\in W|\ ws<w<wt, w\neq s\}$ *and* $\{w\in W|\ wt<w<ws, w\neq sw_0\}$.

(3) *For any total order* $\lhd$ *on* Γ *with* $v\lhd u$, W *has* 5 *two-sided* $\lhd$*-cells* $\{e\}\cup\{w_0\}\cup\{t\}\cup\{tw_0\}\cup W\backslash\{e,t,tw_0,w_0\}$. *The last two-sided cell consists of two left* $\lhd$*-cells* $\{w\in W|\ ws<w<wt, w\neq tw_0\}$ *and* $\{w\in W|\ wt<w<ws, w\neq t\}$.

PROOF. If $u=v$, then the Hecke algebra $\mathcal{H}$ has equal parameters. In this case, the cells of W are Kazhdan-Lusztig cells. The result (1) is known (see, e.g. [**Lu2**]). We only deal with (2). One can prove (3) by switching s to t.

First, we claim $\{s\}$ is a two-sided $\lhd$-cell. Otherwise, there exists $z\in W$ with $z\sim_{LR,\lhd} s$ such that either $s\leqslant_{L,\lhd} z$ or $s\leqslant_{L,\lhd} z^{-1}$. We need only deal with $s\leqslant_{L,\lhd} z$ since we can switch the role between z and z^{-1}.

By the definition of $\leqslant_{L,\lhd}$, either $s=sz>z$ or $M^{s,\lhd}_{s,z}\neq 0$.

In the first case, $z = e$, a contradiction since e and s are not in the same two-sided cell.

In the second case, we have, by [**Xi**, 1.20], $\{s\} = \mathcal{R}(s) \supset \mathcal{R}(z)$, where $\mathcal{R}(w) = \{s \in S \mid ws < w\}$. Notice that $z \neq e$, we have $\mathcal{R}(z) = \{s\}$. On the other hand, $tz < z$. Otherwise, $tz > z$ and $sz > z$ which implies that $z = e$, a contradiction. Consequently, $z = (ts)^i$ for $1 \leqslant i \leqslant [m/2] - 1$. Using induction on $l(y) - l(x)$ and (4.1), one can prove $M_{x,y}^{s,\lhd} = 0$ if $sx < x < y < sy$ (see [**Xi**, 1.21]). In particular, $M_{s,(ts)^i}^{s,\lhd} = 0$, a contradiction. Thus $\{s\}$ is a two-sided $\lhd$-cell.

By the arguments similar to those in [**KL**], one can verify $M_{x,y}^{s,\lhd} + \epsilon_x \epsilon_y M_{yw_0,xw_0}^{s,\lhd} = 0$. Thus, $w \sim_{L,\lhd} y$ if and only if $ww_0 \sim_{L,\lhd} yw_0$. In particular, $\{sw_0\}$ is a two-sided $\lhd$-cell. By [**Xi**, 1.17], we see that $\{e\}$ and $\{w_0\}$ are two distinct two-sided $\lhd$-cells. Finally, we show that all the other elements form a two-sided $\lhd$-cell. By Lemmas 4.1(2) and 4.2(4), the elements $t, st, tst, \dots, w_0 st$ are in the same left $\lhd$-cell. So $tw_0, stw_0, \dots, w_0 stw_0$ are in the same left cell, too. Since $\mathcal{R}(s) \neq \mathcal{R}(t)$, s and t are not in the same left cells (see [**Xi**, 1.20]). Note that $t \sim_{R,\lhd} ts$, $W \backslash \{e, s, sw_0, w_0\}$ forms a two-sided $\lhd$-cell. This proves (2). □

REMARK 4.4. Note that Lemma 4.2 holds if we use $q^{i/2}$ and $q^{j/2}$ with $i < j$ instead of u and v, respectively. Thus, the σ-cells (in unequal parameters case) of a dihedral group are given in Proposition 4.3(2) and (3) for $i < j$ and $i > j$ respectively. They agree with the cells described in [**Lu3**]. Moreover, the Kazhdan-Lusztig cells in $I(m)$ do not coincide with any $\lhd$-cells.

(II) The Weyl group of type $W(B_3)$. Let W be the finite Weyl group of type B_3 with $S = \{s_1, s_2, s_3\}$ as its distinguished generator set. The relations among s_1, s_2, s_3 are $s_1s_2s_1 = s_2s_1s_2$, $s_1s_3 = s_3s_1$, $s_2s_3s_2s_3 = s_3s_2s_3s_2$. Let $u = u_{s_1} = u_{s_2}$ and $v = u_{s_3}$. We assume $u \lhd v \lhd u^2$.

By [**Xi**, 1.17], $\{e\}$ and $\{w_0\}$ are two distinct two-sided $\lhd$-cells of W. Recall [**Xi**, 1.20], $\mathcal{R}(x) = \mathcal{R}(y)$ if $x \sim_{L,\lhd} y$. Therefore, we only need to decompose the subsets with a given $\mathcal{R}$-set into left $\lhd$-cells. On the other hand, since the conjugate map sending $w \in W$ to w_0ww_0 is an automorphism of W, which fixes the Dynkin diagram of $W(B_n)$, we have $w_0sw_0 = s$ for $s \in S$ and $\mathcal{R}(ww_0) = S \backslash \{s\}$ if $\mathcal{R}(w) = \{s\}$. Note that $x \sim_{L,\lhd} y$ if and only if $xw_0 \sim_{L,\lhd} yw_0$ (see the proof of Proposition 4.3). Thus, we only need to deal with the $\mathcal{R}$-sets $\{s_i\}$ for $i = 1, 2, 3$.

PROPOSITION 4.5. *We have*

$$\{w \in W \mid \mathcal{R}(w) = \{s_1\}\} = X_{1,1} \cup X_{1,2},$$
$$\{w \in W \mid \mathcal{R}(w) = \{s_2\}\} = X_{2,1} \cup X_{2,2} \cup X_{2,3} \cup X_{2,4},$$
$$\{w \mid W \mid \mathcal{R}(w) = \{s_3\}\} = X_{3,1} \cup X_{3,2} \cup X_{3,3}$$

where

$X_{1,1} = \{s_1, s_2s_1\}, X_{1,2} = \{s_3s_2s_1, s_2s_3s_2s_1, s_1s_2s_3s_2s_1\},$

$X_{2,1} = \{s_2, s_1s_2\}, X_{2,2} = \{s_1s_3s_2, s_3s_2s_1s_3s_2, s_2s_1s_3s_2\},$

$X_{2,3} = \{s_3s_2, s_2s_3s_2, s_1s_2s_3s_2\}, X_{2,4} = \{s_1s_2s_1s_3s_2, s_1s_3s_2s_1s_3s_2, s_2s_1s_3s_2s_1s_3s_2\},$

$X_{3,1} = \{s_3, s_2s_3, s_1s_2s_3\}, X_{3,2} = \{s_3s_2s_3\},$

$X_{3,3} = \{s_1s_3s_2s_3, s_2s_1s_3s_2s_3, s_3s_2s_1s_3s_2s_3\}.$

PROOF. First, we consider the case $i = 1$. By Lemma 4.1(2), each $X_{1,j}, j = 1, 2$ is in a left $\lhd$-cell . We need to verify $X_{1,1}$ and $X_{1,2}$ are not in the same left $\lhd$-cell.

If it were not the case, then there would be $x \in X_{1,1}$ and $y \in X_{1,2}$ such that either $x = sy > y$ or $M_{x,y}^{s,\lhd} \neq 0$ for some $s \in S$. However, the first result is impossible since $l(x) \leqslant 2$ and $l(y) \geqslant 3$. On the other hand, all possible $M_{x,y}^{s,\lhd}$ for $x \in X_{1,1}, y \in X_{1,2}$ are given as follows, which contradict the fact $M_{x,y}^{s,\lhd} \neq 0$.

$$M_{s_1,s_3s_2s_1}^{s_1} = M_{s_1,s_2s_3s_2s_1}^{s_1,\lhd} = M_{s_2s_1,s_3s_2s_1}^{s_2,\lhd} = M_{s_2s_1,s_1s_2s_3s_2s_1}^{s_2,\lhd} = 0.$$

Consequently, $X_{1,1}$ and $X_{1,2}$ are two distinct left $\lhd$-cells.

Suppose $i = 2$. By a direct computation,

$$M_{s_1s_3s_2,s_3s_2s_1s_3s_2}^{s_1} = M_{s_1s_2s_1s_3s_2,s_2s_1s_3s_2s_1s_3s_2}^{s_1} = u^2v^{-1} + vu^{-2}.$$

Thus

$$s_1s_3s_2 \leqslant_{L,\lhd} s_3s_2s_1s_3s_2 \text{ and } s_1s_2s_1s_3s_2 \leqslant_{L,\lhd} s_2s_1s_3s_2s_1s_3s_2.$$

Using Lemma 4.1(2) again, we see that all the elements in $X_{2,j}$ with $1 \leqslant j \leqslant 4$ are in the same left $\lhd$-cells.

Suppose $i = 3$. Since $M_{s_1s_3s_2s_3,s_3s_2s_1s_3s_2s_3}^{s_1} = u^2v^{-1} + vu^{-2}$, by Lemma 4.1(2), each $X_{3,j}$ is in a left $\lhd$-cell.

One can check, for $i = 2, 3$ and $j \neq k$, $X_{i,j}$ and $X_{i,k}$ are not in the same left $\lhd$-cell by the argument similar to those for $i = 1$. □

One sees easily that all results hold if we specialize u and v to q and $q^{\frac{3}{2}}$, respectively, since we only use the inequality $u \lhd v \lhd u^2$ in the above proof. Thus the left $\lhd$-cells of $W(B_3)$ coincide with the left σ-cells of $W(B_3)$ where $\sigma(u) = q, \sigma(v) = q^{\frac{3}{2}}$.

References

[B] K. Bremke, *On generalized cells in affine Weyl groups,* J. Algebra **191** (1997), 149–173.

[Du1] J. Du, *IC basis and quantum linear groups*, Proc. Symp. in Pure Maths. **56** (1994), 135–148.

[Du2] J. Du, *A Matrix Approach to IC Bases,* CMS Conference Proceedings **14** (1993), 165–174.

[DPS] J. Du, B. Parshall and L. Scott, *Stratifying endomorphism algebras associated to Hecke algebras,* J. Algebra **203** (1998), 169–210.

[H] J.E. Humphreys, *Reflection Groups and Coxeter Groups,* Cambridge University Press, Cambridge, 1990.

[KL] D. Kazhdan and G. Lusztig, *Representations of Coxeter groups and Hecke algebras,* Invent. Math. **53** (1979), 155–174.

[Lu1] G. Lusztig, *Cells in Weyl groups* , Lecture Notes in Mathematics **1024**, 99-111, Berlin, 1983.

[Lu2] G. Lusztig, *Some examples on square integrable representations of semisimple p-adic groups*, Trans. of AMS, **277** (1983), 623–653.

[Lu3] G. Lusztig, *Hecke Algebras with Unequal Parameters,* CRM Monographs, Ser.**18**, Amer. Math. Soc., 2003.

[Xi] N. Xi, *Representations of affine Hecke algebras,* Lecture Notes in Mathematics **1587**, Springer-Verlag, Berlin, 1994.

School of Mathematics, University of New South Wales, Sydney, 2052, Australia
E-mail address: jied@maths.unsw.edu.au

Department of Mathematics, East China Normal University, Shanghai, 200062, China
E-mail address: hbrui@math.ecnu.edu.cn

AMS/IP Studies in Advanced Mathematics
Volume **39**, 2007

On Some Rigidity Problems in Cauchy-Riemann Analysis

Xiaojun Huang and Shanyu Ji

ABSTRACT. This paper is a survey on some rigidity problems in Cauchy-Riemann Geometry. It includes the topics such as holomorphic mappings between balls, the CR version of the Bonnet-Type theorem, super-rigidity and CR transversality for holomorphic mappings between hyperquadrics, rigidity problems for holomorphic maps with symmetry, rigidity problems for holomorphic Segre maps.

1. Introduction

This paper has two parts. We first survey some recent work on various rigidity problems for CR mappings, CR submanifolds embedded into the spheres or hyperquadrics, and some rigidity problems for isolated complex singularities. We then present the computation of the Segre-isomorphisms as a first step to extend the investigation of rigidities for CR maps to holomorphic Segre maps.

We start with a classical result in the conformal geometry. In this setting, we consider open subsets U, V in $\mathbb{R}^n$, equipped with the flat metric ω. We say a smooth map f from U to V is a conformal map if $f^*(\omega) = e^u\omega$. The following result is classical:

THEOREM 1.1. *(Liouville) Assume the above and let $n \geq 3$. Then f is conformal if and only if f is a Mobius transformation: A composition of the following type of transformations: (i) translations, (ii) rotations, (iii) scalings and inversions.*

In CR geometry, the replacement of $\mathbb{R}^n$ is the sphere— the boundary of the unit ball. The conformal transformations should be replaced by CR transformations (or the conjugate of CR transformations), which, roughly speaking, are just the boundary value of holomorphic mappings. (See [BER1] for most notations and definitions). Hence the following result of Poincaré, proved about 100 years ago, can be regarded as a complex version of the Liouville theorem:

THEOREM 1.2. *(Poincaré 1907 [Po] , Tanaka [Ta], Chern-Moser [CM]) Let f be a non-constant holomorphic map from an open piece of the unit sphere $\partial\mathbb{B}^n$ into*

1991 *Mathematics Subject Classification.* Primary 32V40, 32H35; Secondary 32H02, 32V15.

Key words and phrases. Rigidity Problems, Cauchy-Riemann Geometry, holomorphic proper mappings.

The first author was supported in part by NSF-0500626.

the unit sphere ($n \geq 2$). Then f is a linear fractional transformation and extends to a biholomorphic self-map of the unit ball.

One of our concerns here is to address the recent development along these lines of research. Unfortunately, we have to skip many important results in the related fields, where when a local map can be extended to a global object is studied. (See [BER1-2] [KZ] and the references therein).

The second classical result whose CR version we like to pursue here is the so-called Bonnet theorem. In this case, we consider two immersed surfaces M and M' in $\mathbb{R}^3$. Suppose they are parameterized by $U \subset \mathbb{R}^2$ through f and g, respectively. If M and M' are isometric to each other and if the second fundamental form is identical at the corresponding points: $l_{ij}(p) = \langle n(f(p)), f_{ij}(p)\rangle = \widetilde{l}_{ij}(p) = \langle \widetilde{n}(f(p)), g_{ij}(p)\rangle$. Then M and M' are the rigid motion of each other. Namely, there is a Euclidean transformation T such that $T(M) = M'$ (See [Spv]).

In CR geometry, immersed submanifolds are replaced by Levi non-degenerate CR submanifolds. The following result proved by S. Webster in 1977 can be regarded as a generalization of the complex variable version of the classical Bonnet theorem:

THEOREM 1.3. *(Webster [We4]) Let (M_1, p_1) and (M_2, p_2) be two germs of strongly pseudoconvex CR submanifolds of CR codimension 1 in $\partial\mathbb{B}^n$ with $n \geq 5$. If (M_1, p) and (M_2, p_2) are CR diffeomorphic to each other, then they are the rigidity motion of each other. Namely, there is a $T \in Aut(\partial\mathbb{B}^n)$ such that $T(p_1) = p_2$ and $T((M_1, p_1)) = (M_2, p_2)$.*

Acknowledgement: The authors thank D. Zaitsev and N. Mir for many helpful comments and suggestions to this paper

2. Holomorphic Mappings between Balls

We now discuss recent results along the lines of Poincaré's theorem. For simplicity, we restrict ourselves to the global setting, though all results to be stated in this section hold also in the local setting. By the Bochner theorem, any holomorphic (CR) function defined over $\partial\mathbb{B}^n$ extends holomorphically to $\mathbb{B}^n$. Hence, any non constant holomorphic map from $\partial\mathbb{B}^n$ to $\partial\mathbb{B}^n$ extends to a proper holomorphic map from $\mathbb{B}^n$ to $\mathbb{B}^n$. We use the notation Prop$(\mathbb{B}^n, \mathbb{B}^N)$ to denote the collection of all proper holomorphic maps from $\mathbb{B}^n$ into $\mathbb{B}^N$. ($N \geq n > 1$). Now, the aforementioned theorem of Poincaré-Tanaka-Chern-Moser can be stated as follows:

THEOREM 2.1. *(Poincaré [Po]) Let $f \in Prop(\mathbb{B}^n, \mathbb{B}^n)$, that extends holomorphically across $\partial\mathbb{B}^n$. Then $f \in Aut(\mathbb{B}^n)$.*

PROOF. (Sketch) Need only to prove that f is a covering map. For that, it suffices to show that f is a local biholomorphic map. Indeed, if not, then E_f: $= \{z \in \mathbb{B}^n : J_f(z) = 0\}$ is a complex analytic variety of codimension 1. Since $n > 1$, E_f cannot be compact and thus must cut the boundary. Now, an application of the Hopf Lemma shows this is impossible. □

It had been wondered for years that if the boundary regularity assumption in Poincaré's theorem is necessary. That puzzle was answered by H. Alexander in [Alx1-2].

THEOREM 2.2. *(Alexander [Alx1-2]) Let $f \in Prop(\mathbb{B}^n, \mathbb{B}^n)$. Then $f \in Aut(\mathbb{B}^n)$.*

Notice that for an affine complex line L, $L\cap\mathbb{B}^n$ is a complex geodesic in terms of the hyperbolic Kobayashi metric of the ball and an automorphism of $\mathbb{B}^n$ maps an affine line to an affine line. The result of Alexander hence tells that a proper holomorphic self-map of $\mathbb{B}^n$ preserves the complex geodesics of $\mathbb{B}^n$ $(n>1)$. More generally, one says that a map from $\mathbb{B}^n$ into $\mathbb{B}^N$ is a linear map or a totally geodesic embedding if it maps a complex geodesic in $\mathbb{B}^n$ to a complex geodesic in $\mathbb{B}^N$. Using his result on the CR version of the Bonnet theorem (Theorem 1.2), Webster was the first one to look at the geometric structure of proper holomorphic maps between balls in complex spaces of different dimensions. He, in 1979, showed that a proper holomorphic map from $\mathbb{B}^n$ into $\mathbb{B}^{n+1}$ with $n>2$, which is three times differentiable up to the boundary, is a totally geodesic embedding. Subsequently, Cima-Suffridge ([CS1], 1983) reduced the boundary regularity in Webster's theorem to the C^2-regularity. Motivated by a conjecture posed by Cima-Suffridge, Faran in 1986 showed [Fa2] that any proper holomorphic map from $\mathbb{B}^n$ into $\mathbb{B}^N$ with $N<2n-1$, that is analytic up to the boundary, is also a totally geodesic embedding. Forstneric ([Fr1-3]) proved that any proper holomorphic map from $\mathbb{B}^n$ into $\mathbb{B}^N$ is rational, if the map is C^{N-n+1}-regular up to the boundary, which, in particular, reduces the regularity assumption in Faran's linearity theorem to the C^{N-n+1}-smoothness. It had been wondered for years if the super-rigidity holds for proper maps which is C^t-smooth with t independent of the codimension.

On the other hand, the discovery of inner functions had been used in the later 80's to construct proper holomorphic maps from $\mathbb{B}^n$ into $\mathbb{B}^{n+1}$, which cannot be C^2-smooth at any boundary point. (See Alexandroff, Harkim-Sibony [HS], Low, Forstneic[Fr1], Dor [Dor], Stensones, etc)

In 1999 [Hu1] and, subsequently, 2003 [Hu2], we considered two natural questions arising from the above mentioned work. In 1999, it was proved in [Hu1] that any proper holomorphic map which is only C^2-regular up to the boundary must be linear if $N<2n-1$, by applying a different method from the above mentioned work. It is not clear to us if this C^2-regularity is optimal or not for the super-rigidity to hold, the result in [Hu1] gives a first result in which the required regularity is independent of the codimension. Moreover, the basic approaches developed there seem to be quite useful for the study of many other related problems. For instance, it was used in 2005 by Hamada [Ham] to show that all rational proper holomorphic maps from $\mathbb{B}^n$ into $\mathbb{B}^{2n}$ $(n>3)$ are just the D'Angelo family up to the automorphisms.

In [HJ1], it was shown that any proper holomorphic map from $\mathbb{B}^n$ into $\mathbb{B}^N$ with $N=2n-1, n\geq 3$, which is C^2-smooth up to the boundary, is either linear or equivalent to the Whitney map $W: z=(z_1,\cdots,z_n)\rightarrow(z_1,\cdots,z_{n-1},z_nz)$ ([Theorem 1, Theorem 2.3; HJ1]). Since the Whitney map is not an immersion, together with the aforementioned work of Faran [Fa1], this shows that any proper holomorphic embedding from $\mathbb{B}^n$ into $\mathbb{B}^N$ with $N=2n-1$, which is twice continuously differentiable up to the boundary, must be a linear map.

THEOREM 2.3. *(Huang [Hu1], 1999): Let f be a proper holomorphic embedding from $\mathbb{B}^n$ into $\mathbb{B}^N$ with $n\geq 3, N<2n-1$. Assume that f is twice differentiable up to the boundary. Then f is linear.*

THEOREM 2.4. *(Huang-Ji [HJ1], 2001): Let f be a proper holomorphic embedding from $\mathbb{B}^n$ into $\mathbb{B}^N$ with $N=2n-1>2$. Assume that f is twice differentiable up to the boundary. Then f is linear.*

The structure of the maps gets more complicated when $N > 2n-1$. Recall that two proper holomorphic maps f, g from $\mathbb{B}^n$ into $\mathbb{B}^N$ are called equivalent if there are $\sigma \in \mathrm{Aut}(\mathbb{B}^n)$ and $\tau \in \mathrm{Aut}(\mathbb{B}^N)$ such that $g = \tau \circ f \circ \sigma$. It is easy to verify that a map is linear if and only if it is equivalent to the standard big circle embedding $L(z) : z \to (z, 0)$. Faran in 1983 [Fa1] showed that there are four different inequivalent maps from $\mathbb{B}^2$ into $\mathbb{B}^3$, which are C^3-smooth up to the boundary. D'Angelo wrote down a family of holomorphic embeddings from $\mathbb{B}^n$ into $\mathbb{B}^{2n}$ which are mutually inequivalent.

EXAMPLE 2.5. (D'Angelo [DA5][DA2]): Let $z = (z', z_n)$ and

$$F_t = (z', \cos(t)z_n, \sin(t)z_n z)$$

with $t \in (0, \pi/2)$. Then F_t is a proper holomorphic embeddings from $\mathbb{B}^n$ into $\mathbb{B}^{2n}$. Apparently, F_t cannot be linear; for it has degree two. Furthermore, F_t is equivalent to F_s if and only if $t = s$.

When $N \geq 2n$, the linearity breaks down even for embedding. However, by looking at the example of D'Angleo, one sees that when F is restricted to $z_n = const$, it is still linear. That motivated the first author in a recent paper [Hu2] to introduce the following definition:

Let F be a proper holomorphic map from $\mathbb{B}^n$ into $\mathbb{B}^N$. We say that F is κ-linear if for any point $p \in \mathbb{B}^n$, there is an affine complex subspace S_p^κ, that passes through p and is of dimension κ, such that for any affine complex line L contained in S_p^κ, $F(L \cap \mathbb{B}^n)$ is contained in an affine complex line in $\mathbb{C}^N$.

THEOREM 2.6. *(Huang [Hu2], 2003) Let F be a proper holomorphic map from $\mathbb{B}^n$ into $\mathbb{B}^N$, which is C^3-smooth up to the boundary. Write $P(n,\kappa) = \frac{\kappa(2n-\kappa-1)}{2}$. If $1 \leq \kappa \leq n-1$ and $N - n < P(n,\kappa)$, then F is $(n-\kappa+1)$-linear.*

The following example shows that in Theorem 2.6, when $N - n \geq P(n,\kappa)$, one can not expect the $(n-k+1)$-linearity for the map in general.

EXAMPLE 2.7. (Huang [Hu2]): For $k \leq n$, let $W_{n,k} = (\psi_1, \cdots, \psi_k)$ where

$$\psi_1 = (z_1^2, \sqrt{2}z_1z_2, \cdots, \sqrt{2}z_1z_{k-1}, z_1z_k, \cdots, z_1z_n),$$
$$\psi_2 = (z_2^2, \sqrt{2}z_2z_3, \cdots, \sqrt{2}z_2z_{k-1}, z_2z_k, \cdots, z_2z_n),$$
$$\cdots$$
$$\psi_{k-1} = (z_{k-1}^2, z_{k-1}z_k, \cdots, z_{k-1}z_n),$$
$$\psi_k = (z_k, \cdots, z_n).$$

Then $W_{n,k}$ is a proper polynomial map from $\mathbb{B}^n$ into $\mathbb{B}^N$ with $N = n + P(n, k-1)$. Notice that $W_{n,k}$ is not $(n-(k-1)+1)$-linear. (Hence $W_{n,k+1}$, from $\mathbb{B}^n$ into $\mathbb{B}^N$ with $N = n + P(n,k)$, is not $(n-k+1)$-linear.)

When $N < \frac{n(n+1)}{2}$, our result says that at each point in the ball, F has at least two independent directions along which the map is linear. When $N \geq \frac{n(n+1)}{2}$, F usually has no partial linearity. To see this, we just need to notice that the polynomial map H that sends $(z_1, \cdots, z_n)$ to

$$(z_1^2, \sqrt{2}z_1z_2, \cdots, \sqrt{2}z_1z_n, z_2^2, \sqrt{2}z_2z_3, \cdots, \sqrt{2}z_2z_n, \cdots, z_{n-1}z_{n-1}, \sqrt{2}z_{n-1}z_n, z_n^2)$$

is proper from $\mathbb{B}^n$ into $\mathbb{B}^N$ with $N = n + P(n, n-1) = \frac{n(n+1)}{2}$. We mention the interesting similarity between the minimal target dimension N for which the rigidity breaks down in the case we are considering here and the minimal target dimension in the classical Cartan-Janet theorem for which there is no more obstruction to locally isometrically embed an analytic Riemannian manifold of dimension n into $\mathbb{R}^N$.

The above mentioned partial linearity result has applications to some other related problems. For instance, it can be used to prove the following:

THEOREM 2.8. *(Huang-Ji-Xu [HJX1], 2005) Let f be a proper holomorphic map from $\mathbb{B}^n$ into $\mathbb{B}^N$ with $N \leq n(n+1)/2$. Assume that f is three times differentiable up to the boundary. Then f is a rational map.*

The following has been a well-known open question.

PROBLEM 2.9. *Let $f \in Prop(\mathbb{B}^n, \mathbb{B}^N)$. Is there a fixed constant t such that whenever f is t-times differentiable up to the boundary, then f is rational.*

In a more recent paper, the partial linearity in [Hu2], together with the ideas used in [HJ1] and [Ha], has been used to get a new type of gap phenomenon for holomorphic maps:

THEOREM 2.10. *(Huang-Ji-Xu [HJX2], 2005) Let F be a proper holomorphic map from $\mathbb{B}^n$ into $\mathbb{B}^N$, that is three times differentiable up to the boundary. Suppose that $4 \leq n \leq N \leq 3n-4$. Then F is equivalent to*

$$F'_\theta := (F_\theta(z,w), 0, \cdots, 0) = (z, w\cos\theta, z_1 w \sin\theta, \cdots, z_{n-1}w\sin\theta, w^2\sin\theta, 0, \cdots, 0)$$

for some θ with $(0 \leq \theta \leq \frac{\pi}{2})$.

An interesting feature of this result, which is somewhat surprising to us, is that there is no new map added when N runs from $2n$ to $3n-4$. When $N = 3n-3$, the generalized Whitney map $W_{n,3}$ defined in [Example 2.7] properly sends $\mathbb{B}^n$ into $\mathbb{B}^{3n-3}$. Notice that $W_{n,3}$ is not equivalent to F'_θ, for $W_{n,3}$ has geometric rank 2 $(n \geq 3)$. Indeed, [HJX2] contains more results which we discuss as follows.

Denote by $Prop_k(\mathbb{B}^n, \mathbb{B}^N)$ the collection of all proper holomorphic mappings from $\mathbb{B}^n$ into $\mathbb{B}^N$ that are C^k smooth up to the boundary $(k \geq 2)$. By the work in [Hu2], each map $F \in Prop_2(\mathbb{B}^n, \mathbb{B}^N)$ can be associated with an invariant integer $\kappa_0 \in \{0, 1, \cdots, n-1\}$, called its geometric rank (see [Definition 2.2, Hu2] for the precise definition of κ_0). An early result of the first author ([Hu1, Theorem 4.2]) states that F has geometric rank $\kappa_0 = 0$ if and only if F is equivalent to a linear fractional map. When $\kappa_0 \leq n-2$, a map $F \in Prop_3(\mathbb{B}^n, \mathbb{B}^N)$ $(n \geq 3)$ has a very special geometric structure. Indeed, by the results of [Hu2] and [HJX1], such a map must be rational and $(n-\kappa_0)$-linear. At this point, we mention a theorem of Forstneric [Fr2] which states that $Prop_{N-n+1}(\mathbb{B}^n, \mathbb{B}^N) = Rat(\mathbb{B}^n, \mathbb{B}^N)$ for $N \geq n > 1$ (See also a very interesting paper of Mir [Mir] later on a more general situation).

At this point, we would like to give a discussion on a theorem of Forstneric in terms of a formulation of Ebenfelt and Zaitsev [EZ], then present a combining effort with results obtained in [HJX1]. (See already Corollary 2.1 and its proof in [Hu3] when $N = 2n-1$.) Without loss of generality, we need only consider the rationality problem for a CR map F from an open subset of the origin in the Heisenberg hypersurface $\mathbb{H}^n := \{(z,w) \in \mathbb{C}^{n-1} \times \mathbb{C} : \mathrm{Im}(w) = \|z\|^2\}$ into $\mathbb{H}^N \subset \mathbb{C}^N$. $(N \geq n \geq 2)$. Let $L_j = \frac{\partial}{\partial z_j} + 2i\overline{z_j}\frac{\partial}{\partial w}$. For $\alpha = (\alpha_1, \cdots, \alpha_{n-1})$, we define L^α in a standard way. Write $F = (\widetilde{f}, g)$ and define $E_k(p) = \mathrm{span}_{\mathbb{C}}\{L^\alpha \widetilde{f}(p)\}_{\|\alpha\| \leq k}$. Define $r_k(p) = \mathrm{rank} E_k(p)$. Assume that F is C^m-smooth near 0 with $m \geq 3$. Then a result of Forstneric [Fr2] states that if $r_m(p) = N - 1$ for a certain $p \approx 0$, then F is rational. Ebenfelt and Zaitsev [EZ] first noticed that Forstneric in [Fr2] actually proved the following slightly stronger but more technical result than what he stated: If for a certain $k \leq m$, $r_{k-1}(p) \equiv r_k(p)$ for p in a certain open subset of $\mathbb{H}^n$, then F is rational. Namely, as observed by Ebenfelt-Zaitsev [EZ], Forstneric [Fr2] actually showed that, for the above F, either it is rational or, for p in a certain small piece of the Heisenberg hypersurface, $r_k(p) \geq r_{k-1}(p) + 1$ for any $3 \leq k \leq m$. Hence, if F is C^m $(m \geq 3)$, one concludes that either F is rational or $r_m \geq r_2 + (m-2)$. Now, by the work of [HJX1] mentioned above, if the geometric rank of F is not $n-1$ in a neighborhood of 0 in M, then F is rational. On the other hand, by the work in [Hu2], if the geometric rank of F is $n-1$ at some point p close to the origin, then $r_2(p) = \frac{n(n-1)}{2} + (n-1)$. Hence, one sees the proof of the following rationality result, which should be credited to Ebenfelt, Forstneric, Huang, Ji, Xu and Zaitsev:

COROLLARY 2.11. *Let F be a C^m-smooth $(m \geq 3)$ CR map from an open piece of the unit sphere in $\mathbb{C}^n$ into the unit sphere in $\mathbb{C}^N$. If $\frac{n(n+1)}{2} + (m-2) \geq N$, then F is rational.*

In [HJX2], the following general normalization result has been obtained for maps without full rank:

THEOREM 2.12. *Let F be a non-linear proper holomorphic map from $\mathbb{B}^n$ into $\mathbb{B}^N$ with $N \geq n \geq 3$. Assume that F is C^3-smooth up to the boundary and has geometric rank $\kappa_0 \leq n-2$. Then F is equivalent to a proper holomorphic map of the form*

$$H := (z_1, \cdots, z_{k^0}, H_1, \cdots, H_{N-k^0}),$$

where $k^0 = n - \kappa_0$ and $H_j = \sum_{l=k^0+1}^{n} z_l H_{j,l}$ with $H_{j,l}$ holomorphic over $\overline{\mathbb{B}^n}$. Moreover, when $\kappa_0 = 1$, $(H_1, \cdots, H_{N-n+1}) = z_n \cdot h$ with h a rational proper holomorphic map from $\mathbb{B}^n$ into $\mathbb{B}^{N-n+1}$. Both H and h are affine linear maps along each hyperplane defined by $z_n = constant$.

Theorem 2.12 has several immediate applications, which we discuss as follows:

COROLLARY 2.13. *Let $F \in Rat(\mathbb{B}^n, \mathbb{B}^N)$ have geometric rank 1. Assume that $n \geq 3$. Then the degree of F is bounded by $\frac{N-1}{n-1}$.*

The degree estimate in Corollary 2.13 is optimal. Indeed, the Whitney map has degree 2. By letting h in Theorem 2.12 be the Whitney map, we get a proper polynomial map from $\mathbb{B}^n$ into $\mathbb{B}^N$ with $N = 3n-2$, which is of geometric rank 1 and has degree 3. Inductively, we can construct a proper polynomial map from $\mathbb{B}^n$ into $\mathbb{B}^N$ with $N = kn - (k-1)$, which has degree k and geometric rank 1. At this point, we mention a conjecture of D'Angelo which states that the degree of a rational proper map from $\mathbb{B}^n$ into $\mathbb{B}^N$ with $n \geq 3$ is bounded by $\frac{N-1}{n-1}$. Hence, Corollary 2.13 partially provides an affirmative solution to the aforementioned conjecture of D'Angelo.

COROLLARY 2.14. *Let $F \in Prop_3(\mathbb{B}^n, \mathbb{B}^N)$. Suppose $3 \leq n \leq N = 3n-3$ and F has geometric rank 1. Then F is equivalent to*

$$F'_\theta := (F_\theta(z,w), 0, \cdots, 0) = (z, w\cos\theta, z_1 w \sin\theta, \cdots, z_{n-1} w \sin\theta, w^2 \sin\theta, 0, \cdots, 0)$$

for some θ with $(0 \leq \theta \leq \frac{\pi}{2})$.

Notice that by [Lemma 3.1, Hu2], any non-linear map from $\mathbb{B}^n$ into $\mathbb{B}^N$ with $N \leq 3n-4$ has geometric rank one. Hence, Theorem 2.10 is a special case of Corollary 2.14.

3. The CR version of the Bonnet-Type Theorem

Now, we let M be a submanifold of $\mathbb{C}^n$. For any $p \in M$, we can define $T_p^{(1,0)}M$ and $T_p^{(0,1)}M$. If the complex dimension $CR_M(p)$ of $T_p^{(1,0)}M$ is independent of p, we call M a CR submanifold. When the real dimension of M is $2CR_M(p)+1$, we call M a CR submanifold of hypersurface type. For two CR submanifolds M_1 and M_2 with $p_j \in M_j$, we say (M_1,p_1) is CR equivalent to (M_2,p) if there is a smooth dffeomorphism F from (M_1,p_1) to (M_2,p_2) such that $F_*(T^{(1,0)}M_1) = T^{(1,0)}M_2$. Assume that $M \subset \partial\mathbb{B}^n$ be a CR submanifold of hypersurface type. We say that M has CR codimension l if $l = n-1-CR_M$. Webster's theorem gives the rigidity result for CR codimension 1.

For higher codimension, we have the following:

THEOREM 3.1. *(Ebenfelt-Huang-Zaitsev [EHZ1]) Let (M_j,p_j) be CR submanifold of $\partial\mathbb{B}^n$ of CR codimension k. If $k<(n-1)/2$ and (M_1,p_1) is CR equivalent to (M_2,p_2), then there is*

$$T \in Aut(\partial\mathbb{B}^n)$$

such that $T(p_1)=p_2$ and $T((M_1,p_1))=(M_2,p_2)$.

CR submanifolds of the sphere of hypersurface type come naturally as the link of isolated complex singularity.

Let $(V,0) \subset \mathbb{C}^n$ be a germ of complex analytic variety of codimension k with isolated singularity at 0. Let $0<\epsilon<<1$. Then $M_\epsilon = V \cap \partial\mathbb{B}^n_\epsilon$ is called the link of V with radius ϵ. M_ϵ is a CR submanifold of $\partial\mathbb{B}^n_\epsilon$ of hypersurface type of CR codimension k. As an application of the above theorem, one has the following:

THEOREM 3.2. *(Ebenfelt-Huang-Zaitsev [EHZ1]) Let $(V_j,0)$ be germs of complex analytic variety with isolated singularity at 0 of codimension $k<(n-1)/2$. Let $M_{j,\epsilon} = V_j \cap \partial\mathbb{B}^n_\epsilon$ be the corresponding link. If $(M_{1,\epsilon},p_1)$ is CR equivalent to $(M_{2,\epsilon},p_2)$ for certain $p_j \in M_{j,\epsilon}$. Then there is unitary transformation T such that $T(V_1)=V_2$.*

COROLLARY 3.3. *Let $(V,0)$ be the germ of a complex analytic variety with isolated singularity at 0 of codimension $k<(n-1)/2$. Let $M_\epsilon = V \cap \partial\mathbb{B}^n_\epsilon$ be the ϵ-link. If (M_{ϵ_1},p_1) is CR equivalent to (M_ϵ,p_2) for certain $p_j \in M_{\epsilon_j}$. Then there is unitary transformation T such that $T(V)=\frac{\epsilon_2}{\epsilon_1}V$.*

It is not clear to us if $k<(n-1)/2$ in the above theorem is necessary or not. We certainly belive that k cannot be too big with respect to n as in the situation discussed in the previous section. However, we were not able to provide a counter-example for $k=(n-1)/2$.

4. Super-rigidity and CR transversality for holomorphic mappings between hyperquadrics

In this section, we discuss rigidity for holomorphic mappings from a piece of real hyperquadric with positive signature into a hyperquadric in a complex space of larger dimension. Different from the situation for Heisenberg hypersurfaces (hyperquadrics with 0-signature, that is holomorphically equivalent to the unit sphere), the maps always possess strong super-rigidity property no matter what the codimension is. This phenomenon has analogy in holomorphic maps between irreducible bounded symmetric domains of rank at least two, as studied by many authors: Mok, Tsai, Tu, etc... (See the book of Mok [Mok] for results and extended references on the matter.) To state recent results that the first author jointly obtained with S. Baouendi, to we define, for any $0 < \ell < n-1$, the generalized Siegel upper half space $\mathbb{S}^n_\ell$ as follows:

$$\mathbb{S}^n_\ell := \{(z,w) = (z_1,\cdots,z_{n-1},w) \in \mathbb{C}^n : w = u+iv,\ v > -\sum_{j=1}^{\ell}|z_j|^2 + \sum_{j=\ell+1}^{n-1}|z_j|^2\}.$$

Its boundary is the standard hyperquadric

$$\mathbb{H}^n_\ell := \{(z,w) = (z_1,\cdots,z_{n-1},w) \in \mathbb{C}^n : w = u+iv,\ v = -\sum_{j=1}^{\ell}|z_j|^2 + \sum_{j=\ell+1}^{n-1}|z_j|^2\},$$

where ℓ is called the signature of $\mathbb{H}^n_\ell$. For $0 < \ell < n-1$, it is well known that any CR function defined over a connected piece M of $\mathbb{H}^n_\ell$ extends to a holomorphic function in a neighborhood of M in $\mathbb{C}^n$. The main technical result in [BH] is the following Theorem 4.1, which has two parts. The first part is on the super-rigidity property for holomorphic maps and the second part is on the CR transversality for CR mappings along the lines of studies as in Baouendi-Rothschild [BR], etc.

THEOREM 4.1. *(Theorem 1.6 of Baouendi-Huang [BH]) Let M be a small neighborhood of 0 in $\mathbb{H}^n_\ell$ with $0 < \ell < n-1$. Suppose that $F = (f_1,\cdots,f_{N-1},g)$ is a holomorphic map from a neighborhood U of M in $\mathbb{C}^n$ into $\mathbb{C}^N$ with $F(M) \subset \mathbb{H}^N_\ell$, $N \geq n$, and $F(0) = 0$. Suppose either $\ell \leq (n-1)/2$ or F preserves sides in the sense that $F(U \cap \mathbb{S}^n_\ell) \subset \mathbb{S}^N_\ell$. Then the following hold:*
(I). If $\frac{\partial g}{\partial w}(0) \neq 0$, then F is linear fractional. Moreover, there exits $\tau \in Aut_0(\mathbb{H}^N_\ell)$ such that either

$$\tau \circ F(z_1,\cdots,z_{n-1},w) = (z_1,\cdots,z_{n-1},0,\cdots,0,w),\ \textit{or}$$

$$\tau \circ F(z_1,\cdots,z_{n-1},w) = (z_{\ell+1},\cdots,z_{n-1},z_1,\cdots,z_\ell,0,\cdots,0,-w),$$

and the latter can only happen when $\ell = (n-1)/2$.
(II). If $\frac{\partial g}{\partial w}(0) = 0$, then $F(U) \subset \mathbb{H}^N_\ell$. More precisely, there is a constant $\ell \times (N-\ell-1)$ complex matrix V, with $V\,V^t = \mathrm{Id}_\ell$, such that

$$g \equiv 0, \qquad (f_1,\cdots,f_\ell)\cdot V \equiv (f_{\ell+1},\cdots,f_{N-1}). \tag{4.1}$$

We remark that $\mathbb{H}^n_\ell$ is linearly equivalent to $\mathbb{H}^n_{n-1-\ell}$. Also, when $\ell > \frac{n-1}{2}$, the side preserving assumption in Theorem 4.1 is crucial for results to hold. Indeed, let $F = (z_3, z_2^3, z_1, z_2, z_2^3, -w)$. Then F embeds $\mathbb{H}^4_2$ into $\mathbb{H}^6_2$ with $g_w(0) \neq 0$. But F is not linear fractional.

When $N < 2n-1$, Theorem 4.1 (I) already follows from a recent work of Ebenfelt-Huang-Zaitsev [EHZ2].

The above theorem has applications to the study of super-rigidity problems for proper holomorphic mappings between certain classical domains in complex projective space of different dimensions. Write

$$\mathbb{B}^n_\ell := \{[z_0, \cdots, z_n] \in \mathbb{CP}^n : \ |z_0|^2 + \cdots + |z_\ell|^2 > |z_{\ell+1}|^2 + \cdots + |z_n|^2\}.$$

Its automorphism group is a non-compact semi-simple Lie-group of rank $(\ell + 1)$:

$$\mathrm{Aut}(\mathbb{B}^n_\ell) = \{\sigma :\in \mathrm{Hol}(\mathbb{CP}^n, \mathbb{CP}^n) : \sigma([z_0, \cdots, z_n]) = [z_0, \cdots, z_n] \cdot A,$$

$$A \in GL(n+1, \mathbb{C}), \ AE_{\ell+1,n+1}A^t = E_{\ell+1,n+1}.\}$$

where $E_{k,m}$ denotes a diagonal matrix with its first k diagonal elements -1 and the rest $+1$. As a corollary of Theorem 4.1 (I), one has the following:

THEOREM 4.2. *(Theorem 1.1 of [BH]) Let p be a boundary point of $\mathbb{B}^n_\ell$ and let U_p be a small neigborhood of p in $\mathbb{CP}^n$ with $U_p \cap \mathbb{B}^n_\ell$ connected. Assume that F is a holomorphic map from $U_p \cap \mathbb{B}^n_\ell$ into $\mathbb{B}^N_\ell$ for $N \geq n$ and $0 < \ell < n-1$. If for any sequence $\{Z_j\} \subset U_p \cap \mathbb{B}^n_\ell$ with $\lim_{j\to\infty} Z_j \in \partial\mathbb{B}^n_\ell$, $\{F(Z_j)\}$ only has limit points in $\partial\mathbb{B}^N_\ell$. Then F extends to a totally geodesic embedding from $\mathbb{B}^n_\ell$ into $\mathbb{B}^N_\ell$ in the sense that there is a certain $\sigma \in Aut(\mathbb{B}^n_\ell)$ and $\tau \in Aut(\mathbb{B}^N_\ell)$ such that $\tau \circ F \circ \sigma([z_0, \cdots, z_n]) = [z_0, \cdots, z_n, 0, \cdots, 0]$.*

COROLLARY 4.3. *(Corollary 1.2 of Baouendi-Huang [BH]) Any proper holomorphic mapping F from $\mathbb{B}^n_\ell$ into $\mathbb{B}^N_\ell$ for $N \geq n$ and $0 < \ell < n-1$ is a totally geodesic embedding, in the sense that there is a certain $\sigma \in Aut(\mathbb{B}^n_\ell)$ and $\tau \in Aut(\mathbb{B}^N_\ell)$ such that $\tau \circ F \circ \sigma([z_0, \cdots, z_n]) = [z_0, \cdots, z_n, 0, \cdots, 0]$.*

There is no proper holomorphic map from $\mathbb{B}^n_\ell$ into $\mathbb{B}^N_{\ell'}$ for $\ell' < \ell$. One cannot expect, in general, that Corollary 4.3 holds for any $\ell' > \ell$. For instance, one has the following:

EXAMPLE 4.4. Let $F = [z_0^2, \sqrt{2}z_0z_1, z_1^2, z_2^2, \sqrt{2}z_3z_2, z_1^2]$. Then F maps properly $\mathbb{B}^3_1$ into $\mathbb{B}^5_2$. Notice that F is not a linear map.

However, one has the following: (In light of the above example, the target dimension $n + k$ can not be improved. Also, we will use 0_j to denote a zero component at the jth-position in a vector)

THEOREM 4.5. *(Theorem 1.8 of Baouendi-Huang [BH]) Let M be a small neighborhood of 0 in $\mathbb{H}^n_\ell$ with $0 < \ell < n-1$. For $k \geq 0$, let $F = (f_1, \cdots, f_{n+k-1}, g)$ be a holomorphic map from a neighborhood U of M in $\mathbb{C}^n$ into $\mathbb{C}^{n+k}$ with $F(M) \subset \mathbb{H}^{n+k}_{\ell+k}$, and $F(0) = 0$. Suppose that F preserves sides in the sense that $F(U \cap \mathbb{S}^n_\ell) \subset \overline{\mathbb{S}^{n+k}_{\ell+k}}$. Then the following hold:*

(I). If $\dfrac{\partial g}{\partial w}(0) \neq 0$, then F is linear fractional. Moreover, there exits $\tau \in Aut_0(\mathbb{H}^{n+k}_{\ell+k})$ such that either

(4.2)

$$\tau \circ F(z_1, \cdots, z_{n-1}, w) = (z_1, \cdots, z_\ell, 0_{\ell+1}, \cdots, 0_{k+\ell}, z_{\ell+1}, \cdots, z_{n-1}, 0, \cdots, 0, w).$$

(II). If $\dfrac{\partial g}{\partial w}(0) = 0$, then $F(U) \subset \mathbb{H}^{n+k}_{\ell+k}$. More precisely, when $\ell + k \geq n - \ell - 1$, there is an $(n-\ell-1) \times (\ell+k)$ constant complex matrix V, with $V V^t = \mathrm{Id}_{n-\ell-1}$, such that

$$g \equiv 0, \quad (f_1, \cdots, f_{\ell+k}) \equiv (f_{\ell+k+1}, \cdots, f_{n+k-1})V. \tag{4.3}$$

When $\ell + k \le n - \ell - 1$, there is an $(\ell + k) \times (n - \ell - 1)$ constant complex matrix V, with $VV^t = \mathrm{Id}_{\ell+k}$, such that

$$(4.3)' \qquad g \equiv 0, \qquad (f_1, \cdots, f_{\ell+k})V \equiv (f_{\ell+k+1}, \cdots, f_{n+k-1}).$$

COROLLARY 4.6. *(Theorem 4.1 of [BH]) Any proper holomorphic mapping F from $\mathbb{B}^n_\ell$ into $\mathbb{B}^{n+k}_{\ell+k}$ for $N = n + k \ge n$ and $0 < \ell < n - 1$ is a totally geodesic embedding, in the sense that there is a certain $\sigma \in Aut(\mathbb{B}^n_\ell)$ and $\tau \in Aut(\mathbb{B}^N_\ell)$ such that*

$$(4.4) \quad \tau \circ F \circ \sigma([z_0, \cdots, z_n]) = [z_0, \cdots, z_\ell, 0_{\ell+1}, \cdots, 0_{\ell+k}, z_{\ell+k+1}, \cdots, z_n, 0, \cdots, 0].$$

The proof of Theorem 4.1 (II) and Theorem 4.5 (II) is based on the following, which is very useful in some other applicationst (For the definition of $\mathbb{H}^N_{\ell,\ell',n}$ see (2.2) of [BH]):

THEOREM 4.7. *(Lemma 4.1 of Baouendi-Huang [BH]) Let $F = (f, \phi, g)$ be a holomorphic map from a neighborhood M of 0 in $\mathbb{H}^n_\ell$ into $\mathbb{H}^N_{\ell,\ell',n}$, $\ell' \ge \ell > 0, N \ge n > 1$, with $F(0) = 0$. Assume that either $\ell' < n - 1$ or $N - \ell' - 1 < n - 1$. For each $p \in M$, let $F_p = (\widetilde{f_p}, g_p)$ be defined as in [(3.1), BH]. If $(g_p)_w(0) = 0$ for all p sufficiently close to the origin, then $g \equiv 0$ and $F(U) \in \mathbb{H}^N_{\ell,\ell',n}$, where U is a small neighborhood of 0 in $\mathbb{C}^n$. More precisely, when $\ell' \ge (N - \ell' - 1)$, there exists a constant $(N - \ell' - 1) \times \ell'$ complcx matrix V with $VV^t = \mathrm{Id}_{N-\ell'-1}$ such that*

$$(4.5) \quad g \equiv 0, \quad (f_1, \cdots, f_\ell, f_n, \cdots, f_{n+\ell'-\ell-1}) \equiv (f_{\ell+1}, \cdots, f_{n-1}, f_{n+\ell'-\ell}, \cdots, f_{N-1})V;$$

when $\ell' < (N - \ell' - 1)$, there exists a constant $\ell' \times (N - \ell' - 1)$ complex matrix V with $V \cdot V^t = \mathrm{Id}_{\ell'}$ such that

$$(4.5)' \quad g \equiv 0, \quad (f_1, \cdots, f_\ell, f_n, \cdots, f_{n+\ell'-\ell-1}) \cdot V \equiv (f_{\ell+1}, \cdots, f_{n-1}, f_{n+\ell'-\ell}, \cdots, f_{N-1}).$$

The last sentence of [Theorem 1.6 (ii), BH] is to give an equivalent expression of the mathematical content of [Theorem 1.6(ii), BH]). In that sentence, there is a typo which the authors of [BH] would like to correct by making use of this opportunity: Namely, $(f_1, \cdots, f_\ell) \equiv (f_{\ell+1}, \cdots, f_{N-1})V$ should be correcetd to $(f_1, \cdots, f_\ell)V \equiv (f_{\ell+1}, \cdots, f_{N-1})$. (Therefore, the size of the matrix V should be corrected as $\ell \times (N - \ell - 1)$). Similar corections are needed in the last formula of [Theorem 1.8(ii), BH] in the case of $\ell + k < n - \ell - 1$, and in the formula [(4.1), Lemma 4.1, BH] in the case of $\ell' < N - \ell' - 1$. ([Theorem 1.6 (ii), BH], [Theorem 1.8 (ii), BH] and [Lemma 1.6, BH] (with these typos corrected) are copied here as Theorem 4.1(ii), Theorem 4.5 (ii) and Theorem 4.7, respectively.)

One can also study the rigidity for CR submanifolds embedded into $\mathbb{H}^n_\ell$. However, since $\mathbb{H}^n_\ell$ contains complex subspaces, we have to exclude those CR submanifolds embedded in the complex varieties in $\mathbb{H}^n_\ell$. For this, we say M is a CR transversal submanifold of $\mathbb{H}^n_\ell$ of hypersurface type if the complex normal direction of M is also along the complex normal direction of the hyperquadric. Then we have the following:

THEOREM 4.8. *(Ebenfelt-Huang-Zaitsev [EHZ2]) Let M_j be two CR transveral CR submanifolds of hypersurface type. Assume they are Levi non-degenerate with signature ℓ, and of CR codimension k. If $2k < n - 1$ and they are CR equivalent, then they are rigid motion of each other.*

It is an open question if the codiemnsion restriction in the above result can be dropped or not.

5. Rigidity Problems for Holomorphic Maps with Symmetry

Here, we mention a well-known super-rigidity problem in differential geometry: Let G_1 and G_2 be two non-compact semi-simple Lie groups with $dim(G_2) \geq dim(G_1)$. Suppose $\Gamma_j \subset G_j$ are lattices (with certain density or co-compactness properties) and there is a injective homomorphism $\phi : \Gamma_1 \to \Gamma_2$. Can one then extend ϕ to a global homomorphism from G_1 into G_2? This question has been quite well understood when the real ranks of the groups are at least 2 by the work of Mostow, Margulis and Mok-Siu-Yeung, etc. In the rank one case, there are essentially four cases to be considered: The automorphism groups of real, complex and quaternionic hyperbolic spaces, together with that of the hyperbolic Cayley plane. After the work of many people, an important remaining open case is when $G_1 = Aut(\mathbb{B}^n)$ and $G_2 = Aut(\mathbb{B}^N)$ with $1 < n < N, \Gamma_1$ co-compact in G_1 and $\phi(\Gamma_1)$ convex-cocompact. (Notice that when $n = 1$, certain kind of counter-examples have been constructed by Mostow). This problem, after applying the harmonic mapping theory of Siu, is reduced to the following :

Problem 5.1. *(Siu, Mok, etc) Let $\mathbb{B}^n \subset \mathbb{C}^n$ and $\mathbb{B}^N \subset \mathbb{C}^N$ be the unit balls, and let f be a proper holomorphic embedding from $\mathbb{B}^n$ into $\mathbb{B}^N$ $(n, N > 1)$. Suppose that there is a discrete subgroup $\Gamma \subset Aut(\mathbb{B}^N)$ such that Γ fixes $M = f(\mathbb{B}^n)$ and acts co-compactly over M (without fixed point). Is f then a linear embedding?*

This question has been answered in the affirmative by Cao-Mok [CaoM] in case($N \leq 2n - 1$).

Theorem 5.2. *(Cao-Mok [CM], 1990) Let Γ be a lattice of $Aut(\mathbb{B}^N)$ without fixed point. Let F be a proper holomorphic embedding from $\mathbb{B}^n$ into $\mathbb{B}^N$ with $N \leq 2n - 1$ such that Γ stablize and act co-compactly on the image of $F(\mathbb{B}^n)$. Then F is linear.*

Lastly, we mention two results on the bounded symmetric domains:

Theorem 5.3. *(Tsai [Ts], 1993) Let Ω_1 and Ω_2 be two bounded symmetric domains with the same rank at least 2. then any proper holomorphic map from Ω_1 into Ω_2 is a totally geodesic embedding with respect to their Bergman metric.*

Theorem 5.4. *(Tu [Tu], 2002) Let $D^I_{p,p-1}$ and $D^I_{p,p}$ be the standard bounded symmetric domains of the first type with $p - 1 \geq 2$. Then any proper holomorphic map from $D^I_{p,p-1}$ into $D^I_{p,p}$ is a totally geodesic embedding with respect to their Bergman metric.*

6. Segre family of real analytic hypersurfaces and rigidity problems for holomorphic Segre maps

Real analytic CR maps between real analytic CR manifolds in complex spaces naturally induce what we call the holomorphic Segre maps. However, not all holomorphic Segre maps are coming from CR maps. That motivated us to consider the more general rigidity problems for holomorphic Segre maps. This study seems still at a very early stage and we compute in this part of this paper the holomorphic

Segre isomorphisms of the complexification of the Heisenberg hypersurface. We start with some basic notations.

Let M be a real analytic hypersurface in $D \subset \mathbb{C}^n$ with real analytic defining function $r \in C^\omega(D)$. Apparently, for any other local defining function r^* of M, $r^* = s^* r$ with $s^*|_M \neq 0$. Hence we can well define its complexification as the complex submanifold: $\mathcal{M} = \{(z,\xi) \in D \times Conj(D)\ : r(z,\xi) = 0\}$. $\mathcal{M}$ is a complex submanifold of complex codimension 1 in $\mathbb{C}^n \times \mathbb{C}^n$ near $M \times conj(M)$. Here for a set $E \subset \mathbb{C}^n$, $Conj(E) := \{\overline{z} \mid z \in E\}$. For each $\xi \sim Conj(M)$, we can define complex analytic varieties $Q_\xi := \{z \in \mathbb{C}^n : r(z,\xi) = 0\}$ and $\hat{Q}_z := \{\xi \in \mathbb{C}^n : r(z,\xi) = 0\}$. We call Q_ξ and $\hat{Q}_z$ the *Segre variety* of M with respect to ξ and z, respectively. Notice that $\mathcal{M}$ is holomorphically foliated by $\{Q_\xi \times \{\xi\}\}$ and also by $\{\{z\} \times \hat{Q}_z\}$. Due to this reason, $\mathcal{M}$ is called the Segre family associated with M. A fundamental fact for Segre family is its invariant property for holomorphic maps. More precisely, if f is a local holomorphic map from (M,p) to $(\widetilde{M},\widetilde{p})$, then $f(Q_\xi) \subset \widetilde{Q}_{\overline{f}(\xi)}$ and $\overline{f}(\hat{Q}_z) \subset \hat{\widetilde{Q}}_{f(z)}$. Here, for instance, $\widetilde{Q}_{\overline{f}(\xi)}$ is the Segre variety of $\widetilde{M}$ with respect to $\overline{f}(\xi)$. In particular, when f is a holomorphic map from (M,p) to $(\widetilde{M},\widetilde{p})$, f induces a holomorphic map $\mathcal{F} := (f(z), \overline{f}(\xi))$ from $(\mathcal{M},(p,\overline{p}))$ to $(\widetilde{\mathcal{M}},(\widetilde{p},\overline{\widetilde{p}}))$.

Next, let $(\widetilde{M},\widetilde{p})$ be another real analytic hypersurface near $\widetilde{p} \in \mathbb{C}^N$. If there is a holomorphic map f from (M,p) to $(\widetilde{M},\widetilde{p})$, then we have a holomorphic map $(f(z),\overline{f}(\xi))$ from $(\mathcal{M},(p,p))$ to $(\widetilde{\mathcal{M}},(\widetilde{p},\widetilde{p}))$. We say that Φ is a holomorphic Segre map from $(\mathcal{M},A)$ into $(\widetilde{\mathcal{M}},\widetilde{A})$ if Φ is a holomorphic map from $(\mathcal{M},A)$ into $(\widetilde{\mathcal{M}},\widetilde{A})$ such that Φ sends each $Q_\xi \times \{\xi\}$ of $\mathcal{M}$ near A into a certain $Q_{\widetilde{\xi}} \times \{\widetilde{\xi}\}$ of $\widetilde{\mathcal{M}}$ near $\widetilde{A}$ and sends each $\{z\} \times \hat{Q}_z$ into a certain $\{\widetilde{z}\} \times \hat{\widetilde{Q}}_{\widetilde{z}}$. A holomorphic Segre map Φ is called a holomorphic Segre isomorphism if it is also a local biholomorphism. If this is the case, we say $(\mathcal{M},A)$ is Segre-equivalent to $(\widetilde{\mathcal{M}},\widetilde{A})$. In what follows, we only consider the case when Φ is a holmorphic embedding. It can be seen that Φ can be made to take the form: $\Phi(z,\xi) = (\Phi_1(z),\Phi_2(\xi))|_{\mathcal{M}}$. (See Lemma 6.1 below). Apparently, when (M,p) is equivalent to $(\widetilde{M},\widetilde{p})$, then $(\mathcal{M},A)$ is Segre-equivalent to $(\widetilde{\mathcal{M}},\widetilde{A})$ with $A = (p,\overline{p})$ and $\widetilde{A} = (\widetilde{p},\overline{\widetilde{p}})$. We mention that even if M, $\widetilde{M}$ are strongly pseudoconvex, Faran has constructed examples showing that the converse of this statement fails (See [Fa3]).

For simplicity, assume that $p = 0 \in M$ and that M is defined by an equation of the form $r = 2\mathrm{Im}(z_n) + O(|z'| + |\mathrm{Re}(z_n)|)$. On $\mathcal{M}$, there are $(2n-1)$ independent holomorphic one forms

(6.1)
$$\theta^\alpha = dz^\alpha|_{\mathcal{M}},\ \theta_\alpha = d\xi_\alpha|_{\mathcal{M}},\ 1 \le \alpha < n-1, \theta = i d_z r|_{\mathcal{M}} = i r_\alpha dz^\alpha|_{\mathcal{M}} + i r_n dz^n|_{\mathcal{M}}.$$

$\{\theta,\theta^\alpha,\theta_\alpha\}$ is a co-frame for $\mathcal{M}$. In what follows, α,β have range from 1 to $(n-1)$ and the summation convention will be applied. Then $\{\theta,\theta_\alpha\}$ and $\{\theta,\theta^\alpha\}$ are two complete systems, whose corresponding foliations are precisely $\{Q_\xi \times \{\xi\}\}$ and $\{\{z\} \times \hat{Q}_z\}$, respectively.

Also, let $(\widetilde{M},\widetilde{p})$ be another real analytic hypersurface near $\widetilde{p} = 0$ in $\mathbb{C}^N$ with a defining function $\widetilde{r} = 2Im(\widetilde{z}_N) + O(|\widetilde{z}'| + Re(z_N))$. Define similarly the co-frame $\{\widetilde{\theta},\widetilde{\theta}^\alpha,\widetilde{\theta}_\alpha\}$ on $\widetilde{\mathcal{M}}$ near $(\widetilde{p},\overline{\widetilde{p}})$.

LEMMA 6.1. *Suppose Φ is a holomorphic Segre embedding from $(\mathcal{M},0)$ to $(\widetilde{\mathcal{M}},0)$. Then*

$$\begin{cases} \Phi^*(\widetilde{\theta}) = u\theta, \\ \Phi^*(\widetilde{\theta}^\alpha) = u^\alpha\theta + \sum_\beta u^\alpha_\beta \theta^\beta, \\ \Phi^*(\widetilde{\theta}_\alpha) = u_\alpha\theta + \sum_\beta v^\beta_\alpha \theta_\beta. (6.2) \end{cases} \tag{6.2}$$

where $u, u^\alpha, u^\alpha_\beta, v^\beta_\alpha$ are holomorphic near 0 and the holomorphic 1-forms are defined as in (6.1). Moreover $\Phi = (\Phi_1(z), \Phi_2(\xi))|_{\mathcal{M}}$.

PROOF. Suppose that Φ is a holomorphic Segre embedding. Then

$$\{\Phi^*(\widetilde{\theta}), \Phi^*(\widetilde{\theta}^\alpha)\} \text{ and } \{\Phi^*(\widetilde{\theta}), \Phi^*(\widetilde{\theta}_\alpha)\}$$

are complete differential systems over $\mathcal{M}$. Since Φ preserves the Segre varieties, the restriction of $\{\Phi^*(\widetilde{\theta}), \Phi^*(\widetilde{\theta}^\alpha)\}$ (respectively, $\{\Phi^*(\widetilde{\theta}), \Phi^*(\widetilde{\theta}_\alpha)\}$) to the leaves of $\{\theta, \theta^\alpha\}$ (respectively, $\{\theta, \theta_\alpha\}$) must be zero. We conclude the last two formulas in (6.2). The first formula is also obvious because the restriction of $\Phi^*(\widetilde{\theta})$ to leaves of both foliations through the Segre varieties is zero.

Write $\Phi = (\Phi_1, \Phi_2)$. Parametrize $\mathcal{M}$ by (z_α, ξ). Then Φ_2 now can be regarded as a holomorphic function in (z_α, ξ). Since for each fixed ξ, Φ maps $\{Q_\xi \times \{\xi\}\}$ into $\{\widetilde{Q}_{\widetilde{\xi}} \times \{\widetilde{\xi}\}\}$ for a certain $\widetilde{\xi}$, $\Phi_2(z_a, \xi)$ must be constant as z_a varies. That shows that Φ_2 can be expressed as the restriction of a holomorphic function in ξ to $\mathcal{M}$. Similarly, one can verify that Φ_1 can also be regarded as just a holomorphic map in z. □

Now, assume that $\mathcal{M}$ is defined near 0 by a real analytic function of the form: $r = z_n - \rho(z^\alpha, \xi_\alpha, \xi_n)$ with $\rho = O(|z^\alpha| + |\xi_\alpha| + |\xi_n|)$. Assume that M is Levi non-degenerate near 0. Then

$$\begin{cases} z^\alpha = z^\alpha \\ z_n = z_n \\ \rho_\alpha = \rho_\alpha(z^\alpha, \xi_\alpha, \xi_n) \end{cases}$$

can be used to uniquely solve for $(z^\alpha, \xi_\alpha, z_n)$ by the data $(z^\alpha, \rho_\alpha, z^n)$. (See [Ch], [CJ1], etc). Hence, we can use $(z^\alpha, z^n, \rho_\alpha)$ for the coordinates of $\mathcal{M}$ near 0. In the $(z^\alpha, z_n, \rho_\beta)$ coordinates, we define the holomorphic frame $\{\theta^\alpha, \theta_\alpha, \theta\}$ as follows:

$$\begin{cases} \theta = i(dz^n - \rho_\alpha dz^\alpha), \\ \theta^\alpha = dz^\alpha, \\ \theta_\alpha = d\rho_\alpha - \rho_{\alpha\beta}dz^\beta, \quad d\theta = i\theta^\alpha \wedge \theta_\alpha. \end{cases} \tag{6.3}$$

Here $\rho_\alpha = \frac{\partial\rho}{\partial z_\alpha}$, $\rho_{\alpha\beta} = \frac{\partial\rho}{\partial z_\alpha \partial z_\beta}$ are holomorphic functions in (z, ρ_α). Then $\{\theta, \theta_\alpha\}($ $\{\theta, \theta^\alpha\})$ also generates the foliation $\{Q_\xi \times \{\xi\}\}$ ($\{\{z\} \times \hat{Q}_z\}$, respectively).

Next, let $(\widetilde{\mathcal{M}}, 0)$ be the complexification of another real analytic hypersurface $(\widetilde{M}, 0)$. We also choose the same type of the co-frame $(\widetilde{\theta}, \widetilde{\theta}^\alpha, \widetilde{\theta}_\alpha)$ on $(\widetilde{\mathcal{M}}, 0)$ as in (6.3). Suppose that Φ is a holomorphic Segre isomorphism from $(\mathcal{M}, 0)$ to $(\widetilde{\mathcal{M}}, 0)$.

Then by Lemma 6.1, we have

$$\begin{cases} \Phi^*(\widetilde{\theta}) = u\theta, \\ \Phi^*(\widetilde{\theta}^\alpha) = u^\alpha_\beta\theta^\beta + u^\alpha\theta, \\ \Phi^*(\widetilde{\theta}_\alpha) = v^\beta_\alpha\theta_\beta + v_\alpha\theta. \end{cases}$$

with $u, u^\alpha_\beta, v^\alpha_\beta, u^\alpha, v_\beta$ holomorphic near the origin.

Since $d\widetilde{\theta} = i\widetilde{\theta}^\alpha \wedge \widetilde{\theta}_\alpha$, we get $du \wedge \theta + ud\theta = i(u^\alpha_\beta\theta^\beta + u^\alpha\theta) \wedge (v^\beta_\alpha\theta_\beta + v_\alpha\theta)$, from which we get the following

$$\begin{cases} \delta^l_k u = u^\alpha_l v^k_\alpha, \\ du = iu^\alpha_\beta v_\alpha\theta^\beta - iu^\alpha v^\beta_\alpha\theta_\beta + t\theta. \end{cases}$$

Next, we consider the $\mathbb{C}^* := \mathbb{C} - \{0\}$ fiber bundle $\mathcal{E}_0 = \mathcal{M} \times \mathbb{C}^*$ over $\mathcal{M}$, which can be identified with the $\mathbb{C}^*$-fiber bundle over $\mathcal{M}$, whose fiber $\pi^{-1}(P)$ over $P \in \mathcal{M}$ is precisely $\{u\theta|_P\}$ with $u \in \mathbb{C}^*$. Then $\omega = u\theta$ is a tautological global holomorphic 1-form on $\mathcal{E}_0$. Notice that

$$d\omega = ud\theta + du \wedge \theta = iu\theta^\alpha \wedge \theta_\alpha + \omega \wedge (-\frac{du}{u}).$$

Define

$$\begin{cases} \omega^\alpha = u^\alpha\theta + u^\alpha_\beta\theta^\beta, \\ \omega_\alpha = v_\alpha\theta + v^\beta_\alpha\theta_\beta, \end{cases}$$

where $u^\gamma_\beta v^\beta_\kappa = \delta^\gamma_\kappa u$. Then

$$d\omega = i\omega^\alpha \wedge \omega_\alpha + \omega \wedge \left(-\frac{du}{u} - i\frac{u^\alpha}{u}v^\beta_\alpha\theta_\beta + iu^\alpha_\beta\frac{v_\alpha}{u}\theta^\beta \right).$$

Let $\phi = -\frac{du}{u} - i\frac{u^\alpha}{u}v^\beta_\alpha\theta_\beta + iu^\alpha_\beta\frac{v_\alpha}{u}\theta^\beta + t\omega$. Then, the above motivates us to consider the following set of 1-forms:

$$\begin{cases} \omega = u\theta, \\ \omega^\alpha = u^\alpha\theta + u^\alpha_\beta\theta^\beta, \\ \omega_\alpha = v_\alpha\theta + v^\beta_\alpha\theta_\beta, \\ \phi = -\frac{du}{u} - i\frac{u^\alpha}{u}v^\beta_\alpha\theta_\beta + iu^\alpha_\beta\frac{v_\alpha}{u}\theta^\beta + t\theta \\ \delta^l_k u = u^l_\alpha v^\alpha_k. \end{cases} \tag{6.4}$$

A basic property for the above set of 1-forms is the relation:

$$d\omega = i\omega^\alpha \wedge \omega_\alpha + \omega \wedge \phi. \tag{6.5}$$

Choose a special co-frame:

$$\begin{cases} \omega^0 = u\theta, \\ \omega^{0\alpha} = u\theta^\alpha, \\ \omega^0_\alpha = \theta_\alpha, \\ \phi^0 = -\frac{du}{u}. \end{cases}$$

Then, we have

$$\begin{cases} \omega = \omega^0, \\ \omega^\alpha = \frac{u^\alpha}{u}\omega^0 + \frac{u^\alpha_\beta}{u}\omega^{0\beta} \\ \omega_\alpha = \frac{v_\alpha}{u}\omega^0 + v^\beta_\alpha \omega^0_\alpha, \\ \phi = \phi^0 - i\frac{u^\alpha}{u}v^\beta_\alpha\omega^0_\beta + iu^\alpha_\beta\frac{v_\alpha}{u^2}\omega^{0\alpha} + t\omega^0. \end{cases}$$

Hence the space of the set of 1-forms in (6.4) forms a G_1-structure bundle $\mathcal{Y}$ over $\mathcal{E}_0$, where G_1 consists of matrices of the following form:

$$\begin{pmatrix} 1 & 0 & 0 & 0 \\ \frac{u^\alpha}{u} & \frac{u^\alpha_\beta}{u} & 0 & 0 \\ \frac{v_\alpha}{u} & 0 & v^\beta_\alpha & 0 \\ t & i\frac{v^\alpha}{u^2}u^\beta_\alpha & -i\frac{u_\alpha}{u}v^\alpha_\beta & 1 \end{pmatrix}$$

with $u^l_\alpha v^\alpha_k = \delta^l_k u$. Or

$$\begin{pmatrix} 1 & 0 & 0 & 0 \\ u^\alpha & u^\alpha_\beta & 0 & 0 \\ v_\alpha & 0 & v^\beta_\alpha & 0 \\ t & iv^\alpha u^\beta_\alpha & -iu_\alpha v^\alpha_\beta & 1 \end{pmatrix}$$

with $u^l_\alpha v^\alpha_k = \delta^l_k$.

Now, the Segre family $(\mathcal{M}, 0)$ and $(\widetilde{\mathcal{M}}, 0)$ are equivalent if and only if there is a holomorphic map $\mathcal{F}$ from $(\mathcal{E}_0, P_0)$ to $(\widetilde{\mathcal{E}_0}, \widetilde{P_0})$ with $P_0 \in \pi^{-1}(0)$, $\widetilde{P_0} \in \widetilde{\pi}^{-1}(0)$, such that

$$F^* \begin{pmatrix} \widetilde{\omega}^0 \\ \widetilde{\omega}^{0\alpha} \\ \widetilde{\omega}^0_\alpha \\ \widetilde{\phi}^0 \end{pmatrix} (P) = \gamma_{\mathcal{F}}(P) \begin{pmatrix} \omega^0 \\ \omega^{0\alpha} \\ \omega^0_\alpha \\ \phi^0 \end{pmatrix} (P)$$

where $\gamma_{\mathcal{F}}(P)$ is a holomorphic map from a neighborhood of P_0 into G_1.

We consider the G_1-structure bundle $\mathcal{Y}$ over $\mathcal{E}_0$ and lift the above to globally defined holomorphic tautological 1-forms over $\mathcal{Y}$. A fundamental theorem of Chern-Moser and Chern ([CM], [Ch]) states that these forms can be indeed uniquely completed to make the G_1-equivalence problem into an e-equivalence problem over $\mathcal{Y}$. Also, $\mathcal{F}$ is uniquely determined by $\gamma_{\mathcal{F}}(P_0)$ and $\widetilde{P_0}$. (See [Hu4], [Ga]).

We discuss how to determine the holomorphic Segre isomorphisms through their jets at 0. Without loss of generality, we can assume that $\mathcal{M}, \widetilde{\mathcal{M}}$ are defined by the normalized equations as in Chern-Moser [CM]. Then $r = z_n - \xi_n = 2\pi z' \cdot \xi' + O(4)$. Write $F = (f(z), g(z), \phi(\xi), \psi(\xi))$ for a holomorphic Segre isomorphism from $(\mathcal{M}, 0)$ to $(\widetilde{\mathcal{M}}, 0)$. Hence, we have $g(z) - \psi(\xi) = 2\pi i f(z) \cdot \psi(\xi) + O(4)$. Here $z = (z', z_n)$, $\xi = (\xi', \xi_n)$. Write $F^*(\widetilde{\theta}) = k(z, \xi)\theta$, $P_0 = \underline{(0, 1)} \in \pi^{-1}(0)$. Then the lifting of F to $\mathcal{F}$, a holomorphic map from $(\mathcal{E}_0, P_0)$ to $(\widetilde{\mathcal{E}_0}, \widetilde{P_0})$, takes the form: $\mathcal{F} = (F, u/k)$. Notice that $k(0) = g'_{z_n}(0)$. Hence, if $k(0)$ is fixed, then $\widetilde{P}_0 = \mathcal{F}(P_0)$ is fixed too. Next, in the matrix $\gamma_{\mathcal{F}}(P_0)$, $u^\alpha, u^\alpha_\beta, v^\alpha_\beta, v_\alpha$ are determined by the first jets of f, ϕ at 0. It is also clear that t is determined by $\frac{\partial^2 k}{\partial z_n^2}(0)$, which, through a direct computation, can be seen to be determined by $g''_{z_n z_n}(0)$ and $f'_{z_n}(0), \phi'_{\xi_n}(0)$. Therefore, we see that F is completely determined by the first jet of f, ϕ at 0, and $g'_{z_n}(0)$, $g''_{z_n z_n}(0)$.

Write $\mathcal{H}^n := \{(z', w, \xi', \eta) \in \mathbb{C}^{n-1} \times \mathbb{C}^1 \times \mathbb{C}^{n-1} \times \mathbb{C}^1 : w - \eta = 2iz' \cdot \xi\}$ for the complexification of the Heisenberg hypersurface $\mathbb{H}^n$. A natural question one would like to ask is the following:

PROBLEM 6.2. *Let Φ be a holomorphic Segre embedding from $(\mathcal{H}^n, 0)$ into $(\mathcal{H}^N, 0)$ with $N \geq n \geq 2$. What type of rigidity can we get for Φ?*

We say $\Psi \in Aut(\mathcal{H}^n)$ if Ψ is a bimeromorphic map from $\mathcal{H}^n$ to itself, preserving the Segre varieties (both Q and $\hat{Q}$). For any $A_0 = (z'_0, w_0, \xi'_0, \eta_0) \in \mathcal{H}^n$, define $\sigma^0_{A_0}$ by sending $(z', w, \xi', \eta) \in \mathbb{C}^{n-1} \times \mathbb{C}^1 \times \mathbb{C}^{n-1} \times \mathbb{C}^1$ to $(z' + z'_0, w + w_0 + 2iz' \cdot \xi'_0, \xi' + \xi'_0, \eta + \eta_0 - 2i\xi' \cdot z'_0)$. Then $\sigma^0_{A_0} \in Aut(\mathcal{H}^n)$ with $\sigma^0_{A_0}(0) = A_0$. Hence, to understand $Aut(\mathcal{H}^n)$, it suffices to compute $Aut_0(\mathcal{H}^n)$, the set of local self-Segre isomorphisms of $(\mathcal{H}^n, 0)$.

Write $z' = (z_1, \cdots, z_{n-1})$, $\xi' = (\xi_1, \cdots, \xi_{n-1})$, $w, \eta \in \mathbb{C}$, and $z = (z', w)$ $\xi = (\xi', \eta)$ as before. Write $\mathcal{I}_0$ for the group generated by the following type of linear fraction maps: $F(z, \xi) = (S(z), T(\xi)) := (S_1(z), \cdots, S_{n-1}(z), S_n(z), T_1(\xi), \cdots, T_{n-1}(\xi), T_n(\xi)) = (\widetilde{S}(z), S_n(z), \widetilde{T}(\xi), T_n(\xi))$ defined by

$$\widetilde{S}(z) = \frac{\lambda(z' + \vec{a}w)U}{1 - 2i\langle z', \vec{e}\rangle + e_n w}, \quad S_n(z) = \frac{\lambda w}{1 - 2i\langle z', \vec{e}\rangle + e_n w}, \tag{6.6}$$

$$\widetilde{T}(\xi) = \frac{(\xi' + \vec{e}\eta)V}{1 + 2i\langle \xi', \vec{a}\rangle + (e_n + 2i\langle \vec{e}, \vec{a}\rangle)\eta}, \tag{6.7}$$

$$T_n(\xi) = \frac{\lambda\eta}{1 + 2i\langle \xi', \vec{a}\rangle + (e_n + 2i\langle \vec{e}, \vec{a}\rangle)\eta}; \tag{6.8}$$

where U, V are non-singular $(n-1) \times (n-1)$ matrices of complex numbers with $U \cdot V^t = \mathrm{Id}$, $\vec{a} = (a_1, ..., a_{n-1})$, $\vec{e} = (e_1, ..., e_{n-1}) \in \mathbb{C}^{n-1}$, $\lambda \in \mathbb{C}^*$, $e_n \in \mathbb{C}$, $\langle \vec{x}, \vec{y}\rangle = \vec{x} \cdot \vec{y}^t$ for any $\vec{x}, \vec{y} \in \mathbb{C}^{n-1}$. Also, F is uniquely determined by the data $\lambda, \vec{a}, \vec{e}, e_n, A$.

A direct computation shows that $\mathcal{I}_0$ is a subgroup of $Aut_0(\mathcal{H}^n)$ under the action of compositions. From the above arguments, we have the following

THEOREM 6.3. *Let Φ is a local holomorphic Segre self-isomorphism of $(\mathcal{H}_n, 0)$. Then Φ is the restriction of a certain element in $\mathcal{I}_0$. In particular, $Aut_0(\mathcal{H}^n) = \mathcal{I}_0$.*

PROOF. Let $F = (f, g, \phi, \psi) \in Aut_0(\mathcal{H}^n)$. Assume that $g'_w(0) = \sigma$, $f'_{z'} = A$, $\phi'_{\xi'} = B$. Then it is easy to see that $\sigma = A \cdot B^t$. After composing F from the left by a map as in (6.6) (6.7) and (6.8) with $U = (\lambda A)^{-1}, V = B^{-1}, \lambda = \sigma^{-1}, \vec{a}, \vec{e}, e_n = 0$, we can assume that $A = B = \mathrm{Id}, \sigma = 1$. Composing F from the left with a map in (6.6)(6.7) and (6.8) with $\lambda = 1, U, V = \mathrm{id}, e_n = 0, \vec{a} = -f'_w(0), \vec{e} = -\phi'_w(0)$, we can further assume that $f_w = \phi_\eta = 0$. Finally, composing F from the left with $U = V = \mathrm{Id}, \lambda = 1, \vec{a}, \vec{e}, = 0, e_n = \frac{1}{2}g''_{ww}(0)$, we can also make $g''_{ww}(0) = 0$. Now, by the previous arguments on the unique determination of F, we conclude that F is inside the group $\mathcal{I}_0$. □

REMARK 6.4. Let $0 < \ell < n-1$. The Segre family $\mathcal{H}^n_\ell$ of $\mathbb{H}^n_\ell$ is apparently isomorphic to the Segre family of the Heisenberg group through the map

$$(z', w, \xi', \eta) \to (iz_1, \cdots, iz_\ell, z_{\ell+1}, \cdots, w, i\xi_1, \cdots, i\xi_\ell, \xi_{\ell+1}, \cdots, \eta).$$

Hence, we can use Theorem 6.3 to write down the holomorphic Segre self-isomorphisms of $\mathcal{H}^n_\ell$. Namely, $Aut_0(\mathcal{H}^n_\ell)$ consists of the elements of the following form:

$$\widetilde{S}(z) = \frac{\lambda(z' + \vec{a}w)U}{1 - 2i\langle z', \vec{e}\rangle_\ell + e_n w}, \quad S_n(z) = \frac{\lambda w}{1 - 2i\langle z', \vec{e}\rangle_\ell + e_n w},$$
$$\widetilde{T}(\xi) = \frac{(\xi' + \vec{e}\eta)V^t}{1 + 2i\langle \xi', \vec{a}\rangle_\ell + (e_n + 2i\langle \vec{e}, \vec{a}\rangle_\ell)\eta},$$
$$T_n(\xi) = \frac{\lambda\eta}{1 + 2i\langle \xi', \vec{a}\rangle_\ell + (e_n + 2i\langle \vec{e}, \vec{a}\rangle_\ell)\eta};$$

where U, V are non-singular $(n-1)\times(n-1)$ matrices of complex numbers with $U \cdot E_\ell \cdot V^t = E_\ell$ (E_ℓ is the diagonal matrix whose first ℓ diagonal elements are -1 and the other diagonal elements are 1,) $\vec{a} = (a_1, ..., a_{n-1})$, $\vec{e} = (e_1, ..., e_{n-1}) \in \mathbb{C}^{n-1}$, $\lambda \in \mathbb{C}^*$, $e_n \in \mathbb{C}$. Also, $\langle \vec{x}, \vec{y}\rangle_\ell = -\sum_{j=1}^{\ell} x_j y_j + \sum_{j=1+\ell}^{n-1} x_j y_j$ for $\vec{x}, \vec{y} \in \mathbb{C}^{n-1}$.

References

[Alx1] H. Alexander, *Holomorphic mappings from ball and polydisc*, Math. Ann. **209**(1974), 245–256.

[Alx2] H. Alexander, *Proper holomorphic maps in* $\mathbb{C}^n$, Ind. Univ. Math. Journal, **26** (1977), 137–146.

[BER1] S. Baouendi, P. Ebenfelt and L. Rothschild, *Real Submanifolds in Complex Space and Their Mappings, Princeton Mathematical Series*, **47**, Princeton University Press, Princeton, NJ, 1999.

[BER2] S. Baouendi, P. Ebenfelt and L. Rothschild, *Local geometric properties of real submanifolds in complex space*, Bull. Amer. Math. Soc. (N.S.) **37** (2000), no. 3, 309–33.

[BH] S. Baouendi and X. Huang, *Super-Rigidity for holomorphic mappings between hyperquadrtics with positive signature*, Jour. of Diff. Geom. Vol. 69, **2**(2005), 379–398.

[BR] M. S. Baouendi and L. P. Rothschild, *Geometric properties of mappings between hypersurfaces in complex spaces*, J. Differential Geom. **3**(1990), 473–499.

[Car] É. Cartan, *Sur les variétés pseudo-conformal des hypersurfaces de l'espace de deux variables complexes*, Ann. Mat. Pura Appl. (4) **11**(1932), 17–90.

[CD] D. Catlin and J.P. D'Angelo, *A stabilization theorem for Hermitian forms and applications to holomorphic mappings*, Math. Res. Lett. 3, **2**(1996), 149–166.

[CaoM] H. Cao and N. Mok, *Holomorphic immersions between compact hyperbolic space forms*, Invent. Math. 100 (1990), no. 1, 49–61.

[Ch] S. S. Chern, *On the projective structure of a real hypersurface in* $\mathbb{C}^{n+1}$, Math. Scand. **36**(1975), 74–82.

[CJ1] S.-S. Chern and S. Ji, *Projective geometry and Riemann's mapping theorem*, Math. Ann., **302**(1995), 581–600.

[CJ2] S.-S. Chern and S. Ji, *On the Riemann mapping theorem*, Ann. of Math. **144**(1996), 421-439.

[CM] S. S. Chern and J. K. Moser, *Real hypersurfaces in complex manifolds*, Acta Math. **133**(974), 219–271.

[CS1] J. Cima and T. J. Suffridge, *A reflection principle with applications to proper holomorphic mappings*, Math Ann. **265** (1983), 489–500.

[CS2] J.Cima and T. J. Suffridge, *Boundary behavior of rational proper maps*, Duke Math. J. **60**(1990), 135–138.

[CS3] J. Cima and T. Suffridge, *Proper mappings between balls in* $\mathbb{C}^n$, *Complex analysis (University Park, Pa., 1986)*, 66–82, Lecture Notes in Math., **1268**, Springer, Berlin-New York, 1987.

[CKS] J. Cima, S. Krantz, and T. J. Suffridge, *A reflection principle for proper holomorphic mappings of strictly pseudoconvex domains and applications*, Math. Z. **186**(1984), 1–8.

[Cor] K. Corlette, *Archimedean superrigidity and hyperbolic geometry*, Ann. of Math. (2) **135**(1992), no. 1, 165–182.

[DA1] J. D'Angelo, *Real hypersurfaces, orders of contact, and applications*, Ann. Math. **115**(1982), 615–637.

[DA2] J. P. D'Angelo, *Several Complex Variables and the Geometry of Real Hypersurfaces*, CRC Press, Boca Raton, 1993.

[DA3] J. P. D'Angelo, *Polynomial proper holomorphic mappings between balls*, Duke Math J. **57**(1988), 211–219.

[DA4] J. P. D'Angelo, *Proper holomorphic mappings between balls in different complex spaces*, Mich. Math. Jour. **38** (1991), 53–65.

[DA5] J. P. D'Angelo, *Proper holomorphic maps between balls of different dimensions*, Michigan Math. J. **35**(1988), no. 1, 83 – 90.

[Dor] A. Dor, *Proper holomorphic maps between balls in one co-dimension*, Ark. Math., **28**(1990),49–100.

[EHZ1] P. Ebenfelt, X. Huang and D. Zaitsev, *Rigidity of CR-immersions into spheres.* Comm. Anal. Geom. **12**(2004), no. 3, 631–670.

[EHZ2] P. Ebenfelt, X. Huang and D. Zaitsev, *The equivalence problem and rigidity for hypersurfaces embedded into hyperquadrics,* American Jour. of Math. Vol. **127**(2005), 169–191.

[EZ] P. Ebenfelt and D. Zaitsev, Private comminucation, Fall of 2003.

[Fa1] J. Faran, *Maps from the two ball to the three ball*, Invent Math. **68**(1982), 441–475.

[Fa2] J. Faran, *The linearity of proper holomorphic maps between balls in the low codimension case*, J. Differential Geom. **24** (1986), 15–17.

[Fa3] J. Faran, *Segre families and real hypersurfaces*, Invent. Math. **60**(1980), 135–172.

[Fr1] F. Forstneric, *A survey on proper holomorphic mappings*, Proceeding of Year in SCVs at Mittag-Leffler Institute, Math. Notes 38, Princeton, NJ: Princeton University Press , 1992.

[Fr2] F. Forstneric, *Extending proper holomorphic mappings of positive codimension*, Invent. Math. **95**(1989), 31–62.

[Fr3] F. Forstneric, *Complex tangents of real surfaces in complex surfaces*, Duke Math. J. **67** (1992), no. 2, 353–376.

[Ga] R. Gardner, *The method of equivalence and its applications*, CBMS-NSF, regional conference series in applied mathematics, 1989.

[Ham] : Hamada, *Hidetaka Rational proper holomorphic maps from* $\mathbb{B}^n$ *into* $\mathbb{B}^{2n}$, Math. Ann. **331**(2005), no.3, 693–711.

[Ha] M. Hakim, *Applications holomorphes propres continues de domaines strictment pseudoconvexes de* $\mathbf{C}^{\mathbf{n}}$ *dans la boule unité* $\mathbf{C}^{\mathbf{n+1}}$, Duke Math. J. **60** (1990), 115–133.

[HS] M. Hakim and N. Sibony, *Fonctions holomorphes bornes sur la boule unite de* $\mathbb{C}^n$, Invent. Math. **67**(1982), no. 2, 213–222.

[Hu1] X. Huang, *On a linearity problem of proper holomorphic mappings between balls in complex spaces of different dimensions*, J. of Diff. Geom. **51**(1999), 13–33.

[Hu2] X. Huang, *On a semi-linearity property for holomorphic maps*, Asian Jour. of Math. Vol 7 (No. 4)(2003), 463–492.

[Hu3] X. Huang, *On some problems in several complex variables and Cauchy-Riemann Geometry*, Proceedings of ICCM (edited by L. Yang and S. T. Yau), AMS/IP Stud. Adv. Math. **20**(2001), 383–396.

[Hu4] X. Huang, *Lectures on the Local Equivalence Problems for Real Submanifolds in Complex Manifolds*, Lecture Notes in Mathematics **1848** (C.I.M.E. Subseries), Springer-Verlag, 2004, 109–163.

[HJ1] X. Huang and S. Ji, *Mapping* $\mathbb{B}^n$ *into* $\mathbb{B}^{2n-1}$, Invent Math, **145**(2001), 219–250.

[HJ2] X. Huang and S. Ji, *Cartan-Chern-Moser theory on algebraic hypersurfaces and some applications*, Ann. Inst. Fourier (Grenoble) **52**(2002), no. 6, 1793–1831.

[HJX1] X. Huang, S. Ji, and D. Xu, *Several results for holomorphic mappings from* $\mathbb{B}^n$ *into* $\mathbb{B}^N$, Geometric analysis of PDE and several complex variables, 267–292, Contemp. Math., **368**, Amer. Math. Soc., Providence, RI, 2005.

[HJX2] X. Huang, S. Ji and D. Xu, *A new gap phenomenon for proper holomorphic mappings from from* $\mathbb{B}^n$ *into* $\mathbb{B}^N$, preprint, December, 2004. (to appear in Math. Res. Lett. 13(2006), no. 4.)

[JX] S. Ji, and D. Xu, *Maps between* $\mathbb{B}^n$ *and* $\mathbb{B}^N$ *with geometric rank* $k_0 \leq n-2$ *and minimum* N, Asian J. Math. **8** (2004), no. 2, 233–257.

[KZ] W. Kaup and D. Zaitsev, *On local CR transformations of Levi-degenerate group orbits in compact Hermitian symmetric spaces*, Journal of the European Mathematical Society, 2005. (to appear)

[L] L. Lempert, *Imbedding strictly pseudoconvex domains into a ball*, Amer. J. Math. **104** (1982), no. 4,901–904.

[Mir] N. Mir, *Analytic regularity of CR maps into spheres.* Math. Res. Lett. 10 (no.4)(2003), 447–457.

[Mo] N. Mok, *Metric Rigidity Theorems on Hermitian Locally Symmetric Manifolds*, Series in Pure Mathematics, V. 6, World Scientific Publishing Co., 1989.

[MSY] N. Mok, Y. T. Siu and S. K. Yeung, *Geometric superrigidity*, Invent. Math. **113**(1993), 57–83.

[Moo] J. D. Moore, *On extendibility of isometric immersions of spheres*, Duke Math. J. **85**(1996), no. 3, 685–699.

[Mos] J. Moser, *Analytic surfaces in $\mathbb{C}^2$ and their local hull of holomorphy*, Annales Aca - demiæFennicae, Series A.I. Mathematica **10**(1985), 397–410.

[MW] J. Moser and S. Webster, *Normal forms for real surfaces in $\mathbb{C}^2$ near complex tangents and hyperbolic surface transformations*, Acta Math. **150**(1983), 255–296.

[Po] H. Poincaré, *Les fonctions analytiques de deux variables et la représentation conforme*, Rend. Circ. Mat. Palermo, (1907), 185–220.

[Sh] B. Shiffman, *Synthetic projective geometry and Poincaré's theorem on automorphisms of the ball*, Enseign. Math. (2) **41**(1995), no. 3-4, 201–215.

[Spv] M. Spivak, *A Comprehensive Introduction to Differential Geometry*, Volume (3), Publish or Persish, Inc., Wilmington, Delaware, 1979.

[Ste] B. Stensones, *Proper holomorphic mappings between the balls*, J. of Geom. Analysis, **6**(1996), 317–339.

[Ta] N. Tanaka, *On the pseudo-conformal geometry of hypersurfaces of the space of n complex variables*, J. Math. Soc. Japan **14**(1962), 397–429.

[Ts] I. Tsai, *Rigidity of proper holomorphic maps between symmetric domains.* J. Differential Geom. **37**(1993), no. 1, 123–160.

[Tu] Z. Tu, *Rigidity of proper holomorphic mappings between nonequidimensional bounded symmetric domains,* Math. Z. **240**(2002), no. 1, 13–35.

[We1] S. Webster, *On the mapping problem for algebraic real hypersurfaces*, Invent. Math. **43**(1977), 53–68.

[We2] S. Webster, *Biholomorphic mappings and the Bergman kernels off diagonal*, Invent. Math. **51**(1979), 155–169.

[We3] S. Webster, *On mappings an (n+1)-ball in the complex space*, Pac. J. Math. **81**(1979), 267–272.

[We4] S. Webster, *The rigidity of CR hypersurfaces in a sphere*, Indiana Univ. Math. Jour. Vol **28** (1979), 405–416.

[Yue] C. Yue, *Dimension and rigidity of quasi-Fuchsian representations*, Ann. of Math. (2) **143** (1996), no. 2, 331–355.

Department of Mathematics, Rutgers University, New Brunswick, NJ 08903, USA
E-mail address: huangx@math.rutgers.edu

Department of Mathematics, University of Houston, Houston, TX 77204, USA
E-mail address: shanyuji@math.uh.edu

AMS/IP Studies in Advanced Mathematics
Volume **39**, 2007

From CR Geometry to Algebraic Geometry and Combinatorial Geometry

Lixing Jia, Hing Sun Luk, and Stephen S.-T. Yau

ABSTRACT. Strongly pseudoconvex CR manifolds as abstract models of real hypersurfaces have rich geometric and analytic structures. CR geometry was first related to algebraic geometry of singularities in Yau's solution of the Kohn-Rossi conjecture. The purpose of the present paper is to discuss why algebraic geometry plays an important role in studying global CR geometry and why combinatorial geometry plays a role too. Works of Yau and co-workers on several basic questions on compact strongly pseudoconvex CR manifolds in complex Euclidean spaces are explained. Some open problems are also discussed.

1. Introduction

CR manifolds are abstract models of real hypersurfaces in complex spaces. The abstract definition of the boundary as a CR structure on a complex manifold is essentially in Cartan [**Ca**]. For more detail, see [**Ku2**], [**Ku3**]. Strongly pseudoconvex CR manifolds have rich geometric and analytic structures. Namely, there is an intrinsic pseudo conformal geometry for which complete local invariants have been obtained, see for example [**Ch-Mo**], [**Ta1**], [**Fe**], as well as a deep analysis of the $\overline{\partial}_b$ complex, see for example [**Ko-Ro**], [**Fo-St**], [**Ku1**]. The harmonic theory for the $\overline{\partial}_b$ complex on compact strongly pseudoconvex CR manifolds was developed by Kohn [**Ko**]. Using this theory, Boutet de Monvel [**BM**] proved that if X is a compact strongly pseudoconvex CR manifold of dimension $2n-1$, $n \geq 3$, then there exist C^∞ functions $f_1, \ldots, f_N$ on X such that each $\overline{\partial}_b f_j = 0$ and $f = (f_1, \ldots, f_N)$ defines an embedding of X in $\mathbb{C}^N$. Thus, any compact strongly pseudoconvex CR manifold of dimension ≥ 5 can be CR embedded in some complex Euclidean space. On the other hand, 3-dimensional strongly pseudoconvex compact orientable CR-manifolds are not necessarily embeddable. The first example is due to Andreotti according to [**Ro**]. This example also appeared in the list of homogeneous structures of Cartan although the embeddability question was not addressed. Nirenberg [**Ni**] first proved that 3-dimensional CR manifolds might not be locally embeddable. Jacobowitz and Treves [JT 1,2] showed that in fact non-embeddable CR structures

Stephen S.-T. Yau is partially supported by NSF grant and RGC grant. Hing-Sun Luk is supported by Hong Kong RGC grant.

are, in some sense, dense in the space of CR-structures over a 3-dimensional manifold. The theory of harmonic integrals on strongly pseudoconvex CR structures over small balls was due to Kuranishi [**Ku1**]. Using this theory, Kuranishi [**Ku1**] proved that any strongly pseudoconvex CR manifold of dimension $2n-1$ with $n \geq 5$ can be locally CR embedded as a real hypersurface in $\mathbb{C}^n$. For $n = 4$, Akahori [**Ak**] proved that Kuranishi's local embedding theorem is also true. However, the 5-dimensional case of local embeddability of CR manifolds remains open.

Throughout this paper, our CR manifolds are always assumed to be compact orientable and embeddable in some $\mathbb{C}^N$. By a beautiful theorem of Harvey and Lawson [**Ha-La**], these CR manifolds are the boundaries of subvarieties in $\mathbb{C}^N$. This allowed the third author [**Ya1**] to relate CR geometry and algebraic geometry of singularities for the first time. The purpose of this paper is to discuss why algebraic geometry plays an important role in studying CR geometry and why combinatorial geometry also plays a role. Our paper is organized as follows. In section 2, we shall recall the basic notions of CR geometry. In section 3, we use algebraic geometry to study the complex Plateau problem. In section 4, we study the minimal embedding dimension of a compact CR manifold in complex Euclidean space. In section 5, we study invariants of compact strongly pseudoconvex CR manifolds arising from geometry of singularities. Finally in section 6, we explain how combinatorial geometry may play an important role in CR geometry.

2. Preliminary

DEFINITION 2.1. Let X be a connected orientable manifold of real dimension $2n-1$. A CR structure on X is an $(n-1)$-dimensional subbundle S of the complexified tangent bundle $\mathbb{C}TX$ such that

(1) $S \cap \overline{S} = \{0\}$
(2) If L, L' are local sections of S, then so is $[L, L']$.

A manifold with a CR structure is called a CR manifold. There is a unique subbundle $\mathcal{H}$ of the tangent bundle $T(X)$ such that $\mathbb{C}\mathcal{H} = S \oplus \overline{S}$. Furthermore, there is a unique homomorphism $J : \mathcal{H} \longrightarrow \mathcal{H}$ such that $J^2 = -1$ and $S = \{v - iJv : v \in \mathcal{H}\}$. The pair $(\mathcal{H}, J)$ is called the real expression of the CR structure.

DEFINITION 2.2. Let $L_1, \ldots, L_{n-1}$ be a local frame of S. Then $\overline{L}_1, \ldots, \overline{L}_{n-1}$ is a local frame of $\overline{S}$ and one may choose a local section N of TX such that $L_1, \ldots, L_{n-1}$, $\overline{L}_1, \ldots, \overline{L}_{n-1}, N$ is a local frame of $\mathbb{C}TX$. The matrix (c_{ij}) defined by

$$[L_i, \overline{L}_j] = \Sigma a_{ij}^k L_k + \Sigma b_{ij}^k \overline{L}_k + \sqrt{-1} c_{ij} N$$

is Hermitian and is called the Levi form of X.

PROPOSITION 2.1. *The number of non-zero eigenvalues and the absolute value of the signature of the Levi form (c_{ij}) at each point are independent of the choice of $L_1, \ldots, L_{n-1}, N$.*

DEFINITION 2.3. The CR manifold X is called strongly pseudoconvex if the Levi form is definite at each point of X.

THEOREM 2.2. *(Boutet de Monvel [**BM**]). If X is a compact strongly pseudoconvex CR manifold of dimension $(2n-1)$ and $n \geq 3$, then X is CR embeddable in $\mathbb{C}^N$.*

Although there are non-embeddable compact 3-dimensional CR manifolds, in this paper all CR manifolds are assumed to be embeddable in complex Euclidean space. The following theorem links CR geometry and algebraic geometry together.

THEOREM 2.3. *(Harvey-Lawson* [**Ha-La**]*). For any compact connected embeddable CR manifold X, there is a unique complex variety V in $\mathbb{C}^N$ for some N such that the boundary of V is X and V has only normal isolated singularities.*

3. Complex Plateau Problem

Let X be a compact connected CR manifold of dimension $2n-1$ in $\mathbb{C}^N$. The famous complex Plateau problem asks under what conditions on X, X will be a boundary of a complex submanifold in $\mathbb{C}^N$. By a theorem of Harvey and Lawson [**Ha-La**], X is a boundary of a unique complex variety V in $\mathbb{C}^N$. Therefore we need to understand under what conditions on X, V will be a complex submanifold.

In 1963, J.J. Kohn solved the famous $\overline{\partial}$-Neumann problem. Based on this work, Kohn and Rossi [**Ko-Ro**] in 1965 introduced the fundamental CR invariants, the Kohn-Rossi cohomology groups $H^{p,q}_{KR}(X)$. They proved the finite dimensionality of their cohomology groups for $1 \le q \le n-2$ if X is strongly pseudoconvex. Following Tanaka [**Ta2**], we shall recall the definition of Kohn-Rossi cohomology groups as follows.

Let $\{\mathcal{A}^k(X), d\}$ be the De-Rham complex of X with complex coefficients, and let $H^k(X)$ be the De-Rham cohomology groups. There is a natural filtration of the De-Rham complex as follows. For any integer p and k, put $A^k(X) = \Lambda^k(\mathbb{C}T(X)^*)$ and denoted by $F^p(A^k(X))$ the subbundle of $A^k(X)$ consisting of all $\phi \in A^k(X)$ which satisfy the equality

$$\phi(Y_1, \ldots, Y_{p-1}, \overline{Z}_1, \ldots, \overline{Z}_{k-p+1}) = 0$$

for all $Y_1, \ldots, Y_{p-1} \in \mathbb{C}T(X)_x$ and $Z_1, \ldots, Z_{k-p+1} \in S_x, x$ being the origin of ϕ. Then

$$A^k(X) = F^\circ(A^k(X)) \supseteq F^1(A^k(X)) \supseteq \cdots \supseteq F^k(A^k(X)) \supseteq F^{k+1}(A^k(X)) = 0.$$

Setting $F^p(\mathcal{A}^k(X)) = \Gamma(F^p(A^k(X)))$, we have

$$\mathcal{A}^k(X) = F^\circ(\mathcal{A}^k(X)) \supseteq F^1(\mathcal{A}^k(X)) \supseteq \cdots \supseteq F^k(\mathcal{A}^k(X)) \supseteq F^{k+1}(\mathcal{A}^k(X)) = 0.$$

Since clearly $dF^p(\mathcal{A}^k(X)) \subseteq F^p(\mathcal{A}^{k+1}(X))$, the collection $\{F^p(\mathcal{A}^k(X))\}$ gives a filtration of the De-Rham complex.

DEFINITION 3.1. $H^{p,q}_{KR}(X)$, the Kohn-Rossi cohomology group of type (p,q), is defined to be the group $E^{p,q}_1(X)$ of the spectral sequence $\{E^{p,q}_r(X)\}$ associated with the filtration $\{F^p(\mathcal{A}^k(X))\}$.

More explicitly, let

$$A^{p,q}(X) = F^p(A^{p+q}(X)), \qquad \mathcal{A}^{p,q}(X) = \Gamma(A^{p,q}(X))$$
$$C^{p,q}(X) = A^{p,q}(X)/A^{p+1,q-1}(X), \quad \mathcal{C}^{p,q}(X) = \Gamma(C^{p,q}(X)).$$

Since $d : \mathcal{A}^{p,q}(X) \to \mathcal{A}^{p,q+1}(x)$ maps $\mathcal{A}^{p+1,q-1}(X)$ into $\mathcal{A}^{p+1,q}(X)$, it induces an operator $\overline{\partial}_b : \mathcal{C}^{p,q}(X) \to \mathcal{C}^{p,q+1}(X)$. $H^{p,q}_{KR}(X)$ are then the cohomology groups of the complex $\{\mathcal{C}^{p,q}(X), \overline{\partial}_b\}$.

DEFINITION 3.2. $H_h^k(X)$, the holomorphic De-Rham cohomology group of degree k, is defined to be the group $E_2^{k,0}(X)$ of the spectral sequence $\{E_r^{p,q}(X)\}$ associated with the filtration $\{F^p(\mathcal{A}^k(X))\}$.

More explicitly, recall $E_0^{p,q}(X) = \mathcal{C}^{p,q}(X)$ and $d_0 : \mathcal{C}^{p,q}(X) \longrightarrow \mathcal{C}^{p,q+1}(X)$ is the map $\overline{\partial}_b$ above. Note that $E_0^{k,0}(X) = \mathcal{C}^{k,0}(X) = \mathcal{A}^{k,0}(X) \subseteq \mathcal{A}^k(X)$. Next,

$$E_1^{p,q}(X) = \frac{\operatorname{Ker}(d_0 : \mathcal{C}^{p,q}(X) \longrightarrow \mathcal{C}^{p,q+1}(X))}{\operatorname{Im}(d_0 : \mathcal{C}^{p,q-1}(X) \longrightarrow \mathcal{C}^{p,q}(X))}$$

and $d_1 : E_1^{p,q}(X) \longrightarrow E_1^{p+1,q}(X)$ is the naturally induced map. In particular,

$$\begin{aligned} E_1^{k,0}(X) &= \ker(d_0 : \mathcal{C}^{k,0}(X) \longrightarrow \mathcal{C}^{k,1}(X)) \\ &= \{\phi \in \mathcal{A}^{k,0}(X) : d\phi \in \mathcal{A}^{k+1,0}(X)\} \end{aligned}$$

and d_1 is just d on $E_1^{k,0}(X) \subseteq \mathcal{A}^k(X)$. $E_1^{k,0}(X)$ is called the space of holomorphic k-forms on X. Denoting $E_1^{k,0}(X)$ by $\mathcal{S}^k(X)$, we have the holomorphic De Rham complex $\{\mathcal{S}^k(X), d\}$. Then

$$\begin{aligned} E_2^{k,0}(X) &= \frac{\operatorname{Ker}(d : \mathcal{S}^k(X) \longrightarrow \mathcal{S}^{k+1}(X))}{\operatorname{Im}(d : \mathcal{S}^{k-1}(X) \longrightarrow \mathcal{S}^k(x))} \\ &= \frac{\{\text{closed holomorphic } k\text{-forms on } X\}}{\{\text{exact holomorphic } k\text{-forms on } X\}}, \end{aligned}$$

is the holomorphic De Rham cohomology $H_h^k(X)$.

A strongly pseudoconvex complex manifold M is a modification of a Stein space V with isolated singularities. In 1965, Kohn and Rossi [**Ko-Ro**] conjectured that in general, either there is no boundary cohomology of the boundary $X = \partial V$ in degree (p,q) for $q \neq 0,\ n-1$, or it must result from the interior singularities of V. Yau [**Ya1**] solved the Kohn-Rossi conjecture affirmatively in 1981.

THEOREM 3.1. *(Yau [**Ya1**]). Let X be a compact strongly pseudoconvex CR manifold of dimension $2n-1$, $n \geq 3$, which is the boundary of a Stein space V with isolated singularities $x_1, \ldots, x_m$. Then for $1 \leq q \leq n-2$,*

$$H_{KR}^{p,q}(X) \simeq \bigoplus_{i=1}^{m} H_{\{x_i\}}^{q+1}(V, \Omega_V^p),$$

where Ω_V^p is the sheaf of germs of holomorphic p-forms on V. If $x_1, \ldots, x_m$ are hypersurface singularities, then

$$\dim H_{KR}^{p,q}(X) = \begin{cases} 0 & p+q \leq n-2, \quad 1 \leq q \leq n-2 \\ \tau_1 + \cdots + \tau_m & p+q = n-1, n, \quad 1 \leq q \leq n-2 \\ 0 & p+q = n+1, \quad 1 \leq q \leq n-2 \end{cases}$$

where τ_i is the number of moduli of V at x_i.

REMARK 3.1. Let $f : (\mathbb{C}^{n+1}, 0) \longrightarrow (\mathbb{C}, 0)$ be a holomorphic function. Suppose that $V = \{z \in \mathbb{C}^{n+1} : f(z) = 0\}$ has isolated singularity at the origin. Then the local moduli of V is the dimension of the parameter space of the semi universal deformation space of $(V, 0)$. This number is $\tau = \dim \mathbb{C}\{z_0, \ldots, z_n\}/(f, f_{z_0}, \ldots, f_{z_n})$.

As a result of the above theorem, Yau answers the classical complex Plateau problem for some cases.

THEOREM 3.2. *(Yau [**Ya1**]). Let X be a compact connected strongly pseudoconvex CR-manifold of real dimension $2n-1, n \geq 3$, in the boundary of a bounded strongly pseudoconvex domain D in $\mathbb{C}^{n+1}$. Then X is a boundary of a complex submanifold $V \subseteq D - X$ if and only if Kohn-Rossi cohomology groups $H^{p,q}_{KR}(X)$ are zero for $1 \leq q \leq n-2$.*

For $n = 2$ in Theorem 3.2, X is a 3-dimensional CR manifold. The classical complex Plateau problem remains unsolved for over a quarter of a century. The main difficulty is that the Kohn-Rossi cohomology groups are infinite dimensional in this case. Let V be the complex variety with X as its boundary. Then the singularities of V are surface singularities. In order to solve the classical complex Plateau problem for $n = 2$, one would like to ask under what kind of condition on X, V will have only very mild singularities. Our basic observation is the following. Although the Kohn-Rossi cohomology groups are infinite dimensional, we can derive from them the holomorphic De Rham cohomology. Let M be a complex manifold. The k-th holomorphic De Rham cohomology $H^k_h(M)$ of M is defined to be the d-closed holomorphic k-forms quotient by the d-exact holomorphic k-forms. It is well known that if M is a Stein manifold, then the holomorphic De Rham cohomology coincides with the ordinary De Rham cohomology.

DEFINITION 3.3. Let (V, x) be an isolated singularity of dimension n. Let $\pi : (M, A) \longrightarrow (V, x)$ be a resolution of singularity with A as exceptional set. Let

$$s = \dim \Gamma(M - A, \Omega^n)/[d\Gamma(M - A, \Omega^{n-1}) + \Gamma(M, \Omega^n)].$$

s is an invariant of the singularity (V, x). It turns out that the s-invariant plays an important role in the relationship between $H^n_h(M - A)$ and $H^n_h(M)$.

THEOREM 3.3. *(Luk-Yau [**Lu-Ya1**]). Let X be a compact connected $(2n-1)$-dimensional $(n \geq 2)$ strongly pseudoconvex CR manifold. Suppose that X is the boundary of a n-dimensional strongly pseudoconvex complex manifold M which is a modification of a Stein space V with only isolated singularities $\{x_1, \ldots, x_m\}$. Let A be the maximal compact analytic set in M which can be blown down to $\{x_1, \ldots, x_m\}$. Then*

(1) $H^q_h(X) \cong H^q_h(M - A) \cong H^q_h(M)$ for $1 \leq q \leq n-1$.
(2) $H^n_h(X) \cong H^n_h(M - A)$, $\dim H^n_h(M - A) = \dim H^n_h(M) + s$

where $s = s_1 + \cdots + s_m$ and s_i is the s-invariant of the singularity (V, x_i).

REMARK 3.2. The above theorem in particular asserts that up to degree $n-1$, the holomorphic De Rham cohomology can extend across the maximal compact analytic set.

DEFINITION 3.4. A normal surface singularity $(V, 0)$ is Gorenstein if there exists a nowhere vanishing holomorphic 2-form on $V - \{0\}$.

Recall that isolated hypersurface or complete intersection singularities are Gorenstein. It is a natural question to ask for a characterization of Gorenstein surface singularities with vanishing s-invariant.

THEOREM 3.4. *(Luk-Yau [**Lu-Ya1**]). Let $(V, 0)$ be a Gorenstein surface singularity. Let $\pi : M \longrightarrow V$ be a good resolution with $A = \pi^{-1}(0)$ as exceptional set. Assume that M is contractible to A. If $s = 0$, then $(V, 0)$ is a quasi-homogeneous singularity, $H^1(A, \mathbb{C}) = 0$, $\dim H^1(M, \Omega^1) = \dim H^2(A, \mathbb{C}) + \dim H^1(M, \mathcal{O})$, and*

$H_h^1(M) = H_h^2(M) = 0$. *Conversely, if* $(V, 0)$ *is a two dimensional quasi-homogeneous Gorenstein singularity and* $H^1(A, \mathbb{C}) = 0$, *then the s-invariant vanishes.*

Let X be a compact CR manifold with CR-structure S. For any C^∞ functions u, there is a section $\overline{\partial}_b u \in \Gamma(\overline{S}^*)$ defined by $(\overline{\partial}_b u)(\overline{L}) = \overline{L}u$ for any $L \in \Gamma(S)$. This can be generalized as follows:

DEFINITION 3.5. A complex vector bundle E over X is said to be holomorphic if there is a differential operator $\overline{\partial}_E : \Gamma(E) \longrightarrow \Gamma(E \otimes \overline{S}^*)$ such that if $\overline{L}u$ denotes $(\overline{\partial}_E u)(\overline{L})$ for $u \in \Gamma(E)$ and $L \in \Gamma(S)$, then for any $L_1, L_2 \in \Gamma(S)$ and any C^∞ function f on X:

(1) $\overline{L}(fu) = (\overline{L}f)u + f(\overline{L}u)$
(2) $[\overline{L}_1, \overline{L}_2]u = \overline{L}_1\overline{L}_2 u - \overline{L}_2\overline{L}_1 u$.

A solution u of the equation $\overline{\partial}_E u = 0$ is called a holomorphic section.

The vector bundle $\widehat{T}(X) = \mathbb{C}T(X)/\overline{S}$ is holomorphic with respect to the following $\overline{\partial} = \overline{\partial}_{\widehat{T}(X)}$. Let ω be the projection from $\mathbb{C}T(X)$ to $\widehat{T}(X)$. Take any $u \in \Gamma(\widehat{T}(X))$ and express it as $u = \omega(Z)$, $Z \in \Gamma(\mathbb{C}T(X))$. For any $L \in \Gamma(S)$, define $(\overline{\partial}u)(\overline{L}) = \omega([\overline{L}, Z])$. The section $(\overline{\partial}u)(\overline{L})$ of $\widehat{T}(X)$ does not depend on the choice of Z and $\overline{\partial}u$ gives a section of $\widehat{T}(X) \otimes \overline{S}^*$. Further the operator $\overline{\partial}$ satisfies the conditions in Definition 3.5. The resulting holomorphic vector bundle $\widehat{T}(X)$ is called the holomorphic tangent bundle of X.

LEMMA 3.1. *If X is a real hypersurface in a complex manifold M, then the holomorphic tangent bundle $\widehat{T}(X)$ is naturally isomorphic to the restriction to X of the bundle $T^{1,0}(M)$ of all $(1,0)$ tangent vectors to M.*

DEFINITION 3.6. Let X be a compact CR manifold of real dimension $2n-1$. X is said to be a Calabi-Yau CR manifold if there exists a nowhere vanishing holomorphic section in $\Gamma(\Lambda^n \widehat{T}(X)^*)$ where $\widehat{T}(X) = \mathbb{C}T(X)/\overline{S}$ is the holomorphic tangent bundle of X.

REMARK 3.3. (a) Let X be a compact CR manifold of real dimension $2n-1$ in $\mathbb{C}^n$. Then X is a Calabi-Yau CR manifold. (b) Let X be a strongly pseudoconvex CR manifold of real dimension $2n-1$ contained in the boundary of a bounded strongly pseudoconvex domain in $\mathbb{C}^{n+1}$. Then X is a Calabi-Yau manifold.

The following theorem is a fundamental theorem toward the complete solution of the classical complex Plateau problem for 3-dimensional strongly pseudoconvex Calabi-Yau CR manifold in $\mathbb{C}^n$. The theorem is interesting in its own right.

THEOREM 3.5. *(Luk-Yau* [**Lu-Ya1**]*). Let X be a strongly pseudoconvex compact Calabi-Yau CR manifold of dimension 3. Suppose that X is contained in the boundary of a strongly pseudoconvex bounded domain D in $\mathbb{C}^n$. If the holomorphic De Rham cohomology $H_h^2(X) = 0$, then X is a boundary of a complex variety V in D with boundary regularity and V has only isolated singularities in the interior and the normalizations of these singularities are Gorenstein surface singularities with vanishing s-invariant.*

COROLLARY 3.6. *(Luk-Yau* [**Lu-Ya1**]*). Let X be a strongly pseudoconvex compact CR manifold of dimension 3. Suppose that X is contained in the boundary of a strongly pseudoconvex bounded domain D in $\mathbb{C}^3$. If the holomorphic De Rham* *cohomology* $H_h^2(X) = 0$, *then* X *is a boundary of a complex variety* V *in* D *with boundary regularity and* V *has only isolated quasi-homogeneous singularities such that the dual graphs of the exceptional sets in the resolution are star shaped and all the curves are rational.*

4. Minimal embedding dimension of compact CR manifold

Let us first consider a compact strongly pseudoconvex manifold X of dimension $2n-1$ where $n \geq 3$. As mentioned above, X can be CR embedded in some $\mathbb{C}^N$. It is therefore of interest to study the minimal dimensional complex Euclidean space in which X CR embeds. Our starting point is Theorem 3.1 which provides us with obstruction to CR embedding:

THEOREM 4.1. *Let* X *be a compact strongly pseudoconvex CR manifold of dimension* $2n-1, n \geq 3$. *Then* X *cannot be CR embedded in* $\mathbb{C}^n$ *unless all* $H_{KR}^{p,q}(X) = 0$, $1 \leq q \leq n-2$. *Further,* X *cannot be CR embedded in* $\mathbb{C}^{n+1}$ *if one of the following does not hold:*

(1) $H_{KR}^{p,q}(X) = 0$ *for* $p+q \leq n-2$ *and* $1 \leq q \leq n-2$

(2) $\dim H_{KR}^{p,q}(X) = \dim H_{KR}^{p',q'}(X)$ *for* $\left.\begin{matrix} p+q \\ p'+q' \end{matrix}\right\} = n-1, n$ *and* $1 \leq q, q' \leq n-2$

(3) $H_{KR}^{p,q}(X) = 0$ *for* $p+q \geq n+1$ *and* $1 \leq q \leq n-2$.

We next consider an interesting class of CR manifolds.

DEFINITION 4.1. Let X be a CR manifold with structure bundle S. Let α be a smooth S^1-action on X and V be its generating vector field. The S^1-action α is called holomorphic if $\mathcal{L}_V\Gamma(S) \subseteq \Gamma(S)$ where $\mathcal{L}_V$ denotes the Lie derivative. It is called transversal if V is transversal to $S \oplus \overline{S}$ in $\mathbb{C}TX$ at every point of X.

For a CR manifold X which admits a transversal holomorphic S^1-action, the invariant Kohn-Rossi cohomology is defined as follows.

DEFINITION 4.2. With the notation in Definition 4.1, consider first the differential operator on k forms $N : \mathcal{A}^k(X) \to \mathcal{A}^k(X)$ defined by $N\phi = \sqrt{-1}\mathcal{L}_V\phi$, $\phi \in \mathcal{A}^k(X)$. Observe that N leaves invariant the spaces $\mathcal{A}^{p,q}(X)$ and $\mathcal{C}^{p,q}(X)$, and commutes with the operators d and $\overline{\partial}_b$. Hence N acts on the cohomology groups $H_{KR}^{p,q}(X)$. Now define the invariant Kohn-Rossi cohomology by $\widetilde{H}_{KR}^{p,q}(X) = \{c \in H_{KR}^{p,q}(X) : Nc = 0\}$.

For a compact strongly pseudoconvex CR manifold X of dimension $2n-1, n \geq 3$, which admits a transversal holomorphic S^1-action, the invariant Kohn-Rossi cohomology $\widetilde{H}_{KR}^{p,q}(X)$, for $1 \leq p+q \leq 2n-N-1$, are obstructions to CR embedding in $\mathbb{C}^N$. This is implied by the following theorem.

THEOREM 4.2. *(Luk-Yau* [**Lu-Ya2**]*). Let* X *be a compact strongly pseudoconvex CR manifold of dimension* $2n-1, n \geq 3$, *which admits a transversal holomorphic* S^1*-action. Suppose that* X *is CR embeddable in* $\mathbb{C}^N$. *Then* $\widetilde{H}_{KR}^{p,q}(X) = 0$ *for all* $1 \leq p+q \leq 2n-N-1$.

The proof of Theorem contains two main parts. The first part depends heavily on the work of Lawson-Yau [**La-Ya**], which provides us with topological restrictions on X. In particular it can be shown that the De Rham cohomology groups $H^k(X) =$

0 for $1 \leq k \leq 2n - N - 1$. The second part follows Tanaka's differential geometric study on the $\overline{\partial}_b$ cohomology groups [**Ta2**]. The existence of the vector field V with $[V, \Gamma(S)] \subseteq \Gamma(S)$ entails a formalism analogous to Kähler geometry linking the various cohomology groups via harmonic forms. The details of the proof of Theorem 4.2 are contained in [**Lu-Ya2**].

For 3 dimensional compact strongly pseudoconvex CR manifolds, global CR embedding in complex Euclidean space may fail and much work has been done on this phenomenon. See for example [**Bl-Ep**], [**Bu-Ep**], [**Le**]. We only remark that as a consequence of the global invariants to be discussed in the next section, we find obstructions to CR embedding in $\mathbb{C}^3$, assuming that the 3-dimensional strongly pseudoconvex CR manifold is CR embeddable in some $\mathbb{C}^N$ to begin with. These obstructions provide us with numerous examples of such 3-dimensional CR manifolds not CR embeddable in $\mathbb{C}^3$.

REMARK 4.1. It is interesting to note that there are compact strongly pseudoconvex 3 dimensional CR manifolds with arbitrarily large minimal embedding dimensions. For any positive integer N, take any 2-dimensional strongly pseudoconvex complex manifold with maximal compact analytic set A which is a smooth rational curve having self intersection number $-N$. The corresponding weighted dual graph is hence

$$\underset{-N}{\bullet} .$$

On blowing down A, one gets a 2-dimensional rational singularity (V, x). The minimal embedding dimension of (V, x) is $-A \cdot A + 1 = N + 1$. Let X be the intersection of V with a small sphere centered at x. Then the minimal embedding of X is $N + 1$.

5. Global invariants of compact strongly pseudoconvex CR manifolds

As a first step towards the difficult classification problem of compact strongly pseudoconvex CR manifolds, it would be useful to understand the following notion of equivalence which is weaker than CR equivalence.

DEFINITION 5.1. Assume that X_1, X_2 are compact strongly pseudoconvex embeddable CR manifolds of dimension $2n - 1$, $n \geq 2$. By Theorem 2.3, there are unique complex varieties $V_1 \subseteq \mathbb{C}^{N_1}$ and $V_2 \subseteq \mathbb{C}^{N_2}$ such that $\partial V_1 = X_1$, $\partial V_2 = X_2$, V_1 and V_2 have only isolated normal singularities. X_1, X_2 are called algebraically equivalent if V_1 and V_2 have isomorphic singularities Y_1, Y_2, i.e. $(V_1, Y_1) \cong (V_2, Y_1)$ as germs of varieties.

REMARK 5.1. It is not difficult to show that CR equivalence implies algebraic equivalence. Hence all algebro-geometric invariants of the singularities of V are CR invariants of X.

In case a compact strongly pseudoconvex CR manifold X of dimension $2n - 1$ embeds in $\mathbb{C}^{n+1}$, $n \geq 2$, it is the boundary of a complex hypersurface V with isolated singularities $x_1, \ldots, x_m$. In this case, an Artinian algebra can be associated to X as follows.

DEFINITION 5.2. With the above notation, let f_i be a defining function of the germ (V, x_i), $1 \leq i \leq m$. Then the $\mathbb{C}$-algebra $A_i = \mathcal{O}_{n+1}/(f_i, \frac{\partial f_i}{\partial z_0}, \ldots, \frac{\partial f_i}{\partial z_n})$ is a commutative local Artinian algebra called the moduli algebra of (V, x_i). The moduli algebra is independent of the choice of defining function. We associate to the CR manifold X the Artinian algebra $A(X) = \bigoplus_{i=1}^{m} A_i$.

By the work of Mather-Yau [**Ma-Ya**] on isolated hypersurface singularities, it can be shown that the associated Artinian algebras are complete algebraic CR invariants in the following sense.

THEOREM 5.1. *(Luk-Yau [**Lu-Ya4**]). Two compact strongly pseudoconvex real codimension 3 CR manifolds X_1, X_2 are algebraically equivalent if and only if the associated Artinian algebras $A(X_1)$, $A(X_2)$ are isomorphic $\mathbb{C}$-algebras.*

DEFINITION 5.3. With the above notation, let $L(X)$ be the algebra of derivations of $A(X)$. Since $A(X)$ is finite dimensional as $\mathbb{C}$-vector space and $L(X)$ is contained in the endomorphism algebra of $A(X)$, consequently $L(X)$ is a finite dimensional Lie algebra with the obvious Lie algebra structure.

THEOREM 5.2. *(Yau [**Ya2**], [**Ya3**]). With the above notation, $L(X)$ is a finite dimensional solvable Lie algebra.*

We remark that there are Torelli type examples in which the Lie algebras $L(X_t)$ associated to a family of compact strongly pseudoconvex real codimension 3 CR manifolds X_t suffice to distinguish CR equivalence. For example, in the family $X_t = \{(x, y, z) \in \mathbb{C}^3 : x^6 + y^3 + z^2 + tx^4 y = 0 \text{ and } |x|^2 + |y|^2 + |z|^2 = \epsilon^2\}$ where $\epsilon > 0$ is a small fixed number and $t \in \mathbb{C}$ with $4t^2 + 27 \neq 0$, X_{t_1}, X_{t_2} are CR equivalent if and only if $L(X_{t_1}), L(X_{t_2})$ are isomorphic Lie algebras.

QUESTION 5.1. How can one compute $A(X)$ and $L(X)$ directly from X without going through V?

For the rest of this section we consider embeddable 3 dimensional compact strongly pseudoconvex CR manifolds. By taking resolutions of the singularities of the subvariety V bounded by such a CR manifold X in complex Euclidean space, numerical invariants under algebraic equivalence may be defined as follows.

DEFINITION 5.4. Let $\pi : M \to V$ be a resolution of the singularities Y of V such that the exceptional set $A = \pi^{-1}(Y)$ has normal crossing, i.e., the irreducible components A_i of A are nonsingular, they intersect transversally and no three meet at a point. According to Artin [**Ar**], there exists a unique minimal positive divisor Z, called the fundamental cycle, with support on A, such that $Z \cdot A_i \leq 0$ for all A_i. For any positive divisor $D = \Sigma d_i A_i$, let $\mathcal{O}_M(-D)$ be the sheaf of germs of holomorphic functions on M vanishing to order d_i on A_i, let $\mathcal{O}_D = \mathcal{O}_M / \mathcal{O}_M(-D)$ and let $\chi(\mathcal{O}_D) = \sum_{i=0}^{2} (-1)^i \dim H^i(M, \mathcal{O}_D)$. It can be proved that $p_f(X) := 1 - \chi(\mathcal{O}_Z)$, $p_a(X) := \sup(1 - \chi(\mathcal{O}_D))$ where D ranges over all positive divisors with support on A and $p_g(X) := \dim H^1(M, \mathcal{O})$ are defined independent of the resolution π and are invariants of X under algebraic equivalence. The detailed proofs are contained in [**L-Y-Y**]. We refer to $p_f(X), p_a(X)$ and $p_g(X)$ as the fundamental genus, arithmetic genus and geometric genus of X respectively.

The following facts are known:

- $0 \leq p_f(X) \leq p_a(X) \leq p_g(X)$
- $p_f(X) = 0 \Leftrightarrow p_a(X) = 0 \Leftrightarrow p_g(X) = 0$.

Further numerical invariants under algebraic equivalence are given by $m_Z(X) := Z \cdot Z$, $q(X) := \dim H^0(M - A, \Omega^1)/H^0(M, \Omega^1)$, $\chi(X) := K \cdot K + \chi_T(A)$ and $\omega(X) := K \cdot K + \dim H^1(M, \Omega^1)$, where Ω^1 is the sheaf of germs of holomorphic 1-form on M, $\chi_T(A)$ is the topological Euler characteristic of A and K is the canonical divisor on M. These invariants are defined independent of the choice of the resolution π. Since K is a divisor with rational coefficient, $\chi(X)$ and $\omega(X)$ are in general rational numbers.

Using the above invariants, one may attempt a rough algebraic classification of embeddable 3 dimensional compact strongly pseudoconvex CR manifolds.

DEFINITION 5.5. An embeddable 3 dimensional compact strongly pseudoconvex CR manifold X is called a rational (respectively elliptic) CR manifold if $p_a(X) = 0$ (respectively $p_a(X) = 1$).

If X is a rational or an elliptic CR manifold embeddable in $\mathbb{C}^3$ and M_0 is the minimal good resolution of the subvariety V bounded by X in $\mathbb{C}^3$, then the weighted dual graph for the exceptional set of M_0 is completely classified. The same also holds for those X embeddable in $\mathbb{C}^3$ and has $p_g(X) = 1$. With the weighted dual graphs classified, the topology of the embedding of the exceptional set in M_0 is well understood.

As an application, one obtains obstructions to embedding in $\mathbb{C}^3$ for the above three classes of CR manifolds when their weighted dual graphs fail to have the required forms. For example, a rational CR manifolds whose weighted dual graph is not a direct sum of the graphs A_k, D_k, E_6, E_7, E_8 is not embeddable in $\mathbb{C}^3$.

Similarly in view of the following theorem, one obtains numerical obstructions to embedding in $\mathbb{C}^3$ for those CR manifolds failing the conditions in the theorem.

THEOREM 5.3. *(*[**Lu-Ya3**], [**L-Y-Y**]*). Let X be a compact strongly pseudoconvex 3-dimensional CR manifold embeddable in $\mathbb{C}^3$. Then*

(1) *$\chi(X)$ and $\omega(X)$ are integers.*
(2) *$10p_g(X) + \omega(X) \geq 0$*
(3) *If $p_a(X) = 1$, then $\chi(X) \geq -3$*
(4) *If X admits a transversal holomorphic S^1-action, then $6p_g(X)+\chi(X) > 0$.*

We remark that (4) depends on the Durfee conjecture which is solved by Xu and Yau [**Xu-Ya1**].

6. Impacts of Combinatorial geometry in CR geometry

Given a compact strongly pseudoconvex CR manifolds X in $\mathbb{C}^N$, it is a very natural question to ask how many defining equations we need in order to specify X. Also we would like most of these defining functions are holomorphic. Recall that X is a boundary of a Stein variety V. Let $\{f_1, \ldots, f_m\}$ be the set of holomorphic functions which define V. Then X can be defined by $\{f_1, \ldots, f_m, \rho\}$ where ρ is a strictly plurisubharmonic real valued function. Such a set of defining functions of X is called a *natural set of defining functions.*

DEFINITION 6.1. Let X be a compact strongly pseudoconvex $(2n-1)$-dimensional CR manifold in $\mathbb{C}^N$. X is called a complete intersection CR manifold if there exists a natural set of defining functions $\{f_1, \ldots, f_{N-n}, \rho\}$ where $f_1, \ldots, f_{N-n}$ are holomorphic functions.

Clearly not all compact CR manifolds are complete intersection. It is an interesting question to know which CR manifold is a complete intersection.

Now let us restrict ourselves to 3 dimensional compact strongly pseudoconvex CR manifold X. Let x be a normal singularity of the two dimensional Stein space V which bounds X. Let $\pi : M \longrightarrow V$ be a resolution of V such that the irreducible components A_i of $A = \pi^{-1}(x)$ are nonsingular and have only normal crossings. Associated to A is a weighted dual graph Γ (The vertices of Γ correspond to irreducible components A_i. The edges of Γ connecting the vertices corresponding to A_i and A_j, $i \neq j$, correspond to the points of $A_i \cap A_j$. Finally, associated to each A_i is its weight $A_i \cdot A_i$, the topological self-intersection number), which, along with the genera of the A_i, fully describes the topological and differentiable nature of the embedding of A in M. In this way, we have constructed a mapping:

$$\left\{\begin{array}{l}\text{strongly pseudo-}\\ \text{convex 3-dim CR}\\ \text{manifolds}\end{array}\right\} \longrightarrow \left\{\begin{array}{l}\text{2-dim normal}\\ \text{singularities}\end{array}\right\} \longrightarrow \left\{\begin{array}{l}\text{weighted}\\ \text{dual}\\ \text{graphs}\end{array}\right\}$$

$$X \longmapsto (V,x) \longrightarrow \Gamma_X$$

It is natural to ask what kind of Γ_X can one get if X is a complete intersection CR manifold. In other words, we would like to use the combinatorial geometry of Γ_X to understand whether X is a complete intersection CR manifold. Recall that a complete intersection singularity (V,x) is Gorenstein. So there exists an integral cycle $K = \Sigma n_i A_i$ on Γ_X which satisfies the adjunction formula $A_i \cdot K = -A_i \cdot A_i + 2g_i - 2$.

QUESTION 6.1. Describe all graphs Γ_X such that such an integral cycle on Γ_X exists and satisfies the adjunction formula.

THEOREM 6.1. *(Chung-Graham, Xu and Yau* [**C-X-Y**]*). There is an explicit list of all those weighted dual graphs which have only complete intersection 3-dimensional CR manifolds associated with them.*

DEFINITION 6.2. Let X be a $(2n-1)$-dimensional compact strongly pseudoconvex CR manifold in $\mathbb{C}^N$ which is the boundary of a normal variety V with isolated singularities $\{x_1, \ldots, x_k\}$. Then the geometric genus $p_g(X)$ of the CR manifold X is defined as $\dim H^{n-1}(M, \mathcal{O})$ where $\pi : M \to V$ is a resolution of singularities of V.

REMARK 6.1. It can be proved that $p_g(X)$ is an invariant of X. In fact, $p_g(X) = p_g(V, x_1) + \cdots + p_g(V, x_k)$ where $p_g(V, x_i)$ is the geometric genus of the singularity which is equal to $\dim \Gamma(U - \{x_i\}, \Omega^n)/L^2(U - \{x_i\}, \Omega^n)$, where U is a small Stein neighborhood of x_i.

For hypersurface singularity, the geometric genus can be computed in the following manner.

Let $f(z_0, \ldots, z_n) = \Sigma a_\lambda z^\lambda$, where $z^\lambda = z_0^{\lambda_0} \cdots z_n^{\lambda_n}$, be a convergent power series such that $f(0) = 0$ and f has an isolated critical point at the origin.

Newton boundary $\Gamma(f) :=$ union of compact faces of $\Gamma_+(f)$, where

$\Gamma_+(f) :=$ convex hull of the union of the subsets $\{\lambda + (\mathbb{R}^+)^{n+1}\}$ for λ such that $a_\lambda \neq 0$.

Newton Polyhedron $\Gamma_-(f) :=$ cone over $\Gamma(f)$ with cone point at 0.

For each closed face Δ of $\Gamma(f)$, we associate a polynomial $f_\Delta(x) := \sum_{\lambda\in\Delta} a_\lambda x^\lambda$.

DEFINITION 6.3. f is said to be nondegenerate if f_Δ has no critical point in $(\mathbb{C}^*)^{n+1}$ for any $\Delta \in \Gamma(f)$ where $\mathbb{C}^* = \mathbb{C} - \{0\}$.

THEOREM 6.2. *(Merle-Teissier* [**Me-Te**]*). Let $(V, 0)$ be an isolated hypersurface singularity defined by a nondegenerate holomorphic function $f : (\mathbb{C}^{n+1}, 0) \longrightarrow (\mathbb{C}, 0)$. Then the geometric genus $p_g(V, 0) = \#\{p \in \mathbb{Z}_+^{n+1} \cap \Gamma_-(f)\}$.*

DEFINITION 6.4. A polynomial $f(z_0, z_1, \ldots, z_n)$ is said to be weighted homogeneous of type $(w_0, w_1, \ldots, w_n)$, where $w_0, w_1 \ldots, w_n$ are fixed positive rational numbers, if it can be expressed as a linear combination of monomials $z_0^{i_0} z_1^{i_1} \ldots z_n^{i_n}$ for which $i_0/w_0 + i_1/w_1 + \cdots + i_n/w_n = 1$.

As a consequence of the theorem of Merle-Teissier, in case of weighted homogeneous isolated singularity, computing the geometric genus is equivalent to counting the number of positive integral point in a tetrahedron.

Let $a_1 \geq a_2 \geq \ldots \geq a_n \geq 1$ be real numbers. Consider the n-dimensional tetrahedron $\Delta(a_1, \ldots, a_n)$ defined by $\frac{x_1}{a_1} + \ldots + \frac{x_n}{a_n} \leq 1$ and $x_1 \geq 0, \ldots, x_n \geq 0$. Let

$$\begin{aligned} Q(a_1, \cdots, a_n) &= \# \{(x_1, \cdots, x_n) \in (\mathbf{Z}_+ \cup \{0\})^n : \sum_{i=1}^{n} \frac{x_i}{a_i} \leq 1\}, \\ P(a_1, \cdots, a_n) &= \# \{(x_1, \cdots, x_n) \in \mathbf{Z}_+^n : \sum_{i=1}^{n} \frac{x_i}{a_i} \leq 1\}. \end{aligned}$$

Let $a = \frac{1}{a_1} + \ldots + \frac{1}{a_n}$. It is easy to show that

$$P(a_1, \cdots, a_n) = Q(a_1(1-a), \cdots, a_n(1-a)).$$

Counting the number of integral points in a n-dimensional tetrahedron is a classical subject which has attracted a lot of famous mathematicians. For $n = 2$, Hardy and Littlewood in 1922 wrote two famous papers ([**Ha-Li1**], [**Ha-Li2**]) on the number of lattice points of a right-angled triangle because of its relations to their International Congress Talk in 1912 on diophantine approximation ([**Ha-Li3**]). The general result was obtained by D.C. Spencer in 1942 via complex function-theoretic methods (cf. [**Sp1**], [**Sp2**]). The result is as follows. Write $x = a_1 \ldots a_n$. Then

$$Q(a_1, \cdots, a_n) = \frac{1}{n!} a_1 \cdots a_n + \frac{1}{2(n-1)!} \left(\frac{a_n}{a_1} + \cdots + \frac{a_n}{a_{n-1}} + 1\right) a_1 \cdots a_{n-1} + o(x^{n-1}).$$

In analytic number theory, especially in primality testing and in factoring, people are interested in the sharp upper estimate of $Q(a_1, \ldots, a_n)$. However, the above formula of Hardy-Littlewood-Spencer is only asymptotic. So the above result of D.C. Spencer is not sufficient for application in analytic number theory as well as singularity theory.

For motivation from the theory of singularities, people are interested in giving a sharp upper estimate of $P(a_1, \ldots, a_n)$. Recently, Yau and coworkers Xu and Lin have succeeded in giving a sharp polynomial estimate of $P(a_1, \ldots, a_n)$ for $n = 3$, 4 and 5 as follows.

THEOREM 6.3. *(Xu-Yau* [**Xu-Ya2**]*). For $n = 3$, $a_1 \geq a_2 \geq a_3 \geq 2$,*

$$3! P(a_1, a_2, a_3) \leq a_1 a_2 a_3 - (a_1 a_2 + a_1 a_3 + a_2 a_3) + a_1 + a_2$$

and equality is attained iff $a_1 = a_2 = a_3 =$integer.

THEOREM 6.4. *(Xu-Yau [**Xu-Ya3**]). For* $n = 4$, $a_1 \geq a_2 \geq a_3 \geq a_4 \geq 3$,

$$\begin{aligned} & 4!P(a_1, a_2, a_3, a_4) \\ \leq \; & a_1a_2a_3a_4 - \frac{3}{2}(a_1a_2a_3 + a_1a_2a_4 + a_1a_3a_4 + a_2a_3a_4) \\ & + \frac{11}{3}(a_1a_2 + a_1a_3 + a_2a_3) - 2(a_1 + a_2 + a_3) \end{aligned}$$

and equality is attained iff $a_1 = a_2 = a_3 = a_4 =$*integer.*

THEOREM 6.5. *(Lin-Yau [**Li-Ya1**]). For* $n = 5$, $a_1 \geq a_2 \geq a_3 \geq a_4 \geq a_5 \geq 4$,

$$\begin{aligned} & 5!P(a_1, a_2, a_3, a_4, a_5) \\ \leq \; & a_1a_2a_3a_4a_5 - 2(a_1a_2a_3a_4 + a_1a_2a_3a_5 + a_1a_2a_4a_5 + a_1a_3a_4a_5 + a_2a_3a_4a_5) \\ & + \frac{35}{4}(a_1a_2a_3 + a_1a_2a_4 + a_1a_3a_4 + a_2a_3a_4) \\ & - \frac{50}{6}(a_1a_2 + a_1a_3 + a_1a_4 + a_2a_3 + a_2a_4 + a_3a_4) \\ & + 6(a_1 + a_2 + a_3 + a_4) \end{aligned}$$

and equality is attained iff $a_1 = a_2 = a_3 = a_4 = a_5 =$*integer.*

Most recently, Wang and Yau have succeeded in proving the following Theorem for $n \leq 6$.

THEOREM 6.6. *(Wang-Yau [**Wa-Ya**]). Let* $P_n = \#\{(x_1, \ldots, x_n) \in \mathbb{Z}_+^n : \frac{x_1}{a_1} + \cdots + \frac{x_n}{a_n} \leq 1\}$, *where* $a_1 \geq a_2 \geq \cdots \geq a_n \geq n-1$ *are real numbers and* $n \geq 3$. *Let* $b_k^n = \sum_{1 \leq i_1 < \cdots < i_k \leq n-1} i_1 i_2 \ldots i_k$ *for* $1 \leq k \leq n-1$. *Then, for* $n \leq 6$,

$$\begin{aligned} n!P_n \;\leq\; & \prod_{i=1}^{n} a_i + (-1)\frac{b_{n-1}^n}{n}\left(\sum_{k=1}^{n}\frac{1}{a_k}\right)\prod_{i=1}^{n} a_i + (-1)^2\frac{b_{n-2}^n}{\binom{n-1}{1}}\left(\sum_{k=1}^{n-1}\frac{1}{a_k}\right)\prod_{i=1}^{n-1} a_i + \ldots \\ & + (-1)^{k+1}\frac{b_{n-k-1}^n}{\binom{n-1}{k}}\left(\sum_{1 \leq i_1 < \cdots < i_k \leq n-1}\frac{1}{a_{i_1}a_{i_2}\ldots a_{i_k}}\right)\prod_{i=1}^{n-1} a_i + \cdots \\ & + (-1)^{n-1}\frac{b_1^n}{\binom{n-1}{n-2}}(a_1 + a_2 + \cdots + a_{n-1}). \end{aligned}$$

The above sharp estimates not only answer the question raised by people in analytic number theory, but also have helped us to solve the open problem since 1970 of intrinsic characterization of homogeneous polynomials.

THEOREM 6.7. *(Xu-Yau [**Xu-Ya1**]). For* $n = 3$. *Let* $f(x, y, z)$ *be a holomorphic function with isolated singularity at* 0. *Let* $\mu = \dim \mathcal{O}_3/(f_x, f_y, f_z)$ *be the Milnor number,* p_g *be the geometric genus and* ν *be the multiplicity of the singularity. Let* τ *be the dimension of the vector space* $\mathbb{C}\{x, y, z\}/(f, f_x, f_y, f_z)$. *Then, after a biholomorphic change of coordinates,* f *is a homogeneous polynomial iff* $\mu = \tau$ *and* $\mu - \nu + 1 = 6p_g$.

THEOREM 6.8. *(Lin-Yau [**Li-Ya2**], [**Li-Ya3**]). For* $n = 4$. *Let* $f(x, y, z, w)$ *be a holomorphic function with isolated singularity at* 0. *Let* $\mu = \dim \mathcal{O}_4/(f_x, f_y, f_z, f_w)$ *be the Milnor number,* p_g *be the geometric genus and* ν *be the multiplicity of the singularity. Let* τ *be the dimension of the vector space* $\mathbb{C}\{x, y, z, w\}/(f, f_x, f_y, f_z, f_w)$.

Then, after a biholomorphic change of coordinates, f is a homogeneous polynomial iff $\mu = \tau$ and $\mu - (2\nu^3 - 5\nu^2 + 2\nu + 1) = 24p_g$.

More generally, Yau has formulated the following conjecture which gives an intrinsic characterization when a complex hypersurface in $\mathbb{C}^{n+1}$ is a cone over a nonsingular projective variety in $\mathbb{C}P^n$.

Yau Conjecture. *Let $(V, 0)$ be a n-dimensional isolated hypersurface singularity defined by $f(x_0, x_1, \ldots, x_n) = 0$. Let μ be the Milnor number, p_g be the geometric genus, ν be the multiplicity of the singularity and $\tau = \dim \mathbb{C}\{x_0, \ldots, x_n\}/(f, f_{x_0}, \ldots, f_{x_n})$ be the dimension of the semi-universal deformation space of $(V, 0)$. Then after a biholomorphic change of coordinates, f is a homogeneous polynomial if and only if $\mu = \tau$ and $\mu - h(\nu) = (n+1)!p_g$ where $h(\nu) = (\nu-1)^n - \nu(\nu-1)(\nu-2)\cdots(\nu-n+1)$.*

DEFINITION 6.5. A smooth S^1-action on CR manifold X is said to be holomorphic if it preserves the family of subspace $\mathcal{H} \subset TX$ and commutes with J. It is said to be transversal if, in addition, the vector field V which generates the action, is transversal to $\mathcal{H}$ at all point of X.

THEOREM 6.9. *(Lawson-Yau [**La-Ya**]). Let $X \subseteq \mathbb{C}^{n+1}$ be a CR manifold of dimension $2n - 1 > 1$, and suppose X admits a transversal holomorphic S^1-action. Then after a holomorphic change of coordinates in $\mathbb{C}^{n+1}$, M is contained in an affine hypersurface $Y \subset \mathbb{C}^{n+1}$. The hypersurface Y has at most one singular point. It also has a $\mathbb{C}^*$-action and the embedding $X \subset Y$ is S^1-equivariant.*

QUESTION 6.2. Under what conditions on X will Y be a homogeneous hypersurface?

QUESTION 6.3. Given a CR manifold X, how can one compute the geometric genus $p_g(X)$ and Milnor number in terms of X only?

Recall that an embeddable strongly pseudoconvex compact 3-dimensional CR manifold X is said to be rational if $p_g(X) = 0$. For rational 3-dimensional CR manifold, the variety V, which X bounds, has only rational singularities in the interior of V.

THEOREM 6.10. *(Luk-Yau [**Lu-Ya5**]). Let X be a rational 3-dimensional CR manifold. There exist explicit algebraic algorithms to compute the graph invariant Γ in terms of the ring structure of $\bigoplus_{k=0}^{\infty} m^k/m^{k+1}$, where m is the maximal ideal of each singularity.*

QUESTION 6.4. How can one compute $\bigoplus_{k=0}^{\infty} m^k/m^{k+1}$ in terms of the CR manifold X?

References

[Ak] A. Akahori, *The new approach to the local embedding theorem of CR-structures for $n \geq 4$*, Contemp. Math., vol. 49, Amer. Math. Soc., Providence, RI, 1986, pp. 1–10

[Ar] M. Artin, *On isolated rational singularities of surfaces*, Amer. J. Math., **88** (1966), 126–136.

[Bl-Ep] J. Bland, and C. L. Epstein, *Embeddable CR-structures and deformations of pseudoconvex surfaces Part I: Formal deformations*, Jour. of Alg. Geometry, **5** (1996), 277–368.

[BM] L. Boutet de Monvel, *Intégration des éequations de Cauchy-Riemann*, Séminaire Goulaouic-Lions-Schwartz, Ex ré IX, 1974–1975.

[Bu-Ep] D. Burns, and C. L. Epstein, *Embeddability for three dimensional CR-manifolds*, J. Amer. Math. Soc. **3** (1990), 809–841.

[Ca] E. Cartan, *La geometrie pseudo-conforme des hypersurfaces de l'espace de deux variables complexes*, Ann. Mat. IV. Ser. XI, (1932), 17–90.

[Ch-Mo] S. S. Chern, and J. Moser, *Real hypersurfaces in complex manifolds*, Acta Math. **133** (1974), 219–271.

[C-X-Y] F. Chung-Graham, Y.-J. Xu, and S. S.-T. Yau, *Classification of weighted dual graphs admitting only complete intersection singularities structure*, (preprint).

[Fa] E. Falbel, *Non-embeddable CR manifolds and surface singularities*, Invent. Math. **108** (1992), 49–65.

[Fe] C. Fefferman, *Parabolic invariant theory in complex analysis*, Adv. in Math. **31** (1979), 131–262.

[Fo-St] G. B. Folland, and E. M. Stein, *Estimates for the $\overline{\partial}_b$ complex and analysis on the Heisenberg group*, Comm. Pure Appl. Math. **27** (1974), 429–522.

[Ha-Li1] G. H. Hardy, and J. E. Littlewood, *The lattice points of a right-angled triangle*, Proc. London Math. Soc. (2) **20** (1921), 15–36.

[Ha-Li2] G. H. Hardy, and J. E. Littlewood, *The lattice points of a right-angled triangle (seond memoir)*, Hamburg Math. Abh. **1** (1922), 212–249.

[Ha-Li3] G. H. Hardy, and J. E. Littlewood, *Some problems of diophantine approximation*, Proc. 5th Int. Congress of Mathematics, (1912), 223–229.

[Ha-La] R. Harvey, and B. Lawson, *On boundaries of complex analytic varieties I.* Ann. of Math. **102** (1975), 233–290.

[JT1] H. Jacobowitz, and F. Treves, *Non-realizable CR-structures*, Invent. Math. **66** (1982), 231–249.

[JT2] H. Jacobowitz, and F. Treves, *Aberrant CR-structures*, Hokkaiko Math. J. **12** (1983), 276–292.

[Ko] J. J. Kohn, *Boundaries of complex manifolds*, Proc. Conference on Complex Manifolds (Minneapolis), Springer-Verlag, New York, 1965.

[Ko-Ro] J. J. Kohn, and H. Rossi, *On the extension of holomorphic functions from the boundary of a complex manifolds*, Ann. of Math. **81** (1965), 451–472.

[Ku1] M. Kuranishi, *Strongly pseudoconvex CR structures over small balls I, II, III*, Ann. of Math. (2) **115** (1982), 451–500; ibid **116** (1982), 1–64; ibid **116** (1982), 249–330.

[Ku2] M. Kuranishi, *CR geometry and Cartan geometry*, Forum math., **7** (1995), 147–206.

[Ku3] M. Kuranishi, *An approach to the Cartan geometry II*, CR manifolds, Complex Analysis in Several Variables — Memorial Conference of Kiyoshi Oka's Centennial Birthday, Kyoto/Nara 2001, Advanced Studies in Pure Mathematics, **42** (2004), 165–187.

[La-Ya] H. B. Lawson, and S. S.-T. Yau, *Holomorphic symmetries*, Ann. Sci. Ecole Norm. Sup. 4e Série, **20** (1987), 557–577.

[Le] L. Lempert, *Embeddings of three dimensional Cauchy Riemann manifolds*, Math. Ann. **300** (1994), 1–15.

[Li-Ya1] K.-P. Lin, and S. S.-T. Yau, *A sharp upper estimate of the number of integral points in a 5-dimensional tetrahedra*, Journal of Number Theory, **93** (2002), 207–234.

[Li-Ya2] K.-P. Lin, and S. S.-T. Yau, *Classification of affine varieties being cones over nonsingular projective varieties: hypersurface case*, Communication in Analysis and Geometry, vol. 12, no. 5, (2004), 1201–1219.

[Li-Ya3] K.-P. Lin, and S. S.-T. Yau, *Characterization of isolated homogeneous hypersurface singularities in $\mathbb{C}^4$*, (preprint).

[Lu-Ya1] H. S. Luk, and S. S.-T. Yau, *Kohn-Rossi cohomology and its application to the complex Plateau problem II*, (submitted for publication).

[Lu-Ya2] H. S. Luk, and S. S.-T. Yau, *Invariant Kohn-Rossi cohomology and obstructions to embedding of compact real $(2n-1)$-dimensional CR manifolds in $\mathbb{C}^N$*, J. Math. Soc. Japan, **48** (1996), 61–68.

[Lu-Ya3] H. S. Luk, and S. S.-T. Yau, *Obstructions to embedding of real $(2n-1)$-dimensional compact CR manifolds in $\mathbb{C}^{n+1}$*, Proc. of Sympos. Pure Math. **52** (1991), part 3, 261–276.

[Lu-Ya4] H. S. Luk, and S. S.-T. Yau, *Complete algebraic CR invariants of real codimension 3 strongly pseudoconvex CR manifold*, Studies in Advanced Mathematics, vol. 5, Singularities and Complex Geometry, AMS/IP, (1997), 175–182.

[Lu-Ya5] H.-S. Luk, and S. S.-T. Yau, *Explicit construction of graph invariant for strongly pseudoconvex compact 3-dimensional rational CR manifolds*, Compositio Mathematica, **114** (1998), 77–111.

[Lu-Ya6] H.-S. Luk, and S. S.-T. Yau, *Some remarks on compact strongly pseudoconvex CR manifolds*, Advanced Studies in Pure Mathematics **25** (1997), CR-Geometry and Over determined Systems, 237–246.

[L-Y-Y] H. S. Luk, S. S.-T. Yau, and Y. Yu, *Algebraic classification and obstruction to embedding of strongly pseudoconvex compact 3-dimensional CR manifolds in* $\mathbb{C}^3$, Math. Nachr. **170** (1994), 183–200.

[Ma-Ya] Mather, J. and S. S.-T. Yau, *Classification of isolated hypersurfaces singularities by their moduli algebras*, Invent. Math. **69** (1982), 243–251.

[Me-Te] M. Merle, and B. Teissier, *Conditions d'adjonction d'après Du Val*, Séminaire les Singularités des Surfaces (Centre de Math. de l'Ecole Polytechnique, 1976–1977), Lecture Notes in Math., vol. 777, Springer, Berlin, 1980, 229–245.

[Ni] L. Nirenberg, *On a question of Hans Lewy*, Russ. Math. Surv. **29** (1974), 251–262.

[Ro] H. Rossi, *Attaching analytic spaces to an analytic space along a pseudoconcave boundary*, In: Aeppli., A., Calabi, E., Röhrl, H., (eds.), Proc. Conf. Comp. Anal., Minneapolis 1964, Berlin, Heidelberg, New York: Springer 1965.

[Sp1] D. C. Spencer, *On a Hardy-Littlewood problem of diophantine approximation*, Math. Proc. Cambridge Philos. Soc. **35** (1939), 527–547.

[Sp2] D. C. Spencer, *The lattice points of tetrahedra*, J. Math. Phys. **21** (1942), 189–197.

[Ta1] N. Tanaka, *On the pseudo-conformal geometry of hypersurfaces of the space of n complex variables*, J. Math. Soc. Japan, **14** (1962), 397–429.

[Ta2] N. Tanaka, *A differential geometric study on strongly pseudoconvex manifolds*, Lectures in Mathematics, Kyoto University, 9, Kinokuniya Book-Store Co., Ltd., 1975.

[Wa-Ya] X.-J. Wang, and S. S.-T. Yau, *On the GLY conjecture of upper estimate of positive integral points in real tetrahedron*, (preprint).

[Xu-Ya1] Y.-J. Xu, and S. S.-T. Yau, *Durfee conjecture and coordinate free characterization of homogeneous singularities*, J. Differential Geom. **37** (1993), 375–396.

[Xu-Ya2] Y.-J. Xu, and S. S.-T. Yau, *Sharp estimate of number of integral points in a tetrahedron*, J. Reine Angew. Math. **423** (1992), 199–219.

[Xu-Ya3] Y.-J. Xu, and S. S.-T. Yau, *Sharp estimate of number of integral points in a 4-dimensional tetrahedron*, J. Reine Angew. Math. **473** (1996), 1–23.

[Ya1] Stephen S.-T. Yau, *Kohn-Rossi cohomology and its application to the complex Plateau problem, I.* Ann. of Math. **113** (1981), 67–110.

[Ya2] Stephen S.-T. Yau, *Singularities defined by* $sl(2,\mathbb{C})$ *invariant polynomials and solvability of Lie algebras arising from isolated singularities*, Amer. J. Math. **108** (1986), 1215–1240.

[Ya3] Stephen S.-T. Yau, *Solvability of the Lie algebras arising from singularities and non-isolatedness of the singularities defined by the invariant polynomials of* $sl(2,\mathbb{C})$, Amer. J. Math. **113** (1991), 773–778.

Department of Mathematics and Computer Science, Chicago State University, Chicago, IL 60628

E-mail address: `ljia@csu.edu`

Department of Mathematics, Chinese University of Hong Kong, Shatin, Hong Kong

E-mail address: `hsluk@math.cuhk.edu.hk`

Institute of Mathematics, East China Normal University, Shanghai, China

Current address: MSCS, M/C 249, University of Illinois at Chicago, Chicago, IL 60607-7045

E-mail address: `yau@uic.edu`

AMS/IP Studies in Advanced Mathematics
Volume **39**, 2007

Periods of Automorphic Forms

Dihua Jiang

ABSTRACT. In this paper, we discuss periods of automorphic forms from the representation-theoretic point of view. It gives general theory of periods and some special and very interesting cases.

1. Introduction

A classical result of Gelfand and Graev asserts that the Gelfand-Graev models are of multiplicity free over a finite field. For simplicity, we consider here for $\mathrm{GL}_2(\mathbb{F}_q)$, where $\mathbb{F}_q$ is a finite field. Let $B = TU$ be the standard Borel subgroup given by

$$B = \{\begin{pmatrix} a & b \\ 0 & d \end{pmatrix} \in \mathrm{GL}_2\}$$

with the unipotent radical U given by

$$U = \{u(b) = \begin{pmatrix} 1 & b \\ 0 & 1 \end{pmatrix} \in \mathrm{GL}_2\}$$

Fix a nontrivial additive character ψ of $\mathbb{F}_q$, and then define

$$\psi_U(u(b)) := \psi(b).$$

The Gelfand-Graev representation in this case is given by the following induced representation

$$\mathrm{Ind}_{B(\mathbb{F}_q)}^{\mathrm{GL}_2(\mathbb{F}_q)}(\psi_U).$$

It is not hard to check that such an induced representation is of multiplicity free ([**Bmp97**]).

This simple result was generalized to GL_2 over local field by Jacquet and Langlands and played an essential role in their classical work ([**JL70**]). It turned out that it holds for quasi-split reductive groups over local fields. This is the well-known uniqueness property of local Whittaker models ([**Shl74**]), which has been essential in the theory of automorphic L-functions, the theory of Langlands functoriality, the theory of harmonic analysis over local reductive groups, and the spectral theory of automorphic forms.

1991 *Mathematics Subject Classification.* Primary 11F, 22E.
Key words and phrases. Automorphic Forms, L-functions, Periods.

In this notes, we shall explain the periods of automorphic forms as generalization of the theory started from the Gelfand-Graev models. On the other hand, periods of automorphic forms also originated from the systematically study by Y. Manin ([**M72**] and [**M73**]) on modular symbols associated to classical modular forms. Some generalizations of modular symbols related to multiple values of zeta functions has been proposed by Y. Manin recently in [**M05**]. The adelic approach to study periods or modular symbols of automorphic forms was first considered by G. Harder in [**Hd87**] for GL_2. For more general groups and the study of periods or modular symbols of automorphic forms with application to Eisenstein cohomology of arithmetic groups, one may find more interesting discussions in [**Hd03**]. Some work on periods of automorphic forms and modular cycles may be found in [**HLR86**], [**Hr94**], [**Od01**], and [**Kd03**]. More general modular symbols have been studied in [**AB90**].

In the theory of automorphic representations, the periods play the role of automorphic quasi-invariant functionals over the space of automorphic representations. The theory of relative trace formulas, which was first studied by H. Jacquet, stimulated the recent study of periods of automorphic representations, which can be usually used to characterize certain cases of the Langlands functorial liftings of automorphic representations. A well prepared survey paper of H. Jacquet ([**Jcq97**]) provides some detailed discussions on the relative trace formula approach and applications in number theory.

It turns out that periods of automorphic forms are often natural and useful invariants to characterize certain family of automorphic representations, especially in the constructive theory of automorphic forms, such as the theta correspondences in terms of the Weil representations or more generally the automorphic minimal representations, and more recently automorphic descent construction [**GRS01**].

The discussion carried out in this notes is based on the author's own research interests related to this topic. It is clear that some very interesting progress will be omitted without being mentioned. The author's research related to the periods of automorphic forms has been supported in part over years by the NSF research postdoc fellowship, the A. Sloan research fellowship, the McKnight Land-Grant Professorship (UMN), and the regular NSF grants DMS-9896257, DMS-0098003, and DMS-0400414. I would like to thank Professor Stephen Rallis, Professor Ilya Piatetski-Shapiro, Professor Roger Howe, and Professor James Cogdell for their encouragement and support over years and to thank my long term collaborators, Professor David Ginzburg and Professor David Soudry, for stimulating discussions and supports related to the topics in this paper.

Finally, I would like to thank the organizers for inviting me to lecture in the conference. It is my great pleasure to submit this paper to the Proceedings. I spent my time from 1984 to 1989 in the Department of Mathematics, The East China Normal University, starting as a graduate student in Arithmetic Theory of Quadratic Forms, and then working as a junior faculty member. The stimulating research environment in the Department provided me lots of opportunities to learn advanced mathematics during my stay. I am grateful to the teachers in the Research Group of Algebra, especially Professor X.-H. Cao and late Professor Z.-F. Zhu for guiding me during the five years. I would also like to thank Professor H.-W. Lu for his Chinese translation of the well-known expository paper by Professor Stephen Gelbart, *An Elementary Introduction to the Langlands Program* ([**Glb84**]), which attracted my attention for the first time to the modern theory of automorphic forms. The topics included here have been presented in part in lectures at The Joint Number Theory Seminar in New York, The Haar Seminar at The Ohio State University, The Colloquium at The Osaka City University, and The workshop on automorphic forms and automorphic L-functions at RIMS. I would like to thank Professor D. Goldfeld, Professor W.-Z. Luo, and Professor M. Furusawa for invitations.

2. Periods

2.1. Automorphic forms and representations. Let G be a reductive algebraic group defined over a number field k. For simplicity, we may assume that G splits over k. Let $\mathbb{A}$ be the ring of adeles of k. Consider the space of square integrable functions over the quotient $Z_G(\mathbb{A})G(k)\backslash G(\mathbb{A})$, which is denoted by

$$L^2(G) := L^2(Z_G(\mathbb{A})G(k)\backslash G(\mathbb{A})).$$

Note that $Z_G(\mathbb{A})G(k)\backslash G(\mathbb{A})$ has finite volume with respect to the natural Haar measure. Following general spectral theory, one has

$$L^2(G) = L^2_{\text{disc}}(G) \oplus L^2_{\text{cont}}(G),$$

where $L^2_{\text{disc}}(G)$ is the discrete part of the spectrum, which is a Hilbert sum of irreducible subrepresentations with finite multiplicities, and $L^2_{\text{cont}}(G)$ is the continuous part of the spectrum. The discrete part of spectrum can be written as a direct sum of cuspidal spectrum and residual spectrum:

$$L^2_{\text{disc}}(G) = L^2_{\text{cusp}}(G) \oplus L^2_{\text{res}}(G).$$

An irreducible submodule of $L^2_{\text{disc}}(G)$ ($L^2_{\text{cusp}}(G)$, or $L^2_{\text{res}}(G)$, resp.) is called a discrete series (cuspidal, or residual, resp.) automorphic representation of $G(\mathbb{A})$. A general definition of automorphic forms (or representations) can be found in [**BJ79**]. However, the following theorem of Langlands provides a characterization of automorphic representations in terms of cuspidal automorphic representations of Levi subgroups of G, which is also convenient to be taken as an informal definition of general automorphic representations.

THEOREM 2.1 (Langlands [**L79**]). *An irreducible (admissible) representation π of $G(\mathbb{A})$ is automorphic if and only if there exist a parabolic subgroup P with Levi subgroup M, and an irreducible cuspidal automorphic representation σ of $M(\mathbb{A})$ such that π is an irreducible constituent of the parabolically induced representation*

$$\text{Ind}_{P(\mathbb{A})}^{G(\mathbb{A})}(\sigma).$$

A classical example of periods of automorphic forms comes from the Hecke theory. Let π be an irreducible cuspidal automorphic representation of $\text{GL}_2(\mathbb{A})$, and φ_π be a cusp form in π. Consider the following Mellin transform of φ_π

$$\int_{k^\times\backslash\mathbb{A}^\times} \varphi_\pi(\begin{pmatrix} a & 0 \\ 0 & 1 \end{pmatrix})|a|_{\mathbb{A}}^{s-\frac{1}{2}} d^\times a.$$

Since irreducible cuspidal automorphic representations of $\text{GL}_2(\mathbb{A})$ are generic, i.e. having a nonzero Whittaker-Fourier coefficient, and have the following Fourier expansion (The generalization to GL_n can be found in [**Shl74**].)

$$\varphi_\pi(g) = \sum_{\alpha\in k^\times} \mathcal{W}^{\psi}_{\varphi_\pi}(\begin{pmatrix} \alpha & 0 \\ 0 & 1 \end{pmatrix} g)$$

where $\mathcal{W}_{\varphi_\pi}(g)$ is the Whittaker-Fourier coefficient of φ_π with respect to a non-trivial additive character ψ, given by

$$\mathcal{W}^{\psi}_{\varphi_\pi}(g) := \int_{k\backslash\mathbb{A}} \varphi_\pi(\begin{pmatrix}1 & x\\ 0 & 1\end{pmatrix} g)\psi^{-1}(x)dx.$$

Then the Mellin transform can be expressed as

$$\int_{k^\times\backslash\mathbb{A}^\times} \varphi_\pi(\begin{pmatrix}a & 0\\ 0 & 1\end{pmatrix})|a|_{\mathbb{A}}^{s-\frac{1}{2}}d^\times a = \int_{\mathbb{A}^\times} \mathcal{W}^{\psi}_{\varphi_\pi}(\begin{pmatrix}a & 0\\ 0 & 1\end{pmatrix})|a|_{\mathbb{A}}^{s-\frac{1}{2}}d^\times a.$$

By the Jacquet-Langlands version of Hecke theory, there exists a φ_π such that

$$\int_{\mathbb{A}^\times} \mathcal{W}^{\psi}_{\varphi_\pi}(\begin{pmatrix}a & 0\\ 0 & 1\end{pmatrix})|a|_{\mathbb{A}}^{s-\frac{1}{2}}d^\times a = L(s,\pi).$$

Hence at $s = \frac{1}{2}$, the center of the symmetry from the functional equation of $L(s,\pi)$, we have

$$L(\frac{1}{2},\pi) = \int_{k^\times\backslash\mathbb{A}^\times} \varphi_\pi(\begin{pmatrix}a & 0\\ 0 & 1\end{pmatrix})d^\times a.$$

The last integral gives the simplest example of the periods of automorphic forms, which will be discussed in detail in this paper. The general notion of periods of automorphic forms will be a natural generalization of both such simple periods and the Whittaker-Fourier coefficients. The model attached to it will be called the generalized Gelfand-Graev models of automorphic forms, the definition of which is based on the Dynkin-Kostant theory of nilpotent orbits. The generalized Gelfand-Graev representations of reductive groups over finite fields was well discussed in the expository paper by Kawanaka in [**Kwn87**].

2.2. Nilpotent orbits. In order to introduce the general notion of periods of automorphic forms, we have to recall some basic facts of nilpotent orbits. For simplicity, we refer to [**CM93**].

Let $\mathfrak{g}$ be the Lie algebra of G. Consider the theory of nilpotent orbits in $\mathfrak{g}$ over $\mathbb{C}$ or the algebraic closure $\bar{k}$ of k. According to the Dynkin-Kostant theory, for each nilpotent orbit $\mathcal{O}$ in $\mathfrak{g}(\mathbb{C})$, there is a standard parabolic subgroup $P_{\mathcal{O}} = M_{\mathcal{O}}N_{\mathcal{O}}$, whose Lie algebra $\mathfrak{p}_{\mathcal{O}} = \mathfrak{m}_{\mathcal{O}} + \mathfrak{n}_{\mathcal{O}}$ has the following properties. First, the nilpotent radical $\mathfrak{n}_{\mathcal{O}}$ can be written as a graded $\mathfrak{m}_{\mathcal{O}}$-module

$$\mathfrak{n}_{\mathcal{O}} = \mathfrak{n}_1 \oplus \mathfrak{n}_2 \oplus \mathfrak{n}_3 \oplus \cdots,$$

and the intersection of $\mathcal{O}$ with $\mathfrak{n}_2$ is a single Zariski open dense $\mathfrak{m}_{\mathcal{O}}$-orbit. The stabilizer of this Zariski open dense $\mathfrak{m}_{\mathcal{O}}$-orbit is denoted by $\mathfrak{m}^\circ_{\mathcal{O}}$, up to isomorphism. If the nilpotent orbit $\mathcal{O}$ is even, then $\mathfrak{n}_{2i+1} = 0$ for all $i \geq 0$. The algebraic subgroup of G corresponding to $\mathfrak{m}^\circ_{\mathcal{O}}$ is denoted by $M^\circ_{\mathcal{O}}$. For any reductive algebraic subgroup H of $M^\circ_{\mathcal{O}}$, we form a semidirect product

$$J_{\mathcal{O},H} := H \times N_{\mathcal{O}}.$$

When $\mathfrak{n}_1 \neq 0$, the quotient $\mathfrak{n}_{\mathcal{O}}/\mathfrak{n}_3 \oplus \cdots$ has a structure of two-step nilpotent Lie algebra. In this case, we may consider the metaplectic cover of H and $J_{\mathcal{O},H}$, which are denoted by $\widetilde{H}$ and $\widetilde{J}_{\mathcal{O},H}$. In general, we only consider the cases when the subgroup H is relative big, so that the group $J_{\mathcal{O},H}$ has finitely many orbits acting on the flag variety $B\backslash G$ over $\mathbb{C}$ or $\bar{k}$, where B is a Borel subgroup of G. In such cases, the subgroup $J_{\mathcal{O},H}$ is usually called a *spherical subgroup* of G. Some detailed discussion of spherical varieties and spherical subgroups and related topics can be found in ([**Brn95**], [**Hw94**]).

Of course, for precise applications of the above discussion to the theory of automorphic forms, one have to discuss k-rational and k-stable structure of the theory nilpotent orbits and related topics ([**Ktw82**]). A natural formulation of such discussions is to use the theory of Galois cohomology. For this notes, we will avoid such a discussion, instead, we provide an example when $G = \mathrm{SO}(2n+1)$, the k-split odd orthogonal group.

Let k be a field of characteristic zero. Let $(V_{2n+1}, (\cdot,\cdot))$ be a nondegenerate quadratic vector space over k of dimension $2n+1$ with Witt index n. The symmetric bilinear form is given by

$$J_{2n+1}(d) = \begin{pmatrix} & & J_n \\ & d & \\ J_n & & \end{pmatrix} \tag{2.1}$$

where we define $J_n := \begin{pmatrix} & & 1 \\ & J_{n-2} & \\ 1 & & \end{pmatrix}$ inductively. We may choose a basis

$$\{e_1, \cdots, e_n, e_0, e_{-n}, \cdots, e_{-1}\} \tag{2.2}$$

such that

$$(e_i, e_j) = \begin{cases} 1 & \text{if } j = -i, \\ 0 & \text{if } j \neq -i, \\ d & \text{if } j = i = 0. \end{cases}$$

For each $r \in \{1, 2, \cdots, n\}$, we have the following partial polarization

$$V_{2n+1} = X_r \oplus V_{2(n-r)+1} \oplus X_r^* \tag{2.3}$$

where X_r is a totally isotropic subspace of dimension r and X_r^* is the dual of X_r with respect to the non-degenerate bilinear form $(\cdot,\cdot)$, and the subspace $V_{2(n-r)+1}$ is the orthogonal complement of $X_r \oplus X_r^*$. Without loss of generality, we may assume that X_r is generated by $e_1, \cdots, e_r$ and X_r^* is generated by $e_{-r}, \cdots, e_{-1}$.

We recall basic facts on nilpotent orbits of complex classical Lie algebras from [**CM93**]. For $\mathfrak{so}_{2n+1}(\mathbb{C})$, the nilpotent orbits or nilpotent conjugacy classes are in one to one correspondence with the partitions of $2n+1$ with even parts occurring with even multiplicity. Using the notation from [**CM93**], the set of such partitions is denoted by $\mathcal{P}_1(2n+1)$. We call such partitions *odd orthogonal partitions.*

For $r \in \{0, 1, 2, \cdots, n\}$, the partition $[2r+1, 1^{2(n-r)}]$ belongs to $\mathcal{P}_1(2n+1)$. By Theorem 6.3.7, [**CM93**], the partition $[2r+1, 1^{2(n-r)}]$ is special for every $r \in \{0, 1, 2, \cdots, n\}$ and also the involution of partitions stabilizes this family of special partitions. The nilpotent orbit corresponding to $[2r+1, 1^{2(n-r)}]$ is denoted by

$$\mathcal{O}_{n,r} = \mathcal{O}_{[2r+1, 1^{2(n-r)}]}.$$

The orbit corresponding to a special partition is called a special nilpotent orbit. It is easy to see that this family of special nilpotent orbits is linearly ordered. When $r = n$, $\mathcal{O}_{n,n} = \mathcal{O}_{[2n+1]}$ is the regular nilpotent orbit, and when $r = 0$, $\mathcal{O}_{n,0} = \mathcal{O}_{[1^{2n+1}]}$ is the zero nilpotent orbit.

The following discussion is trivial when $r = 0$. Hence we assume that $r \neq 0$. For each $r \in \{1, 2, \cdots, n\}$, there is a standard parabolic Lie subalgebra $\mathfrak{p}_n^r = \mathfrak{m}_n^r \mathfrak{n}_n^r$

(unique up to conjugation) with Levi subalgebra

$$\mathfrak{m}_n^r \cong (\mathfrak{gl}_1)^{\oplus r} \oplus \mathfrak{so}_{2(n-r)+1}$$

and nilpotent radical $\mathfrak{n}_n^r$, whose quotient modulo its derived subalgebra $[\mathfrak{n}_n^r, \mathfrak{n}_n^r]$ is given by

$$\mathfrak{n}_n^r/[\mathfrak{n}_n^r, \mathfrak{n}_n^r] \cong \mathbb{C}^{\oplus(r-1)} \oplus \mathbb{C}^{\oplus(2(n-r)+1)}.$$

Note that if $r = 0$, then $\mathfrak{p}_n^0 = \mathfrak{so}_{2n+1}$. Hence we may assume that the above space is zero when $r = 0$. When $r \neq 0$, the adjoint action of $\mathfrak{m}_n^r$ on $\mathfrak{n}_n^r$ induces an action of $(\mathrm{GL}_1)^r \times \mathrm{SO}_{2(n-r)+1}$ on the affine space $\mathbb{C}^{\oplus(r-1)} \oplus \mathbb{C}^{\oplus(2(n-r)+1)}$ such that for each

$$g = (t_1, t_2, \cdots, t_r; h) \in (\mathrm{GL}_1)^r \times \mathrm{SO}_{2(n-r)+1}$$

and each

$$v = (x_1, x_2, \cdots, x_{r-1}; u) \in \mathbb{C}^{\oplus(r-1)} \oplus \mathbb{C}^{\oplus(2(n-r)+1)},$$

we have

$$g \circ v = (t_1 t_2^{-1} x_1, t_2 t_3^{-1} x_2, \cdots, t_{r_1} t_r^{-1} x_{r-1}; t_r u h^{-1}). \tag{2.4}$$

Moreover, the nilpotent orbit $\mathcal{O}_{[2r+1,1^{2(n-r)}]}$ corresponds to the Zariski open orbit of $(\mathrm{GL}_1)^r \times \mathrm{SO}_{2(n-r)+1}$ on the affine space $\mathbb{C}^{\oplus(r-1)} \oplus \mathbb{C}^{\oplus(2(n-r)+1)}$. It is clear from the action given by (2.4) that we may choose the representatives of the Zariski open orbit to have the following form

$$v = (1, 1, \cdots, 1; u) \in \mathbb{C}^{\oplus(r-1)} \oplus \mathbb{C}^{\oplus(2(n-r)+1)},$$

that is, $x_1 = \cdots = x_{r-1} = 1$. Since the set of such vectors is stable under the the subgroup

$$(\mathrm{GL}_1)^\Delta \times \mathrm{SO}_{2(n-r)+1}$$

of $(\mathrm{GL}_1)^r \times \mathrm{SO}_{2(n-r)+1}$, where $(\mathrm{GL}_1)^\Delta$ is GL_1, diagonally embedded inside $(\mathrm{GL}_1)^r$, this induces further the action of $\mathrm{GL}_1 \times \mathrm{SO}_{2(n-r)+1}$ on the affine space $\mathbb{C}^{\oplus(2(n-r)+1)}$ as follows: for each $(t; h) \in \mathrm{GL}_1 \times \mathrm{SO}_{2(n-r)+1}$ and each $u \in \mathbb{C}^{\oplus(2(n-r)+1)}$, we have

$$(t; h) \circ u = tuh^{-1}. \tag{2.5}$$

Consider the partial polarization of V_{2n+1}

$$V_{2n+1} = X_r \oplus V_{2(n-r)+1} \oplus X_r^*$$

as in (2.3), with $V_{2(n-r)+1} \cong \mathbb{C}^{\oplus(2(n-r)+1)}$. By Witt's theorem, we know that the affine space $V_{2(n-r)+1}$ with the group action of $\mathrm{GL}_1 \times \mathrm{SO}_{2(n-r)+1}$ is a regular prehomogeneous vector space and decomposes over $\mathbb{C}$ (or $\bar{k}$) into three orbits

$$V_{2(n-r)+1} = \{0\} \cup Z_0 \cup Z_1 \tag{2.6}$$

where Z_0 is the set of all nonzero isotropic vectors with respect to the bilinear form and Z_1 is the set of all anisotropic vectors with respect to the bilinear form. It is clear that Z_1 is the Zariski open orbit.

It is important to remark that the above discussion works as well if one replaces the ground field $\mathbb{C}$ by the algebraic closure of a number field or a local field of characteristic zero.

In the sequel, we denote by k a number field or a local field of characteristic zero and by $\bar{k}$ the algebraic closure of k.

We will use the notion of k-stable orbits to discuss the k-rationality of the nilpotent orbits and parameterization. For a detailed discussion of k-stable orbits, we refer to [**Ktw82**]. The k-stable nilpotent orbits associated to the $\bar{k}$-nilpotent orbit $\mathcal{O}_{[2r+1,1^{2(n-r)}]}(\bar{k})$ are in one to one correspondence with the k-stable orbits associated to the $\bar{k}$-orbit $Z_1(\bar{k})$. Note that the k-rational orbit $Z_0(k)$ is k-stable, since Witt's theorem holds for number fields or local fields of characteristic zero.

If $r \neq n$, in the Zariski open $\bar{k}$-orbit $Z_1(\bar{k})$, one has the following parameterization of the k-stable orbits of $\mathrm{GL}_1(k) \times \mathrm{SO}_{2(n-r)+1}(k)$ in terms of the square classes of $k^\times$:

$$Z_1(k) = \cup_{\lambda \in k^\times/(k^\times)^2} Z_1^\lambda, \tag{2.7}$$

where the set Z_1^λ is defined by

$$Z_1^\lambda := \{v \in V_{2(n-r)+1}(k) \mid (v,v) = \lambda \mod (k^\times)^2\}.$$

It is clear that the set Z_1^λ is a single k-rational orbit of $\mathrm{GL}_1(k) \times \mathrm{SO}_{2(n-r)+1}(k)$. Note that $Z_0(k)$ is a single k-rational orbit of $\mathrm{GL}_1(k) \times \mathrm{SO}_{2(n-r)+1}(k)$.

If $r = n$, then $V_{2(n-r)+1}$ is one dimensional and has decomposition

$$V_1(k) = \{0\} \cup k^\times. \tag{2.8}$$

Hence $Z_1(k) = k^\times$ is a single $\mathrm{GL}_1(k)$-orbit. Therefore the k-stable regular nilpotent orbit $\mathcal{O}_{[2n+1]}(k)$ is a single k-rational orbit. Further discussion has been given in [**JS06**]

2.3. Fourier coefficients and periods. Let ψ_0 be a nontrivial additive character of $\mathbb{A}$ which is trivial on k. For each nilpotent orbit $\mathcal{O}$ in the Lie algebra $\mathfrak{g}$ of G, one can define a character $\psi_{\mathcal{O}}$ of $N_2(\mathbb{A})$, which is trivial on $N_2(k)$. When $\mathfrak{n}_1 = 0$, then we have $N_2 = N_{\mathcal{O}}$. In this case, $\psi_{\mathcal{O}}$ is a nontrivial character of $N_{\mathcal{O}}(\mathbb{A})$ and the subgroup $M_{\mathcal{O}}^\circ(\mathbb{A})$ stabilizes the character $\psi_{\mathcal{O}}$. When $\mathfrak{n}_1 \neq 0$, as explained explicitly in [**MW87**], the character $\psi_{\mathcal{O}}$ defines a character, which is denoted also by $\psi_{\mathcal{O}}$, on $N_2'(\mathbb{A})$, where N_2' is a subgroup of $N_{\mathcal{O}}$ such that $N_{\mathcal{O}}/\ker(\psi_{\mathcal{O}})$ is a Heisenberg group with the center $N_2'/\ker(\psi_{\mathcal{O}})$.

We will separate our discussion for these two cases to define Fourier coefficients and periods of automorphic forms φ on $G(\mathbb{A})$.

When $\mathfrak{n}_1 = 0$, this implies that $\mathfrak{n}_{2i+1} = 0$ for all $i \geq 0$, the Fourier coefficient of φ attached to the nilpotent orbit $\mathcal{O}$ or the $\mathcal{O}$-Fourier coefficient of φ is defined by the following integral

$$\mathcal{F}^{\psi_{\mathcal{O}}}(g,\varphi) := \int_{N_{\mathcal{O}}(k)\backslash N_{\mathcal{O}}(\mathbb{A})} \varphi(ng)\psi_{\mathcal{O}}^{-1}(n)dn. \tag{2.9}$$

It is easy to see that the $\mathcal{O}$-Fourier coefficient $\mathcal{F}^{\psi_{\mathcal{O}}}(g,\varphi)$ of φ has left invariant property with respect to the k-rational points $M_{\mathcal{O}}^\circ(k)$ of $M_{\mathcal{O}}^\circ$. Hence it defines an automorphic form on $M_{\mathcal{O}}^\circ(\mathbb{A})$.

Let H be a reductive closed subgroup of $M_{\mathcal{O}}^\circ$. Take an automorphic form ϕ on $H(\mathbb{A})$ and define formally the period integral

$$\mathcal{P}_{\mathcal{O},H}(\varphi,\phi,\psi_{\mathcal{O}}) := \int_{H(k)\backslash H(\mathbb{A})} \mathcal{F}^{\psi_{\mathcal{O}}}(h,\varphi)\phi(h)dh. \tag{2.10}$$

Note that this integral may be divergent. A general problem is to figure out, for a given φ on $G(\mathbb{A})$, what kind of automorphic forms ϕ on $H(\mathbb{A})$ will make the integral nonzero. If the integral is nonzero, one calls that the automorphic form φ is distinguished with respect to the data $(\mathcal{O}, H, \psi_{\mathcal{O}}, \phi)$, or has a generalized Gelfand-Graev model of type $(\mathcal{O}, H, \psi_{\mathcal{O}}, \phi)$.

When $\mathfrak{n}_1 \neq 0$, one define the $\mathcal{O}$-Fourier-Jacobi coefficient of an automorphic forms φ on $G(\mathbb{A})$ by the following integral

$$\mathcal{F}^{\psi_{\mathcal{O}}}(x,\varphi,\Phi) := \int_{N_{\mathcal{O}}(k)\backslash N_{\mathcal{O}}(\mathbb{A})} \varphi(nx)\theta_{\Phi}^{\psi_{\mathcal{O}}}(\bar{n}x)dn \tag{2.11}$$

Some modifications are needed here. It follows also that the $\mathcal{O}$-Fourier-Jacobi coefficient of φ is an automorphic form on $M_{\mathcal{O}}^{\circ}(\mathbb{A})$. Similarly, for an automorphic form ϕ on $H(\mathbb{A})$, one may formally define a period integral

$$\mathcal{P}_{\mathcal{O},H}(\varphi,\phi,\psi_{\mathcal{O}},\Phi) := \int_{H(k)\backslash H(\mathbb{A})} \mathcal{F}^{\psi_{\mathcal{O}}}(h,\varphi,\Phi)\phi(h)dh. \tag{2.12}$$

Assume that the integral converges. If it is nonzero, then we call that the automorphic form φ is distinguished with respect to the data $(\mathcal{O},H,\psi_{\mathcal{O}},\phi)$, or has a generalized Gelfand-Graev model of type $(\mathcal{O},H,\psi_{\mathcal{O}},\phi)$.

3. Some Problems

3.1. Global problems. One of the basic problems in general representation theory is the classification problem. A classification theorem may be very formal. This means that one can classify representations in terms of a certain set of invariants. However, in the modern theory of automorphic forms, representation theory works as a principle to conceptually understand problems from number theory and arithmetic.

Periods of automorphic forms provides the automorphic description of the existence of certain invariant functionals on the space of irreducible automorphic representations. The occurrence of the certain periods may imply certain arithmetic properties of automorphic representations. Jacquet's theory to characterize quadratic base change of irreducible cuspidal automorphic representations of $\mathrm{GL}_n(\mathbb{A}_E)$ from $\mathrm{GL}_n(\mathbb{A}_F)$ is the original example, where E/F is a quadratic extension of fields. By combining the idea of Harder-Langlands-Rapoport ([**HLR86**]), Jacquet's theory asserts that an irreducible cuspidal automorphic representation Π of $\mathrm{GL}_n(\mathbb{A}_E)$ is a base change from $\mathrm{GL}_n(\mathbb{A}_F)$ if and only if Π is distinguished with respect to a certain unitary subgroup of $\mathrm{GL}_n(E)$, that is, the representation Π has a nontrivial period defined over the unitary group ([**Jcq03**], [**JY90**]).

In general, we may propose one qualitative problem and one quantitative problem for periods of automorphic forms defined in the previous section.

PROBLEM 3.1. *Find implications of the existence of nontrivial periods to the structure of the given irreducible discrete automorphic representations in the sense of Langlands functoriality or other types of lifting theories.*

PROBLEM 3.2. *Find explicit formula for a given period as defined in Section 2.3 in terms of special values or the residue of a certain automorphic L-function or other types of invariants attached to the automorphic forms.*

More concretely, one may study the following problem. For a given irreducible discrete series automorphic representation π of $G(\mathbb{A})$, which may not be cuspidal, and for a k-rational nilpotent orbit $\mathcal{O}$, one can define the $\mathcal{O}$-Fourier coefficient $\mathcal{F}^{\psi_{\mathcal{O}}}(g,\varphi_\pi)$ or $\mathcal{O}$-Fourier-Jacobi coefficient $\mathcal{F}^{\psi_{\mathcal{O}}}(x,\varphi_\pi,\Phi)$ of an automorphic form φ_π in the space of π as given in the previous section. It is also a basic problem to understand the spectral decomposition of $\mathcal{F}^{\psi_{\mathcal{O}}}(g,\varphi_\pi)$ or $\mathcal{F}^{\psi_{\mathcal{O}}}(x,\varphi_\pi,\Phi)$ as automorphic representation of $M_{\mathcal{O}}^{\circ}(\mathbb{A})$.

PROBLEM 3.3. *For a given irreducible discrete series automorphic representation π of $G(\mathbb{A})$ and a k-rational nilpotent orbit $\mathcal{O}$, find the discrete part of the spectral decomposition of $\mathcal{F}^{\psi_{\mathcal{O}}}(g, \varphi_\pi)$ or $\mathcal{F}^{\psi_{\mathcal{O}}}(x, \varphi_\pi, \Phi)$ as automorphic representation of $M^\circ_{\mathcal{O}}(\mathbb{A})$.*

Then one may ask how to use the information obtained from Problem 3.3 to characterize the structure of π in the sense of Problem 3.1. More practically, one may choose a particular irreducible discrete series automorphic representation σ of $H(\mathbb{A})$, where H is a big reductive algebraic subgroup of $M^\circ_{\mathcal{O}}$, so that the occurrence of σ in the spectral decomposition of $\mathcal{F}^{\psi_{\mathcal{O}}}(g, \varphi_\pi)$ or $\mathcal{F}^{\psi_{\mathcal{O}}}(x, \varphi_\pi, \Phi)$ when restricted to $H(\mathbb{A})$ should be enough to characterize a certain structure of π. In many cases, the particular representation σ can be chosen to be the trivial representation of $H(\mathbb{A})$ or one-dimensional automorphic character of $H(\mathbb{A})$.

PROBLEM 3.4. *Let π be an irreducible discrete series automorphic representation of $G(\mathbb{A})$. For a given k-rational nilpotent orbit $\mathcal{O}$, find a reductive algebraic subgroup H of $M^\circ_{\mathcal{O}}$ and a particular irreducible automorphic representation σ of $H(\mathbb{A})$, such that the occurrence of σ in the spectral decomposition of $\mathcal{F}^{\psi_{\mathcal{O}}}(g, \varphi_\pi)$ or $\mathcal{F}^{\psi_{\mathcal{O}}}(x, \varphi_\pi, \Phi)$ when restricted to $H(\mathbb{A})$ can be used to characterize a certain structure of π in the sense of Problem 3.1.*

REMARK 3.5. *After we introduce the notion of automorphic L-functions, we will discuss the relations between the periods of automorphic representations and the special values or residues of certain automorphic L-functions. Some connections to arithmetic theory will also be mentioned later.*

3.2. Local problems. One of the main features in the modern theory of automorphic forms and representations is to consider the basic local-global principle from number theory in the content of representations of local compact topological groups. As an important invariant for automorphic representations, it is natural to consider the local version of periods of automorphic forms.

It is a basic theorem that an irreducible automorphic representation π of $G(\mathbb{A})$ can be expressed as a restricted tensor product over all local places v of k of irreducible admissible representations π_v of $G(k_v)$. A period of automorphic representation π produces a certain quasi-invariant functional over the space of the local component π_v for all local places v of k. For simplicity, we consider only here the case when the nilpotent orbit $\mathcal{O}$ is even.

For a given irreducible discrete series automorphic representation π of $G(\mathbb{A})$ and for an even k-rational nilpotent orbit $\mathcal{O}$, if the $\mathcal{O}$-Fourier coefficient $\mathcal{F}^{\psi_{\mathcal{O}}}(g, \varphi_\pi)$ is nonzero, it produces a nontrivial functional in the following space

$$\mathrm{Hom}_{N_{\mathcal{O}}(\mathbb{A})}(\pi, \mathbb{C}_{\psi_{\mathcal{O}}}) \cong \prod_v \mathrm{Hom}_{N_{\mathcal{O}}(k_v)}(\pi_v, \mathbb{C}_{\psi_{\mathcal{O},v}}). \tag{3.1}$$

It follows that for each local place v of k, there exists a nontrivial functional in the space

$$\mathrm{Hom}_{N_{\mathcal{O}}(k_v)}(\pi_v, \mathbb{C}_{\psi_{\mathcal{O},v}}). \tag{3.2}$$

This last space has a natural structure of $M^\circ_{\mathcal{O}}(k_v)$-module. The first local problem can be stated as follows.

PROBLEM 3.6. *Let π_v be an irreducible admissible representation of $G(k_v)$, and let σ_v be an irreducible admissible representation of a big reductive subgroup H of*

$M_{\mathcal{O}}^{\circ}(k_v)$. *Find the dimension of the space*

$$\mathrm{Hom}_{N_{\mathcal{O}}(k_v)}(\pi_v, \mathbb{C}_{\psi_{\mathcal{O},v}})^{(H(k_v),\sigma_v)}$$

which is the $(H(k_v), \sigma_v)$*-isotypic part of* $\mathrm{Hom}_{N_{\mathcal{O}}(k_v)}(\pi_v, \mathbb{C}_{\psi_{\mathcal{O},v}})$.

It is clear that the dimension of the space in Problem 3.6 is independent of the choices of the characters $\psi_{\mathcal{O}}$ attached to the k-stable orbit of $\mathcal{O}$. If a k-rational nilpotent orbit $\mathcal{O}$ has the property that for an irreducible admissible representation σ_v of $H(k_v)$, the dimension of the space in Problem 3.6 is at most one for all irreducible admissible representations π_v of $G(k_v)$, then we say that the nilpotent orbit $\mathcal{O}$ has the uniqueness property of type $(H, \sigma_v$ or has (H, σ_v)-uniqueness. If such a uniqueness property holds for all σ_v, we say that the orbit $\mathcal{O}$ has the H-uniqueness. If such uniqueness holds only for unramified representations and characters, we say that $\mathcal{O}$ has the corresponding unramified uniqueness. Some conjectures of Gross and Prasad for local Bessel models of orthogonal groups were discussed in [**GP92**] and [**GP94**]. Some new evidences and examples for the local Gross-Prasad conjecture can be found in [**GR05**].

PROBLEM 3.7. *Classify all nilpotent orbits* $\mathcal{O}$ *with* $M_{\mathcal{O}}^{\circ}$*-unramified uniqueness, that is, for all unramified irreducible admissible representations* π_v *of* $G(k_v)$*, all unramified irreducible admissible representations of* $M_{\mathcal{O}}^{\circ}(k_v)$*, and for unramified characters* $\psi_{\mathcal{O},v}$*, the dimension of the following space*

$$\mathrm{Hom}_{N_{\mathcal{O}}(k_v)}(\pi_v, \mathbb{C}_{\psi_{\mathcal{O},v}})^{(M_{\mathcal{O}}^{\circ}(k_v),\sigma_v)}$$

is at most one.

Of course, it is an interesting problem to classify the nilpotent orbits with the unramified uniqueness property, but without the uniqueness property in general. From the point of view of the global theory, it is natural to consider the following problem.

PROBLEM 3.8. *For a given* k*-rational nilpotent orbit* $\mathcal{O}$ *with the unramified uniqueness property, and for a given particular irreducible admissible representation* σ_v *of* $M_{\mathcal{O}}^{\circ}(k_v)$*, characterize all irreducible admissible representation* π_v *of* $G(k_v)$ *with the property that the dimension of the space*

$$\mathrm{Hom}_{N_{\mathcal{O}}(k_v)}(\pi_v, \mathbb{C}_{\psi_{\mathcal{O},v}})^{(M_{\mathcal{O}}^{\circ}(k_v),\sigma_v)}$$

is one.

If $\mathcal{O}$ is a k-rational nilpotent orbit with the uniqueness property, then at unramified local place, it is expected that the nontrivial functional attached to $\mathcal{O}$ does not vanish at the unramified vectors of π_v. In this case one says that the unramified vectors are the test vectors for the nontrivial functional attached to the nilpotent orbit $\mathcal{O}$ at the unramified local place. However, at the ramified local places, it is highly desired to develop he theory of local new vectors as test vectors for the nonvanishing of the local functionals. Some significant applications of the theory of local new vectors can be found in the study of special values of automorphic L-functions. See, for instance, the recent development of the Gross-Zagier formula by S.-W. Zhang ([**Z01**]) for some detailed discussion. B. Roberts and R. Schmidt has developed a theory of local new vectors for certain ramified representations of $\mathrm{GSp}_4(k_v)$ ([**RS03**]).

The uniqueness of local Whittaker models has been used by F. Shahidi ([**Sh88**] and [**Sh90**]) to construct his local coefficients, which can be related to the local gamma factors of generic representations and to the local Plancherel measures, and to some other problems in representations of p-adic groups. This work has been generalized by S. Friedberg and D. Goldberg ([**FG99**]) to deal with some nongeneric representations of p-adic groups by introducing the local Bessel models. It is expected that the whole theory should have the same conclusion for the other models attached to the nilpotent orbits with the uniqueness property.

For applications to automorphic L-functions, the uniqueness of Whittaker models has played an indispensable role in the Rankin-Selberg method to establish the basic analytic properties as conjectured by Langlands for families of automorphic L-functions ([**Bmp05**]). The uniqueness of Bessel models attached to the subregular nilpotent orbit of of $\mathfrak{so}_{2n+1}$ was first introduced in [**N75**] to study the standard automorphic L-functions for irreducible generic cuspidal automorphic representations of SO_{2n+1}. This work was generalized in [**G90**] and [**S93**] to the Bessel models attached to nilpotent orbits of $\mathfrak{so}_{2n+1}$ parameterized by partitions $[2r+1, 1^{2(n-r)}]$ with $r = 0, 1, \cdots, n$, for irreducible generic cuspidal automorphic representations of SO_{2n+1}. In [**GPSR97**], the whole theory was established without assuming the irreducible cuspidal automorphic representations being generic. The uniqueness of Bessel models for classical groups has been considered by S. Rallis. Some of his work in this direction can be found in [**GPSR97**].

3.3. Local-global principle. One of the main concerns in the modern theory of automorphic forms is the local-global principle for certain properties which can be formulated for automorphic representations and for representations at all local places.

Let k be a number field and k_v be the local field at a local place v of k. Let π be an irreducible discrete series automorphic representation of $G(\mathbb{A})$. For simplicity, consider an even k-rational nilpotent orbit $\mathcal{O}$ with the uniqueness property, and an irreducible automorphic representation σ of $M_{\mathcal{O}}^{\circ}(\mathbb{A})$. The condition for the nonvanishing of the period $\mathcal{P}_{\mathcal{O}}(\varphi_\pi, \phi_\sigma, \psi_{\mathcal{O}})$ as defined in (2,10) with $H = M_{\mathcal{O}}^{\circ}$ is called a global period condition, and is denoted by **(gpc)**. The nonzero period $\mathcal{P}_{\mathcal{O}}(\varphi_\pi, \phi_\sigma, \psi_{\mathcal{O}})$ defines a nonzero element $\ell_{\mathcal{O}}$ in the following space

$$\mathrm{Hom}_{N_{\mathcal{O}}(\mathbb{A})}(\pi, \mathbb{C}_{\psi_{\mathcal{O}}})^{(M_{\mathcal{O}}^{\circ}(\mathbb{A}),\sigma)}. \tag{3.3}$$

By the uniqueness property for $\mathcal{O}$, the nonzero element $\ell_{\mathcal{O}}$ is unique up to scalar. By (3.1), the nonzero element $\ell_{\mathcal{O}}$ define for each local place v of k a nonzero element $\ell_{\mathcal{O},v}$, which is unique up to scalar, in the following space

$$\mathrm{Hom}_{N_{\mathcal{O}}(k_v)}(\pi_v, \mathbb{C}_{\psi_{\mathcal{O},v}})^{(M_{\mathcal{O}}^{\circ}(k_v),\sigma_v)}. \tag{3.4}$$

The condition for the nonvanishing of the space in (3.4) with the assumption that at all unramified local places v, the element $\ell_{\mathcal{O},v}$ does not vanish at the unramified vectors in π_v, is call a local period condition, and is denoted by **(lpc)**.

PROBLEM 3.9 (Local-Global Principle). *For a k-rational nilpotent orbit $\mathcal{O}$, find a set of conditions depending on π and σ, such that* **(lpc)** *implies* **(gpc)**, *i.e.*

$$\textbf{(lpc)} \oplus \textbf{ what conditions} \cong \textbf{(gpc)}.$$

In reality, it is useful to find the *sufficient conditions* for

$$\textbf{(lpc)} \oplus \textbf{ what conditions} \rightarrow \textbf{(gpc)}.$$

We remark that the conditions which make the equivalence in Problem 3.9 hold are expected to have relation with special values or poles of certain automorphic L-functions. We will discuss some typical cases after we introduce the notion of automorphic L-functions in the next section.

More precise relation between the local periods and the global periods is expected to be an identity which shows that the global period and a certain normalized product of local periods over all the local places are proportional. The proportional constant are expected to be a certain special value or residue of a certain automorphic L-function, which is related to the conditions in the local-global principle. Some interesting examples will be discussed later. It is well-known that the uniqueness of Whittaker models implies that the global Whittaker function attached to an irreducible generic automorphic representation can be expressed as a product of the corresponding local Whittaker functions over all local places.

4. Automorphic L-functions

4.1. L-functions. The simplest L-function is the well-known Riemann zeta function

$$\zeta(s) = \sum_{n=1}^{\infty} \frac{1}{n^s}.$$

It is almost elementary to show that $\zeta(s)$ converges absolutely for the real part of s greater than one, has a meromorphic continuation to the whole complex plane $\mathbb{C}$, and satisfies a functional equation which relates s to $1-s$. Any information about $\zeta(s)$ in the vertical strip $0 \leq Re(s) \leq 1$ will be sensitive, mysterious, important, etc. A recent survey by S. Gelbart and S. Miller ([**GM04**]) provides a good source of discussion on this topics.

R. Langlands introduced the notion of automorphic L-functions attached to automorphic representations as follows. Let ${}^LG$ be the Langlands dual group of the reductive group G. Let π be an irreducible, unitary, cuspidal, automorphic representation of $G(\mathbb{A})$ and ρ be a finite-dimensional complex representation of ${}^LG$. The automorphic L-function attached to the pair (π, ρ) is given by the following Euler product

$$L(s, \pi, \rho) = \prod_v L(s, \pi_v, \rho)$$

if we write $\pi = \otimes_v \pi_v$. For almost all local finite places, the local components π_v are unramified. By Satake's theory, the unramified π are uniquely up to isomorphism determined by a semisimple conjugate class t_{π_v} in the dual group ${}^LG$. At the unramified place v, the local L-factor $L(s, \pi_v, \rho)$ is defined to be

$$L(s, \pi_v, \rho) := \det(I - \rho(t_{\pi_v}) N_v^{-s})^{-1},$$

where N_v is the cardinal of the residual field at v. For the other local places, the local L-factor may be defined via the local Langlands correspondence, or via the Langlands-Shahidi method, or via the Rankin-Selberg method. See [**GSh88**] for more discussions.

THEOREM 4.1 (Langlands ([**GSh88**])). *For any given pair (π, ρ) given as above, the automorphic L-function $L(s, \pi, \rho)$ converges absolutely for the real part of s large.*

CONJECTURE 4.2 (Langlands ([**GSh88**])). *For any given pair (π, ρ) given as above, the automorphic L-function $L(s, \pi, \rho)$ has meromorphic continuation to the whole complex plane $\mathbb{C}$ and satisfies a functional equation which related to s to $1-s$.*

It follows that there is a positive real number S_0, such that $L(s, \pi, \rho)$ will be sensitive, mysterious, and important when s lies in the vertical strip

$$\frac{1}{2} - s_0 \leq Re(s) \leq \frac{1}{2} + s_0.$$

REMARK 4.3. (1) *The Langlands conjecture has been verified for many cases, by either the Langlands-Shahidi method or by the Rankin-Selberg method. See* [**GSh88**] *and* [**Bmp05**] *for detailed discussions.*

(2) *If π is generic, i.e. has a nonzero Whittaker-Fourier coefficient, then one expects that $\frac{1}{2} + s_0 = 1$. Shahidi proved in* [**Sh88**] *for all automorphic L-functions in the Langlands-Shahidi list have the property that the range of the absolute convergence can be made for the real part of s greater than two, which leads a weaker version of the generalized Ramanujan conjecture.*

(3) *The special values or the poles of automorphic L-functions at real value s satisfying*

$$\frac{1}{2} \leq s \leq \frac{1}{2} + s_0$$

should have something to do with the functorial structure of or the arithmetic significance of π.

We will discuss in more detail about the nonvanishing of special values or the existence of poles of automorphic L-functions $L(s, \pi, \rho)$ in terms of the functorial structure of π or the nonvanishing of periods of the automorphic forms in π. One of the most natural links to accomplish this is to use the Langlands theory of Eisenstein series, which was used by Langlands and then by Shahidi to prove the validity of the Langlands conjecture for a big family of automorphic L-functions.

4.2. Residual representations. From Langlands' theory of Eisenstein series ([**L76**] and [**MW95**]), the nonvanishing condition of certain automorphic L-functions is equivalent to the condition for the existence of a residue of certain Eisenstein series.

For simplicity of discussion, we may assume that G is quasi-split or even split over k. Let $P = MN$ be a standard maximal parabolic k-subgroup of G. Let $K = \prod_v K_v$ be a standard maximal compact subgroup of $G(\mathbb{A})$ such that

$$G(\mathbb{A}) = P(\mathbb{A})K$$

is the Iwasawa decomposition of $G(\mathbb{A})$.

Let σ be an irreducible cuspidal automorphic representation of $M(\mathbb{A})$. Consider V_σ-valued smooth functions ϕ_σ over $G(\mathbb{A})$ satisfying following properties:

(1) for a fixed $g \in G(\mathbb{A})$, $k \mapsto \phi_\sigma(gk)$ generates a finite-dimensional representations of K;
(2) for a fixed $g \in G(\mathbb{A})$, $m \mapsto \phi_\sigma(mg)$ is a smooth vector in V_σ and the function ϕ_σ satisfies $\phi_\sigma(ng) = \phi_\sigma(g)$ for all $n \in N(\mathbb{A})$.

Define

$$\Phi(g; s, \phi_\sigma) := \phi_\sigma(g) \exp < s + \rho_P, H_P(g) >$$

where ρ_P is the half of the sum of all positive roots with respect to the center of M, the parameter s is normalized as in [**Sh88**], and H_P is the Harish-Chandra map with respect to P. Then we define Eisenstein series of $G(\mathbb{A})$ attached to the cuspidal datum (P,σ) as

$$E(g;s,\phi_\sigma) := \sum_{\gamma\in P(k)\backslash G(k)} \Phi(\gamma g;s,\phi_\sigma). \tag{4.1}$$

THEOREM 4.4 (Langlands ([**L76**] and [**MW95**]). *For any given cuspidal datum (P,σ), the Eisenstein series $E(g;s,\phi_\sigma)$ converges absolutely for the real part of s large, has meromorphic continuation to the whole complex plane $\mathbb{C}$ (with finitely many poles at the positive real line after suitable normalization), and satisfies a functional equation relating the value s to the value $-s$.*

REMARK 4.5. *It is expected that the structure of the residual representations of $E(g;s,\phi_\sigma)$ at a real number $s_0>0$ should have direct relations with the structure of the cuspidal datum (P,σ). This turns out to be quite useful.*

From the Langlands theory of the constant terms of Eisenstein series, the poles of Eisenstein series can be detected in terms of explicit calculation of the constant terms of the Eisenstein series. The constant term of the Eisenstein series $E(g;s,\phi_\sigma)$ along a standard parabolic subgroup P' is always zero unless $P'=P$ ([**MW95**], II.1.7). In this case the constant term can be expressed as

$$\begin{aligned} E_P(g;s,\phi_\sigma) &= \int_{N(k)\backslash N(\mathbb{A})} E(ug;s,\phi_\sigma)dn \\ &= \Phi(g;s,\phi_\sigma)+\mathcal{M}(w_0,s)(\Phi(\cdot;s,\phi_\sigma))(g) \end{aligned} \tag{4.2}$$

where w_0 the longest Weyl element in the representatives of the double coset decomposition $W_M\backslash W_G/W_M$ of the Weyl groups.

It follows from the Langlands theory of Eisenstein series that $E(g;s,\phi_\sigma)$ has a pole at $s=s_0$ if and only if the term $\mathcal{M}(w_0,s)(\Phi(\cdot;s,\phi_\sigma))$ above has a pole at $s=s_0$ for some holomorphic (or standard) section $\Phi(g;s,\phi_\sigma)$ as defined before. Following the standard argument, if we consider factorizable sections

$$\Phi(\cdot;s,\phi_\sigma)=\otimes_v\Phi_v(\cdot;s,\phi_{\sigma_v}),$$

then we have

$$\mathcal{M}(w_0,s)(\Phi(\cdot;s,\phi_\sigma))=\prod_v \mathcal{M}_v(w_0,s)(\Phi_v(\cdot;s,\phi_{\sigma_v})).$$

By the Langlands-Shahidi theory ([**L71**] and [**Sh88**]) we have

$$\mathcal{M}(w_0,s)=\frac{\prod_l L(ls,\sigma,r_l)}{\prod_l L(ls+1,\sigma,r_l)}\cdot\prod_v \mathcal{N}_v(w_0,s), \tag{4.3}$$

where $\mathcal{N}_v(w_0,s)$ is the normalized intertwining operator

$$\mathcal{N}_v(w_0,s)=\frac{1}{r(s,\pi_v,\sigma_v,w_0)}\cdot\mathcal{M}_v(w_0,s).$$

Here the function $r(s,\pi_v,\sigma_v,w_0)$ is equal to

$$\frac{\prod_l L(ls,\sigma_v,r_l)}{\prod_l L(ls+1,\sigma_v,r_l)\prod_l \epsilon(ls,\sigma_v,r_l,\psi_v)}.$$

By a conjecture of Shahidi [**Sh90**], it is expected that $\mathcal{N}_v(w_0,s)(\Phi_v(\cdot;s,\phi_{\sigma_v}))$ is holomorphic and nonzero for $Re(s)\geq 0$ if σ_v is tempered. It has been checked for many cases when σ_v is tempered and generic, and when G is a classical group ([**CKPSS04**]).

By assuming Shahidi's conjecture on $\mathcal{N}_v(w_0, s)(\Phi_v(\cdot; s, \phi_{\sigma_v}))$, we know that the existence of a pole at $s = s_0$ of $E(g; s, \phi_\sigma)$ is equivalent to the existence of a pole at $s = s_0$ of the product of L-functions

$$\prod_l L(ls, \sigma, r_l).$$

Following from Shahidi ([**Sh88**]), if σ is generic, then the L-function $L(s, \sigma, r_l)$ is holomorphic for $Re(s) > 2$, and by the generalized Ramanujan conjecture, $L(s, \sigma, r_l)$ is holomorphic for $Re(s) > 1$. If we write

$$\prod_l L(ls, \sigma, r_l) = L(s, \sigma, r_1)L(2s, \sigma, r_2) \prod_{l \geq 3} L(ls, \sigma, r_l).$$

It is expected that $\prod_{l \geq 3} L(ls, \sigma, r_l)$ should be holomorphic for $Re(s) \geq \frac{1}{2}$. Then the existence of a pole at $s = s_0$ of the product of L-functions

$$\prod_l L(ls, \sigma, r_l)$$

should be equivalent to either that $L(s, \sigma, r_1)$ has a pole at $s = 1$ (Case (1)) or that $L(s, \sigma, r_1)$ does not vanish at $s = \frac{1}{2}$ and $L(s, \sigma, r_2)$ has a pole at $s = 1$ (Case ($\frac{1}{2}$)). Note that the so called (Case (0)) is the one corresponding to the holomorphicity of the product $\prod_l L(ls, \sigma, r_l)$.

REMARK 4.6. *One expects that Case* (1) *and Case* ($\frac{1}{2}$) *should not occur at the same time, but it is possible for both not to occur (when Case* (0) *occurs. For classical groups and generic σ, this follows from the explicit Langlands functorial lifting from classical groups to the general linear group. In particular, the structures of σ related to these two cases are essentially different.*

4.3. Model-comparison. In the previous subsection, the existence of the residual representation

$$\mathcal{E}_{s_0}(g; \phi_\sigma) := \text{Res}_{s=s_0} E(g; s, \phi_\sigma) \tag{4.4}$$

is expected to be characterized in terms of the existence of a pole at $s = s_0$ of the product of L-functions

$$L(s, \sigma, r_1)L(2s, \sigma, r_2) \tag{4.5}$$

assuming that σ is generic. The question in this subsection is to ask how to describe the structure of the residual representation $\mathcal{E}_{s_0}(g; \phi_\sigma)$ in terms of the structure of the cuspidal support σ, or vise versa. From a representation-theoretic point of view, structures of representations are usually characterized by means of certain invariants attached to the representations. For automorphic representations, it is natural to consider period integrals in order to distinguish automorphic representation from each other. By reciprocity, that an automorphic representation has a certain nonzero period is equivalent to that the automorphic representation has a certain model. For example, it has a Whittaker model if the representation is generic.

When consider two automorphic representations which are related via a certain kind of transfers, say, Langlands functoriality, theta correspondence, or endoscopy lifting, it is natural to expect the two automorphic representations should share certain compatible models. This should be viewed as a version of the Langlands functoriality principle. In reality, one may expect an explicit formula which relates a certain model for one representation to a certain model for the other representation. This is what we called *Model-comparison identity*. Such an idea is not really new. One can find such model-comparison identities in the constructive theory of automorphic forms, such as the Langlands theory of constant terms of Eisenstein series, the Langlands-Shahidi method and the Rankin-Selberg method to study automorphic L-functions, explicit theta correspondences, the descent method (backward liftings) of automorphic forms, relative trace formula method, etc.

We consider in the following the relation between the cuspidal datum (P, σ) and the residual representation $\mathcal{E}_{s_0}(g; \phi_\sigma)$ of $G(\mathbb{A})$. It is the simplest case of Langlands functoriality from (M, σ) to $(G, \mathcal{E}_{s_0}(g; \phi_\sigma))$. The problem is: for a given (M, σ) and the value s_0, to find a suitable spherical subgroup H of G, an automorphic representation τ of $M \cap H$, and an automorphic representation $\mathcal{E}_0(h; \phi_\tau)$ of $H(\mathbb{A})$, such that there exists an identity which relates the period (called the outer period) of $\mathcal{E}_{s_0}(g; \phi_\sigma)$

$$\int_{H(k)\backslash H(\mathbb{A})} \mathcal{E}_{s_0}(h; \phi_\sigma)\mathcal{E}_0(h; \phi_\tau)dh \tag{4.6}$$

and the period (called the inner period) of σ

$$\int_{M\cap H(k)\backslash M\cap H(\mathbb{A})} \phi_\sigma(x)\phi_\tau(x)dx. \tag{4.7}$$

A Theorem one wishes to prove is

THEOREM 4.7. *The nonvanishing of outer period (4.6) is equivalent to the nonvanishing of inner period (4.7).*

REMARK 4.8. *Since the Langlands functoriality from σ to $\mathcal{E}_{s_0}(g; \phi_\sigma)$ is nothing but the 'parabolic induction', it is natural to find the data $(H, \tau, \mathcal{E}_0(h; \phi_\tau))$ through the Frobenius reciprocity law. More precisely, one has to study the orbital decomposition*

$$P\backslash G/H$$

and possible quasi-invariant distributions attached to each of the orbits. Note that the Rankin-Selberg method ([**Bmp05**]*) for automorphic L-functions uses the Zariski open orbit! However, in the model-comparison study, one pays more attention to the other lower dimensional orbits, the closed orbits in particular.*

Of course, one has to regularize the period integral to deal with the convergence problem. One can apply Arthur's truncation method to one of the residual automorphic forms in the integrand. Other methods of regularization may also apply, see [**JLR99**] *and* [**LR04**] *for instance.*

REMARK 4.9. *Some very interesting cases have been treated by this approach and have some striking consequences. A detailed discussion has be carried out in author's recent survey paper* [**Jng06**].

From the above arguments, it follows that the nonvanishing of the inner period in (4.7) implies the existence of the pole at s_0 of the product of L-functions in (4.5). The converse to this statement is known as a version of the conjecture of Gross-Prasad type ([**GP92**] *and* [**GP94**]*). We will discuss this issue in the next section.*

5. Examples

5.1. Special values of the Rankin-Selberg L-functions. In this section we discuss briefly the work joint with D. Ginzburg and S. Rallis on nonvanishing of the central value of the Rankin-Selberg L-functions ([**GJR04**] and [**GJR05**]). The main idea is to use the *Model-comparison identity* approach to link the periods of the generalized Gelfand-Graev type to the central value of the Rankin-Selberg L-functions. More detailed discussions and survey of related work has been written in [**Jng06**].

We may assume that $n \geq r$. We denote that $G_n := \mathrm{SO}_{2n+1}$ and $H_r := \mathrm{SO}_{2r}$. Let $\mathcal{O}_{n,r}$ be the nilpotent orbit of $\mathfrak{so}_{2n+1}$ corresponding to the odd orthogonal partition $[2r+1, 1^{2(n-r)}]$ in $\mathcal{P}_1(2n+1)$. Then we have

$$N_{n,r} = N_{\mathcal{O}_{n,r}} := \{v(u,x,z) = \begin{pmatrix} u & x & z \\ & I_{2r+1} & x^* \\ & & u^* \end{pmatrix}\} \subset G_n$$

where $u \in U_{n-r}$, which is the standard upper triangular unipotent subgroup of GL_{n-r}. Define a character $\psi_{n,r}$ of $N_{n,r}(\mathbb{A})$ by

$$\psi_{n,r}(v) := \psi_0(u_{1,2} + \cdots + u_{r-l-1,r-l})\psi_0(x_{r-l,l+1}).$$

It is clear that $\psi_{n,r}$ is trivial on $N_{n,r}(k)$. Denote the normalizer of $N_{n,r}$ in G_n by

$$M := N_{G_n}(N_{n,r}) \cong \mathrm{GL}_1^{n-r} \times \mathrm{SO}_{2r+1}.$$

Then the connected component of the stabilizer of $\psi_{n,r}$ in M is isomorphic to H_r.

When $n \leq l$, similar notations can be introduced, and we omit the details.

For $\varphi_\sigma \in V_\sigma$ and $\varphi_\tau \in V_\tau$, define $\psi_{n,r}$-Fourier coefficient of φ_σ to be

$$\mathcal{F}^{\psi_{n,r}}(\varphi_\sigma)(h) := \int_{N_{n,r}(k)\backslash N_{n,r}(\mathbb{A})} \varphi_\sigma(vh)\psi_{n,r}(v)^{-1}dv,$$

and the period is defined as follows:

$$\mathcal{P}_{n,r}(\varphi_\sigma, \varphi_\tau, \psi_{n,r}) := \int_{H_r(k)\backslash H_r(\mathbb{A})} \mathcal{F}^{\psi_{n,r}}(\varphi_\sigma)(h)\varphi_\tau(h)dh. \tag{5.1}$$

The following is one of the main results proved in [**GJR04**],[**GJR05**].

THEOREM 5.1 ([**GJR04**],[**GJR05**]). *If the period $\mathcal{P}_{r,l}(\varphi_\sigma, \varphi_\tau, \psi_{r,l})$ is not identically zero on the space $V_\sigma \otimes V_\tau$, then $L(\frac{1}{2}, \pi_1 \times \pi_2)$ is nonzero.*

In order to explain the model-comparison identity approach, we have to introduce the following two Eisenstein series:

(1) the Eisenstein series $E(g; s, \phi_{\pi_1\otimes\tau})$ on SO_{4r+2l} attached to the cuspidal data $\pi_1 \otimes \tau$ of $\mathrm{GL}_{2r} \times \mathrm{SO}_{2l}$;
(2) the Eisenstein series $E(g; s, \phi_{\pi_1\otimes\sigma})$ on SO_{6r+1} attached to the cuspidal data $\pi_1 \otimes \sigma$ of $\mathrm{GL}_{2r} \times \mathrm{SO}_{2r+1}$.

PROPOSITION 5.2 ([**GJR04**],[**GJR05**]). *The Eisenstein series $E(g; s, \phi_{\pi_1\otimes\tau})$ has a possible simple pole at $s = \frac{1}{2}$. The residue, denoted by $\mathcal{E}_{\frac{1}{2}}(g; \phi_{\pi_1\otimes\tau})$, is nonzero if and only if $L(\frac{1}{2}, \pi_1 \times \tau)$ is nonzero. The Eisenstein series $E(g; s, \phi_{\pi_1\times\sigma})$ has a simple pole at $s = 1$, and the residue is denoted by $\mathcal{E}_1(g; \phi_{\pi_1\times\sigma})$.*

Then the 'outer' period is given by

$$\mathcal{P}_{3r,2r+l}(\mathcal{E}_1(\cdot;\phi_{\pi_1\times\sigma}),\mathcal{E}_{\frac{1}{2}}(\cdot;\phi_{\pi_1\otimes\tau}),\psi_{3r,2r+l}).$$

The 'specialization' of Theorem 4.7 to this case is the following theorem.

THEOREM 5.3 ([**GJR04**],[**GJR05**]). *The nonvanishing of outer period*

$$\mathcal{P}_{3r,2r+l}(\mathcal{E}_1(\cdot;\phi_{\pi_1\times\sigma}),\mathcal{E}_{\frac{1}{2}}(\cdot;\phi_{\pi_1\otimes\tau}),\psi_{3r,2r+l})$$

is equivalent to the nonvanishing of the inner period $\mathcal{P}_{r,l}(\varphi_\sigma,\varphi_\tau,\psi_{r,l})$.

More detailed discussions on this Theorem and the relation to the Gross-Prasad conjecture have been given in [**Jng06**]. The analogue for unitary groups is the work in progress of Ginzburg and Jiang. The following example is a special case of the Gross-Prasad conjecture, which was proposed first by H. Jacquet.

5.2. Local-global principle. The most significant lower rank case of the Gross-Prasad conjecture is the case when $G = \mathrm{SO}_4$ and $H = \mathrm{SO}_3$. This can be viewed as the period associated to the zero orbit. In this case, it is a conjecture of Jacquet.

Let π_1, π_2, and π_3 be irreducible cuspidal automorphic representations of $\mathrm{GL}_2(\mathbb{A})$ with the product of the central characters being trivial, i.e.

$$\omega_{\pi_1}\cdot\omega_{\pi_2}\cdot\omega_{\pi_3}=1.$$

The conjecture of Jacquet asserts that the central value of the triple product L-function

$$L(\frac{1}{2},\pi_1\times\pi_2\times\pi_3)$$

is nonzero if and only if there is a unique quaternion algebra D over k such that the trilinear period integral

$$\int_{Z_{\mathrm{D}^\times}(\mathbb{A})\mathrm{D}^\times(k)\backslash\mathrm{D}^\times(\mathbb{A})}\varphi_{\pi_1^{\mathrm{D}}}(x)\varphi_{\pi_2^{\mathrm{D}}}(x)\varphi_{\pi_3^{\mathrm{D}}}(x)dx$$

is nonzero for a certain choice of $\varphi_{\pi_i^{\mathrm{D}}}$, $i=1,2,3$, where π_1^{D} is the image of π_i under the Jacquet-Langlands correspondence.

The conjecture of Jacquet was completely proved in [**HK04**]. In [**Jng98b**] and [**Jng01**], a different approach has been applied to the split period case of Jacquet's conjecture, which was reformulated in the form of the local-global principle. More recently, Boecherer, Furusawa, and Schultze-Pillot proved some special case for the relevant pair $(\mathrm{SO}_5,\mathrm{SO}_4)$ ([**BFSP04**]), and Ichino and Ikeda proved some special case for the relevant pair $(\mathrm{SO}_6,\mathrm{SO}_5)$ ([**II04**]). In those new cases, they obtain precise formula for the special value of the relevant automorphic L-functions.

THEOREM 5.4 ([**HK04**], [**Jng98b**], and [**Jng01**]). *Let* π_i $(i=1,2,3)$ *be irreducible cuspidal automorphic representations of* $GL_2(\mathbf{A})$ *with central character* ω_i $(i=1,2,3)$. *Assume that the product* $\prod_{i=1}^3\omega_i=1$. *Then the following are equivalent.*

(1) *For each local place* v, $Hom_{GL_2(F_v)}(\pi_{1,v}\otimes\pi_{2,v}\otimes\pi_{3,v},1)\neq 0$, *and the central value of the triple product L-function* $L(\frac{1}{2},\pi_1\otimes\pi_2\otimes\pi_3)$ *is nonzero.*

(2) *The (global) trilinear period*

$$\int_{Z(\mathbb{A})GL_2(F)\backslash GL_2(\mathbb{A})} \phi_{\pi_1}(x)\phi_{\pi_2}(x)\phi_{\pi_3}(x)dx$$

is nonzero for certain choice of cusp forms $\phi_{\pi_i} \in \pi_i$, $i = 1, 2, 3$.

This is a perfect example of the local-global principle for periods. It also happens that the conditions we found is sufficient to make **(lpc)** imply **(gpc)**. The following is a good example.

The genericity of automorphic forms has played an indispensable roles in the theory of automorphic forms and L-functions. Generic automorphic forms have nonzero Whittaker-Fourier coefficients, which are the Fourier coefficients associated to the regular nilpotent orbit, and produce the global Whittaker models for the automorphic forms and the automorphic representations. The uniqueness of local Whittaker models proved by J. Shalika [**Shl74**] means that the regular nilpotent orbit has the uniqueness property. The significance of this uniqueness property has been discussed in [**Bmp05**] and [**GSh88**].

It is known that all irreducible cuspidal automorphic representations of $\mathrm{GL}_n(\mathbb{A})$ are generic. The existence of non-generic cuspidal automorphic representations for groups which is not GL_n yields counter-examples to the generalized Ramanujan conjecture ([**HPS79**]). The local-global relation of the genericity has been a very sensitive problem in the theory. From Arthur's conjecture ([**Ar89**], [**Ar04**], and [**Cl**]), it is expected that for a given irreducible cuspidal automorphic representation $\pi = \otimes\pi_v$, if one of the local component π_v is generic, then every local component is generic. Some results for $\mathrm{SO}(2n+1)$ have been proved in [**JS06**].

The local-global principle of genericity asks the condition for an irreducible cuspidal automorphic representation π with all local components π_v being generic to be globally generic, i.e. to have a nonzero Whittaker-Fourier coefficient. The following is a theorem proved for Sp(4).

THEOREM 5.5 ([**KRS92**]). *Let* $\pi = \otimes\pi_v$ *be an irreducible cuspidal automorphic representation of* $\mathrm{Sp}_4(\mathbb{A})$. *Assume that all local components* π_v *are generic, and the partial degree 5 L-function attached to* π, $L(s, \pi, \mathrm{St})$, *has a pole at* $s = 1$. *Then* π *is globally generic.*

In fact, if this is the case, then π is not only just globally generic, but also an image of theta lifting from an even orthogonal groups. More discussions can be found in [**GRS97**].

5.3. Functoriality. The characterization of the image of a certain functorial transfer or lifting in terms of automorphic invariants, i.e. periods of automorphic forms is one of the original motivation of the current study of periods of automorphic forms. The theory of quadratic base change of irreducible cuspidal automorphic representations of GL_n was established by Arthur and Clozel using twisted version of the Arthur-Selberg trace formula ([**AC89**]). By Jacquet's relative trace formula, the theory can be formulated in terms of periods. Let E/k be a quadratic extension of number fields. An irreducible cuspidal automorphic representation Π of $\mathrm{GL}_n(\mathbb{A}_E)$ is a base change from $\mathrm{GL}_n(\mathbb{A}_k)$ if and only if there exists a unitary group U_n in $\mathrm{GL}_n(E)$ such that the period $\mathcal{P}_{\{0\},U_n}(\varphi_\Pi, 1)$ is not identically zero ([**Jcq03**], [**JY90**]).

Another example of characterization of the image of Langlands functorial lifting in terms of periods has been completed via recent work on the Langlands functorial transfer from classical groups to the general linear group. More precisely, let π be an irreducible cuspidal automorphic representation of $\mathrm{GL}_{2n}(\mathbb{A})$. Then the following three statements are equivalent.

(1) π is the image of Langlands functorial lifting from SO_{2n+1};
(2) π has a nonzero Shalike period;
(3) The exterior square L-function $L(s, \pi, \Lambda^2)$ has a simple pole at $s = 1$.

The theory has been established over years through the work of many authors in [**JS90**], [**Bmp90**], [**CKPSS04**], [**GRS01**], [**K02**], [**JS03**].

Some results of this type can also be found in [**GJ00**], [**K00**], [**GJ01**], [**KR94**], [**GRS97**], to mention a few.

6. Final Remarks

In connection with Problem 3.2, one may think of Deligne's conjecture on special values of motivic L-functions ([**D79**]). According the Langlands philosophy, these motivic L-functions are expected to be automorphic. From this point of view, one expects an expression of special values of automorphic L-functions in terms of periods of automorphic forms. If the automorphic forms are motivic, then one expects that the periods of motivic automorphic forms can be expressed as periods in the sense of the Deligne conjecture. In the Rankin-Selberg method approach to automorphic L-functions, one has a natural expression of special values of certain automorphic L-functions in terms of a certain integral. Such integrals may be transferred to periods of automorphic forms, which may be closer to what required by the Deligne conjecture. The well-known Siegel-Weil formula, especially the one extended by Kudla and Rallis, has been the useful tool to obtain the 'right' periods. We refer to [**Hr94**] for detailed discussion. The relative trace formula approach to establish such explicit identities is also very powerful. The generalization of the well-known Waldspurger formula to general cuspidal automorphic forms of GL_2 is obtained in [**BM03**], and the generalization of the Waldspurger formula to high rank group cases are expected by using relative trace formula.

Another aspect connected to Problem 3.2 is to calculate explicitly certain periods of automorphic forms. If the periods can be expressed in terms of something of L-functions, then the analytic method on automorphic L-functions may be used to estimate the periods. Then nice applications to arithmetic may follow. In the recent work [**JLZ06**], one finds that the equidistribution property for certain cycles in the Hilbert modular varieties can be established by explicit calculation of certain periods. It is expected that the method formulated in [**JLZ06**] may be applicable to many other interesting cases.

It is very worthwhile to point out that even the periods of automorphic forms may not have local uniqueness property, one may still obtain a formula which expresses the global period of automorphic forms as a eulerian product of local periods (or functionals) with a natural choice at each local place. An very interesting example has been discussed in [**Jcq01**].

References

[Ar89] Arthur, J. *Unipotent automorphic representations: conjectures.* Orbites et representations II, Asterisque Vol. 171–172 (1989), 13–71.

[Ar04] Arthur, J. *Automorphic representations of* GSp(4). Contributions to automorphic forms, geometry, and number theory, 65–81, Johns Hopkins Univ. Press, 2004.

[AC89] Arthur, James; Clozel, Laurent *Simple algebras, base change, and the advanced theory of the trace formula.* Annals of Mathematics Studies, 120. Princeton University Press, 1989.

[AB90] Ash, A.; Borel, A. *Generalized modular symbols.* Cohomology of arithmetic groups and automorphic forms (Luminy-Marseille, 1989), 57–75, Lecture Notes in Math., 1447, Springer, 1990.

[BM03] Baruch, E.; Mao, Z. *Central value of automorphic L-functions.* Preprint 2003.

[BJ79] Borel, A.; Jacquet, H. *Automorphic forms and automorphic representations.* Proc. Sympos. Pure Math., 33, Automorphic forms, representations and L-functions Part 1, pp. 189–207, Amer. Math. Soc., 1979.

[BFSP04] Boecherer, S.; Furusawa, M.; Schulze-Pillot, R. *On the global Gross-Prasad conjecture for Yoshida liftings.* Contributions to automorphic forms, geometry, and number theory, 105–130, Johns Hopkins Univ. Press, 2004.

[Brn95] Brion, M. *Spherical varieties.* Proceedings of the International Congress of Mathematicians, Vol. 1, 2 (Zurich, 1994), 753–760, Birkhauser, Basel, 1995.

[B61] Bruhat, F. *Distributions sur un groupe localement compact et applications à l'étude des représentations des groupes $\wp$-adiques.* (French) Bull. Soc. Math. France 89 1961 43–75.

[Bmp97] Bump, D. *Automorphic Forms and Representations.* Cambridge Studies in Advanced Mathematics, 55. Cambridge University Press, 1997.

[Bmp05] Bump, D. *The Rankin Selberg Method: An Introduction and Survey.* Automorphic representations, L-functions and applications: progress and prospects, 41–73, Ohio State Univ. Math. Res. Inst. Publ., 11, de Gruyter, Berlin, 2005.

[Bmp90] Bump, D.; Friedberg, S. *The exterior square automorphic L-functions on* $\mathrm{GL}(n)$. Festschrift in honor of I. I. Piatetski-Shapiro on the occasion of his sixtieth birthday, Part II, 47–65, Israel Math. Conf. Proc., 3, 1990.

[Cl] Clozel, L. *Spectral Theory of Automorphic Forms.* Preprint (2001).

[CKPSS04] Cogdell, J. W.; Kim, H. H.; Piatetski-Shapiro, I. I.; Shahidi, F. *Functoriality for the classical groups.* Publ. Math. Inst. Hautes Études Sci. No. 99 (2004), 163–233

[CM93] Collingwood, D.; McGovern, W. *Nilpotent orbits in semisimple Lie algebras.* Van Nostrand Reinhold Mathematics Series. Van Nostrand Reinhold Co., New York, 1993.

[D79] Deligne, P. *Valeurs de fonctions L et périodes d'intégrales.* (French) Proc. Sympos. Pure Math., 33, Automorphic forms, representations and L-functions, Part 2, pp. 313–346, Amer. Math. Soc., 1979.

[FG99] Friedberg, S.; Goldberg, D. *On local coefficients for non-generic representations of some classical groups.* Compositio Math. 116 (1999), no. 2, 133–166.

[FJ93] Friedberg, S.; Jacquet, H. *Linear periods.* J. Reine Angew. Math. 443 (1993), 91–139.

[Glb84] Gelbart, S. *An elementary introduction to the Langlands program.* Bull. Amer. Math. Soc. (N.S.) 10 (1984), no. 2, 177–219.

[GM04] Gelbart, S.; Miller, Stephen D. *Riemann's zeta function and beyond.* Bull. Amer. Math. Soc. (N.S.) 41 (2004), no. 1, 59–112

[GSh88] Gelbart, S.; Shahidi, F. *Analytic properties of automorphic L-functions.* Perspectives in Mathematics, 6. Academic Press, Inc., 1988.

[G90] Ginzburg, D. *L-functions for* $\mathrm{SO}_n \times \mathrm{GL}_k$. J. reine andew. Math. 405(1990), 156–180.

[G95] Ginzburg, D. *On spin L-functions for orthogonal groups.* Duke Math. J. 77 (1995), no. 3, 753–798.

[GJ00] Ginzburg, David; Jiang, Dihua *A Siegel-Weil identity for G_2 and poles of L-functions.* J. Number Theory 82 (2000), no. 2, 256–287.

[GJ01] Ginzburg, D.; Jiang, D. *Periods and liftings: from G_2 to C_3.* Israel J. Math. 123 (2001), 29–59.

[GJR01] Ginzburg, D.; Jiang, D.; Rallis, S., *Nonvanishing of the central critical value of the third symmetric power L-functions.* Forum Math. 13 (2001), no. 1, 109–132.

[GJR02] Ginzburg, D.; Jiang, D.; Rallis, S., *On CAP automorphic representations of a split group of type D_4.* J. Reine Angew. Math. 552 (2002), 179–211.

[GJR04] Ginzburg, D.; Jiang, D.; Rallis, S., *Non-vanishing of the central value of the Rankin-Selberg L-functions. I.*. Journal of AMS. 17 (2004), 679-722.

[GJR05] Ginzburg, D.; Jiang, D.; Rallis, S., *Non-vanishing of the central value of the Rankin-Selberg L-functions. II.*. Automorphic representations, L-functions and applications: progress and prospects, 157–191, Ohio State Univ. Math. Res. Inst. Publ., 11, de Gruyter, Berlin, 2005.

[GPSR97] Ginzburg, D.; Piatetski-Shapiro, I.; Rallis, S. *L functions for the orthogonal group.* Mem. Amer. Math. Soc. 128 (1997), no. 611.

[GRS97] Ginzburg, David; Rallis, Stephen; Soudry, David Periods, *Poles of L-functions and symplectic-orthogonal theta lifts.* J. Reine Angew. Math. 487 (1997), 85–114.

[GRS01] Ginzburg, D.; Rallis, S.; Soudry, D. *Generic automorphic forms on* $\mathrm{SO}(2n+1)$*: functorial lift to* $\mathrm{GL}(2n)$*, endoscopy, and base change.* Internat. Math. Res. Notices 2001, no. 14, 729–764.

[GP92] Gross, B.; Prasad, D. *On the decomposition of a representation of* SO_n *when restricted to* SO_{n-1}. Can. J. Math. Vol. 44 (1992), 974–1002.

[GP94] Gross, B.; Prasad, D. *On irreducible representations of* $\mathrm{SO}_{2n+1}\times\mathrm{SO}_{2m}$. Canad. J. Math. 46 (1994), no. 5, 930–950.

[GR05] Gross, B.: Reeder, M. *From Laplace to Langlands via restriction from* $\mathrm{SO}(2n+1)$ *to* $\mathrm{SO}(2n)$. preprint 2005.

[Hd87] Harder, G. *Eisenstein cohomology of arithmetic groups. The case* GL_2. Invent. Math. 89 (1987), no. 1, 37–118.

[Hd03] Harder, G. *Modular symbils and special values of automorphic L-functions.* preprint (2003)

[HLR86] Harder, G.; Langlands, R. P.; Rapoport, M. *Algebraische Zyklen auf Hilbert-Blumenthal-Flachen.* (German) J. Reine Angew. Math. 366 (1986), 53–120.

[Hr94] Harris, M. *Hodge-de Rham structures and periods of automorphic forms.* Motives (Seattle, WA, 1991), 573–624, PSPM, 55, Part 2, AMS, 1994.

[HK04] Harris, M.; Kudla, S. *On a conjecture of Jacquet.* Contributions to automorphic forms, geometry, and number theory, 355–371, Johns Hopkins Univ. Press, 2004.

[Hw94] Howe, R. *The first fundamental theorem of invariant theory and spherical subgroups.* Algebraic groups and their generalizations: classical methods. 333–346, Proc. Sympos. Pure Math., 56, Part 1, Amer. Math. Soc., 1994.

[HPS79] Howe, R.; Piatetski-Shapiro, I. *A counterexample to the "generalized Ramanujan conjecture" for (quasi-) split groups.* Automorphic forms, representations and L-functions, Part 1, pp. 315–322, Proc. Sympos. Pure Math., 33, Amer. Math. Soc., 1979.

[II04] Ichino, A; Ikeda, T. *On Maass lifts and the central critical values of triple product L-functions.* Preprint 2004

[Jcq97] Jacquet, H. *Automorphic spectrum of symmetric spaces.* Proc. Sympos. Pure Math., AMS, 61(1997), 443–455.

[Jcq01] Jacquet, H. *Factorization of period integrals.* J. Number Theory 87 (2001), no. 1, 109–143.

[Jcq03] Jacquet, H. *Smooth transfer of Kloosterman integrals.* Duke Math. J. 120 (2003), no. 1, 121–152.

[JL70] Jacquet, H.; Langlands, R. *Automorphic forms of* GL_2. Springer Lecture Notes in Mathematics, 114, 1970.

[JLR99] Jacquet, H.; Lapid, E.; Rogawski, J. *Periods of automorphic forms.* J. Amer. Math. Soc. 12 (1999), no. 1, 173–240.

[JR92] Jacquet, H.; Rallis, S. *Symplectic periods.* J. Reine Angew. Math. 423 (1992), 175–197.

[JS90] Jacquet, H.; Shalika, J. *Exterior square L-functions.* Automorphic forms, Shimura varieties, and L-functions, Vol. II, 143–226, Perspect. Math., 11, Academic Press, 1990.

[JY90] Jacquet, H.; Ye, Y. *Une remarque sur le changement de base quadratique.* (French) C. R. Acad. Sci. Paris Sér. I Math. 311 (1990), no. 11, 671–676.

[Jng98a] Jiang, Dihua. G_2*-periods and residual representations.* J. Reine Angew. Math. 497 (1998), 17–46.

[Jng98b] Jiang, D. *Nonvanishing of the central critical value of the triple product L-functions.* Internat. Math. Res. Notices 1998, no. 2, 73–84.

[Jng01] Jiang, D. *On Jacquet's conjecture: the split period case.* Internat. Math. Res. Notices 2001, no. 3, 145–163.

[Jng06] Jiang, D. *Nonvanishing of the central value of the Rankin-Selberg L-functions.* RIMS Kokyuroku 1468, Automorphic Forms and Automorphic L-functions, Feb, 2006.

[JLZ06] Jiang. D.; Li, J.-S.; Zhang, S.-W. *Periods and distribution of cycles on Hilbert modular varieties.* Pure and Applied Mathematics Quarterly, 2(2006)(1), 221–279.

[JMR99] Jiang, D.; Mao, Z; Rallis, S. *A relative Kuznietsov trace formula on* G_2. Manuscripta Math. 99 (1999), no. 3, 411–423.

[JS03] Jiang, D.; Soudry, D. *The local converse theorem for* $\mathrm{SO}(2n+1)$ *and applications.* Ann. of Math. (2) 157 (2003), no. 3, 743–806.

[JS04] Jiang, D.; Soudry, D. *Generic representations and local Langlands reciprocity law for p-adic* SO_{2n+1}. Contributions to automorphic forms, geometry, and number theory, 457–519, Johns Hopkins Univ. Press, 2004.

[JS06] Jiang, D.; Soudry, D. *On the Genericity of Cuspidal Automorphic Forms of* SO_{2n+1}. Accepted by J. Reine Angew. Math. (2006).

[Kwn87] Kawanaka, N. *Shintani lifting and Gelfand-Graev representations.* Proc. Symp. Pure Math. 47, 1987, 147–163.

[K00] Kim, H. *Langlands-Shahidi method and poles of automorphic L-functions. II.* Israel J. Math. 117 (2000), 261–284.

[K02] Kim, H. *Applications of Langlands' functorial lift of odd orthogonal groups.* Trans. Amer. Math. Soc. 354 (2002), no. 7, 2775–2796.

[K] Kim, H. *On local L-functions and normalized intertwining operators.* Canad. J. Math. 57 (2005), no. 3, 535–597.

[Ktw82] Kottwitz, R. *Rational conjugacy classes in reductive groups.* Duke Math. J. 49 (1982), no. 4, 785–806.

[Kd03] Kudla, S. *Special cycles and derivatives of Eisenstein series.* Heegner points and Rankin L-series, 243–270, Math. Sci. Res. Inst. Publ., 49, Cambridge Univ. Press, Cambridge, 2004.

[KR94] Kudla, S.; Rallis, S. *Regularized Siegel-Weil formula: the first term identity.* Ann. of Math. (2) 140 (1994), no. 1, 1–80.

[KRS92] Kudla, S.; Rallis, S.; Soudry, D. *On the degree 5 L-function for* $\mathrm{Sp}(2)$. Invent. Math. 107 (1992), no. 3, 483–541.

[L71] Langlands, R. *Euler products.* Yale Mathematical Monographs, 1. Yale University Press, New Haven, Conn.-London, 1971.

[L76] Langlands, R. *On the functional equations satisfied by Eisenstein series.* Lecture Notes in Mathematics, Vol. 544. Springer-Verlag, 1976.

[L79] Langlands, R. *On the notion of an automorphic representation.* Proc. Sympos. Pure Math., 33, Part 1, pp. 189–207, Amer. Math. Soc., 1979.

[LR04] Lapid, E.; Rallis, S. *A spectral identity between symmetric spaces.* Israel J. Math. 140 (2004), 221–244.

[M72] Manin, Y. *Parabolic points and zeta functions of modular curves.* (Russian) Izv. Akad. Nauk SSSR Ser. Mat. 36 (1972), 19–66.

[M73] Manin, Y. *Periods of cusp forms, and p-adic Hecke series.* (Russian) Mat. Sb. (N.S.) 92(134) (1973), 378–401, 503.

[M05] Manin, Y. *Iterated integrals of modular forms and noncommutative modular symbols.* arXiv:math.NT/0502576

[MW87] Moeglin, C.; Waldspurger, J.-L. *Modéles de Whittaker dégénérés pour des groupes p-adiques.* (French) Math. Z. 196 (1987), no. 3, 427–452.

[MW95] Moeglin, C.; Waldspurger, J.-L. *Spectral decomposition and Eisenstein series* Cambridge Tracts in Mathematics, 113. Cambridge University Press, 1995.

[N75] Novodvorsky, M. *Fontions I pur les groupes orthogonaaux.* (French) C. R. Acad. Sci. Paris. Sér. A 280 (1975), 1421–1422.

[Od01] Oda, T. *Modular cycles on arithmetic quotients of classical domains.* Proc. of Japanese-German Seminar. Hakuba, 2001, 142–152.

[RS03] Roberts, B; Schmidt, R. *New vectors for GSp(4): a conjecture and some evidence.* Research Institute for Mathematical Sciences, Kyoto University, 1338, (2003), 107–121.

[Shl74] Shalika, J. *The multiplicity one theorem for* GL_n. Ann. of Math. (2) 100 (1974), 171–193.

[Sh88] Shahidi, F. *On the Ramanujan conjecture and the finiteness of poles of certain L-functions*, Ann. of Math. (2)(1988)(127), 547-584.

[Sh90] Shahidi, F. *A proof of Langlands' conjecture on Plancherel measures; Complementary series for p-adic groups*. Annals of Math. 132 (1990), 273–330.

[S93] Soudry, D. *Rankin-Selberg convolutions for* $\mathrm{SO}_{2l+1} \times \mathrm{GL}_n$*: local theory.* Mem. AMS. 105, No. 500, 1993.

[Z01] Zhang, S.-W. *Gross-Zagier formula for GL(2).* Asian. J. Math. 5 (2001), no. 2, 183–290.

School of Mathematics, University of Minnesota, Minneapolis, MN 55455, USA
E-mail address: dhjiang@math.umn.edu

AMS/IP Studies in Advanced Mathematics
Volume **39**, 2007

Some problems related to hamiltonian line graphs

Hong-Jian Lai and Yehong Shao

ABSTRACT. Part of this paper summarizes some of the recent developments in the study of hamiltonian line graphs and the related hamiltonian claw-free graphs. The last section of this paper solves some problems on the hamiltonian like indices from a paper by Clark and Wormald in 1983.

1. Definitions and Terminology

Graphs considered here are finite and loopless. Unless otherwise noted, we follow [**2**] for notations and terms. As in [**2**], $\kappa(G)$, $\kappa'(G)$ and $\delta(G)$ represent the *connectivity*, *edge-connectivity*, and the *minimum degree* of a graph G, respectively.

DEFINITION 1.1. A graph G is *nontrivial* if $E(G) \neq \emptyset$.

DEFINITION 1.2. A vertex cut X of G is *essential* if $G - X$ has at least two nontrivial components.

DEFINITION 1.3. For an integer $k > 0$, a graph G is *essentially k-connected* if G does not have an essential cut X with $|X| < k$.

DEFINITION 1.4. An edge cut Y of G is *essential* if $G - Y$ has at least two nontrivial components.

DEFINITION 1.5. For an integer $k > 0$, a graph G is *essentially k-edge-connected* if G does not have an essential edge cut Y with $|Y| < k$.

DEFINITION 1.6. Let G be a graph and let $X \subseteq E(G)$ be an edge subset. The *contraction* G/X is the graph obtained from G by identifying the two ends of each edge in X and then deleting the resulting loops.

For convenience, we use G/e for $G/\{e\}$ and $G/\emptyset = G$; and if H is a subgraph of G, we write G/H for $G/E(H)$.

DEFINITION 1.7. For a graph G, $O(G)$ denotes the set of all vertices of odd degree in G.

DEFINITION 1.8. A graph G is *even* if $O(G) = \emptyset$, is *eulerian* if G is both even and connected, and is *supereulerian* if G contains a spanning eulerian subgraph.

1991 *Mathematics Subject Classification.* Primary 05C45; Secondary 05C38.
Key words and phrases. Line graph, claw-free graph, supereulerian graphs, hamiltonian graphs, hamiltonian index, dominating Eulerian subgraph, essential connectivity.

Definition 1.9. A subgraph H of G is *dominating* if $E(G - V(H)) = \emptyset$.

Definition 1.10. A graph G is *hamiltonian* if G has a spanning cycle. A spanning cycle of G is a *Hamilton cycle* of G.

Definition 1.11. A graph G is *hamiltonian connected* if $\forall u, v \in V(G)$, G has a spanning (u, v)-path.

Definition 1.12. The *line graph* of a graph G, denoted by $L(G)$, has $E(G)$ as its vertex set, where two vertices in $L(G)$ are adjacent if and only if the corresponding edges in G are adjacent.

The following theorem relates hamiltonian line graph $L(G)$ and dominating eulerian subgraph in G.

Theorem 1.13. *(Harary and Nash-Williams* [**10**]*) Let G be a connected graph with $|E(G)| \geq 3$. Then $L(G)$ is hamiltonian if and only if G has a dominating eulerian subgraph.*

Definition 1.14. If $P = v_0 v_1 \cdots v_k$ denotes a trail (or path, respectively), of G, and if $E(P) = \{e_1, e_2, \cdots, e_k\}$ is an edge set such that for $i = 1, 2, \cdots, k$, $e_i = v_{i-1}v_i$, then P is called a (v_0, v_k)-*trail* (or *path*, respectively) of G, and an (e_1, e_k)-trail (or path, respectively) of G. The vertices $v_1, v_2, \cdots, v_{k-1}$ are called the *internal vertices* of P. The edges e_1, e_k are called the *end edges* of P.

Definition 1.15. A trail P of G is *dominating* if every edge of G is incident with an internal vertex of P.

With similar arguments (see page 74 in [**10**]), the following can be proved.

Theorem 1.16. *Let G be a connected graph with $|E(G)| \geq 3$. Then $L(G)$ is hamiltonian connected if and only if for any pair of edges $e_1, e_2 \in E(G)$, G has a dominating (e_1, e_2)-trail.*

Definition 1.17. For a graph G, an induced subgraph H isomorphic to $K_{1,3}$ is called a *claw* of G, and the only vertex of degree 3 of H is the *center* of the claw.

Definition 1.18. A graph G is *claw free* if it does not contain a claw.

Beineke [**1**] and Robertson [**11**] independently proved that a graph G is a line graph if and only if G does not contain a list of 9 graphs as induced subgraphs. Šoltés [**31**], Lai and Šoltés [**15**] indicated for highly connected graphs, this forbidden list can be reduced to fewer graphs. But in any case, a line graph cannot have $K_{1,3}$ as an induced subgraph.

The purpose of this article is to summarize some of the recent development on the study of hamiltonian line graphs, hamiltonian claw-free graphs, and to solve some of the problems from an earlier study by Clark and Wormald [**7**].

2. Sufficient Condition with Local Connectivity

Definition 2.1. A vertex v is *locally connected* if $N(v)$ is connected; and G is *locally connected* if every vertex of G is locally connected.

Theorem 2.2. *(D. J. Oberly and D. P. Sumner* [**26**]*) Every connected, locally connected claw-free graph is hamiltonian.*

DEFINITION 2.3. A graph is *vertex pancyclic* if given any vertex $v \in V(G)$, G has cycles C_i of length i containing v, for each $3 \le i \le |V(G)|$.

THEOREM 2.4. *(L. Clark [**6**], R. H. Shi [**30**], and C.-Q. Zhang [**36**]) Every connected, locally connected claw-free graph is vertex pancyclic.*

DEFINITION 2.5. For a vertex $v \in V(G)$, define

$N_2(v, G)=\{e \in E(G) : e$ has at least one end in $N(v, G)$, but not incident with $v\}$.

DEFINITION 2.6. A vertex v is *locally N_2-connected* in G if $N_2(v, G)$ induces a connected subgraph in G.

DEFINITION 2.7. A graph G is *locally N_2-connected* if every vertex of G is locally N_2-connected.

THEOREM 2.8. *(Ryjáček [**27**]) Let G be a connected, N_2-locally connected claw-free graph without vertices of degree 1, which does not contain an induced subgraph H isomorphic to either G_1 or G_2 (Figure below) such that $N_1(x, G)$ of every vertex x of degree 4 in H is disconnected. Then G is hamiltonian.*

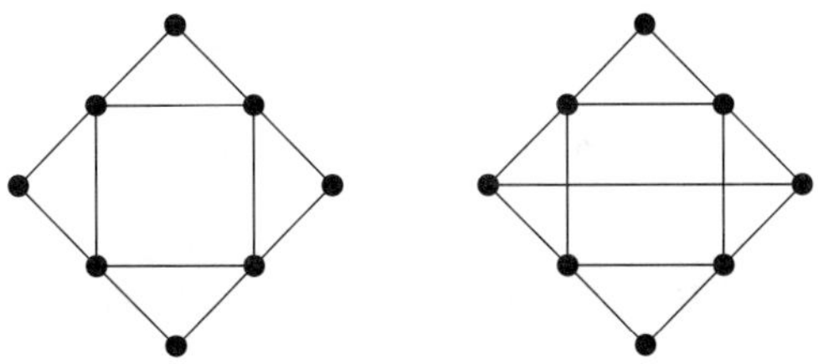

G_1 G_2

Figure 1

The following was conjectured by Ryjáček [**28**].

THEOREM 2.9. *(H.-J. Lai, Y. Shao and M. Zhan [**16**]) Every 3-connected, locally N_2-connected claw-free graph is hamiltonian.*

DEFINITION 2.10. For an integer $k \ge 1$, and a vertex $v \in V(G)$,

$$N^k(v, G) = \{x \in V(G) : dist_G(v, x) \in \{1, 2, \cdots, k\}\}.$$

DEFINITION 2.11. A vertex v is *locally N^k-connected* in G if $N^k(v, G)$ induces a connected subgraph in G.

DEFINITION 2.12. A graph G is *locally N^k-connected* if every vertex of G is locally N^k-connected.

CONJECTURE 2.13. (Li [**22**]) Every 3-connected, locally N^2-connected claw-free graph is hamiltonian.

We now give a counterexample to Conjecture 2.13 as follows:

EXAMPLE 2.14. Let P_{10} denote the Petersen graph. For an integer $k > 0$ and let $P_{10}(k)$ denote the graph obtained from P_{10} by adding k pendant edges at each vertex of P_{10}. Let $G(k) = L(P_{10}(k))$ be the line graph of $P_{10}(k)$. For an example when $k = 3$ see Figure 2.

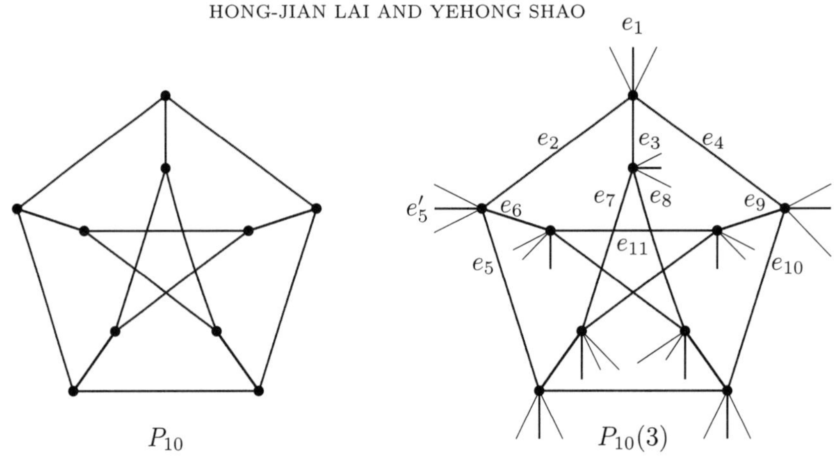

Figure 2 The Petersen graph P_{10} and $P_{10}(3)$

By a well known result of Harary and Nash-Williams [**10**], $G(k)$ does not have a hamilton cycle. On the other hand, we shall check that each $G(k)$ is claw-free 3-connected (hence 3-edge-connected) and N^2-locally connected. Since line graphs are claw-free, $G(k)$ must also be claw-free. Since $P_{10}(k)$ does not have an essential edge cut with size less than 3, $\kappa(G(k)) \geq 3$. It remains to check that each $G(k)$ is N^2-locally connected.

We shall use the notation in Figure 2 to show that $G(k)$ is locally N^2-connected, by the symmetry of the Petersen graph, it suffices to show that both vertices e_1 and e_2 are locally N^2-connected.

Let v_1, v_2 and v_3 denote the vertices in $P_{10}(k)$ that are incident with both e_1 and e_2, both e_2 and e_6, and both e_3 and e_7, respectively. For each vertex $v \in V(P_{10}(k))$, let $K(v)$ denote the complete graph in $G(k)$ induced by the edge incident with v in $P_{10}(k)$.

Since e_1 is a pendant edge in $P_{10}(k)$, it lies in a complete subgraph $K(v_1)$ of $G(k)$ containing e_2, e_3, e_4. Any vertices that are of distance 2 from e_1 in $G(k)$ must be a vertex adjacent to one of e_2, e_3 and e_4. Therefore, e_1 is a locally N^2-connected vertex in $G(k)$, see Figure 3.

It is not difficult to check that e_2 is also a locally N^2-connected vertex in $G(k)$.

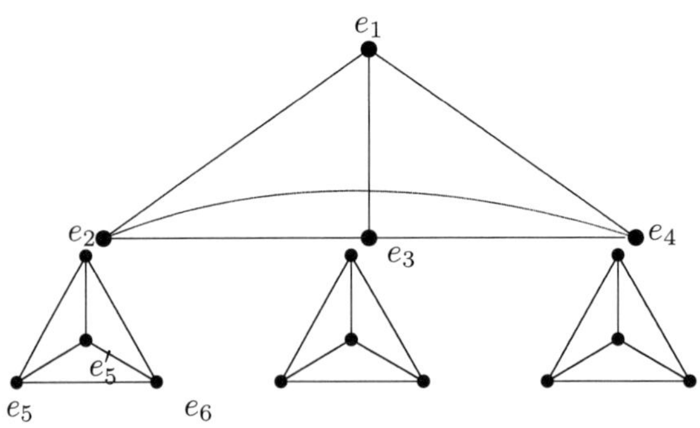

Figure 3 An illustration of the proof that e_1 is locally N^2-connected in $G(k)$

THEOREM 2.15. *(H.-J. Lai, M. Li, Y. Shao and L. Xiong* [**20**]*)*

(1) *Every 2-edge-connected, locally N^2-connected claw-free graph has a spanning Eulerian subgraph with maximum degree at most 4.*
(2) *Every 3-edge-connected, locally N^3-connected claw-free graph has a spanning Eulerian subgraph with maximum degree at most 4.*

DEFINITION 2.16. A graph G is *triangularly connected* if for every pair of edges $e_1, e_2 \in E(G)$, G has a sequence of 3-cycles $C_1, C_2, \cdots, C_l$ such that $e_1 \in C_1, e_2 \in C_l$ and $E(C_i) \cap E(C_{i+1}) \neq \emptyset$, $(1 \leq i \leq l-1)$.

Every connected, locally connected graph is triangularly connected, but triangularly connected graphs may not be locally connected.

THEOREM 2.17. *(H.-J. Lai, L. Miao, Y. Shao and L. Wan* [**29**]*) Every triangularly connected claw-free graph is vertex pancyclic.*

CONJECTURE 2.18. (M. Li, L. Xiong and H.-J. Lai, [**23**]) Every 3-connected 4-cycle connected claw-free graph G with $|E(G)| \geq 3$ is hamiltonian.

We believe that such graphs may even be vertex pancyclic.

3. Degree Conditions

When $\kappa(H) = 2$, Kuipers and Veldman [**13**], and independently Favaron, Flandrin, Li and Ryjáček [**8**], proved that if H is a 2-connected claw-free graph with sufficiently large order ν, and if $\delta(H) \geq \frac{\nu+c}{6}$ (where c is a constant), then H is hamiltonian except a member of ten well-defined families of graphs. When $\kappa(H) = 3$, the following have been proved and proposed.

THEOREM 3.1. *(Kuipers and Veldman*[**13**]*) If H is a 3-connected claw-free simple graph with sufficiently large order ν, and if $\delta(H) \geq \frac{\nu+29}{8}$, then H is hamiltonian.*

THEOREM 3.2. *(Favaron and Fraisse* [**9**]*) If H is a 3-connected claw-free simple graph with order ν, and if $\delta(H) \geq \frac{\nu+37}{10}$, then H is hamiltonian.*

CONJECTURE 3.3. (Kuipers and Veldman [**13**], see also [**9**]) Let H be a 3-connected claw-free simple graph of order ν with $\delta(H) \geq \frac{\nu+6}{10}$. If ν is sufficiently large, then H is hamiltonian.

THEOREM 3.4. *(H.-J. Lai, Y. Shao and M. Zhan* [**18**]*) If H is 3-connected claw-free simple graph with $\nu \geq 196$, and if $\delta(H) \geq \frac{\nu+5}{10}$, then either H is hamiltonian, or $\delta(H) = \frac{\nu+5}{10}$ and $cl(H)$ is the line graph of G obtained from the Petersen graph P_{10} by adding $\frac{\nu-15}{10}$ pendant edges at each vertex of P_{10}.*

4. Matthews, Sumner and Thomassen Conjectures

CONJECTURE 4.1. (Thomassen [**32**]) Every 4-connected line graph is hamiltonian.

CONJECTURE 4.2. (Matthews and Sumner [**24**]) Every 4-connected claw-free graph is hamiltonian.

In 1986, Zhan proved:

THEOREM 4.3. *(Zhan* [**34**]*) If G is a 4-edge-connected graph, then the line graph $L(G)$ is hamiltonian connected.*

THEOREM 4.4. *(Ryjáček* [**28**]*)*

(1) *Conjecture 4.1 and Conjecture 4.2 are equivalent.*
(2) *Every 7-connected claw-free graph is hamiltonian.*

THEOREM 4.5. *(Chen, Lai, Lai, and Weng* [**14**]*) Every 4-connected line graph of a claw free graph is hamiltonian.*

This has been improved by Kriesell.

THEOREM 4.6. *(Kriesell* [**12**]*) Every 4-connected line graph of a claw free graph is hamiltonian connected.*

DEFINITION 4.7. Let C_4 denote a 4-cycle in K_5. The graph $K_5 - E(C_4)$ is called an *hourglass.*

DEFINITION 4.8. A graph G is *hourglass free* if G does not have an induced subgraph isomorphic to $K_5 - E(C_4)$.

THEOREM 4.9. *(Broersma, Kriesell and Ryjáček* [**3**]*) Every 4-connected hourglass free line graph is hamiltonian connected.*

These results suggest that there may be a more general theorem behind them. In fact, the following is recently proved.

THEOREM 4.10. *(Lai, Shao, Yu and Zhan* [**19**]*) Let G be a connected graph with $|V(G)| \geq 4$. The core of this graph G, denoted by G_0, is obtained by deleting all the vertices of degree 1 and contracting exactly one edge xy or yz for each path xyz in G with $d_G(y) = 2$. If every 3-edge-cut of the core G_0 has at least one edge lying in a cycle of length at most 3 in G_0, then the following statements are equivalent.*

(1) *$L(G)$ is hamiltonian connected.*
(2) *$\kappa(L(G)) \geq 3$.*

COROLLARY 4.11. Let G be a graph with $|V(G)| \geq 4$. Suppose that $L(G)$ is hourglass free in which every 3-cut of $L(G)$ is not an independent set. Then $L(G)$ is hamiltonian-connected if and only if $\kappa(L(G)) \geq 3$.

DEFINITION 4.12. A set $B \subset V(G)$ is a *dominating* set if every vertex of G belongs to B or has a neighbor in B.

DEFINITION 4.13. The size of a minimum dominating set of G will be called *dominating number* of G and is denoted by $\gamma(G)$. If $\gamma(G) \leq k$, then G is *k-dominated.*

DEFINITION 4.14. A graph G is *almost claw free* if the vertices that are centers of claws in G are independent and if the neighborhoods of the center of each claw in G is 2-dominated.

Note that every claw free graph is an almost claw free graph and there exist almost claw free graphs that are not claw-free.

COROLLARY 4.15. Every 4-connected line graph of an almost claw free graph is hamiltonian-connected.

We conclude this section with another result and a conjecture in this direction.

THEOREM 4.16. *(Lai, Shao, Wu and Zhou* [**21**]*) Every 3-connected, essentially 11-connected line graph is hamiltonian.*

CONJECTURE 4.17. (H.-J. Lai and L. Šoltés, [**31**]) Every 7-connected claw-free graph is hamiltonian-connected.

5. Problems Related to the Hamiltonian Like Indices

DEFINITION 5.1. For a nontrivial connected graph G, we define $L^0(G) = G$ and for any integer $k > 0$, $L^k(G) = L(L^{k-1}(G))$.

DEFINITION 5.2. The *hamiltonian index* $h(G)$ of G is the smallest positive integer k such that $L^k(G)$ is hamiltonian.

The concept of hamiltonian index was first introduced by Chartrand and Wall [**5**], who showed that (Theorem A of [**5**]) if a connected graph G is not a path, then $L^k(G)$ is defined for any positive integer k. For this reason, we shall assume, throughout this section, that the graph G under discussion is simple, connected and not a path.

Clark and Wormald developed the idea and introduced the hamiltonian like indices.

DEFINITION 5.3. A graph is *edge-hamiltonian* (eh) if each edge lies on a Hamilton cycle.

DEFINITION 5.4. A graph is *pancyclic* (pc) if G has a cycle of length k, for each k with $3 \leq k \leq |V(G)|$.

DEFINITION 5.5. A graph is *vertex pancyclic* (vp) if for every vertex $v \in V(G)$, G has a cycle containing v and of length k, for each k with $3 \leq k \leq |V(G)|$.

DEFINITION 5.6. A graph is *edge pancyclic* (ep) if for every edge $e \in E(G)$, G has a cycle containing e and of length k, for each k with $3 \leq k \leq |V(G)|$.

DEFINITION 5.7. A graph is *hamiltonian connected* (hc) if for every pair of vertices u and v of G, G has a spanning (u, v)-path.

DEFINITION 5.8. For a property $\mathcal{P}$ and a connected nonempty graph G which is not a path, define the $\mathcal{P}$*-index* of G, denoted $\mathcal{P}(G)$, as

$$\mathcal{P}(G) = \begin{cases} \min\{k : L^k(G) \text{ has property } \mathcal{P}\} & \text{if at least one such integer } k \text{ exists} \\ \infty & \text{otherwise} \end{cases}$$

Clark and Wormald [**7**] showed that if a connected, nontrivial graph G is not a path nor a cycle, then the indices $h(G), eh(G), pc(G), vp(G), ep(G), hc(G)$ exist as finite numbers.

The most studied index is $h(G)$, the hamiltonian index. By Theorem 1.13, if a graph G is hamiltonian, then $L(G)$ is also hamiltonian. Therefore, a question arises: for any of the properties $\mathcal{P} \in \{eh, pc, vp, ep, ep, hc\}$, if G has property $\mathcal{P}$, does $L(G)$ also have property $\mathcal{P}$? We in this section will answer this question.

DEFINITION 5.9. For a trail P defined in Definition 1.14, let

$$\partial(P) = \{e \in E(G) : \ e \text{ is incident with an internal vertex of } P\}.$$

If $\partial(P) = E(G)$, then P is a dominating trail. We put the following facts in the form of a lemma, which follows directly from the definition of a line graph.

LEMMA 5.10. *Let $P = v_0v_1 \cdots v_k$ denote a (possibly closed) trail of G, and let $E(P) = \{e_1, e_2, \cdots, e_k\}$ such that for $i = 1, 2, \cdots, k$, $e_i = v_{i-1}v_i$. Suppose that $X \subseteq \partial(P) - E(P)$ can be represented by $X = \{x_1, \cdots, x_{n_1}, x_{n_1+1}, \cdots, x_{n_{k-1}}\}$, such* *that for each* $i = 0, 1, 2, \cdots, k-2$, $x_{n_i+1}, \cdots, x_{n_{i+1}}$ *are incident with the vertex* v_{i+1}, *where we define* $n_0 = 0$. *Then the sequence*

$$e_1, x_1, \cdots, x_{n_1}, e_2, x_{n_1+1}, \cdots, x_{n_{k-1}}, e_k$$

is a path L *in* $L(G)$ *with* $V(L) = E(P) \cup X$.

PROPOSITION 5.11. If G is hamiltonian connected, then $L(G)$ is also hamiltonian connected.

PROOF. By Theorem 1.16, it suffices to show that for any edges $e_1, e_2 \in E(G)$, G has dominating (e_1, e_2)-trail.

We assume that $e_1 = u_1v_1$ and $e_2 = u_2v_2$. Since G is hamiltonian connected, G has a hamiltonian (u_1, u_2)-path P. If $e_1, e_2 \in E(P)$, then since P is a (u_1, u_2)-path, e_1 and e_2 must be the two end edges of the path and so P is a dominating (e_1, e_2)-trail of G. If $e_1, e_2 \notin E(P)$, then $G[E(P) \cup \{e_1, e_2\}]$ is a dominating (e_1, e_2)-trail of G. If $|E(P) \cap \{e_1, e_2\}| = 1$, then we may assume that $e_1 \in E(P)$ and $e_2 \notin E(P)$. Since P is a (u_1, u_2)-path, e_1 must be an end edge of P, and so $G[E(P) \cup \{e_2\}]$ is a dominating (e_1, e_2)-trail of G. □

PROPOSITION 5.12. If G is pancyclic, the $L(G)$ is also pancyclic.

PROOF. For any integer k with $3 \le k \le |E(G)|$, we need to show that $L(G)$ has a cycle of length k.

First assume that $3 \le k \le |V(G)|$. Since G is pancyclic, G has a cycle C with $|E(C)| = k$. By Lemma 5.10, $L(C)$ is a cycle of length k in $L(G)$. Thus we assume that $|V(G)| + 1 \le k \le |E(G)|$. Since G is pancyclic, G has a Hamilton cycle C, and so there exists an edge subset $X \subseteq E(G) - E(C)$ such that $|E(C) \cup X| = k$. It then follows by Lemma 5.10 that $L(G)[E(C) \cup X]$ contains a cycle of length k. Hence $L(G)$ must also be pancyclic. □

PROPOSITION 5.13. If G is edge-hamiltonian, the $L(G)$ is also edge-hamiltonian.

PROOF. Let $f = e_1e_2$ denote an arbitrary edge in $L(G)$, where e_1, e_2 are both adjacent to a vertex u in G.

Suppose first that u has degree 2 in G. Since G is edge-hamiltonian, G has a Hamilton cycle C such that $e_1 \in E(C)$. It follows that C is a dominating cycle of G. Since u has degree 2 in G, $e_2 \in E(C)$. Therefore, e_1 and e_2 are adjacent in C and G has no other edges adjacent to u. It follows by Lemma 5.10 that $L(G)$ has a Hamilton cycle containing the edge $f = e_1e_2$.

Hence we assume that u has degree at least 3 in G. The there exists an edge $e_3 \in E(G) - \{e_1, e_2\}$ such that e_3 is also incident with u in G. Since G is edge-hamiltonian, G has a Hamilton cycle C such that $e_3 \in E(C)$. Since C is a cycle containing e_3 and since e_1, e_2, e_3 are all incident with the same vertex u, we may assume that $e_2 \notin E(C)$. It then follows by Lemma 5.10 that $L(G)$ has a Hamilton cycle containing the edge $f = e_1e_2$. □

PROPOSITION 5.14. If G is vertex pancyclic, the $L(G)$ is also vertex pancyclic.

PROOF. For any integer k with $3 \le k \le |E(G)|$, and for any $e \in V(L(G)) = E(G)$, we need to show that $L(G)$ has a cycle L_k of length k such that $e \in V(L_k)$. Let $e = uv$ such that $deg_G(u) \ge deg_G(v)$. We may assume that G is not a 3-cycle, and since G is vertex pancyclic, $deg_G(u) \ge 3$. Therefore, e lies in a 3-cycle of $L(G)$. We assume that $k \ge 4$.

Since G is vertex pancyclic, G has a cycle C_k with $|E(C_k)| = k$, for $k = 3, 4, \cdots, |V(G)|$, such that $u \in V(C_k)$. If $e \in E(C_k)$, then by Lemma 5.10, $L_k = L(C_k)$ is a cycle of length k containing e in $L(G)$.

Therefore, we assume that $e \notin E(C_k)$. As $k \geq 4$, G has a cycle C_{k-1} of length $k-1$ containing u. Since $deg_G(u) \geq 3$, $|\partial(C_{k-1})| \geq k$ and $e \in \partial(C_{k-1})$. It follows by Lemma 5.10 that $L(G)$ has a cycle of length k containing e.

When $|E(G)| \geq k \geq |V(G)| + 1$, since G has a Hamilton cycle C containing u, a similar argument shows that $L(G)$ has a cycle of length k containing the edge e. Hence $L(G)$ must also be vertex pancyclic. □

PROPOSITION 5.15. If G is edge pancyclic, then $L(G)$ is also edge pancyclic.

PROOF. For any integer k with $3 \leq k \leq |E(G)|$, and for any $e_1, e_2 \in V(L(G)) = E(G)$ with $f = e_1e_2 \in E(L(G))$, we need to show that $L(G)$ has a cycle L_k of length k such that $f \in E(L_k)$. Let u denote the vertex incident to both e_1 and e_2.

We first assume u has degree at least $d \geq 3$ in G, and so that G has an edge $e_3 \in E(G) - \{e_1, e_2\}$. Thus $L(G)$ has a k-cycle containing the edge $f = e_1e_2$, for each $k = 3, 4, \cdots, d$.

Let $k \geq d+1 \geq 4$ be an integer. Assume that $k \leq |V(G)|$. Since G is edge pancyclic, G has a cycle C_{k-1} of length $k-1$ such that $e_3 \in E(C_{k-1})$. If $e_1 \in E(C_{k-1})$, then $e_2 \notin E(C_{k-1})$. Since $e_2 \in \partial(C_{k-1})$, it follows by Lemma 5.10 that $L(G)$ has a cycle of length k containing $f = e_1e_2$. The case when $e_2 \in E(C_{k-1})$ can be similarly proved. Therefore, we assume that $e_1, e_2 \notin E(C_{k-1})$. It follows that $d \geq 4$ and G has an edge $e_4 \in E(G) - \{e_1, e_2, e_3\}$ incident with u in G. Since G is edge-pancyclic, G has a cycle C_{k-2} of length $k-2$ containing e_3. Therefore, $|\partial(C_{k-2})| \geq k$ and $e_1, e_2, e_3, e_4 \in \partial(C_{k-2})$. By Lemma 5.10, $L(G)$ has a cycle of length k containing the edge $f = e_1e_2$. When $|E(G)| \geq k \geq |V(G)| + 1$, since G has a Hamilton cycle C containing e_3, a similar argument shows that $L(G)$ has a cycle of length k containing the edge $f = e_1e_2$.

It remains to prove the case when u has degree 2 in G. In this case, any cycle of G containing e_1 must also contain e_2. Using this fact and the same argument above, we can similarly prove that $L(G)$ has a cycle of length k containing $f = e_1e_2$, for any k with $3 \leq k \leq |E(G)|$. □

There were a few open problems posted at the end of the paper of Clark and Wormald [**7**].

DEFINITION 5.16. For a property $\mathcal{P}$ and integers a, b, with $1 \leq a \leq b$, define

$$\mathcal{P}(a,b) = \begin{cases} \max\{\mathcal{P}(G) : \kappa'(G) \geq a \text{ and } \delta(G) \geq b\} & \text{if the max exists} \\ \infty & \text{otherwise} \end{cases}.$$

Clark and Wormald determined most of the values of $\mathcal{P}(a, b)$ in [**7**]. At the end, they raised this question: What are the values of $h(a, b), eh(a, b)$, and $hc(a, b)$ when $4 \leq a \leq b$?

To answer this question, we shall apply the following theorem by Nash-Williams and Tutte.

THEOREM 5.17. *(Nash-Williams [**25**] and Tutte [**33**]) Let $k > 0$ be an integer. A graph G has k edge-disjoint spanning trees if and only if for any partition $\{V_1, V_2, \cdots, V_t\}$ of $V(G)$, the number of edges in G joining distinct sets of these V_i's is at least $k(t-1)$.*

Using this powerful theorem of Nash-Williams and Tutte, Zhan proved the following.

THEOREM 5.18. *(Zhan [**35**]) If G is 4-edge-connected, then for any edges $e_1, e_2 \in E(G)$, $G - \{e_1, e_2\}$ has two edge-disjoint spanning trees.*

Catlin [**4**] proved the following.

THEOREM 5.19. *(Catlin [**4**]) If G has two edge-disjoint spanning trees, then for any two vertices $u, v \in V(G)$, G has a spanning (u, v)-trail.*

It follows that if $\kappa'(G) \geq \kappa(G) \geq 4$, then by Theorem 5.18 and Theorem 5.19, for any pair of edges $e_1 = u_1v_1, e_2 = u_2v_2 \in E(G)$, the graph $G - \{e_1, e_2\}$ has a spanning (u_1, u_2)-trail C, which can be augmented by adding the edges e_1 and e_2 to result in a dominating (e_1, e_2)-trail of G. By Theorem 1.16, $L(G)$ is hamiltonian connected. This indicates that when $4 \leq a \leq b$, $hc(a, b) \leq 1$. By the definition of these indices, when $4 \leq a \leq b$, we have

$$h(a, b) \leq eh(a, b) \leq hc(a, b) \leq 1.$$

For any integer $m \geq 5$, since $K_{m,4}$ is 4-connected without a Hamilton cycle, $h(4, b) \geq 1$, for any $b \geq 4$. Therefore, when $4 \leq a \leq b$, we have

$$h(a, b) = eh(a, b) = hc(a, b) = 1.$$

This answers the question in [**7**].

References

[1] L. W. Beineke, *Derived graphs and digraphs*, in Beitrdge zur Graphentheorie. Teubner, Leipzig (1968) 17-33.

[2] J. A. Bondy and U. S. R. Murty, *Graph theory with applications*, Macmillan, London and Elsevier, New York, 1976.

[3] H. J. Broersma, M. Kriesell and Z. Ryjáček, *On Factors of 4-Connected Claw-Free Graphs*, J. Graph Theory **37** (2001), 125-136.

[4] P. A. Catlin, *A reduction method to find spanning Eulerian subgraphs*, J. Graph Theory **12** (1988) 29-44.

[5] G. Chartrand and C. E. Wall, *On the hamiltonian index of a graph*, Studia Sci, Math. Hungar. **8** (1973) 43-48.

[6] L. Clark, *Hamiltonian properties of connected locally connected graphs*, Congr. Numer. **32** (1981) 154-176.

[7] L. H. Clark and N. C. Wormald, *Hamiltonian like indices of graphs*, Ars Combinatoria **15** (1983), 131-148.

[8] O. Favaron, E. Flandrin, H. Li and Z. Ryjáček, *Cliques covering and degree conditions for hamiltonicity in claw-free graphs*, Discrete Math., in press.

[9] Odile Favaron and P. Fraisse, *Hamiltonicity and minimum degree in 3-connected claw-free graphs*, J. Combin. Theory Ser. B **82** (2001), 297-305.

[10] F. Harary and C. St. J. A. Nash-Williams, *On eulerian and hamiltonian graphs and line graphs*, Canad. Math. Bull. **8** (1965), 701-709.

[11] F. Harary, *Graph Theory*, Addison-Wesley, Reading, MA (1969).

[12] M. Kriesell, *Every 4-Connected Line Graphs of Claw Free Graphs Are Hamiltonian-Connected*, J. Combin. Theory Ser. B **82** (2001), 306-315.

[13] E. J. Kuipers and H. J. Veldman, *Recognizing claw-free hamiltonian graphs with large minimum degree*, Memorandum 1437, University of Twente, 1998, submitted for publication.

[14] Z. H. Chen, H.-J. Lai, H. Y. Lai, G. Weng, *Jackson's conjecture on eulerian subgraphs*, Combinatorics, Graph Theory, Algorithms and Applications, (eds. by Y. Alavi et al), 53-58, World Scientific, River Edge, NJ (1994).

[15] H.-J. Lai and L. Solés, *Line graphs and forbidden induced subgraphs*, J. Combinatorial Theory, Ser. B **82** (2001), 38-55.

[16] H.-J. Lai, Y. Shao, and M. Zhan, *Hamiltonian N_2-locally Connected Claw-Free Graphs*, J. Graph Theory **48** (2005), 142-146.
[17] H.-J. Lai, X. Li, Y. Ou and H. Poon, *Spanning trails joining given edges*, Graphs and Combinatorics, **21** (2005), 77-88.
[18] H.-J. Lai, Y. Shao and M. Zhan, *Hamiltonicity in 3-connected Claw-Free Graphs*, J. Combinatorial Theory, Ser. B, accepted.
[19] H.-J. Lai, Y. Shao, G. Yu and and M. Zhan, *Hamilton connectedness in 3-connected line graphs*, submitted.
[20] H.-J. Lai, M. Li, Y. Shao, and L. Xiong, *Spanning eulerian subgraphs in N^2-locally connected claw-free graphs*, submitted.
[21] H.-J. Lai, Y. Shao, H. Wu and J. Zhou, *Every 3-connected, essentially 11-connected line graph is hamiltonian*, J. Combinatorial Theory, Ser. B, accepted.
[22] M. Li, *Hamiltonian cycles in N^2-locally connected claw-free graphs*, Ars Combinatoria **62** (2002) 281-288.
[23] M. Li, L. Xiong, H.-J. Lai, Quadrangularly connected claw-free graphs, submitted.
[24] M. M. Matthews and D. P. Sumner, *Hamiltonian results in $K_{1,3}$-free graphs*, J. Graph Theory **8** (1984) 139-146.
[25] C. St. J. A. Nash-Williams, *Edge-disjoint spanning trees of finite graphs*, J. London Math. Soc. **36** (1961) 445-450.
[26] D. J. Oberly and D. P. Sumner, *Every connected, locally connected nontrivial graph with no induced claw is hamiltonian*, J. Graph Theory **3** (1979), 351-356.
[27] Z. Ryjáček, *Hamiltonian circuits in N_2-locally connected $K_{1,3}$-free graphs*, J. Graph Theory **14** (1990), 321-331.
[28] Z. Ryjáček, *On a closure concept in claw-free graphs*, J. Combin. Theory Ser. B **70** (1997), 217-224.
[29] Y. Shao, *Line Graphs and Claw-free Graphs*, Ph.D. Dissertation, West Virginia University, 2005.
[30] R. H. Shi, *Connected and locally connected graphs with no induced claws are vertex pancyclic*, Kexue Tongbao **31** (1986) 427.
[31] L. Soltés, *Forbidden induced subgraphs for line graphs*, Discrete Math. **132** (1994) 391-394.
[32] C. Thomassen, *Reflections on graph theory*, J. Graph Theory **10** (1986) 309-324.
[33] W. T. Tutte, *On the problem of decomposing a graph into n connected factors*, J. London Math. Soc. **36** (1961) 221-230.
[34] S. Zhan, *Hamiltonian connectedness of line graphs*, Ars Combinatoria **22** (1986) 89-95.
[35] S. Zhan, *On hamiltonian line graphs and connectivity*, Discrete Math. **89** (1991) 89-95.
[36] C. Q. Zhang, *Cycles of given length in some $K_{1,3}$-free graphs*, Discrete Math. **78** (1989) 307-313.

Department of Mathematics, West Virginia University, Morgantown, WV 26506
E-mail address: hjlai@math.wvu.edu

Ohio University Southern, Ironton, OH 45638
E-mail address: yehongshao@hotmail.com

AMS/IP Studies in Advanced Mathematics
Volume **39**, 2007

Composition Operators and Isometries on Holomorphic Function Spaces over Domains in $\mathbb{C}^n$

Song-Ying Li

ABSTRACT. The paper gives a short survey on boundedness, compactness and trace ideal for composition operators on Hardy and Bergman spaces. It also gives an easy way to define biholomorphically invariant spaces, and connects these spaces to well-known functions. Some characterization problems on isometries related to compostion operators are posed in end of the paper.

1. Introduction

Let D be a bounded domain in $\mathbb{C}^n$ with C^1 boundary. Many holomorphic function spaces over D have been introduced in last half century, such as Hardy, Bergman, Besov and Sobolev spaces. Properties (boundary behaviour ect.) of functions in those function spaces have been received a great deal of studies. Readers can see parts of them from the following books **[26]** **[29]**, **[66]**, **[71]**, **[75]**, **[76]** and references therein. For any $0 < p \le \infty$, we let $L^p(\partial D)$ be Lebesgue $p-$space with respect to Lebesgue surface measure on ∂D, let $H^p(D)$ be Hardy space of holomorphic functions on D with their boundary value functions belonging to $L^p(\partial D)$. Let $L^p(D)$ be the Lebesgue p-space with respect to the Lebesgue volume measure on D, and let $A^p(D)$ be its holomorphic subspace. It is well-known that $H^2(D)$, $L^2(\partial D)$, $L^2(D)$ and $A^2(D)$ are Hilbert spaces; $H^2(D)$ is a closed subspace of $L^2(\partial D)$ while $A^2(D)$ is a closed subspace of $L^2(D)$. There are two important orthogonal projection operators: $S : L^2(\partial D) \to H^2(D)$ and $P : L^2(D) \to A^2(D)$. S is called Szegö projection and P is called Bergman projection. Both of them can be written as integral operators:

$$(1.1)\quad Sf(z) = \int_{\partial D} S(z,w)f(w)d\sigma(w); \quad Pg(z) = \int_D K(z,w)f(w)dv(w), \quad z \in D$$

1991 *Mathematics Subject Classification.* Primary 32A24.

Key words and phrases. Isometries, composition operators, function spaces.

Partial of this work was done when the author was visiting Wuhan University and Fujian Normal University in China. He would like to thank Professor Hua Chen and Yong-qing Li for their hospitality. The address for those two universities are:
School of Math. and Stat., Wuhan University, Wuhan, Hubei, P.R. China.
Scool of Math. and Computer Sci., Fujian Normal University, Fujian Normal University, Fuzhou, Fujian, P.R. China.

where $S(z, w)$ is Szegö kernel and $K(z, w)$ is the Bergman kernel for D; they are holomorphic in z and anti-holomorphic in w. If one let M_f be the multiplication operator defined as $M_f u = fu$, then the commutator between M_f with S and P are given by $[M_f, S]$ and $[M_f, P]$. Toepliz operator associated with S and P are $T_f^0 = SM_f$ and $T_f = PM_f$. Characterizations for f so that those operators are bounded, compact, or belonging to the Schattern von Neumann class have been obtained by many authors in different function spaces([**2**], [**15**], [**14**], [**8**], [**9**], [**36**], [**37**], etc. and reference therein).

Another interesting and important class of operators is composition operators. Let $\phi : D \to D$ be a holomorphic map. Then the composition operator associated to ϕ is defined as $C_\phi(u)(z) = u(\phi(z))$ for any function u on D. In the past three decades, a great deal of researches have been done on composition operators on some function spaces over D. Most of the works have been focussed on finding characterizations of ϕ so that C_ϕ is boundeded, compact, or in a Schatten class on some function spaces such as the Hardy spaces $H^p(D)$, Bergman spaces $A^p(D)$, Dirichlet spaces and many others (see the book of Cowen and MacCluer [**23**], the survey paper of Russo [**67**], the survey paper of Wogen [**78**], Li [**46**] and the references therein.) When $\phi : D \to D$ is a biholomorphic map, one expects that C_ϕ is a very special operator on some function spacea.

Let $D = B_1$ be the unit disk in complex plane $\mathbb{C}$. Then the holomorphic Besov space is: $B^2(B_1) = \{f : \|f\|_{B^2} < \infty\}$ with $\|f\|_{B^2}^2 = \int_{B_1} |f'(z)|^2 dA(z)$. One can easily show that C_ϕ is an isometry on $B^2(B_1)$ with the semi norm $\|\cdot\|_{B^2}$ for any biholomorphic map or Möbuis transformation ϕ from B_1 to itself. One will see later, C_ϕ is an isometry on many function spaces when $\phi \in \mathrm{Aut}(D)$, the automorphism group of D. On the other hands, one expects that an isometry on many function spaces should be very special. In fact, characterizations for an isometry on Hardy space, Bergman space and little Block space have been obtained by several authors (see [**21**], [**70**], [**41**], [**39, 40**], [**38**] and references therein).

The main purpose of this note is:

1) Give a brief survey on some results of composition operators, which are closely related to works of the author and his collaborators, by viewing composition operators as special Toeplitz operators.

2) Provide several ways to define function spaces so that C_ϕ is an isometry on those function spaces for any $\phi \in \mathrm{Aut}(D)$, and give the equivalent relations between those spaces and the some well-known function spaces.

3) We will introduce some interesting known results on the charaterizations for isometries on some function spaces, and pose some questions for further studies.

2. Composition Operators

It is well-known that C_ϕ is bounded on any holomorphic Hardy space $H^p(D)$ when $D \subset \mathbb{C}$ is bounded domain with C^1 boundary for all $0 < p \le \infty$. In fact, for any $f \in H^p(D)$, one can solve the Dirichlet boundary value problem of Laplacian Δ:

$$(2.1) \qquad \Delta u = 0 \text{ in } D; \text{ and } u(z) = |f(z)|^p, \ z \in \partial D.$$

Since $|f(z)|^p$ is subharmonic, maximum principle shows that $|f(z)|^p \le u(z)$ on D. Since $C_\phi(u)$ solves (2.1) with boundary condition $u = |C_\phi(f)|^p$. Therefore,

$$(2.2). \qquad |C_\phi(f)|^p \le C_\phi(u) \text{ in } D.$$

Let $P(z,w)$ be the Poisson kernel on $D \times \partial D$ for Dirichlet problem (2.1). For any fixed point $z_0 \in D$, there are two constants $0 < a(z_0) \le A(z_0) < \infty$, depending only on D and z_0, so that $a(z_0) \le P(z_0, w) \le A(z_0)$ for $w \in \partial D$. Thus

$$\begin{aligned}
\int_{\partial D} |C_\phi(f)|^p d\sigma(z) &\le \int_{\partial D} u(\phi(z)) d\sigma(z) \\
&\le \frac{1}{a(z_0)} \int_{\partial D} u(\phi(z)) P(z_0, z) d\sigma(z) \\
&= \frac{u(\phi(z_0))}{a(z_0)} \\
&= \frac{1}{a(z_0)} \int_{\partial D} u(z) P(\phi(z_0), z) d\sigma(z) \\
&\le \frac{A(\phi(z_0))}{a(z_0)} \int_{\partial D} |f(z)|^p d\sigma(z).
\end{aligned}$$

Therefore,

$$\|C_\phi(f)\|_{H^p(D)} \le \left[\frac{A(\phi(z_0)}{a(z_0)}\right]^{1/p} \|f\|_{H^p(D)}. \tag{2.3}$$

In particular, when $D = B_1$ the unit disk. By choosing $z_0 = 0$, then $a(0) = A(0) = 1$ and $A(z) \le \frac{1+|z|}{1-|z|}$. Hence

$$\|C_\phi(f)\|_{H^p(B_1)} \le \left[\frac{1 + |\phi(0)|}{1 - |\phi(0)|}\right]^{1/p} \|f\|_{H^p(B_1)}. \tag{2.4}$$

The above argument is no longer true when $n > 1$ since $C_\phi(u)$ may not be harmonic in D if u is merely harmonic in D when $n > 1$. In fact, counterexamples were constructed in [**20**] and [**58**] to show that there is a holomorphic map $\phi = (2z_1z_2, 0) : B_2 \to B_2$ so that C_ϕ is neither bounded on $H^p(B_2)$ nor $A^p(B_2)$ for all $0 < p < \infty$, where B_n is the unit ball in $\mathbb{C}^n$. Let $\phi : D \to D$ be a holomorphic map, we define two pull-bak measures v_ϕ^0 and v_ϕ as follows: For any measurable set $E \subset D$, we let

$$v_\phi^0(E) = \sigma_{\partial D}(\phi^{-1}(E) \cap \partial D\}), \quad v_\phi(E) = v(\phi^{-1}(E)) \tag{2.5}$$

where $\phi^{-1}(E) = \{z \in \overline{D} : \phi(z) \in E\}$. Then one has the following theorem (see [**23**] and [**49**], [**28**], [**8**], etc. for details.)

THEOREM 2.1. *Let D be a bounded strictly pseudoconvex domain in $\mathbb{C}^n$ with C^1 boundary. Let $\phi : D \to D$ be a holomorphic map. Then*

(i) C_ϕ is bounded on $H^p(D)$ if and only if v_ϕ^0 is a Carleson measure;

(ii) C_ϕ is compact on $H^p(D)$ if and only if v_ϕ^0 has vaninshing Carleson measure.

Similar theorem holds if one replaces Hardy spaces by Bergman spaces and replaces v_ϕ^0 by v_ϕ, respectively.

If one let

$$B_{\phi,0}(z)^2 = \int_D \frac{|S(z,w)|^2}{S(z,z)} dv_\phi^0(w), \quad B_\phi(z)^2 = \int_D \frac{|K(z,w)|^2}{K(z,z)} dv_\phi(w) \tag{2.6}$$

Then Theorem 2.1 can be restated as

THEOREM 2.2. *Let D be a bounded strictly pseudoconvex domain in $\mathbb{C}^n$ with C^∞ boundary. Let $\phi : D \to D$ be a holomorphic map. Then*
(i) C_ϕ is bounded on $H^p(D)$ if and only if $B_{\phi,0} \in L^\infty(D)$;
(ii) C_ϕ is compact on $H^p(D)$ if and only if $B_{\phi,0} \in C_0(D)$.
Similar theorem holds by replacing Hardy space by Bergman spaces and by replacing $B_{\phi,0}$ by $B_\phi(z)$.

Compact composition operators on $A^2(B_1)$ were first characterized by Shapiro [**72**] by using Nevanlinna counting function. Hilbert-Schmidt and nuclear composition operators on $H^2(B_1)$ were characterized by Shapiro and Taylor in [**73**]. Leucking and Zhu in [**52**] also used the Nevanlinna counting function to characterize Schatten class composition operators on $H^2(B_1)$ and $A^2(B_1)$. When D is a smoothly bounded strictly pseudoconvex domain, the following result was proved by the author [**44**]:

THEOREM 2.3. *Let D be a smoothly bounded strictly pseudoconvex domain in $\mathbb{C}^n$. Then $C_\phi \in S_p(A^2(D))$ (Schatten Von Neumann $p-$class) if and only if $X_p(\phi) < \infty$, where*

$$X_p(\phi)^p = \int_D \left[K(z,z)^{-1} \int_D |K(z,\phi(w))|^2 dv(w)\right]^{p/2} d\lambda(z) \tag{2.6}$$

for $\frac{2n}{n+1} < p < \infty$, where $d\lambda(z) = K(z,z)dv(z)$.

When D is bounded symmetric domain in $\mathbb{C}^n$, Schattern p-class compositions were characterized by the author and B. Russo [**49**] by the pull-back measure dv_ϕ for $0 < p < \infty$. It is always important and interesting to find a simple and natural condition on the map ϕ which characterizes compact, and Shatten class composition operators C_ϕ. In [**58**], MacCluer and Shapiro proved the following: *C_ϕ is compact on $A^2(B_1)$ if and only if $K(z,z)^{-1}K(\phi(z),\phi(z)) \to 0$ as $z \to \partial B_1$*. In [**52**], Leucking and Zhu proved that: If $D = B_1$ then $Y_p(\phi) \geq \|C_\phi\|_{S_p(A^2(D)}$ for $0 < p \leq 2$; and $\|C_\phi\|_{S_p(A^2(D))} \geq Y_p(\phi)$ for $2 \leq p < \infty$. Here

$$Y_p(\phi)^p = \int_D \left[K(\phi(z),\phi(z))K(z,z)^{-1}\right]^{p/2} d\lambda(z). \tag{2.7}$$

One may extend the latter result to any smoothly bounded domain in $\mathbb{C}^n$ by using their argument (see [**44**].) A natural question was posed in [**52**].

Problem: *Let $2 \leq p < \infty$ and $D = B_1$, is it true $C_\phi \in S_p(A^2(D))$ if and only if $Y_p(\phi) < \infty$?*

Partial results were given in [**44**], in which the author proved that

$$\frac{1}{C(\phi,D)} Y_p(\phi) \leq \|C_\phi\|^p_{S_p(A^2(D))} \leq C(\phi,D) Y_p(\phi), \tag{2.8}$$

where D is a smoothly bounded strictly pseudoconvex domain in $\mathbb{C}^n$ and $C(\phi,D)$ is a constant depending only on D, $\|\phi\|_{C^{n+1}(\overline{D})}$ and $\|C_\phi\|_{S_\infty}$. The smoothness assumption on ϕ here is somewhat natural for $n > 1$ because we know that C_ϕ is not bounded on $A^2(B_2)$ where $\phi(z_1,z_1) = (2z_1z_2, 0)$, a polynomial, but it is not natural for the case $n = 1$. Recently, K. Zhu [**84**] posed another condition for the case $n = 1$. He proved that *(1.3) holds with $C(\phi,B_1) = C(N)$, a positive constant depending on $N = \sup\{\#(\phi^{-1}(w)) : w \in B_1\} < \infty$. Here $\#(E)$ denotes the number of elements in E.* It is clear that the quantity (2.7) is simpler than the quantity (2.6) which characterizes the Shatten class composition operators. It is puzzling and interesting to find out whether $X_p(\phi)$ and $Y_p(\phi)$ are equivalent when $p \neq 2$. In other words, whether the constant $C(N)$ (in the result of Zhu) does really depend on N. In [**45**] (1999, preprint), the author proves that the quantities (1.1) and (1.2) are not equivalent when $D = B_1$ and $p \neq 2$, which answers the question in [**52**] negatively in the sense of the equivalent norms. A precise example was constucted by Jingbo, Xia [**79**], which shows that there is $\phi \in Y_p(B_1)$, but $\|C_\phi\|_{S_p} = \infty$ for $2 < p < \infty$.

Let

$$Z(\phi)^{2p} = \int_D \Big[\int_D |K(\phi(z), \phi(w))|^{p'} dv(w) \Big]^{\frac{p}{p'}} dv(z) \tag{2.9}$$

and

$$W_p^p(\phi) = \int_D \Big[\delta(z)^{n+2} \int_{\partial D} |K(z, \phi(w))|^2 d\sigma(w) \Big]^{p/2} d\lambda(z). \tag{2.10}$$

Another criteria for Schatten class composition operators on $A^2(D)$ in terms of $Z(\phi)$ and criterion for Schatten composition operators on $H^2(D)$ in terms of $W(\phi)$ was also obtained by the author in [**45**], which can be stated as follows:

THEOREM 2.4. *Let D be a smoothly bounded strictly pseudoconvex domain in $\mathbb{C}^n$. Let $\phi : D \to D$ be a holomorphic map.*

(i) For $2 \leq p \leq \infty$, we have

$$\frac{1}{C_p} W_p(\phi) \leq \|C_\phi\|_{S_p(H^2(D))} \leq C_{p,q} W_q(\phi) \tag{2.11}$$

for any $q < p$;

(ii) Let $1 \leq p < \frac{2n}{n-1}$ and let D be either a smoothly bounded strictly pseudoconvex domain or a bounded symmetric domain in $\mathbb{C}^n$. Then

$$\frac{1}{C_p} Z_{2p}(\phi) \leq \|C_\phi\|_{S_{2p}(A^2)} \leq C_p Z_{2p}(\phi). \tag{2.12}$$

where C_p is a positive constant depending only on p and D.

There are many other researches on characterizations for boundedness, compactness of composition on weighted Bergman spaces from A^p to A^q, or other function spaces, we mention a few examples here. For examples, we refer the readers to papers of Z. Cuckovic and R. Zhao [**16**], Luo and Shi and Zhou [**74**], Zhao [**80**], etc. and references therein.

3. Isometric Composition Operators

For a bounded domain D in $\mathbb{C}^n$, we let $\mathrm{Aut}(D)$ be the automorphism group of D. In this section, we will define some function spaces, on which the composition operator C_ϕ is an isometry for any $\phi \in \mathrm{Aut}(D)$. Let

$$B^2(D) = \Big\{ f \in A^2(D) : \|f\|_{B^2}^2 = \int_D |\partial f(z)|^2 dv(z) < \infty \Big\}. \tag{3.1}$$

When D is a bounded domain in the complex plane, it is easy to verify that C_ϕ is an isometry on $B^2(D)$ for any $\phi \in \mathrm{Aut}(D)$. This is no longer true when $n > 1$. It has been proved by Arazy, Fisher, Janson and Peetre [**3**], K. Zhu [**82**] and Peloso [**63**] that C_ϕ is an isomorphism on $B^2(D)$ when D is the unit ball in $\mathbb{C}^n$ and $\phi \in \mathrm{Aut}(D)$. The results for more general function spaces over a more general class of domains were obtained by the author and Luo [**47**]. We will provide a straight forward way to define a general holomorphic Besov spaces so that the following statements hold:

(a) C_ϕ is an isometry on it for any $\phi \in \mathrm{Aut}(D)$;

(b) The new norm is equivalent to the usual norm.

Let $\delta(z)$ be the distance function from z to ∂D. For any $0 < p < \infty$, Besov space over a domain D consisting of all holomorphic functions f on D so that

$$\|f\|^p_{B^p(D)} = \int_D |\partial f(z)|^p \delta(z)^p d\lambda(z) < \infty. \tag{3.2}$$

It is easy prove that $B^p = \mathbb{C}$ when $p \le n$ since $p - n - 1 \le -1$ and $\int_D \delta(z)^{-1} dv(z) = \infty$. In order to avoid that B^p becomes trivial when $0 < p \le n$, one may change the definition for B^p by replacing $|\partial f(z)|\delta(z)$ by $|\partial^k f(z)|\delta(z)^k$ for $pk > n$ (see [**47**] for details). In general, C_ϕ is not an isometry on $B^p(D)$ when $n > 1$. So we introduce here some new norms for new Besov spaces, and prove they are equivalent to old ones.

First, we use a Kähler metric on D to define holomorphic function spaces. Let $g = g_{\alpha\overline{\beta}} dz_\alpha \otimes d\overline{z}^\beta$ be a Kähler metric over D. Then the Laplace operator associated to g is

$$\Delta_g = \sum_{\alpha,\beta=1}^n g^{\alpha\overline{\beta}}(z) \frac{\partial^2}{\partial z_\alpha \partial \overline{z}_\beta}, \quad |\nabla f|^2_g = \sum_{\alpha,\beta} g^{\alpha\overline{\beta}} \Big(\frac{\partial f}{\partial z_\alpha} \overline{\frac{\partial f}{\partial z_\beta}} + \frac{\partial \overline{f}}{\partial z_\alpha} \overline{\frac{\partial \overline{f}}{\partial z_\beta}} \Big) \tag{3.3}$$

Let $O(D)$ be the set of all holomorphic functions over D. For any $0 < p < \infty$, we let $B^p_g(D)$ be the set of all functions $f \in O(D)$ so that

$$\|f\|^p_{g,p} = \int_D |\nabla f(z)|^p_g dv_g(z) < \infty, \quad dv_g(z) = |\det(g_{\alpha\overline{\beta}})|^2 dv(z). \tag{3.4}$$

As we know from a result of Bell [**10**] and Boas and Straube [**13**] that if D is a smoothly bounded domain in $\mathbb{C}^n$ with a smooth plurisubharmonic defining function then ϕ can be extended smoothly up to the boundary of D for any $\phi \in \mathrm{Aut}(D)$. If g is a metric so that there is a constant $C \ge 1$ with

$$\frac{1}{C} g(z) \le g(z) \circ \phi(z) \le C g(z) \tag{3.5}$$

Then C_ϕ is an isomorphic on $B^p_g(D)$ for each $\phi \in \mathrm{Aut}(D)$.

Furthermore, if g is an invariant metric, i.e., g is either Bergman metric or Kähler-Einstein metric, then C_ϕ is an isometry on $B^p_g(D)$ for any $\phi \in \mathrm{Aut}(D)$. In fact, for those two metrics, we have

$$dv_g(z) = dv_g(\phi(z)), \quad \sum_{\alpha,\beta=1}^n g^{\alpha\overline{\beta}}(z) \frac{\partial \phi_k}{\partial z_\alpha} \overline{\frac{\partial \phi_\ell}{dz_\beta}} = g^{k\overline{\ell}}(\phi(z)). \tag{3.6}$$

This implies that

$$\Delta_g |C_\phi f(z)|^2 = (\Delta_g |f|^2)(\phi(z)). \tag{3.7}$$

Notice that if f is holomorphic, then

$$|\nabla f|^2_g = \Delta_g |f(z)|^2. \tag{3.8}$$

Thus

$$\|C_\phi(f)\|_{B^p_g} = \|f\|_{B^p_g}. \tag{3.9}$$

When D is a smoothly bounded strictly pseudoconvex domain in $\mathbb{C}^n$, one may use the result of C. Fefferman [**24**] on asympototic expansion for Bergman kernel function, Cheng and Yau [**19**], Lee and Melrose [**42**] on global regularity for the potential function of Kähler-Einstein metric, as well as the analysis of E. Stein [**76**], to prove $B_g^p(D)$ and $B^p(D)$ are equivalent when $p > 2n$.

Second, we use the Bergman kernel function to define function spaces $Y_p(D)$, which are equivalent to Besov spaces in many cases. Those spaces $Y_p(D)$ were also used a lot in studying the theory of Hankel operators, commutators (see for examples, [**2**], [**3**], [**8**], [**9**], [**14**] and [**43**], ect. and references therein.) For each $z \in D$, we let

$$k_z(w) = K(z,z)^{-1/2}K(w,z), \ w \in D. \tag{3.10}$$

For any $f \in A^2(D)$ we let (Berezin transform)

$$B(f)(z)^2 = \int_D |f(w) - f(z)|^2 |k_z(w)|^2 dv(w). \tag{3.11}$$

The function space $Y_p(D)$ is defined as follows:

$$Y_p(D) = \{f \in A^2(D) : \|f\|_{Y_p(D)} < \infty\}, \quad \|f\|^p_{Y_p(D)} = \int_D B(f)^p(z) d\lambda(z). \tag{3.12}$$

For any $\phi \in \mathrm{Aut}(D)$, since $|k_z(w)|^2 = |k_{\phi(z)}(\phi(w)|^2 |\det \phi'(w)|^2$, we have

$$\begin{aligned} B(C_\phi(f))(z)^2 &= \int_D |C_\phi(f)(w) - C_\phi(f)(z)|^2 |k_z(w)|^2 dv(w) \\ &= \int_D |C_\phi(f)(w) - C_\phi(f)(z)|^2 |k_{\phi(z)}(\phi(w))|^2 |\det \phi'(w)|^2 dv(w) \\ &= \int_D |f(w) - C_\phi(f)(z)|^2 |k_{\phi(z)}(w)|^2 dv(w) \\ &= B(f)^2(\phi(z)) \end{aligned} \tag{3.13}$$

Since $d\lambda(z)$ is invariant measure under biholomorphic map, we have

$$\|f\|_{Y_p} = \|C_\phi(f)\|_{Y_p(D)}, \qquad \text{for all } \phi \in \mathrm{Aut}(D). \tag{3.14}$$

Therefore C_ϕ is an isometry on $Y_p(D)$ for all $\phi \in \mathrm{Aut}(D)$.

Third, for each $0 < p \le \infty$, one can easily see that C_ϕ is an isometry on $L^p(D, d\lambda)$ with norm:

$$\|h\|_{L^p(D,d\lambda)} = \Big[\int_D |h(z)|^p d\lambda(z)\Big]^{1/p}. \tag{3.15}$$

The relations between $B^p(D)$, $B_g^p(D)$ and $Y_p(D)$ can be described as the following theorem.

THEOREM 3.1. *Let D be a smoothly bounded strictly pseudoconvex domain in $\mathbb{C}^n$. Then*

(i) If $0 < p \le n$ then $B^p(D) = B_g^p(D) = Y_p(D) = \mathbb{C}$;

(ii) If $n < p \le 2n$ then $B_g^p(D) = Y_p(D) = \mathbb{C}$, and $B^p(D) \ne \mathbb{C}$ when $p > n$;

(iii) If $2n < p < \infty$, there is a constant $C_p > 0$ so that

$$\frac{1}{C_p}\|f\|_{B^p(D)} \le \|f\|_{B_g^p(D)} \approx \|f\|_{Y_p(D)} \le C_p \|f\|_{B^p(D)} \tag{3.16}$$

for all $f \in B^p(D)$.

Remark: The above theorem remains true on pseudoconvex domain of finite type domain in $\mathbb{C}^2$ or convex domain of finite type in $\mathbb{C}^n$ by replacing 2 (strictly pseudocovex domain is type 2) by type m (see, Li and Luo [**47**] for the details of some arguments.)

There is another space, a little bit different from $B^\infty(B)$ (Bloch space), is BMOA(D). If one defines BMOA-norm (semi-norm) as:

$$A(f)(z)^2 = \int_{\partial D} |f(w) - f(z)|^2 P(z,w) d\sigma(w), \quad P(z,w) = \frac{|S(z,w)|^2}{S(z,z)} \tag{3.17}$$

and

$$\|f\|_{BMOA(D)} = \|A(f)\|_{L^\infty(D)}. \tag{3.18}$$

Then for any $f \in H^2(D)$, we have that $BMO(\partial D)$-norm for f is equivalent to the usual BMOA norm as (3.18) when D is strictly pseudoconvex domain in $\mathbb{C}^n$. With the definition (3.18), we have the following theorem.

THEOREM 3.2. *Let D be a bounded symmetric domain in $\mathbb{C}^n$. Then C_ϕ is an isometry on $BMOA(D)$ for any $\phi \in \mathrm{Aut}(D)$.*

Proof. Let $\phi_z \in \mathrm{Aut}(D)$ so that $\phi_z(0) = z$ and $\phi_z \circ \phi_z = I$. Then

$$\phi_{z_0} \circ \phi_z(w) = \phi_{\phi_{z_0}(\phi_z(0))}(w)$$

and (see, for example, [**68**])

$$\int_{\partial D} g(\phi_z)(w) d\sigma(w) = \int_{\partial D} g(w) P(z,w) d\sigma(w), \quad \phi_z \in \mathrm{Aut}(D).$$

Thus

$$\begin{aligned} \|f\|^2_{BMOA(D)} &= \sup\Big\{ \int_{\partial D} |f(w) - f(z)|^2 P(z,w) d\sigma(w) : z \in D \Big\} \\ &= \sup\Big\{ \int_{\partial D} |f \circ \phi_z(w) - f \circ \phi_z(0)|^2 d\sigma(w) : \phi_z \in \mathrm{Aut}(D) \Big\}. \end{aligned}$$

Let $\phi_{z_0} \in \mathrm{Aut}(D)$ so that $\phi_{z_0}(0) = z_0$. Then

$$\begin{aligned} &\|f \circ \phi_{z_0}\|^2_{BMOA(D)} \\ &= \sup\Big\{ \int_{\partial D} |(f \circ \phi_{z_0}) \circ \phi_z(w) - (f \circ \phi_{z_0})(\phi_z(0))|^2 d\sigma(w) : \phi_z \in \mathrm{Aut}(D) \Big\} \\ &= \sup\Big\{ \int_{\partial D} |f \circ \phi_{\phi_0(\phi_z(0))}(w) - f \circ \phi_{\phi_0(\phi_z(0))}(0)|^2 d\sigma(w) : \phi_z \in \mathrm{Aut}(D) \Big\} \\ &= \sup\Big\{ \int_{\partial D} |f \circ \phi_z(w) - f \circ \phi_z(0)|^2 d\sigma(w) : \phi_z \in \mathrm{Aut}(D) \Big\} \\ &= \|f\|^2_{BMOA}. \end{aligned}$$

Therefore, the proof of the theorem is complete. □

Now what is the relation between $L^p(D, d\lambda)$ and $B^p(D)$? The problem was studied by Zhu [**83**] for bounded symmetric domain, Li and Luo [**47**] for more general pseudoconvex domains in $\mathbb{C}^n$.

Let $d\Lambda(z) = K(z,z)^{-1}dv(z)$ with K being the Bergman kernel function. Let $\tilde{K}$ be the Bergman kernel on D with respect to the measure $d\Lambda(z)$. We define an operator

$$V(f)(z) = K(z)^{-1}\int_D \tilde{K}(z,w)f(w)dv(w). \tag{3.19}$$

Then

$$PV = I_n \text{ on } A^2(D). \tag{3.20}$$

In fact,

$$\begin{aligned}
PV(f)(z) &= \int_D K(z,w)V(f)(w)dv(w) \\
&= \int_D K(z,w)K(w,w)^{-1}\int_D \tilde{K}(w,\xi)f(\xi)dv(\xi)dv(w) \\
&= \int_D\int_D K(z,w)K(w,w)^{-1}\tilde{K}(w,\xi)dv(w)f(\xi)dv(\xi) \\
&= \int_D K(z,\xi)f(\xi)dv(\xi) \\
&= f(z).
\end{aligned}$$

It was proved in [**47**] the following theorem.

THEOREM 3.3. *Let D be a smoothly bounded strictly pseudoconvex in $\mathbb{C}^n$ or pseudoconvex domain of finite type in $\mathbb{C}^n$ or convex domain of finite type in $\mathbb{C}^n$. Then $V : B^p(D) \to L^p(D, d\lambda)$ is bounded. Moreover, $P : L^p(D, d\lambda) \to B^p(D)$ is bounded.*

Note: The above theorem was only true when $n < p \leq \infty$ for usual definition for B^p, it remains true for general p if one modifies the definition for B^p when $p \leq n$.

For any $\phi \in \mathrm{Aut}(D)$, it is easy to show that

$$\tilde{K}(z,w) = \tilde{K}(\phi(z),\phi(w))(\det\phi'(z))^2\overline{\det\phi'(w)}^2 \tag{3.21}$$

If we let

$$E(\phi)(w) = \frac{\det\phi'(w)}{\overline{\det\phi'(w)}}, \tag{3.22}$$

then

PROPOSITION 3.4. *For any biholomorphic map $\phi : D \to D$, we have*

$$VC_\phi(f)(z) = E_\phi(z)\,C_\phi V(PM_{E(\phi^{-1})}f). \tag{3.23}$$

Note that

$$\begin{aligned} V(f\circ\phi)(z) &= K(z,z)^{-1}\int_D \tilde{K}(z,w)f(\phi(w))dv(w) \\ &= K(z,z)^{-1}(\det\phi'(z))^2\int_D \tilde{K}(\phi(z),\phi(w))f(\phi(w))\overline{\det\phi'(w)}^{-2}dv(w) \\ &= E(\phi)(z)K(\phi(z),\phi(z))^{-1}\int_D \tilde{K}(\phi(z),w)f(w)E(\phi^{-1})(w)dv(w) \\ &= E(\phi)(z)K(\phi(z),\phi(z))^{-1}\int_D \tilde{K}(\phi(z),w)P(f(w)E(\phi^{-1}))(w)dv(w) \\ &= E(\phi)(z)C_\phi(V(P(fE(\phi^{-1})))\end{aligned}$$

and

(3.24)

$$\|V(f\circ\phi)\|^p_{L^p(D,d\lambda)} = \|E(\phi)V(fE(\phi))(\phi)\|^p_{L^p(D,d\lambda)} = \|V(P(fE(\phi)))\|^p_{L^p(D,d\lambda)}.$$

Therefore, when $\phi\in \operatorname{Aut}(D)$ with $\phi\in C^1(\overline{D})$, then $PM_{E(\phi^{-1})}: B^p(D)\to B^p(D)$ is bounded. One can see from Theorem 3.3, (3.23) and (3.24) that C_ϕ is an isomorphism on $B^p(D)$ since C_ϕ is an isometry on $L^p(D,d\lambda)$.

4. Isometries

Characterizing an isometry on a given function spaces has been received some studies, some known results along these directions can be summarized as follows:

1) Let D be either the unit ball or unit polydisk in $\mathbb{C}^n$.
If $T: H^p(D)\to H^p(D)$ with $p\in(1,2)\cup(2.\infty]$ is an onto isometry, then

$$T(f)(z) = e^{i\theta}P(\phi^{-1}(0),z)^{-1/p}C_\phi(f)(z)$$

for some $\phi\in\operatorname{Aut}(D)$, where $P(z,w)=S(z,z)^{-1}|S(z,w)|^2$ is Poisson-Szegö kernel.

When D is the unit disk, Statement 1) was proved by Nagasawa in 1959 for $p=\infty$, by deLeeuw, Rudin and Wermer for $p=1$ in 1960 and by Forelli [**25**] for all p.

When D is polydisc, Statement 1) was proved by Schneider in 1971; and when D is the unit ball, Statement 1) was proved by Forelli for $p>2$ and Rudin [**70**] for general p.

The following theorem was proved by Koranyi and Vagi [**41**]

2) Let D be a bounded symmetric domain in $\mathbb{C}^n$. If $T: H^p(D)\to H^p(D)$ with $p\in(1,2)\cup(2.\infty]$ is an into isometry, then

$$T(f)(z) = T(1)C_\phi(f)(z)$$

for some $\phi\in\operatorname{Aut}(D)$.

For Bergman spaces, the following theorem was proved by Kolaski [**39**] and [**40**].

3) Let D be a bounded Runge domain in $\mathbb{C}^n$. Let T be an into isometry on $A^p(D)$ for $p\in(1,)\cup(2,\infty)$. Then $T(f)(z)=T(1)(z)C_\phi(f)(z)$ for some $\phi\in\operatorname{Aut}(D)$.

For the little Bloch space, Cima and Wogen [**21**] for $n=1$ and Krantz and Ma [**38**] for all n prove the following result:

4) Let D be the unit ball in $\mathbb{C}^n$. Let T be an isometry on $B_0(D)$. Then $T(f)(z) = \mu[C_\phi(f)(z) - C_\phi(f)(0)]$ for some $\phi \in \text{Aut}(D)$.

When D is polydisk, some result related 4) was obtained by L. Chen in [**18**].

Viewing above four results, and C_ϕ is an isometry on those Besov space, BMOA constructed in Section 3. One may pose the following question.

Problem: Let D be a smoothly bounded strictly pseudoconvex domain or polydisk in $\mathbb{C}^n$. Let X be one of the following function spaces: $X = Y_p(D)$, $\tilde{B}^p(D)$, $BMOA(D)$, $VMOA(D)$, and let T be an isometry on X. Is it true $Tf = aC_\phi f - bC_\phi(f)(z_0)$ for some $\phi \in Aut(D)$, complex numbers a, b with $|a| = 1$ and some point $z_0 \in D$?

References

[1] P. Ahern and J. Bruna, *Holomorphic functions with absolutely continuous boundary values*, Duke Math. J., **56** (1988), 129–142.

[2] J. Arazy, D. Fisher and J. Peetre, *Hankel operators on weighted Bergman spaces*, Amer. J. Math., **111** (1988), 989–1054.

[3] J. Arazy, D. Fisher, S. Janson and J. Peetre, *Möbius invariance function spaces*, J. Reine Angew., **363** (1985), 110–145.

[4] F. Beatrous, *Boundary continuity of holomorphic function in the ball*, Proceeding of AMS, **97** (1986), 23–29.

[5] F. Beatrous, *Estimates for derivatives of holomorphic functions in pseudoconvex domains*, Math. Z., **191** (1986), 91–116.

[6] F. Beatrous and J, Burbea, *Sobolev spaces of holomorphic functions in the ball*, Dissertations Math., CCLXXVI, Warzawa, 1989.

[7] F. Beatrous and J, Burbea, *Boundary regularity for Harmonic Hardy-Sobolev spaces*, J. London Math Soc., **29** (1989), 160–174.

[8] F. Beatrous, S.-Y. Li, *On the boundedness and compactness of operators of Hankel type*, J. of Functional Analysis, **111** (1993), 350–379.

[9] F. Beatrous, S.-Y. Li, *Trace ideal criteria for operators of Hankel type*, Illinois J. of Math., **39** (1995), 723–754.

[10] S. Bell, *Biholomorphic mappings and the $\bar{\partial}$-problem*, Ann of Math. **114** (1981), 103–113.

[11] A. Bonami, M. M. Peloso and F. Symesak, *Factorization of Hardy spaces and Hankel operators on convex domains in $\mathbb{C}^n$*, preprint, 1999

[12] A. Bonami, M. M. Peloso and F. Symesak, *Powers of the Szegö kernel and Hankel operators on Hardy spaces*, Mich. Math. J., **46** (1999),225–250.

[13] H. Boas and E. Straube, *Sobolev estimates for the $\bar{\partial}$-Neumann operator on domains in $\mathbb{C}^n$ admitting a defining function that is plurisubharmonic on the boundary*, Math. Z. **206** (1991), 81–88.

[14] D. Bekolle, C. Berger, L. Coburn and K. Zhu, *BMO in Bergman metric of bounded symmetric domains*, J. Funct. Anal. **93** (1990), 310–350.

[15] R. Coifman, R. Rochberg and G. Weiss, *Factorization theorem for Hardy spaces in several variables*, Ann. of Math., **103** (1976), 611–635.

[16] Z. Cuckovic, R. Zhao, *Essential Norm Estimates of Weighted Composition Operators between Bergaman Spaces on strongly Pseudoconvex Domains*, preprint.

[17] D. C. Chang, *Estimates of weighted Bergman projections on decoupled domains of finite type*, Nagoya Math. J.

[18] L. Chen, *Operators on the Bloch and Bergman Spaces of Several Complex Variables*, Ph.D. Thesis ay Univ. of California at Irvine, 1994.

[19] S. Y. Cheng and S. T. Yau, *On the existence of a complex Kähler metric on non-compact complex manifolds and the regularity of Fefferman's equation*, Comm. Pure Appl. Math. **33** (1980) 507–544.

[20] A. Cima, and W. Wogen, *Unbounded composition operators on $H^2(B_n)$*, Proc. Amer. Math. Soc. **99** (1987), 477–483.

[21] A. Cima, and W. Wogen, *On isometries of Bloch space*, Illinois J. Math., **24** (1980), 313–316.

[22] C. C. Cowen, *Composition operators on Hilbert spaces of analytic functions; A status report*, Proc. Sympos. Pure Math. **51** (1990), 131–145.

[23] C. C. Cowen, and B. MacCluer, *Composition operators on spaces of analytic functions*, CRC Press, Boca Raton, New York, London, Tokyo, 1995.

[24] C. Fefferman, *The Bergman kernel and biholomorphic mappings of pseudoconvex domaina*, Invent. Math. **26** (1974), 1–65.

[25] F. Forelli, *The isometries of H^p*, Cann. J. Math., **16** (1964), 721–728.

[26] J. Garnett, *Bounded analytic functions*, Academic press, New York, 1981.

[27] L. Hörmander, *L^p-estimates for pluri-subharmonic functions*, Math. Scand. **20** (1967), 65–78.

[28] F. Jafari, *Composition operators in Bergman spaces on bounded symmetric domains*, Cont. Math. **137** (1992), 277-290.

[29] S. G. Krantz, *Function theory of several complex variables*, Wadsworth & Brooks, 2nd edition, 1992.

[30] S. G. Krantz and S-Y. Li, *On factorization theorems for functions in subspaces of L^1 and applications to the Corona problem*, Indiana U. Math. J., **45** (1996), 83–102.

[31] S. G. Krantz, S-Y. Li and R. Rochberg, *Analysis of some function spaces associated to Hankel operators*, Illinois J. of Math., **41** (1997), 398–411.

[32] S. G. Krantz and S-Y. Li, *A Note on Hardy spaces and functions of bounded mean oscillation on domains in $\mathbb{C}^n$*, Mich. Math. J., **41** (1994), 51–72.

[33] S. G. Krantz and S-Y. Li, *Duality theorems on Hardy and Bergman spaces on convex domains of finite type in $\mathbb{C}^n$*, Ann of Fourier Institute, **45** (1995), 1305–1327.

[34] S. G. Krantz and S-Y. Li, *On the decomposition theorems for Hardy spaces in domains in $\mathbb{C}^n$ and application*, J. of Fourier Analysis and Applications, **2** (1995), 65–107.

[35] S. G. Krantz and S-Y. Li, *Area integral characterization of functions in Hardy spacers on domains in $\mathbb{C}^n$*, Complex Variables, **32** (1997), 373–399.

[36] S. G. Krantz, S-Y. Li and R. Rochberg, *Analysis of some function spaces associated to Hankel operators*, Ill. Jour. of Math., **41** (1997), 398–411.

[37] S. G. Krantz, S-Y. Li and R. Rochberg, *The effect of boundary geometry on Hankel operators belonging to the trace ideals of Bergman spaces*, Integr. equ. oper. theory, **28** (1997), 196–213.

[38] S. G. Krantz and D. Ma, *On isometric isomorphisms of the Bloch spaces on the unit ball of $\mathbb{C}^n$*, Mich. Math. J., **36** (1989), 173–180.

[39] C. Kolaski, *Isometries of Bergman space over bounded Runge domains*, Can. J. Math., **33** (1981), 1157–1164.

[40] C. Kolaski, *Isometries of weighted Bergman space*, Can. J. Math., **34** (1982), 910–915.

[41] A. Koranyi and S. Vagi, *On isometries of H^p of bounded symmetruc domains*, Cann. J. Math., **28** (1976), 334–340.

[42] J. Lee and Melrose, J. Lee and R. Melrose, *Boundary behavior of the complex Monge-Ampère equation*, Acta Math. (1982), 159–192.

[43] H. Li, *BMO, VMO, and Hankel operators on Bergmam spaces of strongly pseudoconvex domains*, Jour. of Funct. Anal., 1993.

[44] S-Y. Li, *Trace Ideal criteria for composition operators on Bergman spaces*, Amer. J. of Math, **117** (1995), 1299–1324.

[45] S-Y. Li, *Criterias for Schatten Class Composition Operators on Hardy and Bergman Spaces*, Geometric Function Theory in Several Complex Variables edited by Carl H FitzGerald & Sheng Gong, 2004.

[46] S-Y. Li, *Some function theoritical results on domains in $\mathbb{C}^n$*, Proceeding of ICCM 2001, Editors: C-S Lin, L. Yang and S.T. Yau, International press, 2004

[47] S-Y. Li and W. Luo, *Characterizations for Besov Spaces ans Applications, Part I, J. Math. Anal. Appl.*, 310(2005), 477–491.

[48] S-Y. Li and W. Luo, *Analysis on Besov spaces II: Embedding and Duality Theorems, J. Math. Anal. Appl.*, 2006.

[49] S-Y. Li and B. Russo, *Schatten class composition operators on weighted Bergman spaces of bounded symmetric domains*, Annali di Matematica pura ed applicata, CLXXII (1997), 379–394.

[50] S.-Y. Li and B. Russo, *On the compactness of composition operators in Hardy spaces of several variables*, Proc. Amer. Math. Soc., **123** (1995), 161–171.

[51] D. H. Luecking, *Trace ideal criteria for Toeplitz operators*, J. Funct. Anal. **73** (1987), 345–368.

[52] D. H. Luecking, D, and K. Zhu, *Composition operators belonging to the Schatten ideals*, Amer. J. Math. **114** (1992), 1127–1145.

[53] L. Luo and J. H. Shi, *Composition operators between Bergman spaces on bounded strongly pseudoconvex domains of* $\mathbb{C}^n n$, (Chinese) Chinese Ann. Math. Ser. A **24** (2003), no. 6, 721–728.

[54] B. MacCluer, *Spectra of compact composition operators on* $H^p(B_N)$, Analysis **4** (1984), 87–103.

[55] B. MacCluer, *Composition operators on* $H^p(B_n)$, Mich. Math. J., 32 (1985), 237–248.

[56] B. MacCluer, and J. H. Shapiro, *Angular derivatives and compact composition operators on Hardy and Bergman spaces*, Can. J. Math. **38** (1986), 878–906.

[57] J. McNeal, *Estimates on the Bergman kernels function of convex domains*, Advances in Math., **109** (1994), , 108–139.

[58] J. McNeal and E. M. Stein, *Mapping properties of the Bergman projection on convex domains of finite type*, Duke Math. Jour. **73** (1994), 177–199.

[59] Z. Nehari, *On the bounded bilinear forms*, Ann. of Math., **65** (1957), 153–162.

[60] A. Nagel, J. P. Rosay, E. M. Stein, and S. Wainger, *Estimates for the Bergman and Szegö kernels in* $\mathbb{C}^2$, Ann. Math., **129** (1989), 113-149.

[61] C. Ouyang, W. Yang and R. Zhao, *Möbius invariant* Q_p *spaces associated with the Green's function the unit ball of* $\mathbb{C}^n$, Pacific J. of Math. , **182** (1998), 69–99.

[62] V. V. Peller, *Hankel operators of class* S_p *and their application*, Math. of the USSR Sbornik, **113** (1982), 538–581.

[63] M. M. Peloso, *Möbius Invariant spaces on the ball*, Mich. Math. J., **39** (1992), 509–536.

[64] M. M. Peloso, *Hankel operators on weighted Bergman spaces on strictly pseudoconvex domains*, Ill. J. of Math., **38** (1994), 223–249.

[65] D. H. Phong and E. M. Stein, *Estimates for the Bergman and Szegö projections on strongly pseudoconvex domains*, Duke Math. J, **44** (1977), 695–704.

[66] R. M. Range, *Holomorphic functions and integral representations in several complex variables*, Springer-Verlag, 1986.

[67] B. Russo, *Holomorphic Composition Operators in Several Complex Variables*, Contemporary Mathematics, **213** (1998), 191–212.

[68] B. Russo, *On the Hausdröff-Young theorem for integral operators*, Pacific J. of Math. **68** (1977), 241-253.

[69] R. Rochberg, *Trace ideal criteria for Hankel operators and commutators*, Indiana U. Math. J., **31** (1982), 913–925.

[70] W. Rudin, L^p*-Isometries and Equimeasurability*, Indiana U. Math. J., **3** (1976), 215–228.

[71] W. Rudin, *The Function Theorey in Unit Ball in* $\mathbb{C}^n$, Springer-Verlag, New York, 1980.

[72] J. H. Shapiro, *The essential norm of a composition operator*, Ann. Math. **125** (1987), 375–404.

[73] J. H. Shapiro, and P.D. Taylor, *Compact, nuclear and Hilbert-Schmidt composition operators on* H^2, Indiana U. Math. J., **23** (1973), 471–496.

[74] Z.-H. Zhou, J. H. Shi, *Compactness of composition operators on the Bloch space in classical bounded symmetric domains*, Michigan Math. J. **50** (2002), no. 2, 381–405.

[75] E. M. Stein, *Harmonic Analysis*, Princeton Univ. Press, Princeton, NJ, 1993.

[76] E. M. Stein, *Boundary behavior of holomorphic functions of several complex variables*, Princeton Univ. Press, Princeton, NJ, 1972.

[77] F. Symesak, *Hankel operators on pseudoconvex domains of finite type in* $\mathbb{C}^2$, Canadian J. Math., 1999.

[78] W. R. Wogen, *Composition operators acting on spaces of holomorphic functions on domains in* $\mathbb{C}^n$, Proc. Sym. Pure Math. **51** (1990), 361–366.

[79] J. Xia, *On a proposed characterization of Schatten-class composition operators*, Proc. Amer. Math. Soc. **131** (2003), 2505-2514.

[80] R. Zhao, *Composition Operators from Bloch Type Spaces to Hardy Spaces*, Jour. of Math. Anal. And Appl., **233** (1999), 749–766.

[81] K. Zhu, *Positive Toeplitz operators on weighted Bergman spaces of bounded symmetric domains*, J. Operator Theory, **20** (1988), 329–357.

[82] K. Zhu, *Möbius invariance Hilbert spaces of holomorphic functions in the unit ball of* $\mathbb{C}^n$, Tran. of AMS, **323** (1991), 823–842.

[83] K. Zhu, *Hankel Operators on Bergman spaces of bounded symmetric domains*, Tran. of AMS, **324** (1991), 707–730.

[84] K. Zhu, *Schatten class composition operators on weighted Bergman spaces of the disk*, J. Operator Theory, **46** (2001), 173–181.

DEPARTMENT OF MATHEMATICS, UNIVERSITY OF CALIFORNIA, IRVINE, CA 92697-3875, USA
E-mail address: sli@math.uci.edu

AMS/IP Studies in Advanced Mathematics
Volume **39**, 2007

Localization and String Duality

Kefeng Liu

This paper is dedicated to S.-S. Chern

ABSTRACT. We survey the recent results on Hodge integrals and Gromov-Witten invariants we have obtained through localization techniques. These include the proof of the Mariño-Vafa formula, its two partition generalization and the theory of topological vertex which gives complete and closed formulas for the generating series of all degree and all genera Gromov-Witten invariants of any formal toric Calabi-Yau manifolds. We will also briefly discuss Peng's proof on the Gopakumar-Vafa conjecture and the new recursion formula of Kim.

1. Introduction

According to string theorists, String Theory, as the most promising candidate for the grand unification of all fundamental forces in the nature, should be the final theory of the world, and should be unique. But now there are *five* different looking string theories. As argued by the physicists, these theories should be equivalent, in a way dual to each other. On the other hand all previous theories like the Yang-Mills and the Chern-Simons theory should be parts of string theory. In particular their partition functions should be equal or equivalent to each other in the sense that they are equal after certain transformation. To compute partition functions, physicists use localization technique, a modern version of residue theorem, on infinite dimensional spaces. More precisely they apply localization formally to path integrals which is not well-defined yet in mathematics. In many cases such computations reduce the path integrals to certain integrals of various Chern classes on various finite dimensional moduli spaces, such as the moduli spaces of stable maps and the moduli spaces of vector bundles. The identifications of these partition functions among different theories have produced many surprisingly beautiful mathematical formulas like the famous mirror formula [**22**], as well as the Mariño-Vafa formula [**25**].

The mathematical proofs of these conjectural formulas from the string duality also depend on localization techniques on these various finite dimensional moduli spaces. The purpose of this note is to discuss our recent works on the subject. I

1991 *Mathematics Subject Classification.* 53D30, 57R91, 81T13, 81T30.

The author thanks NSF and NSFC for support during the preparation of this note.

will briefly discuss the proof of the Marinõ-Vafa formula, its generalizations and the related topological vertex theory [**1**]. More precisely we will use localization formulas in various form to compute the integrals of Chern classes on moduli spaces, and to prove those conjectures from string duality. For the proof of the Mariño-Vafa formula and the theory of topological vertex, we note that many aspects of mathematics are involved, such as the Chern-Simons knot invariants, combinatorics of symmetric groups, representations of Kac-Moody algebras, Calabi-Yau manifolds, geometry and topology of moduli space of stable maps, etc.

We remark that localization technique has been very successful in proving many conjectures from physics, see my ICM 2002 lecture [**21**] for more examples. One of our major tools in the proofs of these conjectures is the functorial localization formula which is a variation of the classical localization formula, it transfers computations on complicated spaces to simple spaces, and connects computations of mathematicians and physicists.

Starting from the proof of the Mariño-Vafa formula [**18**], we have proved a series of results about Hodge integrals on the moduli spaces of stable curves, Gromov-Witten invariants for open toric Calabi-Yau manifolds, and their relationship with equivariant indices of elliptic operators on the moduli spaces of framed stable bundles on the projective plane. Here we can only give a brief overview of the results and the main ideas of their proofs. For the details see [**17**], [**15**], [**18**], [**19**], [**20**], [**13**]. While the Marinõ-Vafa formula gives a close formula for the generating series of triple Hodge integrals on the moduli spaces of all genera and any number marked points, the mathematical theory of topological vertex [**13**] gives the most effective ways to compute the Gromov-Witten invariants of any open toric Calabi-Yau manifolds. Recently Pan Peng was able to use our results on topological vertex to give a complete proof of the Gopakumar-Vafa integrality conjecture for any open toric Calabi-Yau manifolds [**27**]. Kim also used our technique to derive new effective recursion formulas for Hodge integrals on the moduli spaces of stable curves [**7**].

The spirit of our results is the duality between gauge theory, Chern-Simons theory and the Calabi-Yau geometry in string theory. One of our observations about the geometric structure of the moduli spaces is the convolution formula which is encoded in the moduli spaces of relative stable maps [**9**], [**10**], and also in the combinatorics of symmetric groups, [**18**], [**13**]. This convolution structure implies the differential equation which we called the cut-and-join equation. The cut-and-join equation arises from both representation theory and geometry. The verification of the cut-and-join equation in combinatorics is a direct computation through character formulas, while its proof in geometry is quite subtle and involves careful analysis of the fixed points on the moduli spaces of relative stable maps, see [**17**]-[**20**] and [**13**] for more details. The coincidence of such kind of equation in both geometry and combinatorics is quite remarkable.

The mathematical theory of topological vertex was motivated by the physical theory as first developed by the Vafa group [**1**], who has been working on string duality for the past several years. Topological vertex theory is a high point of their work starting from their geometric engineering theory and Witten's conjecture that Chern-Simons theory is a string theory [**29**].

The Gopakumar-Vafa integrality conjecture is possibly the most interesting and challenging remaining conjecture in the subject of Gromov-Witten invariants. It is rather surprising that for some cases such invariants can be interpreted as the indices of elliptic operators in gauge theory in [**17**]. A direct proof of the conjecture for open toric Calabi-Yau manifolds was given recently by Peng [**27**] by using the combinatorial formulas for the generating series of all genera and all degree Gromov-Witten invariants of open toric Calabi-Yau. These closed formulas are derived from the theory of topological vertex through the gluing property.

I will start with brief discussions about the Marinõ-Vafa conjecture, then about the other results mentioned above. This note is based on my lecture at the International Conference of Complex Geometry held at the Eastern China Normal University in the summer of 2004. I would like to thank the organizers for the hospitality during my visit. I would also like to thank my collaborators Melissa C.-C. Liu, Jian Zhou and Jun Li for the wonderful experience to solve these conjectures and to develop the theory together.

The great geometer Professor Shiing-Shen Chern passed away on December 3, 2004. I would like to dedicate this article to the memory of Professor Shiing-Shen Chern, from whom I learned not only mathematics, but also the philosophy of life.

2. The Mariño-Vafa Conjecture

Our original motivation to study Hodge integrals was to find a general mirror formula for counting higher genus curves in Calabi-Yau manifolds. To generalize mirror principle to count the number of higher genus curves, we need to first compute Hodge integrals, i.e. the intersection numbers of the λ classes and ψ classes on the Deligne-Mumford moduli space of stable curves $\overline{\mathcal{M}}_{g,h}$. This moduli space is possibly the most famous and most interesting orbifold. It has been studied since Riemann, and by many Fields medalists for the past 50 years, from many different point of views. Still many interesting and challenging problems about the geometry and topology of these moduli spaces remain unsolved. String theory has motivated many fantastic conjectures about these moduli spaces including the famous Witten conjecture which is about the generating series of the integrals of the ψ-classes. We start with the introduction of some notations.

Recall that a point in $\overline{\mathcal{M}}_{g,h}$ consists of $(C, x_1, \dots, x_h)$, a (nodal) curve C of genus g, and n distinguished smooth points on C. The Hodge bundle $\mathbb{E}$ is a rank g vector bundle over $\overline{\mathcal{M}}_{g,h}$ whose fiber over $[(C, x_1, \dots, x_h)]$ is $H^0(C, \omega_C)$, the complex vector space of holomorphic one forms on C. The λ classes are the Chern Classes of $\mathbb{E}$,

$$\lambda_i = c_i(\mathbb{E}) \in H^{2i}(\overline{\mathcal{M}}_{g,h}; \mathbb{Q}).$$

On the other hand, the cotangent line $T^*_{x_i}C$ of C at the i-th marked point x_i induces a line bundle $\mathbb{L}_i$ over $\overline{\mathcal{M}}_{g,h}$. The ψ classes are the Chern classes:

$$\psi_i = c_1(\mathbb{L}_i) \in H^2(\overline{\mathcal{M}}_{g,h}; \mathbb{Q}).$$

Introduce the total Chern class

$$\Lambda^\vee_g(u) = u^g - \lambda_1 u^{g-1} + \cdots + (-1)^g \lambda_g.$$

The Mariño-Vafa formula is about the generating series of the triple Hodge integrals

$$\int_{\overline{\mathcal{M}}_{g,h}} \frac{\Lambda^\vee_g(1)\Lambda^\vee_g(\tau)\Lambda^\vee_g(-\tau-1)}{\prod_{i=1}^h (1-\mu_i\psi_i)},$$

where τ is considered as a parameter here. Later we will see that it actually comes from the weight of the group action, and also from the framing of the knot. Taking

Taylor expansions in τ or in μ_i one can obtain information on the integrals of the Hodge classes and the ψ-classes. The Marinõ-Vafa conjecture asserts that the generating series of such triple Hodge integrals for all genera and any numbers of marked points can be expressed by a close formula which is a *finite* expression in terms of representations of symmetric groups, or Chern-Simons knot invariants.

We remark that the moduli spaces of stable curves have been the sources of many interests from mathematics to physics. Mumford has computed some low genus numbers. The Witten conjecture, proved by Kontsevich, is about the integrals of the ψ-classes.

Let us briefly recall the background of the conjecture. Mariño and Vafa [**25**] made this conjecture based on the large N duality between Chern-Simons and string theory. It starts from the conifold transition. We consider the resolution of singularity of the conifold X defined by

$$\left\{\begin{pmatrix} x & y \\ z & w \end{pmatrix} \in \mathbf{C}^4 : xw - yz = 0\right\}$$

in two different ways:

(1). Deformed conifold T^*S^3

$$\left\{\begin{pmatrix} x & y \\ z & w \end{pmatrix} \in \mathbf{C}^4 : xw - yz = \epsilon\right\}$$

where ϵ a real positive number. This is a symplectic resolution of the singularity.

(2). Resolved conifold by blowing up the singularity, which gives the total space

$$\tilde{X} = \mathcal{O}(-1) \oplus \mathcal{O}(-1) \to \mathbf{P}^1$$

which is explicitly given by

$$\left\{([Z_0, Z_1], \begin{pmatrix} x & y \\ z & w \end{pmatrix}) \in \mathbf{P}^1 \times \mathbf{C}^4 : \begin{array}{l} (x, y) \in [Z_0, Z_1] \\ (z, w) \in [Z_0, Z_1] \end{array}\right\}$$

$$\begin{array}{ccc} \tilde{X} & \subset & \mathbf{P}^1 \times \mathbf{C}^4 \\ \downarrow & & \downarrow \\ X & \subset & \mathbf{C}^4 \end{array}$$

The brief history of the development of the conjecture is as follows. In 1992 Witten first conjectured that the open topological string theory on the deformed conifold T^*S^3 is equivalent to the Chern-Simons gauge theory on S^3. Such idea was pursued further by Gopakumar and Vafa in 1998, and then by Ooguri and Vafa in 2000. Based on the above conifold transition, they conjectured that the open topological string theory on the deformed conifold T^*S^3 is equivalent to the closed topological string theory on the resolved conifold $\tilde{X}$. Ooguri-Vafa only considered the zero framing case. Later Marinõ-Vafa generalized the idea to the non-zero framing case and discovered the beautiful formula for the generating series of the triple Hodge integrals. Recently Vafa and his collaborators systematically developed the theory, and for the past several years, they developed these duality ideas into the most effective tool to compute Gromov-Witten invariants on toric Calabi-Yau manifolds. The high point of their work is the theory of topological vertex. We refer to [**25**] and [**1**] for the details of the physical theory and the history of the development.

Starting with the proof of the Marinõ-Vafa conjecture [**18**], [**19**], we have developed a rather complete mathematical theory of topological vertex [**13**]. Many interesting consequences have been derived for the past year. Now let us see how the string theorists derived mathematical consequence from the above naive idea of string duality. First the Chern-Simons partition function has the form

$$\langle Z(U,V)\rangle = \exp(-F(\lambda,t,V))$$

where U is the holonomy of the $U(N)$ Chern-Simons gauge field around the knot $K \subset S^3$, and V is an extra $U(M)$ matrix. The partition function $\langle Z(U,V)\rangle$ gives the Chern-Simons knot invariants of K.

String duality asserts that the function $F(\lambda,t,V)$ should give the generating series of the open Gromov-Witten invariants of $(\tilde{X}, L_K)$, where L_K is a Lagrangian submanifold of the resolved conifold $\tilde{X}$ canonically associated to the knot K. More precisely by applying the t'Hooft large N expansion, and the "canonical" identifications of parameters similar to mirror formula, which at level k are given by

$$\lambda = \frac{2\pi}{k+N}, \quad t = \frac{2\pi i N}{k+N},$$

we get the partition function of the topological string theory on conifold $\tilde{X}$, and then on $\mathbf{P}^1$. which is just the generating series of the Gromov-Witten invariants. This change of variables is very striking from the point of view of mathematics.

The special case when K is the unknot is already very interesting. In non-zero framing it gives the Mariño-Vafa conjectural formula. In this case $\langle Z(U,V)\rangle$ was first computed in the zero framing by Ooguri-Vafa and in any framing $\tau \in \mathbb{Z}$ by Mariño-Vafa [**25**]. Comparing with Katz-Liu's computations of $F(\lambda,t,V)$, Mariño-Vafa conjectured the striking formula about the generating series of the triple Hodge integrals for all genera and any number of marked points in terms of the Chern-Simons invariants, or equivalently in terms of the representations and combinatorics of symmetric groups. It is interesting to note that the framing in the Mariño-Vafa's computations corresponds to the choice of lifting of the circle action on the pair $(\tilde{X}, L_{\mathrm{unknot}})$ in Katz-Liu's localization computations. Both choices are parametrized by an integer τ which will be considered as a parameter in the triple Hodge integrals. Later we will take derivatives with respect to this parameter to get the cut-and-join equation.

It is natural to ask what mathematical consequence we can have for general duality, that is for general knots in general three manifolds, a first naive question is what kind of general Calabi-Yau manifolds will appear in the duality, in place of the conifold. Some special cases corresponding to the Seifert manifolds are known by gluing several copies of conifolds.

Now we give the precise statement of the Mariño-Vafa conjecture, which is an identity between the geometry of the moduli spaces of stable curves and Chern-Simons knot invariants, or the combinatorics of the representation theory of symmetric groups.

Let us first introduce the geometric side. For every partition $\mu = (\mu_1 \geq \cdots \mu_{l(\mu)} \geq 0)$, we define the triple Hodge integral to be,

$$G_{g,\mu}(\tau) = A(\tau) \cdot \int_{\overline{\mathcal{M}}_{g,l(\mu)}} \frac{\Lambda_g^\vee(1)\Lambda_g^\vee(-\tau-1)\Lambda_g^\vee(\tau)}{\prod_{i=1}^{l(\mu)}(1-\mu_i\psi_i)},$$

where the coefficient

$$A(\tau) = -\frac{\sqrt{-1}^{|\mu|+l(\mu)}}{|\operatorname{Aut}(\mu)|}[\tau(\tau+1)]^{l(\mu)-1}\prod_{i=1}^{l(\mu)}\frac{\prod_{a=1}^{\mu_i-1}(\mu_i\tau+a)}{(\mu_i-1)!}.$$

The expressions, although very complicated, arise naturally from localization computations on the moduli spaces of relative stable maps into $\mathbf{P}^1$ with ramification type μ at ∞.

We now introduce the generating series

$$G_\mu(\lambda;\tau) = \sum_{g\geq 0}\lambda^{2g-2+l(\mu)}G_{g,\mu}(\tau).$$

The special case when $g=0$ is given by

$$\int_{\overline{\mathcal{M}}_{0,l(\mu)}}\frac{\Lambda_0^\vee(1)\Lambda_0^\vee(-\tau-1)\Lambda_0^\vee(\tau)}{\prod_{i=1}^{l(\mu)}(1-\mu_i\psi_i)} = \int_{\overline{\mathcal{M}}_{0,l(\mu)}}\frac{1}{\prod_{i=1}^{l(\mu)}(1-\mu_i\psi_i)}$$

which is known to be equal to $|\mu|^{l(\mu)-3}$ for $l(\mu)\geq 3$, and we use this expression to extend the definition to the case $l(\mu)<3$.

Introduce formal variables $p=(p_1,p_2,\ldots,p_n,\ldots)$, and define

$$p_\mu = p_{\mu_1}\cdots p_{\mu_{l(\mu)}}$$

for any partition μ. These p_{μ_j} correspond to $\operatorname{Tr}V^{\mu_j}$ in the notations of string theorists. The generating series for all genera and all possible marked points are defined to be

$$G(\lambda;\tau;p) = \sum_{|\mu|\geq 1}G_\mu(\lambda;\tau)p_\mu,$$

which encode complete information of the triple Hodge integrals we are interested in.

Next we introduce the representation theoretical side. Let χ_μ denote the character of the irreducible representation of the symmetric group $S_{|\mu|}$, indexed by μ with $|\mu| = \sum_j \mu_j$. Let $C(\mu)$ denote the conjugacy class of $S_{|\mu|}$ indexed by μ. Introduce

$$\mathcal{W}_\mu(\lambda) = \prod_{1\leq a<b\leq l(\mu)}\frac{\sin[(\mu_a-\mu_b+b-a)\lambda/2]}{\sin[(b-a)\lambda/2]}\cdot\frac{1}{\prod_{i=1}^{l(\nu)}\prod_{v=1}^{\mu_i}2\sin[(v-i+l(\mu))\lambda/2]}.$$

This has an interpretation in terms of quantum dimension in Chern-Simons knot theory.

We define the following generating series

$$R(\lambda;\tau;p) =$$

$$\sum_{n\geq 1}\frac{(-1)^{n-1}}{n}\sum_\mu[\sum_{\cup_{i=1}^n\mu^i=\mu}\prod_{i=1}^n\sum_{|\nu^i|=|\mu^i|}\frac{\chi_{\nu^i}(C(\mu^i))}{z_{\mu^i}}e^{\sqrt{-1}(\tau+\frac{1}{2})\kappa_{\nu^i}\lambda/2}\mathcal{W}_{\nu^i}(\lambda)]p_\mu$$

where μ^i are sub-partitions of μ, $z_\mu = \prod_j \mu_j! j^{\mu_j}$ and

$$\kappa_\mu = |\mu| + \sum_i(\mu_i^2 - 2i\mu_i)$$

for a partition μ which is also standard for representation theory of symmetric groups. There is the relation $z_\mu = |\operatorname{Aut}(\mu)|\mu_1\cdots\mu_{l(\mu)}$.

Finally we can give the precise statement of the Mariño-Vafa formula:

Conjecture: *We have the identity*

$$G(\lambda;\tau;p) = R(\lambda;\tau;p).$$

Before discussing the proof of this conjecture, we first give several remarks. This conjecture is a formula: G : Geometry = R : Representations, and the representations of symmetric groups are essentially combinatorics. We note that each $G_\mu(\lambda,\tau)$ is given by a finite and closed expression in terms of the representations of symmetric groups:

$$G_\mu(\lambda,\tau) = \sum_{n\geq 1}\frac{(-1)^{n-1}}{n}\sum_{\cup_{i=1}^n\mu^i=\mu}\prod_{i=1}^n\sum_{|\nu^i|=|\mu^i|}\frac{\chi_{\nu^i}(C(\mu^i))}{z_{\mu^i}}e^{\sqrt{-1}(\tau+\frac{1}{2})\kappa_{\nu^i}\lambda/2}\mathcal{W}_{\nu^i}(\lambda).$$

The generating series $G_\mu(\lambda,\tau)$ gives the values of the triple Hodge integrals for moduli spaces of curves of all genera with $l(\mu)$ marked points. Finally we remark that an equivalent expression of this formula is the following non-connected generating series. In this situation we have a relatively simpler combinatorial expression:

$$G(\lambda;\tau;p)^\bullet = \exp\left[G(\lambda;\tau;p)\right] = \sum_{|\mu|\geq 0}[\sum_{|\nu|=|\mu|}\frac{\chi_\nu(C(\mu))}{z_\mu}e^{\sqrt{-1}(\tau+\frac{1}{2})\kappa_\nu\lambda/2}\mathcal{W}_\nu(\lambda)]p_\mu.$$

According to Mariño and Vafa, this formula gives values for all Hodge integrals up to three Hodge classes. Lu proved that this is right if we combine with some previously known simple formulas about Hodge integrals.

By taking Taylor expansion in τ on both sides of the Mariño-Vafa formula, we have derived various Hodge integral identities in [**20**].

For examples, as easy consequences of the Mariño-Vafa formula and the cut-and-join equation as satisfied by the above generating series, we have unified simple proofs of the λ_g conjecture by comparing the coefficients in τ in the Taylor expansions of the two expressions,

$$\int_{\overline{\mathcal{M}}_{g,n}}\psi_1^{k_1}\cdots\psi_n^{k_n}\lambda_g = \binom{2g+n-3}{k_1,\dots,k_n}\frac{2^{2g-1}-1}{2^{2g-1}}\frac{|B_{2g}|}{(2g)!},$$

for $k_1+\cdots+k_n = 2g-3+n$, and the following identities for Hodge integrals:

$$\int_{\overline{\mathcal{M}}_g}\lambda_{g-1}^3 = \int_{\overline{\mathcal{M}}_g}\lambda_{g-2}\lambda_{g-1}\lambda_g = \frac{1}{2(2g-2)!}\frac{|B_{2g-2}|}{2g-2}\frac{|B_{2g}|}{2g},$$

where B_{2g} are Bernoulli numbers. And

$$\int_{\overline{\mathcal{M}}_{g,1}}\frac{\lambda_{g-1}}{1-\psi_1} = b_g\sum_{i=1}^{2g-1}\frac{1}{i} - \frac{1}{2}\sum_{\substack{g_1+g_2=g\\ g_1,g_2>0}}\frac{(2g_1-1)!(2g_2-1)!}{(2g-1)!}b_{g_1}b_{g_2},$$

where

$$b_g = \begin{cases}1, & g=0,\\ \frac{2^{2g-1}-1}{2^{2g-1}}\frac{|B_{2g}|}{(2g)!}, & g>0.\end{cases}$$

Now let us look at how we proved this conjecture. This is joint work with Chiu-Chu Liu, Jian Zhou.

The first proof of this formula is based on the *Cut-and-Join* equation which is a beautiful match of combinatorics and geometry. The details of the proof is given in [**17**] and [**18**]. First we look at the combinatorial side. Denote by $[s_1, \cdots, s_k]$ a k-cycle in the permutation group. We have the following two obvious operations:

Cut: a k-cycle is cut into an i-cycle and a j-cycle:

$$[s,t]\cdot[s,s_2,\cdots,s_i,t,t_2,\cdots t_j]=[s,s_2,\cdots,s_i][t,t_2,\cdots t_j].$$

Join: an i-cycle and a j-cycle are joined to an $(i+j)$-cycle:

$$[s,t]\cdot[s,s_2,\cdots,s_i][t,t_2,\cdots t_j]=[s,s_2,\cdots,s_i,t,t_2,\cdots t_j].$$

Such operations can be organized into differential equations which we call the cut-and-join equation.

Now we look at the geometry side. In the moduli spaces of stable maps, cut and join have the following geometric meaning: *Cut*: one curve splits into two lower degree or lower genus curves. *Join*: two curves are joined together to give a higher genus or higher degree curve.

The combinatorics and geometry of cut-and-join are reflected in the following two differential equations, which look like heat equation. It is easy to show that such equation is equivalent to a series of systems of linear ordinary differential equations by comparing the coefficients on p_μ. These equations are proved either by easy and direct computations in combinatorics or by localizations on moduli spaces of relative stable maps in geometry. In combinatorics, the proof is given by direct computations and was explored in the combinatorics in the mid 80s and by Zhou [**17**] for this case. The differential operator on the right hand side corresponds to the cut-and-join operations which we also simply denote by (CJ).

LEMMA 2.1.

$$\frac{\partial R}{\partial \tau}=\frac{1}{2}\sqrt{-1}\lambda\sum_{i,j=1}^{\infty}((i+j)p_ip_j\frac{\partial R}{\partial p_{i+j}}+ijp_{i+j}(\frac{\partial R}{\partial p_i}\frac{\partial R}{\partial p_j}+\frac{\partial^2 R}{\partial p_i\partial p_j})).$$

On the geometry side the proof of such equation is given by localization on the moduli spaces of relative stable maps into the the projective line $\mathbf{P}^1$ with fixed ramifications at ∞:

LEMMA 2.2.

$$\frac{\partial G}{\partial \tau}=\frac{1}{2}\sqrt{-1}\lambda\sum_{i,j=1}^{\infty}((i+j)p_ip_j\frac{\partial G}{\partial p_{i+j}}+ijp_{i+j}(\frac{\partial G}{\partial p_i}\frac{\partial G}{\partial p_j}+\frac{\partial^2 G}{\partial p_i\partial p_j})).$$

The proof of the above equation is given in [**17**]. Together with the following **Initial Value**: $\tau=0$,

$$G(\lambda,0,p)=\sum_{d=1}^{\infty}\frac{p_d}{2d\sin\left(\frac{\lambda d}{2}\right)}=R(\lambda,0,p).$$

which is precisely the Ooguri-Vafa formula and which has been proved previously for example in [**30**], we therefore obtain the equality which is the Mariño-Vafa conjecture by the uniqueness of the solution:

THEOREM 2.3. *We have the identity*

$$G(\lambda;\tau;p) = R(\lambda;\tau;p).$$

During the proof we note that the cut-and-join equation is encoded in the geometry of the moduli spaces of stable maps. In fact we later find the convolution formula of the following form, which is a relation for the disconnected version $G^\bullet = \exp G$,

$$G^\bullet_\mu(\lambda,\tau) = \sum_{|\nu|=|\mu|} \Phi^\bullet_{\mu,\nu}(-\sqrt{-1}\tau\lambda) z_\nu K^\bullet_\nu(\lambda)$$

where $\Phi^\bullet_{\mu,\nu}$ is the generating series of double Hurwitz numbers, and z_ν is the combinatorial constant appeared in the previous formulas. Equivalently this gives the explicit solution of the cut-and-join differential equation with initial value $K^\bullet(\lambda)$, which is the generating series of the integrals of certain Euler classes on the moduli spaces of relative stable maps to $\mathbf{P}^1$. See [**14**] for the derivation of this formula, and see [**19**] for the two partition analogue.

The Witten conjecture as proved by Kontsevich states that the generating series of the ψ-class integrals satisfy infinite number of differential equations. The remarkable feature of Mariño-Vafa formula is that it gives a finite close formula. In fact by taking limits in τ and μ_i's one can obtain the Witten conjecture as argued by Okounkov-Pandhrapande. But the combinatorics involved is non-trivial. A much simpler direct proof of the Witten conjecture was obtained recently by Kim and myself. We directly derived the recursion formula which implies both the Virasoro relations and the KdV equations.

The same argument as our proof of the conjecture gives a simple and geometric proof of the ELSV formula for Hurwitz numbers. It reduces to the fact that the push-forward of 1 is a constant in equivariant cohomology for a generically finite-to-one map. See [**18**] for more details.

We would like to briefly explain the technical details of the proof. The proof of the combinatorial cut-and-join formula is based on the Burnside formula and various simple results in symmetric functions. See [**30**], [**15**] and [**18**].

The proof of the geometric cut-and-join formula used the functorial localization formula in [**22**] and [**23**]. Here we only state its simple form for manifolds as used in [**22**], the virtual version of this formula is proved and used in [**23**].

Given X and Y two compact manifolds with torus action. Let $f: X \to Y$ be an equivariant map. Let $F \subset Y$ be a fixed component, and let $E \subset f^{-1}(F)$ denote the fixed components lying inside $f^{-1}(F)$. Let $f_0 = f|_E$, then we have

Functorial Localization Formula: *For $\omega \in H^*_T(X)$ an equivariant cohomology class, we have the identity on F:*

$$f_{0*}\left[\frac{i^*_E\omega}{e_T(E/X)}\right] = \frac{i^*_F(f_*\omega)}{e_T(F/Y)}.$$

This formula, which is a generalization of the Atiyah-Bott localization formula to relative setting, has been applied to various settings to prove many interesting conjectures from physics. It was discovered and effectively used in [**22**]. A virtual version which was first applied to the virtual fundamental cycles in the computations of Gromov-Witten invariants was first proved and used in [**23**].

This formula is very effective and useful because we can use it to push computations on complicated moduli space to simpler moduli space. The moduli spaces used by mathematicians are usually the correct but complicated moduli spaces like the moduli spaces of stable maps, while the moduli spaces used by physicists are usually the simple but the wrong ones like the projective spaces. This functorial localization formula has been used successfully in the proof of the mirror formula [**22**], [**23**], the proof of the Hori-Vafa formula [**16**], and the easy proof of the ELSV formula [**18**]. Our first proof of the Mariño-Vafa formula also used this formula in a crucial way.

More precisely, let $\mathcal{M}_g(\mathbf{P}^1, \mu)$ denote the moduli space of relative stable maps from a genus g curve to $\mathbf{P}^1$ with fixed ramification type μ at ∞, where μ is a fixed partition. We apply the functorial localization formula to the divisor morphism from the relative stable map moduli space to the projective space,

$$\mathrm{Br}: \quad \mathcal{M}_g(\mathbf{P}^1, \mu) \to \mathbf{P}^r,$$

where r denotes the dimension of $\mathcal{M}_g(\mathbf{P}^1, \mu)$. This is similar to the set-up of mirror principle, only with a different linearized moduli space, but in both cases the target spaces are projective spaces.

We found that the fixed points of the target $\mathbf{P}^r$ precisely labels the cut-and-join operations of the triple Hodge integrals. Functorial localization reduces the problem to the study of polynomials in the equivariant cohomology group of $\mathbf{P}^r$. We were able to squeeze out a system of linear equations which implies the cut-and-join equation. Actually we derived a stronger relation than the cut-and-join equation, while the cut-and-join equation we need for the Mariño-Vafa formula is only the very first of such kind of relations. See [**18**] for higher order cut-and-join equations.

As was known in infinite Lie algebra theory, the cut-and-join operator is closely related to and more fundamental than the Virasoro algebras in some sense.

Recently there have appeared two different approaches to the Mariño-Vafa formula. The first one is a direct derivation of the convolution formula which was discovered during our proof of the two partition analogue of the formula [**19**]. See [**14**] for the details of the derivation in this case. The second is by Okounkov-Pandhripande [**26**], they gave a different approach by using the ELSV formula as initial value, and as well as the λ_g conjecture and other recursion relations from localization on the moduli spaces of stable maps to $\mathbf{P}^1$.

3. Two Partition Formula

The two partition analogue of the Mariño-Vafa formula naturally arises from the localization computations of the Gromov-Witten invariants of the open toric Calabi-Yau manifolds, as explained in [**31**].

To state the formula we let μ^+, μ^- be any two partitions. Introduce the Hodge integrals involving these two partitions:

$$G_{\mu^+,\mu^-}(\lambda;\tau) = B(\tau;\mu^+,\mu^-) \cdot \sum_{g\geq 0} \lambda^{2g-2} A_g(\tau;\mu^+,\mu^-)$$

where

$$A_g(\tau;\mu^+,\mu^-) = \int_{\overline{\mathcal{M}}_{g,l(\mu^+)+l(\mu^-)}} \frac{\Lambda_g^\vee(1)\Lambda_g^\vee(\tau)\Lambda_g^\vee(-\tau-1)}{\prod_{i=1}^{l(\mu^+)}(1-\mu_i^+\psi_i)\prod_{j=1}^{l(\mu^-)}\tau\left(\tau-\mu_i^-\psi_{j+l(\mu^+)}\right)}$$

and

$$B(\tau;\mu^+,\mu^-) = -\frac{(\sqrt{-1}\lambda)^{l(\mu^+)+l(\mu^-)}}{|\operatorname{Aut}(\mu^+)||\operatorname{Aut}(\mu^-)|}\left[\tau(\tau+1)\right]^{l(\mu^+)+l(\mu^-)-1}.$$

$$\prod_{i=1}^{l(\mu^+)}\frac{\prod_{a=1}^{\mu_i^+-1}(\mu_i^+\tau+a)}{(\mu_i^+-1)!}\cdot\prod_{i=1}^{l(\mu^-)}\frac{\prod_{a=1}^{\mu_i^--1}(\mu_i^-\frac{1}{\tau}+a)}{(\mu_i^--1)!}.$$

These complicated expressions naturally arise in open string theory, as well as in the localization computations of the Gromov-Witten invariants on open toric Calabi-Yau manifolds.

We introduce two generating series, first on the geometry side,

$$G^\bullet(\lambda;p^+,p^-;\tau) = \exp\left(\sum_{(\mu^+,\mu^-)\in\mathcal{P}^2} G_{\mu^+,\mu^-}(\lambda,\tau)p^+_{\mu^+}p^-_{\mu^-}\right),$$

where $\mathcal{P}^2$ denotes the set of pairs of partitions and $p^\pm_{\mu^\pm}$ are two sets of formal variables associated to the two partitions as in the last section.

On the representation side, we introduce

$$R^\bullet(\lambda;p^+,p^-;\tau) = \sum_{|\nu^\pm|=|\mu^\pm|\geq 0}\frac{\chi_{\nu^+}(C(\mu^+))}{z_{\mu^+}}\frac{\chi_{\nu^-}(C(\mu^-))}{z_{\mu^-}}\cdot e^{\sqrt{-1}(\kappa_{\nu^+}\tau+\kappa_{\nu^-}\tau^{-1})\lambda/2}\mathcal{W}_{\nu^+,\nu^-}p^+_{\mu^+}p^-_{\mu^-}.$$

Here

$$\mathcal{W}_{\mu,\nu} = q^{l(\nu)/2}\mathcal{W}_\mu\cdot s_\nu(\mathcal{E}_\mu(t))$$

$$= (-1)^{|\mu|+|\nu|}q^{\frac{\kappa_\mu+\kappa_\nu+|\mu|+|\nu|}{2}}\sum_\rho q^{-|\rho|}s_{\mu/\rho}(1,q,\dots)s_{\nu/\rho}(1,q,\dots)$$

in terms of the skew Schur functions s_μ [**24**]. They appear naturally in the Chern-Simons invariant of the Hopf link.

THEOREM 3.1. *We have the identity:*

$$G^\bullet(\lambda;p^+,p^-;\tau) = R^\bullet(\lambda;p^+,p^-;\tau).$$

The idea of the proof is similar to that of the proof of the Mariño-Vafa formula. We prove that both sides of the above identity satisfy the same cut-and-join equation of the following type:

$$\frac{\partial}{\partial\tau}H^\bullet = \frac{1}{2}(CJ)^+H^\bullet - \frac{1}{2\tau^2}(CJ)^-H^\bullet,$$

where $(CJ)^{\pm}$ denote the cut-and-join operator, the differential operator with respect to the two set of variables $p^{\pm}$. We then prove that they have the same initial value at $\tau = -1$:

$$G^{\bullet}(\lambda; p^+, p^-; -1) = R^{\bullet}(\lambda; p^+, p^-; -1),$$

which is again given by the Ooguri-Vafa formula [**19**], [**31**].

The cut-and-join equation can be written in a linear matrix form, and such equation follows from the convolution formula of the form

$$K^{\bullet}_{\mu^+,\mu^-}(\lambda) = \sum_{|\nu^{\pm}|=\mu^{\pm}} G^{\bullet}_{\mu^+,\mu^-}(\lambda;\tau) z_{\nu^+} \Phi^{\bullet}_{\nu^+,\mu^+}(-\sqrt{-1}\lambda\tau) z_{\nu^-} \Phi^{\bullet}_{\nu^-,\mu^-}(\frac{-\sqrt{-1}}{\tau}\lambda)$$

where $\Phi^{\bullet}$ denotes the generating series of double Hurwitz numbers, and K_{μ^+,μ^-} is the generating series of certain integrals on the moduli spaces of relative stable maps. For more details see [**19**].

This convolution formula arises naturally from localization computations on the moduli spaces of relative stable maps to $\mathbf{P}^1 \times \mathbf{P}^1$ with the point (∞, ∞) blown up. So it reflects the geometric structure of the moduli spaces. Such convolution type formula was actually discovered during our search for a proof of this formula, both on the geometric and the combinatorial side, see [**19**] for the detailed derivations of the convolution formulas in both geometry and combinatorics.

The proof of the combinatorial side of the convolution formula is again a direct computation. The proof of the geometric side for the convolution equation is to reorganize the generating series from localization contributions on the moduli spaces of relative stable maps into $\mathbf{P}^1 \times \mathbf{P}^1$ with the point (∞, ∞) blown up, in terms of the double Hurwitz numbers. It involves careful analysis and computations.

4. Theory of Topological Vertex

When we worked on the Mariño-Vafa formula and its generalizations, we were simply trying to generalize the method and the formula to involve more partitions, but it turned out that in the three partition case, we naturally met the theory of topological vertex. Topological vertex was first introduced in string theory by Vafa et al in [**1**], it can be deduced from a three partition analogue of the Mariño-Vafa formula in a highly nontrivial way. From this we were able to give a rigorous mathematical foundation for the physical theory. Topological vertex is a high point of the theory of string duality as developed by Vafa and his group for the past several years, starting from Witten's conjectural duality between Chern-Simons and open string theory. It gives the most powerful and effective way to compute the Gromov-Witten invariants for all open toric Calabi-Yau manifolds. In physics it is rare to have two theories agree up to all orders, topological vertex theory gives a very significant example. In mathematics the theory of topological vertex already has many interesting applications. Here we only briefly sketch the rough idea for the three partition analogue of the Mariño-Vafa formula. For its relation to the theory of topological vertex, we refer the reader to [**13**] for the details.

Given any three partitions $\overrightarrow{\mu} = \{\mu^1, \mu^2, \mu^3\}$, the cut-and-join equation in this case, for both the geometry and representation sides, has the form:

$$\frac{\partial}{\partial \tau} F^{\bullet}(\lambda;\tau;\mathbf{p}) = (CJ)^1 F^{\bullet}(\lambda;\tau;\mathbf{p}) + \frac{1}{\tau^2}(CJ)^2 F^{\bullet}(\lambda;\tau;\mathbf{p}) + \frac{1}{(\tau+1)^2}(CJ)^3 F^{\bullet}(\lambda;\tau;\mathbf{p}).$$

The cut-and-join operators $(CJ)^1$, $(CJ)^2$ and $(CJ)^3$ are with respect to the three partitions. More precisely they correspond to the differential operators with respect to the three groups of infinite numbers of variables $\mathbf{p} = \{p^1, p^2, p^3\}$.

The initial value for this differential equation is taken at $\tau = 1$, which is then reduced to the formulas of two partition case. The combinatorial, or the Chern-Simons invariant side is given by $\mathcal{W}_{\vec{\mu}} = \mathcal{W}_{\mu^1,\mu^2,\mu^3}$ which is a combination of the $\mathcal{W}_{\mu,\nu}$ as in the two partition case. See [**13**] for its explicit expression.

On the geometry side,

$$G^\bullet(\lambda;\tau;\mathbf{p}) = \exp(G(\lambda;\tau;\mathbf{p}))$$

is the non-connected version of the generating series of the triple Hodge integral. More precisely,

$$G(\lambda;\tau;\mathbf{p}) = \sum_{\vec{\mu}}[\sum_{g=0}^{\infty} \lambda^{2g-2+l(\vec{\mu})} G_{g,\vec{\mu}}(\tau)] p^1_{\mu^1} p^2_{\mu^2} p^3_{\mu_3}$$

where $l(\vec{\mu}) = l(\mu^1) + l(\mu^2) + l(\mu^3)$ and $G_{g,\vec{\mu}}(\tau)$ denotes the Hodge integrals of the following form,

$$A(\tau)\int_{\overline{\mathcal{M}}_{g,l_1+l_2+l_3}} \frac{\Lambda_g^\vee(1)\Lambda_g^\vee(\tau)\Lambda_g^\vee(-\tau-1)}{\prod_{j=1}^{l_1}(1-\mu_j^1\psi_j)\prod_{j=1}^{l_2}\tau(\tau-\mu_j^2\psi_{l_1+j})} \cdot \frac{(\tau(\tau+1))^{l_1+l_2+l_3-1}}{\prod_{j=1}^{l_3}(\tau+1)(\tau+1+\mu_j^3\psi_{l_1+l_2+j})},$$

where

$$A(\tau) = \frac{-(\sqrt{-1}\lambda)^{l_1+l_2+l_3}}{|\operatorname{Aut}(\mu^1)||\operatorname{Aut}(\mu^2)||\operatorname{Aut}(\mu^3)|} \prod_{j=1}^{l_1} \frac{\prod_{a=1}^{\mu_j^1-1}(\tau\mu_j^1+a)}{(\mu_j^1-1)!} \cdot \prod_{j=1}^{l_2} \frac{\prod_{a=1}^{\mu_j^1-1}((-1-1/\tau)\mu_j^2+a)}{(\mu_j^2-1)!} \prod_{j=1}^{l_3} \frac{\prod_{a=1}^{\mu_j^1-1}(-\mu_j^3/(\tau+1)+a)}{(\mu_j^3-1)!}$$

In the above expression, $l_i = l(\mu^i)$, $i = 1,2,3$. Despite of its complicated coefficients, these triple integrals naturally arise from localizations on the moduli spaces of relative stable maps into the blow-up of $\mathbf{P}^1 \times \mathbf{P}^1 \times \mathbf{P}^1$ along certain divisors. It also naturally appears in open string theory computations [**1**]. See [**13**] for more details.

One of our results in [**13**] states that $G^\bullet(\lambda;\tau;\mathbf{p})$ has a combinatorial expression $R^\bullet(\lambda;\tau;\mathbf{p})$ in terms of the Chern-Simons knot invariants $W_{\vec{\mu}}$, which is a closed combinatorial expression. More precisely it is given by

$$R^\bullet(\lambda;\tau;\mathbf{p}) = \sum_{\vec{\mu}}[\sum_{|\nu^i|=|\mu^i|}\prod_{i=1}^{3}\frac{\chi_{\nu^i}(\mu^i)}{z_{\mu^i}} q^{\frac{1}{2}(\sum_{i=1}^3 \kappa_{\nu^i}\frac{w_{i+1}}{w_i})} W_{\vec{\nu}}(q)] p^1_{\mu^1} p^2_{\mu^2} p^3_{\mu^3}.$$

Here $w_4 = w_1$ and $w_3 = -w_1 - w_2$ and $\tau = \frac{w_2}{w_1}$. Due to the complicated combinatorics in the initial values, the combinatorial expression $W_{\vec{\mu}}$ we obtained is different from the expression $\mathcal{W}_{\vec{\mu}}$ obtained by Vafa et al. Actually our expression is even simpler than theirs in some sense. The expression we obtained is more convenient for mathematical applications such as the proof of the Gopakumar-Vafa conjecture for open toric Calabi-Yau manifolds. It should be possible to identify the two combinatorial expressions by using the classical theory of symmetric functions, as pointed out to us by R. Stanley.

THEOREM 4.1. *We have the equality:*

$$G^{\bullet}(\lambda;\tau;\boldsymbol{p}) = R^{\bullet}(\lambda;\tau;\boldsymbol{p}).$$

The key point to prove the above theorem is still the proof of convolution formulas for both sides which imply the cut-and-join equation. The proof of the convolution formula for $G^{\bullet}(\lambda;\tau;\mathbf{p})$ is much more complicated than the one and two partition cases. See [**13**] for details.

The most useful property of topological vertex is its gluing property induced by the orthogonal relations of the characters of the symmetric group. This is very close to the situation of two dimensional gauge theory. In fact string theorists consider topological vertex as a kind of lattice theory on Calabi-Yau manifolds. By using the gluing formula we can easily obtain closed formulas for generating series of Gromov-Witten invariants of all genera and all degrees, open or closed, for all open toric Calabi-Yau manifolds, in terms of the Chern-Simons knot invariants. Such formulas are always given by finite sum of products of those Chern-Simons type invariants $\mathcal{W}_{\mu,\nu}$'s. The magic of topological vertex is that, by simply looking at the moment map graph of the toric surfaces in the open toric Calabi-Yau, we can immediately write down the closed formula for the generating series for all genera and all degree Gromov-Witten invariants, or more precisely the Euler numbers of certain bundles on the moduli space of stable maps.

Here we only give one example to describe the topological vertex formula for the generating series of the all degree and all genera Gromov-Witten invariants for the open toric Calabi-Yau 3-folds. We write down the explicit close formula of the generating series of the Gromov-Witten invariants for $O(-3) \longrightarrow \mathbf{P}^2$ in terms of the Chern-Simons invariants. In this case the formula is given by

$$\exp\left(\sum_{g=0}^{\infty}\lambda^{2g-2}F_g(t)\right) = \sum_{\nu_1,\nu_2,\nu_3}\mathcal{W}_{\nu_1,\nu_2}\mathcal{W}_{\nu_2,\nu_3}\mathcal{W}_{\nu_3,\nu_1}(-1)^{\sum_{j=1}^{3}|\nu_j|}q^{\frac{1}{2}\sum_{i=1}^{3}\kappa_{\nu_i}}e^{t(\sum_{j=1}^{3}|\nu_j|)}$$

where $q = e^{\sqrt{-1}\lambda}$. The precise definition of $F_g(t)$ will be given in the next section.

For general open toric Calabi-Yau manifolds, the expressions are just similar. They are all given by finite and closed formulas, which are easily read out from the moment map graphs associated to the toric surfaces, with the topological vertex associated to each vertex of the graph.

In [**1**] Vafa and his group first developed the theory of topological vertex by using string duality between Chern-Simons and Calabi-Yau, which is a physical theory. In [**13**] we established the mathematical theory of the topological vertex, and derived various mathematical corollaries, including the relation of the Gromov-Witten invariants to the equivariant index theory as motivated by the Nekrasov conjecture in string duality [**17**].

5. Gopakumar-Vafa Conjecture and Equivariant Indices of Elliptic Operators

Let $N_{g,d}$ denote the so-called Gromov-Witten invariant of genus g and degree d of an open toric Calabi-Yau 3-fold. $N_{g,d}$ is defined to be the Euler number of the obstruction bundle on the moduli space of stable maps of degree $d \in H_2(S, \mathbb{Z})$ from genus g curve into the surface base S. The open toric Calabi-Yau manifold associated to the toric surface S is the total space of the canonical line bundle K_S on S. More precisely

$$N_{g,d} = \int_{[\overline{\mathcal{M}}_g(S,d)]^v} e(V_{g,d})$$

with $V_{g,d} = R^1\pi_* u^* K_S$ a vector bundle on the moduli space induced by the canonical bundle K_S. Here $\pi : U \to \overline{\mathcal{M}}_g(S, d)$ denotes the universal curve and u can be considered as the evaluation or universal map. Let us write

$$F_g(t) = \sum_{d \geq 0} N_{g,d}\, e^{-d \cdot t}.$$

The Gopakumar-Vafa conjecture is stated as follows:

Gopakumar-Vafa Conjecture: *There exists an expression:*

$$\sum_{g=0}^{\infty} \lambda^{2g-2} F_g(t) = \sum_{k=1}^{\infty} \sum_{g,d \geq 0} n_d^g \frac{1}{d} (2 sin \frac{d\lambda}{2})^{2g-2} e^{-kd \cdot t},$$

such that n_d^g are integers, called instanton numbers.

Motivated by the Nekrasov duality conjecture between the four dimensional gauge theory and string theory, we are able to interpret the above integers n_d^g as equivariant indices of certain elliptic operators on the moduli spaces of anti-self-dual connections [**17**]:

THEOREM 5.1. *For certain interesting cases, these n_d^g's can be written as equivariant indices on the moduli spaces of anti-self-dual connections on $\mathbb{C}^2$.*

For more precise statement, we refer the reader to [**17**]. The interesting cases include open toric Calabi-Yau manifolds when S is Hirzebruch surface. The proof of this theorem is to compare fixed point formula expressions for equivariant indices of certain elliptic operators on the moduli spaces of anti-self-dual connections with the combinatorial expressions of the generating series of the Gromov-Witten invariants on the moduli spaces of stable maps. They both can be expressed in terms of Young diagrams of partitions. We find that they agree up to certain highly non-trivial "mirror transformation", a complicated variable change. This result is not only interesting for the index formula interpretation of the instanton numbers, but also for the fact that it gives the first complete examples that the Gopakumar-Vafa conjecture holds for all genera and all degrees.

Recently P. Pan [**27**] has given the proof of the Gopakumar-Vafa conjecture for all open toric Calabi-Yau 3-folds by using the Chern-Simons expressions from the topological vertex. His method is to explore the property of the Chern-Simons expression in great detail with some clever observation about the form of the combinatorial expressions. On the other hand, Kim in [**7**] has derived some remarkable recursion formulas for Hodge integrals of all genera and any number of marked points, involving one λ-classes. His method is to add marked points in the moduli spaces and then follow the localization argument we used to prove the Mariño-Vafa formula.

We strongly believe that there is a more interesting and grand duality picture between Chern-Simons invariants for three dimensional manifolds and the Gromov-Witten invariants for open toric Calabi-Yau manifolds. Our proofs of the Mariño-Vafa formula, and the setup of the mathematical foundation for topological vertex theory and the results of Peng and Kim all together have just opened a small window for a more splendid picture.

References

[1] M. Aganagic, A. Klemm, M. Marino, C. Vafa, *The topological vertex*, preprint, hep-th/0305132.

[2] M. Aganagic, M. Marino, C. Vafa, *All loop topological string amplitudes from Chern-Simons theory*, preprint, hep-th/0206164.

[3] T. Ekedahl, S. Lando, M. Shapiro, A. Vainshtein, *Hurwitz numbers and intersections on moduli spaces of curves*. Invent. Math. **146** (2001), no. 2, 297–327.

[4] I.P. Goulden, D.M. Jackson, A. Vainshtein, *The number of ramified coverings of the sphere by the torus and surfaces of higher genera*, Ann. of Comb. **4** (2000), 27-46.

[5] A. Iqbal, *All genus topological amplitudes and 5-brane webs as Feynman diagrams*, preprint, hep-th/0207114.

[6] S. Katz, C.-C. Liu, *Enumerative geometry of stable maps with Lagrangian boundary conditions and multiple covers of the disc*, Adv. Theor. Math. Phys. **5** (2001), 1-49.

[7] Y. Kim, *Computing Hodge integrals with one lambda-class*, preprint, math-ph/0501018.

[8] M. Kontsevich, *Intersection theory on the moduli space of curves and the matrix Airy function*. Comm. Math. Phys. **147** (1992), no. 1, 1–23.

[9] J. Li, *Stable Morphisms to singular schemes and relative stable morphisms*, J. Diff. Geom. **57** (2001), 509-578.

[10] J. Li, *Relative Gromov-Witten invariants and a degeneration formula of Gromov-Witten invariants*, J. Diff. Geom. **60** (2002), 199-293.

[11] J. Li, *Lecture notes on relative GW-invariants*, preprint.

[12] A.M. Li, G. Zhao, Q. Zheng, *The number of ramified coverings of a Riemann surface by Riemann surface*, Comm. Math. Phys. **213** (2000), no. 3, 685–696.

[13] J. Li, C.-C. Liu, K. Liu, J. Zhou, *A mathematical theory of the topological vertex* , preprint, math.AG/0411247.

[14] C.-C. Liu, *Formulae of one-partition and two-partition Hodge integrals*, preprint, math.AG/0502430.

[15] J. Li, K. Liu, J. Zhou, *Topological string partition functions as equivariant indices*, preprint, math.AG/0412089.

[16] B. Lian, C.-H. Liu, K. Liu, S.-T. Yau, *The S^1-fixed points in quot-schemes and mirror principle computations*, Contemp. Math., 322, 2003, 165-194.

[17] C.-C. Liu, K. Liu, J. Zhou, *On a proof of a conjecture of Mariño-Vafa on Hodge Integrals*, Math. Res. Letters 11, 259-272(2004).

[18] C.-C. Liu, K. Liu, J. Zhou, *A proof of a conjecture of Mariño-Vafa on Hodge Integrals*, J. Differential Geometry, 65(2003) 289-340.

[19] C.-C, Liu, K. Liu, J. Zhou, *A formula of two-partition Hodge integrals*, preprint, math.AG/0310272.

[20] C.-C, Liu, K. Liu, J. Zhou, *Mariño-Vafa formula and Hodge integral identities*, preprint, math.AG/0308015.

[21] K. Liu, *Mathematical results inspired by physics*, Proc. ICM 2002, Vol. III, 457-466.

[22] B. Lian, K. Liu, S.-T. Yau, *Mirror Principle I*, Asian J. Math., 1, 1997, 729-763.

[23] B. Lian, K. Liu, S.-T. Yau, *Mirror Principle III*, Asian J. Math., 3, 1999, 771-800.

[24] I.G. MacDonald, *Symmetric functions and Hall polynomials*, 2nd edition.Claredon Press, 1995.

[25] M. Mariño, C. Vafa, *Framed knots at large N*, Orbifolds in mathematics and physics (Madison, WI, 2001), 185–204, Contemp. Math., 310, Amer. Math. Soc., Providence, RI, 2002.

[26] A. Okounkov, R. Pandharipande, *Hodge integrals and invariants of the unknots*, preprint, math.AG/0307209.
[27] P. Pan, *A simple proof of Gopakumar-Vafa conjecture for local toric Calabi-Yau manifolds*, preprint, math.AG/0410540.
[28] E. Witten, *Quantum field theory and the Jones polynomial*, Commun. Math. Phys. **121** (1989) 351-399.
[29] E. Witten, *Two-dimensional gravity and intersection theory on moduli space.* Surveys in differential geometry (Cambridge, MA, 1990), 243–310, Lehigh Univ., Bethlehem, PA, 1991.
[30] J. Zhou, *Hodge integrals, Hurwitz numbers, and symmetric groups*, preprint, math.AG/0308024.
[31] J. Zhou, *A conjecture on Hodge integrals*, preprint.
[32] J. Zhou, *Localizations on moduli spaces and free field realizations of Feynman rules*, preprint.

CENTER OF MATHEMATICAL SCIENCES, ZHEJIANG UNIVERSITY, AND DEPARTMENT OF MATHEMATICS, UCLA
E-mail address: liu@math.ucla.edu, liu@cms.zju.edu.cn

AMS/IP Studies in Advanced Mathematics
Volume **39**, 2007

Polyhedral and Geometric Convergence of Kleinian Groups

Zhengwu Long, Xiantao Wang, and Yuefei Wang

ABSTRACT. In this paper, we will get a complete generalization of one of Jørgensen and Marden's results about the geometrical and the polyhedral convergence of Kleinian groups.

1. Introduction

Let $\mathrm{M}(\widehat{\mathbb{C}})$ denote the full sense preserving Möbius group acting on the extended complex plane $\widehat{\mathbb{C}}$. Then $\mathrm{M}(\widehat{\mathbb{C}})$ is isomorphic to $\mathrm{PSL}(2,\mathbb{C}) = \mathrm{SL}(2,\mathbb{C})/\{\pm I\}$, where I denotes the unit matrix. We will identify the element in $\mathrm{M}(\widehat{\mathbb{C}})$ with its corresponding one in $\mathrm{PSL}(2,\mathbb{C})$, see [**B**].

It is well known that every element g of $\mathrm{M}(\widehat{\mathbb{C}})$ acting on $\widehat{\mathbb{C}}$ has a natural extension acting on $\mathbb{H}^3$, which is called the Poincaré extension of g and still denoted by g. With this extension, $\mathrm{PSL}(2,\mathbb{C}) \cong Isom^+(\mathbb{H}^3)$, the full group of the orientation preserving isometries of the 3−dimensional hyperbolic space $\mathbb{H}^3$, see [**B, Mas, Ra**] etc. We denote the hyperbolic distance of two given points $x, y \in \mathbb{H}^3$ by $d(x,y)$.

A subgroup $G \subset \mathrm{PSL}(2,\mathbb{C})$ is called discrete if and only if no infinite sequence consisting of distinct elements of G converges to the identity id.

A *Kleinian group* G is a discrete subgroup of $\mathrm{PSL}(2,\mathbb{C})$. Then every Kleinian group acts on $\mathbb{H}^3$ discontinuously.

For a Kleinian group G, let $\Omega(G) \subset \widehat{\mathbb{C}}$ be the discontinuous set of G, i.e., the largest open set of $\widehat{\mathbb{C}}$, where G acts properly discontinuously. The complement $\widehat{\mathbb{C}} - \Omega(G)$ is called the *limit set* of G, denoted by $L(G)$.

For a subgroup $G \subset \mathrm{PSL}(2,\mathbb{C})$, as in [**B**], it is called *elementary* if there exists a finite G-orbit in $\overline{\mathbb{H}}^3 = \mathbb{H}^3 \cup \widehat{\mathbb{C}}$. That is, there is $x \in \overline{\mathbb{H}}^3$ such that the set $\{g(x) : g \in G\}$ is finite. Otherwise G is called *non-elementary*.

When G is a Kleinian group, we know, G is elementary if and only if the cardinality of $L(G)$ is not greater than 2. See [**B**] and [**Mat**] for more details about Kleinian groups.

1991 *Mathematics Subject Classification.* 30F40, 20H10.

Key words and phrases. Kleinian group, Fundamental polyhedron, Geometric convergence, Polyhedral convergence.

The research was partly supported by NSFs of China and Hunan (No. 05JJ10001), Program for NCET (No. 04-1783).

For a non-elementary and torsion-free Kleinian group G, a manifold $(\mathbb{H}^3 \cup \Omega(G))/G$, possibly with boundary, is denoted by M_G and called a *Kleinian manifold.*

As in [**JM**], let Γ be an abstract group and φ_n a sequence of representations (homomorphisms) into $\mathrm{PSL}(2,\mathbb{C})$.

A sequence of representations $\{\varphi_n\}$ is called to converge *algebraically* to G if $\lim \varphi_n(\gamma) = g$ exists as a Möbius transformation for all $\gamma \in \Gamma$ and

$$G = \{g : g = \lim \varphi_n(\gamma), \ \gamma \in \Gamma\}.$$

By [**J**], we know if all φ_n are isomorphisms and all $G_n = \varphi_n(\Gamma)$ are non-elementary Kleinian groups, then G is also a non-elementary Kleinian group. This result also holds in all $m-$dimensional cases, cf. [**WY**] or [**Mar, Wi**].

In [**JK**], Jørgensen and Klein obtained that the algebraic limit group of a sequence of $r-$generator non-elementary Kleinian groups is still a Kleinian group. Martin ([**Mar**]) generalized the mentioned result into the $m-$dimensional cases with the condition "having uniformly bounded torsion". In one of our forthcoming papers, we will get some further generalizations.

The algebraic convergence of a sequence of Kleinian groups does not necessarily imply the convergence of the corresponding Kleinian manifolds. Hence it is necessary for us to introduce the concept of the geometric convergence. The important fact is that the geometric convergence means the convergence of the corresponding hyperbolic manifolds in the geometric sense.

For a sequence $\{G_n\}$ of $\mathrm{PSL}(2,\mathbb{C})$, the group

$$\mathrm{Env}\{G_n\} = \{g \in \mathrm{PSL}(2,\mathbb{C}) : g = \lim g_n, g_n \in G_n\}$$

is called the *envelope* of $\{G_n\}$.

In [**JM**], Jørgensen and Marden proved that

LEMMA 1.1. *Let $\{G_n\}$ be a sequence of Kleinian groups. Then $Env\{G_n\}$ is either a Kleinian group or it is elementary.*

We say a sequence $\{G_n\}$ of $\mathrm{PSL}(2,\mathbb{C})$ converges *geometrically* (to H=Env$\{G_n\}$) if and only if for every subsequence $\{G_{n_j}\}$ of $\{G_n\}$, $\mathrm{Env}\{G_{n_j}\} = \mathrm{Env}\{G_n\}$.

It follows from the definition that (see [**JM**])

LEMMA 1.2. *A sequence $\{G_n\}$ of $PSL(2,\mathbb{C})$ converges geometrically to the group $H = Env\{G_n\}$ if and only if the following two statements are satisfied:*
(a) for every $h \in H$, there exist $g_n \in G_n$ such that $h = \lim g_n$;
(b) if $g_{n_j} \in G_{n_j}$ is such that $\lim g_{n_j} = h$ exists, then $h \in H$.

For the further study of the geometric convergence of Kleinian groups, the following is very important and useful (cf. [**CM, Ta**]). See [**CEG**] or [**BP**] for a proof.

LEMMA 1.3. *A sequence of non-elementary torsion-free Kleinian groups $\{G_n\}$ converges geometrically to a torsion-free Kleinian group H if and only if there exists a sequence $\{(R_n, K_n)\}$ and a sequence of maps $\widetilde{f_n} : B_{R_n} \to \mathbb{B}^3$, where B_R is a ball of radius R centered on the origin in $\mathbb{B}^3$, such that the following hold:*

(1) $R_n \to \infty$ *and* $K_n \to 1$;

(2) *The map* $\widetilde{f_n}$ *is a* K_n*–bilipschitz diffeomorphism onto its image,* $\widetilde{f_n}(0) = 0$*, and for any compact set* A*,* $\widetilde{f_n}|_A$ *converges to the identity;*

(3) *If* $N_n = \mathbb{B}^3/G_n$ *and* $N_\infty = \mathbb{B}^3/H$*, then* $\widetilde{f_n}$ *descends to a map* $f_n : V_n \to N_\infty$*, where* $V_n = B_{R_n}(0)/G_n$ *is a submanifold of* N_n*. Moreover* f_n *is also a* K_n*–bilipschitz diffeomorphism onto its image.*

For a sequence $\{G_n\} \subset \mathrm{PSL}(2,\mathbb{C})$, if its algebraical limit group is G and its geometric limit group is H, then, in general, G is a subgroup of H. But when $G = H$ we call the sequence $\{G_n\}$ is *strongly* convergent.

About when $G = H$ does occure, Jørgensen conjectured that if the algebraic limit and the geometric limit differ, then there must be "new parabolics", that is, there must exist some element $\gamma \in \Gamma$ such that $\lim \varphi_n(\gamma)$ is parabolic, but $\varphi_n(\gamma)$ is parabolic for only finitely many n.

See [**AC1, AC2, AC3, ACM, CM, CT, E, Mat, McM, Oh, Ta, Th**] etc for the discussions about the strong convergence and Jørgensen's conjecture, and the further discussions about the geometric convergence of a sequence of Kleinian groups.

There is another concept of the convergence which is closely related to the geometric convergence—the polyhedral convergence.

For a Kleinian subgroup G of $\mathrm{PSL}(2,\mathbb{C})$ and a given point $O \in \mathbb{H}^3$ which is not fixed by any nontrivial element of G, the (Poincaré or Dirichlet) fundamental polyhedron $P(G)$ of G with the center at O is defined as follows:

$$P(G) = \{x \in \mathbb{H}^3 : \ d(x,O) \le d(x, g(O)), \forall g \in G\}.$$

For any given $r > 0$ and $O \in \mathbb{H}^3$, we denote $B_r(O) = \{x \in \mathbb{H}^3 |\ d(x,O) < r\}$ and $P(G)_r = P(G) \cap B_r(O)$.

A sequence of discrete groups $\{G_n\}$ converges *polyhedrally* to the group H if H is Kleinian and for some point $O \in \mathbb{H}^3$, the fundamental polyhedra $\{P(G_n)\}$ centered at O for G_n converge to that $P(H)$ at O for H, uniformly on compact subsets of $\mathbb{H}^3$.

More precisely, we know (see [**JM**])

LEMMA 1.4. *A sequence of Kleinian groups* $\{G_n\}$ *converges polyhedrally to the Kleinian group* H *if and only if for each* $r > 0$*, the following are satisfied.*

(*i*) *There exists* $N = N(r)$ *such that for each face pairing transformation* h *of* $P(H)_r$*, there exists one face pairing transformation* g_n *of* $P(G_n)_r$ *for all* $n > N$ *such that* $\lim g_n = h$*, and*

(*ii*) *if* g_n *is a face pairing transformation of* $P(G_n)_r$*, then the limit* h *of any convergent subsequence of* $\{g_n\}$ *is a face, edge or vertex pairing transformation of* $P(H)_r$ *(in particular,* $h \ne id$*).*

About the relationship between the geometric convergence and the polyhedral convergence of non-elementary Kleinian groups, Jørgensen and Marden proved in [**JM**] that

THEOREM 1.5. *A sequence* $\{G_n\}$ *of non-elementary Kleinian groups converges geometrically to a Kleinian group if and only if* $\{G_n\}$ *converges polyhedrally to a Kleinian group. The geometric and polyhedral limits of* $\{G_n\}$ *are the same.*

See [**MT, Pa**] etc for the applications of Theorem 1.5.

In this paper,we further generalize this result. By using different methods, we will prove

THEOREM 1.6. *Suppose that* $\{G_n\}$ *is a sequence of Kleinian groups. Then* $\{G_n\}$ *converges polyhedrally to a Kleinian group* G *if and only if*

(a) $\{G_n\}$ *converges geometrically to* G*, and*

(b) there is no sequence $\{S_{n_j}\}, S_{n_j} \in G_{n_j}, S_{n_j} \neq id$*, with* $\lim S_{n_j} = id$*, where* $\{n_j\}$ *is a subsequence of* $\{n\}$*.*

From the following lemma (see [**JM**] or [**MT**]), we know that our result is a complete generalization of Theorem 1.5.

LEMMA 1.7. *Suppose* $\{G_n\}$ *is a sequence of Kleinian groups converging algebraically to Kleinian group* G *which is non-elementary. Then there is no sequence* $\{S_{n_j}\}$, $S_{n_j} \in G_{n_j}$, $S_{n_j} \neq id$*, with* $\lim S_{n_j} = id$.

2. Proof of the main theorem

At first, we introduce several lemmas which will be crucial for us.

LEMMA 2.1. *Suppose that the sequence of Kleinian groups* $\{G_n\}$ *converyes polyhedrally to a Kleinian group* G*. Then* $Env\{G_n\} = G$.

PROOF. Assume that for a given point $O \in \mathbb{H}^3$ the fundamental polyhedra $\{P(G_n)\}$ of G_n centered at O converges to that $P(G)$ at O for G.

First we prove that $\text{Env}\{G_n\} \subset G$.

Given a compact set K in $\mathbb{H}^3$, there exist $r > 0$ and N with the following property: K is covered by the images of the truncated polyhedron $P(G_n)_r$ under all words of length $\leq N$ in the face paring transformations of $P(G_n)_r$ for all large n.

To see that, choose a larger compact set K' containing K in its interior. For large enough numbers r and N, the orbit P_N^* of $P(G)_r$ under words of length $\leq N$ in its face pairing transformations covers K'. When n is large, $P(G_n)_r$ is close to $P(H)_r$ and its orbit P_{nN}^* under words of length $\leq N$ in its face pairing transformations covers K.

Then we can deduce that $\text{Env}\{G_n\} \subset G$. For a given $g \in \text{Env}\{G_n\}$, take $r > d(O, g(O))$ and set $K = \overline{B_r(O)}$. We know $g = \lim g_n, g_n \in G_n$ and for large $n, g_n(O) \in K$. Therefore g_n is a word of length $\leq N$ in the face pairing transformations of $P(G_n)_r$. So $g \in G$.

For the inverse, if $g \in G$ then it is a word in the face pairing transformation of $P(G)$. Obviously the sequence of corresponding words in the face pairing transformations of $P(G_n)$ converges to g. So $g \in \text{Env}\{G_n\}$. □

REMARK 2.2. The methods in the foregoing proof follow from those in the proof of Proposition 3.8 in [**JM**].

LEMMA 2.3. *If G is a Kleinian subgroup of $PSL(2,\mathbb{C})$, then there exist a point $O \in \mathbb{H}^3$ and $\delta > 0$ such that the fundamental polyhedron $P(G)$ of G centered at O contains $B_\delta(O)$.*

PROOF. Since G acts discontinuously in $\mathbb{H}^3$, given a point $O_1 \in \mathbb{H}^3$ and a number $\varepsilon > 0$, we know that the set

$$\{g \in G|\ g(B_\varepsilon(O_1)) \cap B_\varepsilon(O_1) \neq \emptyset, g \neq id\}$$

is finite. Hence we may assume that for all nontrivial $g \in G$, either

(i) $g(O_1) = O_1$ and $g(B_\varepsilon(O_1)) = B_\varepsilon(O_1)$, or

(ii) $g(B_\varepsilon(O_1)) \cap B_\varepsilon(O_1) = \emptyset$.

Then we can choose a point $O \in B_\varepsilon(O_1)$ such that O is not fixed by any non-trivial element of G. It follows that there exists a $\delta > 0$ such that

$$g(B_\delta(O)) \cap B_\delta(O) = \emptyset, \forall g \in G, g \neq id.$$

So $B_\delta(O) \in P(G)$. □

LEMMA 2.4. *Let G be a Kleinian group and $P(G)$ be the fundamental polyhedron centered at $O \in \mathbb{H}^3$. If $r > 0$ and $d(O, g(O)) < r$ for a given $g \in G (g \neq id)$. Then there must exist a face pairing transformation f of $P(G)$ such that $d(O, f(O)) \leq d(O, g(O))$.*

PROOF. We will prove it in $\mathbb{B}^3$. Suppose that O is the origin and L is the segment joining the points O and $g(O)$. Let

$$t_0 = \sup\{t : \forall \tau \in [0, t), \tau g(O) \in P(G)\}.$$

Then the point $x_0 = t_0 g(O)$ must lie in some face of $P(G)$. Let f be the face pairing transformation. Since x_0 is in the bisector of the segment jointed O and $f(O)$, then

$$d(O, f(O)) \leq 2d(O, x_0).$$

Then for every $(t_0 >)\varepsilon > 0$, the point $x_\varepsilon = (t_0 - \varepsilon)g(O) \in P(G)$, so

$$d(O, x_\varepsilon) \leq d(x_\varepsilon, g(O)).$$

Since $d(x_\varepsilon, g(O)) = d(O, g(O)) - d(O, x_\varepsilon)$, we know

$$2d(O, x_\varepsilon) \leq d(O, g(O)).$$

Let $\varepsilon \to 0$. Then $x_\varepsilon \to x_0$ and we have

$$d(O, f(O)) \leq d(O, g(O)).$$

□

The following lemma comes from Proposition 3.5 of [**JM**], see also [**MT**].

LEMMA 2.5. *Let $\{G_n\}$ be a sequence of Kleinian groups. If there exist $O \in \mathbb{H}^3$ and $\delta > 0$ such that*

$$g_n(B_\delta(O)) \cap B_\delta(O) = \emptyset,\ \forall g_n \in G_n, g_n \neq id$$

for all large enough n, then there exists a subsequence of $\{G_n\}$ which converges polyhedrally.

Now we are ready to prove our theorem.

PROOF. First we prove the necessity.

Part (a) follows from Lemma 1.4 and Lemma 2.1.

For the part (b), we assume $O \in \mathbb{H}^3$ such that the fundamental polyhedra $\{P(G_n)\}$ centered at O for $\{G_n\}$ converge to that $P(G)$ of G at O. Then there is some $\delta > 0$ such that $B_{2\delta}(O) \subset P(G)$.

Suppose, to the contrary, that there were a sequence $\{S_{n_j}\}, S_{n_j} \in G_{n_j}, S_{n_j} \neq id$, with $\lim S_{n_j} = id$. Then $S_{n_j}(O) \in B_\delta(O)$ for all large enough j. From Lemma 2.4, for these j, there exist face pairing transformations $g_{n_j} \in G_{n_j}$ satisfying $g_{n_j}(O) \in B_\delta(O)$. We may assume $\lim g_{n_j} = g$. So either $g = id$ or $g \neq id, g \in G$ and $g(O) \in B_{2\delta}(O)$. This is the desired contradiction.

Now we come to prove the sufficiency. We choose $O \in \mathbb{H}^3$ which is not fixed by any nontrivial element of G. Let $P(G)$ be the fundamental polyhedron of G centered at O. Then for some $\delta > 0$, we have $B_{2\delta}(O) \subset P(G)$ and

$$\{g \in G : g(B_{2\delta}(O)) \cap B_{2\delta}(O) \neq \emptyset\} = \{id\}.$$

We claim that when n is large enough,

$$g_n(B_\delta(O)) \cap B_\delta(O) = \emptyset$$

for all $g_n \in G_n, g_n \neq id$.

To the contrary, we suppose that there were a sequence $\{g_{n_j}\}, g_{n_j} \in G_{n_j}, g_{n_j} \neq id$ such that $g_{n_j}(B_\delta(O)) \cap B_\delta(O) \neq \emptyset$. We may assume $\lim g_{n_j} = g$. Then $g \in G$ since $\{G_n\}$ is geometrically convergent. This implies that either $g = id$ or $g \neq id$ and $g(B_{2\delta}(O)) \cap B_{2\delta}(O) \neq \emptyset$. This is impossible.

Hence Lemma 2.5 implies that there is a subsequence $\{G_{n_k}\}$ of $\{G_n\}$ converging polyhedrally to a discrete group H. Then $H = G$.

Now it suffices to prove that conditions (i) and (ii) in Lemma 1.4 are satisfied by $\{G_n\}$.

If h is a face pairing transformation, then there exists a sequence $\{g_n\}$, $g_n \in G_n$ such that $\lim g_n = h$. We claim that for all large n, g_n is a face pairing transformation of $P(G_n)$. For if it is not, then there exists a subsequence $\{g_{n_l}\}$ of $\{g_n\}$ such that every g_{n_l} is not a face pairing transformation of $P(G_{n_l})$ and $\lim g_{n_l} = h$. For the sequence $\{G_{n_l}\}$, there is a subsequence (denoted by $\{G_{n_l}\}$ still) converges polyhedrally to G. The observation that if $\lim g_n = \lim h_n$ then we have $g_n = h_n$ for all large n shows that for all large l, g_{n_l} is a face pairing transformation. This contradiction completes the proof of condition (i).

For the condition (ii), let g_n be a face pairing transformation of $P(G_n)$ and let h be the limit of the subsequence of $\{g_{n_j}\}$. Since $\{G_{n_j}\}$ has a polyhedrally convergent subsequence we know h must be a face, edge or vertex pairing transformation of $P(H)$.

This implies that $\{G_n\}$ itself converges polyhedrally to G. □

3. An example

The following example shows that an element of infinite order in a Kleinian group can have more than one rth root in the group.

EXAMPLE 3.1. Let

$$g(z) = 6z, \quad h(z) = -z, \quad G_1 =< g, h >,$$
$$D_1 = \mathbb{H}^2 \cap \{z||z| > 1\} \cap \{z||z| < 6\}$$

and let

$$f(z) = \frac{(1+4i)z + 16}{z + 1 - 4i}, \quad G_2 =< f >,$$
$$D_2 = \{z||z - (1+4i)| > 1\} \cap \{z||z - (-1+4i)| > 1\}.$$

Then D_i is a G_i-packing ($i = 1, 2$), respectively.

Assume that $G =< G_1, G_2 >$. Then, by Theorem 5.3.15 in [**B**], G is a Kleinian group. But

$$g^2(z) = 36z$$

and

$$(gh)^2 = 36z.$$

References

[AC1] J. M. Anderson and R. D. Canary, *Cores of hyperbolic* 3*−manifolds and limits of Kleinian groups*, Amer. J. Math., 118(1996), 745-779.

[AC2] J. M. Anderson and R. D. Canary, *Cores of hyperbolic* 3*−manifolds and limits of Kleinian groups II*, J. London Math. Soc., 61(2000), 489-505.

[AC3] J. M. Anderson and R. D. Canary, *Algebraic limits of Kleinian groups which rearrange the pages of a book*, Invent. Math., 196(1996), 205-218.

[ACM] J. M. Anderson, R. D. Canary and D. McCullough, *The topology of deformation spaces of Kleinian groups*, Ann. Math., 152(2000), 693-741.

[B] A. F. Beardon, *The geometry of discrete groups*, Graduate text in Mathematics, vol. 91, Springer-Verlag, 1983.

[BP] R. Benedetti and C. Petronio, *Lectures on hyperbolic geometry*, Springer-Verlag, 1992.

[CEG] R. D. Canary, D. B. A. Epstein and P. Green, *Notes on notes of Thurston*, in Analytical and geometric Aspects of Hyperbolic spaces, Cambridge University Press, 1987, 3-92.

[CM] R. D. Canary and Y. N. Minsky, *On the limits of tame hyperbolic* 3*−manifolds*, J. Diff. Geom., 43(1996), 1-41.

[CT] R. D. Canary and E. C. Taylor, *Hausdorff dimension and limits of Kleinian groups*, GAFA, 9(1999), 283-297.

[E] R. Evans, *Weakly Type-Preserving Sequences and Strong Convergence*, Geom. Dedicata 108 (2004), 71-92.

[J] T. Jørgensen, *On discrete groups of Möbius transformations*, Amer. J. Math., 98(1976), 739-749.

[JK] T. Jørgensen and P. Klein, *Algebraic convergence of finitely generated Kleinian groups*, Quart. J. Math. Oxford, 33(1982), 325-332.

[JM] T. Jørgensen and A. Marden, *Algebraic and Geometric Convergence of Kleinian Groups*, Math. Scan. 66(1990), 47-72.

[Mas] B. Maskit, *Kleinian groups*, Springer-Verlag, New York, 1988.

[Mar] G. J. Martin, *On discrete Möbius groups in all dimensions*, Acta Math., 163(1989), 253-289.

[Mat] K. Matsuzaki, *Convergence of the Hausdorff dimension of the limit sets of Kleinian groups*, Contemp. Math., 256(2000), 243-254.

[MT] K. Matsuzaki and M. Taniguchi, *Hyperbolic manifolds and Kleinian groups*, Oxford Math. Monographs, Oxford, 1998.

[McM] C. T. McMullen, *Hausdorff dimension and conformal dynamics I: Strong convergence of Kleinian groups*, J. Diff. Geometry, 51(1999), 471-515.

[Oh] K. Ohshika, *Strong convergence of Kleinian groups and Carathéodory convergence of domains of discontinuity*, Math. Proc. Camb. Phil. Soc., 112(1992), 297-307.

[Pa] J. Parkkonen, *Geometric limits of cyclic groups and the shape of Schottky space*, Math. Proc. Camb. Phil. Soc., 137(2004), 55-68.

[Ra] J. G. Ratcliffe, *Foundations of Hyperbolic Manifolds*, Graduate text in Mathematics, vol. 149, Springer-Verlag, 1994.

[Ta] E. C. Taylor, *Geometric finiteness and the convergence of Kleinian groups*, Communications in Analysis and Geometry, 5(1997), 497-533.

[Th] W. P. Thurston, *The geometry and topology of 3-manifolds*, Princeton University, 1977.

[WY] X. Wang and W. Yang, *Discreteness criteria of Möbius groups of high dimensions and convergence theorem of Kleinian groups*, Adv. in Math., 159 (2001), 68-82.

[Wi] N. J. Wielenberg, *Discrete Möbius groups: Fundamental polyhedra and convergence*, Amer. J. Math., 99(1977), 861-877.

Academy of Mathematics and Systems Science and Graduate School of Chinese Academy of Sciences, Beijing 100080, P. R.China

E-mail address: zwlong@amss.ac.cn

Department of Mathematics, Hunan Normal University, Changsha, Hunan 410081, P. R. China

E-mail address: xtwang@hunnu.edu.cn

Academy of Mathematics and Systems Science, Chinese Academy of Sciences, Beijing 100080, P. R. China

E-mail address: wangyf@math.ac.cn

AMS/IP Studies in Advanced Mathematics
Volume **39**, 2007

Rigidity Problems on Compact Quotients of Bounded Symmetric Domains

Ngaiming Mok

ABSTRACT. We consider in this article rigidity problems on Hermitian locally symmetric spaces of the noncompact type, i.e., quotients of bounded symmetric domains by torsion-free discrete groups of biholomorphic automorphisms. On the one hand we give an introduction to differential-geometric techniques on rigidity problems, on the other hand we quickly lead the reader to more recent rigidity results. Topics discussed include local rigidity of complex structure, metric rigidiy, extremal bounded holomorphic functions, and gap rigidity, which is related to the characterization of compact totally geodesic complex submanifolds. In relation to local rigidity we adopt a proof which exploits the structure of zeros of holomorphic bisectional curvatures of the Bergman metric. This paves the way to the proof of various forms of metric rigidity, which in its optimal form can be applied to the study of extremal problems on bounded holomorphic functions on irreducible bounded symmetric domains Ω of rank ≥ 2 with applications to proving embedding theorems. As regards gap rigidity, we sample a number of available techniques for proving differential-geometric or algebro-geometric versions of gap rigidity, formulating at the same time a related problem on almost holomorphic G-structures modelled on irreducible bounded symmetric domains of rank ≥ 2.

Rigidity problems are of fundamental importance in Complex Geometry. They have been widely studied, especially in the case of model manifolds, including compact Hermitian locally symmetric spaces. There are a wide range of rigidity problems on such manifolds. Some concern only the complex structure, such as local rigidity, rigidity under deformation, and uniqueness results on complex structures. For the treatment of such problems on model manifolds often one makes use of canonical metrics on the model manifolds or associated holomorphic vector bundles, and derive the rigidity results using differential-geometric techniques. Other rigidity problems are by their very formulation differential-geometric in nature. This includes problems on metric rigidity and on the characterization of compact totally geodesic complex submanifolds. They may however have consequences which concern primarily the complex structure, such as rigidity results on holomorphic mappings, which follow from metric rigidity.

2000 *Mathematics Subject Classification.* 53C35, 53C55.

Partially supported by a research grant of the Research Grants Council, Hong Kong.

In this article we consider exclusively rigidity problems on Hermitian locally symmetric spaces of the noncompact type, i.e., quotients of bounded symmetric domains by torsion-free discrete groups of biholomorphic automorphisms. It is our purpose to give on the one hand an introduction to differential-geometric techniques on rigidity problems on such model manifolds, and on the other hand to quickly lead the reader to more recent rigidity results concerning holomorphic mappings and extremal bounded holomorphic functions (Mok [Mok4,5]), and on the characterization of compact totally geodesic complex submanifolds (Mok [Mok4, 2002], Eyssidieux-Mok [EM2, 2004]). We will start with a proof of local rigidity of compact quotients of irreducible bounded symmetric domains of dimension ≥ 2, a special case of a celebrated result proven by Calabi-Vesentini [CV, 1960]) in the case of classical domains and by Borel ([Bo1, 1960]) in the case of exceptional domains. Then we establish Hermitian metric rigidity (Mok [Mok1, 1987]) and derive a generalization applicable to continuous complex Finsler metrics ([Mok4,5]), which in its optimal form allows the author to study extremal bounded holomorphic functions ([Mok5, 2004]) and deduce an embedding theorem for liftings to universal covers of nonconstant holomorphic mappings into quotients of bounded (not necessarily symmetric) domains. In the last section we study the gap phenomenon for pairs of bounded symmetric domains. In its original form in Eyssidieux-Mok ([EM1, 1995]) gap rigidity is said to hold for a pair of bounded symmetric domains $\Omega' \subset \Omega$ if any compact complex submanifold $S \subset X := \Omega/\Gamma$ locally resembling $\Omega' \subset \Omega$ is necessarily totally geodesic. This may be termed gap rigidity in the complex topology, which amounts to a strong form of pinching theorem for certain compact complex submanifolds of quotients of bounded symmetric domains. There is also in some cases a stronger form of gap rigidity, in terms of intersection theory (Eyssidieux [Eys1,2], Mok [Mok4], Eyssidieux-Mok (EM2]), which we call gap rigidity in the Zariski topology. We give in this section a survey of techniques of proving gap rigidity and explain also the construction of a counter-example to the gap phenomenon as given in [EM2].

A few words on the organization and in the choice of materials are in order. For rigidity results concerning the complex structure we give in §1 a proof of the most basic result of local rigidity, but completely leave aside strong rigidity of compact Kähler manifolds as developed by Siu ([Siu1, 1980; Siu2, 1981]) using harmonic maps (for which a number of surveys ([Siu3,4], [Mok3]) are available). In the proof of local rigidity we adopt however the proof of the $\partial\overline{\partial}$-Bochner-Kodaira formula as given in [Siu3], the idea of which originated from Siu's work on harmonic maps. For the proof of local rigidity in the locally irreducible case of rank ≥ 2, in place of the computations of Calabi-Vesentini [CV] and Borel [Bo1] of eigenvalues of curvature operators we give a proof basing on the study of the structure of zeros of holomorphic bisectional curvature in Hermitian symmetric spaces of the noncompact type. This derivation, while applied only to the proof of local rigidity, is also applicable to prove vanishing theorems for higher cohomology groups in a more general setting (Siu [Siu3]). Here the variation of proof of local rigidity in the Hermitian locally symmetric case paves the way for the proof of Hermitian and Finsler metric rigidity in §2, for which the structure of the zeros of holomorphic bisectional curvatures play a crucial role. In §2 on metric rigidity we restrict our attention to compact quotients of irreducible bounded symmetric domains of rank ≥ 2. We start with a formulation of Finsler metric rigidity in the smooth case, viz., the fact that lengths of minimal characteristic vectors (cf. (2.1)) with respect to a smooth complex Finsler metric of nonpositive curvature must agree up to a fixed multiplicative constant with those given by the Kähler-Einstein metric. We give a deduction of Hermitian metric rigidity and its consequences by a polarization argument already introduced in §1 in connection with local rigidity. We then strengthen Finsler metric rigidity in two senses, requiring the complex Finsler metric only to be continuous or assuming a definition of the metric and nonpositivity of its curvature only on leaves of a canonical foliation of the minimal characteristic bundle (cf. (2.1)). This strengthening of metric rigidity is then applied to the study of holomorphic mappings into quotients of bounded domains, where we construct such "partial" continuous Finsler metrices on the domain manifold by means of the given holomorphic mappings. The use of such metrics brings in the role of extremal bounded holomorphic functions for which we sketch some ideas introducing the reader to the original article ([Mok5]). The embedding theorem in the latter article leads then to natural unresolved questions of rigidity when target manifolds are quotients of bounded homogeneous domains. We also give an application of the embedding theorem to study holomorphic mappings onto normal projective spaces, using a result of Margulus [Mar], to show that they must be unramified covering maps unless the target manifolds have finite fundamental groups. In §3 on gap rigidity it is our intention to provide some motivation on the problem, and to explore different strategies of proof, with the hope to shed some light on the general problem, which remains largely unexplored. For these reasons we sketch the original proof of gap rigidity in the complex topology in the Siegel modular case, even though the result itself has now been superseded. This is done through the study of solutions of certain second-order elliptic equations. In this light we also sketch a differential-geometric proof of a result of Shioda's [Sh, 1972] in relation to finiteness of Mordell-Weil groups over modular curves. The proof makes use of eigensection equations which we relate to Eichler's automorphic forms (cf. Silverberg [Sil, 1985]). We also include generalizations along this line to general Kuga families as given in Mok-To [MT, 1993]. A counter-example to the general Gap Phenomenon in the complex topology ([EM2]) is given, motivating the formulation of a more restricted conjectural version of gap rigidity. We present in some details, together with an introduction to the algebraic background on bounded symmetric domains, the verification of gap rigidity in the Zariski topology for pairs (Ω, Ω') obtained from diagonal embeddings (cf. (3.6)) through the use of the Poincaré-Lelong equation, leaving aside in the case of higher-dimensional submanifolds discussions involving Geometric Invariant Theory to the original article [EM2]. We explain a link between metric rigidity and gap rigidity as given by an application to the study of holomorphic mappings ([Mok4]), and another link as given by a proof of gap rigidity in relation to quadric structures ([Mok4]), where metric rigidity and results on Kähler-Einstein metrics are used. This latter approach leads to a natural question involving almost holomorphic geometric structures modelled on bounded symmetric domains which we formulate at the end of the article.

Table of Contents

1. Local rigidity of complex structure under deformation

(1.1) Rigidity problems are of fundamental importance in Complex Geometry. In this article we discuss a number of rigidity problems on Hermitian locally symmetric spaces of the noncompact type X, i.e., quotients of bounded symmetric domains by torsion-free discrete groups of biholomorphic automorphisms. These manifolds are of particular importance as space forms in differential geometry and often as moduli spaces for certain algebro-geometric classification problems, e.g., Siegel modular varieties as parameter spaces for polarized Abelian varieties. In this section we will be solely concerned with the case where X is compact. We start with the classical local rigidity theorem of Calabi-Vesentini [CV] and Borel [Bo1]. For a compact complex manifold X, local rigidity of complex structure follows from the vanishing of $H^1(X, T_X)$ for the holomorphic tangent bundle T_X. In differential-geometric terms vanishing theorems most often arise from special curvature properties of canonical metrics. Let (V, h) be a Hermitian holomorphic vector bundle over a complex manifold X and denote by $\Theta_{v\overline{w}\alpha\overline{\beta}}$ the curvature tensor of (V, h); $v, w \in V_x$; $\alpha, \beta \in T_x(X)$; $x \in X$. We have at $x \in X$ the Hermitian bilinear form Q_x on $V_x \otimes \overline{T_x(X)}$ defined by $Q_x(v \otimes \overline{\beta}, w \otimes \overline{\alpha}) = \Theta_{v\overline{w}\alpha\overline{\beta}}$ and extended by Hermitian bilinearity. We say that (V, h) is of nonpositive (strictly negative) curvature in the dual sense of Nakano if and only if Q_x is negative semidefinite (resp. negative definite) at each point $x \in X$. We have

THEOREM 1. *Let (X, g) be a compact Kähler manifold of complex dimension at least 2, and (V, h) be a Hermitian holomorphic vector bundle on X of strictly negative curvature in the dual sense of Nakano. Then, $H^1(X, V) = 0$.*

The usual proof of Theorem 1 is by means of a Bochner-Kodaira formula. After Siu [Siu1] obtained a $\partial\overline{\partial}$-Bochner formula for proving strong rigidity of compact Kähler manifolds using harmonic maps, he realized that the same approach could have yielded vanishing theorems for Hermitian holomorphic vector bundles. There is the conceptual advantage that by this approach, in that one knows *a priori* that the resulting integral formula will only involve the curvature of the Hermitian holomorphic vector bundle (V, h) and not the curvature of (X, g). We will present this proof here.

PROOF OF THEOREM. Let u be a V-valued harmonic $(0,1)$-form with respect to (X, g) and the Hermitian metric h on V, i.e., $\overline{\partial} u = \overline{\partial}^* u = 0$, where $\overline{\partial}^*$ stands for the adjoint operator of $\overline{\partial}$ with respect to the given metrics. Write

$$u = \sum_{i,\alpha} u^\alpha_{\overline{i}} e_\alpha \otimes d\overline{z^i} \ . \tag{1}$$

Associated to u we have a real (1,1)-form η and a real (2,2)-form ξ defined by

$$\xi = \sqrt{-1} \sum_{i,j,\alpha,\beta} h_{\alpha\overline{\beta}} u^\alpha_{\overline{j}} \overline{u^\beta_{\overline{i}}} dz^i \wedge d\overline{z^j} \ ; \qquad \eta = \sqrt{-1}\partial\overline{\partial}\xi \ . \tag{2}$$

Let $x \in X$. Choose holomorphic fiber coordinates adapted to (V, h) at x so that $h_{\alpha\overline{\beta}}(x) = \delta_{\alpha\beta}$ and $dh_{\alpha\overline{\beta}}(x) = 0$. Taking derivatives and adopting the Einstein convention for summations, we have

$$
\begin{aligned}
-\eta = {} & (\partial_k \partial_{\overline{\ell}} h_{\alpha\overline{\beta}}) u^{\alpha}_{\overline{j}} \overline{u^{\beta}_{\overline{i}}} dz^k \wedge d\overline{z^\ell} \wedge dz^i \wedge d\overline{z^j} \ \ldots \text{(I)} \\
& + h_{\alpha\overline{\beta}} (\partial_k u^{\alpha}_{\overline{j}}) (\partial_{\overline{\ell}} \overline{u^{\beta}_{\overline{i}}}) dz^k \wedge d\overline{z^\ell} \wedge dz^i \wedge d\overline{z^j} \ \ldots \text{(II)} \\
& + h_{\alpha\overline{\beta}} (\partial_{\overline{\ell}} u^{\alpha}_{\overline{j}}) (\partial_k \overline{u^{\beta}_{\overline{i}}}) dz^k \wedge d\overline{z^\ell} \wedge dz^i \wedge d\overline{z^j} \ \ldots \text{(III)} \\
& + h_{\alpha\overline{\beta}} (\partial_k \partial_{\overline{\ell}} u^{\alpha}_{\overline{j}}) \overline{u^{\beta}_{\overline{i}}} dz^k \wedge d\overline{z^\ell} \wedge dz^i \wedge d\overline{z^j} \ \ldots \text{(IV)} \\
& + h_{\alpha\overline{\beta}} u^{\alpha}_{\overline{j}} (\partial_k \partial_{\overline{\ell}} \overline{u^{\beta}_{\overline{i}}}) dz^k \wedge d\overline{z^\ell} \wedge dz^i \wedge d\overline{z^j} \ \ldots \text{(V)}
\end{aligned} \tag{3}
$$

We assert that all terms on the right-hand side vanish except for the first two. Regarding (III), from $\overline{\partial} u = 0$ it follows that $\partial_{\overline{\ell}} u^{\alpha}_{\overline{j}} = \partial_{\overline{j}} u^{\alpha}_{\overline{\ell}}$, i.e., we have symmetry in the pair (j, ℓ) of indices. On the other hand,

$$
dz^k \wedge d\overline{z^\ell} \wedge dz^i \wedge d\overline{z^j} = -dz^k \wedge d\overline{z^j} \wedge dz^i \wedge d\overline{z^\ell} \ . \tag{4}
$$

Summing up over j, ℓ we observe the vanishing of (III). Regarding (IV), observing the symmetry of $\partial_k \partial_{\overline{\ell}} u^{\alpha}_{\overline{j}}$ in (j, ℓ) the same argument applies. As to (V), it suffices to write $\partial_k \partial_{\overline{\ell}} \overline{u^{\beta}_{\overline{i}}} = \partial_{\overline{\ell}} \overline{\partial_{\overline{k}} u^{\beta}_{\overline{i}}}$ and use the symmetry in (k, i) to conclude its vanishing. It follows from (3) that

$$
-\eta \wedge \omega^{n-2} = [\text{(I)} + \text{(II)}] \wedge \omega^{n-2} = \text{Curvature term} + \text{Gradient term} \ . \tag{5}
$$

At $x \in X$ choose a system of complex geodesic coordinates (z_i) such that $g_{i\overline{j}}(x) = \delta_{ij}$ and $dg_{i\overline{j}}(x) = 0$, so that the Riemann-Christoffel symbols Γ^k_{ij} vanish at x. Write

$$
\omega = \sqrt{-1} \sum_{i,j} g_{i\overline{j}} dz^i \wedge d\overline{z^j} \tag{6}
$$

for the Kähler form ω of (X, g). From the harmonicity of u we have

$$
0 = \overline{\partial}^* u = -\sum_{i,j} \left(g^{i\overline{j}} \nabla_i u^{\alpha}_{\overline{j}} \right) e_\alpha \ , \tag{7}
$$

which at x translates into

$$
0 = \sum_{i,\alpha} (\nabla_i u^{\alpha}_{\overline{i}}) e_\alpha = \sum_{i,\alpha} (\partial_i u^{\alpha}_{\overline{i}}) e_\alpha \ . \tag{8}
$$

For the expression involving (II) in (5) we have at x

$$
\begin{aligned}
\text{(II)} \wedge \omega^{n-2} = {} & \sum_{k,j} \left(\partial_k u^{\alpha}_{\overline{j}} \overline{\partial_k u^{\alpha}_{\overline{j}}} dz^k \wedge d\overline{z^k} \wedge dz^j \wedge d\overline{z^j} \right) \wedge \omega^{n-2} \\
& - \sum_{k,j} \left(\partial_k u^{\alpha}_{\overline{k}} \overline{\partial_j u^{\alpha}_{\overline{j}}} dz^k \wedge d\overline{z^k} \wedge dz^j \wedge d\overline{z^j} \right) \wedge \omega^{n-2} \\
= {} & (n-2)! \sum_{k,j} \left| \partial_k u^{\alpha}_{\overline{j}} \right|^2 dz^1 \wedge d\overline{z^1} \wedge \cdots \wedge dz^n \wedge d\overline{z^n} \ ,
\end{aligned} \tag{9}
$$

where we have simplified the expression using (8). Therefore,

$$
\int_X -\eta \wedge \frac{\omega^{n-2}}{(n-2)!} = \int_X \text{Curvature term} + \int_X \|\nabla u\|^2 \frac{\omega^n}{n!} \ . \tag{10}
$$

Write Θ for the curvature form of (V,h) so that with respect to the chosen base and fiber holomorphic coordinate systems we have $\Theta_{\alpha\overline{\beta}k\overline{\ell}}(x) = -\partial_k\partial_{\overline{\ell}}h_{\alpha\overline{\beta}}(x)$. At x we have

$$(I) \wedge \frac{\omega^{n-2}}{(n-2)!} = \sum_{k,\ell,\alpha,\beta} \left(\Theta_{\alpha\overline{\beta}k\overline{\ell}} u^{\alpha}_{\overline{k}} \overline{u^{\beta}_{\overline{\ell}}} - \Theta_{\alpha\overline{\beta}k\overline{k}} u^{\alpha}_{\overline{\ell}} \overline{u^{\beta}_{\overline{\ell}}}\right) \frac{\omega^n}{n!} \tag{11}$$

Since $\eta = \sqrt{-1}\partial\overline{\partial}\xi$, hence $d\eta = 0$, dividing (5) by $(n-2)!$ and integrating over X we obtain by Stokes' Theorem

$$0 = \int_X -\eta \wedge \frac{\omega^{n-2}}{(n-2)!} = \int_X \left(\sum_{k,\ell,\alpha,\beta} \left(\Theta_{\alpha\overline{\beta}k\overline{\ell}} u^{\alpha}_{\overline{k}} \overline{u^{\beta}_{\overline{\ell}}} - \Theta_{\alpha\overline{\beta}k\overline{k}} u^{\alpha}_{\overline{\ell}} \overline{u^{\beta}_{\overline{\ell}}}\right) + \|\nabla\eta\|^2 \right) \frac{\omega^n}{n!}\,, \tag{12}$$

Recall the Bochner-Kodaira formula for the (1,0)-gradient

$$\int_X \left\|\partial u\right\|^2 + \left\|\overline{\partial}^* u\right\|^2 = \int_X K(u,u) + \left\|\nabla\eta\right\|^2 , \tag{13}$$

where in the integrals the volume form $\frac{\omega^n}{n!}$ is implicit. (12) is simply the Bochner-Kodaira formula (13) in the case where u is harmonic. Write $\Theta_{k\overline{\ell}} := \sum_{\alpha,\beta} h^{\alpha\overline{\beta}}\Theta_{\alpha\overline{\beta}k\overline{\ell}}$. We have $(\Theta_{k\overline{l}}) < 0$ whenever (V,h) is of strictly negative curvature in the dual sense of Nakano. When $V = L$ is a holomorphic line bundle, taking at $x \in X$ $\alpha = e_1$ to be of unit length, writing $u_{\overline{k}}$ for $u^1_{\overline{k}}$, and choosing furthermore local holomorphic coordinates (z_k) such that at x the Hermitian matrix $\left(\Theta_{k\overline{\ell}}\right)$ is diagonalized, $\Theta_{k\overline{\ell}} = -c_k\delta_{k\ell}$, we have

$$K(u,u) = \Big(\sum_k c_k\Big)\Big(\sum_l |u_{\overline{\ell}}|^2\Big) - \sum_k c_k|u_{\overline{k}}|^2 = \sum_k (c - c_k)|u_{\overline{k}}|^2 \tag{14}$$

where $-c = -\sum_k c_k > 0$ is the scalar curvature. This gives Kodaira's Vanishing Theorem for negative line bundles, which holds true more generally for (L,h) on an n-dimensional compact Kähler manifold, $n \geq 2$, such that at each point the sum of any $n-1$ eigenvalues of the curvature form is negative. In the general case of a Hermitian holomorphic vector bundle (V,h) of strictly negative curvature in the dual sense of Nakano, it remains to prove that

$$(\dagger) \qquad K(u,u) = \sum_{k\neq\ell}\Big(-\sum_{\alpha,\beta}\Theta_{\alpha\overline{\beta}k\overline{k}} u^{\alpha}_{\overline{\ell}}\overline{u^{\beta}_{\overline{\ell}}} + \sum_{\alpha,\beta}\Theta_{\alpha\overline{\beta}k\overline{\ell}} u^{\alpha}_{\overline{k}}\overline{u^{\beta}_{\overline{\ell}}}\Big) > 0$$

whenever $(u^{\alpha}_{\overline{i}}) \neq 0$. Pick any pair of indices (k,ℓ), $k \neq \ell$. Consider the matrix $(A^{\gamma\overline{i}}) = (A^{\gamma\overline{i}}_{k\ell})$, where

$$A^{\gamma\overline{k}} = u^{\gamma}_{\overline{\ell}},\ A^{\gamma\overline{\ell}} = -u^{\gamma}_{\overline{k}};\ A^{\gamma\overline{i}} = 0 \ \text{ whenever } \ i \neq k,\ell\,. \tag{15}$$

Then,

$$
\begin{aligned}
0 \leq & \sum_{\alpha,\beta,i,j} -\Theta_{\alpha\overline{\beta}i\overline{j}} A^{\alpha\overline{j}}\, \overline{A^{\beta\overline{i}}} \\
&= -\sum_{\alpha,\beta} \Theta_{\alpha\overline{\beta}k\overline{k}} A^{\alpha\overline{k}}\, \overline{A^{\beta\overline{k}}} - \sum_{\alpha,\beta} \Theta_{\alpha\overline{\beta}\ell\overline{\ell}} A^{\alpha\overline{\ell}}\, \overline{A^{\beta\overline{\ell}}} \\
&\quad - \sum_{\alpha,\beta} \Theta_{\alpha\overline{\beta}k\overline{\ell}} A^{\alpha\overline{\ell}}\, \overline{A^{\beta\overline{k}}} - \sum_{\alpha,\beta} \Theta_{\alpha\overline{\beta}\ell\overline{k}} A^{\alpha\overline{k}}\, \overline{A^{\beta\overline{\ell}}} \\
&= -\sum_{\alpha,\beta} \Theta_{\alpha\overline{\beta}k\overline{k}} u^{\alpha}_{\overline{\ell}} \overline{u^{\beta}_{\overline{\ell}}} - \sum_{\alpha,\beta} \Theta_{\alpha\overline{\beta}\ell\overline{\ell}} u^{\alpha}_{\overline{k}} \overline{u^{\beta}_{\overline{k}}} \\
&\quad + \sum_{\alpha,\beta} \Theta_{\alpha\overline{\beta}k\overline{\ell}} u^{\alpha}_{\overline{k}} \overline{u^{\beta}_{\overline{\ell}}} + \sum_{\alpha,\beta} \Theta_{\alpha\overline{\beta}\ell\overline{k}} u^{\alpha}_{\overline{\ell}} \overline{u^{\beta}_{\overline{k}}} \\
&= 2\Big(-\sum_{\alpha,\beta} \Theta_{\alpha\overline{\beta}k\overline{k}} u^{\alpha}_{\overline{\ell}} \overline{u^{\beta}_{\overline{\ell}}} + \sum_{\alpha,\beta} \Theta_{\alpha\overline{\beta}k\overline{\ell}} u^{\alpha}_{\overline{k}} \overline{u^{\beta}_{\overline{\ell}}} \Big) .
\end{aligned} \tag{16}
$$

Summing up over (k,ℓ), $k \neq \ell$, it follows from (16) that $K(u,u) \geq 0$ and that equality holds if and only if each $\big(A^{\gamma\overline{i}}_{k\ell}\big)$ vanishes for each choice of (k,ℓ), i.e., if and only if $u = 0$. We have thus verified (†) and shown that $H^1(X,V) = 0$ on an n-dimensional compact Kähler manifold X, $n \geq 2$, whenever (V,h) is of strictly negative curvature in the dual Nakano sense, proving Theorem 1.

(1.2) We will now give a proof of the local rigidity of compact quotients of irreducible bounded symmetric domains of complex dimension ≥ 2. More precisely, we have

THEOREM 2. *Let Ω be an irreducible bounded symmetric domain of complex dimension ≥ 2 and $\Gamma \subset \mathrm{Aut}(\Omega)$ be a torsion-free cocompact lattice, $X := \Omega/\Gamma$. Then, $H^1(X, T_X) = 0$. In particular, local rigidity holds for X.*

Theorem 2 is a special case of a vanishing theorem on cohomology groups originally proved by Calabi and Vesentini [CV] for classical bounded symmetric domains and by Borel [Bo2] for exceptional domains. Their proofs involved the computation of eigenvalues of certain Hermitian bilinear forms associated to the curvature tensor. We will now give a proof basing on Theorem 1 by considering the kernel of the Hermitian bilinear form K in the Bochner-Kodaira formula discussed in the above.

PROOF OF THEOREM 2 FOR TYPE I DOMAINS. The tangent bundle T_X, equipped with the Hermitian metric given by the Bergman metric g on X, is of nonpositive curvature in the dual sense of Nakano. Fix $x \in X$ and let Q be the curvature operator on X regarded as a Hermitian bilinear form on $T_x \otimes \overline{T_x}$, i.e., Q is defined by $Q(\xi \otimes \overline{\eta}; \xi' \otimes \overline{\eta'}) = R_{\xi\overline{\eta}\eta'\overline{\xi'}}$ for decomposable elements; $\eta, \eta', \xi, \xi' \in T_x$; and extended to $T_x \otimes \overline{T_x}$ by Hermitian sequilinearity. We consider now first of all the special case where Ω is a bounded symmetric domain of Type I and of dimension > 1, given by

$$\Omega = D^I_{m,n} = \big\{ m \times n \text{ matrices } Z : I - \overline{Z}^t Z > 0 \big\} \ , \quad m \geq n \ ; \quad m > 1 \ .$$

When $n = 1$ we have the m-ball B^m which is of strictly negative curvature in the dual sense of Nakano, so that Theorem 1 applies immediately. For the purpose of proving Theorem 2 for bounded symmetric domains of Type I we will restrict ourselves to $n \geq 2$ in what follows. Up to a scalar multiple the Kähler form for the

Bergman metric of $D^I_{m,n}$ is given by $\omega = \sqrt{-1}\partial\overline{\partial}\varphi$, $\varphi = -\log\det(I - \overline{Z}^t Z)$. Note that the Euclidean coordinate system is a complex geodesic coordinate system at the origin o of Ω. A direct computation at o using the potential function φ shows that

$$R_{ij,\overline{k\ell},pq,\overline{rs}} = -\delta_{ik}\delta_{pr}\delta_{js}\delta_{lq} - \delta_{ir}\delta_{pk}\delta_{j\ell}\delta_{qs}$$

which gives

$$\sum R_{ij,\overline{k\ell},pq,\overline{rs}} A^{ij,\overline{k\ell}}\ \overline{A^{rs,\overline{pq}}} = -\left(\sum A^{ij,\overline{i\ell}}\ \overline{A^{pj,\overline{p\ell}}} + \sum A^{ij,\overline{kj}}\ \overline{A^{iq,\overline{kq}}}\right) \leq 0 \quad (1)$$

showing the nonpositivity of Q, i.e., that (X,g) is of nonpositive curvature in the dual sense of Nakano. We note in particular that

$$\begin{aligned}\mathrm{Bisect}(X,Y) &= R_{X\overline{X}Y\overline{Y}}\\ &= Q(X\otimes\overline{Y}, X\otimes\overline{Y}) \qquad (2)\\ &= -\|X\overline{Y}^t\|^2 - \|X^t\overline{Y}\|^2\end{aligned}$$

As an example if we let

$$X_o = \begin{bmatrix} * & 0\\ 0 & 0\end{bmatrix} \quad \text{and} \quad Y_o = \begin{bmatrix} 0 & 0\\ 0 & *\end{bmatrix}, \qquad (3)$$

then

$$X_o\overline{Y_o}^t = X_o^t\overline{Y_o} = 0 \qquad (4)$$

showing that $\mathrm{Bisect}(X_o, Y_o) = 0$. We note that in general from the nonpositivity of Q it follows that

$$X\otimes\overline{Y} \in \mathrm{Ker}(Q) \quad \text{whenever} \quad \mathrm{Bisect}(X,Y) = 0\ . \qquad (5)$$

Let now

$$u = \sum_{i,j,k,\ell} u^{k\ell}_{\overline{ij}} \frac{\partial}{\partial z_{k\ell}} \otimes d\overline{z^{ij}} \qquad (6)$$

be a T_X-valued harmonic (0,1)-form. By the proof of Theorem 1, $K(u,u) = 0$ everywhere on X, and we have

$$u\left(\frac{\partial}{\partial\overline{z_{ij}}}\right)\otimes\frac{\partial}{\partial z_{k\ell}} - u\left(\frac{\partial}{\partial\overline{z_{k\ell}}}\right)\otimes\frac{\partial}{\partial z_{ij}} \in \mathrm{Ker}(Q)\ . \qquad (7)$$

Writing $u_{\overline{ij}} = u\big(\frac{\partial}{\partial\overline{z_{ij}}}\big)$, etc. we assert that in the special case where $i=j$, $k=\ell$, $i\neq k$, we have

$$(\sharp) \qquad u_{\overline{ii}}\otimes\frac{\partial}{\partial z_{kk}};\quad u_{\overline{kk}}\otimes\frac{\partial}{\partial z_{ii}} \in \mathrm{Ker}(Q)\ .$$

From the identity

$$Q(A+B, A+B) = Q(A,A) + Q(B,B) + 2\mathrm{Re}Q(A,B)$$

to deduce $(\sharp)$ from (7) it suffices to show that $Q\big(u_{\overline{ii}}\otimes\frac{\partial}{\partial z_{kk}}, u_{\overline{kk}}\otimes\frac{\partial}{\partial z_{ii}}\big) = 0$. This is indeed the case because $\mathrm{Bisect}\big(\frac{\partial}{\partial z_{ii}}, \frac{\partial}{\partial z_{kk}}\big) = 0$, so that $\frac{\partial}{\partial z_{ii}}\otimes\frac{\overline{\partial}}{\partial z_{kk}} \in \mathrm{Ker}(Q)$ by (5), which implies

$$Q\left(u_{\overline{ii}}\otimes\frac{\partial}{\partial z_{kk}}, u_{\overline{kk}}\otimes\frac{\partial}{\partial z_{ii}}\right) = Q\left(\frac{\partial}{\partial z_{ii}}\otimes\overline{\frac{\partial}{\partial z_{kk}}}, u_{\overline{kk}}\otimes\overline{u_{\overline{ii}}}\right) = 0\ . \qquad (8)$$

Let α,ζ be unit vectors in $T_o(\Omega)$ such that $\mathrm{Bisect}(\alpha,\zeta) = 0$. Representing α resp. ζ as $m\times n$-matrices X resp. Y, there exist unitary matrices $U\in U(m)$, $V\in U(n)$

such that UXV resp. UYV can be put in the form X_o resp. Y_o as in (3) and (4). It follows from the argument giving ($\sharp$) that

$$u(\overline{\alpha}) \otimes \overline{\zeta} \in \mathrm{Ker}(Q) \quad \text{whenever} \quad R_{\alpha\overline{\alpha}\zeta\overline{\zeta}} = 0 \ .$$

For any $\alpha \in T_o(\Omega)$, we define $\mathcal{N}_\alpha$ to consist of all $\beta \in T_o(\Omega)$ such that $R_{\alpha\overline{\alpha}\beta\overline{\beta}} = 0$. A vector $\alpha \in T_o(\Omega)$ is said to be a minimal characteristic vector at o if and only if it corresponds to a matrix of rank 1. Given a complex vector space V and $\pi : V - \{0\} \to \mathbb{P}V$ the canonical projection, for any $A \subset \mathbb{P}V$ we write $\widetilde{A}$ for $\pi^{-1}(A)$, the homogenization of A. Denote by $\mathcal{M}_o \subset \mathbb{P}T_o(X)$ the projective submanifold consisting of projectivizations of minimal characteristic vectors. For $\alpha \in \widetilde{\mathcal{M}_o}$ of unit length, up to unitary transformations $U \in U(m)$ and $V \in U(n)$ we may assume α to be $\frac{\partial}{\partial z_{11}}$. From this it is immediate to check that $\bigcap_{\beta \in \mathcal{N}_\alpha} \mathcal{N}_\beta = \mathbb{C}\alpha$. It follows that

$$u(\overline{\alpha}) = \lambda(\alpha)\alpha \tag{9}$$

for some smooth function λ on $\widetilde{\mathcal{M}_o}$. We assert that (9) implies that $u = 0$. To see this let $\alpha_o \in \widetilde{\mathcal{M}_o}$, $\eta \in T_o(\Omega)$ be a nonzero vector tangent to $\widetilde{\mathcal{M}_o}$ at α_o and not proportional to α_o. From (9) it follows that at α_o we have

$$0 = (\partial_\eta \lambda)\alpha_o + \lambda(\alpha_o)\eta \tag{10}$$

which is impossible unless $\lambda(\alpha_o) = 0$. This implies that λ and hence u vanishes identically on $\widetilde{\mathcal{M}_o}$, as asserted.

(1.3) To prove Theorem 2 for the general case of an irreducible bounded symmetric domain Ω of rank ≥ 2 the proof given above for Type I domains works equally well. We take this occasion to consider irreducible bounded symmetric domains Ω in general and discuss properties of the curvature operator shared by Bergman metrics on all such domains. To start with we have the following complete list of the set of all irreducible bounded symmetric domains. Here we denote the Euclidean space of all $m \times n$-matrices with complex coefficients by $M(m, n; \mathbb{C})$. Furthermore, given a simple Lie algebra $\mathfrak{g}$ over $\mathbb{C}$, we say that an irreducible bounded symmetric domain Ω is of type $\mathfrak{g}$ to mean that $\mathrm{Aut}_o(\Omega) := G$ is a Lie group whose Lie algebra is a real form of $\mathfrak{g}$. The following is a complete list of irreducible bounded symmetric domains:

Irreducible Classical Symmetric Domains

$$\begin{aligned}
D^I_{m,n} &:= \{Z \in M(m,n;\mathbb{C}) : I - \overline{Z}^t Z > 0\} \ , \quad m, n \geq 1; \\
D^{II}_n &:= \{Z \in D^I_{n,n} : Z^t = -Z\} \ , \quad n \geq 2; \\
D^{III}_n &:= \{Z \in D^I_{n,n} : Z^t = Z\} \ , \quad n \geq 1; \\
D^{IV}_n &:= \Big\{(z_1, \ldots, z_n) \in \mathbb{C}^n : \|z\|^2 < 2 \ ; \\
&\qquad \|z\|^2 < 1 + \Big|\frac{1}{2}\sum_{i=1}^n z_i^2\Big|^2\Big\} \ , \quad n \geq 3 \ .
\end{aligned}$$

Exceptional Domains

$$D^V \ , \quad \text{of type } E_6, \quad \dim_{\mathbb{C}}\left(D^V\right) = 16 \ ;$$

$$D^{VI} \ , \quad \text{of type } E_7, \quad \dim_{\mathbb{C}}\left(D^{VI}\right) = 27 \ .$$

There is some duplication in the listing, viz. $D^I_{1,1} \cong D^{II}_2 \cong D^{III}_1 \cong \Delta$, the unit disk, $D^{II}_3 \cong D^I_{3,1} \cong B^3$, the 3-dimensional unit ball, $D^{III}_2 \cong D^{IV}_3$, $D^I_{2,2} \cong D^{IV}_4$ and $D^{II}_4 \cong D^{IV}_6$. Sometimes it is more natural to make use of an unbounded realization of a Hermitian symmetric space of the noncompact type. This is the case of the Siegel upper half-plane $\mathcal{H}_n$ of genus n defined by

$$\mathcal{H}_n := \left\{\tau \in M(n,n;\mathbb{C}) : \tau^t = \tau;\ \mathrm{Im}(\tau) > 0\right\} ,$$

the natural parameter space for principally polarized n-dimensional Abelian varieties, where each τ represents the lattice $L_\tau \subset \mathbb{C}$ spanned by the unit vectors and the column vectors of the matrix τ. The conditions of symmetry and positivity of the imaginary part are the Riemann bilinear relations which guarantee that $\mathbb{C}/L_\tau$ admits by a positive line bundle E such that $\frac{(c_1(E))^n}{n!} = 1$. The Siegel upper half-plane is biholomorphic to the bounded symmetric domain D^{III}_n, as given by the Cayley transform $Z = (\tau - iI_n)(\tau + iI_n)^{-1}$, where I_n stands for the identity $n \times n$ matrix.

(1.4) Write now $G = \mathrm{Aut}_o(\Omega)$ for the identity component of the automorphism group of Ω, $\mathfrak{g}$ for its Lie algebra, and by $K \subset G$ the isotropy group at $o = eK$, so that $\Omega = G/K$. Denote by $B_\mathfrak{g}$ the Killing form on the simple Lie algebra $\mathfrak{g}$. Let g_0 be any G-invariant Kähler metric on Ω, which is necessarily Kähler-Einstein and just a constant multiple of the Bergman metric. The structure of (Ω, g_0) as a Riemannian symmetric manifold is defined by a Cartan involution σ on $\mathfrak{g}$ whose fixed point set is precisely $\mathfrak{k}$. Denote by θ the involution on G such that $d\theta = \sigma$. Let $\mathfrak{g} = \mathfrak{k} + \mathfrak{m}$ be the Cartan decomposition of $\mathfrak{g}$ with respect σ. $\mathfrak{k}$ is reductive with a one-dimensional centre $\mathfrak{z}$ corresponding to a circle group $Z \subset K$, and the almost complex structure on Ω is defined by an element ι of order 4 in $\mathfrak{z}$ such that $\iota^2 = \theta$. Let $\mathfrak{m}^\mathbb{C} := \mathfrak{m} \otimes_\mathbb{R} \mathbb{C} = \mathfrak{m}^+ \oplus \mathfrak{m}^-$ be the eigenspace decomposition of $\mathfrak{m}^\mathbb{C}$ into eigenspaces corresponding to the eigenvalues i resp. $-i$. Then, we can identify $\mathfrak{m}^+$ with $T^{1,0}_o(\Omega)$ and $\mathfrak{m}^-$ with $T^{0,1}_o(\Omega)$. We will identify canonically the (1,0)-component of the complexified tangent bundle T_Ω of Ω with the holomorphic tangent bundle T_Ω, so that the (0,1)-component $T^{0,1}_\Omega$ is canonically identified with $\overline{T_\Omega}$. An element of $T_o(\Omega) \otimes \overline{T_o(\Omega)}$ corresponds to an element A of $\mathfrak{m}^+ \otimes \mathfrak{m}^-$. The Lie bracket on $\mathfrak{g}$, extended to its complexification $\mathfrak{g}^\mathbb{C}$, yields a complex linear map: $[\cdot,\cdot] : \mathfrak{m}^+ \otimes \mathfrak{m}^- \to \mathfrak{k}^\mathbb{C}$. ¿From a standard formula for the Riemannian curvature tensor on a Riemannian symmetric manifold (cf. Wolf [Wo, Thm.(841), p.245-246]) for the curvature operator Q on $T_o(\Omega) \otimes \overline{T_o(\Omega)}$ we have

$$Q(A,A) = -c\left\|[A,\overline{A}]\right\|^2$$

where $\|\cdot\|$ on $\mathfrak{k}^\mathbb{C}$ is induced by the Killing form $B_\mathfrak{g}$, and $c > 0$ is a constant. This shows that for any (not necessarily irreducible) bounded symmetric domain Ω equipped with the Kähler-Einstein metric g_0, (Ω, g_0) is of nonpositive curvature in the dual sense of Nakano. In particular, it is of nonpositive holomorphic bisectional curvature. For the proof of Theorem 2 in general we will need to study the structure of the zeros of bisectional curvature. Since Q is negative semi-definite, $R_{\alpha\overline{\alpha}\zeta\overline{\zeta}} = 0$ implies $Q(\alpha \otimes \overline{\zeta}; \alpha \otimes \overline{\zeta}) = 0$, so that $\alpha \otimes \overline{\zeta}$ lies on $\mathrm{Ker}(Q)$. We have the analogue of the notion of the minimal characteristic subvariety $\mathcal{M}_x \subset \mathbb{P}T_x(X)$ (cf. (2.1)). Let now $E \subset T_x(X) \otimes \overline{T_x(X)}$ be defined by

$$E = \{\alpha \otimes \overline{\zeta} : \alpha \in \widetilde{\mathcal{M}}_x, R_{\alpha\overline{\alpha}\zeta\overline{\zeta}} = 0\} .$$

Then, $E = \big(\widetilde{\mathcal{M}}_x \otimes \overline{T_x(X)}\big) \cap \mathrm{Ker}(Q)$ is an analytic subvariety of $(T_x(X) - \{o\}) \times \overline{T_x(X)}$. In particular, as α varies holomorphically in $\widetilde{\mathcal{M}}_x$, $\overline{\zeta}$ varies holomorphically in $\overline{T_x(X)}$, i.e., ζ varies anti-holomorphically in $T_x(X)$. If we take a smooth point (α_o, β_o) of A the orbit of α_o resp. ζ_o must generate $T_o(\Omega)$ since K acts on $T_o(\Omega)$ irreducibly. To complete the proof of Theorem 2 we will use the following general lemma on polarization.

LEMMA 1 (THE POLARIZATION ARGUMENT). *Let V and W be finite-dimensional complex vector spaces. Denote by W' the complex vector space which agrees with W as a real vector space but is endowed with the conjugate complex structure of that of W. (In other words, denoting by J_W resp. $J_{W'}$ the almost complex structures of W resp. W' , we have $J_{W'} = -J_W$.) Let $E \subset V \times W'$ a connected complex-analytic submanifold of some open subset of $V \times W'$. Regarding E as a subset of $V \times W$ and denoting by $pr_V : E \to V$, $pr_W : E \to W$ the canonical projections, assume that $pr_V(E)$ resp. $pr_W(E)$ spans V resp. W as a complex vector space. Write $E^{\#} \subset V \otimes W$ for the subset $\{\alpha \otimes \beta : (\alpha, \beta) \in V \times W\}$. Then, $E^{\#}$ spans the tensor product $V \otimes W$ as a complex vector space.*

PROOF. The complex linear span of E is completely determined by its germ at any base point. Pick now $(\alpha_0, \beta_0) \in E$ so that $pr_V|_E : E \to V \times W$, $pr_{W'}|_E : E \to W'$ are of maximal rank at (α_0, β_0). Shrinking E if necessarily we may assume that $pr_V|_E$ is a holomorphic submersion onto a locally closed complex submanifold $\Sigma \subset V$. Let now α range over a sufficiently small open neighborhood U of α_0 in Σ and $(\alpha, \beta(\alpha)) \in E$, $\beta(\alpha_0) = \beta_0$, be chosen such that $\beta : U \to W'$ is a holomorphic map, i.e., $\beta : U \to W$ is an anti-holomorphic map. Then, $\mathrm{Span}(E^\sharp) \supset \mathrm{Span}\{\alpha \otimes \beta(\alpha) : \alpha \in U\}$. Hence, $\mathrm{Span}(E^\sharp)$ contains $\alpha_0 \otimes \beta_0$ and any partial derivative of $\alpha \otimes \beta(\alpha)$ at α_0 with respect to a choice of holomorphic coordinates (z_k) on U. In particular, taking partial derivatives in the $(1,0)$ direction and using the fact that β is anti-holomorphic we conclude that $\mathrm{Span}(E^\sharp)$ contains every $\gamma \otimes \beta_0$ where γ is a partial derivative of α at α_0 in the $(1,0)$ direction. For instance

$$\frac{\partial}{\partial z_k}(\alpha \otimes \beta(\alpha))\Big|_{\alpha=\alpha_0} = \frac{\partial \alpha}{\partial z_k}(\alpha_0) \otimes \beta_0.$$

But the complex linear span of α_0 and all such γ agrees with that of E, which is by assumption the same as V. Thus, $\mathrm{Span}(E^\sharp) \supset V \otimes \{\beta_0\}$. Varying now β_0 and using the assumption that $pr_W(E)$ spans W we conclude that $\mathrm{Span}(E^\sharp) = V \otimes W$, as desired.

PROOF OF THEOREM 2 CONTINUED. It suffices to consider the case where Ω is an irreducible bounded symmetric domain of rank ≥ 2. In this case the curvature operator Q as a Hermitian form on $T_o(\Omega) \otimes \overline{T_o(\Omega)}$ negative semi-definite with a nonzero kernel. By the same argument in the case of $\Omega = D^I_{m,n}$, $m \geq n > 1$, we have the analogue of $(\sharp'')$ there, viz., for any harmonic T_X-valued (0,1)-form on X we have

$$u(\overline{\alpha}) \otimes \zeta \in \mathrm{Ker}(Q) \quad \text{whenever} \quad R_{\alpha\overline{\alpha}\zeta\overline{\zeta}} = 0\ . \qquad (\sharp'')$$

By considering the set $E = \big(\widetilde{\mathcal{M}}_x \otimes \overline{T_x(X)}\big) \cap \mathrm{Ker}(Q)$ and applying the polarization argument (Lemma 1) we conclude that $u(\overline{\xi}) \otimes \eta \in \mathrm{Ker}(Q)$ for any $\xi, \eta \in T_o(\Omega)$. This implies that $\mathrm{Im}(u) \subset T_o(\Omega)$ consists of vectors μ such that $R_{\mu\overline{\mu}\eta\overline{\eta}} = 0$ for every $\eta \in T_o(\Omega)$, which contradicts with the fact that (Ω, g_0) is of constant negative Ricci curvature unless $u \equiv 0$. □

2. Metric rigidity and extremal bounded holomorphic functions on arithmetic varieties of rank ≥ 2

(2.1) Let Ω be an irreducible bounded symmetric domain. When Ω is of rank 1, i.e., Ω is an n-dimensional complex unit ball B^n, any two unit (1,0) tangent vectors are equivalent to each other under the automorphism group of Ω. This is the case because the isotropy group at the origin $o \in B^n$ is the unitary group. In the case of Ω of higher rank this is no longer the case. In the notations of (1.3) for $\Omega = D^I_{m,n}$; $m, n \geq 1$; $SU(m) \times SU(n)$ acts (with a finite kernel) as the full isotropy group at $o \in D^M_{m,n}$ by the action $(A, B) \cdot X = AXB$ where $A \in SU(m)$, $B \in SU(n)$ and $X \in M(m, n; \mathbb{C})$. Thus, in particular the rank of the matrix X, which represents a (1,0) tangent vector at the origin, is unchanged under the action of the isotropy group. As in (1.5) we have the Harish-Chandra and Borel embeddings $D^I_{m,n} \subset M(m, n; \mathbb{C}) \cong \mathbb{C}^{mn} \subset G(n, m)$, where $G(n, m)$ stands for the Grassmannian of complex n-planes in $\mathbb{C}^{m+n}$. In this case $SL(m + n; \mathbb{C})$ acts (with a finite kernel) as the full group $G^{\mathbb{C}}$ of biholomorphisms so that the parabolic (isotropy) subgroup $P \subset G^{\mathbb{C}}$ at $o \in G(n, m)$ is the subgroup generated by the reductive group $L \subset GL(mn; \mathbb{C})$ which is the image of $GL(m, \mathbb{C}) \times GL(n, \mathbb{C})$ under the action $(A, B) \cdot Z = AZB$, and by the unipotent subgroup $U \cong M(m, n; \mathbb{C})$ where $C \in U \cong M(m, n; \mathbb{C})$ acts on $G(n, m)$ by $C \cdot Z = Z(I_n + C^t Z)^{-1}$. Elements of $U \subset P$ are characterized by the fact that their Jacobian matrices at the fixed point o are the identity matrix. ¿From this description it follows that the rank of $X \in M(m, n; \mathbb{C})$ is actually invariant under the parabolic subgroup P. This observation leads to the construction of $G^{\mathbb{C}}$-invariant subvarieties of the projectivized tangent bundle of $G(n, m)$. With the proper definition of the rank of a (1,0) tangent vector the construction applies to any bounded symmetric domain Ω and its dual M, which is a Hermitian symmetric manifold of the compact type. When restricted to Ω and passing to quotient manifolds they lead to holomorphic bundles, to be called minimal characteristic bundles, which served in [Mok1,2,4,5] as the source of various forms of metric rigidity with applications to rigidity phenomena for holomorphic mappings. In what follows we will give a general description of the construction of characteristic bundles. This relies on some basic facts about Lie algebras relevant to the study of bounded symmetric domains, as can be found for instance in [Mok2, Chapter 3 and 5]. Geometrically we have the following basic statement about totally geodesic polydisks on a bounded symmetric domain.

POLYDISK THEOREM (cf. Wolf [Wo, p.280]). *Let Ω be a bounded symmetric domain of rank r, equipped with the Kähler-Einstein metric g. Then, there exists an r-dimensional totally geodesic complex submanifold Π biholomorphic to the polydisk Δ^r. Moreover, the identity component $\mathrm{Aut}_o(\Omega)$ of $\mathrm{Aut}(\Omega)$ acts transitively on the space of all such polydisks.*

In our study of metric rigidity we make use of ergodic-theoretic properties of the action of an irreducible lattice $\Gamma \subset \mathrm{Aut}_o(\Omega) := G$ on right homogeneous spaces G/H where $H \subset G$ is a noncompact closed subgroup. For instance, by the Polydisk Theorem the space of maximal polydisks on a bounded symmetric domain Ω of rank ≥ 2 can be identified as such a right-homogeneous space G/H. A basic result in this context is given by

MOORE'S ERGODICITY THEOREM. (CF. ZIMMER [ZI, THM. (2.2.6), P.19]) *Let G be a semisimple real Lie group and Γ be an irreducible lattice on G, i.e., $\Gamma \backslash G$*

is of finite volume in the left invariant Haar measure. Suppose $H \subset G$ is a closed subgroup. Consider the action of H on $\Gamma \setminus G$ by multiplication on the right. Then, H acts ergodically if and only if H is noncompact.

For other related results in Ergodic Theory relevant to metric rigidity we refer the reader to [Mok5, (3.1)] and the references cited there from Zimmer [Zim].

Characteristic bundles

Write G for $\mathrm{Aut}_o(\Omega)$ and let $K \subset G$ be the isotropy subgroup at a point $o \in \Omega$, so that $\Omega = G/K$. Ω is a Hermitian symmetric space with respect to the Bergman metric. K acts faithfully on the real tangent space $T_o^{\mathbb{R}}(\Omega)$. Let $\mathfrak{a}$ be a maximal abelian subspace of the real tangent space, whose dimension r is the rank of Ω as a Riemannian symmetric space. Then, $T_o^{\mathbb{R}}(\Omega) = \bigcup_{k\in K} k\mathfrak{a}$. We have $\mathfrak{a} \cap J\mathfrak{a} = 0$ for the canonical complex structure J on Ω. The complexification $(\mathfrak{a} + J\mathfrak{a}) \otimes_{\mathbb{R}} \mathbb{C}$ decomposes into $\mathfrak{a}^+ \oplus \overline{\mathfrak{a}^+}$, where $\mathfrak{a}^+ \subset T_o(\Omega)$, the holomorphic tangent space at o, so that $T_o(\Omega) = \bigcup_{k\in K} k\mathfrak{a}^+$. Thus, any (1,0) tangent vector ξ at o is equivalent under the action of the isotropy group to some $\eta \in \mathfrak{a}^+$. By the Polydisk Theorem, there exists a totally geodesic (holomorphic) polydisk $\Pi \cong \Delta^r$ passing through o, $\Pi \subset \Omega$, such that $T_o(\Pi) = \mathfrak{a}^+$. With respect to Euclidean coordinates on $\Pi \cong \Delta^r$, we can write $\eta = (\eta_1, ..., \eta_r)$. For $\eta \neq 0$ we will say that η is of rank k, $1 \leq k \leq r$, if and only if exactly k of the coefficients η_j are non-zero. The automorphisms of Π extend to global automorphisms of Ω belonging to G. Thus, any $\eta \in T_o(\Omega)$ is equivalent under K to a unique vector $\eta = (\eta_1, ..., \eta_r)$, such that each coefficient is real and $\eta_1 \geq \eta_2 \cdots \geq \eta_r \geq 0$. We call η the normal form of ξ under K.

By means of the Borel embedding we consider now Ω as an open subset of its compact dual M, which is a Hermitian symmetric manifold of the compact type. For instance, for $\Omega = D^I_{m,n}$, the compact dual M is the Grassmannian manifold $G(n,m)$, and for $\Omega = D^{IV}_n$, M is the n-dimensional hyperquadric $\mathbf{Q}^n$. Write $G^{\mathbb{C}}$ for the identity component of the automorphism group of M and $P \subset G^{\mathbb{C}}$ for the isotropy subgroup at o. $G^{\mathbb{C}} \supset G$ is a complexification of G. Consider the action of P on $\mathbb{P}T_o(M)$. Let L be a Levi subgroup of P, i.e., $L \subset P$ is a maximal reductive subgroup. L can be taken to be a complexification of K. We write $L := K^{\mathbb{C}}$. We have $P = K^{\mathbb{C}} \cdot M^-$, where M^- is the unipotent radical of P. M^- is abelian and acts trivially on $T_o(M) = T_o(\Omega)$. Let H_o be the identity component of the automorphism group of the polydisk D, $H \subset G$. Then, $H \cong SU(1,1)^r$. Its complexification $H^{\mathbb{C}}$ inside $\mathbb{C}$, $H \cong SL(2,\mathbb{C})^r$, acts transitively on a polysphere $\Sigma \cong \mathbb{P}_1^r$ such that $(\Pi;\Sigma), \Pi \subset \Sigma$, is a dual pair of Hermitian symmetric spaces. Since H contains $\mathbb{C}^* \times \cdots \times \mathbb{C}^*$ (r times), any $\eta = (\eta_1, ..., \eta_r)$ in the normal form under K must be equivalent under $K^{\mathbb{C}}$ to a vector of the form $\eta^{(k)} = (1, ...1; 0, ...0)$, with exactly the first k entries being equal to 1, where k is the rank of η. In particular, two non-zero vectors of the same rank in $T_o(\Omega) = T_o(M)$ are equivalent under $K^{\mathbb{C}}$ and hence under P. Moreover, using the action of $(\mathbb{C}^*)^r$, for $k < r$ any non-zero vector of rank k is a limit of vectors of rank $k+1$. Furthermore, any two vectors in $T_o(M)$ of different ranks are not equivalent under P. In the case of $M = G(m,n)$ this follows from taking the rank to mean the usual rank of a matrix. In general, this can be checked using Lie algebras.

For the action of P on $\mathbb{P}T_o(M)$ it follows from the above that there are precisely r orbits $\mathcal{O}_k, 1 \leq k \leq r$, such that the topological closures $\overline{\mathcal{O}_k}$ form an ascending chain of subvarieties of $\mathbb{P}T_o(M)$, with $\overline{\mathcal{O}_r} = \mathbb{P}T_o(M)$. Here $\mathcal{O}_k = P[\eta]$ for any $[\eta] \in \mathbb{P}T_o(M)$ of rank k. Given any η of rank k, $k < r$, the closure of the G-orbit of its projectivization $[\eta]$ defines a holomorphic bundle of projective subvarieties $\mathcal{S}_k(M) \subsetneq \mathbb{P}T_M$, called the k-th characteristic bundle, where $\mathcal{S}_{k,o}(M) = \overline{\mathcal{O}}_k$. The restrictions $\mathcal{S}_k(\Omega)$ to Ω are invariant under the action of G. Their quotients under $\Gamma \subset G$ will be denoted by $\mathcal{S}_k(X)$.

A (1,0)-vector η is said to be a *generic* vector if and only if it is of rank r. A (1,0)-vector η of rank k will be called a k-th characteristic vector, and a nonzero vector will be called a characteristic vector whenever it is a k-th characteristic vector for some $k, 1 \le k < r$. (This is a deviation from the usage in earlier articles, where the term 'characteristic vector' was used only for the case $k = 1$). We write $\mathcal{S}(X)$ for $\mathcal{S}_{k-1}(X)$, etc., and call $\mathcal{S}(X)$ the maximal characteristic bundle. (It was called the highest characteristic bundle in [Mok4]). It consists precisely of projectivizations of characteristic vectors η. A first characteristic vector is also called a minimal characteristic vector. We write $\mathcal{M}(X)$ for $\mathcal{S}_1(X)$, etc. and also call it the minimal characteristic bundle. $\mathcal{M}(M)$ consists precisely of projectivizations of (1,0)-vectors tangent to minimal rational curves on M, i.e., lines on M with respect to the first canonical projective embedding of M. $\mathcal{M}(\Omega)$ corresponds on the other hand to tangents of minimal disks on Ω.

The minimal characteristic bundle as a foliated manifold

Let g_0 be a canonical Kähler-Einstein metric on Ω, and denote by g the induced Kähler-Einstein metric on the quotient manifold X. Write ω for the Kähler form of (X, g). We denote by R the curvature tensor of (Ω, g_0) or that of (X, g). Write $\mathcal{M} := \mathcal{M}(X)$ for the minimal characteristic bundle on X. We assume now that Ω is of rank ≥ 2. There is a canonical foliation $\mathcal{N}$ on $\mathcal{M}$ whose leaves can be described as follows. For any $[\eta] \in \mathbb{P}T_o(\Omega)$ let $N_\eta = \{\zeta \in T_o(\Omega) = R_{\eta\overline{\eta}\zeta\overline{\zeta}} = 0\}$, the null-space of η. Write $q = \dim(\Omega_o)$. Let $\triangle \subset \Omega$ be the unique minimal disk passing through o such that $T_o(\triangle) = \mathbb{C}\alpha$. Then, there exists a unique totally geodesic complex submanifold Ω_o passing through o such that $T_o(\Omega_o) = N_\alpha$. Moreover, $\mathbb{C}\alpha \oplus N_\alpha$ is tangent to a unique totally geodesic $(q+1)$-dimensional complex submanifold which can be identified with $\triangle \times \Omega_o$. The above description is clear for the case of Type-I domains $D^I_{m,n}$, $m \ge n \ge 2$, in which case, taking α without loss of generality to be the matrix E_{11} (the matrix with $(1,1)$ entry equal to 1 and all other entries equal to 0), N_α is nothing other than the complex vector space of matrices with vanishing first row and first column, thus $N_\alpha = T_o(\Omega_o)$, where Ω_o can be identified with $D^I_{m-1,n-1}$, and $\Delta \times \Omega_o$ sits in Ω as a totally geodesic complex submanifold. The same type of verification works out for classical domains, and a verification applicable to any Ω can be obtained from root space decompositions.

Identify $\{o\} \times \Omega_o$ with Ω_o. For every $z \in \Omega_o$ write $[\alpha(z)] := \mathbb{P}T_z(\triangle \times \{z\}) \in \mathcal{M}_z(\Omega)$. As z runs over Ω_o, this defines a lifting of Ω_o to a complex submanifold $F \subset \mathcal{M}(\Omega)$ which is precisely the leaf of the lifting of $\mathcal{N}$ to $\mathcal{M}(\Omega)$ passing through $[\alpha]$. Note that G acts transitively on $\mathcal{M}(\Omega)$. Let $H \subset G$ be the closed subgroup which fixes Ω_o as a set. The leaf space of the lifted foliation on Ω can be identified as the homogeneous space G/H. Set-theoretically the leaf space of $\mathcal{N}$ is then given by $\Gamma \setminus G/H$.

There is a description of the distribution $[\alpha] \to \mathcal{N}_{[\alpha]}$ which does not involve curvature. For a point z on the compact dual M of Ω let $\alpha \in \widetilde{\mathcal{M}}_z$ be a minimal characteristic vector, and $C \subset M$ be the unique minimal rational curve (line) on M passing through z and tangent to α. Write $T_M|_C \cong \mathcal{O}(2) \oplus \mathcal{O}(1)^p \oplus \mathcal{O}^r$. Let P_α stand for the fiber of the positive part $\mathcal{O}(2)\oplus\mathcal{O}(1)^p$ at z, which is well-defined independent of the choice of the Grothendieck decomposition. ¿From the deformation theory of rational curves we have $T_{[\alpha]}(\mathcal{M}_o) = P_\alpha/\mathbb{C}\alpha$, so that P_α varies holomorphically as α varies over $\widetilde{\mathcal{M}}(M)$. Suppose now $z \in \Omega \subset M$. Then, $r = q$ and the null-space N_α of α is nothing other than the orthogonal complement of P_α in $T_z(\Omega)$ with respect to the Kähler-Einstein metric g_0, While leaves of the foliation $\mathcal{N}$ are holomorphic, from the above description it is easy to see that the foliation itself is not holomorphic. In fact, denoting by $\pi : \mathcal{M} \to X$ the canonical projection, the assignment $[\alpha] \to N_{[\alpha]} = d\pi(\mathcal{N}_{[\alpha]})$ already fails to be holomorphic when restricted to one fiber $\mathcal{M}_x$.

<u>The foliated minimal characteristic bundle with a transverse measure</u>

The foliation $\mathcal{N}$ on $\mathcal{M}$ as defined is not holomorphic, while all leaves are holomorphic. A typical example of such foliations arises from the nontrivial kernel of a nonpositive (1,1)-form which is of constant rank as a Hermitian bilinear form. We will see that this is indeed the case.

The holomorphic tangent bundle T_X defines the tautological line bundle L over $\mathbb{P}T_X$, and the Hermitian metric on T_X defined by g then corresponds to a Hermitian metric $\widehat{g}$ on L, and we call $(L, \widehat{g})$ the Hermitian tautological line bundle. Since (T_X, g) is of nonpositive curvature in the sense of Griffiths, $(L, \widehat{g})$ is of nonpositive curvature (cf. [Mok2, (4.2), Proposition 1]) so that $-c_1(L, \widehat{g}) \geq 0$. with a kernel at $[\alpha] \in \mathcal{M}$, to be denoted $A_{[\alpha]} \subset T_{[\alpha]}(X)$, such that $d\pi(A_{[\alpha]}) = N_\alpha$ (cf. [Mok2, loc. cit.]). Thus $A_{[\alpha]}$ is a lifting to $\mathcal{M}_x$ of N_α which can be determined, as follows. Choose a holomorphic coordinate system at x which is complex geodesic at x with respect to the Kähler metric g. Then, $A_{[\alpha]} \subset T_{[\alpha]}(X)$ is 'horizontal' with respect to the local holomorphic trivialization of $\mathbb{P}T_X$ on a neighborhood of x defined by the chosen holomorphic coordinate system. Now the Euclidean coordinate system for $\Omega \subset \mathbb{C}^n$ given by the Harish-Chandra realization serves as a complex geodesic coordinate system at $o \in \Omega$. This is the case because the symmetry at o is given by the mapping $z \to -z$ in terms of Harish-Chandra coordinates, implying that the Taylor expansion at o of the Kähler metric g has no odd-order terms, in particular $dg_{i\bar{j}}(o) = 0$. Now the 'horizontal' lifting of $N_{[\alpha]}$ with respect to Harish-Chandra coordinates agrees precisely with $\mathcal{N}_{[\alpha]}$, since the totally geodesic complex submanifold $\Omega_o \subset \Omega$ (which passes through $o \in \Omega$) used in the definition of $\mathcal{N}$ sits in Ω as the intersection of Ω with a complex linear subspace, implying therefore that $A_{[\alpha]} = \mathcal{N}_{[\alpha]}$. In particular, we have demonstrated that $-c_1(L, \widehat{g})|_{\mathcal{M}}$ is positive semi-definite with precisely a q-dimensional kernel at each point, and the foliation $\mathcal{N}$ on X is nothing other than the foliation defined by the smooth d-closed (1,1)-form $-c_1(L, \widehat{g})|_{\mathcal{M}}$.

The closed (1,1)-form $\lambda := -c_1(L, \widehat{g})|_{\mathcal{M}}$ is a real 2-form on the real $2(n+p)$-dimensional underlying smooth manifold of the minimal characteristic bundle $\mathcal{M}$. As a skew-symmetric bilinear form on $\mathcal{M}$, λ is of constant rank $4p+2$. The foliation $\mathcal{N}$ is precisely defined by the distribution $Ker(\lambda)$, which is integrable because λ is d-closed. For the corresponding foliation $\widetilde{\mathcal{N}}$ on $\mathcal{M}(\Omega)$, the leaves are closed, and the leaf space can be given the structure of a smooth real $(4p+2)$-dimensional manifold of G/H. The real skew-symmetric bilinear form λ corresponds to some $\widetilde{\lambda}$ on $\mathcal{M}(\Omega)$. G/H is then endowed with a quotient skew-symmetric bilinear form, to be denoted by $\overline{\lambda}$. $\overline{\lambda}$ is G-invariant by definition non-degenerate everywhere on G/H, and $\wedge^{4p+2}\overline{\lambda} = d\mu$ is a G-invariant volume form on the homogeneous space

G/H. Since Γ acts ergodically on G/H the leaf space $\Gamma\backslash G/H$ of $\mathcal{N}$ on $\mathcal{M}$ does not carry the structure of a smooth manifold. In this sense $\overline{\lambda}$ does not descend to the leaf space of $\mathcal{N}$. However, the structure of $\mathcal{M}$ as a foliated manifold in the small lifts to $\mathcal{M}(\Omega)$, and as far as integration on $\mathcal{M}$ is concerned we can sometimes make use of the volume form $d\mu$ on local pieces of $\mathcal{M}$ as will be seen in (2.2).

(2.2) For a locally irreducible compact Hermitian locally symmetric manifold $X := \Omega/\Gamma$ of rank ≥ 2 we will now make use of the structure of its minimal characteristic bundle $\mathcal{M}$ as a foliated manifold to establish results on metric rigidity. To start with we prove a form of metric rigidity implicit in [Mok 2, (3.1)] for smooth complex Finsler metrics. The original result on Hermitian metric rigidity [Mok 1] follows readily from Finsler metric rigidity by the polarization argument ([(1.4), Lemma 1]). For applications to rigidity problems related to bounded holomorphic functions a key ingredient is a generalization of Finsler metric rigidity, which we will be discuss in (2.3). We have

THEOREM 3. *Let Ω be an irreducible bounded symmetric domain of rank ≥ 2 and $\Gamma \subset \mathrm{Aut}(\Omega)$ be a torsion-free cocompact lattice, $X := \Omega/\Gamma$. Let g be a canonical Kähler-Einstein metric on X, and h be any smooth complex Finsler metric on X of nonpositive curvature. Denote by $\|\cdot\|_g$ resp. $\|\cdot\|_h$ lengths of vectors measured with respect to g resp. h. Then, there exists a positive constant c such that for any minimal characteristic vector $\alpha \in T(X)$ we have $\|\alpha\|_h = c\|\alpha\|_g$.*

PROOF. Let ν be a Kähler form on the minimal characteristic bundle $\pi : \mathcal{M} \to X$. For instance, we may take $\nu = -c_1(L,\widehat{g}) + \pi^*\omega$. $\pi : \mathcal{M} \to X$ is of fiber dimension p and of total dimension $n+p = 2n-q-1$. We have

$$\begin{aligned} &\int_{\mathcal{M}} -c_1(L,h) \wedge (-c_1(L,\widehat{g}))^{2n-2q-1} \wedge \nu^{q-1} \\ &= \int_{\mathcal{M}} (-c_1(L,\widehat{g}))^{2n-2q} \wedge \nu^{q-1} = 0. \end{aligned} \tag{1}$$

The integrand of the left-hand side of is then a nonnegative, and (1) forces the identical vanishing

$$-c_1(L,h) \wedge \big(-c_1(L,\widehat{g})\big)^{2n-2q-1} \equiv 0 \quad \text{on} \quad \mathcal{M}. \tag{2}$$

Write $h = e^u g$ on $\mathcal{M}$. Then, $-c_1(L,h) = -c_1(L,\widehat{g}) + \frac{\sqrt{-1}}{2\pi}\partial\overline{\partial}u$. It follows from (2) that

$$\sqrt{-1}\partial\overline{\partial}u \wedge \big(-c_1(L,\widehat{g})\big)^{2n-2q-1} \equiv 0 \quad \text{on} \quad \mathcal{M}. \tag{3}$$

Mutiplying the left-hand side of (3) by u and taking exterior product with ν^{q-1}, we conclude by integrating by part that

$$\int_{\mathcal{M}} \sqrt{-1}\partial u \wedge \overline{\partial} u \wedge (-c_1(L,\widehat{g}))^{2n-2q-1} \wedge \nu^{q-1} = 0. \tag{4}$$

forcing the function u to be constant on every local leaf of $\mathcal{N}$. Thus, we may regard u as a function on the leaf space of $\mathcal{N}$, which can be identified with $\Gamma\backslash G/H$, where H is noncompact. By Moore's Ergodicity Theorem Γ acts ergodically on G/H, i.e., any Γ-invariant measurable subset of G/H is either of full or zero measure on G/H with respect to the measure induced by the Haar measure on G. (Here by full measure we mean that the complement is of measure zero.) Write $\mathcal{M}(\Omega) = G/E$ for some compact subgroup $E \subset K$, and write $\varphi : G \to G/E = \mathcal{M}(\Omega)$ for the canonical projection. If the (continuous) function u were nonconstant, some sublevel set

$\{a < (\pi \circ \varphi)^* u < b\}$ would give an open subset of G which is of neither full nor zero measure, contradicting Moore's Ergodicity Theorem. In other words, we have shown that u must be a constant on $\mathcal{M}$, proving that for any $\alpha \in \widetilde{\mathcal{M}}$, $\|\alpha\|_h = c\|\alpha\|_g$ for some global constant $c > 0$, as desired.

We now deduce Hermitian metric rigidity in its original form, together with its corollary giving a rigidity result for holomorphic mappings.

THEOREM 4 ([MOK1, 1987]). *Let Ω be an irreducible bounded symmetric domain of rank ≥ 2 and $\Gamma \subset \mathrm{Aut}(\Omega)$ be a torsion-free cocompact lattice, $X := \Omega/\Gamma$. Let g be the canonical Kähler-Einstein metric on X and h be any smooth Hermitian metric of nonpositive curvature in the sense of Griffiths. Denote by $\|\cdot\|_g$ resp. $\|\cdot\|_h$ lengths of vectors measured with respect to g resp. h. Then, there exists a positive constant c such that for any $\eta \in T(X)$, we have $\|\eta\|_h = c\|\eta\|_g$. As a consequence, if $f : X \to N$ is a nonconstant holomorphic mapping into a complex manifold N endowed with a Kähler metric s of nonpositive holomorphic bisectional curvature, then $f : (x, g) \to (N, s)$ is up to a normalizing constant a totally geodesic isometric immersion.*

PROOF. Denote by $g(\cdot, \bar{\cdot})$ resp. $h(\cdot, \bar{\cdot})$ the Hermitian inner products with resp. to the Hermitian metrics g resp. h. By Theorem 3 there exists a constant $c > 0$ such that $h(\alpha, \overline{\alpha}) = cg(\alpha, \overline{\alpha})$ for any minimal characteristic vector α at x, i.e. for $\alpha \in \widetilde{\mathcal{M}_x}$. At $x \in X$ regard both g_x and h_x as complex bilinear functions on $T_x(X) \times \overline{T_x(X)}$, i.e., we may regard g_x, h_x as elements of the complex vector space $T_x^*(X) \otimes \overline{T_x^*(X)}$. Now $\widetilde{\mathcal{M}}_x(X) = \widetilde{\mathcal{S}}_{1,x}(X)$ is complex analytic. We have $(h_x - cg_x)(A) = 0$ for any $A \in T_x(X) \otimes \overline{T_x(X)}$ lying in the complex linear span of $\alpha \otimes \alpha, \alpha \in \widetilde{\mathcal{M}}_x(X)$. When η varies holomorphically on $\widetilde{\mathcal{M}}_x(X)$, $\overline{\eta}$ varies anti-holomorphically. Since $\widetilde{\mathcal{M}}_x$ is linearly non-degenerate in $T_x(X)$ by the polarization argument [(1.4), Lemma 1] we conclude that $\alpha \otimes \overline{\alpha}$ spans $T_x(X) \otimes \overline{T_x(X)}$ as a complex vector space, so that $h(\xi, \overline{\eta}) = cg(\xi, \overline{\eta})$ for any $\xi, \eta \in T_x(X)$. Since $x \in X$ is arbitrary we have $h \equiv cg$ proving the first half of Theorem 4.

For the second half let now $f : X \to N$ be a nonconstant holomorphic mapping. Then $h = g + f^*s$ is a Hermitian metric of nonpositive curvature in the sense of Griffiths. By Theorem 4 it follows that there exists some constant $c > 1$ such that $h = cg$, i.e. $f^*s = c'g$; $c' = c - 1 > 0$; so that f is necessarily an isometric immersion up to a normalizing constant. Without loss of generality we may take $c' = 1$, so that f is an isometric immersion. Now suppose furthermore that (N, s) is Kähler, identifying (X, g) locally as a Kähler submanifold of N we have by the Gauss equation $R^X_{\xi\overline{\xi}\eta\overline{\eta}} = R^N_{\xi\overline{\xi}\eta\overline{\eta}} - \|\sigma(\xi, \eta)\|^2$, where $x \in X$ is identified with $f(x) \in N$, and $\xi, \eta \in T_x(X) \subset T_x(N)$, R^X resp. R^N denote the curvature tensors of (X, g) resp. (N, s), and $\sigma : S^2T_X \to f^*T_N/T_X$ denotes the (2,0) part of the second fundamental form of the holomorphic isometric immersion $X \hookrightarrow N$. Write now R for R^X. Let (α, ζ) be a zero of holomorphic bisectional curvatures of (X, g), i.e. $R_{\alpha\overline{\alpha}\zeta\overline{\zeta}} = 0$. Since $R^N_{\alpha\overline{\alpha}\zeta\overline{\zeta}} \leq 0$ it follows from the Gauss equation that $R^N_{\alpha\overline{\alpha}\zeta\overline{\zeta}} = \|\sigma(\alpha, \zeta)\|^2 = 0$, hence $\sigma(\alpha, \zeta) = 0$. As in the proof of [(1.4), Theorem 2] for the case of rank ≥ 2 we know again by the polarization argument [(1.4), Lemma 1] that such decomposable vectors $\alpha \otimes \zeta$ span $T_x(X) \otimes T_x(X)$ as a complex vector space. Thus, $\sigma(\xi, \eta) = 0$ for all $\xi, \eta \in T_x(X)$. As $x \in X$ is arbitrary the holomorphic isometric embedding $f : (X, g) \to (N, s)$ is actually totally geodesic, as desired.

REMARKS.

(1) One of the motivations for proving Hermitian metric rigidity was to give an explanation, in terms of Hermitian geometry, of an early result of Matsushima's [Ma] according to which the first Betti number of a compact Kähler manifold (X, g) uniformized by an irreducible bounded symmetric domain of rank ≥ 2 must necessarily vanish. This results from Hermitian metric rigidity, since one can construct a non-canonical Kähler metric of nonpositive curvature on X from a nontrivial (closed) holomorphic 1-form ν by taking the Kähler form to be $\sqrt{-1}\nu \wedge \overline{\nu} + \omega$, ω being the Kähler form of (X, g).

(2) In the second half of Theorem 4, in the event that the target manifold (N, s) is actually complete and of nonpositive *Riemannian* sectional curvature, then the lifting $F : \Omega \to \widetilde{N}$ to universal covering spaces is actually up to a normalizing constant a totally geodesic isometric embedding, in view of the Cartan-Hadamard Theorem.

(2.3) While the rigidity result in Theorem 4 for holomorphic mappings applies in particular to holomorphic mappings between Hermitian locally symmetric manifolds of the noncompact type when the domain manifold X is compact and of rank ≥ 2, it is nonetheless difficult to construct in general Hermitian metrics of nonpositive curvature on target manifolds. Believing that rigidity for holomorphic mappings on X should hold under much more general conditions of nonpositivity of curvature for target manifolds, we study in [Mok5, 2004] holomorphic mappings of X into quotients of bounded domains in Stein manifolds. For such domains there is an ample supply of bounded holomorphic functions, and we have on them Carathéodory metrics, which are continuous complex Finsler metrics of nonpositive curvature in the generalized sense. A key ingredient in our study is a generalization of Finsler metric rigidity (Theorem 3) for smooth complex Finsler metrics, in two senses. First of all, we need to take care of complex Finsler metrics which are only continuous and of nonpositive curvature in the generalized sense, where we have in mind the Carathéodory metric, which are actually Lipschitz. Secondly, which turns out to be crucial, we need to consider new 'partially defined' complex Finsler metrics which are of nonpositive curvature only along leaves of the foliation $\mathcal{N}$ as defined in (2.1). As it turns out, the 'partially defined' Finsler metrics are continuous Hermitian metrics of the Lipschitz class on the tautological line bundle L defined only on the minimal characteristic bundle $\mathcal{M}$. In relation to Theorem 3 and in terms of the description of $\mathcal{M}$ as a foliated manifold with a transverse measure as given in (2.1), we formulate our rigidity result in a general form, as follows.

PROPOSITION 1. *Let (Z, ω) be an m-dimensional compact Kähler manifold, and θ be a smooth closed nonnegative (1,1)-form on Z such that $Ker(\theta)$ is of constant rank $q > 0$ everywhere on Z. Denote by $\mathcal{K}$ the foliation on Z with holomorphic leaves defined by the distribution $Re(Ker(\theta))$. Let $u : Z \to \mathbb{R}$ be a continuous function whose restriction to every leaf $\mathcal{L}$ of the foliation $\mathcal{K}$ is plurisubharmonic. Then, the restriction of u to every leaf $\mathcal{L}$ is pluriharmonic. If in addition du is locally integrable, then u is constant on every leaf of $\mathcal{K}$. If there is furthermore a dense leaf of $\mathcal{K}$, then u is constant on Z.*

A continuous Hermitian metric h on a holomorphic line bundle E is said to be of nonpositive curvature in the generalized sense if and only if its curvature current is a closed positive current. h is said to be of the Lipschitz class if $h = e^u g$ for some smooth Hermitian metric g such that du is bounded. A continuous complex Finsler metric h on a holomorphic vector bundle V is said to be of nonpositive curvature in the generalized sense if and only if the induced continuous Hermitian metric $\widehat{h}$ on the tautological line bundle L over the $\mathbb{P}V$ is of nonpositive curvature in the generalized sense.

COROLLARY 1. *The analogue of Theorem 3 on Finsler metric rigidity holds true for continuous complex Finsler metrics h of nonpositive curvature in the generalized sense.*

PROOF. With the analogous notations as in Theorem 3 and its proof, we write $h = e^u\widehat{g}$. The first Chern current of h is given by

$$c_1(L,h) = -\frac{\sqrt{-1}}{2\pi}\partial\overline{\partial}u + c_1(L,\widehat{g}).$$

By assumption $-c_1(L,h) \geq 0$ as a current. Since $c_1(L,\widehat{g})$ is smooth, we conclude that u is almost plurisubharmonic in the sense that locally $u+\varphi$ is plurisubharmonic for some smooth function φ. Since the gradient of a plurisubharmonic function is locally integrable, we conclude that du is locally integrable. Hence Proposition 1 is applicable in the present context in which Z is the minimal characteristic bundle $\mathcal{M}$ on X, θ is $-c_1(L,\widehat{g})|_{\mathcal{M}}$ and $\mathcal{K}$ is the canonical foliation $\mathcal{N}$ on $\mathcal{M}$. Finally, since Γ acts ergodically on G/H by Moore's Ergodicity Theorem, there exists a dense leave Λ on $\mathcal{M}$, and we conclude from Proposition 1 that u is a constant, i.e., $h = cg$ for some constant $c > 0$, as desired.

Proposition 1 incorporates [Mok5, Proposition 5], where we further assumed u to be a Lipschitz function, and an argument of [Mok4, Proposition 3] in which we require only du to be integrable. [Mok5, loc. cit.] involves the use of closed currents which are leaf-wise positive. We give here the detailed justification which was omitted in [Mok5].

PROOF OF PROPOSITION 1. Consider

$$T = \sqrt{-1}\partial u \wedge \theta^{m-q} \wedge \omega^{q-1}, \qquad S = dT. \tag{1}$$

Then, S is an (m,m)-current, thus acting on smooth functions. If u is smooth by by Stokes' Theorem we have

$$S(1) = \int_Z \sqrt{-1}\partial\overline{\partial}u \wedge \theta^{m-q} \wedge \omega^{q-1} = 0. \tag{2}$$

The integrand on the right-hand side is nonnegative by the leaf-wise plurisubharmonicity of u, and thus it vanishes identically on Z implying that u is pluriharmonic on each leaf of $\mathcal{K}$. The integration on the right-hand side can be interpreted in terms of the foliated structure of $\mathcal{K}$ in analogy to that of $\mathcal{M}$ in (2.1), as follows. Denote by $B^n_{\mathbb{C}}(r)$ (resp. $B^n_{\mathbb{R}}(r)$) the n-dimensional complex (resp. real) Euclidean open ball of radius $r > 0$, and write $B^n_{\mathbb{C}} := B^n_{\mathbb{C}}(1)$, etc. Let $x \in X$. There exists a coordinate neighborhood U^o_x of x in X which admits a diffeomorphism Φ_x onto $V^o_x \times B^q_{\mathbb{C}}(2)$ for some open subset V^o_x diffeomorphic to and identified with $B^{2s}_{\mathbb{R}}(2)$, where $s = m - q$, such that $\Phi_x^{-1}|_{\{t\}\times B^q_{\mathbb{C}}}$ is a biholomorphism onto a leaf of $\mathcal{K}|_{U^o_x}$. Shrinking U^o_x we obtain $U_x \Subset U^o_x$ such that U_x corresponds under Φ_x to $V_x \times B^q_{\mathbb{C}}$, where $V_x \Subset V^o_x$ corresponds to $B^{2s}_{\mathbb{R}} \Subset B^{2s}_{\mathbb{R}}(2)$. We call V_x constructed this way a priviledged $\mathcal{K}$-box. Z is now covered by a finite number of priviledged $\mathcal{K}$-boxes U_i.

We can choose open subsets $W_i \subset U_i$ such that $\{W_i\}$ are mutually disjoint, and such that the complement of $\bigcup W_i$ is a set of measure zero on Z. In the smooth case the integral over Z in (2) can be decomposed into the sum of the integrals over the finite number of open subset W_i, over which we can write

$$\int_{W_i} \sqrt{-1}\partial\overline{\partial} u \wedge \theta^{m-q} \wedge \omega^{q-1} = \int_{V_i} \left(\int_{L_t \cap W_i} \sqrt{-1}\partial\overline{\partial} u \wedge \omega^{q-1} \right) d\mu(t). \tag{3}$$

Here U_i is an open set of the form U_x, and V_i corresponds to V_x, L_t is the leaf of $\mathcal{K}|_{U_x}$ given by $\Phi_x^{-1}(\{t\} \times B^q_{\mathbb{C}})$, and $d\mu(t)$ is the transverse measure on a local $\mathcal{K}$-leaf space obtained in analogy with the local transverse measure (with the same notation) on the minimal characteristic bundle $\mathcal{M}$ by means of a quotient symplectic form $\overline{\lambda}$, as given in (2.1). In what follows we will drop the subscript i in the notations.

We proceed to justify the analogue of (3) in the case where u is only assumed to be continuous and leaf-wise plurisubharmonic. It suffices to verify that for any smooth function ρ of compact support on U we have

$$(\dagger) \qquad S(\rho) = \int_V \left(\int_{L_t} \rho \sqrt{-1}\partial\overline{\partial} u_t \wedge \omega^{q-1} \right) d\mu(t)\ ,$$

Here u_t means the restriction of u to the local $\mathcal{K}$-leaf L_t. The right-hand side of ($\dagger$) has a well-defined meaning, as follows. Each u_t being plurisubharmonic on L_t, $P_t := \sqrt{-1}\partial\overline{\partial} u_t$ is a closed positive current. It is thus a current whose coefficients are complex measures, and the integral over L_t is the integral of ρ against a positive multiple of the trace of P_t with respect to ω, which is a nonnegative measure. Because the (continuous) functions $u_t, t \in V$, are uniformly bounded, integrating by part we conclude that P_t are of uniformly bounded total mass. The leaf-wise integrals on the right-hand side of ($\dagger$) are uniformly bounded independent of t, and the integral in ($\dagger$) with respect to $d\mu(t)$ makes sense. Thus $P_t \wedge \omega^{q-1}$ defines a positive (q,q)-current, i.e., a positive measure $d\tau_t$ on L_t, and the integral against $d\mu(t)$ defines a measure $d\tau$ for which we have by definition the Fubini's Theorem $\int_U f d\tau = \int_V \left(f d\tau_t\right) d\mu(t)$ for any bounded measurable function f, and in particular for $f = \rho$ a smooth function of compact support on U as given in ($\dagger$).

To justify ($\dagger$) we approximate the function u by smooth functions. Since u is continuous on X by local smoothing with respect to standard kernels we obtain a sequence of smooth functions u_n which converge uniformly on $\overline{V}$ to u. Thus, writing $u_{n,t} := u_n\big|_{L_t}$ we have

$$\begin{aligned} \left| \int_{L_t} \rho \sqrt{-1}\partial\overline{\partial} u_{n,t} \wedge \omega^{q-1} - \int_{L_t} \rho d\tau_t \right| &= \left| \int_{L_t} (u_{n,t} - u_t) \sqrt{-1}\partial\overline{\partial} \rho \wedge \omega^{q-1} \right| \\ &\leq C(\rho) \cdot \epsilon_n \end{aligned} \tag{4}$$

for some positive constant $C(\rho)$ depending only on ρ and for $\epsilon_n = \sup\left\{|u_n(x) - u(x)| : x \in \overline{V}\right\}$, which tends to 0 as $n \to \infty$. It follows that

$$\int_U \rho \sqrt{-1}\partial\overline{\partial} u \wedge \theta^{m-q} \wedge \omega^{q-1} = \int_V \left(\int_{L_t} \rho \sqrt{-1}\partial\overline{\partial} u_t \wedge \omega^{q-1} \right) d\mu(t) \tag{5}$$

holds true for any ρ of compact support on U. This shows that on U, $\sqrt{-1}\partial\overline{\partial} u \wedge \theta^{m-q} \wedge \omega^{q-1}$ is an (m,m)-current of order 0, more precisely the nonnegative measure

$d\tau$. In particular, applying to the characteristic function χ_W we obtain the analogue of (3) for the continuous function u in the hypothesis of Proposition 1.

Under the additional assumption that du is locally integrable we proceed to deduce that u is constant on each leaf. By local smoothing and partition of unity there exists a sequence (u_n) of smooth functions on Z such that u_n converges to u uniformly, du_n converges to du in L^1 and $\partial\overline{\partial}u_n$ converges to $\partial\overline{\partial}u$ as distributions of order 0 (i.e., their coefficients converge as measures). Write on Z

$$T_n = \sqrt{-1}\;\overline{\partial}u_n \wedge \theta^{m-q} \wedge \omega^{q-1} \tag{6}$$

Then T_n converges to T in L^1 and dT_n converges to $dT = S = 0$ as distributions of order 0. Consider now

$$\sqrt{-1}d(u_nT_n) = \sqrt{-1}du_n \wedge T_n + \sqrt{-1}u_n dT_n \ . \tag{7}$$

Integrating over Z, by Stokes' Theorem

$$\int_Z \sqrt{-1}du_n \wedge T_n = \int_Z \sqrt{-1}\partial u_n \wedge \overline{\partial}u_n \wedge \theta^{m-q} \wedge \omega^{q-1} \to 0 \ . \tag{8}$$

Define on Z the Hermitian bilinear form B on smooth $(1,0)$-forms φ by

$$B(\varphi,\psi) = \int_S \sqrt{-1}\varphi \wedge \overline{\psi} \wedge \theta^{m-q} \wedge \omega^{q-1} \ . \tag{9}$$

B is positive semi-definite. Note that $B(\varphi,\psi)$ can also be defined for φ of class L^1 and ψ smooth. Fix any smooth $(1,0)$-form ψ on Z. By the Cauchy-Schwarz inequality $B(\partial u_n,\psi)^2 \le B(\partial u_n,\partial u_n)B(\psi,\psi)$. By (8), $B(\partial u_n,\partial u_n) \to 0$ as $n \to \infty$. It follows that $B(\partial u_n,\psi) \to 0$ as $n \to \infty$. Since $\partial u_n \to \partial u$ as distributions we have $B(\partial u,\psi) = 0$ for any smooth ψ. Hence the L^1 differential form ∂u vanishes almost everywhere on leaves of $\mathcal{K}$. As a consequence, u is constant on almost every leaf of $\mathcal{K}$, and thus on every leaf by continuity. Finally, u must be constant on Z whenever there exists a dense leaf, again by continuity, proving Proposition 1. □

(2.4) We now make use of results on metric rigidity as given by [(2.3), Proposition 1] to study holomorphic mappings into complex manifolds whose universal covering spaces admit nontrivial bounded holomorphic functions, as given in [Mok5]. As an example, we proved

THEOREM 5 (MOK [MOK5, 2004]). *Let Ω be an irreducible bounded symmetric domain of rank ≥ 2 and $\Gamma \subset Aut(\Omega)$ be a torsion-free cocompact lattice, $X := \Omega/\Gamma$. Let D be a bounded domain in some Stein manifold, $\Phi \subset Aut(D)$ be a torsion-free discrete group of automorphisms, $N := D/\Phi$. Let $f : X \to N$ be a nonconstant holomorphic mapping and $F : \Omega \to D$ be its lifting to universal covering spaces. Then, $F : \Omega \to D$ is a holomorphic embedding.*

REMARKS.

(1) In [(2.2), Theorem 4] the lifted holomorphic mapping to universal covering spaces is only known to be an immersion. The only instance in which injectivity was known requires nonpositivity of *Riemannian* sectional curvatures (cf. Remark (2) there).
(2) The analogue of Theorem 5 holds for any target manifold N provided that there exists a single bounded holomorphic function h on $\widetilde{N}$ such that F^*h is nonconstant.

(3) Theorem 5 generalizes to the case where Ω is reducible and $\Gamma \subset \mathrm{Aut}(\Omega)$ is a torsion-free irreducible (not necessarily cocompact) lattice.

To prove Theorem 5, we make use of the Carathéodory metric and a 'partially defined' complex Finsler metrics on $X := \Omega/\Gamma$, which correspond to a continuous Hermitian metric of the Lipschitz class on the minimal characteristic bundle $\mathcal{M}$. For the constructions we consider first of all the following general situation. Let $\mathcal{H}$ be the set of all holomorphic mappings of Ω into the unit disk Δ endowed with the the Poincaré metric ds^2_Δ of constant curvature -1. The norm $\|\cdot\|$ on T_Δ will be understood to be in terms of ds^2_Δ. Let $\mathcal{G} \subset \mathcal{H}$ be a subset satisfying the following conditions.

(a) For any $g \in \mathcal{F}$ and any $\gamma \in \Gamma$ the composite map $g \circ \gamma : \Omega \to \Delta$ lies on $\mathcal{F}$.
(b) If g_n are elements of $\mathcal{G}$ such that g_n converges uniformly on compact subsets to some $g \in \mathcal{H}$, then $g \in \mathcal{G}$.

Define now a length function $\kappa(\mathcal{G})$ on (0,1)-vectors η by $\|\eta\|_{\kappa(\mathcal{G})} = \sup\{\|dg(\eta)\| : g \in \mathcal{G}\}$. When $\mathcal{G} = \mathcal{H}$ this length function defines precisely the Carathéodory metric κ, which is a continuous complex Finsler metric on Ω of nonpositive curvature in the generalized sense. We will call $\kappa(\mathcal{G})$ the $\mathcal{G}$-Carathéodory metric on Ω. ¿From Cauchy estimates it follows that $\kappa(\mathcal{G})$ is of the Lipschitz class. κ is by construction invariant under $\mathrm{Aut}(\Omega)$. In general, under the assumption (a), $\|\eta\|_{\kappa(\mathcal{G})}$ is Γ-invariant. Under the assumption (b), for any (1,0)-vector η, there exists $g \in \mathcal{G}$ such that $\|\eta\|_{\kappa(\mathcal{G})} = \|dg(\eta)\|$. We call g a $\kappa(\mathcal{G})$-extremal (bounded holomorphic) function for η. The Poincaré metric ds^2_Δ is of negative curvature. The length function $\kappa(\mathcal{G})$ can be locally represented as the (continuous) supremum of a family of (smooth) log-plurisubharmonic functions, and as such $\kappa(\mathcal{G})$ is of nonpositive curvature in the generalized sense.

To apply the above construction to the study of the holomorphic mapping $f : X \to N$ consider now $\mathcal{F} = F^*(\mathcal{H}_D)$, where $\mathcal{H}_D$ stands for the set of holomorphic maps of the universal covering domain D of N into the unit disk Δ. By Montel's Theorem, given any $g_n := F^*h_n \in \mathcal{F}$ with $h_n \in \mathcal{H}$ which converges to some $g \in \mathcal{F}$, there exists a subsequence of h_n which converges uniformly on compact subsets to some $h \in \mathcal{H}_D$, as a consequence of which $g = F^*h \in \mathcal{F}$, so that (b) is satisfied. The $\mathcal{F}$-Carathéodory metric $\kappa(\mathcal{F})$ reflects properties of the holomorphic mapping. In fact, $\kappa(\mathcal{F}) = F^*\kappa_D$, and the $\mathcal{F}$-Carathéodory metric is degenerate at a (1,0)-vector η if and only if $\eta \in \mathrm{Ker}(dF)$. The same set-up as in Theorem 5 applies to holomorphic mappings $F : \Omega \to D$ which are Γ-equivariant, i.e, inducing a homomorphism $F_* : \Gamma \to \mathrm{Aut}(D)$, and an analogue of Theorem 5 holds for such mappings $F : \Omega \to N$ without assuming that F descends to some $f : X \to N$, as given in [Mok5, Theorem 5]. From [(2.3), Corollary 1] we concude that for any minimal characteristic vector α, we have $df(\alpha) \neq 0$. A generalization of Corollary 1 applied to holomorphic mappings in the proof of [Mok4, Theorem 3] allows one to conclude that $df(\eta) \neq 0$ for any characteristic vector η (i.e., k-characteristic vector with $1 \leq k < r$), but the argument fails for generic vectors η. Instead we will make use of the following result on $\mathcal{F}$-Carathéodory extremal functions.

Proposition 2. *Let $P \subset \Omega$ be a maximal polydisk, $P \cong \Delta^r$, and use Euclidean coordinates of the latter as coordinates for P. Let $x \in P, x = (x_1; x')$ and denote by $P' \subset P$ the polydisk corresponding to $\{x_1\} \times \Delta^{r-1}$. Let α be a minimal characteristic vector at x tangent to the minimal disk Δ_α corresponding to $\Delta \times \{x'\}$ and denote* *by s a $\kappa(\mathcal{F})$-extremal function at x for α. Then $s(x_1; z_2, \cdots, z_r) = s(x_1)$ for any $(z_2, \cdots, z_r) \in P'$.*

PROOF. By (2.1) we have a totally geodesic complex submanifold $\Delta \times \Omega_o \subset \Omega$, such that Ω_o is biholomorphic to an irreducible bounded symmetric domain of rank $r-1$. P' sits in Ω_o as a maximal polydisk. We may assume x to be the origin and take α to be $\frac{\partial}{\partial z_1}$ at 0. By Proposition 1, $\|\frac{\partial}{\partial z_1}\|_{\kappa(\mathcal{F})}$ is constant on $\Delta \times \Omega_o$. For any $z \in P$ write α_z for $\frac{\partial}{\partial z_1}$ at z. At $z = (0, z')$ we have (2.1).

$$\|\alpha_z\|_{\kappa(\mathcal{F})} \geq \|\alpha_z\|_s = \|ds(\alpha_z)\| = \frac{|ds(\alpha_z)|}{1 - |s(0, z')|^2} := e^{\psi(z')} ,$$

$$\|\alpha_o\|_{\kappa(\mathcal{F})} = \frac{|ds(\alpha_o)|}{1 - |s(o)|^2} = ds(\alpha_o) = e^{\psi(o)} .$$

By Proposition 1, $\|\alpha_z\|_{\kappa(\mathcal{F})}$ is constant on P', so that $\psi(z')$ attains its maximum on P' at the origin. On the other hand, by Lemma 2, $\psi(z')$ is plurisubharmonic in $z' \in P'$. It follows that $\psi \equiv C$ for some constant $C > 0$, so that

$$|ds(\alpha_z)| = C(1 - |s(0, z')|^2),$$

which violates the maximum principle for the plurisubharmonic function $|ds(\alpha_z)|$ in z' unless $s(0, z') = 0$ for any $z' \in P'$. In other words, we have proven that $s(0, z') = s(0)$, as desired.

We proceed to derive from Proposition 2 that F is an immersion by a streamlining of the proof given in [Mok5].

PROOF THAT F IS AN IMMERSION IN THEOREM 5. Let $x \in \Omega$ and $\eta \in T_x(\Omega)$ such that $dF(\eta) = 0$. Assuming that $\eta \neq 0$ we are going to derive a contradiction. Endow $T_x(\Omega)$ with the Hermitian inner product defined by g_0. Let $\alpha \in \widetilde{\mathcal{M}_x}$ be the point at a minimal distance from η. Then $\eta = \alpha + \zeta$, where $\zeta \in N_\alpha$, the null-space of ζ. There is a maximal polydisk $P \cong \Delta^r = \Delta \times \Delta^{r-1}$ passing through x such that $\eta \in T_x(P)$ and α is tangent to the first direct factor. Let s be a $\kappa(\mathcal{F})$-extremal function for η. By Proposition 2 and in the notations there we have $s(x_1; z_2, \cdots, z_r) = s(x_1)$, so that $ds(\zeta) = 0$. $0 = ds(\eta) = ds(\alpha) + ds(\zeta) = ds(\alpha)$. But as s is an extremal function for α, by Proposition 1 $ds(\alpha) \neq 0$, a contradiction. This proves that $F : \Omega \to \widetilde{N}$ is an immersion, as desired. □

To prove injectivity of $F : \Omega \to \widetilde{N}$ it is not enough to consider $\mathcal{F}$-Carathéodory metrics. Instead we will resort to the construction of a 'partial' continuous complex Finsler metric which only measures the lengths of minimal characteristic vectors, thus equivalently a continuous Hermitian metric on the minimal characteristic bundle $\mathcal{M}(\Omega)$. The metric is Γ-invariant and descends thus to $\mathcal{M}$. Like the Carathéodory metric it is defined by bounded holomorphic functions, and it turns out to be *a priori* of nonpositive curvature only along leaves of the canonical foliation on $\mathcal{M}$ as defined in (2.1). The construction goes as follows.

Let $\mathcal{G}$ be a Γ-invariant family of holomorphic functions satisfying the conditions (a) and (b) as in the definition of the $\mathcal{G}$-Carathéodory metric. Let α be a minimal characteristic vector at $x \in \Omega$, Δ_α be a the unique minimal disk passing through x tangent to α, and $\Omega_o \subset \Omega$ be the unique totally geodesic complex submanifold passing through x tangent to the null space N_α, so that $\Delta \times \Omega_o \subset \Omega$ is a totally geodesic complex submanifold, as in (2.1). Fix an arbitrary $\epsilon > 0$. Denote by $S_{\alpha,\epsilon}$ the circle on Δ_α centred at x of radius ϵ with respect to the Kähler-Einstein metric g_0. We define a length function $e(\mathcal{G})$ on $\widetilde{\mathcal{M}}$, as follows. Let $\alpha(\zeta)$ be a smooth vector field on the geodesic circle $S_{\alpha,\epsilon}$ of constant length equal to $\|\alpha\|$. For $s \in \mathcal{G}$, let $\|\alpha\|'_s$ to be the average of $\|ds(\alpha(\zeta))\|$ over $\zeta \in S_{\alpha,\epsilon}$ with respect to an S^1-invariant metric on $S_{\alpha,\epsilon}$, i.e., with respect to polar angles when the latter is identified with the unit circle S^1 in the usual way. We define $\|\alpha\|_{e(\mathcal{G})}$ to be $\sup\{\|\alpha\|'_s : s \in \mathcal{G}\}$. Taking the zero vector to be of length 0, the length function $e(\mathcal{G})$ corresponds to a continuous Hermitian metric on the tautological line bundle L over the minimal characteristic bundle $\widetilde{\mathcal{M}}$. By Cauchy estimates we see that $e(\mathcal{G})$ is of the Lipschitz class. However, as explained in [Mok5, (3.2)] the procedure of averaging over geodesic circles and taking suprema does not lead *a priori* to a Hermitian metric of nonpositive curvature in the generalized sense.

By Γ-invariance $e(\mathcal{G})$ descends to $\mathcal{M}$. Depending on the context we will use the same notation $e(\mathcal{G})$ both on Ω and on X. We may call $e(\mathcal{G})$ the $\mathcal{G}$-Carathéodory (S^1-averaging) characteristic metric. Strictly speaking, its construction depends on the choice of some $\epsilon > 0$, and we may write $e(\mathcal{G}) = \kappa(\epsilon, \mathcal{G})$, the latter notation suggesting its relationship to the Carathéodory metric $\kappa(\mathcal{G})$. In fact, $\kappa(\epsilon, \mathcal{G})$ converges (decreases) uniformly on $\widetilde{\mathcal{M}}$ to $\kappa(\mathcal{G})|_{\widetilde{\mathcal{M}}}$. In the sequel as in the case of the Carathéodory metric we will be making use of $e(\mathcal{G})$ for $\mathcal{G} = F^*\mathcal{H}_D$.

Although $e(\mathcal{G})$ is not *a priori* of nonpositive curvature, it turns out to be of nonpositive curvature in the generalized sense when restricted to leaves of the canonical foliation $\mathcal{N}$ on $\mathcal{M}$. This observation, which is crucial for our purpose, follows readily from the following lemma on plurisubharmonic functions.

LEMMA 2. *Let $U \subset \mathbb{C}^n$ be an open subset, and $a, b \in \mathbb{R}; a < b$. Let $u : [a, b] \times U \to \mathbb{R}$ be a continuous function such that for any $t \in [a, b]$, writing $u_t(z) := u(t, z)$, $u_t : U \to \mathbb{R}$ is plurisubharmonic. Define $\varphi : U \to R$ by $\varphi(z) := \log \int_a^b e^{u_t(z)} dt$. Then, φ is plurisubharmonic. Moreover, $e^\varphi \sqrt{-1}\partial\overline{\partial}\varphi \geq \int_a^b e^{u_t} \sqrt{-1}\partial\overline{\partial}u_t dt$ in the sense of currents.*

The basic principle underlying Lemma 2 is the fact that the space of Hermitian metrics of nonpositive curvature is preserved under taking sums. We sketch a proof of Lemma 2. For a more detailed geometric proof in terms of Hermitian metrics and the Gauss equation, we refer the reader to [Mok5, Lemma 4].

SKETCH OF PROOF OF LEMMA 2. By standard smoothing arguments we may assume without loss of generality that u is actually smooth. The function φ is defined as an integral, and can thus be approximated by Riemann sums. Expressing φ as a uniform limit of Riemann sums, it suffices to prove the analogue of the lemma for the sum of a finite number of smooth plurisubharmonic functions in place of an integral over $[a, b]$. Thus, we have $u_i : [a, b] \times U \to \mathbb{R}$ smooth and plurisubharmonic for $1 \leq i \leq N$, and, defining a new $\varphi(z) = \log(e^{u_1(z)} + \cdots + e^{u_N(z)})$, we have to prove that φ is plurisubharmonic, and that $e^\varphi \sqrt{-1}\partial\overline{\partial}\varphi \geq \sum e^{u_k} \sqrt{-1}\partial\overline{\partial}u_k$ as smooth (1,1)-forms. The case of $N = 2$ follows from

$$(e^{u_1} + e^{u_2})\sqrt{-1}\partial\overline{\partial}\log(e^{u_1} + e^{u_2})$$

$$= e^{u_1}\sqrt{-1}\partial\overline{\partial}u_1 + e^{u_2}\sqrt{-1}\partial\overline{\partial}u_2 + \frac{e^{u_1+u_2}}{e^{u_1} + e^{u_2}}\sqrt{-1}(\partial u_1 - \partial u_2) \wedge \overline{(\partial u_1 - \partial u_2)}\,.$$

The case of general N follows by induction. □

In analogy with Proposition 3 for the Carathéodory metric, which plays a crucial role in the proof that F is an immersion in Theorem 5, we have the following result $e(\mathcal{F})$-extremal functions, which plays a crucial role in the proof that F is injective.

PROPOSITION 3. *Let $P \subset \Omega$ be a maximal polydisk, $P \cong \Delta^r$, and use Euclidean coordinates of the latter as coordinates for P. Let $x \in P, x = (x_1; x')$ and denote by $P' \subset P$ the polydisk corresponding to $\{x_1\} \times \Delta^{r-1}$. Let α be a minimal characteristic vector at x tangent to the minimal disk Δ_α corresponding to $\Delta \times \{x'\}$ and denote by s an $e(\mathcal{F})$-extremal function at x for α. Then $s(z_1; z_2, \cdots, z_r) = s(z_1)$ for any $(z_2, \cdots, z_r) \in P'$.*

PROOF. We use the notations in the proof of Proposition 2. Let $\delta > 0$ be such that $S_{\alpha,\epsilon} = \{(\delta e^{i\theta}; o) : \theta \in \mathbb{R}\}$ for the geodesic circle $S_{\alpha,\epsilon} \subset \Delta_\alpha$. Then, for $z \in P' \subset N$, $z = (0, z')$, and a suitable normalizing constant $c > 0$ we have

$$\|\alpha_z\|_{e(\mathcal{F})} \geq \|\alpha_z\|'_s = c \int_0^{2\pi} \|ds(\delta e^{i\theta}; z')\| d\theta = c \int_0^{2\pi} \frac{|ds(\delta e^{i\theta}, z')|}{1 - |s(\delta e^{i\theta}, z')|^2} d\theta := e^{\varphi(z')} ,$$

$$\|\alpha_o\|_{e(\mathcal{F})} = c \int_0^{2\pi} \frac{|ds(\delta e^{i\theta}, o)|}{1 - |s(\delta e^{i\theta}, o)|^2} = e^{\varphi(o)} .$$

By Proposition 1, $\|\alpha_z\|_{e(\mathcal{F})}$ is constant on P'. As a consequence $\varphi(z') \geq \varphi(o)$, so that $\varphi(z')$ attains its maximum on P' at the origin. On the other hand, by Lemma 2, $\varphi(z')$ is plurisubharmonic in $z' \in P'$. It follows that $\varphi \equiv C$ on P' for some constant $C > 0$. Write $e^{u_\theta(z')}$ for the integrand in the definition of $e^{\varphi(z)}$and note that u_θ is plurisubharmonic in z. Again by Lemma 2,

$$e^\varphi \sqrt{-1}\partial\bar{\partial}\varphi \geq c \int_0^{2\pi} e^{u_\theta} \sqrt{-1}\partial\bar{\partial}u_\theta \; d\theta \geq 0$$

in the sense of currents. It follows that for almost all $\theta \in [0, 2\pi]$, u_θ is pluriharmonic. However, $\sqrt{-1}\partial\bar{\partial}u_\theta$ is the pull-back of the Kähler form of (Δ, ds^2_Δ) by $\sigma_\theta : P' \to \Delta$, given by $\sigma_\theta(y') = s(\delta e^{i\theta}; y')$, so that σ_θ must be constant for almost all $\theta \in [0, 2\pi]$, hence for all θ by continuity. Thus, the $e(\mathcal{F})$-extremal function s must be of the form $s(z_1; z_2, \cdots, z_r) = s(z_1)$ when restricted to the polydisk P, as desired.

For the proof of Theorem 5 the final step is to derive injectivity of $F : \Omega \to D$ from Proposition 3. This step involves a substantial use of Ergodic Theory, applied in our context to the action of Γ on the left homogeneous spaces G/H for certain noncompact $H \subset G$. Primarily we make use of Moore's Ergodicity Theorem and consequently the density of 'generic' leaves of such Γ-actions. We sketch here an outline of our argument and refer the reader to [Mok5, (2.2) and (3.4)] for details.

SKETCH OF PROOF OF INJECTIVITY OF F IN THEOREM 5. Let $x, y \in \Omega$ be two distinct points. Then $F(x) \neq F(y)$ if and only if $\mathcal{F}$ separates x and y. There exists a maximal polydisk $P = \Delta \times P'$ containing both x and y such that the projections to the unit disk Δ separates x and y. From Proposition 4 it is known that extremal functions s, when restricted to P, are of the form $s(z_1; z') = s(z_1)$, and we will have shown that $\mathcal{F}$ separates points if we can take the function $s(z_1)$ on Δ to be injective. Actually, it suffices to derive from such extremal functions some $\sigma \in \mathcal{F}$ satisfying $\sigma(z_1; z') = \sigma(z_1) = \lambda z_1$ on P for some $\lambda \neq 0$. We call such a function σ a special function, which need not be an $e(\mathcal{F})$-extremal function.

Given a holomorphic function g on the unit disk, by averaging $e^{-i\theta}g(e^{i\theta}z_1)$ over $e^{i\theta} \in S^1$, the unit circle, we obtain a linearization of g at $0 \in \Delta$, which is precisely $g'(0)z_1$. Identifying $\mathrm{Aut}_o(P) \cong SU(1,1)^r$ in a canonical way as a subgroup of $\mathrm{Aut}_o(\Omega) = G$, for any $\theta \in \mathbb{R}$, the biholomorphism $\rho_\theta(z_1; z') := (e^{i\theta} z_1; z')$ on the maximal polydisk P extends to an automorphism τ_θ of Ω. Choosing $\epsilon > 0$ sufficiently small in the definition of $e(\mathcal{F}) = \kappa(\epsilon, \mathcal{F})$ we derive from Finsler metric rigidity and Cauchy estimates that for any $e(\mathcal{F})$-extremal function s as in Proposition 4, $s(z_1; z') = s(z_1)$, we have $|s'(0)| > c > 0$ for some constant c independent of the maximal polydisk P. Thus, we would be able to produce a special function $\sigma \in \mathcal{F}$ if $s \circ \tau_\theta \in \mathcal{F}$, which is however not known.

We note that $s \circ \gamma \in \mathcal{F}$ for any $\gamma \in \Gamma$. While $\Gamma \subset G$ is discrete by Moore's Ergodicity Theorem, its left action on G/H is ergodic for any noncompact closed subgroup $H \subset G$. As a consequence, the orbit under Γ of $\nu H \in G/H$ is dense in G/H, provided that νH lies outside a certain null set $E \subset G/H$. Adopting the notations of Proposition 4, we fix a triple $(P, P'; \alpha)$, $P = \Delta \times P'$, and consider the subgroup $H \subset G$ consisting of $\mu \in G$ such that $\mu(P) = P, \mu(P') = P'$ and such that $d\mu(\alpha)$ projects to the same vector as α under the canonical projection $\pi : P \to \Delta$. We may say that the extremal function s in Proposition 4 is adapted to $(P, P'; \alpha)$. Since $s(z_1, z') = s(z_1)$, $s|_P$ is invariant under the group H. Suppose we choose $\gamma_i \in \Gamma$ such that $\gamma_i H$ converges to $\tau_\theta H$. Then, by Lemma 3 below, $s \circ \gamma_i^{-1}\big|_P$ converges to $s \circ \tau_{-\theta}\big|_P$, and the S^1-averaging argument applies to produce a special function σ adapted to $(P, P'; \alpha)$.

It remains to take care of the the null set E. For instance, there may in fact be a maximal polydisk P such that its orbit under Γ gives a discrete set of maximal polydisks on Ω. Then, completing P to a triple $(P, P'; \alpha)$, the latter corresponds to an element of G/H whose orbit under Γ is discrete, and the argument above to produce special functions by S^1-averaging fails. However, from the estimate $|s'(0)| > c > 0$ for the $e(\mathcal{F})$-extremal function S the S^1-averaging argument produces a special function σ for which $|\lambda| = |S'(0)|$ is bounded from below independent of (P, P', α), which allows us to take care of the 'exceptional' by taking limits to obtain special functions for *every* triple $(P, P'; \alpha)$. This proves that $F : \Omega \to D$ is injective and completes the proof of Theorem 5. □

We include now the lemma on convergence of extremal functions referred to in the preceding sketch of proof of Theorem 5, which supplements [Mok5, (3.4), proof of Theorem 1 in the cocompact case] (where on Page 22, line 16, $\gamma_i^* s$ should have read $(\gamma_i^{-1})^* s$, i.e., $s \circ \gamma_i^{-1}$).

LEMMA 3. *Suppose $\gamma_i \in \Gamma$ are such that $\gamma_i H$ converges to $\tau_\theta H$ in G/H. Then, $s \circ \gamma_i^{-1}$ converges to $s \circ \tau_{-\theta}$ on P, i.e., $s\big(\gamma_i^{-1}(z; z')\big)$ converges to $s(e^{-i\theta} z; z')$ uniformly on compact subsets of P.*

PROOF. Write $\gamma_i = \lambda_i \tau_\theta h_i$, where $h_i \in H$ and $\lambda_i \in G$ converges to the identity element $e \in G$. Then for $(z; z') \in P$

$$\begin{aligned}(s \circ \gamma_i^{-1})\big(\lambda_i(z; z')\big) &= s\big(\gamma_i^{-1}(\lambda_i(z; z'))\big) = s\big(h_i^{-1}\tau_\theta^{-1}\lambda_i^{-1}(\lambda_i(z; z'))\big) \\ &= s\big(h_i^{-1}\tau_{-\theta}(z; z')\big) = s\big(h_i^{-1}(e^{-i\theta} z; z')\big) = s\big(e^{-i\theta} z; \mu_i(z')\big)\end{aligned} \tag{1}$$

for some $\mu_i \in \mathrm{Aut}(P')$. Here we make use of the fact that any $h \in H$ preserves P, and that $h|_P$ is necessarily of the form $h(z; z') = \big(z, \nu(z')\big)$, where $\nu \in \mathrm{Aut}(P')$. By Proposition 3, we conclude that

$$(s \circ \gamma_i^{-1})\big(\lambda_i(z; z')\big) = s(e^{-i\theta} z; z') = (s \circ \tau_{-\theta})(z, z') \,. \tag{2}$$

Fix an arbitrary compact subset $Q \subset P$. Then there exists a compact subset $Q' \subset \Omega$ such that $\lambda_i(Q) \subset Q'$ for any i. On the other hand, $s \circ \gamma_i^{-1} : \Omega \to \Delta$, so that by Cauchy estimates

$$\left|(s \circ \gamma_i^{-1})\big(\lambda_i(z; z')\big) - (s \circ \gamma_i^{-1})(z; z')\right| \le C(Q')\|\lambda_i(z, z') - (z; z')\| , \qquad (3)$$

where $C(Q')$ is a constant depending only on Q' (and independent of i), and $\|\cdot\|$ denotes the Euclidean norm. Since λ_i converges to $e \in G$, we conclude that the right hand side of (3) converges to 0. It follows from (2) that

$$\lim_{i\to\infty} \|(s \circ \gamma_i^{-1}) - (s \circ \tau_{-\theta})\|_Q = 0$$

for every compact subset $Q \subset P$, where $\|\cdot\|_Q$ denotes the supremum norm for continuous functions on Q. In other words, $s \circ \gamma_i^{-1}$ converges uniformly on compact subsets of P to $s \circ \tau_{-\theta}$, as desired.

As pointed out by Qi-Keng Lu, Theorem 5 applies in particular to the case where D is a bounded homogeneous domain, where in general holomorphic bisectional curvatures with respect to the Bergman metric need not be nonpositive. Theorem 5 raises the following very interesting question.

QUESTION. If in the statement of Theorem 5 we assume furthermore that D is a bounded homogeneous domain, is the holomorphic mapping $F : \Omega \to D$, which is now known to be an embedding, necessarily a totally geodesic isometric embedding up to a normalizing constant with respect to the Bergman metrics on Ω and on D?

It should be noted that, when restricted to minimal characteristic vectors, df (and hence dF) preserves lengths up to a fixed normalizing constant with respect to the Carathéodory metric, by Proposition 1, but the argument there does not apply in general to the Bergman metric on bounded homogeneous domains.

THEOREM 6. *In the notations of Theorem 5 write $X = \Omega/\Gamma$. Let Z be a normal projective variety, and $f : X \to Z$ be a surjective holomorphic map. Then, either the fundamental group $\pi_1(Z)$ is finite, or $f : X \to Z$ is an unramified covering map.*

SKETCH OF PROOF. Let $\tau : Z' \to Z$ be an covering of Z corresponding to the subgroup $f_*(\Gamma) \subset \pi_1(Z)$. Then, f lifts to a map $f' : X \to Z'$. Since Z is normal, Z' is normal and irreducible, and $f' : X \to Z'$ must be surjective by the Proper Mapping Theorem, implying that $f_*(\Gamma) \subset \pi_1(Z)$ must be of finite index. By a result of Margulis [Mar, Chapter VIII, Theorem A, p.258ff.], any normal subgroup of Γ is either finite or of finite index in Γ. If $\pi_1(Z)$ is infinite, then $f_*(\Gamma) \subset \pi_1(Z)$ must also be infinite, so that by Margulis' result $\mathrm{Ker}(f_*)$ must be finite, and the lifting $F : \Omega \to \widetilde{Z}$ to universal covering spaces must be a finite proper surjective map. We can now construct bounded holomorphic functions h on $\widetilde{Z}$ from bounded holomorphic functions on the bounded symmetric domain Ω by taking elementary symmetric functions of values of h on general fibers of F and by extension using the normality of $\widetilde{Z}$. The Γ-invariant complex Finsler metric θ on Ω constructed by pulling back bounded holomorphic functions on $\widetilde{Z}$ must be such that $\|\eta\|_\theta$ is nonzero for every nonzero (1,0)-vector η on Ω, by a slight generalization of Theoroem 5 with the same proof. But at the ramification locus of F this would contradict with the construction of θ, proving that $F : \Omega \to \widetilde{Z}$, and hence $f : X \to Z$, must be unramified. □

3. Gap rigidity for pairs of bounded symmetric domains

(3.1) In this section we consider a rigidity phenomenon on quotients of bounded symmetric domains which we will call gap rigidity. In oversimplified form it can be formulated as follows. Let Ω be a bounded symmetric domain and $\Gamma \subset \mathrm{Aut}(\Omega)$ be a discrete group of automorphisms acting without fixed points. Writing $X := \Omega/\Gamma$ and denoting by $S \subset X$ a *compact* complex submanifold of sufficiently small second fundamental form, we ask whether $S \subset X$ is necessarily totally geodesic. An affirmative answer to this question under certain geometric hypotheses to be made precise will be called a gap rigidity theorem.

To put things in persepcetive, we consider first a special case of the problem which remains up to this point completely open. Let Ω be the n-ball B^n, $n \geq 2$. Let $S \subset B^n/\Gamma := X$ be a compact complex submanifold such that the second fundamental form is *everywhere* sufficiently small. The question is whether $S \subset X$ is necessarily a totally geodesic submanifold. Because the Bergman metric on B^n is of constant negative holomorphic sectional curvature, a pointwise bound on the second fundamental form means exactly a pointwise pinching condition on the holomorphic sectional curvatures of the complex submanifold S. Here by saying that the second fundamental form is everywhere "sufficiently small", we mean that its pointwise norm is bounded by some absolute constant ϵ, independent of Γ and thus depending only on n. We note that the problem is only meaningful when we consider all compact complex submanifolds S of a fixed dimension, especially we should not assume any bound on the volume of S. For instance, in the case of $n = 2$ and $\dim(S) = 1$, the integral of the square of the norm of the second fundemental form, suitably normalized, represents the first Chern class, hence the degree, of some holomorphic line bundle. If the integral is sufficiently small it has to vanish since the degree here must be a nonnegative integer. Thus, if ϵ is chosen such that $\epsilon^2 \cdot \mathrm{Volume}(S) < 2\pi$ we conclude readily that S must be totally geodesic. This question on gap rigidity on B^n raised in the above turns out to be a difficult problem. For a discussion on some special cases of this unsolved problem we refer the reader to Eyssidieux-Mok [EM2,§3].

We explain here on the other hand a first positive result together with the original motivations for studying gap rigidity. It concerns the case where Ω is the bounded symmetric domain of Type III, which is biholomorphic to the Siegel upper half-plane, and where $S \subset X := \Omega/\Gamma$ is a compact holomorphic curve. As will be explained in (3.5) this first result has now been superseded by a different method. However, it is altogether possible that the original line of reasoning remains useful to attack other cases of the problem.

In Mok [Mok3, 1991] we were studying among other things ramifications of Hermitian metric rigidity. A natural question arises in the case of the Siegel upper half-plane $\mathcal{H}_n$ of genus n, which parametrizes polarized Abelian varieties of dimension n. Let $n \geq 2$. Consider the moduli space $X_\Gamma := \mathcal{H}_n/\Gamma$ of principally polarized Abelian varieties of dimension n with a certain level structure so that the corresponding discrete group Γ of symplectic transformations is torsion-free. Then, we have over X_Γ a universal family $\pi : \mathcal{A}_\Gamma \to X_\Gamma$ of such Abelian varieties. The universal family admits a projective compactification, which can be regarded as a geometric model of the universal polarized Abelian variety over the function field of X_Γ. It was known (Shioda [Sh] for $n = 1$, Silverberg [Sil] in general) that the Mordell-Weil group of this universal Abelian variety is finite, which means that every holomorphic section of the universal family is a section of finite order. In [Mok3] our perspective was to find a differential-geometric explanation of this in terms of invariant metrics, thereby to obtain a generalization of a number of known results at that point to the general case of Kuga families of polarized Abelian varieties without constant parts over Shimura varieties, which are arithmetic quotients of certain bounded symmetric domains. In Mok-To [MT, 1993] we resolved the last question, by resorting to considering eigensection equations satisfied by differential forms arising from global holomorphic sections of the universal family. In Mok [Mok3] we prove a certain gap rigidity theorem for compact holomorphic curves on quotients of the Siegel upper half-plane by making use of invariant Kähler metrics on universal families of polarized Abelian varieties, and in Eyssidieux-Mok [EM1, 1995] we resolved the same problem with effective constants by making use of perturbations of eigensection equations. To put our argument in concrete terms we sketch first of all a differential-geometric proof of Shioda's result, as follows:

THEOREM 7 (SHIODA [SH, 1972]). *Let $\mathcal{H} \subset \mathbb{C}$ denote the upper half-plane, $\Gamma \subset \mathbb{P}SL(2,\mathbb{Z})$ be a torsion-free subgroup of finite index, and write $X_\Gamma = \mathcal{H}/\Gamma$ for the quotient Riemann surface. Denote by $\pi : \mathcal{A}_\Gamma \to X_\Gamma$ the universal family, and let $\overline{\pi} : \overline{\mathcal{A}}_\Gamma \to \overline{X}_\Gamma$ be a projective compactification as an elliptic surface, which is a geometric model for the associated modular elliptic curve A_Γ over the function field $\mathbb{C}(\overline{X}_\Gamma)$. Then,* $\mathrm{rank}_{\mathbb{Z}}(A_\Gamma(\mathbb{C}(\overline{X}_\Gamma)) = 0$ *for the Mordell-Weil group $A_\Gamma(\mathbb{C}(\overline{X}_\Gamma))$. In other words, there are only a finite number of holomorphic sections of $\overline{\pi} : \overline{\mathcal{A}}_\Gamma \to \overline{X}_\Gamma$ over $\overline{X}_\Gamma$.*

The original proof of Shioda's result was obtained by methods of Algebraic Geometry which involves a determination of the Néron-Severi group of the elliptic surface, noting that a holomorphic section of the latter of infinite order leads to an element of $\mathrm{Pic}(\overline{X}_\Gamma) \otimes_{\mathbb{Z}} \mathbb{Q}$ linearly independent from standard elements of the latter vector space. Silverberg ([Sil, 1985]) generalized the result among other things to the case of the Siegel upper half-plane. In her proof she made use of Eichler automorphic forms. We will sketch a differential-geometric proof of Shioda's result here. In the proof we associate to each section of the elliptic surface an eigensection, corresponding to a negative eigenvalue, of some elliptic operator which is in fact related to Eichler automorphic forms. To start with we discuss the notion of Eichler automorphic forms in the case of $n = 1$ (cf. Gunning [Gu]).

Eichler automorphic forms associated to a rational point of A_Γ

A holomorphic section of s of $\pi : \mathcal{A}_\Gamma \to X_\Gamma$ lifts to a holomorphic function $f : \mathcal{H} \mapsto \mathbb{C}$ satisfying the functional equation

$$f(\gamma\tau) = \frac{f(\tau)}{c_\gamma\tau + d_\gamma} + A_\gamma\Big(\frac{a_\gamma\tau + b_\gamma}{c_\gamma\tau + d_\gamma}\Big) + B_\gamma, \tag{$\sharp$}$$

where $\gamma(\tau) = \frac{a_\gamma\tau+b_\gamma}{c_\gamma\tau+d_\gamma}$, $\gamma \in \Gamma$. We have

$$\frac{f'(\gamma\tau)}{(c_\gamma\tau + d_\gamma)^2} = -\frac{c_\gamma}{(c_\gamma\tau + d_\gamma)^2}f(\tau) + \frac{f'(\tau)}{(c_\gamma\tau + d_\gamma)} + \frac{A_\gamma}{(c_\gamma\tau + d_\gamma)^2} .$$

Hence

$$f'(\gamma\tau) = -c_\gamma f(\tau) + (c_\gamma\tau + d_\gamma)f'(\tau) + A_\gamma ,$$

$$\frac{f''(\gamma\tau)}{(c_\gamma\tau + d_\gamma)^2} = -c_\gamma f'(\tau) + c_\gamma f'(\tau) + (c_\gamma\tau + d_\gamma)f''(\tau) ;$$

yielding

$$f''(\gamma\tau) = (c_\gamma \tau + d)^3 f''(\tau) \ . \tag{$\flat$}$$

Thus from the functional equation ($\sharp$) we obtain the transformation rule ($\flat$) for the ordinary second derivative f'' under Deck transformations $\gamma \in \Gamma$. In other words f'' defines an Eichler automorphic form $\alpha = \alpha(f)$ on the modular curve X_Γ.

We will associate to rational points of A_Γ smooth sections of some holomorphic line bundle which are related to Eichler automorphic forms with respect to covariant differentiation arising from canonical metrics.

Sketch of a differential-geometric proof of Theorem 7. In what follows we assume that $K_{X_\Gamma}^{1/2}$ exists, which is the case whenever $-1 \notin \Gamma$. The Eichler automorphic form α is an element of $\Gamma(X_\Gamma, K_{X_\Gamma}^{3/2})$. Such automorphic forms may exist, and the question is whether they can arise from a section σ of $\overline{\pi} : \overline{\mathcal{A}_\Gamma} \to \overline{X_\Gamma}$. Over the universal covering space $\mathcal{H} \times \mathbb{C}$ there is the notion of a horizontal section, viz., the graph of a holomorphic function h on $\mathcal{H}$ such that $h(\tau) = A\tau + B$ for some real numbers A and B. There is then a canonical decomposition of $T_{\mathcal{H}\times\mathbb{C}}$ into a direct sum $W \oplus H$, where W stands for the bundle of vertical vectors, i.e., W is the relative tangent bundle of the canonical projection $\rho : \mathcal{H} \times \mathbb{C} \to \mathcal{H}$, and at a point $x \in \mathcal{H} \times \mathbb{C}$, H_y stands for the tangent space of the unique horizontal section passing through y. Thus H defines a foliation with 1-dimensional holomorphic leaves on $\mathcal{H} \times \mathbb{C}$ whose leaves are precisely the horizontal sections. Given a local holomorphic section s of $\pi : \mathcal{A}_\Gamma \to X_\Gamma$, over $U \subset X_\Gamma$, a point $x \in U$, and $\xi \in T_x(U)$, we can decompose $ds(\xi)$ into vertical and horizontal components according to the decomposition $T(\mathcal{H} \times \mathbb{C}) = W \oplus H$, leading to the vertical component $\eta = \eta_s$ of ds, so that η vanishes if and only if s is a horizontal section. η is a smooth V-valued (0,1)-form, where V stands for the universal line bundle over X_Γ. When we have actually a global holomorphic section s of $\overline{\pi} : \overline{\mathcal{A}_\Gamma} \to \overline{X_\Gamma}$, the restriction of s to X_Γ gives $\eta : T_{X_\Gamma} \mapsto T_{X_\Gamma}^{1/2}$, noting that the universal line bundle V is a square root of the tangent bundle. Thus, $\eta \in \mathcal{C}^\infty(X_\Gamma, K_{X_\Gamma}^{-1/2})$. From the fact that s is a holomorphic section over $\overline{X_\Gamma}$, not just X_Γ, one can deduce that η is square-integrable with respect to the canonical metrics on V ad X_Γ (cf. Mok-To [MT]), which allows us to perform integration by parts as if we were dealing with a compact base manifold. Denote by ∇ the (1,0)-component of covariant differentiation on X_Γ arising from canonical metrics on V and T_{X_Γ}. From the definitions of the Eichler automorphic form it is easy to check that $\nabla\eta = c\alpha$ for some $c \neq 0$. From the holomorphicity of α we have

$$\begin{aligned} \overline{\partial}\alpha = 0 \Rightarrow \overline{\partial}\nabla\eta = 0 \\ \Rightarrow \overline{\partial}\,\overline{\partial}^*\eta = 0 \Rightarrow \overline{\partial}^*\overline{\partial}\eta = -\eta \ . \end{aligned}$$

Integrating by parts

$$\int_{X_\Gamma} \langle \overline{\partial}^*\overline{\partial}\eta, \eta \rangle = -\int_{X_\Gamma} \langle \eta, \eta \rangle \ , \quad \text{i.e.}\ ,$$

$$\int_{X_\Gamma} \|\overline{\partial}\eta\|^2 = -\int_{X_\Gamma} \|\eta\|^2,$$

which forces η to vanish identically on X_Γ. Hence $\eta_s \equiv 0$ for every holomorphic section s of $\overline{\pi} : \overline{\mathcal{A}_\Gamma} \to \overline{X_\Gamma}$. This forces s to be horizontal, and hence of finite order, proving Theorem 7. □

Along the same line of argument of using eigensection equations of elliptic differential operators as explained, we established

THEOREM 8 (MOK-TO [MT, 1993]). *Let $\pi : \mathcal{A}_\Gamma \to X_\Gamma$ be a Kuga family of polarized Abelian varieties without locally constant parts and $\overline{\pi} : \overline{\mathcal{A}_\Gamma} \to \overline{X_\Gamma}$ be a projective compactification. Then,* $\text{rank}_{\mathbb{Z}}(A_\Gamma(\mathbb{C}(\overline{X_\Gamma})) = 0$ *for the Mordell-Weil group $A_\Gamma(\mathbb{C}(\overline{X_\Gamma}))$. In other words, there are only a finite number of holomorphic sections of $\overline{\pi} : \overline{\mathcal{A}_\Gamma} \to \overline{X_\Gamma}$ over $\overline{X_\Gamma}$.*

SKETCH OF PROOF. Consider first of all the case of a Siegel modular variety, where $X_\Gamma = \Omega/\Gamma$, $\Omega \cong \mathcal{H}_n$, and Γ a torsion-free subgroup of finite index in $Sp(n, \mathbb{Z})$. We have over X_Γ a universal holomorphic vector bundle V of nonpositive curvature in the dual sense of Nakano and a universal family of polarized Abelian varieties $\mathcal{A}_\Gamma$. For a local holomorphic section of $\mathcal{A}_\Gamma$ over an open subset U of X_Γ we can associate a smooth V-valued (0,1)-form η which satisfies an eigensection equation for some locally homogeneous elliptic operator. In the case where $n = 1$ the equation is of the form (†) $\overline{\partial}^*\overline{\partial}\eta = -\eta$ as explained in the above. For n in general it is of the form (†′) $\overline{\partial}^*\overline{\partial}\eta = -A\eta$, where A is a self-adjoint nonnegative operator. There is a decomposition $\eta = \eta' + \eta''$, corresponding to a canonical decomposition $V \otimes \overline{T} = P' \oplus P''$, such that A is positive on P' and vanishes on P''. As a consequence, the integral of $\langle -A\eta, \eta\rangle$ over X is positive unless $\eta' = 0$, which in turn implies that $\eta = 0$ [MT, §4]. In the general case of a Kuga family, up to a finite unramified cover we may assume $X_\Gamma \subset X'_{\Gamma'}$, where $X'_{\Gamma'}$ is a Siegel modular variety. We have analogously a canonical decomposition $V \otimes \overline{T} = P' \oplus P''$, $\eta = \eta' + \eta''$; such that $A|_{P'} > 0$ and $A|_{P''} \equiv 0$. When $V|_{X_\Gamma}$ has no locally flat component, the argument that $\eta' = 0$ forces $\eta = 0$ remains valid. In general, V_{X_Γ} may have flat factors even when the Kuga family $\pi : \mathcal{A}_\Gamma \to X_\Gamma$ does not have locally constant parts. By a simple algebraic argument [MT, §6] involving taking traces of algebraic numbers the argument in general can be reduced to the former case. □

For a formulation of the result of [EM1] on gap rigidity mentioned in (3.1) note that on the bounded symmetric domain D_n^{III} there is a maximal totally geodesic polydisk of dimension n. An immersed totally geodesic curve on a manifold uniformized by $D_n^{III} \cong \mathcal{H}_n$ is said to be of the diagonal type if it can be lifted to the diagonal of a maximal polydisk. We have

PROPOSITION 4 (EYSSIDIEUX-MOK [EM1, 1995]). *Let $n \geq 2$ be an integer. Let $\Gamma \subset \text{Aut}_o(\mathcal{H}_n)$ be a torsion-free discrete group of automorphsms and write $X_\Gamma := \mathcal{H}_n/\Gamma$. Normalize the canonical Kähler-Einstein metric on $\mathcal{H}_n$ so that a totally geodesic curve of the diagonal type is of constant holomorphic sectional curvature -1. Let $f : C \subset X_\Gamma$ be an immersed compact holomorphic curve such that*

$$-\left(1 + \frac{1}{4n}\right) < \textit{Gauss curvature of } C \ (\leq -1)$$

Then, C is an immersed totally geodesic curve of the diagonal type.

SKETCH OF PROOF. With reference to the proof in Theorem 8, in the case where C is a totally geodesic holomorphic curve of the diagonal type, (†′) is exactly (†) as in the proof of Shioda's result except that η is vector-valued. More precisely, the universal vector bundle V splits over C as the direct sum of n Hermitian holomorphic line bundles isomorphic to each other, and accordingly η decomposes as

$\eta = \eta_1 + \cdots + \eta_n$ such that $\overline{\partial}^*\overline{\partial}\eta_i = -\eta_i$ for $1 \le i \le n$. In general, for any holomorphic curve $C \subset X_\Gamma$ and for V-valued (0,1)-form any η obtained from a holomorphic section of the universal family $\pi : \mathcal{A}_\Gamma \to X_\Gamma$ over C we have a perturbed second order elliptic differential equation ([EM1, §3, Eqn.(8)])

$$\overline{\partial}^*\overline{\partial}\eta = \Phi\eta + S\eta \,. \qquad (\dagger'')$$

Denote by Θ the End(V)-valued curvature (1,1) form of V. Then, $\Theta = \Phi \otimes \omega$ where ω is the Kähler form on C, and S is defined by the second fundamental form σ of C in X_Γ. Φ is always nonpositive, and is strictly negative when C is locally approximable by a totally geodesic holomorphic curve of the diagonal type. When the second fundamental form is sufficiently small, integrating ($\dagger''$) over X still yields a contradiction unless $\eta = 0$. From this one concludes that any holomorphic section of $\mathcal{A}_\Gamma$ over C is of finite order. The argument is actually stronger. Let $\pi : \mathcal{H}_n \to X_\Gamma$ be the universal covering and denote by $\widetilde{C} \subset \mathcal{H}_n$ a connected component of $\pi^{-1}(C)$. Any $\gamma \in \Gamma$ is of the form $\begin{bmatrix} A_\gamma & B_\gamma \\ C_\gamma & D_\gamma \end{bmatrix} \in Sp(n, \mathbb{Z})$ where $A_\gamma, B_\gamma, C_\gamma$ and D_γ are $n \times n$-matrices. A global holomorphic section over C can be interpreted as an n-tuple of global holomorphic functions $F : \widetilde{C} \mapsto \mathbb{C}^n$ satisfying the functional equations

$$F(\gamma\tau) = \left[(C_\gamma\tau + D_\gamma)^t\right]^{-1} F(\tau) + \left(A_\gamma\tau + B_\gamma\right)\left(C_\gamma\tau + D_\gamma\right)^{-1} P_\gamma + Q_\gamma \,, \qquad (\sharp)$$

where we only require the column vectors P_γ and Q_γ to have *real* coefficients. When the coefficients of P_γ and Q_γ are integers F leads to a *bona fide* holomorphic section s of $\mathcal{A}_\gamma$ over C. In the general case s is only 'multi-valued', but from the definition of η as in the proof of Shioda's result, the smooth tensor η thus associated remains globally defined on C. The number μ of linearly independent solutions to the perturbed eigensection equation ($\dagger''$) can be interpreted as a Hodge-theoretic invariant. In the case where $C \subset X_\Gamma$ is a compact holomorphic curve locally approximable by a totally geodesic holomorphic curve of the diagonal type, this number is computable using the Riemann-Roch Theorem, and is shown to be positive unless S is totally geodesic. This is a *non-vanishing theorem*, which contradicts the vanishing theorem on μ obtained from the perturbed eigensection equation, proving by contradiction that any compact holomorphic curve locally approximable by a totally geodesic holomorphic curve of the diagonal type must necessarily be totally geodesic. The main result of [EM1] was obtained along this line of argument and the pinching constants were obtained by an explicit determination and estimates of the perturbation term in the equation ($\dagger'$). For details of the argument we refer the reader to [EM1]. □

PROBLEM AND REMARKS. It will be very interesting to obtain analogues of Proposition 3 for certain pairs of bounded symmetric domains (Ω, Ω') by exploiting the approach of proving both vanishing and non-vanishing theorems for a hypothetical compact almost totally geodesic complex submanifold which is not totally geodesic. To look for non-vanishing theorems, in place of the geometric perspective of considering multi-valued sections of the universal family one should use cohomological interpretations in terms of harmonic forms associated to local systems, and look for vanishing theorems for higher cohomology groups which are *stable* in the sense that they remain valid under a small perturbation arising from the second fundamental form.

Taking the above result as a *prototype*, we formulated in [EM1] a conjectural gap phenomenon. As a starting point, by standard estimates on solutions of ordinary linear differential equations we proved.

LEMMA 4. *Let $\Omega \Subset \mathbf{C}^N$ be a bounded symmetric domain. Fix $x_0 \in \Omega$ and let $B(r) \subset \Omega$ denote the geodesic ball (with respect to the Bergman metric) of radius r and centered at x_0. For $\delta > 0$ sufficiently small ($\delta < \delta_0$) there exists $\epsilon > 0$ such that the following holds:*

> *For any ϵ-pinched connected complex submanifold $V \subset B(x_0; 1)$, $x_0 \in V$, there exists a unique equivalence class of totally geodesic complex submanifold on Ω, to be represented by $j : \Omega' \hookrightarrow \Omega$, and a totally geodesic complex submanifold $\Xi \subset B(1)$ modelled on $(\Omega, \Omega'; j)$ such that the Hausdorff distance between $\Xi \cap B(\frac{1}{2})$ and $V \cap B(\frac{1}{2})$ is less than δ.*

Here we say that V is ϵ-pinched to mean that the norm of the second fundamental form is less than ϵ everywhere on V. The distance between a point p and a set F on a metric space (M, d), to be denoted $\mathrm{dist}(p, E)$, is the infimum of distances from p to points q on E, and the Hausdorff distance between two subsets $E, F \subset M$ is the supremum of $\mathrm{dist}(p, F)$ as p ranges over points on E.

DEFINITION (GAP PHENOMENON). Let $\Omega \Subset \mathbf{C}^N$ be a bounded symmetric domain and $j : \Omega' \hookrightarrow \Omega$ be a totally geodesic complex submanifold. We say that the gap phenomenon holds for $(\Omega, \Omega'; j)$ if and only if there exists $\epsilon < \epsilon(\delta_0)$ (δ_0 as in Proposition) for which the following holds:

> For any torsion-free discrete group $\Gamma \subset \mathrm{Aut}(\Omega)$ of automorphisms and any ϵ-pinched immersed compact complex submanifold $S \hookrightarrow \Omega/\Gamma$ modelled on $(\Omega, \Omega'; j)$, S is necessarily totally geodesic.

Proposition 3 can be interpreted as an effective version of the gap phenomenon for the special case of $(\mathcal{H}_n, \mathcal{H}; j)$, where $j : \mathcal{H} \to \mathcal{H}_n$ denotes the 'diagonal' embedding defined by $j(\tau) = \tau I_n \in \mathcal{H}_n$. Very recently, we realized that the Gap Phenomenon in its general form does not always hold. Indeed, it fails for the case of $(\Delta^2, \Delta \times \{0\})$, as follows.

THEOREM 9 (EYSSIDIEUX-MOK [EM2, 2004]). *There exist sequences of compact Riemann surfaces S_k, T_k of genus ≥ 2; and branched double covers $f_k : S_k \to T_k$ such that, writing ds^2_C for the Poincaré metric of Gaussian curvature -2 on a compact Riemann surface C, and defining*

$$\mu_k := \sup \left\{ \frac{f_k^* ds^2_{T_k}(x)}{ds^2_{S_k}(x)} : x \in S_k \right\} ,$$

we have

$$\lim_{k \to \infty} \mu_k = 0 .$$

As a consequence, gap rigidity fails for $(\Delta^2, \Delta \times \{0\})$.

The last statement on the failure of gap rigidity follows readily when we consider the graphs Γ_k of f_k, as subvarieties of $S_k \times T_k$, which are uniformized by the bi-disk. The ratio of metrics appearing in the definition of μ_k is nothing other than $\|df_k\|^2$, measured with respect to Poincaré metrics of Gaussian curvature -2 on

S_k resp. T_k. A uniform bound on $\|df_k\|^2$ implies a uniform bound on the second fundamental form of Γ_k in $S_k \times T_k$, which tends to zero as μ_k tends to zero.

In our construction, which is elementary, the target compact Riemann surface can be taken to be the same for all k. $f : S_k \to T$ will be constructed as double covers branched over sets of points on T to be made precise whose cardinalities grow to infinity. To put our construction in perspective we consider first of all any surjective holomorphic map $f : S \to T$ between compact Riemann surfaces of genus ≥ 2. For a compact Riemann surface C we write $g(C)$ for its genus. By the Riemann-Hurwicz formula,we have

$$2g(S) - 2 = r(2g(T) - 2) + e \ ,$$

where r denotes the sheeting number and e denotes the cardinality of the ramification divisor. By the Gauss-Bonnet Theorem on a compact Riemann surface C of genus ≥ 2 we have

$$\int_C -2ds_C^2 = 4\pi(1 - g(C)); \text{ i.e., } \frac{1}{\pi}\int_C ds_C^2 = 2g(C) - 2.$$

On the other hand,

$$\frac{1}{\pi}\int_S f^* ds_T^2 = \frac{r}{\pi}\int_T ds_T^2 = r(2g(T) - 2)$$
$$\frac{1}{\pi}\int_S ds_S^2 = 2g(S) - 2 \ .$$

As a consequence, with respect to the Poincaré metric on S we have

$$\text{Average}\Big(\frac{f^* ds_T^2}{ds_S^2}\Big) = r\Big(\frac{2g(T) - 2}{2g(S) - 2}\Big) := \nu.$$

In our construction T is fixed, of genus 2, and $r = 2$. Denoting by e_k the cardinality of the ramification divisor of $f_k : S_k \to T$, and by ν_k the ratio ν for the map f_k, by the Riemann-Hurwitz formula ν_k is roughly $\frac{4}{e_k}$. A crucial point in our construction is to find $f_k : S_k \to T$ such that f_k is as uniformly area-decreasing as possible. Obviously $\|df_k\|$ has to vanish at the ramification points, and 'uniformity' is relative. In fact, as our example will show, μ_k does not tend to zero at the same rate as ν_k. Beyond uniformity, our construction has to be such that the invariant metrics can be compared at least asymptotically. We do this by taking T to be a suitable double cover of an elliptic curve E, and by constructing double branched covers $f_k : S_k \to T$ in such a way that the Poincaré metrics on S_k and T all descend to singular Hermitian metrics of constant negative curvature with prescribed poles and pole orders. A comparision between Poincaré metrics is then deduced from a comparison between very special singular metrics on E which are related to each by homomorphisms of E as an Abelian group.

PROOF OF THEOREM 9. Let $L \subset \mathbb{C}$ be a lattice and write $E = \mathbb{C}/L$ for the corresponding elliptic curve. Let $\tau \in E$ be a nonzero 2-torsion point and $h : T \to E$ be a double cover branched over $\{0, \tau\}$. For a positive integer m we write $\Phi_m : E \to E$ for the homomorphism $\Phi_m(x) = mx$. For a positive integer m we define $D_m := \Phi_m^{-1}(\{0, \tau\})$. Note that $\text{Card}(D_m) = 2m^2$, $D_1 = \{0, \tau\}$. For a positive integer $m = 2k - 1$, we have $m\tau = 2k\tau - \tau \equiv -\tau \equiv \tau \bmod L$, hence $D_m \supset D_1$. Since the cardinality of $D_m - D_1$ is even, there exists a double cover $f_k : S_k \to T$

branched over $h^{-1}(D_m - D_1)$. We claim that the sequence of holomorphic maps $f_k : S_k \to T$ furnishes an example satisfying the statement of Theorem 9.

Since $h : T \to E$ is a double cover, there is an involution switching the two distinct points $h^{-1}(q)$ for q outside the branch locus of $D_1 = \{0, \tau\}$ of h. The Poincaré metric ds^2_T is invariant under the involution and descends to a Hermitian metric on $E - D_1$ with a singularity of order $\frac{1}{2}$ at 0 and at τ, which can be regarded as a continuous Hermitian metric on $T_E \otimes [D_1]^{-\frac{1}{2}}$. Likewise for each k the double cover $f_k : S_k \to T$ is invariant under an involution and descends to a continuous Hermitian metric on $T_E \otimes [D_m]^{-\frac{1}{2}}$. ¿From the uniqueness of singular Hermitian metrics of curvature -2 with prescribed orders of poles, we conclude that

$$(h \circ f_k)_* ds^2_{S_k} = \Phi^*_m(h_* ds^2_T) \tag{1}$$

Near 0, we have

$$\Phi^*_m\Big(\frac{|dz|^2}{|z|}\Big) = \frac{m^2|dz|^2}{|mz|} = m\frac{|dz|^2}{|z|} \ ,$$

and an analogous statement holds true at τ. On the other hand, outside of two given small disks centred at 0 resp. τ we have $h_* ds^2_T \geq \epsilon \cdot ds^2_E$ for some $\epsilon > 0$, where ds^2_E denotes the Euclidean metric on E, so that

$$\Phi^*_m(h_* ds^2_T) \geq m^2 \epsilon \cdot ds^2_E \ , \tag{2}$$

¿From (1) and (2) it follows that for some positive constant C, we have $\mu_k \leq \frac{C}{k} \to 0$ as $k \to \infty$, as desired. □

REMARKS. It is clear from the proof that $\mu_k \geq \frac{c}{k}$ for some $c > 0$. Recall that $\nu_k := 2\big(\frac{2g(T)-2}{2g(S)-2}\big)$ is the average of $\|df_k\|^2$ over S with respect to the Poincaré metric $ds^2_{S_k}$. Since the cardinality of D_k is of the order of k^2, ν_k is of the order $\frac{1}{k^2}$, and μ_k decreases to zero at a rate comparable to the square root of ν_k.

In view of Theorem 9, gap rigidity does not always hold. Taking into consideration such counter-examples to the gap phenomenon, we have the following conjectural formulation of a sufficient curvature condition for the validity of gap rigidity for a pair (Ω, Ω') of bounded symmetric domains.

CONJECTURE. *Let Ω be a bounded symmetric domain, and $\Omega' \subset \Omega$ be a totally geodesic complex submanifold. Denote by N the normal bundle of Ω' in Ω, equipped with the Hermitian metric h induced by the Bergman metric g on Ω, so that (N, h) is a homogeneous holomorphic vector bundle on Ω'. Suppose (N, h) does not contain a nontrivial flat direct summand, i.e., an isometric direct summand of zero curvature. Then, gap rigidity in the Zariski topology holds for (Ω, Ω').*

The Conjecture above should more properly be regarded as a Working Hypothesis, and its resolution remains difficult. For instance, the case of (B^2, Δ) discussed in Eyssidieux-Mok [EM2, Section 32] remains unresolved. We will further discuss some interesting special cases in (3.7).

(3.3) Proposition 3 of (3.2) is an example of a characterization of certain compact totally geodesic complex submanifolds in terms of a pinching on holomorphic sectional curvatures, and as such is by its very formulation differential-geometric in nature. This gives an example of gap rigidity in the complex topology. As it turns out, there is a stronger notion of gap rigidity, which says that certain compact complex submanifolds of quotients of bounded symmetric domains are totally geodesic, provided that their tangent spaces at all points satisfy a genericity condition in algebro-geometric terms. We call this gap rigidity in the Zariski topology. More precisely, we have

DEFINITION (GAP RIGIDITY IN THE ZARISKI TOPOLOGY). Let Ω, Ω' be bounded symmetric domains, $j : \Omega' \hookrightarrow \Omega$ be a holomorphic totally geodesic embedding and identify Ω' with $j(\Omega') \subset \Omega$. Let G be the identity component of the automorphism group of Ω. We say that $(\Omega, \Omega'; j)$; $\dim(\Omega) = n$, $\dim(\Omega') = n'$; exhibits gap rigidity in the Zariski topology if and only if there exists a G-invariant complex-analytic subvariety $\mathcal{Z}_\Omega \subset \mathbb{G}_\Omega :=$ Grassmann bundle of n'-planes, which descends to $\mathcal{Z}_X \subset \mathbb{G}_X := \mathbb{G}_\Omega/\Gamma$ for any $X = \Omega/\Gamma$, such that the following holds:

(a) $[T_o(\Omega')] \notin \mathcal{Z}_{\Omega,o}$;
(b) For any compact complex n'-dimensional immersed submanifold $f : S \to X = \Omega/\Gamma$ such that $\big[df(T_x(S))\big] \notin \mathcal{Z}_{X,x}$ for all $x \in S$, S must be totally geodesic.

REMARKS. We note that in (b) we do not require that S must be modelled on $(\Omega, \Omega'; j)$. There are examples for which there exist two n'-dimensional bounded symmetric domains Ω'_1, Ω'_2 not biholomorphic to each other, together with holomorphic totally geodesic embeddings $j_1 : \Omega'_1 \hookrightarrow \Omega$, $j_2 : \Omega'_2 \hookrightarrow \Omega$; and a G-invariant complex-analytic subvariety $\mathcal{Z}_\Omega \subset \mathbb{G}_\Omega$, such that (a) holds true for both Ω'_1 and Ω'_2. Nonetheless, given a bounded symmetric domain Ω, up to holomorphic isometries there are only at most a finite number of (n'-dimensional) complex totally geodesic submanifolds, and (b) says in particular that S is modelled on one of the finitely many possibilities $(\Omega, \Omega'_k; j_k)$ up to holomorphic isometries for which in addition (a) holds true. In particular, if $(\Omega, \Omega'; j)$ exhibits gap rigidity in the Zariski topology, it also exhibits gap rigidity in the complex topology.

When the embedding j is understood, we will just say that (Ω, Ω') exhibits gap rigidity in the Zariski topology. To start with we have the simple example of gap rigidity in the Zariski topology, where the ambient domain Ω is reducible.

PROPOSITION 5. *Let D be an irreducible bounded symmetric domain, and $\Omega = D \times \cdots \times D$ be the Cartesian product of k copies of Ω, where $k > 1$. Denote by Ω' the diagonal of Ω. Then, (Ω, Ω') exhibits gap rigidity in the Zariski sense.*

PROOF. Let $\Gamma \subset \mathrm{Aut}_o(\Omega)$ be a torsion-free discrete subgroup. Write $n' = \dim(\Omega') = \dim(D)$. Call an n'-plane generic if and only if its projection to each individual factor Ω is a local biholomorphism. If $S \subset X = \Omega/\Gamma$ is such that $T_x(S)$ is generic for every $x \in S$, $\dim(S) = n'$, then we obtain Kähler-Einstein metrics by projection onto each of the k individual factors. Proposition 4 follows readily from the uniqueness of Kähler-Einstein metrics, noting that a holomorphic isometry between two non-empty open subsets of D is the restriction of an automorphism of D.

(3.4) Eyssidieux [Eys1, 1997] studied compact Kähler manifolds underlying variations of Hodge structures by considering Euler-Poincaré characteristics of assoicated differential complexes. Let $(S, \mathbb{V})$ be a polarized variation of Hodge structures on an m-dimensional compact Kähler manifold S with immersive period map. Eyssidieux proved in [Eys1] the Lefschetz-Gromov vanishing theorem for L^2-cohomology with coefficients in $\mathbb{V}$ on an appropriate normal cover $\widetilde{S}$ of S in degrees $\neq m$. From this he deduced Chern number inequalities called Arakelov inequalities. More precisely, when one replaces S by a suitable tower of finite covers $\pi_n : S \to S$, $S_1 = S$, asymptotically the Euler-Poincaré characteristic χ_n of some differential complex arising from $\mathbb{V}$ is concentrated in dimension m. This yields $(-1)^m \lim_{n\to\infty} \frac{\chi_n}{[\pi_1(X):\pi_1(X_n)]} \geq 0$ and hence $(-1)^m\chi \geq 0$, $\chi = \chi_1$, by the multiplicative nature of the Euler-Poincaré characteristic. The cases of equality leads to characterization of certain totally geodesic compact complex submanifolds of Ω/Γ. We remark that these Chern class inequalities are in general *not* local, and the characterization of holomorphic geodesic cycles does not follow from the proof of Arakelov inequalities.

Accompanying the Arakelov inequalities and the characterization of certain totally geodesic holomorphic cycles in the case of equality, Eyssidieux [Eys1] obtained the first examples where gap rigidity in the Zariski topology holds for pairs (Ω, Ω') where Ω is irreducible. His results are for the more general context of period domains. In Eyssidieux [Eys2] he produced tables of lists of pair (Ω, Ω') of bounded symmetric domains where the methods of [Eys1] apply. Gap rigidity in the Zariski topology is obtained as follows. The relevant Euler-Poincaré characteristics can be computed using Gauss-Manin complexes, which are complexes of homomorphisms of holomorphic vector bundles. When the holomorphic tangent space at each point is in some precise sense generic, we have an exact sequence of holomorphic vector bundles, yielding the vanishing of the Euler-Poincaré characteristic in question, and implying that $S \subset \Omega/\Gamma$ is totally geodesic, by the characterization of the equality case of Arakelov inequalities.

The original Arakelov inequality is the case where Ω is the Siegel upper half-plain $\mathcal{H}_n$. In this case the holomorphic tangent bundle $T_{\mathcal{H}_n}$ is canonically isomorphic to S^2V, where V stands for the universal vector bundle on $\mathcal{H}_n$. For a local holomorphic curve $E \subset \mathcal{H}_n$ at $x \in K$ we have naturally a linear map $T_x(E) \otimes V_x^* \to V_x$, i.e., $V_x^* \to V_x \otimes K_x$, where K denotes the canonical line bundle of E. For a torsion-free discrete subgroup $\Gamma \subset \mathrm{Aut}(\mathcal{H}_n)$ and a holomorphic curve $C \subset X := \mathcal{H}_n/\Gamma$ we obtain thus a homomorphism $\varphi : V_C^* \to V_C \otimes K_C$ where V_C stands for the induced universal vector bundle on C, and K_C is the canonical line bundle. On the Siegel space $\mathcal{H}_n$ at $x \in \mathcal{H}_n$ the tangent space can be identified with the space of symmetric n-by-n matrices with complex coefficients. The rank of the matrix is invariant under the action of $\mathrm{Aut}(\mathcal{H}_n)$ on $\mathcal{H}_n$. As it turns out $0 \to V_C^* \xrightarrow{\varphi} V_C \otimes K_C \to 0$ is precisely the Gauss-Manin complex, and this complex is an exact sequence if and only if each tangent space $T_x(C)$ is generated by a vector of rank n.

(3.5) While the approach of Eyssidieux [Eys1, 2] provides an ample supply of examples and yields at the same time the deeper characterization of totally geodesic holomorphic cycles by the equality case of Arakelov inequalities, there are also many examples which do not fall into the Hodge-theoretic setting given there. In Mok [Mok4], partly motivated by a rigidity problem for holomorphic mappings ([3.6, Theorem 1] here) we considered gap rigidity in the Zariski topology in the setting of intersection theory using the Poincaré-Lelong equation. For Ω an irreducible bounded symmetric domain and $\Omega' \subset \Omega$ of dimension 1 we completely determined the cases where gap rigidity in the Zariski topology holds for (Ω, Ω').

DEFINITION (CHARACTERISTIC CODIMENSION). Let Ω be an irreducible bounded symmetric domain of rank $r \geq 2$, G be the identity component of the automorphism group of Ω, and o be any reference point on Ω. Denote by $\mathcal{S}_o \subset \mathbb{P}T_o(\Omega)$

the $(r-1)-th$ characteristic subvariety of Ω, also called the maximal (highest) characteristic subvariety, as given in (2.1). The positive integer $\kappa(\Omega) :=$ codim($\mathcal{S}_o$ in $\mathbb{P}T_o(\Omega)$) will be called the characteristic codimension of Ω.

In order to apply the Poincaré-Lelong equation to deduce gap rigidity for curves, we require the ambient domain Ω to be of characteristic codimension 1. We enumerate here such bounded symmetric domains.

Complete list of Ω with $\kappa(\Omega) = 1$:

(1) Ω of Type $\mathrm{I}_{m,n}$ with $m = n > 1$;
(2) Ω of Type II_n with n even, $n \geq 4$;
(3) Ω of Type III_n, $n \geq 3$;
(4) Ω of Type IV_n, $n \geq 3$;
(5) Ω of Type VI (the 27-dimensional exceptional domain pertaining to E_7).

As it turns out, from the classification given above, Ω is of characteristic codimension 1 if and only if it is a tube domain. In [Mok4] we proved

THEOREM 10 (MOK [MOK4, 2002]). *Let Ω be an irreducible bounded symmetric domain, and $\Gamma \subset \mathrm{Aut}(\Omega)$ be torsion-free discrete subgroup, $X := \Omega/\Gamma$. Denote by $\mathcal{S}_x \subset \mathbb{P}T_x(\Omega)$ the maximal characteristic subvariety and $\widetilde{\mathcal{S}_x} \subset T_x(\Omega) - \{0\}$ its homogenization. Let $f : C \to X$ be an immersed compact holomorphic curve. Suppose $q(\Omega) = 1$ and, for every point $x \in C$, the tangent space $T_x(C)$ is spanned by some generic vector η, i.e. $\eta \notin \widetilde{\mathcal{S}_x}$. Then, $f : C \to X$ is totally geodesic.*

REMARKS. For irreducible bounded symmetric domains Ω, Theorem 1 is optimal for the characterization of $(\Omega, D; j)$ exhibiting gap rigidity in the Zariski topology in the case where D is 1-dimensional. In fact, gap rigidity in the Zariski topology holds true *if and only if* $q(\Omega) = 1$ and D is the diagonal of a maximal polydisk.

PROOF OF THEOREM 10. Assume that $q(\Omega) = 1$. Then, there exists a locally homogeneous divisor $\mathcal{S} \subset \mathbb{P}T_X$ corresponding to the set of projectivizations of non-generic tangent vectors. We have tautologically $\mathcal{S} = \{s = 0\}$ for some $s \in \Gamma(X, [\mathcal{S}])$. Denote by $\pi : \mathbb{P}T_X \to X$ the canonical projection, and by L the tautological line bundle over $\mathbb{P}T_X$. Let M be the compact dual of Ω and identify Ω as a domain on M by the Borel embedding. On M we have analogously $\mathcal{S}_M \subset \mathbb{P}T_M$. Denote by $\rho : \mathbb{P}T_M \to M$ the canonical projection, and by $\mathcal{O}(1)$ the positive generator of the $\mathrm{Pic}(M) \cong \mathbb{Z}$. We have $\mathrm{Pic}(\mathbb{P}T_M) \cong \mathbb{Z}^2$, generated by $\rho^*\mathcal{O}(1)$ and the tautological line bundle L_M over $\mathbb{P}T_M$, and a curvature computation in [Mok4] shows that in fact

$$[\mathcal{S}_M] \cong L_M^{-r} \otimes \rho^*\mathcal{O}(2) . \tag{1}$$

Denote by E the negative locally homogeneous line bundle on X dual to $\mathcal{O}(1)$ on M and write r for rank(Ω). By duality we have

$$[\mathcal{S}] \cong L^{-r} \otimes \pi^* E^2 . \tag{2}$$

Let $f : C \to X$ be a compact holomorphic curve and denote by $\widehat{C} \subset \mathbb{P}T_M$ the tautological lifting of C. Observe that if $C \subset X$ is totally geodesic and of diagonal type, then $[T_x(C)] \notin \mathcal{S}_x$ for any $x \in C$, so that $\mathcal{S} \cap \widehat{C} = \emptyset$ and $[\mathcal{S}] \cdot \widehat{C} = 0$. If $[T_x(C)] \notin \mathcal{S}_x$ for a generic $x \in C$, then,

$$[\mathcal{S}] \cdot \widehat{C} \geq 0 . \tag{3}$$

On the other hand, the intersection number can be computed from the Poincaré-Lelong equation

$$\sqrt{-1}\partial\overline{\partial}\log\|s\|^2 = rc_1(L,\hat{g}_0) - 2\pi^*c_1(E,h_0) + [\mathcal{S}]\ , \tag{4}$$

so that

$$[\mathcal{S}]\cdot\widehat{C} = r\int_{\widehat{C}} c_1(L,\hat{g}_0) - 2\int_C c_1(E,h_0) = \frac{r}{2\pi}\int_C \mathrm{Ric}(C,g_0|_C) - 2\int_C c_1(E,h_0)\ . \tag{5}$$

When $f : C \to X$ is an immersed totally geodesic holomorphic curve of diagonal type, the Gauss curvature is everywhere equal to $\frac{-2}{r}$. In general, for $f : C \to X$ an immersed holomorphic curve, by the Gauss equation we have

$$\text{Gauss curvature} \leq \frac{-2}{r}\ , \tag{6}$$

where equality holds if and only if (a) C is tangent to a local totally geodesic holomorphic curve of diagonal type; (b) the second fundamental form vanishes. Under the assumption of Theorem 10, $\mathcal{S}\cap\widehat{C} = \emptyset$, and hence $[\mathcal{S}]\cdot\widehat{C} = 0$. From (3) and (4) it follows that C must be totally geodesic of diagonal type. □

REMARKS. The divisor $[\mathcal{S}] \subset \mathbb{P}T_X$ is in general not numerically effective. Let $C \subset X$ be a totally geodesic curve descending from a minimal disk (i.e., C is dual to a minimal rational curve). Then, $[\mathcal{S}]\cdot\widehat{C} > 0$. On the other hand, let $C^\flat$ be a holomorphic lifting of C such that for $[\beta] \in C^\flat$ lying over x with $T_x(C) = \mathbb{C}\alpha$, we have $R_{\alpha\overline{\alpha}\beta\overline{\beta}} = 0$. Then, $L|_{C^\flat} \cong \mathcal{O}$, $[\mathcal{S}]\big|_{C^\flat} \cong \pi^*E^2$, and hence $[\mathcal{S}]\cdot C^\flat < 0$.

THEOREM 11 (MOK [MOK4, 2002]). *Let Ω be an irreducible bounded symmetric domain of rank $r \geq 2$ and of characteristic codimension 1. Let Γ be a torsion-free cocompact discrete group of biholomorphic automorphisms of Ω and write $X := \Omega/\Gamma$. Let Z be a complex manifold carrying a continuous complex Finsler metric of nonpositive curvature. Then, any nonconstant holomorphic map $f : X \to Z$ is necessarily an immersion at some point.*

PROOF. Proposition 1 of (2.3) includes in particular the generalization of Finsler metric rigidity [(2.2), Theorem 3] for continuous complex Finsler metrics of nonpositive curvature. When the argument of Finsler rigidity is applied to k-th characteristic bundles, $1 \leq k \leq r-1$, $r = \text{rank}\ (\Omega)$, the arguments in [Mok4, Proposition 4, last paragraphs] shows that, in the notations of Theorem 3 here, $h = e^u g$ with u being constant on the image of every G-orbit of a non-generic vector under the natural projection $\pi : \mathbb{P}T_\Omega \to \mathbb{P}T_X$. Denote by $F : \Omega \to \widetilde{Z}$ the lifting of $f : X \to Z$ to universal covering manifolds. If $dF(\gamma) = 0$ for some non-generic nonzero vector γ. Then dF vanishes on the G-orbit of γ, and hence on the $G^{\mathbb{C}}$-orbit of γ, since the kernel of dF is complex-analytic. This forces a contradiction since $\widetilde{\mathcal{M}}(X)$ is in the topological closure of each stratum $\mathcal{S}_k(X) - \mathcal{S}_{k-1}(X), 1 \leq k \leq r-1$. We have thus proven that $\mathbb{P}(\mathrm{Ker}(df)) \cap \mathcal{S} = \emptyset$.

Let now $X = \Omega/\Gamma$ where Ω is of characteristic codimension 1, $\Gamma \subset \mathrm{Aut}(\Omega)$ is torsion-free and cocompact, and $f : X \to Z$ be a nonconstant holomorphic mapping into a complex manifold Z equipped with a continuous complex Finsler metric θ of nonpositive curvature. Suppose $f : X \to Z$ is of maximal rank $< \dim(X)$. Then, for a general point y of $f(X) := Y$, the fiber $f^{-1}(y)$ is a smooth p-dimensional manifold for some $p \geq 1$, $p = \dim(X) - \dim(Y)$. For $x \in X$, let F^x be the fiber $f^{-1}(f(x))$. For $x \in X$ generic, $\mathbb{P}T_x(F^x) \cong \mathbb{P}^{p-1}$ must be disjoint from

$\mathcal{S}_x \subset \mathbb{P}T_x$. Since $\mathcal{S}_x \subset \mathbb{P}T_x(X)$ is a hypersurface, we must have $p = 1$, so that each irreducible component F_k^x of F^x must lift tautologically to $\widehat{F}_k^x \subset \mathbb{P}T(X)$ such that $\widehat{F}_k^x \cap \mathcal{S} = \emptyset$. By Theorem 9, $F_k^x \subset X$ is a compact totally geodesic holomorphic curve of the diagonal type. For such a compact holomorphic curve the normal bundle $N_{F_k^x|X}$ is strictly negative. Varying x we obtain a positive-dimensional holomorphic family of such compact holomorphic curves. On the other hand, since $N_{F_k^x|X}$ is strictly negative, $F_k^x \subset X$ is an exceptional curve, and must be the unique compact holomorphic curve on some tubular neighborhood U of F_k^x in X, a plain contradiction. This means that $f : X \to Z$ must be of maximal rank $= \dim(X)$ at some point, i.e., f is generically finite onto its image, as desired.

(3.6) In order to formulate results on gap rigidity in the Zariski topology for higher-dimensional complex submanifolds we discuss here some further basic facts on holomorphic embeddings between bounded symmetric domains. To start with, we collect some basic facts, notions and notations on bounded symmetric domains beyond those given in (1.4), as follow.

Bounded Symmetric Domains

Let Ω be a bounded symmetric domain, not necessarily irreducible. As in (1.4) we write $\Omega = G/K$ and use the same notations and conventions there, except that Ω may be reducible. Let $\mathfrak{g} = \mathfrak{k} \oplus \mathfrak{m}$ be the Cartan decomposition. Choose $H_0 \in \mathfrak{z} :=$ Centre $(\mathfrak{k})$ such that $ad(H_0)^2 = \theta$, where $ad(H_0)$ defines an integrable almost complex structure on Ω. We have the decomposition $\mathfrak{m}^{\mathbb{C}} = \mathfrak{m}^+ \oplus \mathfrak{m}^-$ into $\pm i$-eigenspaces of $ad(H_0)$, $\mathfrak{m}^{\mathbb{C}} := \mathfrak{m} \otimes_{\mathbb{R}} \mathbb{C}$. Write $o = eK$. We call $(\mathfrak{g}, H_0)$ a semisimple Lie algebra of the Hermitian and noncompact type. We have the following basic notions regarding embeddings between bounded symmetric domains.

Embedding of Bounded Symmetric Domains

Let $(\mathfrak{g}', H_0')$, $(\mathfrak{g}, H_0)$ be semisimple Lie algebras of the Hermitian and noncompact type, and $\rho : \mathfrak{g}' \to \mathfrak{g}$ be a Lie algebra homomorphisms

- We say that $\rho : (\mathfrak{g}', H_0') \to (\mathfrak{g}, H_0)$ is an (H_1)-homomorphism if and only if
$$ad(H_0) \circ \rho = \rho \circ ad(H_0') \ .$$
- We say that $\rho : (\mathfrak{g}', H_0') \to (\mathfrak{g}, H_0)$ is an (H_2)-homomorphism if and only if
$$\rho(H_0') = H_0 \ .$$

An (H_1)-homomorphism induces a totally geodesic holomorphic embedding between bounded symmetric domains associated to $(\mathfrak{g}', H')$ resp. $(\mathfrak{g}, H_0)$. (H_2)-homomorphisms are (H_1) (cf. Satake [Sa, pp 83-88]). In 1965 Satake [Sa] classified all (H_2)-embeddings into classical domains. In 1967 Ihara [Iha] obtained the full classification of (H_2)-embeddings.

For $\Omega = G/K$ as in the above, a G-invariant Kähler metric g_0 can be determined on Ω by the Killing form. When Ω is irreducible, g_0 is Kähler-Einstein, and the Einstein constant is fixed. When Ω is irreducible, $\dim(\Omega) = n$, writing $\{e_1, \ldots, e_n\}$ for an orthonormal basis of $\mathfrak{m}^+ = T_0(\Omega)$, and $\sum(\mathfrak{m}^+) = \sqrt{-1}\sum_{i=1}^{n}[e_i, \overline{e}_i]$, we have $\sum(\mathfrak{m}^+) = \sqrt{-1}c_\Omega H_0$ for some $c_\Omega \in \mathbb{R}$. When Ω is reducible, we slightly modify (H_2)-homomorphisms to introduce the notion of (H_3)-homomorphisms, as follows.

(H_3)-*Embeddings*

Let $\rho : (\mathfrak{g}', H_0') \to (\mathfrak{g}, H_0)$ be an (H_1)-embedding corresponding to a totally geodesic holomorphic embedding $j : \Omega' \to \Omega$. Write $\Omega' = \Omega_1' \times \cdots \times \Omega_a'$ for the decomposition of Ω' into a Cartesian product of irreducible bounded symmetric domains Ω_k', $1 \leq k \leq a$. Define positive constants $d_{\Omega_k',\Omega}$ by $g_0^{\Omega}|_{\Omega_k'} = d_{\Omega_k',\Omega} \cdot g_0^{\Omega'}$

- We say that ρ is an (H_3)-embedding if and only if

$$\rho\Big(\sum_{k=1}^{N} c_{\Omega_k'} d_{\Omega_k',\Omega} H_{0k}'\Big) \in \mathbb{R}H_0 \ .$$

If $\rho : (\mathfrak{g}', H_0') \to (\mathfrak{g}, H_0)$ is an (H_3)-embedding, we also call $j : \Omega \to \Omega'$ an (H_3)-embedding, or a totally geodesic holomorphic embedding of diagonal type. The lemma below follows readily from the definitions (cf. [EM2, Lemma 3]).

LEMMA 5. *(H_3)-embeddings are (H_2). An (H_2)-embedding is (H_3) if and only if $g_0^{\Omega}|_{\Omega'}$ is Einstein.*

In order to generalize Theorem 9 to the case where $\Omega' \subset \Omega$ is of higher dimension, we will need to make use of results from geometric invariant theory (GIT), especially involving the moment map. We refer the reader to [EM2, (2.1)-(2.3)] for details on the arguments involving Geometric Invariant Theory, and to the standard reference Mumford-Fogarty-Kirwan [MFk] for the necessary background.

THEOREM 12 (EYSSIDIEUX-MOK [EM2, 2004]. *Let Ω be an irreducible bounded symmetric domain. Let $j : \Omega' \to \Omega$ be a totally geodesic holomorphic embedding of the diagonal type, $\dim(\Omega') = n'$, $\dim(\Omega) = n$. Then, there exists a nonempty K-invariant hypersurface $\mathcal{H}_o \subset \mathrm{Gr}(n', \mathbb{C}^n)$ for which the following holds.*

(1) *$[T_o(\Omega')] \notin \mathcal{H}_o$.*
(2) *Write $\mathcal{H} \to X = \Omega/\Gamma$ for the corresponding locally homogeneous holomorphic subbundle of $\pi : \mathbb{P}T_X \to X$. Then, for any n'-dimensional immersed compact complex submanifold $f : S \to X$ such that for $x \in S$, $[T_x(S)] \notin \mathcal{H}_x$, the compact complex manifold $S \subset X$ is totally geodesic.*

SKETCH OF PROOF. Denote by $\mathfrak{l} \subset \mathfrak{k}$ the semisimple part of $\mathfrak{k}$, $\mathfrak{l} = [\mathfrak{k}, \mathfrak{k}]$. Denote by $\kappa : \mathfrak{k} \to \mathfrak{l}^*$ the complex linear map induced by the Killing form. For any $E \in \mathrm{Gr}(n', T_0(\Omega)) = \mathbb{G}$, choose an orthonormal basis $\{e_i\}$ and set

$$\mu(E) = \kappa\Big(- \sum_i [e_i, \bar{e}_i]\Big) \ ,$$

where $\{e_i\}_{1 \leq i \leq n}$ stands for an orthonormal basis of E. When $j : \Omega' \to \Omega$ is a totally geodesic holomorphic embedding of the diagonal type, we have $\mu([T_o(\Omega')]) = 0$. The moment map of the adjoint action of $U(n)$ on $M_n(C)$ is given by $A \mapsto [A, \overline{A}^t]$. From this one deduces that μ is the moment map for the Hamiltonian action of K on the Kähler manifold $\mathbb{G}$ (cf. [EM2, Section 2.3]). The Hamiltonian action extends to a linearizable action of $K^{\mathbb{C}}$ on $\mathbb{G}$. The existence of a K-invariant hypersurface $\mathcal{H}_o \subset \mathrm{Gr}(n', \mathbb{C}^n)$ satisfying (1) amounts to saying that $[T_o(\Omega')]$ is a semistable point of the $K^{\mathbb{C}}$ action on $\mathbb{G}$ in the sense of Geometric Invariant Theory (GIT). GIT-semistables point of $\mathbb{G}$ are points whose $K^{\mathbb{C}}$-orbits meet $\mu^{-1}(0)$. In particular, $\mu^{-1}(0)$ are GIT-semistable. In other words, there exists a K-invariant hypersurface $\mathcal{Z}_o \subset \mathrm{Gr}(n', T_o(\Omega))$ such that $[T_o(\Omega')] \notin \mathcal{Z}_o$.

As in the proof of Theorem 10 we denote by M the compact dual of Ω and identify Ω as an open subset of M by the Borel embedding. Denote by $\mathbb{G}_M$ the

Grassmann bundle of n'-planes on M, $\rho : \mathbb{G}_M \to M$ the canonical projection, and by L_M the tautological line bundle on $\mathbb{G}_M$. Let $\mathcal{Z}_M \subset \mathbb{G}_M$ be the $G^{\mathbb{C}}$-invariant hypersurface, corresponding to $\mathcal{Z}_o \subset \mathrm{Gr}(n', T_o(\Omega))$. ¿From $\mathrm{Pic}(\mathbb{G}_M) \cong \mathbb{Z}^2$ the divisor line bundle $[\mathcal{Z}_M]$ must be of the form $L_M^{-m} \otimes \rho^*\mathcal{O}(\ell)$ for some positive integers m and ℓ, where $\mathcal{O}(1)$ denotes the positive generator of $\mathrm{Pic}(M)$. Thus, $\mathcal{Z}_M$ is the zero set of some $s \in \Gamma(\mathbb{G}_M, L_M^{-m} \otimes \rho^*\mathcal{O}(\ell))$ which is a $G^{\mathbb{C}}$-invariant sections. On $\Omega \subset M$, s is G-invariant. Denote by E_o the holomorphic negative line bundle on Ω dual to $\mathcal{O}(1)$ on M and h_o be a canonical Hermitian metric on E_o. Normalize the canonical Kähler metric g_o on Ω so that $c_1(E, h) = \omega$ for the induced Kähler form ω on X. In the notations of the proof of Theorem 10, we have

$$\frac{\sqrt{-1}}{2\pi}\partial\bar{\partial}\log\|s\|^2 = mc_1(L, \hat{g}) - \ell c_1(\pi^*E, \pi^*h) + [\mathcal{Z}_\Omega]\ . \tag{1}$$

As opposed to the case of Theorem 10 for curves, it is not clear what m and ℓ are. For our purpose it is the ratio between m and ℓ and not their precise values which concerns us. By Borel [Bo2, 1963], there exists $\Gamma' \subset \mathrm{Aut}(\Omega')$ such that $S_0 = \Omega'/\Gamma'$ is compact. Since $[T_x(S_0)] \notin \mathcal{Z}_{X,x}$ for any $x \in S_0$, integrating over the lifting $\widehat{S}_0$ of S_0 to $\mathbb{G}_X\big|_{S_0}$, we have

$$\begin{aligned} 0 &= \int_{\widehat{S}_0} \big(mc_1(L, \hat{g}) - \ell c_1(\pi^*E, \pi^*h)\big) \wedge (\pi^*\omega)^{n'-1} \\ &= \int_{S_0} \big(mc_1(K_{S_0}^{-1}, \det(g|_{S_0})) - \ell c_1(E, h)\big) \wedge \omega^{n'-1} \\ &= \int_{S_0} \Big(\frac{m}{2\pi}\mathrm{Ric}(g|_{S_0}) - \ell c_1(E, h)\Big) \wedge \omega^{n'-1} \\ &= \int_{S_0} \Big(\frac{m}{2\pi n'}K(g|_{S_0}) + \ell\Big)\omega^{n'}\ , \end{aligned} \tag{2}$$

where K denotes scalar curvature. By local homogeneity the last integrand vanishes identically on S_0. In other words,

$$\frac{m}{2\pi n'}K(g_0|_{\Omega'}) + \ell \equiv 0\ . \tag{3}$$

Suppose now $S \subset X = \Omega/\Gamma$ as in the hypothesis. We have $\widehat{S} \cap \mathcal{Z}_X = \emptyset$, so that

$$\int_S \Big(\frac{m}{2\pi n'}K(g|_S) + \ell\Big)\omega^{n'} = 0\ . \tag{4}$$

Define $\Sigma : \mathrm{Gr}\big(n', T_0(\Omega)\big) \to \mathfrak{k}$ by

$$\Sigma(E) = \sqrt{-1}\sum_{i=1}^{n'}[e_i, \bar{e}_i]\ , \tag{5}$$

where $\{e_i\}$ is any orthonoromal basis. $\|\Sigma(E)\|$ is a minimum if $\Sigma(E) \in \mathfrak{z}$, thus whenever $E = T_0(\Omega')$, where $\Omega' \hookrightarrow \Omega$ is (H_3). Now

$$K\big(g_0|_{\Omega'}\big) = -C\big\|\Sigma(T_0(\Omega'))\big\|^2 \tag{6}$$

for a universal constant C. For every $x \in S$ by the Gauss equation

$$K(g|_S)_x = -C\big\|\Sigma(T_xS)\big\|^2 - \|\sigma_x\|^2 \leq K\big(g_0\big|_{\Omega'}\big) \tag{7}$$

where σ is the second fundamental form. Comparing with (1) and (2) we get

$$K(g|_S)_x = K\big(g_0\big|_{\Omega'}\big) , \quad \sigma_x \equiv 0 . \tag{8}$$

In particular, $f : S \to X$ is a totally geodesic immersion, as desired. □

Tables of pairs of bounded symmetric domain (Ω, Ω') exhibiting gap rigidity in the Zariski topology and arising from (H_3)-embeddings i.e., embedding of the diagonal type, can be read from [EM2, (2.6)]. They are based on Satake [Sa] and Ihara [Iha]. Explicit examples of gap pairs (Ω, Ω') in which a $K^{\mathbb{C}}$-invariant hypersurface $\mathcal{H}_o$ in the Grassmannian $\mathbb{G}$ is determined, are given in [EM2, (2.5)]. While such hypersurfaces are not in general unique, it is interesting to note that in these examples $\mathcal{H}_o$ is a hypersurface of degree equal to rank(Ω). While in Theorem 11 the arguments leading to GIT-semistability are existential, it is interesting to answer the following question.

QUESTION. Let Ω be an irreducible bounded symmetric domain, and (Ω, Ω') be a gap pair in the Zariski topology arising from an embedding of the diagonal type.

(1) Is it possible, in terms of Killing forms or other Lie-algebraic data, to determine an explicit example of a $K^{\mathbb{C}}$-invariant hypersurface $\mathcal{H}_o \subset \mathbb{G}$ such that $[T_o(\Omega')] \notin \mathcal{H}_o$.
(2) Can one always find $\mathcal{H}_o \subset \mathbb{G}$, either by construction as in (1) or by existential arguments from Geometric Invariant Theory, to be of degree equal to rank(Ω)?

(3.7) In (3.3)for any bounded symmetric domain D we have the example $(D^k, D; \delta)$, $k > 1$, where δ stands for the diagonal embedding, for which gap rigidity holds in the Zariski topology. In this case the proof makes use of the uniqueness of Kähler-Einstein metrics on compact Kähler manifolds with ample canonical line bundle. There is a more interesting situation, where the amibient domain is irreducible, in which we can deduce gap rigidity in the Zariski topology using Kähler-Einstein metrics. This is the case concerning holomorphic quadric structures, which corresponds to the pair (D_n^{IV}, D_k^{IV}), $k \geq 3$, for which gap rigidity in the Zariski topology was established in [Mok4] using Hermitian metric rigidity ([Mok1]) and the existence of Kähler-Einstein metrics, as applied in Kobayashi-Ochiai [KO] in the context of G-structures modelled on irreducible bounded symmetric domains of rank ≥ 2. This case for Type-IV domains actually subsumes under the method using the Poincaré-Lelong equation as explained in (3.6), but the original method of proof in [Mok2] may be relevant to more general situations. We will sketch here the argument in the quadric case. Recall that for an n-dimensional complex manifold Z, and a linear subgroup G of $GL(n; \mathbb{C})$, by a smooth (resp. holomorphic) G-structure we mean a smooth (resp. holomorphic) reduction of the holomorphic frame bundle, with structure group $GL(n; \mathbb{C})$, to G. In the case of the Grassmannian $G(p, q)$ or quotients of its noncompact duals, the Type-I domains $D^I_{p,q}$, we have G-structures corresponding to the holomorphic splitting of the tangent bundle $T = U \otimes V$ into a tensor product of universal bundles U resp. V, of rank p resp. q, which is a nontrivial restriction when $p, q \geq 2$. In the case of G-structures modelled on the hyperquadric a holomorphic G-structure is nothing other than a holomorphic conformal structure, also called a holomorphic quadric structure, which is equivalently given by a holomorphic section θ of $S^2 T^*_Z \otimes L$, where L is some holomorphic line bundle, such that θ defines (up to scalar multiples) a non-degenerate symmetric biliner form on each tangent space $T_x(Z)$.

Argument using G-structures in the quadric case

Consider the case of (D_n^{IV}, D_k^{IV}), $k \geq 3$. Let $S \subset D_n^{IV}/\Gamma$ be a k-dimensional compact complex manifold. D_m^{IV}, $m \geq 3$, is an irreducible bounded symmetric domain of rank 2 and of characteristic codimension 1. Its dual manifold is $\mathbf{Q}^m$, the m-dimensional hyperquadric on which the (minimal) characteristic subbundle $\mathcal{M}(\mathbf{Q}^m) \subset \mathbb{P}T_{\mathbf{Q}^m}$ consists precisely of projectivizations of null vectors. Restricting to the domain D_m^{IV} and in the notations of [(3.5), Proof of Theorem 9], there is an $\mathrm{Aut}(D_m^{IV})$-invariant section $s \in \Gamma(L^{-2} \otimes \pi^* E^2)$, which can be interpreted as a twisted holomorphic quadratic form on the tangent bundle. In the situation $S \subset X := D_n^{IV}/\Gamma$ we are considering taking $m = n$ and taking s to be the $\mathrm{Aut}(D_n^{IV})$-invariant section, by descending to X and restricting to S we obtain a twisted holomorphic quadratic form θ on S. If we require this θ to be non-degenerate on S, we obtain a holomorphic quadric structure on S. Since S carries by restriction from X a Kähler metric of strictly negative curvature, its canonical bundle is ample, and by Yau [Yau] there exists on X a Kähler-Einstein metric. By Kobayashi-Ochiai [KO], any compact Kähler-Einstein manifold carrying a holomorphic G-structure modelled on an irreducible bounded symmetric domain D of rank ≥ 2 must be biholomorphic to a quotient of D. (We say that the holomorphic G-structure is integrable.) Thus, S is biholomorphically isomorphic to some quotient Z of D_k^{IV}, $k \geq 3$, and the inclusion $S \subset X$ is then a holomorphic embedding of Z into X. By Hermitian metric rigidity it follows that S is totally geodesic in X, proving gap rigidity.

The proof of gap rigidity in relation to quadric structures suggests that there may be more general situations about gap rigidity concerning G-structures modelled on irreducible bounded symmetric domains of rank ≥ 2. For instance, when (Ω, Ω') is a pair of irreducible bounded symmetric domains such that Ω' and hence Ω are of rank ≥ 2, and when the normal bundle of Ω' in Ω does not have a nontrivial flat direct summand (as in [(3.4), Conjecture]) one may ask whether gap rigidity holds for (Ω, Ω'). The case of $(D_n^{\mathrm{IV}}, D_k^{IV})$; $n > k \geq 3$; belongs to this situation, where we have given a proof of gap rigidity in the Zariski topology, both by using the Poincaré-Lelong equation and by using holomorphic quadric structures. However, in many cases, e.g., $(D_{m,n}^I, D_{p,n}^I)$, where $m \geq n \geq 2$, and $m > p \geq 2$, the geometric assumption as in the Conjecture is satisfied while it is not possible to apply the Poincaré-Lelong equation. In these cases it is tempting to make use of the approach using G-structures modelled on Ω'. In this situation we are content with considering gap rigidity in the complex topology, i.e., we consider almost geodesic compact complex submanifolds S of quotients X of Ω by discrete torsion-free subgroups Γ of automorphisms, where S is locally modelled on $\Omega' \hookrightarrow \Omega$.

In the case of $(D_n^{\mathrm{IV}}, D_k^{IV})$, $n > k \geq 3$, the canonical quadric structures from the ambient domain restrict and descend to the compact complex manifold S to give a holomorphic quadric structure, and we resort to the use of Kähler-Einstein metrics (Kobayashi-Ochiai [KO]) to conclude that S is itself biholomorphically isomorphic to a quotient of D_k^{IV}. In the situation of pairs (Ω, Ω') that we now consider, in general it does not make sense to restrict holomorphic G-structures. For a bounded symmetric domain $\Omega = G/K$ (in standard notations) and hence its quotient manifolds we have canonical $K^{\mathbb{C}}$-structures. Write also $\Omega' = G'/K'$ (in standard notations), and from now on we write G for $K'^{\mathbb{C}}$. (G is just a default notation and has nothing to do with G or G'. When $S \subset X = \Omega/\Gamma$ is almost geodesic and modelled on Ω, Ω', lifting to the universal covering domain locally we can make use of approximating totally geodesic complex submanifolds conjugate to Ω' to define holomorphic G-structures. Although *a priori* this cannot be done globally the local G-structures can be pieced together to yield smooth G-structures which are in some sense almost holomorphic. One way of piecing the G-structures together is as follows. Write $N = \dim(\Omega), N' = \dim(\Omega')$. Consider on X the Grassmann bundle $\mathcal{G}$ of complex N'-planes in the holomorphic tangent spaces $T_x(X) \cong \mathbb{C}^N$. In $\mathcal{G}$ there is a smooth locally homogeneous subbundle $\mathcal{C} \subset \mathcal{G}$ whose fiber $\mathcal{C}_x$ over each $x \in \Omega$ consists of tangent N'-planes which lift to tangent spaces of totally geodesic complex submanifolds $\Sigma \subset \Omega$, where $\Sigma = \gamma(\Omega')$ for some $\gamma \in G$. Here and in what follows 'locally homogeneous' refers to the fact that the lifting to the universal covering domain is invariant under G. There is a locally homogeneous tubular neighborhood $\mathcal{U}$ of $\mathcal{C}$ in $\mathcal{G}$ which admits a smooth retraction ρ onto $\mathcal{C}$ respecting the canonical projection of $\mathcal{U}$ onto X. More precisely, for $\delta > 0$ let $\mathcal{U}_x \subset \mathcal{G}_x$ be the open subset of all points at a distance less than δ to $\mathcal{C}_x$, where distances are measured with respect to the Kähler metric on the Grassmannian $\mathcal{G}_x$ induced by the canonical Kähler-Einstein metric on X. For an N'-plane in $T_x(X)$, we define $\rho([E])$ to be the point on $\mathcal{C}_x$ at a shortest distance from $[E]$. ρ is well-defined and is a smooth retraction provided that $\delta > 0$ is sufficiently small. When $S \subset X$ is ϵ-geodesic and modelled on (Ω, Ω'), for ϵ sufficiently small the tangent bundle of S gives a holomorphic section of $\mathcal{G}|_S$ lying inside $\mathcal{U}|_S$. The smooth retraction then defines a smooth section μ of $\mathcal{C}|_S$ which is almost holomorphic. In a sense to be made more precise below, μ can then be used to define a smooth G-structure which is almost holomorphic.

Almost-holomorphic G-structures

It remains to formulate properly the notion of ϵ-holomorphic G-structures in such a way that when $S \subset X := \Omega/\Gamma$ is ϵ-geodesic and modelled on (Ω, Ω'), then it admits an ϵ'-holomorphic geodesic structure, where ϵ' tends to 0 as ϵ tends to 0. For the formulation we describe first of all some basic facts associated to holomorphic G-structures coming from irreducible bounded symmetric domains Ω. G $\subset GL(n;\mathbb{C})$ is reductive, and its highest weight orbit defines in the case of Ω and its dual manifold M precisely the minimal characteristic bundle as a subvariety of the projectivized tangent bundle. Since this definition is purely representation-theoretic, we can define the minimal characteristic bundle $\mathcal{M}(Z) \subset \mathbb{P}T_Z$ on any complex manifold Z admitting a holomorphic G-structure. The second symmetric power $S^2\mathbb{C}^n$ splits as a direct sum of precisely 2 irreducible G-representation spaces (cf. Borel [Bo1]), which on Z corresponds to a holomorphic splitting $S^2T_Z = H \oplus J$, where H_x is spanned by $\alpha \circ \alpha$ as α ranges over highest weight vectors at x. A smooth G-structure on a complex manifold Z is defined equivalently by a smooth subbundle $\mathcal{M}_Z \subset \mathbb{P}T_Z$ where each fiber $\mathcal{M}_x \subset \mathbb{P}T_x(Z)$ over $x \in Z$ corresponds to the highest weight orbit of the G-represenation on $\mathbb{C}^n$, while the variation of these minimal characteristic subvarieties is only smooth. We still have in this case the splitting $S^2T_Z = H \oplus J$ as a smooth complex vector bundle. This splitting is easily seen to be holomorphic if and only if the minimal characteristic bundle $\mathcal{M}_Z \subset \mathbb{P}T_Z$

is holomorphic. Consider the canonical projection map $\mu : S^2T_Z \to H \subset S^2T_Z$ as a smooth endomorphism of the holomorphic vector bundle S^2T_Z. We can now measure the lack of holomorphicity of the G-structure by the failure of ν to be holomorphic, as follows.

DEFINITION. Let Z be a compact complex manifold with ample canonical line bundle, and denote by h the canonical Kähler-Eintein metric on Z of Ricci curvature -1. We say that Z admits an ϵ-holomorphic G-structure if it admits a smooth G-structure, such that, for the smooth decompositionm $S^2T_Z = H \oplus J$, and the canonical bundle endomorphism $\nu : S^2T_Z \to H \subset S^2T_Z$, we have $\|\overline{\partial}\nu\| < \epsilon$ everywhere on Z.

In the case of ϵ-geodesic compact Kähler manifolds $S \subset X := \Omega/\Gamma$ modelled on (Ω, Ω'), where we use the notation S for the submanifold in place of the abstract manifold Z, and the notation Ω' to stand for a subdomain in the context of gap rigidity in place of the notation D, we can make use of the smooth section μ of $\mathcal{C}|_S$ over S to define a smooth G-structure, as follows. At each $x \in S$, $[T_x(S)] := \tau(x) \in \mathcal{C}_x$, and define $\mu(x) := \rho(\tau(x))$ for the smooth retraction $\rho : \mathcal{U} \to \mathcal{C}$ constructed. To define a smooth G-structure is equivalently to define a smooth minimal characteristic bundle on S modelled on Ω'. This can be done by pulling back the minimal characteristic subvarieties at $\rho(\tau(x))$, provided that we have at the same time a linear isomorphism λ_x from $T_x(S)$ to the vector space $E_x \subset T_x(X)$ defined by $\rho(\tau(x))$. Since $T_x(S)$ is sufficiently close to E_x, we can define $\lambda_x(\eta)$ to be the orthogonal projection of η to E_x, defined in terms of the canonical Kähler-Einstein on X. Finally, our notion of ϵ-holomorphic G-structure has the desired property that an ϵ-geodesic compact complex submanifold $S \subset X := \Omega/\Gamma$ modelled on (Ω, Ω') admits necessarily an ϵ'-holomorphic G-structure, where $\epsilon' = \epsilon'(\epsilon)$ tends to 0 as ϵ tend to 0, by an analogue of the convergence argument of [EM1, (1.1), Proposition 1]. (The latter is stated as [(3.2), Lemma 3] in the current article.)

REMARKS.

(1) In the case of quadric structures in the canonical decompostion $S^2T_Z = H \oplus J$, J is a complex line bundle. For holomorphic quadric structures T_Z is holomorphically isomorphic to a $T_Z^* \otimes L$ for some holomorphic line bundle L, and $J \subset S^2T_Z$ corresponds under the latter isomorphism to a holomorphic line subbundle of $S^2T_Z^* \otimes L^2$, which is spanned at every point by the twisted holomorphic non-degenerate quadratic form defining the quadric structure.

(2) The direct sum decomposition $S^2T_Z = H \oplus J$ is an isometric decomposition when the background metric h is Kähler-Einstein (cf Siu [Siu4, Proposition (1.6)]), which forces H to be invariant under holonomy. Consequently, the set of tangent vectors whose squares lie in H is invariant under holonomy. But this set corresponds precisely to the minimal characteristic bundle, proving the invariance of the latter under holonomy, and hence restricting the holonomy group at any $o \in Z$ to be contained in the subgroup of $GL(T_o(D))$ preserving $\mathcal{M}_o$, which corresponds precisely to G. It follows readily from Berger's characterization of Riemannian symmetry that Z must be uniformized by D if h is of negative Ricci curvature. This proves the result of Kobayashi-Ochiai [KO] mentioned in the above.

We have the following general conjecture regarding compact complex manifolds with ample canonical line bundles admitting almost holomorphic G-structures.

CONJECTURE. *Let D be an n-dimensional irreducible bounded symmetric domain of rank ≥ 2. Let $o \in D$ be any reference point and denote by $K \subset Aut(D)$ the isotropy subgroup of holomorphic isometries at o. The map $\Phi : K \to GL((T_o(D))$ defined by $\Phi(\gamma) = d\gamma(o)$ being injective, we identify K as a subgroup of $GL(T_o(D))$. Fixing an identification of $T_o(D)$ as a complex vector space with $\mathbb{C}^n$ we write* G $\subset$ *$GL(n;\mathbb{C})$ for the subgroup corresponding to the complexification $K^{\mathbb{C}} \subset GL(T_o(D))$. Let now Z be an n-dimensional compact complex manifold with ample canonical line bundle and write h for the unique Kähler-Einstein metric on Z of constant Ricci curvature -1. Let $\epsilon > 0$ and suppose Z admits an ϵ-holomorphic* G*-structure. Then, if ϵ is sufficiently small, Z admits a holomorphic* G*-structure and is hence uniformized by the bounded symmetric domain D.*

A positive resolution of the Conjecture would imply that (Ω, D) is a gap pair for any ambient bounded symmetric domain Ω, not necessarily irreducible, which contains a totally geodesic complex submanifold biholomorphically isomorphic to and identified with D. This is the case because of Hermitian metric rigidity [(2.1), Theorem 4]. For the verification of gap rigidity of (Ω, D) in the complex topology it is in fact sufficient to resolve the Conjecture in the affirmative under the additional assumptions that the curvature tensor is *a priori* bounded from above by a prescribed, fixed but arbitrarily small positive number, and that Z admits some finite covering space of injectivity radius at least equal to 1. For the case of $S \subset X = \Omega/\Gamma$ at hand the latter assumption is valid from the residual finiteness of the fundamental groups Γ and the pinching condition on second fundamental forms on Z, while the former assumption follows from solving the Monge-Ampere equation by the continuity method with estimates, given that S is ϵ-geodesic for a sufficiently small ϵ and modelled on (Ω, D), and the injectivity radius can be taken to be at least 1. We also note that while in [(3.2), Conjecture] we consider only the case of pairs of bounded symmetric domains (Ω, D) where the Hermitian holomorphic normal bundle of D in Ω admits no trivial isometric direct summand, without which the Gap Phenomenon has been shown to fail in general, it is not clear whether this condition is necessary when D is an irreducible bounded symmetric domain of rank ≥ 2. For instance, if we take Ω to be a product of $D \times U$, where U is an arbitrary bounded symmetric domain, and identify D with $D \times \{o\}$ for any reference point $o \in U$, then a compact complex manifold S modelled on (Ω, D) admits an integrable holomorphic G-structure (arising from D) by lifting to the universal cover and taking the canonical projection from $D \times U$ onto D, and must therefore be itself biholomorpically isometric to a compact quotient of the bounded symmetic domain D, as a consequence of which $S \subset \Omega/\Gamma$ must be totally geodesic by Hermitian metric rigidity.

References

[Bo1] A. Borel, *On the curvature tensor of the Hermitian symmetric manifolds.* Ann. Math., **71** (1960), 508–521.

[Bo2] A. Borel, *Compact Clifford-Klein forms of symmetric spaces*, Topology, **2** (1963), 111–122.

[CV] E. Calabi and E. Vesentini, *On compact locally symmetric Kähler manifolds*, Ann. Math., **71** (1960), 472–507.

[Eys1] P. Eyssidieux, *La caractéristique d'Euler du complexe de Gauss-Manin*, J. Reine Angew. Math., **490** (1997), 155–212.

[Eys2] P. Eyssidieux, *Kähler hyperbolicity and variations of Hodge structures*, in New Trends in Algebraic Geometry, Cambridge 1999, pp.71–92.

[EM1] P. Eyssidieux and N. Mok, *Characterization of certain holomorphic geodesic cycles on Hermitian locally symmetric manifolds of the noncompact type*, Ann. Math. Studies, **138**, Princeton 1995, pp.85–118.

[EM2] P. Eyssidieux and N. Mok, *On the validity or failure of gap rigidity for certain pairs of bounded symmetric domains*, Asian J. Math., **8** (2004), 773–794.

[Gu] R.C. Gunning, *The Eichler cohomology groups and automorphic forms*, Trans. Amer. Math. Soc., **100** (1961), 44–62.

[He] S. Helgason, *Differential Geometry, Lie Groups and Symmetric Spaces*, Academic Press, New York, 1978.

[Iha] S. Ihara, *Holomorphic imbeddings of symmetric domains*, J. Math. Soc. Japan, **19** 261-302; Supplement, ibid., 543–544.

[KO] S. Kobayashi and T. Ochiai, *Holomorphic structures modeled after compact hermitian symmetric spaces*, in Hano et al. (eds), Manifolds and Lie Groups, Papers in honor of Yozo Matsushima, Progr. Math. 14, Birkhäuser, Basel 1981, pp.207–221.

[Ma] Y. Matsushima, *On the first Betti number of compact quotient spaces of higher-dimensional symmetric spaces*, Ann. Math., **75** (1962), 312–330.

[Mar] G.A. Margulis, *Discrete Subgroups of Semisimple Lie Groups*, Ergebnisse der Mathematik und ihrer Grenzgebiete, 3. Folge, Band 17, Springer, Berlin-Heidelberg-New York 1991.

[Mok1] N. Mok, *Uniqueness theorems of Hermitian metrics of seminegative curvature on locally symmetric spaces of negative Ricci curvature*, Ann. Math., **125** (1987), 105–152.

[Mok2] N. Mok, *Metric Rigidity Theorems on Hermitian Locally Symmetric Manifolds*, Series in Pure Mathematics, World Scientific Singapore-New Jersey-London-Hong Kong, 1989.

[Mok3] N. Mok, *Aspects of Kähler geometry on arithmetic varieties*, Proceedings of Symposia in Pure Mathematics, **52** (1991), 335–396.

[Mok4] N. Mok, *Characterization of certain holomorphic geodesic cycles on quotients of bounded symmetric domains in terms of tangent subspaces*, Compos. Math., **132** (2002), 289–309.

[Mok5] N. Mok, *Extremal bounded holomorphic functions and an embedding theorem for arithmetic varieties of rank* ≥ 2, Invent. Math., **158** (2004), 1–31.

[MFK] D. Mumford, J. Fogarty and F. Kirwan, *Geometric Invariant Theory*, 3rd edition, Ergebnisse der Mathematik und ihrer Grenzgebiete, 3. Folge, Band 34, Springer, Berlin-Heidelberg-New York 1994.

[MT] N. Mok and W.-K. To, *Eigensections on Kuga families of abelian vareities and finiteness of their Mordell-Weil groups*, J. Reine Angew. Math., **444** (1993), 29–78.

[Sa] I. Satake, *Holomorphic imbeddings of symmetric domains into a Siegel space*, Amer. J. Math., **90** (1965), 425–461.

[Sh] T. Shioda, *On elliptic modular surfaces*, J. Math. Soc. Japan, **24** (1972), 20–59.

[Siu1] Y.-T. Siu, *The complex analyticity of harmonic maps and the strong rigidity of compact Kähler manifolds*, Ann. Math., **112** (1980), 73–111.

[Siu2] Y.-T. Siu, *Strong rigidity of compact quotients of exceptional bounded symmetric domains*, Duke Math. J., **48** (1981), 857–871.

[Siu3] Y.-T. Siu, *Complex-analyticity of harmonic maps, vanishing and Lefschetz theorems*, J. Diff. Geom., **17** (1982), 55–138.

[Siu4] Y.-T. Siu, *Uniformization in several complex variables*, in Contemporary Geometry, Univ. Ser. Math., Plemum, New York, 1991, pp.95–130.

[Siu5] Y.-T. Siu, *Lectures on Hermitian-Einstein Metrics for Stable Bundles and Kähler-Einstein Metrics*, DMV Seminar, Band 8, Birkhäuser, Basel-Boston 1987.

[Sil] A. Silverberg, *Mordell-Weil groups of generic abelian varieties*, Invent. Math., **81** (1985), 71–108.

[To] W.-K. To, *Hermitian metrics of seminegative curvature on quotients of bounded symmetric domains*, Invent. Math., **95** (1989) 559–578.

[Wo] J.A. Wolf, *Fine structure of Hermitian symmetric spaces*, in Geometry of Symmetric Spaces, ed. Boothby-Weiss, Marcel Dekker, New York 1972, pp.271–357.

[Zim] R.J. Zimmer, *Ergodic Theory and Semisimple Groups*, Monographs in Mathematics, Birkhäuser, Boston-Basel-Stuttgart, 1984.

[Yau] S.-T. Yau, *On the Ricci curvarture of a compact Káhler manifold and the complex Monge-Ampère equation, I*, Comm. Pure Appl. Math., **31**(1978), 339–411.

Department of Mathematics, The University of Hong Kong, Pokfulam Road, Hong Kong
E-mail address: nmok@hkucc.hku.hk

AMS/IP Studies in Advanced Mathematics
Volume **39**, 2007

The Second Main Theorem with Hypersurfaces over Function Fields

Min Ru

ABSTRACT. Let C be an irreducible nonsingular projective algebraic curve of genus g defined over an algebraically closed field k of characteristic 0. Let K be the function field of C. Let $f_0, \dots, f_n \in K$ be algebraically independent over k (i.e. there exists no homogeneous polynomial Q in $n+1$ variables with coefficients in k such that $Q(f_0, \dots, f_n) \equiv 0$). Given homogeneous polynomials $Q_1, \dots, Q_q$ in $n+1$ variables with coefficients in k, located in general position. Assume that $q \geq n+2$. In this paper, we establish an upper bound for the height function $h(f_0, \dots, f_n)$ in terms of the number of the zeros of $Q_1(f_0, \dots, f_n), \dots, Q_q(f_0, \dots, f_n)$.

1. Introduction

Let C be an irreducible nonsingular projective algebraic curve of genus g defined over an algebraically closed field k of characteristic 0. Let K be the function field of C. Let $f \in K$ (it can be regarded as a map $f : C \to \mathbb{P}^1$). Then we know that $\deg(f)$ equals the number of zeros (or poles) of f counted with multiplicities. In other words, the growth of f can be measured either in terms of the degree or in terms of the number of zeros of $f - a$ for $a \in k$. On the other hand, rather than the algebraic case above, Nevanlinna theory deals the transcendental case. It studies the holomorphic maps $f : \mathbb{C} \to \mathbb{P}^1(\mathbb{C})$. In particular, Nevanlinna's Second Main Theorem gives an upper bound of the characteristic(height) function of f(compare to the degree of f) in terms of the number of zeros of $f - a_j$ for $a_j \in \mathbb{C} \cup \{\infty\}$, $j = 1, \dots, q$, with $q \geq 3$, in the disk of radius r. In 1933, H. Cartan [Ca] generalized Nevanlinna's theorem to holomorphic curves into the complex projective space $\mathbb{P}^n(\mathbb{C})$ and established a defect relation for linearly nondegenerate holomorphic curves intersecting hyperplanes. Recently, the author [Ru2] extended H. Cartan's result and established a defect relation for algebraically nondegenerate holomorphic curves $f : \mathbb{C} \to \mathbb{P}^n(\mathbb{C})$ intersecting curvilinear hypersurfaces. This

---

1991 *Mathematics Subject Classification.* Primary 11j13, 11j25; Secondary 32H30.

Key words and phrases. Algebraic curves, Function field, Riemann-Hurwitz theorem, Second Main Theorem.

The author is supported in part by NSA grant H98230-05-1-0042. The United State Government is authorized to reproduce and distribute reprints notwithstanding any copyright notation hereon.

completely settles a long-standing conjecture of B. Shiffman. The purpose of this paper is to consider the analogy of above result in the function field case.

This paper is organized as follows. In section 2, we first establish a second main theorem for $f \in K$. The result is indeed a trivial consequence of the Riemann-Hurwitz theorem. In section 3, we extend this result to $f = (f_0, \ldots, f_n)$ where $f_0, \ldots, f_n \in K$. In section 4, we establish a general second main theorem for algebraic curves intersecting hyperplanes, located in general position. In section 5, we we establish a second main theorem for algebraic curves intersecting hypersurfaces, located in general position.

2. The Riemann-Hurwitz theorem and its consequence

Let C be an irreducible nonsingular projective algebraic curve of genus g defined over an algebraically closed field k of characteristic 0. Let K be the function field of C. Let $f \in K$ and let $a_1, \ldots, a_q \in k \cup \{\infty\}$ be distinct points. Let $S = f^{-1}(\{a_1, \ldots, a_q\}) \subset C$. Define the *ramification set*

$$r(S) := \sum_{x \in S} (\text{mult}_f(x) - 1),$$

where $\text{mult}_x(f)$ is the multiplicity of f at the point x. By the definition, we trivially have the following Lemma.

LEMMA 2.1. $q \deg(f) = |S| + r(S)$, *where* $\deg(f)$ *is the degree of* f.

The following simple version of The Second Main Theorem over function fields is a consequence of the Riemann-Hurwitz theorem.

THEOREM 2.2. (**The Second Main Theorem over function fields**) *Let C be an irreducible nonsingular projective algebraic curve of genus g defined over an algebraically closed field k of characteristic 0. Let K be the function field of C. Let $f \in K$ and let $a_1, \ldots, a_q \in k \cup \{\infty\}$ be distinct points. Let $S = f^{-1}(\{a_1, \ldots, a_q\}) \subset C$. Then*

$$(q-2)\deg(f) \le |S| + 2(g-1)$$

PROOF. Applying the Hurwitz theorem to the mapping $f : C \to k \cup \{\infty\}$ gives $r(S) \le 2\deg(f) + 2(g-1)$. Combining it with Lemma 2.1, we have $(q-2)\deg(f) \le |S| + 2(g-1)$. □

3. The theory of algebraic curves

In this section, we extend the above result to algebraic curves in $\mathbb{P}^n$.

Let C be an irreducible nonsingular projective algebraic curve of genus g defined over an algebraically closed field k of characteristic 0. Let K be the function field of C. Let $f : C \to \mathbb{P}^n$ be a linearly nondegenerate algebraic curve. We can lift f locally to k^{n+1}– that is, in a neighborhood of any point $P \in C$ we can find vector-valued regular function $\mathbf{f} = (f_0(z), ..., f_n(z))$ to k^{n+1} such that $f(z) = [f_0(z), ..., f_n(z)]$. Conversely, for $\mathbf{f} : C \to k^{n+1}$ any vector-valued function, the map $f(z) = [f_0(z), ..., f_n(z)]$ is well-defined even if $\mathbf{f}$ at isolated points. To see this, simply let z be a local coordinate centered around a zero P of $\mathbf{f}$ and let $e_P = -\min\{\text{ord}_P(f_i)\}$, then the map $\tilde{f}(z) = [z^{e_P} f_0(z), ... z^{e_P} f_n(z)]$ is well-defined and extends f.

Let $f : C \to \mathbb{P}^n$ be a linearly nondegenerate algebraic curve and let f be given locally by the vector function $\mathbf{f}(z) = (f_0(z), ..., f_n(z))$. We define the ith associate curve of f:

$$f^i : C \to \mathbb{P}(\wedge^{i+1}k^{n+1})$$

by

$$f^i(z) = [\mathbf{f}(z) \wedge \mathbf{f}'(z) \wedge \cdots \wedge \mathbf{f}^{(i)}(z)].$$

Let $P \in C$ and let z be a local coordinate centered around P with $z(P) = 0$. Write $f = [\mathbf{f}(z)] = [f_0(z), ..., f_n(z)]$ with $f_0(0) \neq 0$. Making a linear change of coordinates in k^{n+1}, we may take $\mathbf{f}(0) = (1, 0, ..., 0)$. We have $f_1(0) = \cdots = f_n(0) = 0$. Write $(f_1(z), ..., f_n(z)) = z^{\delta_1}(f_1^1(z), ..., f_n^1(z))$ with $(f_1^1(0), ..., f_n^1(0)) \neq 0$. Now make a linear change of the last n coordinates in k^{n+1} so that $(f_1^1(0), ..., f_n^1(0)) = (1, 0, ..., 0)$. Write $(f_2^1(z), ..., f_n^1(z)) = z^{\delta_2 - \delta_1}(f_2^2(z), ..., f_n^2(z))$ with $(f_2^2(0), ..., f_n^2(0)) \neq 0$. Now make a change of the last $n-1$ coordinates in k^{n+1} so that $(f_2^2(0), ..., f_n^2(0)) = (1, 0, ..., 0)$, and continuing in this way we end up with a system of coordinates for k^{n+1} in terms of which

$$\mathbf{f}(z) = (z^{\delta_0} + \cdots, z^{\delta_1} + \cdots, ..., z^{\delta_n} + \cdots), \tag{3.1}$$

where $0 = \delta_0 < \delta_1 < \cdots < \delta_n$, and where z is a local coordinate centered around P with $z(P) = 0$. The integers

$$\nu_i = \delta_{i+1} - \delta_i - 1, \ 0 \leq i \leq n-1 \tag{3.2}$$

are called the *stationary indices of order* i at the point $z = 0$. The stationary point, that is, the points with non-zero stationary index, are isolated and hence are finite in number. Let

$$\sigma_i = \sum_{x \in C} \nu_i(x). \tag{3.3}$$

Let d_i be the degree of the i-th associate curve f^i of f. Then we have the following Plücker formula

LEMMA 3.1. **(Plücker formula)**

$$2d_i - d_{i+1} - d_{i-1} = 2g - 2 - \sigma_i, \ 1 \leq i \leq n-1.$$

Plücker formula is a generalization of Hurwitz's theorem. For the proof of Plücker formula, see ([GH], p. 270).

By Lemma 3.1, we have

$$\sum_{0 \leq i \leq n-1} (n-i)\sigma_i = (n+1)d_0 + n(n+1)(g-1). \tag{3.4}$$

THEOREM 3.2. **(A special form of SMT over function fields)** *Let C be an irreducible nonsingular projective algebraic curve of genus g defined over an algebraically closed field k of characteristic 0. Let $f : C \to \mathbb{P}^n$ be a linearly nondegenerate algebraic curve(i.e. its image is not contained any hyperplanes in $\mathbb{P}^n(k)$). Let $L_1, \ldots, L_q$ be linear forms in $n+1$ variables with coefficients in k, located in general position, i.e., any $n+1$ of them are linearly independent over k. Let $H_1, \ldots, H_q$ be the hyperplanes defined by L_i's. Let $S = \cup_{j=1}^q f^{-1}(H_j)$. Then*

$$(q - (n+1)) \deg(f) \leq \frac{1}{2} n(n+1)\{2(g-1) + |S|\}.$$

PROOF. Write $S = \{P_1, \ldots, P_r\}$. First of all, if H_j intersects $f(C)$ at certain point $p_l \in C$ with some multiplicity μ_{jl}, then, by the definition,

$$\sum_{1 \le l \le r} \mu_{jl} = d. \tag{3.5}$$

Secondly, since $H_1, \ldots, H_q$ are in general position, at most n hyperplanes can intersect $f(C)$ at P_l, hence there exists subset $A \subset \{1, 2, \ldots, q\}$ with $\#A = n$ such that

$$\sum_{1 \le j \le q} \mu_{jl} \le \sum_{i \in A} \mu_{il}. \tag{3.6}$$

Take a parameter z for C such that $z(P_l) = 0$, let the equations of $f(C)$ at P_l are as in (3.1). At P_l the maximum possible value of $\mu_{il}, i \in A$ is $\delta_n(P_l)$, and then for the unique hyperplane $\zeta_n = 0$. A second hyperplane can intersect $f(C)$ at P_l with multiplicities at most $\delta_{n-1}(P_l), \ldots$. It follows that

$$\sum_{i \in A} \mu_{il} \le \delta_1(P_l) + \cdots + \delta_n(P_l). \tag{3.7}$$

By (3.2),

$$\delta_1(P_l) + \cdots + \delta_n(P_l) = \sum_{0 \le i \le n-1} (n - i)\nu_i(P_l) + \frac{1}{2}n(n+1). \tag{3.8}$$

Combining (3.8) with (3.3), (3.4), (3.5), (3.6) and (3.7), we get

$$q \deg(f) \le (n+1) \deg(f) + n(n+1)(g-1) + \frac{1}{2}n(n+1)|S|.$$

This finishes the proof. □

4. The general second main theorem over function fields

In this section, we derive a general Second Main Theorem. Let C be an irreducible nonsingular projective algebraic curve of genus g defined over an algebraically closed field k of characteristic 0. Let K be the function field of C. let $f : C \to \mathbb{P}^n$ be an algebraic curve. Let $P \in C$ and let f be given locally, around P, by the vector function $\mathbf{f}(z) = (f_0(z), ..., f_n(z))$, where $f_0, \ldots, f_n \in K$, without common zeros. We define the height(compare to the degree) of f as

$$h_K(f) := \sum_{P \in C} -\min\{v_P(f_0), \ldots, v_P(f_n)\}, \tag{4.1}$$

where $v_P(f_j)$ is the order of f_j at P. Note that the definition of $h_K(f)$ is independent of the choice of the reduced representation $\mathbf{f}(z) = (f_0(z), ..., f_n(z))$, and since $f_0, \ldots, f_n$ do not have common zeros,

$$\min\{v_P(f_0), \ldots, v_P(f_n)\} \le 0$$

for all $P \in C$. Hence, $h_K(f) \ge 0$. For a nonzero element $f \in K$, we define

$$h_K(f) := \sum_{P \in C} -\min\{0, v_P(f)\},$$

where $v_P(f)$ is the order of f at P.

THEOREM 4.1. (**The SMT over function fields**) *Let C be an irreducible nonsingular projective algebraic curve of genus g defined over an algebraically closed field k of characteristic 0. Let K be the function field of C. Let S be a finite set of points of C. For each point $P \in S$, let $L_{P,j}, 1 \le j \le q$, be linear forms in $n+1$ variables with coefficients in k. Assume that, for every $P \in S$, the linear forms $L_{P,j}, 1 \le j \le q$, are in general position, i.e. all collections of up to $n+1$ those linear forms are linearly independent over k. Then*

$$\sum_{j=1}^{q}\sum_{P\in S}(v_P(L_{P,j}(f_0,\dots,f_n)) - \min_{0\le i\le n}\{v_P(f_i)\})$$
$$\le (n+1)h_K(f) + \frac{1}{2}n(n+1)\{2(g-1)+|S|\}.$$

for every linearly nondegenerate algebraic curve $f : C \to \mathbb{P}^n$. Here, at $P \in S$, we locally write $f = [f_0, \dots, f_n]$ with $f_0, \dots, f_n \in K$.

PROOF. To prove this theorem, we only need to modify a little bit of the proof given in section 2. The inequality trivially holds for $q \le n+1$. So we assume that $q \ge n+1$. Denote by $l_{P,i} = L_{P,i}(f_0, \dots, f_n)$. For each $P \in S$, there exists a distinct subset $\{l_{P,i_1}, \dots, l_{P,i_{n+1}}\} \subset \{l_{P,1}, \dots, l_{P,q}\}$ which are linearly independent over k, such that

(4.2)
$$v_P(l_{P,i_1}) \ge v_P(l_{P,i_2}) \ge \cdots \ge v_P(l_{P,i_{n+1}}) \ge v_P(l_{P,j}), \text{for any } j \ne i_1, \dots, i_{n+1}.$$

Since $L_{P,i_1}, \dots, L_{P,i_{n+1}}$ are linearly independent over k, there exists b_{ij} such that $f_i = \sum_{j=1}^{n+1} b_{ij} l_{P,i_j}$. Hence $v_P(f_i) \ge \min_{1\le j\le n+1}\{v_P(l_{P,i_j})\}$ for $0 \le i \le n$. On the other hand, $l_{P,j_i} = \sum_{l=0}^{n} a_{j_i,l} f_l$ with $a_{j_i,l} \in k$, therefore

$$v_P(l_{P,i_j}) \ge \min_{0\le i\le n}\{v_P(f_i)\}.$$

Hence, we have

$$\min_{0\le i\le n}\{v_P(f_i)\} = \min_{1\le j\le n+1}\{v_P(l_{P,i_j})\}. \tag{4.3}$$

Similar to (4.4), if we consider $\{l_{P,i_1}, \dots, l_{P,i_n}, l_{P,j}\}$ where $j \ne i_1, \dots, i_{n+1}$ and noticing (4.3), we have, for any $j \ne i_1, \dots, i_{n+1}$,

$$\min_{0\le i\le n}\{v_P(f_i)\} = \min\{v_P(l_{P,i_1}), \dots, v_P(l_{P,i_n}), v_P(l_{P,j})\} = v_P(l_{P,j}). \tag{4.4}$$

Let $P \in C$ and write f locally around P as $f = [f_0, \dots, f_n]$ where $f_0, \dots, f_n \in K$ and are linearly independent over k. We then have

$$f(P) = ((t^{e_P} f_0)(P), \dots, (t^{e_P} f_n)(P))$$

where $e_P := -\min\{v_P(f_0), \dots, v_P(f_n)\}$ and t is a local parameter of C at P. Assume that $v_P(f_0) \le v_P(f_1) \le \cdots \le v_P(f_n)$, then, writing it in the form of (3.1), we found that, for $0 \le i \le n$,

$$\delta_i = v_P(f_i) + e_P. \tag{4.5}$$

Now, since, for every $P \in S$, $L_{P,i_1}, \dots, L_{P,i_{n+1}}$ are linearly independent over k, after k-linear change of coordinates, we may assume that $f = (l_{P,i_1}, \dots, l_{P,i_{n+1}})$. Hence, by (4.5), while writing f in the form of (3.1)

$$\delta_i = v_P(l_{P,j_{i+1}}) + e_P \tag{4.6}$$

for $0 \le i \le n$. Hence, for $P \in S$, by (3.2) and (4.6),

$$\begin{aligned}\sum_{0\le i\le n-1}(n-i)\nu_i(P) &= \sum_{i=0}^{n}(\delta_i(P)-i)\\ &= \sum_{j=1}^{n+1}(v_P(l_{P,i_j})+e_P)-\frac{n(n+1)}{2}\\ &= \sum_{j=1}^{q}(v_P(l_{P,j})+e_P)-\frac{n(n+1)}{2}.\end{aligned}$$

The above inequality together with (3.4) give

$$\begin{aligned}\sum_{j=1}^{q}\sum_{P\in S}(v_P(l_{P,j})+e_P) &\le \sum_{0\le i\le n-1}(n-i)\sum_{P\in S}\nu_i(P)+\frac{n(n+1)}{2}|S|\\ &\le \sum_{0\le i\le n-1}(n-i)\sigma_i+\frac{n(n+1)}{2}|S|\\ &= (n+1)h_K(f)+n(n+1)(g-1)+\frac{n(n+1)}{2}|S|\\ &= (n+1)h_K(f)+\frac{1}{2}n(n+1)\{2(g-1)+|S|\}.\end{aligned}$$

Hence

$$\begin{aligned}&\sum_{j=1}^{q}\sum_{P\in S}(v_P(L_{P,j}(f_0,\dots,f_n))-\min_{0\le i\le n}\{v_P(f_i)\})\\ &\le (n+1)h_K(f)+\frac{1}{2}n(n+1)\{2(g-1)+|S|\}.\end{aligned}$$

This completes the proof of the theorem. □

5. The Second Main Theorem with Hypersurfaces over Function Fields

In this section, we extend Theorem 4.1 to hypersurfaces. Let C be an irreducible nonsingular projective algebraic curve defined over an algebraically closed field k of characteristic zero. Let K be the function field of C. Let $Q_1,\dots,Q_q$ be homogeneous polynomials in $n+1$ variables with coefficients in k. The zero sets of $Q_j, 1\le j\le q$, define hypersurfaces $D_1,\dots,D_q$ in $\mathbb{P}^n$. The hypersurfaces $D_1,\dots,D_q$, $q>n$, in $\mathbb{P}^n$ are said to be *in general position* if $\bigcap_{k=1}^{n+1}\mathrm{Supp}(D_{j_k})=\emptyset$ for any distinct $j_1,\dots,j_{n+1}$, where $\mathrm{Supp}(D)$ means the support of the divisor D. We prove the following theorem.

THEOREM 5.1. **(The SMT with hyperpsurfaces over function fields)** *Let C be an irreducible nonsingular projective algebraic curve defined over an algebraically closed field k of characteristic zero. Let K be the function field of C. Let S be a finite set of points of C. Let $Q_1,\dots,Q_q$ be homogeneous polynomials of degree $d_1,\dots,d_q$ respectively, in $n+1$ variables with coefficients in k. Assume that*

$Q_1, \dots, Q_q$ are in general position. Then, for any $\epsilon > 0$,

$$\sum_{j=1}^{q} \sum_{P \in S} d_j^{-1}(v_P(Q_j(f_0, \dots, f_n)) - \min_{0 \le i \le n}\{v_P(f_i)\})$$
$$\le (n+1+\epsilon)h_K(f) + C(\epsilon)$$

for every algebraic curve $f : C \to \mathbb{P}^n$ which is algebraically independent over k (i.e. there does not exist non-constant homogeneous polynomial Q in $n+1$ variables with coefficients in k, such that $Q(f) \equiv 0$), where, at $P \in S$, we locally write $f = [f_0, \dots, f_n]$ with $f_0, \dots, f_n \in K$, and where $C(\epsilon)$ is a constant which depend only on ϵ, K, S and the coefficients of $Q_1, \dots, Q_q$ and can be determined effectively.

PROOF. First of all, by replacing Q_j by Q_j^{d/d_j} if necessary, where d is the l.c.m. of $d_j's$, we can assume that $\deg(Q_1) = \deg(Q_2) = \dots = \deg(Q_q) = d$. For a fixed big integer N, which will be chosen later, denote by V_N the space of homogeneous polynomials in $\mathbb{C}[X_0, \dots, X_n]$ of degree N. Fix a basis $\phi_1, \dots, \phi_u$ where $u = \dim V_N$. We write locally as $f = [f_0, \dots, f_n]$ where $f_0, \dots, f_n \in K$ which are algebraically independent over k, we consider

(5.1) $$F = [\phi_1(f_0, \dots, f_n), \dots, \phi_u(f_0, \dots, f_n)] : C \to \mathbb{P}^{u-1}$$

Note that functions $\phi_1(f_0, \dots, f_n), \dots, \phi_u(f_0, \dots, f_n)$ are linearly independent over k.

Arrange, by the lexicographic order, the n-tuples $(\mathbf{i}) = (i_1, \dots, i_n)$ of non-negative integers such that $\sigma(\mathbf{i}) := \sum_j i_j \le N/d$. Pick n distinct polynomials $\gamma_1, \dots, \gamma_n \in \{Q_1, \dots, Q_q\}$. Define the spaces $W_{(\mathbf{i})} = W_{N,(\mathbf{i})}$ by

$$W_{(\mathbf{i})} = \sum_{(\mathbf{e}) \ge (\mathbf{i})} \gamma_1^{e_1} \cdots \gamma_n^{e_n} V_{N - d\sigma(\mathbf{e})}.$$

Plainly $W_{(0,\dots,0)} = V_N$ and $W_{(\mathbf{i})} \supset W_{(\mathbf{i}')}$ if $(\mathbf{i}') \ge (\mathbf{i})$, so the $W_{(\mathbf{i})}$ in fact define a filtration of V_N. We first investigate quotients between consecutive spaces in the filtration. Suppose that $(\mathbf{i}')$ follows $(\mathbf{i})$ in the ordering. We recall the following lemma in [Ru2]

LEMMA 5.2. *There exists an integer N_0 dependent only on $\gamma_1, \dots, \gamma_n$ such that, in the above notation,*

$$\Delta_{(\mathbf{i})} := \dim \frac{W_{(\mathbf{i})}}{W_{(\mathbf{i}')}} = d^n$$

provided $d\sigma(\mathbf{i}) < N - N_0$. Also, for the remaining n-tuples $(\mathbf{i})$, $\dim \frac{W_{(\mathbf{i})}}{W_{(\mathbf{i}')}}$ is bounded (by $\dim V_{N_0}$.)

For the proof, see Lemma 3.3 in [Ru2].

We now chose a suitable basis $\{\psi_1, \dots, \psi_u\}$ for V_N is the following way. We start with the last nonzero $W_{(\mathbf{i})}$ and pick any basis of it. Then we continue inductively as follows: suppose $(\mathbf{i}') > (\mathbf{i})$ are consecutive n-tuples such that $d\sigma(\mathbf{i}), d\sigma(\mathbf{i}') \le N$ and assume that we have chosen a basis of $W_{(\mathbf{i}')}$. It follows directly from the definition that we may pick representatives in $W_{(\mathbf{i})}$ for the quotient space $W_{(\mathbf{i})}/W_{(\mathbf{i}')}$, of the form $\gamma_1^{i_1} \cdots \gamma_n^{i_n}\gamma$, where $\gamma \in V_{N - d\sigma(\mathbf{i})}$. We extend the previously constructed basis in $W_{(\mathbf{i}')}$ by adding these representatives. In particular, we have obtained a basis for $W_{(\mathbf{i})}$ and our inductive procedure may go on unless $W_{(\mathbf{i})} = V_N$, in which case we stop. In this way, we obtain a basis $\{\psi_1, \dots, \psi_u\}$ for V_N.

Let ψ be an element of the basis, constructed with respect to $W_{(\mathbf{i})}/W_{(\mathbf{i}')}$, so we may write $\psi = \gamma_1^{i_1} \cdots \gamma_n^{i_n}\gamma$, where $\gamma \in V_{N-d\sigma(\mathbf{i})}$. Then,

$$\gamma_1^{i_1}(f_0, \dots, f_n) \cdots \gamma_n^{i_n}(f_0, \dots, f_n) = \psi(f_0, \dots, f_n) \cdot \gamma^{-1}(f_0, \dots, f_n).$$

Hence, for each $P \in S$, note that $\gamma \in V_{N-d\sigma(\mathbf{i})}$,

$$\begin{aligned} & i_1 v_P(\gamma_1(f_0, \dots, f_n)) + \cdots + i_n v_P(\gamma_n(f_0, \dots, f_n)) \\ &= v_P(\psi(f_0, \dots, f_n)) - v_P(\gamma(f_0, \dots, f_n)) \\ &= v_P(\psi(f_0, \dots, f_n)) - v_P(\gamma(f_0 t^{e_P}, \dots, f_n t^{e_P})) + (N - d\sigma(\mathbf{i}))e_P \\ &\leq v_P(\psi(f_0, \dots, f_n)) + (N - d\sigma(\mathbf{i}))e_P, \end{aligned}$$

where $e_P = -\min_{0 \leq i \leq n}\{v_P(f_i)\}$. Therefore

$$\begin{aligned} & i_1[v_P(\gamma_1(f_0, \dots, f_n)) + de_P] + \cdots + i_n[v_P(\gamma_n(f_0, \dots, f_n)) + de_P] \\ &\leq v_P(\psi(f_0, \dots, f_n)) + Ne_P. \end{aligned}$$

Observe that there are precisely $\Delta_{(\mathbf{i})}$ such functions ψ in our basis. Hence, taking the sum over all functions in the basis,

$$\begin{aligned} & \sum_{(\mathbf{i})} \Delta_{(\mathbf{i})} \{i_1[v_P(\gamma_1(f_0, \dots, f_n)) + de_P] + \cdots + i_n[v_P(\gamma_n(f_0, \dots, f_n)) + de_P]\} \\ & \leq \sum_{t=1}^{u} v_P(\psi_t(f_0, \dots, f_n)) + Nue_P. \end{aligned} \tag{5.2}$$

We now estimate the sums. First,

$$u = \frac{(N+n)!}{N!n!} = \frac{N^n}{n!} + O(N^{n-1}). \tag{5.3}$$

Second, since the number of non-negative integer m-tuples with sum $\leq T$ is equal to the number of non-negative integer $(m+1)$-tuples with sum exactly $T \in \mathbb{Z}$, which is $\binom{T+m}{m}$, and since the sum below is independent of j, we have that, for N divisible by d and for every j,

$$\begin{aligned} \sum_{(\mathbf{i})} i_j &= \frac{1}{n+1} \sum_{(\hat{\mathbf{i}})} \sum_{j=1}^{n+1} i_j = \frac{1}{n+1} \sum_{(\hat{\mathbf{i}})} \frac{N}{d} \\ &= \frac{1}{n+1} \binom{N/d+n}{n} \frac{N}{d} = \frac{N^{n+1}}{d^{n+1}(n+1)!} + O(N^n), \end{aligned} \tag{5.4}$$

where the sum $\sum_{(\hat{\mathbf{i}})}$ is taken over the non-negative integer $(n+1)$-tuples with sum exactly N/d. Combining (5.4) and Lemma 5.2, we have, for every $1 \leq j \leq n$,

$$\sum_{(\mathbf{i})} i_j \Delta_{(\mathbf{i})} = \frac{N^{n+1}}{d(n+1)!} + O(N^n), \tag{5.5}$$

where again the summations are taken over the n-tuples with sum $\leq N/d$; also the various constants in the "O" terms depend only on the original data, $\gamma_1, \dots, \gamma_n$, hence only on $Q_1, \ldots, Q_q$, not on f and P. (5.2), (5.3) and (5.5) yield

$$\left\{\sum_{j=1}^{n}[v_P(\gamma_j(f_0,\ldots,f_n)) + de_P]\right\}\frac{N^{n+1}}{d(n+1)!}(1+O(N^{-1})) \tag{5.6}$$
$$\leq \sum_{t=1}^{u} v_P(\psi_t(f_0,\ldots,f_n)) + uNe_P.$$

We write $\{\psi_1, \ldots, \psi_u\}$ as linear forms $L_1, \ldots, L_u$ in $\phi_1, \ldots, \phi_u$ so that $\psi_t(\mathbf{f}_\lambda) = L_t(\mathbf{F}_\lambda), t = 1, \ldots, u$. The linear forms $L_1, \ldots, L_u$ are linearly independent. Hence, (5.6) becomes

$$\sum_{j=1}^{n}[v_P(\gamma_j(f_0,\ldots,f_n)) + de_P] \tag{5.7}$$
$$\leq \frac{d(n+1)!}{N^{n+1}(1+O(N^{-1}))}\left[\sum_{t=1}^{u} v_P(\psi_t(f_0,\ldots,f_n)) + uNe_P\right].$$

Now, for every $P \in S$, there exists a renumbering $\{Q_{P,i_1}, \ldots, Q_{P,i_q}\} = \{Q_1, \ldots, Q_q\}$ such that

(5.8)
$$v_P(Q_{P,i_1}(f_0,\ldots,f_n)) \geq v_p(Q_{P,i_2}(f_0,\cdots,f_n)) \geq \cdots \geq v_P(Q_{P,i_q}(f_0,\cdots,f_n)).$$

Similar to (4.4) and (4.5), we claim that

$$d \min_{0\leq i\leq n}\{v_P(f_i)\} = \min_{1\leq j\leq n}\{v_P(Q_{i_j}(f_0,\ldots,f_n))\}, \tag{5.9}$$

and, for any $j \neq i_1, \ldots, i_n$,

$$d \min_{0\leq i\leq n}\{v_P(f_i)\} = v_P(Q_j(f_0,\ldots,f_n)). \tag{5.10}$$

In fact, first of all, since Q_{i_j} is a homogeneous polynomial of degree d with coefficients in k, we trivially have

$$v_P(Q_{i_j}(f_0,\ldots,f_n)) \geq d \min_{0\leq i\leq n}\{v_P(f_i)\}, \tag{5.11}$$

hence $\min_{1\leq j\leq n+1}\{v_P(Q_{i_j}(f_0,\ldots,f_n))\} \geq d\min_{0\leq i\leq n}\{v_P(f_i)\}$. On the other hand, by Hilbert's Nullstellensatz, that for any integer $l, 0 \leq l \leq n$, there is an integer $m_l \geq d$ such that

$$x_l^{m_l} = \sum_{j=1}^{n+1} b_{jl}(x_0,\ldots,x_n)Q_{i_j}(x_0,\ldots,x_n),$$

where $b_{jl}, 1 \leq j \leq n+1, 0 \leq l \leq n$, are the homogeneous forms with coefficients in $\mathbb{C}$ of degree $m_l - d$. Hence, for $0 \leq l \leq n$,

$$f_l^{m_l} = \sum_{j=1}^{n+1} b_{jl}(f_0,\ldots,f_n)Q_{i_j}(f_0,\ldots,f_n).$$

So, for $0 \leq l \leq n$,

$$m_l v_P(f_l) \geq (m_l - d)\min_{0\leq i\leq n}\{v_P(f_i)\} + \min_{1\leq j\leq n+1}\{v_P(Q_{i_j}(f_0,\ldots,f_n))\}.$$

Therefore,

$$d \min_{0\leq i\leq n}\{v_P(f_i)\} \geq \min_{1\leq j\leq n+1}\{v_P(Q_{i_j}(f_0,\ldots,f_n))\}. \tag{5.12}$$

Combining (5.11) and (5.12) gives

$$d \min_{0\le i\le n}\{v_P(f_i)\} \ge \min_{1\le j\le n+1}\{v_P(Q_{i_j}(f_0,\dots,f_n))\} = \min_{1\le j\le n}\{v_P(Q_{i_j}(f_0,\dots,f_n))\}.$$

This proves (5.9). The proof of (5.10) is similar to (4.5). (5.9) and (5.10) imply that

$$\begin{aligned}(5.13)\qquad &\sum_{j=1}^{q}\sum_{P\in S}[v_P(Q_j(f_0,\dots,f_n)) - d\min_{0\le i\le n}\{v_P(f_j)\}]\\ &= \sum_{P\in S}\sum_{j=1}^{n}[v_P(Q_{P,i_j}(f_0,\dots,f_n)) - d\min_{0\le i\le n}\{v_P(f_j)\}].\end{aligned}$$

Now take $\gamma_1 = Q_{P,i_1},\dots,\gamma_n = Q_{P,i_n}$, then the linear forms $L_1,\dots,L_u$ obtained as above depends on $Q_{P,i_1},\dots,Q_{P,i_n}$, hence we write them as $L_{P,1},\dots,L_{P,u}$. From (5.7) and (5.13), we have

$$\begin{aligned}&\sum_{j=1}^{q}\sum_{P\in S}[v_P(Q_j(f_0,\dots,f_n)) - d\min_{0\le i\le n}\{v_P(f_j)\}]\\ &= \sum_{P\in S}\sum_{j=1}^{n}[v_P(Q_{P,i_j}(f_0,\dots,f_n)) - d\min_{0\le i\le n}\{v_P(f_j)\}]\\ &= \sum_{P\in S}\sum_{j=1}^{n}[v_P(Q_{P,i_j}(f_0,\dots,f_n)) + de_P]\\ (5.14)\qquad &\le \frac{d(n+1)!}{N^{n+1}(1+O(N^{-1}))}\left[\sum_{j=1}^{u}\sum_{P\in S} v_P(L_{P,j}(\mathbf{F})) + uN\sum_{P\in S} e_P\right]\\ &= \frac{d(n+1)!}{N^{n+1}(1+O(N^{-1}))}\left\{\sum_{j=1}^{u}\sum_{P\in S}[v_P(L_{P,j}(\mathbf{F})) + e_P(F)]\right.\\ &\left. + uN\sum_{P\in S} e_P - u\sum_{P\in S} e_P(F)\right\},\end{aligned}$$

where $e_P(F) = \min_{1\le j\le u}\{v_P(\phi_j(f_0,\dots,f_n))\}$ and

$$\mathbf{F} = (\phi_1(f_0,\dots,f_n),\dots,\phi_u(f_0,\dots,f_n)).$$

Similar to the proof of (5.9), we have $e_P(F) = Ne_P$. Hence $h_K(F) = Nh_K(f)$ and (5.14) becomes

$$\begin{aligned}&\sum_{j=1}^{q}\sum_{P\in S}[v_P(Q_j(f_0,\dots,f_n)) - d\min_{0\le i\le n}\{v_P(f_j)\}]\\ (5.15)\qquad &\le \frac{d(n+1)!}{N^{n+1}(1+O(N^{-1}))}\left\{\sum_{j=1}^{u}\sum_{P\in S}[v_P(L_{P,j}(\mathbf{F})) + e_P(F)]\right\}.\end{aligned}$$

Applying (4.2) to the map F and $L_{P,j}$ gives

$$\sum_{j=1}^{u}\sum_{P\in S}[v_P(L_{P,j}(\mathbf{F})) + e_P(F)] \le uh_K(F) + \frac{1}{2}u(u+1)\{2(g-1)+|S|\}.$$

Combining it with (5.15), we have

$$\begin{aligned}
&\sum_{j=1}^{q}\sum_{P\in S}[v_P(Q_j(f_0,\dots,f_n))-d\min_{0\le i\le n}\{v_P(f_j)\}]\\
&\le \frac{d(n+1)!}{N^{n+1}(1+O(N^{-1}))}\left[uh_K(F)+\frac{1}{2}u(u+1)\{2(g-1)+|S|\}\right]\\
&\le \frac{d(n+1)!}{N^{n+1}(1+O(N^{-1}))}\left[uNh_K(f)+\frac{1}{2}u(u+1)\{2(g-1)+|S|\}\right].
\end{aligned}$$

Since, by (5.3),

$$u=\frac{(N+n)!}{N!n!}=\frac{N^n}{n!}+O(N^{n-1}),$$

for any $\epsilon>0$, we choose N such that

$$\frac{d(n+1)!uN}{N^{n+1}(1+O(N^{-1}))}=d(n+1)(1+O(N^{-1}))<d(n+1)+\epsilon.$$

Using such N we obtain that

$$\begin{aligned}
&\sum_{j=1}^{q}\sum_{P\in S}[v_P(Q_j(f_0,\dots,f_n))-d\min_{0\le i\le n}\{v_P(f_j)\}]\\
&\le (d(n+1)+\epsilon)h_K(f)+C(\epsilon),
\end{aligned}$$

where $C(\epsilon)$ is a constant which depend only on ϵ, K, S and the coefficients of the hypersurfaces $Q_1,\dots,Q_q$ and can be determined effectively. This proves the theorem. □

References

[Ca] H. Cartan, *Sur les zeros des combinaisions linearires de p fonctions holomorpes donnees*, Mathematica(Cluj) **7**(1933), 80-103.

[GH] P. Griffiths and J. Harris, *Principles of algebraic geometry*, Wiley, 1994.

[Ru1] M. Ru, *Nevanlinna theory and its relation to Diophantine approximation*, World Scientific Publishing, 2001.

[Ru2] M. Ru, *A defect relation for holomorphic curves intersecting hypersurfaces*, Amer. J. Math., **126**(2004), 215-226.

[Vo] J.F. Voloch, *Diagonal equations over function fields*, Bol. Soc. Bras. Mat., **16**(1985), 29-39.

[V] P. Vojta, *Diophantine approximations and value distribution theory*, Springer Verlag, New York, 1987

[W] J.T.-Y. Wang, *The truncated second main theorem of function fields*, J. Number Theory, **58**(1996), 139-157.

Department of Mathematics, University of Houston, Houston, TX 77204

E-mail address: minru@math.uh.edu

AMS/IP Studies in Advanced Mathematics
Volume **39**, 2007

Presentations for finite complex reflection groups

Jian-yi Shi

ABSTRACT. We survey our achievements on the classification of congruence classes of presentations for the finite complex reflection groups. The classification is described in terms of certain graphs for the imprimitive groups, and is with the help of the computer programmes for the primitive groups.

Shephard and Todd classified all the finite complex reflection groups in their paper [**10**]. A finite complex reflection group G can be presented by generators and relations just as that for a Coxeter group. We already have one presentation for each irreducible finite complex reflection group (see [**1**]). However, such a presentation is not unique for G in general. Different presentations of G may reveal various properties of G (see [**5**] for example). Then it is desirable to define an equivalence relation, called *congruence relation*, among the presentations of G and then to classify all the presentations of G into congruence classes.

Finite complex reflection groups are divided into two main classes: primitive and imprimitive. Any imprimitive complex reflection group has the form $G(m, p, n)$ for some positive integers m, p, n with $p|m$ (reading " p divides m "), $m > 2$, $n > 1$, and $(m, p, n) \neq (m, m, 2)$ (see [**2**]). The imprimitive complex reflection groups form an infinite series. There are 28 primitive complex reflection groups in total, 12 of them has exactly one congruence class of presentations since they can be generated by only two reflections (see [**1**, Appendix 2]).

I completed the classification of the presentations (S, P) for the groups $G(m, p, n)$ according to their congruence (see [**8**] [**9**]). The classification was made separately in the cases of $p = 1$, $p = m$ and $1 < p < m$. We established a bijection between the set of congruence classes of presentations (S, P) of the group $G(m, m, n)$ (resp., $G(m, 1, n)$, $G(m, p, n)$ with $1 < p < m$) and the set of isomorphism classes of certain graphs Γ_S (resp., of certain rooted graphs Γ_S^r). The relation set P was chosen to be the set P_S of the basic relations on S, which was defined separately in the cases of $p = 1$, $p = m$ and $1 < p < m$. The latter can now be treated in a unified way (see Section 4).

I, together with my students L. Wang and P. Zeng, found the number of congruence classes of the presentations for all the primitive complex reflection groups

Key words and phrases. Complex reflection groups, presentations, congruence classes.

Supported by Nankai Univ., the 973 Project of MST of China, the NSF of China, the SF of the Univ. Doctoral Program of ME of China, the Shanghai Priority Academic Discipline, and the CST of Shanghai.

G except for the group G_{34}. We also found a representative set for all the congruence classes of presentations for 10 primitive complex reflection groups: G_7, G_{11}, G_{12}, G_{15}, G_{19}, G_{24}, G_{25}, G_{26}, G_{27}, G_{32} (the notation is due to Shephard-Todd, see [**10**]), each of which is generated by at least three reflections (see [**6**] [**7**] [**11**] [**12**]). We achieved these results with the help of computer programmes.

In the present paper, we give a survey on the above achievements.

1. Preliminaries

1.1. Let V be an n-dimensional complex vector space with a hermitian form $(\ ,\ )$. A *reflection* s on V is by definition an invertible linear transformation on V with $o(s) < \infty$ and $\dim V^s = n-1$, where $o(s)$ denotes the order of s and $V^s := \{v \in V \mid s \cdot v = v\}$. Any reflection has the form $s_{\alpha,\zeta}$ for some non-zero vector $\alpha \in V$ and some root of unity ζ, where $s_{\alpha,\zeta}$ is defined by

$$s_{\alpha,\zeta}(v) = v + (\zeta - 1)(v, \alpha)\alpha \qquad \text{for all } v \in V.$$

We also write $s_{a,d}$ for $s_{a,\zeta}$ if $\zeta = e^{2\pi i/d}$. A *reflection group* G on V is a finite group generated by reflections on V.

A reflection group G on V is called a *real* group or a *Coxeter group* if there is a G-invariant $\mathbb{R}$-subspace V_0 of V such that the canonical map $\mathbb{C} \otimes_{\mathbb{R}} V_0 \to V$ is bijective. If this is not the case, then G is called *complex*. (Note that, according to this definition, a real reflection group is not complex.)

1.2. A reflection group G in V is *imprimitive*, if there exists a decomposition $V = V_1 \oplus ... \oplus V_r$ into a direct sum of proper subspaces $V_1, ..., V_r$ such that G permutes $\mathbf{V} := \{V_i \mid 1 \leq i \leq r\}$ ($\mathbf{V}$ is called an *imprimitive system* of G in V). Otherwise G is *primitive*.

1.3. Let S_n be the symmetric group over n numbers $1, 2, ..., n$. For $\sigma \in S_n$, denote by $[(a_1, ..., a_n)|\sigma]$ the $n \times n$ monomial matrix with non-zero entries a_i in the $(i, (i)\sigma)$-positions. For $p|m$ in $\mathbb{N}$, set

$$G(m,p,n) = \left\{ [(a_1, ..., a_n)|\sigma] \,\middle|\, a_i \in \mathbb{C}, a_i^m = 1, \sigma \in S_n, \left(\prod\nolimits_j a_j\right)^{m/p} = 1 \right\}.$$

Any imprimitive complex reflection group G on V has the matrix form $G(m,p,n)$ with respect to a basis $e_1, e_2, ..., e_n$ for some $m, p, n \in \mathbb{N}$ with $m > 2$, $n > 1$, $p|m$ and $(m,p,n) \neq (m,m,2)$. In particular, the imprimitive system $\{V_1, ..., V_r\}$ of G in V consists of one-dimensional subspaces.

1.4. It is well known that to each finite complex reflection group G, we can associate a root system (R, f), where R is a finite G-invariant set of unit vectors in the vector space V and $f : R \to \mathbb{N} \setminus \{1\}$ is a function, constant on any G-orbit, such that G is generated by the reflection set $\{s_{\alpha,f(\alpha)} \mid \alpha \in R\}$. As a root system for G, (R, f) is determined by G up to scalar factors (see [**3**, Subsection 1.9]). One can define a simple root system (B, w) in (R, f), where $B \subset R$ has the minimal cardinality with the following properties: $G \cdot B = R$, G is generated by $\{s_{\alpha,w(\alpha)} \mid \alpha \in B\}$, and $w = f|_B$. A simple root system B for R is not uniquely determined by G. In [**6**], we introduce an equivalence relation on simple root systems: Two simple root systems (B, w) and (B', w') of (R, f) are *equivalent*, written $B \sim B'$, if there exists a bijection $\phi : B \longrightarrow B'$ such that for any $\alpha, \beta \in B$,

(1) $w(\alpha) = w'(\phi(\alpha))$ and,

(2) $\langle s_{\alpha,w(\alpha)}, s_{\beta,w(\beta)}\rangle \cong \langle s_{\phi(\alpha),w'(\phi(\alpha))}, s_{\phi(\beta),w'(\phi(\beta))}\rangle$, where the notation $\langle x,y\rangle$ stands for the subgroup of G generated by the elements $x, y \in G$.

1.5. For a reflection group G, *a presentation of G by generators and relations* (or *a presentation* in short) is by definition a pair (S, P), where

(1) S is a finite set of reflection generators for G with minimal possible cardinality.

(2) P is a finite relation set on S, and any other relation on S is a consequence of the relations in P.

Clearly, for any simple system (B, w) for G, $S = \{s_{\alpha,w(\alpha)} \mid \alpha \in B\}$ forms a generator set of a presentation of G. Call (B, w) *the associated simple system* of the presentation.

1.6. Two presentations (S, P) and (S', P') for G are *congruent*, if there exists a bijection $\eta : S \longrightarrow S'$ such that for any $s, t \in S$,(*) $\quad \langle s,t\rangle \cong \langle \eta(s), \eta(t)\rangle$,

Note that when a generator set S of the group G is given, we assume that all the relations on S are known. Thus by the definition, we see that the congruence of a presentation (S, P) of G is determined entirely by the generator set S, the relation set P plays no role.

We see that two presentations of G are congruent if and only if their associated simple root systems are equivalent.

1.7. For a given G, one way to calculate the number of congruence classes of presentations is to find the number of equivalence classes of simple root systems in the root system of G. The latter can be done by calculation of the groups $\langle s_\alpha, s_\beta\rangle$ for all pairs of reflections $s_\alpha \neq s_\beta$ in G with respect to $\alpha, \beta \in R$, and also by calculation of all the permutations of R arising from the action of the reflections s_α, $\alpha \in R$. This has been done when G is primitive (see 5.1).

2. Graphs associated to reflection sets of $G(m, p, n)$

2.1. We have $G(m, m, n) \subseteq G(m, p, n) \subseteq G(m, 1, n)$ for any $1 \leq p \leq m$ with $p|m$. There exists two kinds of reflections in $G(m, 1, n)$ as follows.(i) $s(i, j; k) := [(1, ..., 1, \zeta_m^{-k}, 1, ..., 1, \zeta_m^k, 1, ..., 1)|(i, j)]$, where $\zeta_m := e^{2\pi i/m}$, the numbers ζ_m^{-k}, ζ_m^k are the ith, resp. jth components of the n-tuple and (i, j) is the transposition of i and j for some $i < j$. Call $s(i, j; k)$ *a reflection of type I*. Set $s(j, i; k) = s(i, j; -k)$.(ii) $s(i; k) := [(1, ..., 1, \zeta_m^k, 1, ..., 1)|1]$ for some $k \in \mathbb{Z}$ with $m \mid k$, where ζ_m^k is the ith component of the n-tuple. Call $s(i; k)$ *a reflection of type II*. $s(i; k)$ has the order $m/\gcd\{m, k\}$.

All the reflections of type I lie in the subgroup $G(m, m, n)$.

2.2. Let $X = \{s(i_h, j_h; k_h) \mid h \in J\}$ be a set of reflections of type I in $G(m, 1, n)$ for some finite index set J. We associate to X a digraph $\overline{\Gamma}_X = (V, E)$ as follows. Its node set V is $[n] := \{1, 2, ..., n\}$, and its arrow set E consists of all the ordered pairs (i, j), $i < j$, with labels k for any $s(i, j; k) \in X$ (hence, if $s(i, j; k) \in X$ and $i > j$, then $\overline{\Gamma}_X = (V, E)$ contains an arrow (j, i) with the label $-k$). Denote by Γ_X the underlying graph of $\overline{\Gamma}_X$, i.e., Γ_X is obtained from $\overline{\Gamma}_X$ by replacing each labelled arrow by an unlabelled edge.

Clearly, the graph Γ_X has no loop but may have multi-edges between two nodes.

The above definition of a graph can be extended: to any set X of reflections of $G(m, 1, n)$, we define a graph Γ_X to be $\Gamma_{X'}$, where X' is the subset of X consisting of all the reflections of type I. When X contains exactly one reflection of type II (say $s(i;k)$), we define another graph, denoted by Γ_X^r, which is obtained from Γ_X by rooting the node i, i.e., Γ_X^r is a rooted graph with the rooted node i. Sometimes we denote Γ_X^r by $([n], E, i)$.

2.3. Example. Let $n = 6$.

(1) If $X = \{s(1,2;4), s(3,4;2), s(4,6;0), s(3,4;3)\}$, then $\overline{\Gamma}_X$ is as in Fig. 1 (a).

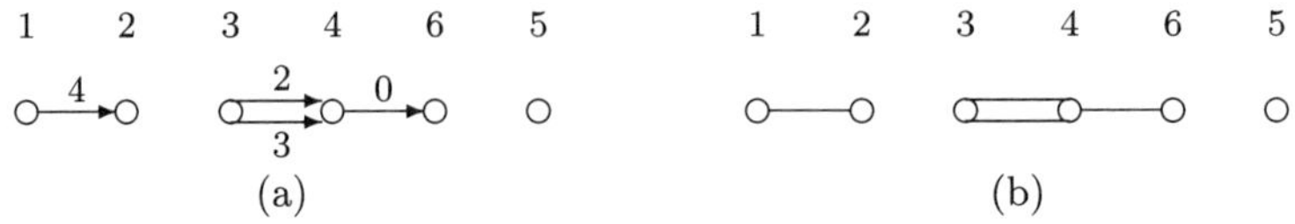

Fig. 1

and Γ_X is as in Fig. 1 (b).

(2) Let $Y = X \cup \{s(6;3)\}$ with X as in (1). Then Γ_Y^r is as in Fig. 2.

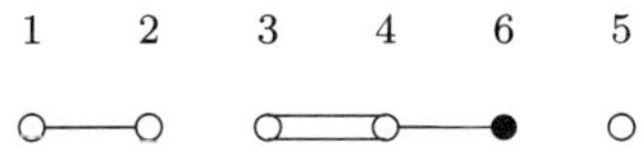

Fig. 2

where the node labelled by 6 is rooted.

2.4. Two graphs (N, E) and (N', E') are *isomorphic*, if there exists a bijection $\eta : N \to N'$ such that for any $v, w \in N$, $\{v, w\}$ is in E if and only if $\{\eta(v), \eta(w)\}$ is in E'.

Two rooted graphs (N, E, i) and (N', E', i') are *isomorphic*, if there exists a bijection $\eta : N \to N'$ with $\eta(i) = i'$ such that for any $v, w \in N$, $\{v, w\}$ is in E if and only if $\{\eta(v), \eta(w)\}$ is in E'.

2.5. In [**8**, Lemma 2.1], we showed that the generator set S of a presentation (S, P) of the group $G(m, 1, n)$ consists of $n-1$ reflections of type I and one reflection of order m. Then we showed the following

2.6. Theorem. (see [**8**, Theorem 2.8]) *Let S be a subset of $G(m, 1, n)$ consisting of $n-1$ reflections of type I and one reflection of order m ($m > 2$ as assumed). Then S is the generator set of a presentation for $G(m, 1, n)$ if and only if the graph Γ_S^r is a rooted tree.*

2.7. Assume that X is a reflection set of $G(m, 1, n)$ with Γ_X connected and having exactly one circle. Then there is some integer r, $2 \leq r \leq n$ such that X contains the reflections $s(a_h, a_{h+1}; k_h)$, $1 \leq h \leq r$, for some integers k_h (the subscripts are modulo r). Denote by $\delta(X)$ the absolute value of $\sum_{h=1}^{r} k_h$.

By [**8**, Lemma 2.7] and [**1**, Appendix 2], we see that the generator set S of a presentation (S, P) of the group $G(m, m, n)$ consists of n reflections of type I such that the graph Γ_S is connected (hence contains exactly one circle). Then we showed the following

2.8. Theorem. (see [**8**, Theorem 2.19]) *Let S be a subset of $G(m,m,n)$ consisting of n reflections of type I with Γ_S connected. Then S is the generator set of a presentation of $G(m,m,n)$ if and only if the value $\delta(S)$ is coprime to m.*

2.9. Next we consider the group $G(m,p,n)$ for any $m,p,n \in \mathbb{N}$ with $p|m$ and $1 < p < m$. By [**9**, Lemma 2.2], we know that for $1 < p < m$ with $p|m$, the generator set S in a presentation (S,P) of the group $G(m,p,n)$ consists of n reflections of type I and one reflection of order m/p and type II such that the graph Γ_S is connected (hence containing exactly one circle). Then we get the following

2.10. Theorem. (see [**9**, Theorem 2.4]) *Assume that S is a subset of $G(m,p,n)$ consisting of n reflections of type I and one reflection of order m/p and type II such that Γ_S is connected. Then S is the generator set of a presentation of $G(m,p,n)$ if and only if* $\gcd\{p,\delta(S)\} = 1$.

3. The classification of presentations of $G(m,p,n)$

Let $\Sigma(m,p,n)$ be the set of the presentations (S,P) of $G(m,p,n)$ and let $\widetilde{\Sigma}(m,p,n)$ be the set of congruence classes of $\Sigma(m,p,n)$. In the present section, we shall describe the set $\widetilde{\Sigma}(m,p,n)$ in the cases of $p = 1$, $p = m$ and $1 < p < m$, separately.

3.1. It is known by [**8**, Subsection 3.1] that $S, S' \in \Sigma(m,1,n)$ are congruent if and only if $\Gamma_S^r \cong \Gamma_{S'}^r$. Also, it is known by [**8**, Subsection 3.3] that $S, S' \in \Sigma(m,m,n)$ are congruent if and only if $\Gamma_S \cong \Gamma_{S'}$. So we get the following two theorems concerning the classification of congruence classes of presentations for the groups $G(m,1,n)$ and $G(m,m,n)$.

3.2. Theorem. (see [**8**, Theorem 3.2]) *The map $(S,P) \to \Gamma_S^r$ induces a bijection from the set $\widetilde{\Sigma}(m,1,n)$ to the set of isomorphism classes of rooted trees with n nodes.*

3.3. Theorem. (see [**8**, Theorem 3.4]) *The map $(S,P) \to \Gamma_S$ induces a bijection from the set $\widetilde{\Sigma}(m,m,n)$ to the set of isomorphism classes of connected graphs with n nodes and n edges (or equivalently with n nodes and exactly one circle).*

3.4. Example. Let $n = 4$.
(1) There are 4 isomorphic classes of rooted trees of nodes 4 (see Fig. 3).

Fig. 3

Hence $G(m,1,4)$ has 4 congruence classes of presentations.
(2) There are 5 isomorphic classes of connected graphs with exactly one circle (see Fig 4).

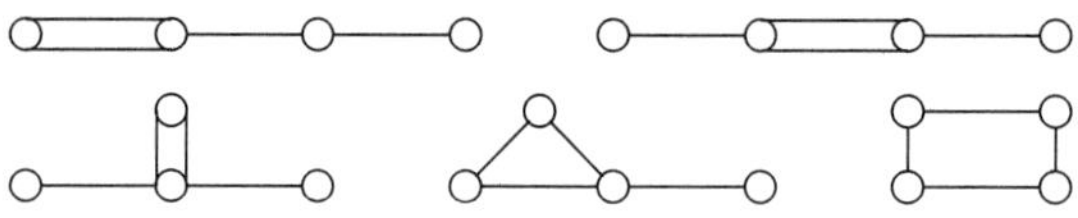

Fig. 4

Hence $G(m, m, 4)$ has 5 congruence classes of presentations.

In the remaining part of the section, we always assume $1 < p < m$ and $p|m$.

3.5. It is known by [**9**, Lemma 2.7] that $S, S' \in \Sigma(m, p, n)$ are congruent if and only if one of the following conditions holds:

(1) the circle of Γ_S^r contains more than two nodes and $\Gamma_S^r \cong \Gamma_{S'}^r$ (see 2.4);

(2) the circle of Γ_S^r contains exactly two nodes, $\Gamma_S^r \cong \Gamma_{S'}^r$ and $\gcd\{m, \delta(S)\} = \gcd\{m, \delta(S')\}$.

3.6. Denote by $\Lambda(m, p)$ the set of all the numbers $d \in \mathbb{N}$ such that $d|m$ and $\gcd\{d, p\} = 1$. Let $\Gamma(m, p, n)$ be the set of all the connected rooted graphs with n nodes and n edges. Let $\Gamma_1(m, p, n)$ be the set consisting of all the rooted graphs in $\Gamma(m, p, n)$ each of which contains a two-node circle. Let $\Gamma_2(m, p, n)$ be the complement of $\Gamma_1(m, p, n)$ in $\Gamma(m, p, n)$. Denote by $\widetilde{\Gamma}(m, p, n)$, resp., $\widetilde{\Gamma}_i(m, p, n)$ the set of the isomorphism classes in the set $\Gamma(m, p, n)$, resp., $\Gamma_i(m, p, n)$ for $i = 1, 2$ (see 2.4).

The following result describes all the congruence classes of presentations for $G(m, p, n)$ in terms of rooted graphs.

3.7. Theorem. (see [**9**, Theorem 2.9]) *(1) The map* $\psi : S \mapsto \Gamma_S^r$ *from* $\Sigma(m, p, n)$ *to* $\Gamma(m, p, n)$ *induces a surjection* $\widetilde{\psi}$: $\widetilde{\Sigma}(m, p, n) \twoheadrightarrow \widetilde{\Gamma}(m, p, n)$. *(2) Let* $\widetilde{\Sigma}_i(m, p, n) = \widetilde{\psi}^{-1}(\widetilde{\Gamma}_i(m, p, n))$ *for* $i = 1, 2$. *Then the map* $\widetilde{\psi}$ *gives rise to a bijection:* $\widetilde{\Sigma}_2(m, p, n) \rightarrowtail\!\!\!\twoheadrightarrow \widetilde{\Gamma}_2(m, p, n)$*; also,* $S \mapsto (\Gamma_S^r, \gcd\{m, \delta(S)\})$ *induces a bijection:* $\widetilde{\Sigma}_1(m, p, n) \rightarrowtail\!\!\!\twoheadrightarrow \widetilde{\Gamma}_1(m, p, n) \times \Lambda(m, p)$.

3.8. Example. Let $n = 4$, $m = 6$ and $p = 2$. Then $\Lambda(6, 2) = \{1, 3\}$. There exist 13 isomorphic classes of rooted connected graphs with exactly one circle, 9 of them contain a two-node circle (see Fig. 5). So $G(6, 2, 4)$ has $22 = 9 \times 2 + 4$ congruence classes of presentations.

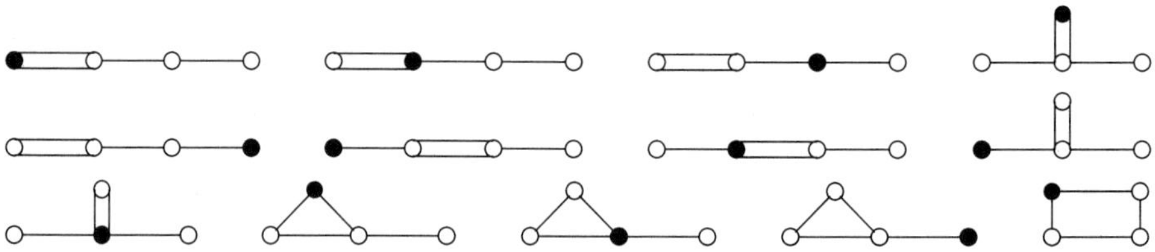

Fig. 5

4. The relation set of a presentation for $G(m,p,n)$

We always assume $1 \leq p \leq m$, $m > 2$, $n > 1$, $p|m$ and $(m,p,n) \neq (m,m,2)$ in the section. In [**8**, Section 4] and [**9**, Section 4], we defined the basic relations on the generator set S of a presentation (S,P) for the groups $G(m,p,n)$ in the cases of $p=1$, $p=m$ and $1<p<m$ separately. In the present section, we shall give a uniform treatment for these relations.

4.1. Let $S \in \Sigma(m,p,n)$. By the results stated in the previous sections, we can write

(4.1) $$S = \{s = s(a,k), t_h \mid h \in J\},$$

where $\gcd\{m,k\} = \gcd\{m,p\}$ (Thus, if $p=m$ then $s=1$, which can be removed from S), J is an index set with $|J| = n-1$ if $p=1$ and $|J| = n$ if $1<p\leq m$, and all the reflections t_h, $h \in J$, are of type I. The graph Γ_S is always connected. When $1 \leq p < m$, we have the rooted graph Γ_S^r with the node a rooted; when $1 < p \leq m$, the graph Γ_S contains exactly one circle.

4.2. Now assume $1 < p \leq m$. Take any node x of Γ_S. Call a sequence of nodes $\xi_x : a_0 = x, a_1, ..., a_r = x$ in Γ_S *a generalized circle sequence of* Γ_S^r *at the node* x if S contains reflections $t_h = s(a_{h-1}, a_h; k_h)$ for $1 \leq h \leq r$ with some integers k_h, where $t_l \neq t_{l+1}$ for $1 \leq l < r$. Since the graph Γ_S is connected and contains a unique circle, the sequence ξ_x always exists. ξ_x contains all the nodes on the circle of Γ_S and is uniquely determined by the set S and the node x up to an orientation of the circle.

When $x = a$ is the rooted node of Γ_S^r (this is the case only when $1<p<m$), ξ_x is also called *a root-circle sequence of* Γ_S^r.

Call $s_{hj} := t_h t_{h+1} ... t_{j-1} t_j t_{j-1} ... t_h$ *a path reflection of* Γ_S *in* ξ_x for any $1 \leq h < j \leq r$ (see Fig. 6).

Fig. 6

Let c_x, c'_x be the smallest, resp., the largest integer with the node a_{c_x}, resp., $a_{c'_x}$ lying on the circle of Γ_S. Then x is on the circle of Γ_S if and only if $c_x = 0$ and $c'_x = r$.

4.3. Example.

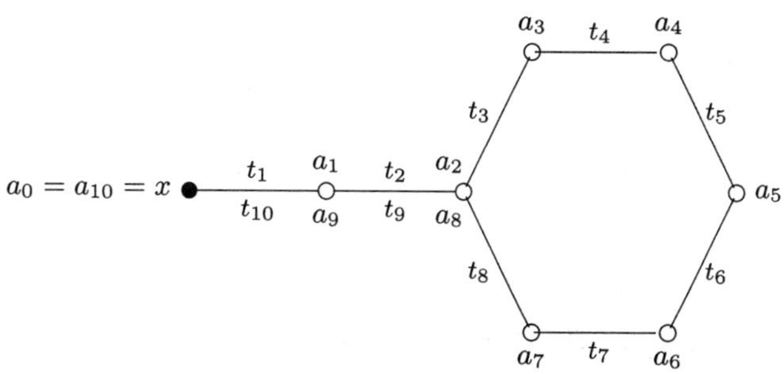

Fig. 7

Here the generalized circle sequence ξ_x at the node x is $a_0 = x, a_1, ..., a_{10} = x$, which is also a root-circle sequence. Hence $r = 10$, $c_x = 2$, $c'_x = 8$, $a_1 = a_9$, $a_2 = a_8$ and $s_{1,4} = t_1t_2t_3t_4t_3t_2t_1$. We have $s_{1,2} = s_{9,10}$ since $t_1 = t_{10}$, $t_2 = t_9$ and $t_1t_2t_1 = t_2t_1t_2$.

4.4. Let S be as in (4.1). By [**8**, Theorems 4.17 and 4.20] and [**9**, Theorem 6.2], we see that (S, P) is a presentation of $G(m,p,n)$ if P consists of relations (A)–(M) as follow. Here the path reflections s_{1j}, $s_{j+1,r}$ are with respect to the generalized circle sequence ξ_x in 4.2 and satisfy $c_x < j < c'_x$ whenever it is applicable. $\{x, a_j\}$ is called *an admissible node pair* of Γ_S, at which we are allowed to talk about relations (J)-(L) on S, where x is required to be the rooted node a of Γ_S^r for relations (K)-(L). $u, v \in S$ in (M) satisfy that the edges $e(u)$, $e(v)$ are incident to ξ_x at the nodes x, a_j respectively.

(A) $s^{m/p} = 1$.

(B) $t_i^2 = 1$ for $i \in J$.

(C) $t_it_j = t_jt_i$ if the edges $e(t_i)$ and $e(t_j)$ have no common end node.

(D) $t_it_jt_i = t_jt_it_j$ if the edges $e(t_i)$ and $e(t_j)$ have exactly one common end node.

(E) $st_ist_i = t_ist_is$ if a is an end node of $e(t_i)$.

(F) $st_i = t_is$ if a is not an end node of $e(t_i)$.

(G) $(t_it_j)^{m/d} = 1$ if $t_i \neq t_j$ with $e(t_i)$ and $e(t_j)$ having two common end nodes, where $d = \gcd\{m, \delta(S)\}$.

(H) $t_i \cdot t_jt_lt_j = t_jt_lt_j \cdot t_i$ for any triple $X = \{t_i, t_j, t_l\} \subseteq S$ with Γ_X having a branching node.

(I) $s \cdot t_it_jt_i = t_it_jt_i \cdot s$, if $e(t_i)$ and $e(t_j)$ have exactly one common end node a.

(J) $(s_{1j}s_{j+1,r})^{\frac{m}{\gcd\{m,\delta(S)\}}} = 1$.

(K) $ss_{1j}s_{j+1,r} = s_{1j}s_{j+1,r}s$.

(L) $(s_{j+1,r}s_{1j})^{p-1} = s^{-\delta(S)}s_{1j}s^{\delta(S)}s_{j+1,r}$.

(M) (a) $us_{1j}u \cdot vs_{j+1,r}v = vs_{j+1,r}v \cdot us_{1j}u$.

(b) $us_{1j}s_{j+1,r}us_{1j}s_{j+1,r} = s_{1j}s_{j+1,r}us_{1j}s_{j+1,r}u$.

(c) $vs_{1j}s_{j+1,r}vs_{1j}s_{j+1,r} = s_{1j}s_{j+1,r}vs_{1j}s_{j+1,r}v$.

Here any of the above relations involving the reflection s (i.e., any of the relations (A), (E), (F), (I), (K) and (L)) is applicable only in the case of $1 \leq p < m$; while any of the above relations involving a circle (i.e., any of the relations (G), (J), (K), (L) and (M)) is applicable only in the case of $1 < p \leq m$.

4.5. In 4.4, call (A)-(B) the *order relations*, (C)-(G) the *braid relations* on S. Call (A)-(F) the *order-braid relations* (or *o.b. relations* in short) on S (note that relation (G) is not included in the o.b. relations on S). Call (H) the *branching relations*, (I) the *root-braid relations*, (H) the *branching relations*, (I) the *root-braid relations*, (G), (J) the *circle relations*, (K) the *root-circle relations*, (L) the *circle-root relations*, and (M) the *branching-circle relations* on S.

Call all the relations (A)-(M) above the basic relations on S. We have the following

4.6. Theorem. (see [**8**, Theorems 4.17 and 4.20] and [**9**, Theorem 6.2]) *Let $S \in \Sigma(m,p,n)$ and let P_S be the set of all the basic relations on S. Then (S, P_S) forms a presentation of $G(m,p,n)$.*

4.7. A presentation (S, P) of a reflection group G is *essential* if (S, P_0) is not a presentation of G for any proper subset P_0 of P.

Let $S \in \Sigma(m, p, n)$ be as in (4.1) with $1 \leq p \leq m$ and let P_S be the set of all the basic relations (A)-(M) on S. Then the presentation (S, P_S) of $G(m, p, n)$ is not essential in general.

Let us take the case of $1 < p < m$ as an example,

(a) Let (B′) be any relation in (B). Then (B′) is equivalent to (B) under the assumption of (D).

(b) Let (K′) (resp., (L′)) be a relation in (K) (resp., (L)) at any one admissible node pair. Then (K′) (resp., (L′)) is equivalent to (K) (resp., (L)) under the assumption of the o.b. relations on S.

(c) For any branching node v of Γ_S, fix some $t_v \in S$ of type I with $e(t_v)$ incident to v. Set

(H′) The relation $t_v \cdot tt't = tt't \cdot t_v$ holds for any $t \neq t'$ in $S \setminus \{t_v\}$ of type I with $\Gamma_{\{t_v, t, t'\}}$ having v as a branching node.

Then (H′) is a subset of and is equivalent to (H) under the assumption of the o.b. relations on S.

(d) Let a be the rooted node in Γ_S^r. Fix some $t_a \in S$ of type I with $e(t_a)$ incident to a. Set (I′) $s \cdot t_a t t_a = t_a t t_a \cdot s$ for any $t \in \Gamma_S \setminus \{t_a\}$ of type I with $e(t)$, $e(t_a)$ having just one common end node a.

Then (I′) is a subset of and is equivalent to (I) under the assumption of the o.b. relations and the branching relations on S. (e) (G) is a special case of (J), while (J) is a consequence of the o.b. relations and the relations (K)-(L) on S.

(f) Let (M′) be the relations (M) if Γ_S^r has a two-node circle and be the empty set of relations if otherwise.

(g) Assume that Γ_S^r has a two-node circle with the rooted node on the circle and not adjacent to any node outside the circle and that $\gcd\{\delta(S), m\} = p$. Then relation (E) is a consequence of (L) and the other o.b. relations on S. Let (E′) be the empty set of relations in this case and be the relation (E) in any other case.

Let P_S' be the collection of relations (A), (B′), (C), (D), (E′), (F), (H′), (I′), (K′), (L′), (M′). Then (S, P_S') is again a presentation of $G(m, p, n)$ (see [**9**, Remark 6.9 (1)]).

One may ask if the presentation (S, P_S') is always essential. The answer is still negative. Recently, Liu and Shi have showed that each of the relations (H′), (I′), (M′) could be further reduced (see [**4**]).

4.8. Among all the presentations (S, P_S) of $G(m, 1, n)$, the relation set P_S has a simpler form when Γ_S^r is a string with the rooted node at one end (see [**1**, Appendix 2]). Among all the presentations (S, P_S) of $G(m, p, n)$, $1 < p \leq m$, we single out two kinds of presentations whose relation sets have simpler forms:

(i) One is when Γ_S is a string with a two-node circle at one end, and with the rooted node on the circle, not incident to any node outside the circle if $1 < p < m$ (see [**1**, Appendix 2]);

(ii) The other is when Γ_S is a circle. In this case, if $p = m$, then the relation set P_S can only consist of some o.b. relations and one circle relation (see [**5**, Proposition 3.3]); if $1 < p < m$, then P_S can only consist of some o.b. relations, one root-braid relation, one root-circle relation and one circle-root relation (see [**9**, Remark 6.9 (2)]).

5. Presentations for the primitive complex reflection groups

5.1. In Table 1, we record results of L. Wang, P. Zeng and J. Y. Shi on the numbers $N(G)$ of congruence classes of presentations for the primitive complex reflection groups G (see 1.7), where the numbers $N(G_i)$ for $i = 12, 24, 25, 26$ were obtained by Shi (see [**6**]), for $i = 13, 22, 27, 29, 31, 33$ by Wang (see [**11**]), and for $i = 7, 11, 15, 19, 32$ by Zeng (see [**12**]).

G	$N(G)$	G	$N(G)$
G_7	2	G_{11}	4
G_{12}	5	G_{13}	4
G_{15}	4	G_{19}	6
G_{22}	18	G_{24}	3
G_{25}	2	G_{26}	2
G_{27}	6	G_{29}	9
G_{31}	61	G_{32}	5
G_{33}	14		

Table 1

Since any G generated by ≤ 2 reflections satisfies $N(G) = 1$, G_{34} is the only primitive complex reflection group with $N(G_{34})$ unknown.

5.2. We list *a representative set for the congruence classes of presentations* (or *r.c.p.* for brevity) of the primitive complex reflection groups G_{12}, G_{24}, G_{27}, G_{25}, G_{26}, G_7, G_{15}, G_{19}, G_{11}, G_{32} (see 5.3–5.12), where the groups G_{12}, G_{24}, G_{25}, G_{26}, G_{19}, G_{11}, G_{32} were done by Shi (see [**6**] [**7**]), G_{27} by Wang (see [**11**]), and G_7, G_{15} by Zeng (see [**12**]). The first presentation for each group was given in [**1**, Appendix 2]). According to the Shephard-Todd's classification (see [**10**]), the groups G_{25}, G_{26}, G_7, G_{15}, G_{19}, G_{11} and G_{32} form the full set of such primitive complex reflection groups each of which is generated by more than two reflections and contains some reflections of order > 2.

5.3. r.c.p. for the group G_{12}. (see [**6**, Propositions 3.3–3.7]):

(1) $G_{12} = \langle s, u, t \mid s^2 = u^2 = t^2 = 1, suts = utsu = tsut\rangle$.

(2) $G_{12} = \langle s, u, w \mid s^2 = u^2 = w^2 = 1, susus u = ususus, sws = wsw, swus = uswu\rangle$.

(3) $G_{12} = \langle s, u, x \mid s^2 = u^2 = x^2 = 1, suxs = usux, xusx = uxus\rangle$.

(4) $G_{12} = \langle s, u, y \mid s^2 = u^2 = y^2 = 1, usuyu = susuy, ysuyu = suyus\rangle$.

(5) $G_{12} = \langle s, x, y \mid s^2 = x^2 = y^2 = 1, sxsx = xsxs, sys = ysy, xyx = yxy, sxsxy = ysxsx\rangle$.

Here $w = ututu$, $x = utu$, $y = tst$, $t = sususuw = uxu = uyusy = yxsxy$ and $u = sxyxs$.

5.4. r.c.p. for the group G_{24}. (see [**6**, Propositions 4.3–4.5]):

(1) $G_{24} = \langle s, u, t \mid s^2 = u^2 = t^2 = 1, stst = tsts, sus = usu, utut = tutu, tusutu = usutus\rangle$.

(2) $G_{24} = \langle s, t, x \mid s^2 = t^2 = x^2 = 1, stst = tsts, sxs = xsx, txt = xtx, stsxstsx = xstsxsts\rangle$.

(3) $G_{24} = \langle s, t, y \mid s^2 = t^2 = y^2 = 1, stst = tsts, tyty = ytyt, tstyts = ytstyt\rangle$.

Here $x = sus$, $y = tut$ and $u = sxs = tyt$.

5.5. r.c.p. for the group G_{27}. (see [**11**, Propositions 3.1.1–3.1.6]):

(1) $G_{27} = \langle s,u,t \mid s^2 = u^2 = t^2 = 1, utu = tut, stst = tsts, susus = ususu, usutus = tusutu\rangle$.

(2) $G_{27} = \langle s,u,w \mid s^2 = u^2 = w^2 = 1, susus = ususu, uwu = wuw, sws = wsw, suwusuwu = uwusuwus\rangle$.

(3) $G_{27} = \langle s,t,w \mid s^2 = t^2 = w^2 = 1, stst = tsts, wtw = twt, sws = wsw, swtwswtwsw = wtwswtwswt\rangle$.

(4) $G_{27} = \langle s,u,x \mid s^2 = u^2 = x^2 = 1, susus = ususu, sxsx = xsxs, usxsus = sxsusx, uxsxux = xsxuxs\rangle$.

(5) $G_{27} = \langle s,t,y \mid s^2 = t^2 = y^2 = 1, stst = tsts, sysys = ysysy, ysytys = tysyty, tsysts = systsy\rangle$.

(6) $G_{27} = \langle s,w,y \mid s^2 = w^2 = y^2 = 1, sws = wsw, sysys = ysysy, wysywy = ysywys, wsyswsys = syswsysw\rangle$.

Here $w = utu$, $x = sts$, $y = susus$, $t = uwu = sxs = ysywysy$ and $u = ysy$.

5.6. r.c.p. for the group G_{25}. (see [**6**, Propositions 5.3–5.4]):

(1) $G_{25} = \langle t,u,v \mid t^3 = u^3 = v^3 = 1, tut = utu, uvu = vuv, tv = vt\rangle$.

(2) $G_{25} = \langle t,u,x \mid t^3 = u^3 = x^3 = 1, tut = utu, uxu = xux, xtux = uxtu\rangle$.

Here $x = u^2vu$ and $v = uxu^2$.

Each of the above presentations for G_{12}, G_{24}, G_{27} and G_{25} becomes essential after removing any two of the order relations.

5.7. r.c.p. for the group G_{26}. (see [**6**, Propositions 6.3–6.4]):

(1) $G_{26} = \langle t,u,v \mid t^2 = u^3 = v^3 = 1, tutu = utut, uvu = vuv, tv = vt\rangle$.

(2) $G_{26} = \langle t,u,x \mid t^2 = u^3 = x^3 = 1, tutu = utut, uxu = xux, uxtu = xtux\rangle$.

Here $x = u^2vu$ and $v = uxu^2$.

Each presentation becomes essential after removing any one of the order 3 relations.

5.8. r.c.p. for the group G_7. (see [**12**, Propositions 4.1.1–4.1.2]):

(1) $G_7 = \langle t,u,s \mid t^2 = u^3 = s^3 = 1, tus = ust = stu\rangle$.

(2) $G_7 = \langle t,x,s \mid t^2 = x^3 = s^3 = 1, sxt = txs, txtsts = ststxt\rangle$.

Here $x = tsus^2t$ and $u = s^2txts$.

5.9. r.c.p. for the group G_{15}. (see [**12**, Propositions 4.2.1–4.2.4]):

(1) $G_{15} = \langle t,u,s \mid t^2 = u^2 = s^3 = 1, tus = ust, stusu = tusus\rangle$.

(2) $G_{15} = \langle t,x,s \mid t^2 = x^2 = s^3 = 1, tsxs^2ts = sxs^2tst, xstsxs^2t = stsxstsx\rangle$.

(3) $G_{15} = \langle t,y,s \mid t^2 = y^2 = s^3 = 1, tsy = syt, ytsys = tsysy\rangle$.

(4) $G_{15} = \langle t,z,s \mid t^2 = z^2 = s^3 = 1, tzts = ztst, sztstz = ztszts\rangle$.

Here $x = s^2tuts$, $y = s^2us$ and $z = tut$. Hence $u = tsxs^2t = sys^2 = tzt$.

5.10. r.c.p. for the group G_{19}. (see [**7**, (2.1.1) and Propositions 2.4–2.8]) :

(1) $G_{19} = \langle t,u,s \mid t^2 = u^3 = s^5 = 1, tus = ust = stu\rangle$

(2) $G_{19} = \langle t,u,w \mid t^2 = u^3 = w^5 = 1, wuwu = uwuw, u^2t \cdot w^2 = w^2 \cdot utu\rangle$

(3) $G_{19} = \langle t,v,w \mid t^2 = v^3 = w^5 = 1, wv^2wv = vwv^2w, vw \cdot wvt = wvt \cdot vw\rangle$

(4) $G_{19} = \langle t,x,z \mid t^2 = x^3 = z^5 = 1, xzxz = zxzx, zxz^2 \cdot zt = tz \cdot zxz^2\rangle$

(5) $G_{19} = \langle t,y,w \mid t^2 = y^3 = w^5 = 1, ywyw = wywy, yw^2y \cdot wt = wt \cdot w^2y^2\rangle$

Here $v = s^3u^2s^2$, $x = utsu^2s^4tu^2$, $y = x^2 = utsus^4tu^2$, $w = su^2s^2us^4$ and $z = s^2$. Hence $s = w^2tu^2tuw = tvw^3v^2t = tyw^3y^2t = z^3$ and $u = s^2v^2s^3 = w^3v^2w^2 = w^2yw^3 = z^2tz^2xztx$.

5.11. r.c.p. for the group G_{11}. (see [**7**, (3.1.1) and Propositions 3.3–3.5]):
(1) $G_{11} = \langle t, u, s \mid t^2 = u^3 = s^4 = 1, tus = ust = stu\rangle$
(2) $G_{11} = \langle t, u, z \mid t^2 = u^3 = z^4 = 1, tuz = zut, tuzu = uzut\rangle$
(3) $G_{11} = \langle r, u, s \mid r^2 = u^3 = s^4 = 1, usus = susu, usu \cdot rs = rs \cdot su^2\rangle$
(4) $G_{11} = \langle r, u, w \mid r^2 = u^3 = w^4 = 1, wuwu = uwuw, w \cdot w^2r \cdot u = u \cdot w^2r \cdot w\rangle$
Here $w = u^2su$, $z = u^2s^3u$ and $r = us^3tsu^2$. Hence $s = uwu^2 = uz^3u^2$ and $t = su^2rus^3 = w^3rw = zrz^3$.

All the above presentations for the groups G_7, G_{15}, G_{19} and G_{11} are essential.

5.12. r.c.p. for the group G_{32}. (see [**7**, (4.1.1) and Propositions 4.3–4.6]):

(1) $G_{32} = \langle t, u, v, w \mid t^3 = u^3 = v^3 = w^3 = 1, tut = utu, uvu = vuv, vwv = wvw, tv = vt, tw = wt, uw = wu\rangle$

(2) $G_{32} = \langle t, u, v, x \mid t^3 = u^3 = v^3 = x^3 = 1, tut = utu, tv = vt, tx = xt, vuxv = xvux = uxvu\rangle$

(3) $G_{32} = \langle r, u, v, x \mid r^3 = u^3 = v^3 = x^3 = 1, uvu = vuv, rur = uru, vxv = xvx, uvru = vruv, uxvu = vuxv, vrxv = xvrx\rangle$

(4) $G_{32} = \langle t, s, w, y \mid t^3 = s^3 = w^3 = y^3 = 1, tw = wt, tyt = yty, ywy = wyw, sws = wsw, stys = ysty, syws = wsyw\rangle$

(5) $G_{32} = \langle t, u, z, m \mid t^3 = u^3 = z^3 = m^3 = 1, uzu^2m^2tm = m^2tmuzu^2, tz = zt, um = mu, zmz = mzm, tut = utu, uzu = zuz, tmt = mtm\rangle$

Here $x = vwv^2$, $r = utu^2$, $s = uvu^2$, $y = u^2vu$, $z = wvw^2$ and $m = t^2u^2vut$. Hence $t = u^2ru$, $u = sys^2$, $v = s^2ys = utmt^2u^2$ and $w = v^2xv = utmt^2u^2zutm^2t^2u^2$.

Any of the presentations (1)–(4) of the group G_{32} becomes essential after removing any three of the four order relations. However, in order to make presentation (5) of G_{32} essential, one needs to remove any one of the relations $zmz = mzm$, $tut = utu$, $uzu = zuz$ and $tmt = mtm$ in addition.

References

[1] M. Broué, G. Malle and R. Rouquier, *Complex reflection groups, braid groups, Hecke algebras*, J. Reine Angew. Math., **500** (1998), 127–190.

[2] A. M. Cohen, *Finite complex reflection groups*, Ann. Scient. Éc. Norm. Sup. **4**$^{\mathbf{e}}$ série t. **9** (1976), 379-436.

[3] R. B. Howlett and J. Y. Shi, *On regularity of finite reflection groups*, Manuscripta Mathematica, **102(3)** (2000), 325–333.

[4] X. L. Liu and J. Y. Shi, in preparation.

[5] J. Y. Shi, *Certain imprimitive reflection groups and their generic versions*, Trans. Amer. Math. Soc., **354** (2002), 2115–2129.

[6] J. Y. Shi, *Simple root systems and presentations for certain complex reflection groups*, to appear in Comm. in Algebra.

[7] J. Y. Shi, *Presentations for three complex reflection groups*, preprint.

[8] J. Y. Shi, *Congruence classes of presentations for the complex reflection groups $G(m,1,n)$ and $G(m,m,n)$*, to appear in Indag. Math.

[9] J. Y. Shi, *Congruence classes of presentations for the complex reflection groups $G(m,p,n)$*, J. Algebra, **284**(1) (2005), 392–414.

[10] G. C. Shephard and J. A. Todd, *Finite unitary reflection groups*, Canad. J. Math., **6** (1954), 274–304.

[11] L. Wang, *Simple root systems and presentations for the primitive complex reflection groups generated by involutive reflections*, Master Dissertation in ECNU, 2003.

[12] P. Zeng, *Simple root systems and presentations for the primitive complex reflection groups containing reflections of order* > 2, Master Dissertation in ECNU, 2003.

Department of Mathematics, East China Normal University, Shanghai, 200062, China and Center for Combinatorics, Nankai University, Tianjin, 300071, China

E-mail address: jyshi@math.ecnu.edu.cn

AMS/IP Studies in Advanced Mathematics
Volume **39**, 2007

Representations of Finite Lie Algebras and Geometry of Reductive Lie Algebras

Bin Shu

ABSTRACT. In the paper [**10**], the authors established a connection between representations of a finite Lie algebra $\mathfrak{g}_0$ over a finite field $\mathbb{F}_q$ and their algebraic extension $\mathfrak{g} = \mathfrak{g}_0 \otimes_{\mathbb{F}_q} k$ for $k = \bar{\mathbb{F}}_q$, which particularly describes irreducible representations of $\mathfrak{g}_0$ through irreducible representations of $\mathfrak{g}$. In this process, the so-called $F_{\mathfrak{g}}$-stable modules play a crucial role, which arise from the Frobenius-Lie morphism $F_{\mathfrak{g}}$ on $\mathfrak{g}$, a defining mapping of $\mathfrak{g}_0$ in $\mathfrak{g}$. In particular, fix $\mathfrak{g} = \mathrm{Lie}(G)$ for a connected reductive algebraic group G over k, defined over $\mathbb{F}_q$, with corresponding Frobenius map F. Then the corresponding Frobenius-Lie morphism $F_{\mathfrak{g}}$ naturally arises. One has a finite reductive Lie algebra $\mathfrak{g}_0 = \mathfrak{g}^{F_{\mathfrak{g}}}$ over $\mathbb{F}_q$. In [**26**], it was proved that a closed conical subvariety X in the restricted nilpotent cone of $\mathfrak{g}$ admits an $\mathbb{F}_q$-rational structure if and only if X is the support variety of a stable $\mathfrak{g}$-module. As an application of this result, we show that the Zariski closures of nilpotent orbits in Lie algebras of classical groups admits $\mathbb{F}_q$-rational structures under some restrictions on p, the characteristic of k.

1. Preliminaries

Throughout this paper, let $q = p^d$ be an arbitrary given power of a prime p, $\mathbb{F}_q$ the finite field of q elements and k the algebraic closure $\bar{\mathbb{F}}_q$ of $\mathbb{F}_q$. For any $n \geq 1$, let $\mathbb{F}_{q^n}$ denote the unique extension field of $\mathbb{F}_q$ of degree n contained in $k = \bar{\mathbb{F}}_q$. All vector spaces are assumed to be finite dimensional.

Let G be a connected reductive algebraic group over k which is defined over $\mathbb{F}_q$, with corresponding (geometric) Frobenius map F. Let $\mathfrak{g} = \mathrm{Lie}(G)$. Consider $A := k[G]$ and $F^* : A \to A$. Set $\sigma : A \to A$ to be defined via $\sigma(f) = (F^*)^{-1}(f^q)$ for all $f \in A$. Then σ is an arithmetic Frobenius map satisfying the condition that σ is a bijective $\mathbb{F}_q$-algebra homomorphism with $\sigma(af) = a^q\sigma(f)$, $a \in k, f \in A$ and that there exists some $m \geq 1$ such that $\sigma^m(f) = f$. Denote by A^σ the set of fixed points

2000 *Mathematics Subject Classification.* 17B20, 17B35, 20G15.

Key words and phrases. Nilpotent orbits, support varieties, Frobenius maps, $\mathbb{F}_q$-rational structures.

This work is done during the author's visit to Institute of Mathematical Sciences, University of Virginia (2004-2005). He would like to express great thanks to the department for the hospitality and the financial support. This work is supported partially by the NSF of China (Grant: 10271047), CSC, a project of science and technology of Shanghai, and the Shanghai principal project to Math. Dept. ECNU.

of σ in A, which is often simply denoted by A_0. Then A_0 is an $\mathbb{F}_q$-algebra, satisfying $A \cong A_0 \otimes_{\mathbb{F}_q} k$. Set $\mathfrak{g}_0 = \mathfrak{g} \cap \mathrm{Der}_{\mathbb{F}_q}(A_0) = \{D \in \mathrm{Der}_{\mathbb{F}_q}(A_0) \mid (1 \otimes D) \circ \Delta = \Delta \circ D\}$ for the co-multiplication Δ on Hopf algebra A. Then $\mathfrak{g} \cong \mathfrak{g}_0 \otimes_{\mathbb{F}_q} k$ (cf. [**2**, I.3.4]). With such an $\mathbb{F}_q$-structure of $\mathfrak{g}$, we have a natural "Frobenius" map $F_{\mathfrak{g}}$ on $\mathfrak{g}$, called "Frobenius-Lie morphism". Frobenius-Lie morphisms can also be defined for an arbitrary Lie algebra over k. A finite Lie algebra $\mathfrak{g}_0$ over $\mathbb{F}_q$ may be regarded as an $\mathbb{F}_q$-Lie algebra of fixed-point set of a Frobenius-Lie morphism $F_{\mathfrak{g}}$ corresponding to $\mathfrak{g} := \mathfrak{g}_0 \otimes_{\mathbb{F}_q} k$. This viewpoint enables us to establish a close relation between representations of $\mathfrak{g}_0$ and $\mathfrak{g}$, similar to [**8**]. In the setup of establishing such a relation, $F_{\mathfrak{g}}$-stable modules play a very important role which is a class of $\mathfrak{g}$-modules coming up from a $\mathfrak{g}_0$-module structure. The result in [**26**], announced here, shows that $F_{\mathfrak{g}}$-stable modules completely reflect $\mathbb{F}_q$-rational property of a nilpotent variety.

2. Representations of Finite Lie algebras

2.1. We first give the definition of Frobenius mappings and $\mathbb{F}_q$-structures on vector spaces.

DEFINTION 2.1. (1) Let V and W be two vector space over k. A mapping f from V to W is called a q semi-linear if it satisfies the following conditions:

(a) f is $\mathbb{F}_q$-linear;

(b) $f(av) = a^q f(v)$ for all $v \in V$ and $a \in k$.

Furthermore, f is called a q-semi-linear isomorphism if it's bijective. If furthermore $V = W$, then f is called a q semi-linear automorphism.

(2) A q semi-linear automorphism F on V such that for any $v \in V$, $F^n(v) = v$ for some $n > 0$ is called a q-Frobenius mapping (a Frobenius mapping for short if the q is clear in the context). The smallest r satisfying $F^r(v) = v$ is called a period of v (under F), denoted by $p_F(v)$.

Obviously, it's the same to give a Frobenius mapping F on V as to give an $\mathbb{F}_q$-structure V_0 on V which means by definition that the natural homomorphism $V_0 \otimes_{\mathbb{F}_q} k \to V$ is an isomorphism. They are related via $V_0 = V^F := \{v \in V \mid F(v) = v\}$. We shall always identify V with $V_0 \otimes k$ in the sequel.

2.2. For further discussion, we introduce the intertwining operator $F_{(V,W)}$. Let $\mathcal{V}_{k,\mathbb{F}_q}$ be the category whose objects are vector spaces over k with *fixed* $\mathbb{F}_q$-structures and whose morphisms are linear maps $f : V \to W$ defined over $\mathbb{F}_q$, namely, $f(V_0) \subseteq W_0$. Clearly, if F_V and F_W are the Frobenius maps on V and W defining V_0 and W_0, respectively, then f is defined over $\mathbb{F}_q$ if and only if $F_W \circ f = f \circ F_V$. In particular, if F and F' are two Frobenius maps on V, then any $\mathbb{F}_q$-linear isomorphism from V^F to $V^{F'}$ induces a k-linear isomorphism θ on V such that $\theta \circ F = F' \circ \theta$. Thus, up to isomorphism in $\mathcal{V}_{k,\mathbb{F}_q}$, the $\mathbb{F}_q$-structure on V is unique. Note that $\mathcal{V}_{k,\mathbb{F}_q}$ is an abelian category.

We shall represent the fixed $\mathbb{F}_q$-structure on a k-space V by a Frobenius map F_V. Thus, the objects V in $\mathcal{V}_{k,\mathbb{F}_q}$ are, sometimes, written as (V, F_V), and

$$\mathrm{Hom}_{\mathcal{V}_{k,\mathbb{F}_q}}(V, W) = \{f \in \mathrm{Hom}_k(V, W) \mid F_W \circ f = f \circ F_V\}.$$

For $(V, F_V), (W, F_W) \in \mathrm{Ob}(\mathcal{V}_{k,\mathbb{F}_q})$, there is an induced $\mathbb{F}_q$-linear map

$$F_{(V,W)} : \mathrm{Hom}_k(V, W) \to \mathrm{Hom}_k(V, W); f \mapsto F_{(V,W)}(f) = F_W \circ f \circ F_V^{-1}.$$

We have the following Lemma.

LEMMA 2.2. *Maintain the notation introduced above and let* $F = F_{(V,W)}$. *We have an* $\mathbb{F}_q$*-space isomorphism*

$$\mathrm{Hom}_{\mathbb{F}_q}(V^{F_V}, W^{F_W}) \cong \mathrm{Hom}_{\mathcal{V}_{k,\mathbb{F}_q}}(V, W) = \mathrm{Hom}_k(V, W)^F.$$

In particular, if V *is finite dimensional, then* F *is a Frobenius map on* $\mathrm{hom}_k(V, W)$

2.3. Frobenius-Lie morphisms. Let $\mathfrak{g}$ be a finite-dimensional Lie algebra over k. A mapping $F : \mathfrak{g} \to \mathfrak{g}$ is called a *Frobenius-Lie morphism* on $\mathfrak{g}$ if it is a Frobenius mapping on the k-space $\mathfrak{g}$, and if it preserves the Lie product: $F[x, y] = [F(x), F(y)]$ for all $x, y \in \mathfrak{g}$. Clearly, a Frobenius-Lie morphism on $\mathfrak{g}$ is also a Lie algebra $\mathbb{F}_q$-automorphism. Also, if $\mathfrak{g} = A^-$ is the Lie algebra obtained from an associative algebra A with the Lie product $[x, y] = xy - yx$ for all $x, y \in A$, then any Frobenius morphism on A induces a Frobenius-Lie morphism on A^-.

The set of all Frobenius-Lie morphisms on $\mathfrak{g}$ is denoted by $\mathfrak{F}_q(\mathfrak{g})$. For any $\sigma \in \mathrm{Aut}(\mathfrak{g})$, it's easy to see by direct verification that $\sigma_* F := \sigma^{-1} F \sigma$ is still a Frobenius-Lie morphisms on $\mathfrak{g}$. Two Frobenius-Lie morphism F_1 and F_2 on $\mathfrak{g}$ are called being equivalent, denoted by $F_1 \sim F_2$, if $\mathfrak{g}^{F_1} \cong \mathfrak{g}^{F_2}$ as Lie $\mathbb{F}_q$-algebras. The following Lemma is clear.

LEMMA 2.3. *Let* F_1 *and* F_2 *be two Frobenius Lie morphisms on* $\mathfrak{g}$.

(a) $F_1 = F_2$ *if and only if* $g^{F_1} = g^{F_2}$*;*

(b) $F_1 \sim F_2$ *if and only if* $F_1 = \sigma_*(F_2)$ *for some* $\sigma \in \mathrm{Aut}(\mathfrak{g})$.

2.4. Frobenius-Lie morphisms on reductive Lie algebras. Now we suppose the Lie algebra $\mathfrak{g}$ is endowed with a p-mapping $[p]$. Suppose F is a Frobenius-Lie morphism. Call F a restricted Frobenius morphism if it satisfies an additional condition $F(x^{[p]}) = F(x)^{[p]}$. We have the following fact.

LEMMA 2.4. *Suppose* F *is a Frobenius-Lie morphism on a restricted Lie algebra* $(\mathfrak{g}, [p])$. *Then*

(a) F *is a restricted morphism if and only if* $\mathfrak{g}^F$ *is a restricted subalgebra of* $\mathfrak{g}$ *over* $\mathbb{F}_q$.

(b) *If* F *is a restricted morphism, then the* (q^n-)*Frobenius-Lie morphism* F^n *is also restricted.*

In the following, we consider $\mathfrak{g} = \mathrm{Lie}(G)$ for a simple algebraic group G over k with characteristic $p > 3$. Set $\mathrm{Aut}(\mathfrak{g})$ by $\mathfrak{G}$. Fix a Cartan subalgebra $\mathfrak{h}$ of $\mathfrak{g}$. Let $\mathfrak{g} = \mathfrak{h} + \sum_\alpha \mathfrak{g}_\alpha$ be the Cartan decomposition relative to $\mathfrak{h}$. Let $\alpha_1, \cdots, \alpha_r$ be a fundamental system of roots relative to $\mathfrak{h}$. For each root $\alpha \neq 0$, let $0 \neq e_\alpha \in \mathfrak{g}_\alpha$. Recall that for any $\phi \in \mathfrak{G}$, ϕ can be written uniquely as $s = \gamma \cdot \sigma$ where γ is a graph automorphism and $\sigma \in E = E(\mathfrak{h})$, where $E = E(\mathfrak{h})$ is the Chevalley group constructed relative to $\mathfrak{h}$. Note that E is independent of the choice of $\mathfrak{h}$ as k is algebraically closed [**27**, IV.1.2]. Furthermore, $\mathfrak{G}$ is a semi-direct product of the graph automorphism group Γ and E, and E is invariant [**27**, III.5.1].

Suppose $\mathfrak{g}_0$ is a split classical Lie algebra over $\mathbb{F}_q$, with respect to the Cartan subalgebra $\mathfrak{h}$ and $\mathfrak{g} = \mathfrak{g}_0 \otimes_{\mathbb{F}_q} k$. Denote the corresponding Frobenius morphism by F_0, that is to say $\mathfrak{g}_0 = \mathfrak{g}^{F_0}$. Obviously, F_0 is a restricted Frobenius morphism, called a split Frobenius morphism related to $\mathfrak{h}$. With an application of Lang-Steinberg Theorem we have

THEOREM 2.5. *(*[**10**, 4.4]*) Let* $\mathfrak{g} = \mathrm{Lie}(G)$ *for a simple algebraic group* G *over an algebraically closed field of characteristic* $p > 3$.

(a) *For any Frobenius morphism F on $\mathfrak{g}$, there exist a unique element $\gamma \in \Gamma$ and an element $\sigma \in E$ such that $F = \gamma \cdot \sigma^{-1} F_0 \sigma$. Furthermore, F is equivalent to $F' = \gamma' \cdot F_0$, where $\gamma' = \gamma_0^{-1}\gamma\gamma_0$ for some $\gamma_0 \in \Gamma$.*

(b) *The equivalence class of F is uniquely determined by the conjugacy class of γ in Γ. Precisely, for two Frobenius morphisms F_1 and F_2 and writing both as $F_1 = \gamma_1 \cdot F_0$ and $F_2 = \gamma_2 \cdot F_0$, up to equivalence as in (a), then $F_1 \sim F_2$ if and only if there exists a graph automorphism $\gamma_0 \in \Gamma$ such that $\gamma_1 = \gamma_0^{-1}\gamma_2\gamma_0$.*

2.5. Results on $F_{\mathfrak{g}}$-stable module cases. Now we turn to an arbitrary given finite-dimensional Lie algebra $\mathfrak{g}$ over k. Let V be a finite dimensional $\mathfrak{g}$-module and let $\rho = \rho_V : \mathfrak{g} \to \mathfrak{gl}(V)$ be the corresponding representation. We say that V is *$F_{\mathfrak{g}}$-stable* (or simply F-stable) if there is an $\mathbb{F}_q$-structure V_0 of V such that ρ induces a representation $\rho_0 : \mathfrak{g}^F \to \mathfrak{gl}_{\mathbb{F}_q}(V_0)$ of $\mathfrak{g}^F$. By the definition of an $\mathbb{F}_q$-structure, V is F-stable if and only if there is a Frobenius map $F_V : V \to V$ such that

$$F_V(xv) = F_{\mathfrak{g}}(x)F_V(v), \quad \text{for all } x \in \mathfrak{g}, v \in V. \tag{2.1}$$

In terms of the corresponding representation ρ, the F-stability of V simply means that there is a Frobenius map F_V on V such that $\rho \circ F_{\mathfrak{g}} = F_{(V,V)} \circ \rho$. In the following arguments, we often denote both $F_{\mathfrak{g}}$ and F_V simply as F if the context is clear. For example, we sometimes use g^F and V^F in the same time, but obviously with different meanings of F. Thus, if V is an F-stable $\mathfrak{g}$-module with respect to F_V, then $V = V^F \otimes_{\mathbb{F}_q} k$, $F_V(v \otimes a) = v \otimes a^q$ for all $v \in V^F$, $a \in k$, and V^F is an $\mathfrak{g}^F$-module.

Up to isomorphism, it doesn't matter which Frobenius maps (or $\mathbb{F}_q$-structures) on V we shall work with when considering F-stable modules. This is because we have the following lemma which directly follows from Noether-Deuring Theorem (c.f. [**7**, p.139]).

LEMMA 2.6. *Let (V_1, F_1) and (V_2, F_2) be two F-stable $\mathfrak{g}$-modules. Then $V_1^{F_1} \cong V_2^{F_2}$ as $\mathfrak{g}^F$-modules if and only if $V_1 \cong V_2$ as $\mathfrak{g}$-modules. In particular, if $V_1 = V_2 = V$, then $V^{F_1} \cong V^{F_2}$ as $\mathfrak{g}^F$-modules.*

In the following, we sometimes use the notation (V, F_V) for an F-stable module V. Let $\mathbf{mod}^F$-$\mathfrak{g}$ denote the category whose objects are F-stable modules $V = (V, F_V)$. The morphisms from (V, F_V) to (W, F_W) are given by homomorphisms $\theta \in \hom_{\mathfrak{g}}(V, W)$ such that $\theta \circ F_V = F_W \circ \theta$, that is, $\mathfrak{g}$-module homomorphisms compatible with their $\mathbb{F}_q$-structures. It is easy to see that $\mathbf{mod}^F$-$\mathfrak{g}$ is an abelian $\mathbb{F}_q$-category. We have the following basic theorem which allows us to embed a module category defined over a finite field into a category defined over the algebraic closure of the finite field.

THEOREM 2.7. *([**10**, 2.4]) The abelian category $\mathbf{mod}^F$-$\mathfrak{g}$ is equivalent to the category* mod $\mathfrak{g}^F$ *of finite dimensional $\mathfrak{g}^F$-modules. In particular, there is a one-to-one correspondence between isoclasses of irreducible (indecomposable, resp.) $\mathfrak{g}^F$-modules and isoclasses of irreducible (indecomposable, resp) F-stable $\mathfrak{g}$-modules.*

2.6. Results on $F_{\mathfrak{g}}$-twisted module cases. Keep the notations in the previous arguments. In general, a $\mathfrak{g}$-module may not be $F_{\mathfrak{g}}$-stable. Let V be a $\mathfrak{g}$-module. Let F_g be a Frobenius-Lie morphism of $\mathfrak{g}$, and F_V be a Frobenius map of V with respect to $\mathbb{F}_q$-structures. Then obviously F_g^n is also a Frobenius-Lie morphism of $\mathfrak{g}$, and F_V^n also a Frobenius map of V with respect to $\mathbb{F}_{q^n}$-structure for some positive integer n.

DEFINTION 2.8. A $\mathfrak{g}$-module V is called F-periodic if there is a positive integer n so that (V, F^n) is F_g^r-stable. In this case, let r be the smallest of such n, then r is called the period of F, denoted by $p(V) = r$.

The obvious fact is that if V is F_g^n-stable, then r divides n. Furthermore, we have the following facts.

PROPOSITION 2.9. ([**10**, 2.7]) Let $\mathfrak{g}$ be finite dimensional. Then every $\mathfrak{g}$-module is F-periodic.

There is another description of being periodic, with the aid of "twist" arising from Frobenius morphisms.

Let V be a $\mathfrak{g}$-module and $F_V : V \to V$ a Frobenius map on the space V. Note that V is not necessarily F-stable. We define its Frobenius twist (with respect to the Frobenius map F_V) to be the $\mathfrak{g}$-module $V^{[F_V]}$ such that $V^{[F_V]} = V$ as vector spaces with F-twisted action

$$x * v := F_V\big(F_{\mathfrak{g}}^{-1}(x)F_V^{-1}(v)\big) \text{ for all } x \in \mathfrak{g}, v \in V.$$

If $\rho : \mathfrak{g} \to \mathfrak{gl}(V)$ and $\rho^{[F_V]} : \mathfrak{g} \to \mathfrak{gl}(V^{[F_V]})$ denote the corresponding representations, then

$$\rho^{[F_V]}(x) = F_{(V,V)}(\rho(F_{\mathfrak{g}}^{-1}(x))) = F_V \circ \rho(F_{\mathfrak{g}}^{-1}(x)) \circ F_V^{-1} \text{ for all } x \in \mathfrak{g},$$

where $F_{(V,V)}$ is the induced Frobenius map on $\mathfrak{gl}(V)$.

Observe that any two Frobenius maps F_V and F'_V on the $\mathfrak{g}$-module V give rise to the linear isomorphism $F'_V \circ F_V^{-1}$, which is clearly a $\mathfrak{g}$-module isomorphism from $V^{[F_V]}$ to $V^{[F'_V]}$. Hence, up to isomorphism, the Frobenius twist $V^{[F_V]}$ is independent of the choice of the Frobenius map F_V on V. Thus, we may denote $V^{[F_V]}$ and $\rho^{[F_V]}$ by $V^{[1]}$ and $\rho^{[1]}$ respectively. Similarly, we define $V^{[-1]}$ to be the $\mathfrak{g}$-module given by $\rho^{[-1]} : \mathfrak{g} \to \mathfrak{gl}(V)$ where

$$\rho^{[-1]}(x) = F_V^{-1} \circ \rho(F_{\mathfrak{g}}(x)) \circ F_V \text{ for all } x \in \mathfrak{g}. \tag{2.2}$$

Inductively, for each integer $s > 1$, we define $V^{[s]} = (V^{[s-1]})^{[1]}$ and $V^{[-s]} = (V^{[-s+1]})^{[-1]}$ with respect to the *same* given F_V. Furthermore, for a $\mathfrak{g}$-module homomorphism $f : V \to W$, $f^{[1]} := F_{(V,W)}(f)$ is also a $\mathfrak{g}$-module homomorphism from $V^{[1]}$ to $W^{[1]}$. Thus we have a functor $(-)^{[1]}$ from $\operatorname{mod} \mathfrak{g}$ to $\operatorname{mod} \mathfrak{g}$. Similarly, we have a functor $(-)^{[-1]}$. It's readily verified that these two functors are inverses to each other. Hence, the Frobenius twist $(\ )^{[1]}$ defines a category equivalence from $\mathbf{mod}^F$-$\mathfrak{g}$ onto itself. In particular, if V and W are two $\mathfrak{g}$-modules, then for any given integer s, $V \cong W$ if and only if $V^{[s]} \cong W^{[s]}$.

PROPOSITION 2.10. ([**10**, 2.10]) Let V be an $\mathfrak{g}$-module with a Frobenius map F_V and $V^{[1]}$ denote the Frobenius twist of V with respect to F_V.

(a) (V, F_V) is F-stable if and only if, as $\mathfrak{g}$-modules, $V^{[1]} = V$.

(b) $V^{[1]} \cong V$ if and only if there exists a Frobenius map F'_V on V such that V is F-stable with respect to F'_V, that is, $V^{[F'_V]} = V$ as $\mathfrak{g}$-modules.

(c) V is F_V-periodic if and only if $V^{[r]} \cong V$ for some integer r. More precisely, $V^{[r]} \cong V$ if and only if there exists a Frobenius map F'_V such that V is $F^r_{\mathfrak{g}}$-stable with respect to $(F'_V)^r$.

Thus the F-period $p(V)$ of V is just the minimal integer r such that $V^{[r]} \cong V$. And if $V^{[s]} \cong V$, then $p(v) \mid s$. Also, $p(V)$ is independent of the choice of the Frobenius map on V.

Let V be F-periodic with F-period $r = p(V)$. Then, there is a Frobenius map F_V on V such that $V^{[r]} = V$ as $\mathfrak{g}$-modules. Thus, $V, V^{[1]}, \cdots, V^{[r-1]}$ are pairwise non-isomorphic. Let

$$\tilde{V} = V \oplus V^{[1]} \oplus \cdots \oplus V^{[r-1]} \tag{2.3}$$

and define a Frobenius map $F_{\tilde{V}} : \tilde{V} \to \tilde{V}$ by

$$F_{\tilde{V}}(v_0, v_1, \cdots, v_{r-1}) = (F_V(v_{r-1}), F_V(v_0), \cdots, F_V(v_{r-2})). \tag{2.4}$$

Then $(\tilde{V}, F_{\tilde{V}})$ is F-stable. Thus, $\tilde{V}^F := \tilde{V}^{F_{\tilde{V}}}$ is an $\mathfrak{g}^F$-module. By Lemma 2.6 we know that, up to isomorphism, $\tilde{V}^F$ is independent of the choice of F_V.

THEOREM 2.11. ([**10**, 2.14]) *Let V be an F-periodic indecomposable (resp. irreducible) $\mathfrak{g}$-module with F-period r. Then $(\tilde{V}, F_{\tilde{V}})$ is indecomposable (resp. irreducible) in* $\textbf{mod}^F$*-$\mathfrak{g}$. Moreover, every indecomposable (resp. irreducible) $\mathfrak{g}^F$-module is isomorphic to a module of the form $\tilde{V}^F$ for some F-periodic indecomposable (resp. irreducible) module V.*

Thus, for a finite-dimensional Lie algebra $\mathfrak{g}$, there is a one-to-one correspondence between the isoclasses of indecomposable (resp. irreducible) $\mathfrak{g}^F$-modules and the F-orbits of the isoclasses of indecomposable (resp. irreducible) $\mathfrak{g}$-modules.

3. $\mathbb{F}_q$-rational structure on support varieties

3.1. $\mathbb{F}_q$-rational structure of an affine verity.

DEFINTION 3.1. Let (X, A) be an affine variety over $k = \bar{\mathbb{F}}_q$. We say that (X, A) is defined over $\mathbb{F}_q$, if there is a morphism $F : X \to X$ such that the following conditions hold for the algebra homomorphism $F^* : A \to A$:

(a) F^* is injective and $F^*(A) = A^q$.
(b) For each $f \in A$, there exists some $m \geq 1$ such that $(F^*)^m(f) = f^{q^m}$.

In this case, we also say that (X, A) admits an $\mathbb{F}_q$-rational structure, and that F is the corresponding Frobenius map.

Furthermore, we set $X^F := \{x \in X \mid F(x) = x\}$ for the fixed-point set of F. We also write this set as $X(\mathbb{F}_q)$ and call it the set of $\mathbb{F}_q$-rational points in X.

In the above definition, since $F^*(A) = A^q$ and F^* is injective, we obtain a well-defined map $\sigma : A \to A$ such that $\sigma(f) = (F^*)^{-1}(f^q)$ for all $f \in A$. It is just an arithmetic Frobenius map satisfying the following:

(a) The map $\sigma : A \to A$ is a bijective $\mathbb{F}_q$-algebra homomorphism such that $\sigma(af) = a^q\sigma(f)$, $a \in k$, $f \in A$.
(b) For each $f \in A$, there exists some $m \geq 1$ such that $\sigma^m(f) = f$.

PROPOSITION 3.2. ([**16**, 4.1.3 and 4.1.5], [**9**, 3.3]) Retain the above notations. Set $A_0 := \{f \in A \mid \sigma(f) = f\} = \{f \in A \mid F^*(f) = f^q\}$. The following statements hold.

(a) A_0 is a finitely generated $\mathbb{F}_q$-subalgebra of A such that the natural map $A_0 \otimes_{\mathbb{F}_q} k \to A$ is an algebra isomorphism.
(b) For any closed subset $V \subset X$, the following conditions are equivalent.
(1) $F(V) \subset V$.
(2) $F(V) = V$.
(3) The vanishing ideal $\mathbf{I}_A(V)$ of V is generated by elements in A_0.
(4) The ideal $\mathbf{I}_A(V)$ has the form $I_0 \otimes k$ where I_0 is an ideal of A_0.
(5) $V = \mathbf{V}_X(S)$ for some subset $S \subset A_0$, where $\mathbf{V}_X(S)$ is the algebraic set defined by S.

Thus, if a closed subvariety V is F-stable, then V is defined over $\mathbb{F}_q$, thereby V admits an $\mathbb{F}_q$-rational structure.

EXAMPLE 3.3. Consider the standard Frobenius map $F : k^n \to k^n$. Let $A = k[x_1, \cdots, x_n]$. Let $V \subset k^n$ be an algebraic set. The A_0 defined in Proposition 3.2 is given by $A_0 = \mathbb{F}_q[x_1, \cdots, x_n]$. If V is F-stable, then set $A_0[V] := \mathbb{F}_q[x_1, \cdots, x_n]/I_0$, where $I_0 = \mathbf{I}(V) \cap \mathbb{F}_q[x_1, \cdots, x_n]$. Then $A_0[V]$ is precisely the subalgebra defined in Proposition 3.2 for $(V, A[V])$.

3.2. Support varieties. Next, we recall some facts about support varieties (cf. [**12**], [**13**] and [**14**]). For any restricted Lie algebra $\mathfrak{g}$, the cohomology variety $|\mathfrak{g}|$ of $\mathfrak{g}$ is the affine variety associated to $H^{\mathrm{ev}}(u(\mathfrak{g}), k)^{(-1)}$. Here and below, the $H^{\mathrm{ev}}(u(\mathfrak{g}), k)^{(-1)}$ with superscript (-1) means the twisted ring. Let V be a finite-dimensional $\mathfrak{g}$-module, the support variety $|\mathfrak{g}|_M$ is the support in $|\mathfrak{g}|$ of the $H^{\mathrm{ev}}(u(\mathfrak{g}), k)^{(-1)}$-module $H^\bullet(u(\mathfrak{g}), M \otimes V^*)^{(-1)}$, $|\mathfrak{g}|_V := \mathrm{Supp}(H^\bullet(u(\mathfrak{g}), V \otimes V^*)^{(-1)})$ (cf. [**14**],[**15**]). Denote by $\Phi : |\mathfrak{g}| \to \mathfrak{g}$ the morphism of varieties associated with the natural map $\Phi^* : S^\bullet(\mathfrak{g}^*) \to H^{\mathrm{ev}}(u(\mathfrak{g}), k)^{(-1)}$, where $S^\bullet(\mathfrak{g}^*)$ is the symmetric algebra on the dual of $\mathfrak{g}$. We may identify $\Phi(|\mathfrak{g}|_V)$ with the support of the graded $S^\bullet(\mathfrak{g}^*)$-module $H^\bullet(u(\mathfrak{g}), V \otimes V^*)^{(-1)}$. So defined, $\Phi(|\mathfrak{g}|_V)$ is a conical closed subvariety of the restricted nilpotent cone $\mathcal{N}_1 := \{x \in \mathfrak{g} \mid x^{[p]} = 0\}$ and $\Phi(|\mathfrak{g}|_V) = \{x \in \mathcal{N}_1 \mid V$ is not projective as a restricted $\langle x \rangle$-module $\} \cup \{0\}$ (cf. [**19**] and [**13**]).

3.3. Let G be a connected reductive algebraic group over k, which defined over $\mathbb{F}_q$, with corresponding Frobenius map F on G and Frobenius-Lie morphism $F_{\mathfrak{g}}$ on $\mathfrak{g} = \mathrm{Lie}(G)$. By application of Theorem 2.5, we have

PROPOSITION 3.4. ([**26**]) All restricted projective modules for the Lie algebra $\mathfrak{g} = \mathrm{Lie}(G)$ are $F_{\mathfrak{g}}$-stable under Frobenius-Lie morphism $F_{\mathfrak{g}}$. Furthermore, if (V, F_V) and (W, F_W) are two finite-dimensional restricted stable $\mathfrak{g}$-modules with $V_0 = V^{F_V}$ and $W_0 = W^{F_W}$ and if denote by $\mathrm{Ext}^n(V, W)$ (resp. $\mathrm{Ext}_0^n(V_0, W_0)$) the n-th extension between restricted $\mathfrak{g}$-modules V and W over $u(\mathfrak{g})$ (resp. the n-th extension for $u(\mathfrak{g}_0)$-modules V_0 and W_0), then Then $\mathrm{Ext}^n(V, W)^{\mathfrak{F}_n} \cong \mathrm{Ext}_0^n(V_0, W_0)$ for $n = 0, 1, 2, \cdots$, where $\mathfrak{F}_n$ is a Frobenius map induced by F (for detail, cf [**26**]).

The following fact is clear.

LEMMA 3.5. *Suppose G is a connected reductive algebraic groups over k, defined over $\mathbb{F}_q$ with corresponding Frobenius map F on G and Frobenius-Lie morphism $F_{\mathfrak{g}}$ on $\mathfrak{g}$. Then, $\Phi(|\mathfrak{g}|)$ is defined over $\mathbb{F}_q$ such that the associated $\mathbb{F}_q$-rational structure is given by the restriction of $F_{\mathfrak{g}}$ in $\Phi(|\mathfrak{g}|)$.*

3.4. $\mathbb{F}_q$-rational structure on support varieties.

THEOREM 3.6. ([**26**, 4.4]) *Maintain the notations as in Lemma 3.5. If V is a restricted $F_\mathfrak{g}$-stable $\mathfrak{g}$-module, then there is an $\mathbb{F}_q$-rational structure on $\Phi(|\mathfrak{g}|_V)$. Conversely, if X is a closed conical subvariety of $\Phi(|\mathfrak{g}|)$ which admits an $\mathbb{F}_q$-rational structure arising from $F_\mathfrak{g}$, then there exists a restricted $F_\mathfrak{g}$-stable $\mathfrak{g}$-module V such that $X = \Phi(|\mathfrak{g}|_V)$.*

For the proof, we need to construct a $F_\mathfrak{g}$-stable module corresponding to the $F_\mathfrak{g}$-stable subvariety X. The construction comes from Calson's module L_ξ (cf. [**3**] and [**1**]). Here we give some necessary argument for this construction.

Suppose V_0 is a restricted $\mathfrak{g}_0$-module. Recall the construction of a minimal projective resolution $\cdots \to P_0^{(n)} \to \cdots P_0^{(0)} \to V_0$ of V_0 is as follows. Set $P_0^{(0)} = P(V_0)$, the projective cover of V_0, and define the *syzygy* module $\Omega(V_0)$ as the kernel of the given map $P(V_0) \to V_0$. Inductively, set $P_0^{(n)} = P(\Omega^n(V_0))$ and $\Omega^n(V_0) = \Omega(\Omega^{n-1}(V_0)$. When $V_0 = \mathbb{F}_q$, then $H^n(u(\mathfrak{g}_0), \mathbb{F}_q) \cong \mathrm{Hom}_{u(\mathfrak{g}_0)}(\Omega^n(\mathbb{F}_q), \mathbb{F}_q)$. Suppose $\zeta \in H^{2n}(u(\mathfrak{g}_0), \mathbb{F}_q)$. Then such a ζ can be represented by a map $\zeta : \Omega^{2n}(\mathbb{F}_q) \to \mathbb{F}_q$ whose kernel is the $u(\mathfrak{g}_0)$-module $\mathbb{L}_\zeta$. In the meantime, we have for $V = V_0 \otimes_{\mathbb{F}_q} k$, $P^{(n)} = P(\Omega^n(V))$ and $\Omega^n(V) = \Omega(\Omega^{n-1}(V))$. By Proposition 3.4 and the exactness of the field extension functor from the $u(\mathfrak{g}_0)$-module category to the $u(\mathfrak{g})$-module category, $P^{(n)} \cong P_0^{(n)} \otimes_{\mathbb{F}_q} k$, and $\Omega^n(V) \cong \Omega^n(V_0) \otimes_{\mathbb{F}_q} k$. Especially, $\Omega^{2n}(k) \cong \Omega^{2n}(\mathbb{F}_q) \otimes_{\mathbb{F}_q} k$. Then Propositions 3.2 and 3.4 enable us to construct the desired Carlson's module.

We call a rational G-module stable if it is $F_\mathfrak{g}$-stable as a $\mathfrak{g}$-module. By [**13**, Theorem 1.2], we have a corollary to the above theorem.

COROLLARY 3.7. Suppose V is a finite-dimensional rational stable G-module V, then $\Phi(|\mathfrak{g}|_V)$ is a closed conical G-invariant variety defined over $\mathbb{F}_q$.

The inverse statement of the above is a version of an old Friedlander-Parshall's conjecture ([**13**]): any closed conical G-invariant subvariety of $\Phi(|\mathfrak{g}|)$ is the support variety of a finite-dimensional rational stable G-module V. We will give a positive answer in the case when the subvarieties are the closures of nilpotent orbits in the next section (Reference [**24**] gives a more general argument. Thanks to [**24**], we may give an application of Theorem 3.6 in the next section).

4. Application

4.1. In this section, we always suppose that G is a connected reductive algebraic group over $k = \bar{\mathbb{F}}_q$, which is split over $\mathbb{F}_q$. Suppose T is a given maximal torus of G, and $X(T)$ the character group of T, and $\mathfrak{g} = \mathrm{Lie}(G)$. Set Π the simple root system of G and Φ (resp. Φ^+) the corresponding root system (resp. the positive root system). And set W be the Weyl group of G. Furthermore, we always fix $F_\mathfrak{g}$ to be the *split Frobenius-Lie morphism* related to $\mathfrak{h} = \mathrm{Lie}(T)$. Then we have the following Lemma.

LEMMA 4.1. *All Weyl modules $V(\lambda)$ and dual Weyl modules $H^0(\lambda)$ of G are $F_\mathfrak{g}$-stable, as $\mathfrak{g}$-modules. Here $\lambda \in X(T)_+$, the dominant weight set in $X(T)$.*

PROOF. It can be seen directly from the constructions of both modules (cf. [**18**] and [**11**]). □

4.2. $\mathbb{F}_q$-rational structures on Richardson nilpotent orbits. Before giving the following theorem, we just recall some related facts. For a given standard parabolic subgroup $P_J = L_J U_J$ with $J \subset \Pi$, the Levi subgroup L_J and the unipotent radical U_J inside P_J, there is a unique nilpotent orbit C in $\mathfrak{g}$ under the G-Adjoint action such that $C \cap u_J$ is an open dense subset of $u_J := \mathrm{Lie}(U_J)$ (cf. [**7**, Theorem 5.2.3]). This C is called a Richardson orbit, and $\bar{C} = G \cdot u_J$. In [**20**], Jantzen conjectured that under the condition that p is good, the support variety of $\mathfrak{g}$ over $H^0(\lambda)$ for $\lambda \in X(T)_+$ coincides with the closure of a Richardson orbit represented by $J \in \Pi$ which is determined by a $w \in W$ satisfying $w\Phi_\lambda = \Phi_J$ for $\Phi_\lambda := \{\alpha \in \Phi \mid \langle \lambda + \rho, \alpha^\vee \rangle \in \mathbb{Z}p\}$ (cf. [**20**, 2.2]). Here ρ is a half of the sum of all positive roots. This conjecture was proved in the case $\mathfrak{gl}(m)$ in the same paper. The conjecture for the general cases was proved to be true by D. Nakano, B. Parshall and D. Vella ([**23**, 6.2]). In the following we can reformulate it from the viewpoint of the $\mathbb{F}_q$-rational structure (cf. Theorem 3.6).

THEOREM 4.2. *Let $\mathcal{O}$ be a Richardson nilpotent orbit in $\mathfrak{g}$ under G-adjoint action. Then there exists a rational stable G-module V such that the Zariski closure $\bar{\mathcal{O}}$ of $\mathcal{O}$ equals to $\Phi(|\mathfrak{g}|_V)$. Hence $\bar{\mathcal{O}}$ admits an $\mathbb{F}_q$-rational structure.*

PROOF. Now that $\mathcal{O}$ is a Richardson nilpotent orbit, then we can set $\bar{\mathcal{O}} = G \cdot u_J$ for $J \in \Pi$, and $\mathcal{O}$ is the unique nilpotent orbit in $\mathfrak{g}$ such that $\mathcal{O} \cup u_J$ is an open dense subset of u_J. Choose $\lambda_J \in X(T)_+$ as follows: $\lambda_J = (2p-1)\rho + \sum_{i \in \Pi \setminus J} \varpi_i$ where the ϖ_i's are fundamental weights. Then $\Phi_{\lambda_J} = \Phi_J$. By Theorem ([**23**, 6.2]), $\Phi(|\mathfrak{g}|_{H^0(\lambda_J)}) = \bar{\mathcal{O}}$. The second statement is just implied by the above argument together with Theorem 3.6. □

4.3. The rational structures on generic nilpotent orbits for G of type $A_n = SL(n+1,k)$, $B_n = SO(2n+1,k)$ $C_n = SP(2n+1,k)$ and $D_n = SO(2n,k)$. Maintain the notations as the previous sections. Suppose G is a classical general groups in $GL(m,k)$. Since $F_\mathfrak{g}$ is a split Frobeniu-Lie morphism, it's just the restriction in $\mathfrak{g}$ of the standard Frobenius map from $\mathfrak{gl}(m)$ to $\mathfrak{gl}(m)$ by raising all matrix entries to the qth power $F_q : (a_{ij}) \to (a_{ij}^q)$ for $(a_{ij}) \in \mathfrak{gl}(m)$. So the following lemma is clear.

LEMMA 4.3. *Maintain the notations as above: $\mathfrak{g} = \mathrm{Lie}G$ for G a classical linear group, and $F_\mathfrak{g}$ the standard Frobenius map on $\mathfrak{g}$ and also $\mathfrak{gl}(m)$. Let V be a restricted $\mathfrak{gl}(m)$-module. The following statements hold.*

(a) *If V is a $F_\mathfrak{g}$-stable module of $\mathfrak{gl}(m)$, then V is also a F_q-stable module.*
(b) *(cf.* [**14**] *and* [**19**, 3.4]*)* $\Phi(|\mathfrak{g}|_V) = \Phi(|\mathfrak{gl}(m)|_V) \cap \mathfrak{g}$.

Now we turn to our aim. First, make some conventions that for a given m, $\mathcal{O}(\mathfrak{gl})$ stands for a nilpotent orbit in $\mathfrak{gl}(m)$ under $GL(m,k)$-Adjoin action, and that $\mathcal{O}(\mathfrak{g})$ stands for a nilpotent orbit in $\mathfrak{g}$ under G-adjoint action.

PROPOSITION 4.4. (cf. [**21**, §1] and [**6**, 5.1]) (1) Let $\mathcal{O}$ be a nilpotent orbit in $\mathfrak{gl}(m)$. Then $\mathcal{O}$ is parameterized by a unique partition $\mathbf{d} = (d_1 \geq d_2 \geq \cdots \geq d_r) \vdash m$ such that there is a unique representative in $\mathcal{O}$ with a standard Jordan form admitting ordered Jordan Blocks $(J_{d_1} \cdots, J_{d_r})$. (Note: $\mathbf{d}$ is also written as $\mathbf{d} = [1^{r_1}2^{r_2}3^{r_3}\cdots]$ if r_i is the number of i with $d_i = i$)

(2) Let $\mathcal{O}(\mathfrak{g})$ be a nilpotent orbit in $\mathfrak{g}$. Then $\mathcal{O}$ is parameterized by a unique partition $\mathbf{d} = (d_1 \geq \cdots \geq d_r) = [1^{r_1}2^{r_2}3^{r_3}\cdots]$ with the same meaning as in (1) and we have:

(a) For type A, $\mathbf{d}$ any possible partition of m;
(b) For type B, r_i is even for all even i;
(c) For type C, r_i is even for all odd i; Conversely, such a partition corresponds to a unique nilpotent orbit in $\mathfrak{g}$ for the above types.
(d) For type D, r_i is even for all even i. If $\mathbf{d}$ has only even parts, each having even multiplicity, which is called a very even partition, then such a partition $\mathbf{d}$ corresponds to two orbits, denoted by $\mathcal{O}^I(\mathfrak{g})_{\mathbf{d}}$ and $\mathcal{O}^{II}(\mathfrak{g})_{\mathbf{d}}$. A partition which is not very even corresponds to a unique nilpotent orbit in $\mathfrak{g}$.

We denote the nilpotent orbit in $\mathfrak{g}$ by $\mathcal{O}(\mathfrak{g})_{\mathbf{d}}$ for all cases except for $\mathbf{d}$ very even. Thus we have the following proposition.

PROPOSITION 4.5. ([**22**] and [**29**]) Let h =Coxeter number of G, $\mathcal{O}(\mathfrak{g})_{\mathbf{d}}$ be a nilpotent orbit in $\mathfrak{g}$ with $\mathbf{d}$ not being very even. Assume the following conditions are satisfied:

(a) For type B_n, assume $p > h+1 = 2n+1$; Set $m = 2n+1$.
(b) For type C_n, assume $p > h = 2n$; Set $m = 2n$.
(c) For type D_n, assume $p > h+2 = 2n$. Set $m = 2n$.

Then there is $\mathbf{a} \vdash m$ such that $\bar{\mathcal{O}}(\mathfrak{g})_{\mathbf{d}} = \overline{\mathcal{O}(\mathfrak{gl}(m))_{\mathbf{a}}} \cap \mathfrak{g}$.

THEOREM 4.6. *Maintain the above notations. Assume the conditions (a)(b)(c) in the above proposition are satified. Then the Zariski closures of all nilpotent orbits in $\mathfrak{g}$ admit an $\mathbb{F}_q$-rational structure.*

PROOF. First suppose $\mathcal{O} = \mathcal{O}(\mathfrak{g})^I_{\mathbf{d}}$ or $\mathcal{O}(\mathfrak{g})^{II}_{\mathbf{d}}$ with $\mathbf{d}$ very even. Then those orbits are Richardson. Theorem 4.2 implies that the statement is true in this case.

In the following argument, we suppose $\mathbf{d}$ is not very even. By Proposition 4.5, we have $\overline{\mathcal{O}(\mathfrak{g})_{\mathbf{d}}} = \overline{\mathcal{O}(\mathfrak{gl})_{\mathbf{a}}} \cap \mathfrak{g}$ for an $\mathbf{a} \vdash m$. Notice that all nilpotent orbits in $\mathfrak{gl}(m)$ are Richardson, hence by Theorem 4.2 we have $\overline{\mathcal{O}(\mathfrak{gl})_{\mathbf{a}}} = \Phi(|\mathfrak{gl}(m)|_{H(\lambda)})$ where $H(\lambda) = H^0_{GL(m,k)}(\lambda)$ for $\lambda \in X(T(GL(m,k)))$ where $T(GL(m,k))$ stands for the maximal torus of $GL(m,k)$. By Lemma 4.3(b), $\Phi(|\mathfrak{g}|_{H(\lambda)}) = \Phi(|\mathfrak{gl}(m)|_{H(\lambda)} \cap \mathfrak{g} = \overline{\mathcal{O}(\mathfrak{g})_{\mathbf{d}}}$. Thus, from Lemma 4.3(a) it follows that the closure of a not-very-even nilpotent orbit $\mathcal{O}$ is defined over $\mathbb{F}_q$. The proof is completed. □

References

[1] Bensen, D., *Modular representation theory: new trends and methods*, Springer-Verlag, LNM 1081, 1984.
[2] Borel, A., *Linear algebraic groups*, Springer-Verlag, GTM 129, 1991.
[3] Calson, J., *The variety of an indecomposalbe module is connected*, Inven. Math. 77, 291-299 (1984).
[4] Carlson, J. F., Lin, Z. and Nakano, D. *Support varieties for modules over Chevalley groups and classical Lie algebras*, preprint.
[5] Carter, R.W., *Finite groups of Lie type*, Wiley-Interscience, 1985.
[6] Collingwood, D.H. and McGovern, W.M., *Nilpotent orbits in semisimple Lie algebras*, Van Nostrand Reinhold, New York, 1993.
[7] Curtis, C.W. and Reiner, I., *Methods of representation theory. With applications to finite groups and orders*, Vol.I, A Wiley-Interscience Publication, New York, 1981.
[8] Deng, B. and Du, J., *Frobenius morphisms and representations of algebras*, Trans. Amer.Math. Soc. (to appear).
[9] Digne, F. and Michel, J., *Representaiotions of finite grouops of Lie type*, London Math. Soc., Student Texts, 21. Cambridge University Press, Cambridge, 1991.

[10] Du, J. and Shu B., *On finite Lie algebras and their representations*, preprint (2004).
[11] Cao,X.H. and Wang, J.P., *Introduction to representations of linear algebraic groups*, Science Publishing House, Beijing, 1987. (in Chinese)
[12] Fridelander, E. M. and Parshall, B. J., *Cohomlolgy of infinitesimal and discrete groups*, Math. Ann. 273 (1986), 353-374.
[13] Friedlander, E. M. and Parshall, B. J., *Support varieties for restricted Lie algebras*, Inv. Math. 86 (1986), 553-562.
[14] Friedlander, E. M. and Parshall, B. J., *Geometry of p-unipotent Lie algebras*, J. Algebra 109 (1987), 25-45.
[15] Friedlander, E. M. and Parshall, B. J., *Modular representations theroy of Lie algebras*, Amer. J. Math. 110 (1988), 117-126.
[16] Geck, M., *An introduction to algebraic geometry and algebraic groups*, Oxford University Press, 2003.
[17] Humphreys, J.E., *Introduction to Lie algebras and representation theroy*, Springer-Verlag, 1972
[18] Jantzen, J.C., *Representations of algebraic groups*, Academic Press, 1987.
[19] Jantzen, J.C., *Restricted Lie alegbra cohomology*, pp. 91-108, in *Algebraic groups: Proceedings of a conference in honor of T. A. Springer*, Lecture Notes in Mathematics 1271, Springer-Verlag, 1987.
[20] Jantzen, J.C., *Support varieties of Weyl modules*, Bull. LMS 19 (1987), 238-244.
[21] Jantzen, J.C. *Nilpotent orbits in representation theory*, pages 1-221, in "Lie Theory, Lie algebras and representations", Progress in Mathematics 228, Birhauser, Boston.Basel.Berlin, 2004.
[22] Kempken, B., *Induced conjugacy classes in classical Lie-algebras*, Abh. Math. Sem. Univ. Hamb. 53(1983), 53-83.
[23] Nakano, D., Parshall, B. and Vella, D., *Support varieties for algebraic groups*, J. Reine Angew. Math. 547(2002), 15-49.
[24] Li, W., *Support varieties and algebraic groups*, Doctoral Disseration, University of Virginia, 2001.
[25] Lin, Z. and Nakano, D., *Complexity for modules over finite Chevalley groups and classical Lie algebras*, Inv. Math. 138(1999), 85-101.
[26] Parshall,B. and Shu,B., On $\mathbb{F}_q$-rational structure of nilpotent cones in reductive Lie algebras, preprint (2005).
[27] Seligmen, G.B., *Modular Lie algebras*, Springer-Verlag, Berlin Heidelberg New York, 1967.
[28] Strade, H. and Farnsteiner, R., *Modular Lie algebras and their representations*, Marcel Dekker, New York, 1988.
[29] Springer, T.A. and Steinberg, R., *Conjugacy class*, pages 167-266, in "Seminar on algebraic groups and related finite groups, Lecture Notes in Mathematics 131, Springer-Verlag, Berlin, 1970.

Department of Mathematics, East China Normal University, Shanghai 200062, China.
E-mail address: `bshu@euler.math.ecnu.edu.cn`

AMS/IP Studies in Advanced Mathematics
Volume **39**, 2007

Perspectives On Geometric Analysis

Shing-Tung Yau

This essay grew from a talk I gave on the occasion of the seventieth anniversary of the Chinese Mathematical Society. I dedicate the lecture to the memory of my teacher S. S. Chern who had passed away half a year before (December 2004).

During my graduate studies, I was rather free in picking research topics. I [**727**] worked on fundamental groups of manifolds with non-positive curvature. But in the second year of my studies, I started to look into differential equations on manifolds. However, at that time, Chern was very much interested in the work of Bott on holomorphic vector fields. Also he told me that I should work on Riemann hypothesis. (Weil had told him that it was time for the hypothesis to be settled.) While Chern did not express his opinions about my research on geometric analysis, he started to appreciate it a few years later. In fact, after Chern gave a course on Calabi's works on affine geometry in 1972 at Berkeley, S. Y. Cheng told me about these inspiring lectures. By 1973, Cheng and I started to work on some problems mentioned in Chern's lectures. We did not realize that the great geometers Pogorelov, Calabi and Nirenberg were also working on them. We were excited that we solved some of the conjectures of Calabi on improper affine spheres. But soon after we found out that Pogorelov [**559**] published his results right before us by different arguments. Nevertheless our ideas are useful in handling other problems in affine geometry, and my knowledge about Monge-Ampère equations started to broaden in these years. Chern was very pleased by my work, especially after I [**732**] solved the problem of Calabi on Kähler Einstein metric in 1976. I had been at Stanford, and Chern proposed to the Berkeley Math Department that they hire me. I visited Berkeley in 1977 for a year and gave a course on geometric analysis with emphasis on isometric embedding.

Chern nominated me to give a plenary talk at the International Congress in Helsinki. The talk [**733**] went well, but my decision not to stay at Berkeley did not quite please him. Nevertheless he recommended me for a position on the faculty at the Institute for Advanced Study. Before I accepted a faculty position at the Institute, I organized a special year on geometry in 1979 at the Institute at the invitation of Borel. That was an exciting year because most people in geometric analysis came.

To appear in *Survey in Differential Geometry*, Vol. **X**, 2005.
This research is supported by NSF grants DMS-0244464, DMS-0354737 and DMS-0306600.

In 1979, I visited China at the invitation of Professor L. K. Hua. I gave a series of talks on the bubbling process of Sacks-Uhlenbeck [**577**]. I suggested to the Chinese mathematicians that they apply similar arguments for a Jordan curve bounding two surfaces with the same constant mean curvature. I thought it would be a good exercise for getting into this exciting field of geometric analysis. The problem was indeed picked up by a group of students of Professor G. Y. Wang [**360**]. But unfortunately it also initiated some ugly fights during the meeting of the sixtieth anniversary of the Chinese Mathematical Society. Professor Wang was forced to resign, and this event hampered development of this beautiful subject in China in the past ten years.

In 1980, Chern decided to develop geometric analysis on a large scale. He initiated a series of international conferences on differential geometry and differential equations to be held each year in China. For the first year, a large group of the most distinguished mathematicians was gathered in Beijing to give lectures (see [**147**]). I lectured on open problems in geometry [**735**]. It took a much longer time than I expected for Chinese mathematicians to pick up some of these problems.

To his disappointment, Chern's enthusiasm about developing differential equations and differential geometry in China did not stimulate as much activity as he had hoped. Most Chinese mathematicians were trained in analysis but were rather weak in geometry. The goal of geometric analysis for understanding geometry was not appreciated. The major research center on differential geometry came from students of Chern, Hua and B. C. Su. The works of J. Q. Zhong (see, e.g., [**751, 523, 524**]) were remarkable. Unfortunately he died about twenty years ago. Q. K. Lu studied the Bergman metric extensively. C. H. Gu [**294**] studied gauge theory and considered harmonic map where the domain is $R^{1,1}$. J. X. Hong (see, e.g., [**343, 316**]) did some interesting work on isometric embedding of surfaces into $\mathbb{R}^3$. In the past five years, the research center at the Chinese University of Hong Kong, led by L. F. Tam and X. P. Zhu, has produced first class work related to Hamilton's Ricci flow (see, e.g., [**124, 125, 128, 129, 112**]).

In the hope that it will advance Chern's ambition to build up geometric analysis, I will explain my personal view to my Chinese colleagues. I will consider this article to be successful if it conveys to my readers the excitement of developments in differential geometry which have been taking place during the period when it has been my good fortune to contribute. I do not claim this article covers all aspects of the subject. In fact, I have given priority to those works closest to my personal experience, and, consequently, I have given insufficient space to aspects of differential geometry in which I have not participated. In spite of these shortcomings, I hope that my readers, particularly those too young to know the origins of geometric analysis, will be interested to learn how the field looks to someone who was there. I would like to thank comments given by R. Bryant, H. D. Cao, J. Jost, H. Lawson, N. C. Leung, T. J. Li, Peter Li, J. Li, K. F. Liu, D. Phong , D. Stroock , X. W. Wang, S. Scott, S. Wolpert and S. W. Zhang. I am also grateful to J. X. Fu, especially for his help of tracking down references for the major part of this survey. When Fu went back to China, this task was taken up by P. Peng and X. F. Sun to whom I am grateful also.

In this whole survey, I follow the following:

Basic Philosophy:

Functions, tensors and subvarieties governed by natural differential equations provide deep insight into geometric structures. Information about these objects will give a way to construct a geometric structure. They also provide important information for physics, algebraic geometry and topology. Conversely it is vital to learn ideas from these fields.

Behind such basic philosophy, there are basic invariants to understand how space is twisted. This is provided by Chern classes [148], which appear in every branch of mathematics and theoretical physics. So far we barely understand the analytic meaning of the first Chern class. It will take much more time for geometers to understand the analytic meaning of the higher Chern forms. The analytic expression of Chern classes by differential forms have opened up a new horizon for global geometry. Professor Chern's influence on mathematics is forever.

An old Chinese poem says:

The reeds and rushes are abundant,

and the white dew has yet to dry.

The man whom I admire is on the bank of the river.

I go against the stream in quest of him,

But the way is difficult and turns to the right.

I go down the stream in quest of him,

and Lo! He is on the island in the midst of the water.

May the charm and beauty be always the guiding principle of geometry!

Contents

1. History and contributors of the subject

1.1. Founding fathers of the subject. Since the whole development of geometry depends heavily on the past, we start out with historical developments. The following are samples of work before 1970 which provided fruitful ideas and methods.

- **Fermat's principle of calculus of variation** (Shortest path in various media).
- **Calculus (Newton and Leibnitz)**: Path of bodies governed by law of nature.
- **Euler, Lagrange**: Foundation for the variational principle and the study of partial differential equations. Derivations of equations for fluids and for minimal surfaces.
- **Fourier, Hilbert**: Decomposition of functions into eigenfunctions, spectral analysis.
- **Gauss, Riemann**: Concept of intrinsic geometry.
- **Riemann, Dirichlet, Hilbert**: Solving Dirichlet boundary value problem for harmonic function using variational method.
- **Maxwell**: Electromagnetism, gauge fields, unification of forces.
- **Christoffel, Levi-Civita, Bianchi, Ricci**: Calculus on manifolds.
- **Riemann, Poincaré, Koebe, Teichmüller**: Riemann surface uniformization theory, conformal deformation.
- **Frobenius, Cartan, Poincaré**: Exterior differentiation and Poincaré lemma.
- **Cartan**: Exterior differential system, connections on fiber bundle.
- **Einstein, Hilbert**: Einstein equation and Hilbert action.
- **Dirac**: Spinors, Dirac equation, quantum field theory.
- **Riemann, Hilbert, Poincaré, Klein, Picard, Ahlfors, Beurling, Carlsson**: Application of complex analysis to geometry.
- **Kähler, Hodge**: Kähler metric and Hodge theory.
- **Hilbert, Cohn-Vossen, Lewy, Weyl, Hopf, Pogorelov, Efimov, Nirenberg**: Global surface theory in three space based on analysis.
- **Weierstrass, Riemann, Lebesgue, Courant, Douglas, Radó, Morrey**: Minimal surface theory.
- **Gauss, Green, Poincaré, Schauder, Morrey**: Potential theory, regularity theory for elliptic equations.
- **Weyl, Hodge, Kodaira, de Rham, Milgram-Rosenbloom, Atiyah-Singer**: de Rham-Hodge theory, integral operators, heat equation, spectral theory of elliptic self-adjoint operators.
- **Riemann, Roch, Hirzebruch, Atiyah-Singer**: Riemann-Roch formula and index theory.
- **Pontrjagin, Chern, Allendoerfer-Weil**: Global topological invariants defined by curvature forms.
- **Todd, Pontrjagin, Chern, Hirzebruch Grothendieck, Atiyah**: Characteristic classes and K-theory in topology and algebraic geometry.
- **Leray, Serre**: Sheaf theory.
- **Bochner-Kodaira**: Vanishing of cohomology groups based on the curvature consideration.

- **Birkhoff, Morse, Bott, Smale**: Critical point theory, global topology, homotopy groups of classical groups.
- **De Giorgi-Nash-Moser**: Regularity theory for the higher dimensional elliptic equation and the parabolic equation of divergence type.
- **Kodaira, Morrey, Grauert, Hua, Hörmander, Bergman, Kohn, Andreotti-Vesentini**: Embedding of complex manifolds, $\bar{\partial}$-Neumann problem, L^2 method, kernel functions.
- **Kodaira-Spencer, Newlander-Nirenberg**: Deformation of geometric structures.
- **Federer-Fleming, Almgren, Allard, Bombieri, De Giorgi, Giusti**: Varifolds and minimal varieties in higher dimensions.
- **Eells-Sampson, Al'ber**: Existence of harmonic maps into manifolds with non-positive curvature.
- **Calabi**: Affine geometry and conjectures on Kähler Einstein metric.

1.2. Modern Contributors. The major contributors can be roughly mentioned in the following periods:

I. 1972 to 1982: M. Atiyah, R. Bott, I. Singer, E. Calabi, L. Nirenberg, A. Pogorelov, R. Schoen, L. Simon, K. Uhlenbeck, S. Donaldson, R. Hamilton, C. Taubes, W. Thurston, E. Stein, C. Fefferman, Y. T. Siu, L. Caffarelli, J. Kohn, S. Y. Cheng, M. Kuranishi, J. Cheeger, D. Gromoll, R. Harvey, H. Lawson, M. Gromov, T. Aubin, V. Patodi, N. Hitchin, V. Guillemin, R. Melrose, Colin de Verdière, M. Taylor, R. Bryant, H. Wu, R. Greene, Peter Li, D. Phong, S. Wolpert, J. Pitts, N. Trudinger, T. Hildebrandt, S. Kobayashi, R. Hardt, J. Spruck, C. Gerhardt, B. White, R. Gulliver, F. Warner, J. Kazdan.

Highlights of the works in this period include a deep understanding of the spectrum of elliptic operators, introduction of self-dual connections for four manifolds, introduction of a geometrization program for three manifolds, an understanding of minimal surface theory, Monge-Ampére equations and the application of the theory to algebraic geometry and general relativity.

II. 1983 to 1992: In 1983, Schoen and I started to give lectures on geometric analysis at the Institute for Advanced Study. J. Q. Zhong took notes on the majority of our lectures. The lectures were continued in 1985 in San Diego. During the period of 1985 and 1986, K. C. Chang and W. Y. Ding came to take notes of some part of our lectures. The book *Lectures on Differential Geometry* was published in Chinese around 1989 [**602**]. It did have great influence for a generation of Chinese mathematicians to become interested in this subject. At the same time, a large group of my students made contributions to the subject. This includes A. Treibergs, T. Parker, R. Bartnik, S. Bando, L. Saper, M. Stern, H. D. Cao, B. Chow, W. X. Shi, F. Y. Zheng and G. Tian.

At the same time, J. Bismut, C. S. Lin, N. Mok, J. Q. Zhong, J. Jost, G. Huisken, D. Jerison, P. Sarnak, K. Fukaya, T. Mabuchi, T. Ilmanen, C. Croke, D. Stroock, Price, F. H. Lin, S. Zelditch, D. Christodoulou, S. Klainerman, V. Moncrief, C. L. Terng, Michael Wolf, M. Anderson, C. LeBrun, M. Micallef, J. Moore, John Lee, A. Chang, N. Korevaar were making contributions in various directions. One should also mention that in this period important work was done by the authors in the first group. For example, Donaldson, Taubes [**651**] and Uhlenbeck [**684, 685**] did spectacular work on Yang-Mills theory of general manifolds which led Donaldson [**194**] to solve the outstanding question on four manifold topology.

Donaldson [**195**], Uhlenbeck-Yau [**687**] proved the existence of Hermitian Yang-Mills connection on stable bundles. Schoen [**586**] solved the Yamabe problem.

III. 1993 to now: Many mathematicians joined the subject. This includes P. Kronheimer, B. Mrowka, J. Demailly, T. Colding, W. Minicozzi, T. Tao, R. Thomas, Zworski, Y. Eliashberg, Toth, Andrews, L. F. Tam, N. C. Leung, Y. B. Ruan, W. D. Ruan, R. Wentworth, A. Grigor'yan, L. Saloff-Coste, J. X. Hong, X. P. Zhu, M. T. Wang, A. K. Liu, K. F. Liu, X. F. Sun, T. J. Li, X. J. Wang, J. Loftin, H. Bray, J. P. Wang, L. Ni, P. F. Guan, N. Kapouleas, P. Ozsváth, Z. Szabó and Y. I. Li. The most important event is of course the first major breakthrough of Hamilton [**313**] in 1995 on the Ricci flow. I did propose to him in 1982 to use his flow to solve Thurston's conjecture. But after this paper by Hamilton, it is finally realized that it is feasible to solve the full geometrization program by geometric analysis. (A key step was the estimates on parabolic equations initiated by Li-Yau [**443**] and accomplished by Hamilton for Ricci flow [**310, 311**].) In 2002, Perelman [**547, 548**] brought in fresh new ideas to solve important steps that are remained in the program. Many contributors, including Colding-Minicozzi [**172**], Shioya-Yamaguchi [**612**] and Chen-Zhu [**128**], [**129**] have helped in filling important gaps in the arguments of Hamilton-Perelman. Cao-Zhu has just finished a long manuscript which gives a complete detail account of the program. The monograph will be published by International Press. In the other direction, we see the important development of Seiberg-Witten theory [**717**]. Taubes [**657, 658, 659, 660**] was able to prove the remarkable theorem for counting pseudo-holomorphic curves in terms of his invariants. Kronheimer-Mrowka [**400**] were able to solve the Thom conjecture that holomorphic curves provide the lowest genus surfaces in representing homology in algebraic surfaces. (Ozsváth-Szabó had a symplectic version [**544**].)

2. Construction of functions in geometry

The following is the basic principle [**733**]:

> **Linear or non-linear analysis is developed to understand the underlying geometric or combinatorial structure. In the process, geometry will provide deeper insight of analysis. An important guideline is that space of special functions defined by the structure of the space can be used to define the structure of this space itself.**

Algebraic geometers have defined the Zariski topology of an algebraic variety using ring of rational functions. In differential geometry, one should extract information about the metric and topology of the manifolds from functions defined over it. Naturally, these functions should be defined either by geometric construction or by differential equations given by the underlying structure of the geometry. (Integral equations have not been used extensively as the idea of linking local geometry to global geometry is more compatible with the ideas of differential equations.) A natural generalization of functions consists of the following: differential forms, spinors, and sections of vector bundles.

The dual concepts of differential forms or sections of vector bundles are submanifolds or foliations. From the differential equations that arise from the variational principle, we have minimal submanifolds or holomorphic cycles. Naturally the properties of such objects or the moduli space of such objects govern the geometry of the underlying manifold. A very good example is Morse theory on the space of loops on a manifold (see [**514**]).

I shall now discuss various methods for constructing functions or tensors of geometric interest.

2.1. Polynomials from ambient space. If the manifold is isometrically embedded into Euclidean space, a natural class of functions are the restrictions of polynomials from Euclidean space. However, isometric embedding in general is not rigid, and so functions constructed in such a way are usually not too useful.

On the other hand, if a manifold is embedded into Euclidean space in a canonical manner and the geometry of this submanifold is defined by some group of linear transformations of the Euclidean space, the polynomials restricted to the submanifold do play important roles.

2.1.1. *Linear functions being the harmonic function or eigenfunction of the submanifold.* For minimal submanifolds in Euclidean space, the restrictions of linear functions are harmonic functions. Since the sum of the norm square of the gradient of the coordinate functions is equal to one, it is fruitful to construct classical potentials using coordinate functions. This principle was used by Cheng-Li-Yau [**139**] in 1982 to give a comparison theorem for a heat kernel for minimal submanifolds in Euclidean space, sphere and hyperbolic space. Li-Tian [**437**] also considered a similar estimate for complex submanifolds of $\mathbb{C}P^n$. But this follows from [**139**] as such submanifolds can be lifted to a minimal submanifold in S^{2n+1}.

Another very important property of a linear function is that when it is restricted to a minimal hypersurface in a sphere S^{n+1}, it is automatically an eigenfunction. When the hypersurface is embedded, I conjectured that the first eigenfunction is linear and the first eigenvalue of the hypersurface is equal to n (see [**735**]). While this conjecture is not completely solved, the work of Choi-Wang [**154**] gives strong support. They proved that the first eigenvalue has a lower bound depending only on n. Such a result was good enough for Choi-Schoen [**152**] to prove a compactness result for embedded minimal surfaces in S^3.

2.1.2. *Support functions.* An important class of functions that are constructed from the ambient space are the support functions of a hypersurface. These are functions defined on the sphere and are related to the Gauss map of the hypersurface. The famous Minkowski problem reduces to solving some Monge-Ampère equation for such support functions. This was done by Nirenberg [**536**], Pogorelov [**556**], Cheng-Yau [**143**]. The question of prescribed symmetric functions of principal curvatures has been studied by many people: Pogorelov [**560**], Caffarelli-Nirenberg-Spruck [**91**], P. F. Guan and his coauthors (see [**296, 295**]), Gerhardt [**248**], etc. It is not clear whether one can formulate a useful Minkowski problem for higher codimensional submanifolds.

The question of isometric embedding of surfaces into three space can also be written in terms of the Darboux equation for the support function. The major global result is the Weyl embedding theorem for convex surfaces, which was proved by Pogorelov [**557, 558**] and Nirenberg [**536**]. The rigidity part was due to Cohn-Vossen and an important estimate was due to Weyl himself. For local isometric embeddings, there is work by C. S. Lin [**453, 454**], which are followed by Han-Hong-Lin [**316**]. The global problem for surfaces with negative curvature was studied by Hong [**343**]. In all these problems, infinitesimal rigidity plays an important role.

Unfortunately they are only well understood for a convex hypersurface. It is intuitively clear that generically, every closed surface is infinitesimally rigid. However, significant works only appeared for very special surfaces. Rado studied the set of surfaces that are obtained by rotating a curve around an axis. The surfaces constructed depend on the height of the curve. it turns out that such surfaces are infinitesimally rigid except on a set of heights which form part of a spectrum of some Sturm-Liouville operator.

2.1.3. *Gradient estimates of natural functions induced from ambient space.* A priori estimates are the basic tools for nonlinear analysis. In general the first step is to control the ellipticity of the problem. In the case of the Minkowski problem, we need to control the Hessian of the support function. For minimal submanifolds and other submanifold problem, we need gradient estimates which we shall discuss in Chapter 4. In 1974 and 1975, S. Y. Cheng and I [**142, 146**] developed several gradient estimates for linear or quadratic polynomials in order to control metrics of submanifolds in Minkowski spacetime or affine space. This kind of idea can be used to deal with many different metric problems in geometry.

The first theorem concerns a spacelike hypersurface M in the Minkowski space $\mathbb{R}^{n,1}$. The following important question arose: Since the metric on $\mathbb{R}^{n,1}$ is $\sum(dx^i)^2 - dt^2$, the restriction of this metric on M need not be complete even though it may be true for the induced Euclidean metric. In order to prove the equivalence of these two concepts for hypersurfaces whose mean curvatures are controlled, Cheng and I proved the gradient estimate of the function

$$\langle X, X\rangle = \sum_i (x^i)^2 - t^2$$

restricted on the hypersurface.

By choosing a coordinate system, the function $\langle X, X\rangle$ can be assumed to be positive and proper on M. For any positive proper function f defined on M, if we prove the following gradient estimate

$$\frac{|\nabla f|}{f} \leq C$$

where C is independent of f, then we can prove the induced metric on M is complete. This is obtained by integrating the inequality to get

$$|\log f(x) - \log f(y)| \leq Cd(x,y)$$

so that when $f(y) \to \infty$, $d(x,y) \to \infty$. Once we knew the metric was complete, we proved the Bernstein theorem which says that maximal spacelike hypersurface must be linear. Such work was then generalized by Treibergs [**681**], C. Gerhardt [**247**] and R. Bartnik [**40**] for hypersurfaces in more general spacetime. (It is still an important problem to understand the behavior of a maximal spacelike hypersurface foliation for general spacetime when we assume the spacetime is evolved by Einstein equation from a nonsingular data set.)

Another important example is the study of affine hypersurfaces M^n in an affine space A^{n+1}. These are the improper affine spheres

$$\det(u_{ij}) = 1$$

where u is a convex function or the hyperbolic affine spheres

$$\det(u_{ij}) = \left(-\frac{1}{u}\right)^{n+2}$$

where u is convex and zero on $\partial\Omega$ and Ω is a convex domain. Note that these equations describe hypersurfaces where the affine normals are either parallel or converge to a point.

For affine geometry, there is an affine invariant metric defined on M which is

$$(\det h_{ij})^{-\frac{1}{n+2}} \sum h_{ij} dx^i dx^j$$

where h_{ij} is the second fundamental form of M. A fundamental question is whether this metric is complete or not.

A coordinate system in A^{n+1} is chosen so that the height function is a proper positive function defined on M. The gradient estimate of the height function gives a way to prove completeness of the affine metric. Cheng and I [**146**] did find such an estimate which is similar to the one given above.

Once completeness of the affine metric is known, it is straight forward to obtain important properties of the affine spheres, some of which were conjectured by Calabi. For example we proved that an improper affine sphere is a paraboloid and that every proper convex cone admits a foliation of hyperbolic affine spheres. The statement about improper affine sphere was first proved by Jörgens [**362**], Calabi [**93**] and Pogorelov [**559**]. Conversely, we also proved that every hyperbolic affine sphere is asymptotic to a convex cone. (The estimate of Cheng-Yau was reproduced again by a Chinese mathematician who claimed to prove the result ten years afterwards.) Much more recently, Trudinger and X. J. Wang [**683**] solved the Bernstein problem for an affine minimal surface, thereby settling a conjecture by Chern. They found a counterexample for dim≥ 10. These results are solid contributions to fourth order elliptic equations.

The argument of using gradient estimates for some naturally defined function was also used by me to prove that the Kähler Einstein metric constructed by Cheng and myself is complete for any bounded pseudo-convex domain [**144**]. (It appeared in my paper with Mok [**522**] who proved the converse statement which says that if the Kähler Einstein metric is complete, the domain is pseudo-convex.)

It should be noted that in most cases, gradient estimates amount to control of ellipticity of the nonlinear elliptic equation.

> **Comment**: To control a metric, find functions that are capable of describing the geometry and give gradient or higher order estimates for these functions.

2.2. Geometric construction of functions.

2.2.1. *Distance function and Busemann function.* When manifolds cannot be embedded into the linear space, we construct functions adapted to the metric structure. Obviously the distance function is the first major function to be constructed. A very important property of the distance function is that when the Ricci curvature of the manifold is greater than the Ricci curvature of a model manifold which is spherical symmetric at one point, the Laplacian of the distance function is not greater than the Laplacian of the distance function of the model manifold in the sense of distribution. This fact was used by Cheeger-Yau [**123**] to give a sharp lower estimate of the heat kernel of such manifolds. An argument of this type was also used by Perelman in his recent work.

Gromov [**282**] developed a remarkable Morse theory for the distance function (a preliminary version was developed by K. Grove and K. Shiohama [**293**]) to compare the topology of a geodesic ball to that of a large ball, thereby obtaining a bound on the Betti numbers of compact manifolds with nonnegative sectional curvature. (He can also allow the manifolds to have negative curvature. But in this case the diameter and the lower bound of the curvature will enter into the estimate.)

We can also take the distance function from a hypersurface and compute the Hessian of the distance function. In general, one can prove comparison theorems, and the principle curvatures of the hypersurface will come into the estimates. However the upper bound of the Laplacian of the function depends only on the Ricci curvature of the ambient manifold and the mean curvature of the hypersurface. This kind of calculation was used in the sixties by Penrose and Hawking to study the focal locus of a closed surface under the assumption that the surface is "trapped" which means the mean curvatures are negative in both the ingoing and the outgoing null directions. This information allowed them to prove the first singularity theorem in general relativity (which demonstrates that the black hole singularity is stable under perturbation). The distance to hypersurfaces can be used as barrier functions to prove the existence of a minimal surface as was shown by Meeks-Yau [**503**], [**504**]. T. Frankel used the idea of minimizing the distance between two submanifolds to detect the topology of minimal surfaces. In particular, two maximal spacelike hypersurfaces in spacetime which satisfy the energy condition must be disjoint if they are parallel at infinity.

Out of the distance function, we can construct the Busemann function in the following way:

Given a geodesic ray $\gamma : [0, \infty) \to M$ *so that*

$$\textit{distance}(\gamma(t_1), \gamma(t_2)) = t_2 - t_1,$$

where $\| \frac{d\gamma}{dt} \| = 1$, *one defines*

$$B_\gamma(x) = \lim_{t \to \infty} (d(x, \gamma(t)) - t).$$

This function generalizes the notion of a linear function. For a hyperbolic space form, its level set defines horospheres. For manifolds with positive curvature, it is concave. Cohn-Vossen (for surface) and Gromoll-Meyer [**277**] used it to prove that a complete noncompact manifold with positive curvature is diffeomorphic to $\mathbb{R}^n$.

A very important property of the Busemann function is that it is superharmonic on complete manifolds with nonnegative Ricci curvature in the sense of distribution. This is the key to prove the splitting principle of Cheeger-Gromoll [**118**]. Various versions of this splitting principle have been important for applications to the structure of manifolds. When I [**732**] proved the Calabi conjecture, the splitting principle was used by me and others to prove the structure theorem for Kähler manifolds with a nonnegative first Chern Class. (The argument for the structure theorem is due to Calabi [**92**] who knew how to handle the first Betti number. Kobayashi [**385**] and Michelsohn [**512**] wrote up the formal argument and Beauville [**45**] had a survey article on this development.)

In 1974, I was able to use the Busemann function to estimate the volume of complete manifolds with nonnegative Ricci curvature [**730**]. After long discussions with me, Gromov [**283**] realized that my argument of Busemann function amounts to compare volumes of geodesic balls. The comparison theorem of Bishop-Gromov had been used extensively in metric geometry.

If we consider $\inf_\gamma B_\gamma$, where γ ranges from all geodesic rays from a point on the manifold, we may be able to obtain a proper exhaustion of the manifold. When M is a complete manifold with finite volume and its curvature is pinched by two negative constants, Siu and I [**630**] did prove that such a function gives a concave exhaustion of the manifold. If the manifold is also Kähler, we were able to prove that one can compactify the manifold by adding a point to each end to form a compact complex variety. In the other direction, Schoen-Yau [**599**] was able to use the Busemann function to construct a barrier for the existence of minimal surfaces to prove that any complete three dimensional manifold with positive Ricci curvature is diffeomorphic to Euclidean space.

The Busemann function also gives a way to detect the angular structure at infinity of the manifold. It can be used to construct the Poisson kernel of hyperbolic space form. For a simply connected complete manifold with bounded and strongly negative curvature, it is used as a barrier to solve the Dirichlet problem for bounded harmonic functions, after they are mollified at infinity. This was achieved by Sullivan [**643**] and Anderson [**8**]. Schoen and Anderson [**9**] obtained the Harnack inequality for a bounded harmonic function and identified the Martin boundary of such manifolds. W. Ballmann [**27**] then studied the Dirichlet problem for manifolds of non-positive curvature. Schoen and I [**602**] conjectured that nontrivial bounded harmonic function exists if the manifold has bounded geometry and a positive first eigenvalue. Many important cases were settled in [**602**]. Lyons-Sullivan [**483**] proved the existence of nontrivial bounded harmonic functions using the non-amenability of groups acting on the manifold.

The abundance of bounded harmonic functions on the universal cover of a compact manifold should mean that the manifold is "hyperbolic". Hence if the Dirichlet problem is solvable on the universal cover, one expects the Gromov volume of the manifold to be greater than zero.

The Martin boundary was studied by L. Ji and MacPherson (see [**301, 359**]) for the compactification of various symmetric spaces. For product of manifolds with negative curvature, it was determined by A. Freire [**231**]. For rank one complete manifolds with non-positive curvature, work has been done by Ballman-Ledrappier [**28**] and Cao-Fan-Ledrappier [**104**]. It should be nice to generalize the work of L. K. Hua on symmetric spaces with higher rank to general manifolds with non-positive curvature. Hua found that bounded harmonic functions satisfy extra equations (see [**346**]).

2.2.2. *Length function defined on loop space.* If we look at the space of loops in a manifold, we can take the length of each loop and thereby define a natural function on the space of loops. This is a function for which Morse theory found rich application. Bott [**70**] made use of it to prove his periodicity theorem. Bott [**67, 69**] and Morse also developed formula for computing index a geodesic. Bott showed that the index of a closed geodesic and its linearized Poincaré map determine the indexes of iterates of this geodesic. Starting from the famous works of Poincaré, Birkhoff, Morse and Ljusternik-Shnirel'man, there has been extensive work on proving the existence of a closed geodesic using Morse theory on the space of loops. Klingenberg and his students developed powerful tools (see [**384**]). Gromoll-Meyer [**278**] did important work in which they proved the existence of infinitely many closed geodesics assuming the Betti number of the free loop space of the manifold grows unboundedly. They used the results of Bott [**69**], Serre and some version of degenerate Morse theory. There was also later work by Ballmann, Ziller, G. Thorbergsson, Hingston and Kramer (see, e.g., [**30, 326, 399**]), who improved the Gromoll-Meyer theorem to give a low estimate of the growth of the number of geometrically distinct closed geodesics of length $\leq t$. In most cases, they grow at least as fast as the prime numbers. The classical important question that every metric on S^2 supports an infinite number of closed geodesics was also solved affirmatively by Franks [**227**], Bangert [**35**] and Hingston [**327**]. An important achievement was made by Vigué-Poirrier and D. Sullivan [**691**] who proved that the Gromoll-Meyer condition for the existence of infinite numbers of closed geodesics is satisfied if and only if the rational cohomology algebra of the manifold has at least two generators. They made use of Sullivan's theory of the rational homotopic type. When the metric is Finsler, the most recent work of Victor Bangert and Yiming Long [**36**] showed the existence of two closed geodesics on the two dimensional sphere. (Katok [**375**] produced an example which shows that two is optimal.) Length function is a natural concept in Finsler geometry. In the last fifty years, Finsler geometry has not been popular in western world. But under the leadership of Chern, David Bao, Z. Shen, X. H. Mo and M. Ji did develop Finsler geometry much further (see, e.g., [**37**]).

A special class of manifolds, all of whose geodesics are closed, have occupied quite a lot of interest of distinguished geometers. It started from the work of Zoll (1903) for surfaces where Guillemin did important contributions. Bott [**68**] has determined the cohomology ring of these manifolds. The well known Blaschke conjecture was proved by L. Green [**268**] for two dimension and by M. Berger and J. Kazdan (see [**51**]) for higher dimensional spheres. Weinstein [**709**] and C. T. Yang [**723, 724, 725**] made important contributions to the conjecture for other homotopic types.

2.2.3. *Displacement functions.* When the manifold has negative curvature, the length function of curves is related to the displacement function defined in the following way:

If γ is an element of the fundamental group acting on the universal cover of a complete manifold with non-positive curvature, we consider the function $d(x, \gamma(x))$: The study of such a function gives rise to properties of compact manifolds with non-positive curvature. For example, in my thesis, I generalized the Preissmann theorem to the effect that every solvable subgroup of the fundamental group must be a finite extension of an Abelian group which is the fundamental group of a totally geodesic flat sub-torus [**727**]. Gromoll-Wolf [**279**] and Lawson-Yau [**410**] also proved that if the fundamental group of such a manifold has no center and splits as a product, then the manifold splits as a metric product. Strong rigidity result for discrete group acting on product of manifolds irreducibly was obtained by Jost-Yau [**370**] where they proved that these manifolds are homogeneous if the discrete group also appears as fundamental group of compact manifolds with nonpositive curvature.

When the manifold has bounded curvature, Margulis studied those points where $d(x, \gamma(x))$ is small and proved the famous Margulis lemma which was used extensively by Gromov [**280**] to study the structure of manifolds with non-positive curvature.

Comment: The lower bound of sectional curvature (or Ricci curvature) of a manifold gives upper estimate of the Hessian (or the Laplacian) of the distance functions. Since most functions constructed in geometry come from distance functions, we have partial control of the Hessian of these functions. The information provides us with basic tools to construct barrier functions for harmonic analysis or to produce convex functions. The Hessian of distance functions come from computations of second variation of geodesics. If we consider the second variation of closed geodesic loops, we get information about the Morse index of the loop, which enable us to link global topology to the existence of many closed geodesics or curvatures of the manifold.

We always look for canonical objects through geometric constructions and deform them to find their global properties.

2.3. Functions and tensors defined by linear differential equations. Direct construction of functions or tensors based on geometric intuitions alone is not rich enough to handle the very complicated geometric world. One should produce global geometric objects based on global differential equations. Often the construction depends on the maximal principle, integration by part, or the method of contradictions, and they are not necessarily geometric intuitive. On the other hand, basic principle of global differential equations does fit well with modern geometry in relating local data to global behavior. In order for the theory to be effective, the global differential operator has to be constructed from a geometric structure naturally.

The key to understanding any self-adjoint linear elliptic differential operator is to understand its spectral resolution and the detail of the structure of objects in the process of the resolution: eigenvalues or eigenfunctions are particularly important for their relation to geometry. Low eigenvalues and low eigenfunctions give deep information about global geometry such as topology or isoperimetric inequalities. High eigenvalues and high eigenfunctions are related to local geometry such as curvature forms or characteristic forms. Semiclassical analysis in quantum physics give a way to relate these two ends. This results in using either the heat equation or the hyperbolic equation.

There are many important first order differential operators: d, δ, $\bar{\partial}$, Dirac operator. All these operators have contributed to a deeper understanding of geometry. They form systems of equations. Our understanding of them is not as deep as the Laplacian acting on functions. The future of geometry will rest on an understanding of global systems of equations and their relation to global topology. The index theorem is the most important contribution. It provides information about the kernel (or cokernel). We still need to have a deeper understanding of the spectrum of these operators.

2.3.1. *Laplacian.*

(a). Harmonic functions. The spectral resolution of the Laplacian gives rise to eigenfunctions. Harmonic functions are therefore the simplest functions that play important roles in geometry.

If the manifold is compact, the maximum principle shows that harmonic functions are constant. However, when we try to understand the singularities of compact manifolds, we may create noncompact manifolds by scaling and blowing up processes, at which point harmonic functions can play an important role.

The first important question about harmonic functions on a complete manifold is the Liouville theorem. I started my research on analysis by understanding the right formulation of the Liouville theorem. In 1971, I thought that it is natural to prove that for complete manifolds with a non-negative Ricci curvature, there is no nontrivial harmonic function [**728**]. I also thought that in the opposite case, when a complete manifold has strongly negative curvature and is simply connected, one should be able to solve Dirichlet problem for bounded harmonic functions.

The gradient estimates [**728**] that I derived for a positive harmonic function come from a suitable interpretation of the Schwarz lemma in complex analysis. In fact, I generalized the Ahlfors Schwarz lemma before I understood how to work out the gradient estimates for harmonic functions. The generalized Schwarz lemma [**734**] says that holomorphic maps, from a complete Kähler manifold with Ricci curvature bounded from below to a Hermitian manifold with holomorphic bisectional curvature bounded from above by a negative constant, are distance decreasing with constants depending only on the bound on the curvature. This generalization has since found many applications such as the study of the geometry of moduli spaces by Liu-Sun-Yau [**467, 468**]. They used it to prove the equivalence of the Bergman metric with the Kähler-Einstein metric on the moduli space. They also proved that these metrics are equivalent to the Teichmüller metric and the McMullen metric.

The classical Liouville theorem has a natural generalization: Polynomial growth harmonic functions are in fact polynomials. Motivated by this fact and several complex variables, I asked whether the space of polynomial growth harmonic functions with a fixed growth rate is finite dimension with the upper bound depending only on the growth rate [**737**]. This was proved by Colding-Minicozzi [**167**] and generalized by Peter Li [**435**]. (Functions can be replaced by sections of bundles). In a beautiful series of papers (see, e.g., [**438, 439**]), P. Li and J. P. Wang studied the space of harmonic functions in relation to the geometry of manifolds. In the case when harmonic functions are holomorphic, they form a ring. I am curious about the structure of this ring. In particular, is it finitely generated when the manifold is complete and has a nonnegative Ricci curvature? A natural generalization of such a question is to consider holomorphic sections of line bundles, especially powers of canonical line bundles. This is part of Mori's minimal model program.

(b). Eigenvalues and eigenfunctions. Eigenvalues reflect the geometry of manifolds very precisely. For domains, estimates of them date back to Lord Rayleigh. Hermann Weyl [**707**] solved a problem of Hilbert's on the asymptotic behavior of eigenvalues in relation to the volume of the domain and hence initiated a new subject of spectral geometry. Pólya-Szegö, Faber, Krahn and Levy gave estimates of eigenvalues of various geometric problems. On a general manifold, Cheeger [**113**] was the first person to relate a lower estimate of the first eigenvalue with the isoperimetric constant (now called the Cheeger constant). One may note that many questions on the eigenvalue for domains are still unsolved. The most noted one is the Pólya conjecture which gave a sharp lower estimate of the Dirichlet problem in terms of volume. Li-Yau [**442**] did settle the average version of the Pólya conjecture.

The gradient estimate that I found for harmonic functions can be generalized to cover eigenfunctions and Peter Li [**434**] was the first one to apply it to finding estimates for eigenvalues for manifolds with positive Ricci curvature. (If the Ricci curvature has a positive lower bound, this is due to Lichnerowicz.) Li-Yau [**440**] then solved the well-known problem of estimating eigenvalues of manifolds in terms of their diameter and the lower bound on their Ricci curvature. Li-Yau conjectured the sharp constant for their estimates, and Zhong-Yang [**751**] were able to prove this conjecture by sharpening Li and Yau's arguments. A probabilistic argument was developed by Chen and Wang [**132**] to derive these inequalities. The precise upper bound for the eigenvalue was first obtained by S. Y. Cheng [**136**] also in terms of diameter and lower bound of the Ricci curvature. Cheng's theorem gives a very good demonstration of how the analysis of functions provides information about geometry. As a corollary of his theorem, he proved that if a compact manifold M^n has a Ricci curvature $\geq n-1$ and the diameter is equal to π, then the manifold is isometric to the sphere. He used a lower estimate for eigenvalues based on the work of Lichnerowicz and Obata. Colding [**166**] was able to use functions with properties close to those of the first eigenfunction to prove a pinching theorem which states that: When the Ricci curvature is bounded below by $n-1$ and the volume is close to that of the unit sphere, the manifold is diffeomorphic to the sphere. There is extensive work by Colding-Cheeger [**115, 116, 117**] and Perelman (see, e.g., [**87**]) devoted to the understanding of Gromov's theory of Hausdorff convergence for manifolds. The tools they used include the comparison theorem, the splitting theorem of Cheeger and Gromoll, and the ideas introduced earlier by Colding.

A very precise estimate of eigenvalues of the Laplacian has been important in many areas of mathematics. For example, the idea of Szegö [**647**]-Hersch [**325**] on the upper bound of the first eigenvalue in terms of the area alone was generalized by me to the higher genus in joint works with P. Yang [**726**] and P. Li [**441**]. For genus one, this was Berger's conjecture, as I was informed by Cheng. After Cheng showed me the paper of Hersch, I realized how to create trial functions by taking the branched conformal cover of S^2. While the constant in the paper of Yang-Yau [**726**] for torus is not the best possible, the recent work of Jakobson, Levitin, Nadirashvili, Nigam and Polterovich [**356**] demonstrated that the constant for a genus two surface may be the best possible and may be achieved by Bolza's surface.

Shortly afterwards, I applied the argument of [**726**] to prove that a Riemann surface defined by an arithmetic group must have a relative high degree when it is branched over the sphere. This observation of using Selberg's estimate coupled with Li-Yau [**441**] was made in 1985 when I was in San Diego, where I also used similar idea to estimate genus of mini-max surface in a three dimensional manifolds and also to prove positivity of Hawking mass. After I arrived in Harvard, I discussed the idea with my colleague N. Elkies and B. Mazur. The paper was finally written up and published in 1995 [**741**]. In the mean while, ideas of using my work on eigenvalue coupled with Selberg's work to study congruence subgroup was generalized by D. Abramovich [**1**] (my idea was conveyed by Elkies to him) and by P. Zograf [**754**] to the case where the curve has cusps. Most recently Ian Agol [**2**] also used similar idea to study arithmetic Kleinian reflection groups.

In a beautiful article, N. Korevaar [**395**] gave an upper bound, depending only on genus and n, for the n-th eigenvalue λ_n of a Riemann surface. His result answered a challenge of mine (see [**735**]) when I met him in Utah in 1989. Grigor'yan, Netrusov and I [**276**] were able to give a simplified proof and apply the estimate to bound the index of minimal surfaces. There are also works by P. Sarnak (see, e.g., [**582, 355**]) on understanding eigenfunctions for such Riemann surfaces. Iwaniec-Sarnack [**355**] showed that the estimate of the maximum norm of the n-th eigenfunction on an arithmetic surface has significant interest in number theory. Wolpert [**721**] analyzes perturbation stability of embedded eigenvalues and applies asymptotic perturbation theory and harmonic map theory to show that stability is equivalent to the non-vanishing of certain standard quantities in number theory. There was also the work of Schoen-Wolpert-Yau [**591**] on the behavior of eigenvalues $\lambda_1, \cdots, \lambda_{2g-3}$ for a compact Riemann surface of genus g. These are eigenvalues that may tend to zero for metrics with curvature -1. However, λ_{2g-2}, λ_{2g-1}, $\cdots$, λ_{4g-1} always appear in $[c_g, \frac{1}{4}]$ where $c_g > 0$ depends only on g. It will be nice to find the optimal c_g.

In this regard, one may mention the very deep problem of Selberg on lower estimate of λ_1 for surfaces defined by an arithmetic group. Selberg proved that it is grater than $\frac{3}{16}$ and it was later improved by Luo-Rudnick-Sarnak [**478**]. For a higher dimensional locally symmetric space, there is a similar question of Selberg and results similar to Selberg's were found by J. S. Li [**423**] and Cogdell-Li-Piatctski-Shapiro-Sarnak [**165**]. Many researchers attempt to use Kazdhan's property T for discrete groups to study Selberg's problem.

There are many important properties of eigenfunctions that were studied in the seventies. For example, Cheng [**137**] found a beautiful estimate of multiplicities of eigenvalues of Riemann surfaces based only on genus. The idea was used by Colin de Verdière [**174**] to embedded graphes into $\mathbb{R}^3$ when they satisfy nice combinatorial properties. The connectivity and the topology of nodal domains are very interesting questions. Melas [**506**] did prove that for a convex planar domain, the nodal line of second eigenfunctions must intersect the boundary in exactly two points. Very little is known about the number of nodal domains except the famous theorem of Courant that the number of nodal domains of the m-th eigenfunction is less than m.

There are several important questions related to the size of nodal sets and the number of critical points of eigenfunctions. I made a conjecture (see [**735**]) about the area of nodal sets, and significant progress toward its resolution was made by Donnelly-Fefferman [**206**], Dong [**205**] and F. H. Lin [**456**]. The number of critical points of an eigenfunction is difficult to determine. I [**742**] managed to prove the existence of a critical point near the nodal set. Jakobson and Nadirashvili [**357**] gave a counterexample to my conjecture that the number of critical points of the n-th eigenfunction is unbounded when n tends to infinity. I believe the conjecture is true for generic metrics and deserves to be studied extensively. Nadirashvili and his coauthors [**342, 317**] were also the first to show that the critical sets of eigenfunctions in n-dimensional manifold have a finite H^{n-2}-Hausdorff measure. Afterwards, Han-Hardt-Lin [**315**] gave an explicit estimate.

When there is potential V, the eigenvalues of $-\triangle + V$ are also important. When V is convex, with Singer, Wong and Stephen Yau, I applied the argument that I had with Peter Li to estimate the gap $\lambda_2 - \lambda_1$ [**620**]. When V is arbitrary,

I [**743**] observed how this gap depends on the lower eigenvalue of the Hessian of $-\log\psi$, where ψ is the ground state. The method of continuity was used by me in 1980 to reprove the work of Brascamp-Lieb [**78**] on the convexity of $-\log\psi$ when V is convex (This work appeared in the appendix of [**620**]). When V is the scalar curvature, this was studied by Schoen and myself extensively. In fact, in [**600**], we found an upper estimate of the first Dirichlet eigenvalue of the operator $-\triangle+\frac{1}{2}R$ in terms of $\frac{3\pi^2}{2r^2}$ where r is a certain concept of radius related to loops in a three dimensional manifold. (If we replace loops with higher dimensional spheres, one can define a similar concept of radius. It will be nice if such a concept can be related to eigenvalues of differential forms.) This operator is naturally related to conformal deformation, stability of minimal surfaces, etc. (The works of D. Fischer-Colbrie and Schoen [**222**], Micallef [**508**], Schoen-Yau [**593, 599**] on stable minimal surfaces all depend on an understanding of spectrum of this operator.) The parabolic version appears in the recent work of Perelman.

If there is a closed non-degenerate elliptic geodesic in the manifold, Babič [**25**], Guillemin and Weinstein [**300**] found a sequence of eigenvalues of the Laplacian which can be expressed in terms of the length, the rotation angles and the Morse index of the geodesic.

> **Comment**: It is important to understand how harmonic functions or eigenfunctions oscillate. Gradient estimate is a good tool to achieve this. Gradient estimate for the log of the eigenfunction can be used to prove the Louville theorem or give a good estimate of eigenvalues. For higher eigenfunctions, it is important to understand its zero set and its growth. By controlling this information, one can estimate the dimension of these functions. A good upper estimate for eigenvalues depends on geometric intuition which may lead to construction of trial functions that are more adaptive to geometry. It should be emphasized that a clean and sharp estimate for the linear operator is key to obtaining good estimates for the nonlinear operator.

(c). Heat kernel. Most of the work on the heat kernel over Euclidean space can be generalized to those manifolds where Sobolev and Poincaré inequalities hold. (It should be noted that Aubin [**22, 24**] and Talenti [**648**] did find best constant for various Sobolev inequalities on Euclidean space.) These inequalities are all related to isoperimetric inequalities. C. Croke [**177**] was able to follow my work [**729**] on Poincaré inequalities to prove the Sobolev inequality depending only on volume, diameter and the lower bound of Ricci curvature. Arguments of John Nash were then used by Cheng-Li-Yau [**138**] to give estimates of the heat kernel and its higher derivatives. In this paper, an estimate of the injectivity radius was derived and this estimate turns out to play a role in Hamilton's theory of Ricci flow. A year later, Cheeger-Gromov-Taylor [**121**] made use of the wave kernel to reprove this estimate. D. Stroock (see [**534**]) was able to use his methods from Malliavin's calculus to give remarkable estimates for the heat kernel at a pair of points where one point is at the cut locus of another point.

The estimate of the heat kernel was later generalized by Davies [**184, 185**], Saloff-Coste [**578**] and Grigor'yan [**274, 275**] to complete manifolds with polynomial volume growth and volume doubling property. Since these are quasi-isometric invariants, their analysis can be applied to graphs or discrete groups. See Grigor'yan's survey [**275**] and Saloff-Coste's survey [**579**].

On the other hand, the original gradient estimate that I derived is a pointwise inequality that is much more adaptable to nonlinear theory. Peter Li and I [**443**] were able to find a parabolic version of it in 1984. We observed its significance for estimates on the heat kernel and its relation to the variational principle for paths in spacetime. Coupled with the work of Cheeger-Yau [**123**], this gives a very precise estimate of the heat kernel. Such ideas turn out to provide fundamental estimates which play a crucial role in the analysis of Hamilton's Ricci flow [**310, 311**].

Not much is known about the heat kernel on differential forms or differential forms with twisted coefficients. The fundamental idea of using the heat equation to prove the Hodge theory came from Milgram-Rosenbloom. The heat kernel for differential forms with twisted coefficients does play an important role in the analytic proof of the index theorem, as was demonstrated by Atiyah-Bott-Patodi [**13**]. It is the alternating sum that exhibits cancellations and gives rise to index of elliptic operators. When t is small, the alternating sum reduces to a calculation of curvature forms. When t is large, it gives global information on harmonic forms. Since the index of the operator is independent of t, we can relate the index to characteristic forms.

If a compact manifold is the quotient of a non-compact manifold by a discrete group and if the heat kernel of the non-compact manifold can be computed explicitly, it can be averaged to give the heat kernel of the quotient manifold. Since the integral of the later kernel on the diagonal can be computed by the spectrum to be $\sum e^{-t\lambda_i}$, one can relate the displacement function of the discrete group to the spectrum. This is the Selberg trace formula relating length of closed geodesics to the spectrum of the Laplacian.

> **Comment**: Understanding the heat kernel is almost the same as understanding the heat equation. However, heat kernel satisfies semi-group properties, which enables one to give a good estimate of the maximum norm or higher order derivative norms as long as the Sobolev inequality can be proved. It is useful to look at the heat equation in spacetime where the Li-Yau gradient estimate is naturally defined. The estimate provides special pathes in spacetime for the estimate of the kernel. However, the effects of closed geodesics have not been found in the heat equation approach. A sharp improvement of the Li-Yau estimate may lead to such information.

(d). Isoperimetric inequalities. Isoperimetric inequality is a beautiful subject. It has a long history. Besides its relation to eigenvalues, it reviews the deep structure of manifolds. The best constant for the inequality is important. Pólya-Szegö [**561**], G. Faber (1923), E. Krahn (1925) and P. Lévy (1951) made fundamental contributions. Gromov generalized the idea of Lévy to obtain a good estimate of Cheeger's constant (see [**287**]). C. Croke [**178**] and Cao-Escobar [**103**] have worked on domains in a simply connected manifold with non-positive curvature. It is assumed that the inequality holds for any minimal subvariety in Euclidean space. But it is not known to be true for the best constant. Li-Schoen-Yau [**436**] did prove it in the case of a minimal surface with a connected boundary, and E. Lutwak, Deane Yang and G. Y. Zhang did some beautiful work in the affine geometry case (see,

e.g., [**480, 481**]). In Hamilton's proof of Ricci flow convergent to the round metric on S^2, he demonstrated that the isoperimetric constant of the metric is improving. One sees how the constant controls the geometry of the manifold.

> **Comment**: The variational principle has been the most important method in geometry since the Greek mathematicians. Fixing the area of the domain and minimizing the length of the boundary is the most classical form of isoperimetric inequality. This principle has been generalized to much more general situations in geometry and mathematical physics. In most cases, one tries to prove existence of the extremal object and establish isoperimetric inequalities by calculating corresponding quantities for the extremal object. There is also the idea of rearrangement or symmetrization to prove isoperimetric inequalities. In the other direction, there is the duality principle in the alculus of variation: instead of minimizing the length of the boundary, one can fix it and maximize the area it encloses. The principle can be effective in complicated variational problems.

(e). Harmonic analysis on discrete geometry. There are many other ideas in geometric analysis that can be discretized and applied to graph theory. This is especially true for the theory of spectrum of graphs. Some of these were carried out by F.Chung, Grigor'yan and myself (see the reference of Chung's survey [**163**]). But the results in [**163**] are far away from being exhaustive. On the other hand, Margulis [**487**] and Lubotzky-Phillips-Sarnak [**476**] were able to make use of discrete group and number theory to construct expanding graphes. Methods to construct and classify these expanding graphs are important for application in computer science. It should be noted that Kazhdan's property (T) [**378**] did play an important role in such discussions. It is also important to see how to give a good decomposition of any graph using the spectral method.

The most important work for the geometry of a finitely presented group was done by Gromov [**282**]. He proved the fundamental structure theorem of groups where volume grows at most polynomially. These groups must be virtually nilpotent. Geometric ideas were developed by Varopoulos and his coauthors [**689, 38**] on the precise behaviors of the heat kernel in terms of volume growth. As an application of the theory of amenable groups, R. Brooks [**80**] was able to prove that if a manifold covers a compact set by a discrete group Γ, then it has positive eigenvalue if and only if Γ is non-amenable.

Gromov [**281**] also developed a rich theory of hyperbolic groups using concepts of isoperimetric type inequalities. It would be nice to characterize these groups that are fundamental groups of compact manifolds with non-positive curvature or locally symmetric spaces.

> **Comment**: The geometry of a graph or complex can be used as a good testing ground for geometric ideas. They can be important in understanding smooth geometric structures. Many rough geometric concepts such as isoperimetric inequalities, can be found on graphs and in fact they play some roles in computer network theory. On the other hand, many natural geometric concepts should be generalized to graphs. For example, the concept of the fiber bundle, bundle theory over graphs and harmonic forms. It is likely

one needs to have a good way to define the concept of equivalence between such objects. When we approximate a smooth manifold by a graph or complex, we only care about the limiting object and therefore some equivalence relations should be allowed. In the case of Cayley graph of a finitely generated group, it depends on the choice of the generating set, and properties independent of this generating set are preferable if we are only interested in the group itself. In the other direction, computer networks and other practical subjects have independent interest in graph theory. A close collaboration between geometer and computer scientists would be fruitful.

(f). Harmonic analysis via hyperbolic operators. There are important works of Fefferman, Phong, Lieb, Duistermaat, Guillemin, Melrose, Colin de Verdier, Taylor, Toth, Zelditch and Sarnak on understanding the spectrum of the Laplacian from the point of view of semi-classical analysis (see, e.g. [**220, 209, 323, 583**]). Some of their ideas can be traced back to the geometric optics analysis of J. Keller. The fundamental work of Duistermaat and Hörmander [**208**] on propagation of singularities was also used extensively. There has been a lot of progress on the very difficult question of determining when one "Can hear the shape of a drum" by, among others, Melrose (see [**507**]), Guillemin [**297**] and Zelditch [**749**]. (Priori to this, Guillemin and Kazhdan [**298**] proved that no negatively curved closed surface can be isospectrally deformed.) The first counterexample for closed manifolds was given by J. Milnor [**515**] on a 16 dimensional torus. The idea was generalized by Sunada [**645**], Gordon-Wilson [**261**]. For domains in Euclidean spaces, there were examples by Urakawa in three dimensions. Two dimensional counterexamples were given by Gordon-Webb-Wolpert [**260**], Wilson and Szabó [**646**]. Most of the ideas for counterexamples are related to the Selberg trace formula discussed in the section of heat kernel. The semi-classical analysis based on the hyperbolic operator also gives a very precise estimate or relation between the geodesic and the spectrum. The support of the singularities of the trace of the wave kernel $\sum e^{\sqrt{-1}t\sqrt{\lambda_i}}$ is a subset of the set of the lengths of closed geodesics. It is difficult to achieve such results by elliptic theory. However, most results are asymptotic in nature. It will be remarkable if both methods can be combined.

Comment: Fourier expansion has been a very powerful tool in analysis and geometry. Practically any general theorem in classical Fourier analysis should have a counterpart in analysis of the spectrum of the Laplacian. The theory of geometric optics and the propagation of a singularity gives deep understanding of the singularity of a wave kernel. Geodesic and closed geodesic becomes an important means to understand eigenvalues. However, the theory has not been fruitful for the Laplacian acting on differential forms. Should areas of minimal submanifolds play a role? In the case of Kähler manifolds, holomorphic cycles or the volume of special Lagrangian cycles should be important, as the length of close geodesics appear in the exponential decay term of the heat kernel. It would be useful to sharpen the heat equation method to capture this lower order information.

(g). Harmonic forms. Natural generalizations of harmonic or holomorphic functions are harmonic or holomorphic sections of bundles with connections. The most important bundles are the exterior power of cotangent bundles. Using the Levi-Civita connection, harmonic sections are harmonic forms which, by the theory of de Rham and Hodge, give canonical representation of cohomology classes. The major research on harmonic forms comes from Bochner's vanishing theorem [**58**]. But our understanding is still poor except for 1-forms or when the curvature operator is positive, in which case, the Bochner argument proved the manifold to be a homology sphere. If there is any nontrivial operator which commutes with the Laplacian, the eigenforms split accordingly. Making use of special structures of such splitting, the Bochner method can be more effective. For example, when the manifold is Kähler, differential forms can be decomposed further to (p, q)-forms and the Kodaira vanishing theorem [**388**] yields much more powerful information, when the (p, q) forms are twisted with a line bundle or vector bundles. Similar arguments can be applied to manifolds with a special holonomy group depending on the representation theory of the holonomy group. When the complex structure moves holomorphically, the subbundles of (p, q) forms in the bundle of $(p+q)$ forms do not necessarily deform holomorphically. The concept of Hodge filtration is therefore introduced. When we deform the complex structure around a point where the complex structure degenerates, there is a monodromy group acting on the Hodge filtration. The works of Griffiths-Schmid [**273**] and Schmid's $SL_2(\mathbb{R})$ theorem [**584**] give powerful control on the degeneration of the Hodge structure. Deligne's theory of mixed Hodge structure [**186**] plays a fundamental role for studying singular algebraic varieties. The theory of variation of Hodge structures is closely related to the study of period of the differential forms. This theory also appears in the subject of mirror symmetry. It is desirable to give a precise generalization of these works to higher dimensional moduli spaces where Kaplan-Cattani-Schmid made important contributions.

Harmonic forms give canonical representation to de Rham cohomology. However the wedge product of harmonic forms need not be harmonic. The obstruction comes from secondary cohomology cooperation. K. T. Chen [**131**] studied the case of 1-forms and Sullivan [**642**] studied the general case and gave a minimal model theory for a rational homotopic type of a manifold. Using $\partial\bar{\partial}$-lemma of Kähler manifolds, Deligne-Griffiths-Morgan-Sullivan [**187**] showed that the rational homotopic type is formal for Kähler manifolds.

The importance of harmonic forms is that they give canonical representation to the de Rham cohomology which is isomorphic to singular cohomology over real numbers. It gives a powerful tool to relate local geometry to global topology. In fact the vanishing theorem of Bochner-Kodaira-Lichnerowicz allows one to deduce from sign of curvature to vanishing of cohomology. This has been one of the most powerful tools in geometry in the past fifty years.

The idea of harmonic forms came from fluid dynamics and Maxwell equations. The non-Abelian version is the Yang-Mills theory. Most of the works on Yang-Mills theory have been focused on these gauge fields where the absolute minimum is achieved by some (topological) characteristic number. (These are called BPS state in physics literature.) When the dimension of the manifold is four, the star operator maps two form to two form and it makes sense to require the curvature form to be self-dual or anti-self-dual depending whether the curvature form is invariant or anti-invariant under the star operator. These curvature forms can be interpreted as non-Abelian harmonic forms. The remarkable fact is that when the metric is Kähler, the anti-self-dual connections give rise to holomorphic bundles. The moduli space of such bundles can often be computed using tools from algebraic geometry.

If we take the product space $M \times \mathcal{M}$ where M is the four dimensional manifold and $\mathcal{M}$ is the moduli space of anti-self-dual connections, there is a universal bundle V over $M \times \mathcal{M}$. By studying the slant product and the Chern classes of V, we can construct polynomials on the cohomology of M that are invariants of the differentiable structure of M. These are Donaldson polynomials (see [**203**]). In general $\mathcal{M}$ is not compact and Donaldson has to construct cycles in $\mathcal{M}$ for such operations. Donaldson invariants are believed to be equivalent to Seiberg-Witten invariants, where the vanishing theorem can apply and powerful geometric consequences can be found. Kronheimer and Mrowka [**400**] built an important concept of simple type for Donaldson invariants. It is believed that Donaldson invariants of algebraic surfaces of general type are of simple type.

If the manifold is symplectic, we can look at the moduli space of pseudo-holomorphic curves. (These are J-invariant maps from Riemann surfaces to the manifold. J is an almost complex structure that is tame to the symplectic form.) Symplectic invariants can be created and they are called Gromov-Witten invariants. Y. Ruan [**575**] has observed that they need not be diffeomorphic invariants. It may still be interesting to know whether Gromov-Witten invariants are invariants of differentiable structures for Calabi-Yau manifolds.

De Rham cohomology can only capture the non-torsion part of the singular cohomology. Weil [**706**] and Allendoerfer-Eells [**5**] attempted to use differential forms with poles to compute cohomology with integer coefficients. Perhaps one should study Chern forms of a complex bundle with a connection that satisfies the Yang-Mills equation and whose curvature is square integrable. The singular set of the connection may be allowed to be minimal submanifolds. The moduli space of such objects may give information about integral cohomology. It should be noted that Cheeger-Simons [**122**] did develop a rich theory of differential character with values in $\mathbb{R}/\mathbb{Z}$. It depends on the connections of the bundle. Witten managed to integrate the Chern-Simons forms [**151**] on the space of connections to obtain the knot invariants of Jones [**361**].

When we look for different operators acting on different forms, we may have to look into different kinds of harmonic forms. For example, if we are interesting in $\partial\bar{\partial}$ cohomology, we may look for the operator $\left(\partial\bar{\partial}\right)^* \partial\bar{\partial} + \partial\partial^* + \bar{\partial}\bar{\partial}^*$. It would be interesting to see how super-symmetry may generalize the concept of harmonic forms.

> **Comment**: The theory of harmonic form is tremendously powerful because it provides a natural link between global topology, analysis, geometry, algebraic geometry and arithmetic geometry. However, our analytic understanding of high degree forms is poor. For one forms, we can integrate along paths. For two forms, we can take an interior product with a vector field to create a moment map. For closed $(1,1)$-forms in a Kähler manifold, we can express them locally as $\partial\bar{\partial}f$. However, we do not have good ways to reduce a high degree form to functions which are easier to understand. Good estimates of higher degree forms will be very important.

2.3.2. *$\bar{\partial}$-operator.* Construction of holomorphic functions or holomorphic sections of vector bundles and holomorphic curves are keys to understanding complex manifolds.

In order to demonstrate the idea behind the philosophy of determining the structure of manifolds by function theory, I was motivated to generalize the uniformization theory of a Riemann surface to higher dimensions when I was a graduate student. During this period, I was influenced by the works of Greene-Wu [**271**] in formulating these conjectures. Greene and Wu were interested in knowing whether the manifolds are Stein or not.

When the complete Kähler manifold is compact with positive bisectional curvature, this is the Frankel conjecture, as was proved independently by Mori [**526**] and Siu-Yau [**629**]. Both arguments depend on the construction of rational curves of low degree. Mori's argument is stronger, and it will be good to capture his result by the analytic method. When the manifold has nonnegative bisectional curvature and positive Ricci curvature, Mok-Zhong [**523**] and Mok [**519**], using ideas of Bando [**31**] in his thesis on Hamilton's Ricci flow, proved that the manifold is Hermitian symmetric.

When the complete Kähler manifold is noncompact with positive bisectional curvature, I conjectured that it must be biholomorphic to $\mathbb{C}^n$ (see [**735**]). Siu-Yau [**628**] made the first attempt to prove such a conjecture by using the L^2-method of Hörmander [**345**] to construct holomorphic functions with slow growth. (Note that Hörmander's method goes back to Kodaira, which was also generalized by Calabi-Vesentini [**94**].) Singular weight functions were used in this paper and later much more refined arguments were developed by Nadel [**531**] and Siu [**626**] using what is called the multiplier ideal sheaf method. Siu found important applications of this method in algebraic geometry and also related the idea to the powerful work of J. Kohn on weakly pseudo-convex domain.

This work of Siu-Yau was followed by Siu-Mok-Yau [**520**] and Mok [**517, 518**] under assumptions about the decay of curvature and volume growth. Shi [**607, 608, 609**] introduced Hamilton's Ricci flow to study my conjecture, and his work is fundamental. This was followed by beautiful works of Cao [**100, 101**], Chen-Zhu [**126, 127**] and Chen-Tang-Zhu [**124**]. Assuming the manifold has maximal Euclidean volume growth and bounded curvature, Chen-Tang-Zhu [**124**] (for complex dimension two) and then Ni [**535**] (for all higher dimension) were able to prove the manifold can be compactified as a complex variety. Last year, Albert Chau and Tam [**112**] were finally able to settle the conjecture assuming maximal Euclidean volume growth and bounded curvature. An important lemma of L. Ni [**535**] was used, where a conjecture of mine (see [**738**] or the introduction of [**535**]) was proved. The conjecture says that maximal volume growth implies scalar curvature decays quadratically in the average sense.

While we see great accomplishments for Kähler manifolds with positive curvature, very little is known for Kähler manifolds, which are complete simply connected with strongly negative curvature. It is conjectured to be a bounded domain in $\mathbb{C}^n$. (Some people told me that Kodaira considered a similar problem. But I cannot find the appropriate reference.) The major problem is to construct bounded holomorphic functions.

The difficulty of construction of bounded holomorphic functions is that the basic principle of the L^2-method of Hörmander comes from Kodaira's vanishing theorem. It is difficult to obtain elegant results by going from weighted L^2 space to bounded functions. In this connection, I was able to show that non-trivial bounded holomorphic functions do not exist on a complete manifold with non-negative Ricci curvature [**734**].

If the manifold is the universal cover of a compact Kähler manifold M which has a homotopically nontrivial map to a compact Riemann surface with genus > 1, then one can construct a bounded holomorphic function, using arguments of Jost-Yau [**368**]. In particular, if M has a map to a product of Riemann surfaces with genus > 1 with nontrivial topological degree, the universal cover should have a good chance to be a bounded domain.

Of course, this kind of construction is based on the fact that holomorphic functions are harmonic. Certain rigidity based on curvature forced the converse to be true. For functions, the target space has no topology and rigidity is not expected. Bounded holomorphic functions can not be constructed by solving the Dirichlet problem unless some boundary condition is assumed. This would make good sense if the boundary has a nice CR structure. Indeed, for odd dimensional real submanifold in $\mathbb{C}^n$ which has maximal complex linear subspace on each tangent plane, Harvey-Lawson [**319, 320**] proved the remarkable theorem that they bound complex submanifolds. Unfortunately the boundary of a complete simply connected manifold with bounded negative curvature does not have a smooth boundary. It will be nice to define a CR structure on such a singular boundary. One may mention the remarkable work of Kuranishi [**403, 404, 405**] on embedding of an abstract CR structure.

Historically a motivation for the development of the $\bar{\partial}$ operator came from the Levi problem, which was solved by Morrey, Grauert and greatly improved by Kohn and Hörmander. Their methods are powerful in studying pseudoconvex manifolds.

In this regard, one may mention the conjecture of Shafarevich that the universal cover of an algebraic manifold is pseudoconvex. Many years ago, I conjectured that if the second homotopy group of the manifold is trivial, its universal cover can be embedded into a domain of some algebraic manifold where the covering transformations act on the domain by birational transformations. One may also mention the work of S. Frankel [**226**] on proving that an algebraic manifold is Hermitian symmetric if the universal cover is a convex domain in complex Euclidean space.

> **Comment**: The $\bar{\partial}$ operator is the fundamental operator in complex geometry. Classically it was used to solve the uniformization theorem, the Levi problem and the Corona problems. We have seen much progress on the higher dimensional generalizations of the first two problems. However, due to poor understanding of the construction of bounded holomorphic functions, we are far away from understanding the Corona problem in a higher dimension and many related geometric questions.

2.3.3. *Dirac operator.* A very important bundle is the bundle of spinors. The Dirac operator acting on spinors is the most mysterious but major geometric operator. Atiyah-Singer were the first mathematicians to study it in geometry and by thoroughly understanding the Dirac operator, they were able to prove their celebrated index theorem [**20**]. On a Kähler manifold, the Dirac operator can be considered as a $\bar{\partial} + \bar{\partial}^*$ operator acting on differential forms with coefficients on the square root of the canonical line bundle. Atiyah-Singer's original proof can be traced back to the celebrated Riemann-Roch-Hirzebruch formula and the Hirzebruch index formula. The formulas of Gauss-Bonnet-Chern and Atiyah-Singer-Hirzebruch should certainly be considered as the most fundamental identities in geometry. The vanishing theorem of Lichnerowicz [**451**] on harmonic spinors over spin manifolds with positive scalar curvature gives strong information. Through the Atiyah-Singer index theorem, it gives the vanishing theorem for the $\hat{A}$-genus and the α invariants for spin manifolds with positive scalar curvature. The method was later sharpened by Hitchin [**329**] to prove that every Einstein metric over $K3-$surfaces must be Kähler and Ricci flat. An effective use of Lichnerowicz formula for a $\text{spin}_{\mathbb{C}}$ structure for a four dimensional manifold is important for Seiberg-Witten theory, which couples the Dirac operator with a complex line bundle. Lawson-Yau [**411**] were able to use Lichnerowicz's work coupled with Hitchin's work to prove a large class of smooth manifolds have no smooth non-Abelian group action and, by using modular forms, K. F. Liu proved a loop space analogue of the Lawson-Yau's theorem for the vanishing of the Witten genus in [**465**].

On the basis of the surgery result of Schoen-Yau [**593, 596**] and Gromov-Lawson [**288, 289**], one expects that a suitable converse to Lichnerowicz's theorem exists. The chief result is that surgery on spheres with codimension ≥ 3 preserves a class of metrics with positive scalar curvature. Once geometric surgery is proved, standard works on cobordism theory allow one to deduce existence results for simply connected manifolds with positive scalar curvature. The best work in this direction is due to Stolz [**637**] who gave a complete answer in the case of simply connected manifolds with dimension greater then 4. I also suggested the possibility of performing surgery on an asymptotic hyperbolic manifold with conformal boundary whose scalar curvature is positive. This is related to the recent work of Witten-Yau [**718**] on the connectedness of the conformal boundary.

The study of metrics with positive scalar curvature is the first important step in understanding the positive mass conjecture in general relativity. Schoen-Yau [**594, 598**] gave the first proof using ideas of minimal surfaces. Three years later, Witten [**712**] gave a proof using harmonic spinors. Both approaches have been fundamental to questions related to mass and other conserved quantities in general relativity. In the other direction, Schoen-Yau [**596**] generalized their argument in 1979 to find topological obstructions for higher dimensional manifolds with positive scalar curvature. Subsequently Gromov-Lawson [**288, 289**] observed that the Lichnerowicz theorem can be coupled with a fundamental group and give topological obstructions for a metric with positive scalar curvature. This work was related to the Novikov conjecture where many authors, including Lusztig [**479**], Rosenberg [**567**], Weinberger [**708**] and G. L. Yu [**747**] made contributions.

Besides its importance in demonstrating the stability of Minkowski spacetime, the positive mass conjecture was used by Schoen [**586**] in a remarkable manner to finish the proof of the Yamabe problem where Trudinger [**682**] and Aubin [**21**] made substantial contributions.

> **Comment**: The Dirac operator is perhaps one of the most mysterious operators in geometry. When it is twisted with other bundles, it gives the symbol of all first order elliptic operators. When it couples with a complex line bundle it gives the Seiberg-Witten theory which provides powerful information for four manifolds. On

> the other hand, there were two different methods to study metrics with positive scalar curvature. It should be fruitful to combine both methods: the method of Dirac operator and the method of minimal submanifolds.

2.3.4. *First order operator twisted by vector fields or endomorphisms of bundles.* Given a vector field X on a manifold, we can consider the complex of differential forms ω so that $L_X\omega = 0$. On such complex, $d + \iota_X$ defines a differential and the resulting cohomology is called equivariant cohomology.

During the seventies, Bott [**71**] and Atiyah-Bott [**12**] developed the localization formula for equivariant cohomology. Both the concepts of a moment map and equivariant cohomology have become very important tools for computations of various geometric quantities, especially Chern numbers of natural bundles. The famous work of Atiyah, Guillemin-Sternberg on the convexity of the image of the moment map gives a strong application of equivariant cohomology to toric geometry. The formula of Duistermaat-Heckman [**207**] played an important role in motivation for evaluation of path integrals. These works have been used by Jeffrey and Kirwan [**358**] and by K. F. Liu and his coauthors on several topics: the mirror principle (Lian-Liu-Yau [**447, 448, 449, 450**]), topological vertex (Li-Liu-Liu-Zhou [**428**]), etc. The idea of applying localization to enumerative geometry was initiated by Kontsevich [**391**] and later by Givental [**255**] and Lian-Liu-Yau [**447**] independently. (Lian-Liu-Yau [**447**] formulated a functorial localization formula which has been fundamental for various calculations in mirror geometry.) These works solve the identities conjectured by Candelas et al [**98**] based on mirror symmetry, and provide good examples of the ways in which conformal field theory can be a source of inspiration when looking at classical problems in mathematics.

If we twist the $\bar\partial$ operator with an endomorphism valued holomorphic one form s so that $s \circ s = 0$, it gives rise to a complex $\bar\partial + s$. This was the Higgs theory initiated by Hitchin [**330**] and studied extensively by Simpson [**617**]. There is extensive work of Zuo Kang and Jost-Zuo (see [**755**]) on Higgs theory and representation of fundamental groups of algebraic manifold.

In string theory, there is a three form H and the cohomology of $d^c + H$ has not been well understood. It would be interesting to develop a deeper understanding of such twisted cohomology and its localization.

> **Comment**: The idea of deforming a de Rham operator by twisting with some other zero order operators has given powerful information to geometry. Witten's idea of the analytic proof of Morse theory is an example. Equivariant cohomology is another example. We expect to see more works in such directions.

2.3.5. *Spectrum and global geometry.* Weyl made a famous address in the early fifties. The title of his talk was *The Eigenvalue Problem Old and New.* He was excited by the work of Minakshisundaram and Pleijel which asserts that the zeta function $\zeta(s) = \sum_\lambda \lambda^{-s}$, where λ are eigenvalues of the Laplacian, not only makes sense for Re's large, but also has meromorphic extension to the whole complex s-plane, the position of whose poles could be described explicitly. In particular, it is analytic near $s = 0$. Formally $\frac{d\zeta(s)}{ds}|_{s=0}$ can be viewed as $-\log\det(\Delta)$. This gives a definition of determinant of Laplacian which entered into the fundamental work of Ray-Singer relating Reidemeister's combinational invariant of a manifold with analytic torsion defined by determinants of the Laplacians acting on differential forms of various degrees. Other application of zeta function expressed in terms of kernel is the calculation of the asymptotic growth of eigenvalues in terms of volume of the manifold. Tauberian type theorem is needed.

This initiated the subject of finding formula to relate spectrum of manifolds with their global geometry. Atiyah and Singer [**20**] were the most important contributors to this beautiful subject. Atiyah-Bott-Patodi [**13**] applied the heat kernel expansion to a proof of the local index theorem. Atiyah-Patodi-Singer [**17, 18, 19**] initiated the study of spectrum flow and gave important global spectral invariants on odd dimensional manifolds. These global invariants become boundary terms for the L^2-index theorem developed by Atiyah-Donnelly-Singer [**14**] and Mark Stern [**636**]. (A method of Callias [**95**] has been used for such calculations.) Witten [**713, 714**] has introduced supersymmetry and analytic deformation of the de Rham complex to Morse theory, and thereby revealed a new aspect of the connection between global geometry and theoretical physics. Witten's work has been generalized by Demailly [**189**] and Bismut-Zhang [**55, 56**] to study the holomorphic Morse inequality and analytic torsion. Novikov [**537**] also studied Morse theory for one forms. Witten's work on Morse theory inspired the work of Floer (see, e.g., [**223, 224, 225**]) who used his ideas in Floer cohomology to prove Arnold's conjecture in case where the manifold has vanishing higher homotopic group. Floer's theory is related to knot theory (through Chern-Simon's theory [**151**]) on three manifolds. Atiyah, Donaldson, Taubes , Dan Freed, P. Braam, and others (see, e.g., [**10, 654, 76, 228**]) all contributed to this subject. Fukaya-Ono [**240**], Oh [**540**], Kontsevich [**392**], Hofer-Wysocki-Zehnder [**339**], G. Liu-Tian [**464**], all studied such a theory in symplectic geometry. Some part of Arnold's conjecture on fixed points of groups acting on symplectic manifolds was claimed to be proven. But a completely satisfactory proof has not been forthcoming.

One should also mention here the very important work of Cheeger [**114**] and Müller [**529**] in which they verify the conjecture of Ray-Singer equating analytic torsion with the combinational torsion of the manifold. The fundamental idea of Ray-Singer [**563**] on holomorphic torsion is still being vigorously developed. It appeared in the beautiful work of Vafa et al [**50**]. Many more works on analytic torsion were advanced by Quillen, Todorov, Kontsevich, Borcherds , Bismut, Lott, Zhang and Z. Q. Lu (see [**54**] and it's reference, [**363**], [**63, 64**]). The local version of the index theorem by Atiyah-Bott-Patodi [**13**] was later extended in an sophisticated way by Bismut [**53**] to an index theorem for a family of elliptic operators.(The local index argument dates back to the foundational work of McKean-Singer [**495**] where methods were developed to calculate coefficients of heat kernel expansion.) The study of elliptic genus by Witten [**715**], Bott-Taubes [**72**], Taubes [**653**], K. F. Liu [**466**] and M. Hopkins [**344**] has built a bridge between topology and modular form.

> **Comment**: The subject of relating the spectrum to global topology is extremely rich. It is likely that we have only touched part of this rich subject. The deformation of spectrum associated with the deformation of geometric structure is always a fascinating subject. Global invariants are created by spectral flows. Determinants of elliptic operators are introduced to understand measures of infinite dimensional space. Geometric invariants that are created by asymptotic expansion of heat or wave kernels are in general not

well understood. It will be a long time before we have a much better understanding of the global behavior of spectrum.

3. Mappings between manifolds and rigidity of geometric structures

There is a need to exhibit a geometric structure in a simpler space: hence we embed algebraic manifolds into complex projective space, we isometrically embed a Riemannian manifold into Euclidean space and we classify structures such as bundles by studying maps into Grassmannian.

We are also interested in probing the structure of a manifold by mapping Riemann surfaces inside the manifold, an important example being holomorphic curves in algebraic manifolds. Of course, we are also interested in maps that can be used to compare the geometric structures of different manifolds.

3.1. Embedding.

3.1.1. *Embedding theorems.* Holomorphic sections of holomorphic line bundles have always been important in algebraic geometry. The Riemann-Roch formula coupled with vanishing theorems gave very powerful existence results for sections of line bundles. The Kodaira embedding theorem [**389**] which said that every Hodge manifold is projective has initiated the theory of holomorphic embedding of Kähler manifolds. For example, Hirzebruch-Kodaira [**328**] proved that every odd (complex) dimensional Kähler manifold diffeomorphic to projective space is biholomorphic to projective space. (I proved the same statement for even dimensional Kähler manifolds based on Kähler Einstein metric.)

Given an orthonormal basis of holomorphic sections of a very ample line bundle, we can embed the manifold into projective space. The induced metric is the Bergman metric associated with the line bundle. Note that the original definition of the Bergman metric used the canonical line bundle and L^2-holomorphic sections.

In the process of understanding the relation between stability of a manifold and the existence of the Kähler Einstein metric, I [**737**] proposed that every Hodge metric can be approximated by the Bergman metric as long as we allow the power of the line bundle to be large. Following the ideas of the paper of Siu-Yau [**628**], Tian [**672**] proved the C^2 convergence in his thesis under my guidance. My other student W. D. Ruan [**571**] then proved C^∞ convergence in his thesis. This work was followed by Lu [**475**], Zelditch [**748**] and Catlin [**109**] who observed that the asymptotic expansion of the kernel function follows from some rather standard expressions of the Szegö kernel, going back to Fefferman [**219**] and Boutet de Monvel-Sjöstrand [**74**] on the circle bundle associated with the holomorphic line bundle over the Kähler manifold. Recently, Dai, Liu, Ma and Marinescu [**181**] [**484**] obtained the asymptotic expansion of the kernel function by using the heat kernel method, and gave a general way to compute the coefficients, thus also extended it to symplectic and orbifold cases.

Kodaira's proof of embedding Hodge manifolds by the sufficiently high power of a positive line bundle is not effective. Matsusaka [**489**] and later Kollár [**390**], Siu [**624**] were able to provide effective estimate of the power. Demailly [**191, 192**] and Siu [**624, 626**] made a remarkable contribution toward the solution of the famous Fujita conjecture [**236**] (see also Ein and Lazarsfeld [**212**]). Siu's powerful method also leads to a proof of the deformation invariance of plurigenera of algebraic manifolds [**625**]. It should be noticed that the extension theorem of Ohsawa-Takegoshi played an important role in this last work of Siu.

Comment: The idea of embedding a geometric structure is clearly important as once they are put in the same space, we can compare them and study the moduli space of the geometric structure much better. For example, one can define Chow coordinate of a projective manifold and we can study various concepts of geometric stability of these structures. At this moment, there is no natural universal space of Kähler manifolds or complex manifolds as we may not have a positive holomorphic line bundle over such manifolds to embed into complex projective space. In a similar vein, it will be nice to find a universal space for symplectic manifolds.

3.1.2. *Compactification.* Problem of compactification of the manifold dates back to Siegel, Satake, Baily-Borel [**26**] and Borel-Serre [**66**]. They are important for representation theory, for algebraic geometry and for number theory.

For geometry of non-compact manifolds, we like to control behavior of differentiable forms at infinity. A good exhaustion function is needed.

Construction of a proper exhaustion function with a bounded Hessian on a complete manifold with a bounded curvature was achieved by Schoen-Yau [**602**] in 1983 in our lectures in Princeton. Based on this exhaustion function, M. Dafermos [**179**] was able to reprove a theorem of Cheeger-Gromov [**120**] that such manifolds admit an exhaustion by compact hypersurfaces with bounded second fundamental form. Such exhaustions are useful to understand characteristic forms on noncompact manifolds as the boundary term can be controlled by the second fundamental form of the hypersurfaces.

After my work with Siu [**630**] on compactification of a strongly, negatively curved Kähler manifold with finite volume, I proposed that every complete Kähler manifold with bounded curvature, finite volume and finite topology should be compactifiable to be a compact complex variety. I suggested this problem to Mok and Zhong in 1982 who did significant work [**524**] in this direction. (The compactification by Mok-Zhong is not canonical and it is desirable to find an algebraic geometric analogue of Borel-Baily compactification [**26**] so that we can study the L^2-cohomology in terms of the intersection cohomology of the compactification.) Recall that the important conjecture of Zucker on identifying L^2-cohomology with the intersection cohomology of the Borel-Baily compactification was settled by Saper-Stern [**581**] and Looijenga [**472**]. (Intersection cohomology was introduced by Goresky-MacPherson [**263, 264**]. It is a topological concept and hence the Zucker conjecture gives a topological meaning of the L^2-cohomology.) It would be nice to find compactification for algebraic varieties so that suitable form of intersection cohomology can be used to understand L^2 cohomology. Goresky-Harder-MacPherson [**262**] and Saper [**580**] have contributed a lot toward this kind of question. For moduli space of bundles, or polarized projective structures, compactification means degeneration of these structures in a suitable canonical manner. For algebraic curves, there is Deligne-Mumford compactification [**188**] which has played a fundamental role in understanding algebraic curves. Geometric invariant theory (see [**530**]) gives a powerful method to introduce the concept of stable structures. Semi-stable structures can give points at infinity. The compactification based on the geometric invariant theory for moduli space of surfaces of the general type was done by Gieseker [**253**]. For a higher dimension, this was done by Viehweg [**690**]. Detailed analysis of the divisors at infinity is still missing.

Comment: Compactification of a manifold is very much related to the embedding problem. One needs to construct functions or sections of bundles near infinity. For the moduli space of geometric structures, it amounts to degenerately the structures canonically. It will be important to study the degeneration of Hermitian Yang-Mills connections and Kähler Einstein metrics.

3.1.3. *Isometric embedding.* Given a metric tensor on a manifold, the problem of isometric embedding is equivalent to find enough functions $f_1, \cdots, f_N$ so that the metric can be written as $\sum(df_i)^2$. Much work was accomplished for two dimensional surfaces as was mentioned in section 2.1.2. Isometric embedding for the general dimension was solved in the famous work of J. Nash [**532, 533**]. Nash used his famous implicit function theorem which depends on various smoothing operators to gain derivatives. In a remarkable work, Günther [**305**] was able to avoid the Nash procedure. He used only the standard Hölder regularity estimate for the Laplacian to reproduce the Nash isometric embedding with the same regularity result. In his book [**285**], Gromov was able to lower the codimension of the work of Nash. He called his method the h-principle.

When the dimension of the manifold is n, the expected dimension of the Euclidean space for the manifold to be isometrically embedded is $\frac{n(n+1)}{2}$. It is important to understand manifolds isometrically embedded into Euclidean space with this optimal dimension. Only in such a dimension does it make sense to talk about rigidity questions. It remains a major open problem whether one can find a nontrivial smooth family of isometric embeddings of a closed manifold into Euclidean space with an optimal dimension. Such a nontrivial family was found for a polyhedron in Euclidean three space by Connelly [**175**]. For a general manifold, it is desirable to find a canonical isometric embedding into a given Euclidean space by minimizing the L^2 norm of its mean curvature within the space of isometric embeddings.

Chern told me that he and H. Lewy studied local isometric embedding of a three manifold into six dimensional Euclidean space. But they didn't wrote any manuscript on it. The major work in this subject was done by E. Berger, Bryant, Griffiths and Yang [**84**] [**47**]. They showed that a generic three dimensional embedding system is strictly hyperbolic, and the generic four dimensional system is a real principal type. Local existence is true for a generic metric using a hyperbolic operator and the Nash-Moser implicit function theorem.

If the target space of isometric embedding is a linear space with indefinite metric, it is possible that the problem is easier. For example, by a theorem of Pogorelov [**557, 558**], any metric on the two dimensional sphere can be isometrically embedded into a three dimensional hyperbolic space-form (where the sectional curvature may be a large negative constant). Hence it can always be embedded into the hyperboloid of the Minkowski spacetime. This statement may also be true for surfaces with higher genus. The fundamental group may cause obstruction, hence the first step should be an attempt to canonically embed any complete metric (with bounded curvature) on a simply connected surface into a three dimensional hyperbolic space form. It should be also very interesting to study the rigidity problem of a space-like surface in Minkowski spacetime. Besides requesting the metric to be the induced metric, we shall need one more equation. Such an equation should be related to the second fundamental form. A candidate appeared in the work of M. Liu-Yau [**462, 463**] on the quasi-local mass in general relativity.

In the other direction, Calabi found the condition for a Kähler metric to be isometrically and holomorphically embedded into Hilbert space with an indefinite signature. In the course of his investigation, he introduced some kind of distance function that can be defined by the Kähler potential and enjoys many interesting properties. Calabi's work in this direction which should be relevant to the flat coordinate appeared in the recent works of Vafa et al [**50**].

> **Comment**: The theory of isometric embedding is a classical subject. But our knowledge is still rather limited, especially in dimension greater than three. Many difficult problems are related to nonlinear mixed type equation or hyperbolic differential systems over a closed manifold.

3.2. Rigidity of harmonic maps with negative curvature. One can define the energy of maps between manifolds and the critical maps are called harmonic maps. In 1964, Eells-Sampson [**211**] and Al'ber [**3**] independently proved the existence of such maps in their homotopy class if the image manifold has a non-positive curvature.

When I was working on manifolds with non-positive curvature, I realized that it is possible to use harmonic map to reprove some of the theorems in my thesis. I was convinced that it is possible to use harmonic maps to study rigidity questions in geometry such as Mostow's theorem [**527**]. In 1976, I proved the Calabi conjecture and applied the newly proved existence of the Kähler Einstein metric and the Mostow rigidity theorem to prove uniqueness of a complex structure on the quotient of the ball [**731**]. Motivated by this theorem, I proposed to use the harmonic map to prove the rigidity of a complex structure for Kähler manifolds with strongly negative sectional curvature. I proposed this to Siu who carried out the idea when the image manifold satisfies a stronger negative curvature condition [**621**]. Jost-Yau [**366**] proved that for harmonic maps into manifolds with non-positive curvature, the fibers give rise to holomorphic foliations even when the map is not holomorphic. Such a work was found to be used in the work of Corlette, Simpson et al.

A further result was obtained by Jost-Yau [**369**] and Mok-Siu-Yeung [**521**] on the proof of the superrigidity theorem of Margulis [**486**], improving an earlier result of Corlette [**176**] who proved superrigidity for a certain rank one locally symmetric space. Complete understanding of superrigidity for the quotient of a complex ball is not yet available. One needs to find more structures for harmonic maps which reflect the underlying structure of the manifold. The analytic proof of super-rigidity was based on an argument of Matsushima [**491**] as was suggested by Calabi. (This was a topic discussed by Calabi in the special year on geometry in the Institute for Advanced Study.)

The discrete analogue of harmonic maps is also important. When the image manifold is a metric space, there are works by Gromov-Schoen [**290**], Korevaar-Schoen [**397**] and Jost [**365**]. Margulis knew that the super-rigidity for both the continuous and the discrete case is enough to prove Selberg's conjecture for the arithmeticity of lattices in groups with rank ≥ 2. Unfortunately, the analytic argument mentioned above only works if the lattices are cocompact as it is difficult to find a degree one smooth map with finite energy for non-cocompact lattices. Harmonic maps into a tree have given interesting applications to group theory. When the domain manifold is a simplicial complex, there are articles by Ballmann-Śwwiątkowski [**29**] and M. T. Wang [**699, 700**], where they introduce maps from complices which are generalizations of buildings. They also generalized the work of H. Garland [**244**] on the vanishing of the cohomology group for p-adic buildings.

Using the concept of the center of gravity, Besson-Courtois-Gallot [**52**] give a metric rigidity theorem for rank one locally symmetric space. They also proved a rigidity theorem for manifolds with negative curvature: if the fundamental group can be split as a nontrivial free product over some other group C, the manifold can be split along a totally geodesic submanifold with the fundamental group C.

> **Comment**: The harmonic map gives the first step in matching geometric structures of different manifolds. Eells-Sampson derived it from the variational principle. One can also use different elliptic operators to define maps which satisfy elliptic equations. Higher dimensional applications are mostly based on the assumption that the image manifold has a metric with non-positive curvature. In such a case, existence is easier and uniqueness (as shown by Hartman) is also true. Up to now, significant results on higher dimensional harmonic maps are based on such assumptions. Generalization to kähler manifold should be reasonable. The second homotopic group should play a role as one may look at it as a generalization of the work of Sacks-Uhlenbeck. It may be possible to use harmonic maps to study the moduli of geometric structure on a fixed manifold as was done by Michael Wolf for Riemann surfaces. It will also be nice to see how a harmonic map can be used to compare graphs.

3.3. Holomorphic maps. The works of Liouville, Picard, Schwarz-Pick and Ahlfors show the importance of hyperbolic complex analysis. Grauèrt-Reckziegel [**266**] generalized this kind of analysis to higher dimensional complex manifolds. Kobayashi [**386**] and H. Wu [**722**] put this theory in an elegant setting. Kobayashi introduced the concept of hyperbolic complex manifolds. Its elegant formulation has been influential. An important application of the negative curvature metric is the extension theorem for holomorphic maps, as was achieved by the work of Griffiths-Schmid [**273**] on maps to a period domain and by the extension theorem of Borel [**65**] on compactification of Hermitian symmetric space. A major question was Lang's conjecture: on an algebraic manifold of a general type, there exists a proper subvariety such that the image of any holomorphic map from $\mathbb{C}$ must be a subset of this subvariety. It has deep arithmetic geometric meaning. In terms of the Kobayashi metric, it says that the Kobayashi metric is nonzero on a Zariski open set. Many works were done towards subvarieties of Abelian variety by Bloch, Green-Griffiths, Kobayashi-Ochiai, Voitag and Faltings. For generic hypersurfaces in $\mathbb{C}P^n$, there is work by Siu [**627**]. They developed the tool of jet differentials and meromorphic connections. For algebraic surfaces with $C_1^2 > 2C_2$, Lu-Yau [**473**] proved Lang's conjecture, based on the ideas of Bogomolov.

> **Comment**: Holomorphic maps have been studied for a long time. There is no general method to construct such maps based on the knowledge of topology alone, except the harmonic map approach proposed by me and carried out by Siu, Jost-Yau and others. But the approach is effective only for manifolds with negative curvature. For rigidity questions, the most interesting manifolds are

Kähler manifolds with non-positive Ricci curvature, which give the major chunk of algebraic manifolds of a general type. The Kähler-Einstein metric should provide tools to study such problems. Is there any intrinsic way, based on the metric, to find the largest subvariety where the image of all holomorphic maps from the complex line lie? Deformation theory of such a subvariety should be interesting. There is also the question of when the holomorphic image of the complex line will intersect a divisor. Cheng and I did find good conditions for the complement of a divisor to admit the complete Kähler-Einstein metric. For such a geometry, the holomorphic line should either intersect the divisor or a subset of some subvarieties. This kind of questions are very much related to arithmetic questions if the manifolds are defined over number fields.

3.4. Harmonic maps from two dimensional surfaces and pseudoholomorphic curves. Harmonic maps behave especially well for Riemann surface. Morrey was the first one who solved the Dirichlet problem for energy minimizing harmonic map into any Riemannian manifold.

Another major breakthrough was made by Sacks-Uhlenbeck [**577**] in 1978 where they constructed minimal spheres in Riemannian manifolds representing elements in the second homotopy group using a beautiful extension theorem of a harmonic map at an isolated point. By pushing their method further, Siu-Yau [**629**] studied the bubbling process for the harmonic map and made use of it to prove a stable harmonic map must be holomorphic under curvature assumptions. As a consequence, they proved the famous conjecture of Frankel that a Kähler manifold with positive bisectional curvature is $\mathbb{C}P^n$, as was discussed in section [2.3.2].

Gromov [**284**] then realized that a pseudoholomorphic curve for an almost complex structure can be used in a similar way to prove rigidity of a symplectic structure on $\mathbb{C}P^n$. The bubbling process mentioned above was sharpened further to give compactification of the moduli space of pseudoholomorphic maps by Ye [**745**] and Parker-Wolfson [**545**]. Based on these ideas, Kontsevich [**391**] introduced the concept of stable maps and the compactification of their moduli spaces.

The formal definitions of Gromov-Witten invariants and quantum cohomology were based on these developments and the ideas of physicists. For example, quantum cohomology was initiated by Vafa (see, e.g., [**688**]) and his coauthors (the name was suggested by Greene and me). Associativity in quantum cohomology was due to four physicists WDVV [**716, 193**]. The mathematical treatment (done by Ruan [**575**] and subsequently by Ruan-Tian [**576**]) followed the gluing ideas of the physicists. Ruan-Tian made use of the ideas of Taubes [**652**]. But important points were overlooked. A. Zinger [**752, 753**] has recently completed these arguments.

In close analogy with Donaldson's theory, one needs to introduce the concept of virtual cycle in the moduli space of stable maps. The algebraic setting of such a concept is deeper than the symplectic case and is more relevant to the development for algebraic geometry. The major idea was conceived by Jun Li who also did the algebraic geometric counterpart of Donaldson's theory (see [**424, 429**]). (The same comment applies to the concept of the relative Gromov-Witten invariant, where Jun Li made the vital contribution in the algebraic setting [**426, 427**].) The symplectic version of Li-Tian [**430**] ignores difficulties, many of which were completed recently by A. Zinger [**752, 753**].

Sacks-Uhlenbeck studied harmonic maps from higher genus Riemann surfaces. Independently, Schoen-Yau [**597**] studied the concept of the action of an L_1^2 map on the fundamental group of a manifold. It was used to prove the existence of a harmonic map with prescribed action on the fundamental group. Jost-Yau [**367**] generalized such action on fundamental group to a more general setting which allows the domain manifold to be higher dimensional. Recently F. H. Lin developed this idea further [**458**]. He studied extensively geometric measure theory on the space of maps (see, e.g., [**455, 457**]). The action on the second homotopy group is much more difficult to understand. I think there should exist a harmonic map with nontrivial action on the second homotopic group if such a continuous map exists. Such an existence theorem will give interesting applications to Kähler geometry.

There is a supersymmetric version of harmonic maps studied by string theorists. This is obtained by coupling the map with Dirac spinors in different ways (which corresponds to different string theories). While this kind of world sheet theory is fundamental for the development of string theory, geometers have not paid much attention to the supersymmetric harmonic map. Interesting applications may be found. The most recent paper of Chen, Jost, Li and Wang [**133**] does address to a related problem where they studied the regularity and energy identities for Dirac-Harmonic maps.

> **Comment**: Maps from circle or Riemann surfaces into a Riemannian manifold give a good deal of information about the manifold. The capability to construct holomorphic or pseudo-holomorphic maps from spheres with low degree was the major reason that Mori, Siu-Yau and Taubes were able to prove the rigidity of algebraic or symplectic structures on the complex projective space. It will be desirable to find more ways to construct such maps from low genus curves to manifolds that are not of a rational type. Their moduli space can be used to produce various invariants. An outstanding problem is to understand the invariants on counting curves of a higher genus which appeared in the fundamental paper of Vafa et al [**50**].

3.5. Morse theory for maps and topological applications. The energy functional for maps from S^2 into a manifold does not quite give rise to Morse theory. But the perturbation method of Sacks-Uhlenbeck did provide enough information for Micallef-Moore [**509**] to prove some structure theorem for manifolds with positive isotropic curvature. (Micallef and Wang [**510**] then proved the vanishing of second Betti number in the even dimensional case. If the manifold is irreducible, has non-negative isotropic curvature and non-vanishing second Betti number, then they proved that its second Betti number equals to one and it is Kähler with positive first Chern class.)

If the image manifold has negative curvature, the theorem of Eells-Sampson [**211**] says that any map can be canonically deformed by the heat flow to a unique harmonic map. Hence the topology of the space of maps is given by the space of homomorphism between the fundamental groups of the manifolds. This gives some information of the topology of manifolds with negative curvature. Farrell and Jones

[**217**] have done much deeper analysis on the differentiable structure of manifolds with negative curvature.

Schoen-Yau [**597**] exploited the uniqueness theorem for harmonic maps to demonstrate that only finite groups can act smoothly on a manifold which admits a non-zero degree map onto a compact manifold with negative curvature. The size of the finite group can also be controlled. If the image manifold has non-positive curvature, then the only compact continuous group actions are given by the torus.

The topology of the space of maps into Calabi-Yau manifolds should be very interesting for string theory. Sullivan [**644**] has developed an equivariant homology theory for loop space. It will be interesting to link such a theory with quantum cohomology when the manifold has a symplectic structure.

> **Comment**: Morse theory has been one of the most powerful tools in geometry and topology as it connects local to global information. One does not expect full Morse theory for harmonic maps as we have difficulty even proving their existence. However, if their existence can be proven, the perturbation technique may be used and powerful conclusions may be drawn.

3.6. Wave maps. In early eighties, C. H. Gu [**294**] studied harmonic maps when the domain manifold is the two dimensional Minkowski spacetime. They are called wave maps. Unfortunately, good global theory took much longer to develop as there were not many good a priori estimates. This subject was studied extensively by Christodoulou , Klainerman, Tao, Tataru and M. Struwe (see, e.g., [**160, 383, 649, 650, 606**]). It is hoped that such theory may shed some light on Einstein equations.

> **Comment**: The geometric or physical meaning of wave maps should be studied. The problem of vibrating membrane gives a good motivation to study time-like minimal hypersurface in a Minkowski spacetime. One can study the vibration of submanifold by looking into the minimal time-like hypersurface with the boundary given by the submanifold. It is a mystery how such vibrations can be related to the eigenvalues of the submanifold.

3.7. Integrable system. Classically, Bäcklund (1875) was able to find a nonlinear transformation to create a surface with constant curvature in $\mathbb{R}^3$ from another one. The nonlinear equation behind it is the Sine-Gordon equation. Then in 1965, Kruskal and Zabusky (see [**401**]) discovered solitons and subsequently in 1967, Gardner, Greene, Kruskal and Miura [**243**] discovered the inverse scatting method to solve the KdV equations. The subject of a completely integrable system became popular.

Uhlenbeck [**686**] used techniques from integrable systems to construct harmonic maps from S^2 to $U(n)$, Bryant [**82**] and Hitchin [**332**] also contributed to related constructions using twistor theory and spectral curves. These inspired Burstall, Ferus, Pedit and Pinkall [**88**] to construct harmonic maps from a torus to any compact symmetric space. In a series of papers, Terng and Uhlenbeck [**664, 665**] used loop group factorizations to solve the inverse scattering problem and to construct Bäcklund transformations for soliton equations, including Schrödinger maps from $\mathbb{R}^{1,1}$ to a Hermitian symmetric space. There are recent attempts by Martin Schmidt [**585**] to use an integrable system to study the Willmore surface.

The integrable system also appeared naturally in several geometric questions such as the Schottky problem (see Mulase [**528**]) and the Witten conjecture on Chern numbers of bundles over moduli space of curves.

Geroch found the Backlund transformation for axially symmetric stationary solutions of Einstein equations. It will be nice to find such nonlinear transformations for more general geometric structures.

> **Comment**: It is always important to find an explicit solution to nonlinear problem. Hopefully an integrable system can help us to understand general structures of geometry.

3.8. Regularity theory. The major work on regularity theory of harmonic maps in higher dimensions was done by Schoen-Uhlenbeck [**588, 589**]. (There is a weaker version due to Giaquinta-Giusti [**251**] and also the earlier work of Ladyzhenskaya-Ural'ceva and Hildebrandt-Kaul-Widman where the image manifolds for the maps are more restrictive.) Leon Simon (see [**616**]) made a deep contribution to the structure of harmonic maps or minimal subvarieties near their singularity. This was followed by F. H. Lin [**457**]. The following is still a fundamental problem: Are singularities of harmonic maps or minimal submanifolds stable when we perturb the metric of the manifolds? Presumably some of them are. Can we characterize them? How big is the codimension of generic singularities?

In the other direction Schoen-Yau [**592**] also proved that degree one harmonic maps are one to one if the image surface has a non-positive curvature. Results of this type work only for two dimensional surfaces. It will be nice to study the set where the Jacobian vanishes.

> **Comment**: There is a very rich theory of stable singularity for smooth maps. However, in most problems, we can only afford to deform certain background geometric structures, while the extremal objects are still constrained by the elliptic variational problem. Understanding this kind of stable singularity should play fundamental roles in geometry.

4. Submanifolds defined by variational principles

4.1. Teichmüller space. The totality of the pair of polarized Kähler manifolds with a homotopic equivalence to a fixed manifold gives rise to the Teichmüller space. For an Algebraic curve, this is the classical Teichmüller space. This space is important for the construction of the mapping problem for minimal surfaces of a higher genus.

In fact, given a conformal structure on a Riemann surface Σ, a harmonic map from Σ to a fixed Riemannian manifold may minimize energy within a certain homotopy class. However, it may not be conformal and may not be a minimal surface. In order to obtain a minimal surface, we need to vary the conformal structure on Σ also. Since the space of conformal structures on a surface is not compact, one needs to make sure the minimum can be achieved.

If the map f induces an injection on the fundamental group of the domain surface, Schoen-Yau [**593**] proved the energy of the harmonic map is proper on the moduli space of conformal structure on this surface by making use of a theorem of Linda Keen [**379**]. Based on a theory of topology of the L_1^2 map, they proved the existence of incompressible minimal surfaces. As a product of this argument, it is possible to find a nice exhaustion function for the Teichmüller space. Michael Wolf [**719**] was able to use harmonic maps to give a compactification of Teichmüller space which he proved to be equivalent to the Thurston compactification. S. Wolpert studied extensively the behavior of the Weil-Petersson metric (see Wolpert's survey [**720**]). A remarkable theorem of Royden [**570**] says that the Teichmüller metric is the same as the Kobayashi metric. C. McMullen [**498**] introduced a new Kähler metric on the moduli space which can be used to demonstrate that the moduli space is hyperbolic in the sense of Gromov [**286**]. The great detail of comparison of various intrinsic metrics on the Teichmüller space had been a major problem [**737**]. It was accomplished recently in the works of Liu-Sun-Yau [**467, 468**]. Actually Liu-Sun-Yau introduced new metrics with bounded negative curvature and geometry and found the stability of the logarithmic cotangent bundle of the moduli spaces. Recently L. Habermann and J. Jost [**306, 307**] also study the geometry of the Weil-Petersson metric associated to the Bergmann metric on the Riemann surface instead of the Poincaré metric.

> **Comment**: For a conformally invariant variational problem, Teichmüller space plays a fundamental role. It covers the moduli space of curves and in many ways behaves like a Hermitian symmetric space of noncompact type. Unfortunately, there is no good canonical realization of it as a pseudo-convex domain in Euclidean space. For example, we do not know whether it can be realized as a smooth domain or not.
>
> There is also Teichmüller space for other algebraic manifolds, such as Calabi-Yau manifolds. It is an important question in understanding their global behavior.

4.2. Classical minimal surfaces in Euclidean space. There is a long and rich history of minimal surfaces in Euclidean space. Recent contributions include works by Meeks, Osserman, Lawson, Gulliver, White, Hildebrandt, Rosenberg, Collin, Hoffman, Karcher, Ros, Colding, Minicozzi, Rodríguez, Nadirashvili and others (see the reference in Colding and Minicozzi's survey [**173**]) on embedded minimal surfaces in Euclidean space. They come close to classifying complete embedded minimal surfaces and a good understanding of complete minimal surface in a bounded domain. For example, Meeks-Rosenberg [**499**] proved that the plane and helicoid are the only properly embedded simply connected minimal surfaces in $\mathbb{R}^3$.

Calabi also initiated the study of isometric embedding of Riemann surfaces into S^N as minimal surfaces. The geometry of minimal spheres and minimal torus was then pursued by many geometers [**106**], [**149**], [**82**], [**332**], [**408**].

> **Comment**: This is one of the most beautiful subjects in geometry where Riemann made important contributions. Classification of complete minimal surface is nearly accomplished. However a similar problem for compact minimal surfaces in S^3 is far from being solved. It is also difficult to detect which set of disjoint Jordan curves can bound a connected minimal surface. The classification of moduli space of complete minimal surfaces with finite total curvature should be studied in detail.

4.3. Douglas-Morrey solution, embeddedness and application to topology of three manifolds. In a series of papers started in 1978, Meeks-Yau [**501, 502, 503, 504**] settled a classical conjecture that the Douglas solution for the Plateau problem is embedded if the boundary curve is a subset of a mean convex boundary. (One should note that Osserman [**542**] had already settled the old problem of non-existence of branched points for the Douglas solution while Gulliver [**304**] proved non-existence of false branched points.) We made use of the area minimizing property of minimal surfaces to prove these surfaces are equivariant with respect to the group action. Embedded surfaces which are equivariant play important roles for finite group actions on manifolds. Coupling with a theorem of Thurston, we can then prove the Smith conjecture [**744**] for cyclic groups acting on the spheres: that the set of fixed points is not a knotted curve.

The Douglas-Morrey solution of the Plateau problem is obtained by fixing the genus of the surfaces. However, it is difficult to minimize the area when the genus is allowed to be arbitrary large. This was settled by Hardt-Simon [**318**] by proving the boundary regularity of the varifold solution of the Plateau problem. In the other direction, Almgren-Simon [**7**] succeeded in minimizing the area among embedded disks with a given boundary in Euclidean space. The technique was used by Meeks-Simon-Yau [**500**] to prove the existence of embedded minimal spheres enclosing a fake ball. This theorem has been important to prove that the universal covering of an irreducible three manifold is irreducible. They also gave conditions for the existence of embedding minimal surfaces of a higher genus. This work was followed by topologists Freedman-Hass-Scott [**230**]. Pitts [**553**] used the mini-max argument for varifolds to prove the existence of an embedded minimal surfaces. Simon-Smith (unpublished) managed to prove the existence of an embedded minimax sphere for any metric on the three sphere. J. Jost [**364**] then extended it to find four mini-max spheres. Pitts-Rubinstein (see, e.g., [**554**]) continued to study such mini-max surfaces. Since such mini-max surfaces have Morse index one, I was interested in representing such minimal surface as a Heegard splitting of the three manifolds. I estimated its genus based on the fact that the second eigenvalue of the stability operator is nonnegative. This argument (dates back to Szego-Hersch) is to map the surface conformally to S^2. Hence we can use three coordinate functions, orthogonal to the first eigenfunction, to be trial functions. The estimate gave an upper bound of the genus for mini-max surfaces in compact manifolds with a positive scalar curvature. About twenty years ago, I was hoping to use such an estimate to control a Heegard genus as a way to prove Poincaré conjecture. While the program has not materialized, three manifold topologists did adapt the ideas of Meeks-Yau to handle combinational type minimal surfaces and gave applications in three manifold topology.

The most recent works of Colding and Minicozzi [**168, 169, 170, 171**] on lamination by minimal surfaces and estimates of minimal surfaces without the area bound are quite remarkable. They [**172**] made contributions to Hamilton's Ricci flow by bounding the total time for evolution. Part of the idea came from the above mentioned inequality.

> **Comment**: The application of minimal surface theory to three manifold topology is a very rich subject. However, one needs to have a deep understanding of the construction of minimal surfaces. For example, if minimal surfaces are constructed by the method of

mini-max, one needs to know the relation of their Morse index to the dimension of the family of surfaces that we use to perform the procedure of mini-max. A detail understanding may lead to a new proof of the Smale conjecture, as we may construct a minimal surface by a homotopic group of embeddings of surfaces. Conversely, topological methods should help us to classify closed minimal surfaces.

4.4. Surfaces related to classical relativity. Besides minimal surfaces, another important class of surfaces are surfaces with constant mean curvature and also surfaces that minimize the L^2-norm of the mean curvature. It is important to know the existence of such surfaces in a three dimensional manifold with nonnegative scalar curvature, as they are relevant to question in general relativity.

The existence of minimal spheres is related to the existence of black holes. The most effective method was developed by Schoen-Yau [**600**] where they [**595**] proved the existence theorem for the equation of Jang. It should be nice to find new methods to prove existence of stable minimal spheres. The extremum of the Hawking mass is related to minimization of the L^2 norm of mean curvature. Their existence and behavior have not been understood.

For surfaces with constant mean curvature, we have the concept of stability. (Fixing the volume it encloses, the second variation of area is non-negative.) Making use of my work on eigenvalues with Peter Li, I proved with Christodoulou [**161**] that the Hawking mass of such a surface is positive. (This was part of my contribution to the proposed joint project with Christodoulou-Klainerman which did not materialize.) This fact was used by Huisken and me [**352**] to prove uniqueness and the existence of foliation by constant mean curvature spheres for a three dimensional asymptotically flat manifold with positive mass. (We initiated this research in 1986. Ye studied our work and proved existence of similar foliations under various conditions, see [**746**].)

This foliation was used by Huisken and Yau [**352**] to give a canonical coordinate system at infinity. It defines the concept of center of gravity where important properties for general relativity are found. The most notable is that total linear momentum is equal to the total mass multiple with the velocity of the center of the mass. One expects to find good asymptotic properties of the tensors in general relativity along these canonical surfaces. We hope to find a good definition of angular momentum based on this concept of center of gravity so that global inequality like total mass can dominate the square norm of angular momentum.

The idea of using the foliation of surfaces satisfying various properties (constant Gauss curvature, for example) to study three manifolds in general relativity is first developed by R. Bartnik [**41**]. His idea of quasi-spherical foliation gives a good parametrization of a large class of metrics with positive scalar curvature.

Some of these ideas were used by Shi-Tam [**610**] to study quantities associated to spheres which bound three manifolds with positive scalar curvature. Such a quantity is realized to be the quasi-local mass of Brown-York [**81**]. At the same time, Melissa Liu and Yau [**462, 463**] were able to define a new quasi-local mass for general spacetimes in general relativity, where some of the ideas of Shi-Tam were used. Further works by M. T. Wang and myself generalized Liu-Yau's work by studying surfaces in hyperbolic space-form.

My interest in quasi-local mass dates back to the paper that I wrote with Schoen [**600**] on the existence of a black hole due to the condensation of matter. It is desirable to find a quasi-local mass which includes the effect of matter and the nonlinear effect of gravity. Hopefully one can prove that when such a mass is larger than a constant multiple of the square root of the area, a black hole forms. This has not been achieved.

> **Comment**: When surfaces theory appears in general relativity, we gain intuitions from both geometry and physics together. This is a fascinating subject.

4.5. Higher dimensional minimal subvarieties. Higher dimensional minimal subvarieties are very important for geometry. There are works by Federer-Fleming [**218**], Almgren [**6**] and Allard [**4**]. The attempt to prove the Bernstein conjecture, that minimal graphs are linear, was a strong drive for its development. Bombieri, De Giorgi and Giusti [**62**] found the famous counterexample to the Bernstein problem. It initiated a great deal of interest in the area minimizing cone (as a graph must be area minimizing). Schoen-Simon-Yau [**587**] found a completely different approach to the proof of Bernstein problem in low dimensions. This paper on stable minimal hypersurfaces initiated many developments on curvature estimates for the codimension one stable hypersurfaces in higher dimension. There are also works by L. Simon with Caffarelli and Hardt [**90**] on constructing minimal hypersurfaces by deforming stable minimal cones. Recently N. Wickramasekera [**710, 711**] did some deep work on stable minimal (branched) hypersurfaces which generalizes Schoen-Simon-Yau.

Michael-Simon [**511**] proved the Sobolev inequality and mean valued inequalities for such manifolds. This enables one to apply the classical argument of harmonic analysis to minimal submanifolds. For a minimal graph, Bombieri-Giusti [**61**] used ideas of De Giorgi-Nash to prove gradient estimates of the graph. N. Korevaar [**394**] was able to reprove this gradient estimate based on the maximal principle.

The best regularity result for higher codimension was done by F. Almgren [**6**] when he proved that for any area minimizing variety, the singular set has the codimension of at least two. How such a result can be used for geometry remains to be seen.

It was observed by Schoen-Yau [**593**] that for a closed stable minimal hypersurface in a manifold with positive scalar curvature, the first eigenfunction of the second variational operator can be used to conformally deform the metric so that the scalar curvature is positive. This provides an induction process to study manifolds with a positive scalar curvature. For example, if the manifold admits a nonzero degree map to the torus, one can then construct stable minimal hypersurfaces inductively until we find a two dimensional surface with higher genus which cannot support a metric with positive scalar curvature. At this moment, the argument encounters difficulty for dimensions greater than seven as we may have problems of singularity. In any case, we did apply the argument to prove the positive action conjecture in general relativity. The question of which type of singularities for minimal subvariety are generic under metric perturbation remains a major question for the theory of minimal submanifolds.

Perhaps the most important possible application of the theory of minimal submanifolds is the Hodge conjecture: whether a multiple of a (p,p) type integral cohomology class in a projective manifold can be represented by an algebraic cycle. Lawson made an attempt by combining a result of Lawson-Simons [**409**] and work of J. King [**381**] and Harvey-Shiffman [**322**]. (Lawson-Simons proved that currents in $\mathbb{C}P^n$ which are minimizing with respect to the projective group action are complex subvarieties.) The problem of how to use the hypothesis of (p,p) type has been difficult. In general, the algebraic cycles are not effective. This creates difficulties for analytic methods. The work of King [**381**] and Shiffman [**611**] on complex currents may be relevant.

Perhaps one should generalize the Hodge conjecture to include general (p,q) classes, as it is possible that every integral cycle in $\bigoplus_{i=-k}^{k} H^{p-i,p+i}$ is rationally homologous to an algebraic sum of minimal varieties such that there is a $p-k$ dimensional complex space in the tangent space for almost every point of the variety: it may be important to assume the metric to be canonical, e.g. the Kähler Einstein metric.

A dual question is how to represent a homology class by Lagrangian cycles which are minimal submanifolds also. When the manifold is Calabi-Yau, these are special Lagrangian cycles. Since they are supposed to be dual to holomorphic cycles, there should be an analogue of the Hodge conjecture. For example if $\dim_{\mathbb{C}} M = n$ is odd, any integral element in $\bigoplus_{i+j=n} H^{i,j}$ should be representable by special Lagrangian cycles up to a rational multiple provided the cup product of it with the Kähler class is zero.

A very much related question is: if the Chern classes of a complex vector bundle are of (p,p) type, does the vector bundle, after adding a holomorphic vector bundle, admit a holomorphic structure? If the above generalization of the Hodge conjecture holds, there should be a similar generalization for the vector bundle. It should also be noted that Voisin [**692**] observed that Chern classes of all holomorphic bundles do not necessarily generate all rational (p,p) classes. On the other hand, the Kähler manifold that she constructed is not projective.

These questions had a lot more success for four dimensional symplectic manifolds by the work of Taubes both on the existence of pseudoholomorphic curves [**661**] and on the existence of anti-self-dual connections [**651, 652**]. On a Kähler surface, anti-self-dual connections are Hermitian connections for a holomorphic vector bundle. In particular, Taubes gave a method to construct holomorphic vector bundles over Kähler surfaces. Unfortunately this theorem does not provide much information on the Hodge conjecture as it follows from Lefschetz theorem in this dimension.

Another important class of minimal varieties is the class of special Lagrangian cycles in Calabi-Yau manifolds. Such cycles were first developed by Harvey-Lawson [**321**] in connection to calibrated geometry. Major works were done by Schoen-Wolfson [**590**], Yng-Ing Lee [**415**] and Butscher [**89**]. One expects Lagrangian cycles to be mirror to holomorphic bundles and special Lagrangian cycles to be mirror to Hermitian-Yang-Mills connections. Hence by the Donaldson-Uhlenbeck-Yau theorem, it is related to stability. The concept of stability for Lagrangian cycles was discussed by Joyce and Thomas. Since the Yang-Mills flow for Hermitian connection exists for all time, Thomas-Yau [**667**] suggested an analogy with the mean curvature flow for Lagrangian cycles. For stable Lagrangian cycles, mean curvature flow should converge to special Lagrangian cycles. See M. T. Wang [**701, 702**], Smoczyk [**632**] and Smoczyk-Wang [**633**]. The geometry of mirror symmetry was explained by Strominger-Yau-Zaslow in [**639**] using a family of special Lagrangian tori. There are other manifolds with special holonomy group. They have similar calibrated submanifolds. Conan Leung has contributed to studies of such manifolds and their mirrors (see, e.g. [**418, 419**]).

Submanifolds of space forms are called isoparametric if the normal bundle is flat and the principal curvatures are constants along parallel normal fields. These were studied by E. Cartan [**107**]. Minimal submanifolds, with constant scalar curvature are believed to be isoparametric surfaces. There is work done by Lawson [**407**], Chern-de Carmo-Kobayashi [**150**] and Peng-Terng [**546**]. Recently there has been extensive work by Terng and Thorbergsson (see Terng's survey [**663**] and Thorbergsson [**668**]). Terng [**662**] related isometric embedded hyperbolic spaces in Euclidean space to soliton theory. A nice theory of Lax pair and loop groups related to geometry has been developed.

> **Comment**: The theory of higher dimensional minimal submanifolds is one of the deepest subjects in geometry. Unfortunately our knowledge of the subject is not mature enough to give applications to solve outstanding problems in geometry, such as the Hodge conjecture. But the future is bright.

4.6. Geometric flows. The major geometric flows are flows of submanifold driven by mean curvature, gauss curvature, inverse mean curvature. Flows that change geometric structures are Ricci flows and Einstein flow.

Mean curvature flow for varifolds was initiated by Brakke [**77**]. The level set approach was studied by many people: S. Osher, L. Evans, Giga, etc (see [**541, 216, 135**]). Huisken [**347, 348**] did the first important work when the initial surface is convex. His recent work with Sinestrari [**350, 351**] on mean convex surfaces is remarkable and gives a good understanding of the structure of singularities of mean curvature flow. Mean curvature flow has many geometric applications. For example, the work of Huisken-Yau mentioned in 4.4 was achieved by mean curvature flow. Mean curvature flow for spacelike hypersurfaces in Lorentzian manifolds should be very interesting. Ecker [**210**] did interesting work in this direction. It will be nice to find the Li-Yau type estimate for such flows.

The inverse mean curvature was proposed by Geroch [**249**] to understand the Penrose conjecture relating the mass with the area of the black hole. Such a procedure was finally carried out by Huisken-Ilmanen [**349**] when the scalar curvature is non-negative. There was a different proof by H. Bray [**79**] subsequently.

Ricci flow has had spectacular successes in recent years. However, not much progress has been made on the Calabi flow (see Chang's survey [**110**]) for Kähler metrics. They are higher order problems where the maximal principle has not been effective. An important contribution was made by Chrúsciel [**162**] for Riemann surface. Inspired by the concept of the Bondi mass in general relativity, Chruściel was able to give a new estimate for the Calabi flow. Unfortunately, a higher dimensional analogue had not been found.

Natural higher order elliptic problems are difficult to handle. Affine minimal surfaces and Willmore surfaces are such examples. L. Simon [**615**] made an important contribution to the regularity of the Willmore surfaces. The corresponding flow problem should be interesting.

The dynamics of Einstein equations for general relativity is a very difficult subject. The Cauchy problem was considered by many people: A. Lichnerowicz, Y.

Choquet-Bruhat, J. York, V. Moncrief, H. Friedrich, D. Christodoulou, S. Klainerman, H. Lindblad, M. Dafermos (see, e.g., [**452**], [**157**], [**156**], [**233**], [**158**], [**382**], [**459**], [**180**]). But the global behavior is still far from being understood. The major unsolved problem is to formulate and prove the fundamental question of Penrose on Cosmic censorship. I suggested to Klainerman and Christodoulou to consider small initial data for the Einstein system. The treatment of stability of Minkowski spacetime was accomplished by Christodoulou-Klainerman [**159**] under small perturbation of flat spacetime and fast fall off conditions. Recently Lindblad and Rodnianski [**459**] gave a simpler proof. A few years ago, N. Zipser (Harvard thesis) added Maxwell equation to gravity and still proved stability of Minkowski spacetime. There is remarkable progress on the problem of Cosmic censorship by M. Dafermos [**180**]. He made an important contribution for the spherical case. Stability for Schwarzschild or Kerr solutions is far from being known. Finster-Kamran-Smoller-Yau [**221**] had studied decay properties of Dirac particles with such background. The work does indicate the stability of these classical spacetimes.

The no hair theorem for stationary black holes is a major theorem in general relativity. it was proved by W. Israël [**353**], B. Carter [**108**], D. Robinson [**566**] and S. Hawking [**324**]. But the proof is not completely rigorous for the Kerr metric. In any case, the existing uniqueness theorem does assume regularity of the horizon of the black hole. It is not clear to me whether a nontrivial asymptotically flat solitary solution of a vacuum Einstein equation has to be the Schwarzschild solution. There is a possibility that the Killing field is spacelike. In that case, there may be a new interesting vacuum solution.

There is extensive literature on spacelike hypersurfaces with constant mean curvature. The foliation defined by them gives interesting dynamics of Einstein equation. These surfaces are interesting even for $\mathbb{R}^{n,1}$. A. Treiberges studied it extensively [**681**]. Li, Choi-Treibergs [**153**] and T. Wan [**695**] observed that the Gauss maps of such surfaces give very nice examples of harmonic maps mapping into the disk. Recently Fisher and Moncrief used them to study the evolution equation of Einstein in $2+1$ dimension.

> **Comment**: The dynamics of submanifolds and geometric structures reveal the true nature of these geometric objects deeply. In the process of arriving at a stationary object or a solitary solution, it encounters singularities. Understanding the structures of such singularity will solve many outstanding conjectures in topology such as Shoenflies conjecture.

5. Construction of geometric structures on bundles and manifolds

A fundamental question is how to build geometric structures over a given manifold. In general, the group of topological equivalences that leaves this geometric structure invariant should be a special group. With the exception of symplectic structures, these groups are usually finite dimensional. When the geometric structure is unique (up to equivalence), it can be used to produce key information about the topological structure.

The study of special geometric structures dates back to Sophus Lie, Klein and Cartan. In most cases, we like to be able to parallel transport vectors along paths so that we can define the concept of holonomy group.

5.1. Geometric structures with noncompact holonomy group. When the holonomy group is not compact, there are examples of projective flat structure, affine flat structure and conformally flat structure. It is not a trivial matter to determine which topological manifolds admit such structures. Since the structure is flat, there is a unique continuation property and hence one can construct a developing map from a suitable cover of the manifold to the real projective space, the affine space and the sphere respectively. The map gives rise to a representation of the fundamental group of the manifold to the real projective group, the special linear group and the Möbius group respectively. This holonomy representation gives a great deal of information for the geometric structure. Unfortunately, the map is not injective in general. In the case where it is injective, the manifold can be obtained as a quotient of a domain by a discrete subgroup of the corresponding Lie group. In this case, a lot more can be said about the manifold as the theories of partial differential equations and discrete groups can play important roles.

5.1.1. *Projective flat structure.* If a projective flat manifold can be projectively embedded as a bounded domain, Cheng-Yau [**144**] were able to construct a canonical metric from the real Monge-Ampère equation which generalizes the Hilbert metric. When the manifold is two dimensional, there are works of C. P. Wang [**696**] and J. Loftin [**469**] on how to associate such metrics to a conformal structure and a holomorphic section of the cubic power of a canonical bundle. This is a beautiful theory related to the hyperbolic affine sphere mentioned in chapter one.

There are fundamental works by S. Y. Choi, W. Goldman (see the reference of Choi-Goldman [**155**]), N. Hitchin [**333**] and others on the geometric decomposition and the moduli of projective structures on Riemann surfaces. It should be interesting to extend them to three or four dimensional manifolds.

5.1.2. *Affine flat structure.* It is a difficult question to determine which manifolds admit flat affine structures. For example, it is still open whether the Euler number of such spaces is zero, although great progress was made by D. Sullivan [**640**]. W. Goldman [**257**] has also found topological constraints on three manifolds in terms of fundamental groups. The difficulty arises as there is no useful metric that is compatible with the underlining affine structure. This motivated Cheng-Yau [**145**] to define the concept of affine Kähler metric.

When Cheng and I considered the concept of affine Kähler metric, we thought that it was a natural analogue of Kähler metrics. However, compact nonsingular examples are not bountiful. Strominger-Yau-Zaslow [**639**] proposed the construction of mirror manifolds by constructing the quotient space of a Calabi-Yau manifold by a special Lagrangian torus. At the limit of the large Kähler class, it was pointed out by Hitchin [**334**] that the quotient space admits a natural affine structure with a compatible affine Kähler structure. But in general, we do expect singularities of such structure. It now becomes a deep question to understand what kind of singularity is allowed and how we build the Calabi-Yau manifold from such structures. Loftin-Yau-Zaslow [**470**] have initiated the study of the structure of a "Y" type singularity. Hopefully one can find an existence theorem for affine structures over compact manifolds with prescribed singularities along codimension two stratified submanifolds.

5.1.3. *Conformally flat structure.* Construction of conformally flat manifolds is also a very interesting topic. Similar to projective flat or affine flat manifolds, there are simple constraints from curvature representation for the Pontrjagin classes.

The deeper problem is to understand the fundamental group and the developing map. When the structure admits a conformal metric with positive scalar curvature, Schoen-Yau [**601**] proved the rather remarkable theorem that the developing map is injective. Hence such a manifold must be the quotient of a domain in S^n by a discrete subgroup of Möbius transformations. It would be interesting to classify such manifolds. In this regard, the Yamabe problem as was solved by Schoen [**586**] did provide a conformal metric with constant scalar curvature. One hopes to be able to use such metrics to control the conformal structure. Unfortunately the metric is not unique and a deep understanding of the moduli space of conformal metrics with constant scalar curvature should be important.

Kazdan-Warner [**377**] and Korevaar-Mazzeo-Pacard-Schoen [**396**] developed a conformal method to understand Nirenberg's problem on prescribed scalar curvature. It was followed by Chen-Lin [**130**], Chang-Gursky-Yang [**111**]. Chen-Lin has related this problem to mean field theory. Their computations in the relevant degree theory involves deep analysis. One should generalize their works to functions which are sections of a flat line bundle because it is related to the previously mentioned work of Loftin-Yau-Zaslow [**470**]. In any case, the integrability condition of Kazdan and Warner is still not fully understood.

It is curious that while bundle theory was used extensively in Riemannian geometry, it has not been used in the study of these geometries. One can construct real projective space bundles, affine bundles or sphere bundles by mapping coordinate charts projectively, affinely or conformally to the corresponding model spaces (possible with dimensions different from the original manifold) and gluing the target model spaces together to form natural bundles. Perhaps one may study their associated Chern-Simons forms [**151**].

Many years ago, H. C. Wang [**698**] proved the theorem that if a compact complex manifold has trivial holomorphic tangent bundle, it is covered by a complex Lie group. It will be nice to generalize and interpret such a theorem in terms of Hermitian connections on the manifold with a special holonomy group and torsion.

This program was discussed in my paper [**740**] on algebraic characterization of locally Hermitian symmetric spaces. For a holomorphic stable vector bundle V, we can form a stable vector bundles from V by taking irreducible representation of $GL(n,\mathbb{C})$ from decomposition of the tensor product representation $\bigotimes_p V \bigotimes_q V^*$. By twisting with powers of canonical line bundle, we can form irreducible stable bundles with trivial determinant line bundle. In general, such bundles may not have holomorphic sections. If they do, the section must be parallel with respect to the Hermitian-Yang-Mills connection on the bundle, and the structure group of V can be reduced to a smaller group. Hermitian-Yang-Mills connections with reduced holonomy group have good geometric properties. We may formulate a principle: For stable holomorphic bundles, existence of nontrivial holomorphic invariants implies the existence of parallel tensors and therefore the reduction of structure group. If the holonomy group is reduced to discrete group, the bundle will provide representations of the fundamental group into unitary group. This should compare with Wang's theorem when the bundle is the tangent bundle.

Comment: Geometric structures with a noncompact holonomy group is less intuitive than Riemannian geometry. Perhaps we need to deepen our intuitions by relating them to other geometric structures, especially those structures that may carry physical meaning.

5.2. Uniformization for three manifolds. An important goal of geometry is to build a canonical metric associated to a given topology. Besides the uniformization theorem in two dimensions, the only (spectacular) work in higher dimensions is the geometrization program of Thurston (see [**670**]).

W. Thurston made use of ideas from Riemann surface theory, W. Haken's work [**308**] on three-manifolds, G. Mostow's rigidity [**527**] to build his manifolds. Many mathematicians have contributed to the understanding of this program of Thurston's. (e.g., J. Morgan [**525**], C. McMullen [**496, 497**], J. Otal [**543**], J. Porti [**562**].) Thurston's orbifold program was finally proved by M. Boileau, B. Leeb and J. Porti [**60**]. However, all this work has to assume that an incompressible surface (or corresponding surface in case of orbifold) exists. When R. Hamilton [**309**] had his initial success on his Ricci flow, I suggested (around 1981) to him to use his flow to break up the manifold and prove Thurston's conjecture. His generalization of the theory of Li-Yau [**443**] to Ricci flow [**310, 311**] and his seminal paper in 1996 [**313**] on breaking up the manifold mark cornerstone of the remarkable program. Perelman's recent idea [**547, 548**] built on these two works and has gone deeply into the problem. Detailed discussions have been pursued by Hamilton, Colding-Minicozzi, Shioya-Yamaguchi, Zhu, Cao and Huisken in the past two years. Hopefully it may lead to the final settlement of the geometrization program. This theory of Hamilton and Perelman should be considered as a crowning achievement of geometric analysis in the past thirty years. Most ideas developed in this period by geometric analysts are used.

Let me now explain briefly the work of Hamilton and Perelman.

In early 90's, Hamilton [**311, 312, 313**] developed methods and theorems to understand the structure of singularities of the Ricci flow. Taking up my suggestions, he proved a fundamental Li-Yau type differential inequality (now called the Li-Yau-Hamilton estimate) for the Ricci flow with non-negative curvature in all dimensions. He gave a beautiful interpretation of the work of Li-Yau and observed the associated inequalities should be equalities for solitary solutions. He then established a compactness theorem for (smooth) solutions to the Ricci flow, and observed (also independently by T. Ivey[**354**]) a pinching estimate for the curvature for three-manifolds. By imposing an injectivity radius condition, he rescaled the metric to show that each singularity is asymptotic to one of the three singularity models. For type I singularities in dimension three, Hamilton established an isoperimetric ratio estimate to verify the injectivity radius condition and obtained spherical or neck-like structures. Based on the Li-Yau-Hamilton estimate, Hamilton showed that any type II model is either a Ricci soliton with a neck-like structure or the product of the cigar soliton with the real line. Similar characterization for type III model was obtained by Chen-Zhu [**125**]. Hence Hamilton had already obtained the canonical neighborhood structures (consisting of spherical, neck-like and cap-like regions) for the singularities of three-dimensional Ricci flow.

But two obstacles remained: one is the injectivity radius condition and the other is the possibility of forming a singularity modelled on the product of the cigar soliton with a real line which could not be removed by surgery. Recently, Perelman [**547**] removed these two stumbling blocks in Hamilton's program by establishing a local injectivity radius estimate (also called "Little Loop Lemma" by Hamilton in [**312**]). Perelman proved the Little Loop Lemma in two ways, one with an entropy functional he introduced in [**547**], the other with a reduced distance function based on the same idea as Li-Yau's path integral in obtaining their inequality [**547**]. This reduced distance question gives rise to a Gaussian type integral which he called reduced volume. The reduced volume satisfies monotonicity property. Furthermore, Perelman [**548**] developed a refined rescaling argument (by considering local limits and weak limits in Alexandrov spaces) for singularities of the Ricci flow on three-manifolds to obtain a uniform and global version of the canonical neighborhood structure theorem.

After obtaining the canonical neighborhoods for the singularities, one performs geometric surgery by cutting off the singularities and continue the Ricci flow. In [**313**], Hamilton initiated such a surgery procedure for four-manifolds with a positive isotropic curvature. Perelman [**548**] adapted Hamilton's geometric surgery procedure to three-manifolds. The most important question is how to prevent the surgery times from accumulations and make sure there are only a finite number of surgeries on each finite time interval. When one performs the surgeries with a given accuracy at each surgery time, it is possible that the errors may add up which causes the surgery times to accumulate. Hence at each step of surgery one is required to perform the surgery more accurate than the former one. In [**549**], Perelman presented a clever idea on how to find "fine" necks, how to glue "fine" caps and how to use rescaling arguments to justify the discreteness of the surgery times. In the process of rescaling for surgically modified solutions, one encounters the difficulty of how to use Hamilton's compactness theorem, which works only for smooth solutions. The idea to overcome such difficulty consists of two parts. The first part, due to Perelman [**548**], is to choose cutoff radius (in neck-like regions) small enough to push the surgical regions far away in space. The second part, due to Chen-Zhu [**129**] and Cao-Zhu [**102**], is to show that the solutions are smooth on some uniform small time intervals (on compact subsets) so that Hamilton's compactness theorem can be used.

Once surgeries are known to be discrete in time, one can complete Schoen-Yau's classification [**599**] for three-manifolds with positive scalar curvature. For simply connected three manifolds, if one can show solution to the Ricci flow with surgery extincts in finite time, Poincaré conjecture will be proved. Recently, such a finite extinction time result was proposed by Perelman [**549**] and a proof appeared in Colding-Minicozzi [**172**].

For the full geometrization program, one still needs to find the long time behavior of surgically modified solutions. In [**314**], Hamilton studied the long time behavior of the Ricci flow on a compact three-manifold for a special class of (smooth) solutions called "nonsingular solutions". Hamilton proved that any (three-dimensional) nonsingular solution either collapses or subsequently converges to a metric with constant curvature on the compact manifold, or at large time it admits a thick-thin decomposition where the thick part consists of a finite number of hyperbolic pieces and the thin part collapses. Moreover, by adapting Schoen-Yau's minimal surface arguments to a parabolic version, Hamilton showed that the boundary of hyperbolic pieces are incompressible tori. Then by combining with the collapsing results of Cheeger-Gromov [**119**], any nonsingular solution to the Ricci flow is geometrizable.

In [**547, 548**], Perelman modified Hamilton's arguments to analyze the long-time behavior of arbitrary solutions to the Ricci flow and solutions with surgery in dimension three. Perelman also argued by showing a thick-thin decomposition, except that he can only show the thin part has (local) lower bound on sectional curvature. For the thick part, based on Li-Yau-Hamilton estimate, Perelman established a crucial elliptic type Harnack estimate to conclude the thick part consisting of hyperbolic pieces. For the thin part, he announced a new collapsing result which states that if a three-manifold collapses with a (local) lower bound on the sectional curvature, it is then a graph manifold. However, the proof of the new collapsing result has not been published. Shioya and Yamaguchi [**612, 613**] offered a proof for compact manifolds. Very recently, Cao-Zhu claimed to have a complete proof for compact manifolds based only on the Shioya-Yamaguchi's collapsing result.

Hopefully all these arguments can be checked thoroughly in the near future. It should also be interesting to see whether other famous problems in three manifold can be settled by analysis: Does every three dimensional hyperbolic manifold admit a finite cover with nontrivial first Betti number?

Hyperbolic metrics have been used by topologists to give invariants for three dimensional manifolds. Thurston [**669**] observed that the volume of a hyperbolic metric is an important topological invariant. The associated Chern-Simons [**151**] invariant, which is defined by mod integers, can be looked upon as a phase for such manifolds. These invariants appeared later in Witten's theory of $2+1$ dimensional gravity [**715**] and S. Gukov [**302**] was able to relate them to fundamental questions in knot theory.

> **Comment**: This is the most spectacular development in the last thirty years. Once the three manifold is hyperbolic, Ricci flow does not give much more information. Happily one may obtain further information by performing reduction from four dimension Ricci flow to three dimension by circle action. Is there any effective way to understand the totality of all hyperbolic manifolds with finite volume by constructing flows that may break up topology?

5.3. Four manifolds. The major accomplishment of Thurston, Hamilton, Perelman et al is the ability to create a canonical structure on three manifolds. Such a structure has not even been conjectured for four manifolds despite the great success of Donaldson invariants and Seiberg-Witten invariants. Taubes [**655**] did prove a remarkable existence theorem for self-dual metrics on a rather general class of four dimensional manifolds. Unfortunately their moduli space is not understood and their topological implication is not clear at this moment. Since the twistor space of Taubes metric admits integrable complex structure, ideas from complex geometry may be helpful. Prior to the construction of Taubes, Donaldson-Friedman [**204**] and LeBrun [**412**] have used ideas from twistor theory to construct self-dual metrics on the connected sum of $\mathbb{C}P^2$.

The problem of four manifold is the lack of good diffeomorphic invariants. Donaldson or Seiberg-Witten provide such invariants. But they are not powerful enough to control the full structure of the manifold. A true understanding of four manifolds probably should come from understanding the question of existence of the integrable complex structures. The Riemann-Roch-Hirzebruch formula has been the basic tool to find the integrability condition. In the last twenty-five years, there are nonlinear methods from Käher-Einstein metrics, harmonic maps, anti-self-dual connections and Seiberg-Witten invariants. However, one needs an existence theorem to find a canonical way to deform an almost complex structure to an integrable complex structure. What kind of obstructions do we expect? The work of Donaldson [**198, 200**] and Gompf [**259**] gave a good characterization of symplectic manifolds in terms of Lefschetz fibration. It may be useful to know under which condition such fibration will give rise to complex structures. I did ask several of my students to work on it. But no definite answer is known. J Jost and I [**368**] studied the rigidity part: if a Kähler surface has a topological map to a Riemann surface with higher genus, it can be deformed to be a holomorphic map by changing the complex structure of the Riemann surface. One can derive from the work of Griffiths [**272**] that every algebraic surface has a Zariski open set which admits a complete Kähler-Einstein metric with finite volume and is covered by a contractible pseudo-convex domain. Perhaps one can classify these manifolds by topological means.

While the Donaldson invariant gave the first counterexample to the h-cobordism theorem and irreducibility (nontrivial connected sum with manifolds not homotopic to $\mathbb{C}P^2$) of four manifolds, the Seiberg-Witten invariant gave the remarkable result that an algebraic surface of general type can not be diffeomorphic to rational or elliptic surfaces. It also solves the famous Thom conjecture that holomorphic curves realize the lowest genus for embedded surface in a Kähler surface (Kronheimer-Mrowka [**400**] and Ozsváth-Szabó [**544**]). One wonders whether one can construct a diffeomorphic invariant based on metrics which are a generalization of Kähler-Einstein metrics.

> **Comment**: A good conjectural statement need to be made on the topology of four manifolds that may admit an integrable complex structure. Pseudo-holomorphic curve and fibration by Riemann surfaces should provide important information. Geometric flows may still be the major tool.

5.4. Special connections on bundles. In the seventies, theoretic physicists were very much interested in the theory of instantons: self-dual connections on four manifolds. Singer was able to communicate the favor of this excitement to the mathematical community which soon led to his paper with Atiyah and Hitchin [**16**] and also the complete solution of the problem over the four sphere by Atiyah-Hitchin-Drinfel'd-Manin [**15**] using twistor technique of Penrose.

While the paper of Atiyah-Hitchin-Singer [**16**] laid the algebraic and geometric foundation for self-dual connections, the analytic foundation was laid by Uhlenbeck [**684, 685**] where she established the removable singularity theorem and compactness theorem for Yang-Mills connections. This eventually led to the fundamental works of Taubes [**652**] and Donaldson [**194**] which revolutionized four manifold topology.

In the other direction, Atiyah-Bott [**11**] applied Morse theory to the space of connections over Riemann surface. They solved important questions on the moduli space of holomorphic bundles which was studied by Narasimhan, Seshadri, Ramanathan, Newstead and Harder. In the paper of Atiyah-Bott, Morse theory, moment map and localization of equivariant cohomology were introduced on the subject of vector bundle. It laid the foundation of works in last twenty years.

The analogue of anti-self dual connections over Kähler manifolds are Hermitian Yang-Mills connections, which was shown by Donaldson [**195**] for Kähler surfaces and Uhlenbeck-Yau [**687**] for general Kähler manifolds to be equivalent to the polystability of bundles. (That polystability of bundle is a consequence of existence of Hermitian Yang-Mills connection was first observed by Lüber [**482**]. Donaldson [**196**] was able to make use of the theorem of Mehta-Ramanathan [**505**] and ideas of above two papers to prove the theorem for projective manifold). It was generalized by C. Simpson [**617**], using ideas of Hitchin [**331**], to bundles with Higgs fields. It has important applications to the theory of variation of a Hodge structure [**618, 619**]. G. Daskalopoulos and R. Wentworth [**183**] studied such a theory for moduli space of vector bundles over curves. Li-Yau [**431**] generalized the existence of Hermitian Yang-Mills connections to non-Kähler manifolds. (Buchdahl [**86**] subsequently did the same for complex surfaces.) Li-Yau-Zheng [**433**] then used the result to give a complete proof of Bogomolov's theorem for class VII_0 surfaces. The only missing parts of the classification of non-Kähler surfaces are those complex surfaces with a finite number of holomorphic curves. It is possible that the argument of Li-Yau-Zheng can be used. One may want to use Hermitian Yang-Mills connections with poles along such curves. I expect more applications of Donaldson-Uhlenbeck-Yau theory to algebraic geometry.

It should be noted that the construction of Taubes [**655**] on the anti-self-dual connection is achieved by singular perturbation after gluing instantons from S^4. The method is rather different from Donaldson-Uhlenbeck-Yau. While it applies to arbitrary four manifolds, it does require some careful choice of Chern classes for the bundle. It will be nice to find a concept of stability for a general complex bundle so that a similar procedure of Donaldson-Uhlenbeck-Yau can be applied. The method of singular perturbation has an algebraic geometric counterpart as was found by Gieseker-Li [**254**] and O'Grady [**538**] who proved that moduli spaces of algebraic bundles with fixed Chern classes over algebraic surfaces are irreducible. Li [**425**] also obtained information about Betti number of such moduli space. Not many general theorems are known for bundles over algebraic manifolds of a higher dimension. It will be especially useful for bundles over Calabi-Yau manifolds.

D. Gieseker [**252**] developed the geometric invariant theory for the moduli space of bundles and introduced the Gieseker stability of bundles. Conan Leung [**417**] introduced the analytic counterpart of such bundles in his thesis under my guidance. While it is a natural concept, there is still an analytic problem to be resolved. (He assumed the curvature of the bundles to be uniformly bounded.)

There were attempts by de Bartolomeis-Tian [**43**] to generalize Yang-Mills theory to symplectic manifolds and also by Tian [**675**] to manifolds with a special holonomy group, as was initiated by the work of Donaldson and Thomas. However, the arguments for both papers are not complete and still need to be finished.

For a given natural structure on a manifold, we can often fix a structure and linearize the equation to obtain a natural connection on the tangent bundle. Usually we obtain Yang-Mills connections with the extra structure given by the holonomy group of the original structure. It is interesting to speculate whether an iterated procedure can be constructed to find an interesting metric or not. In any case, we can draw analogous properties between bundle theory and metric theory. The concept of stability for bundles is reasonably well understood for the holomorphic category. I believed that for each natural geometric structure, there should be a concept of stability. Donaldson [**196**] was able to explain stability in terms of moment map, generalizing the work of Atiyah-Bott [**11**] for bundles over Riemann surfaces. It will be nice to find moment maps for other geometric structures.

> **Comment**: Bundles with anti-self-dual connections or Hermitian Yang-Mills connections have been important for geometry. However, we do not have good estimates of their curvature of such connections. Such an estimate would be useful to handle important problems such as the Hartshorne question (see, e.g., [**39**]) on the splitting of rank two bundle over high dimension complex projective space.

5.5. Symplectic structures. Symplectic geometry had many important breakthroughs in the past twenty years. A moment map was developed by Atiyah-Bott [**12**], Guillemin-Sternberg [**299**] who proved the image of the map is a convex polytope. Kirwan and Donaldson had developed such a theory to be a powerful tool. The Marsden-Weinstein [**488**] reduction has become a useful method in many branches of geometry. At around the same time, other parts of symplectic topology were developed by Donaldson [**199**], Taubes [**656**], Gompf [**258**], Kronheimer-Mrowka [**400**] and others.

The phenomenon of symplectic rigidity is manifested by the existence of symplectic invariants measuring the 2–dimensional size of a symplectic manifold. The first such invariant was discovered by Gromov [**284**] via pseudo-holomorphic curves. Hofer [**338**] then developed several symplectic invariants based on variational methods and successfully applied them to Weinstein conjectures. Ekeland-Hofer [**213**] introduced a concept of symplectic capacity and used it to provide a characterization of a symplectomorphism not involving any derivatives. The C^0-closed property of the symplectomorphism group as a subgroup of the diffeomorphism group then follows, which was independently established by Y. Eliashberg [**214**] via wave front methods. Hofer-Zehnder [**341**] introduced another capacity and discovered the displacement-energy on $\mathbb{R}^{2n}$. By relating the two invariants with the energy-capacity inequality, Hofer [**337**] found a bi-invariant norm on the infinite dimensional group of Hamiltonian symplectomorphisms of $\mathbb{R}^{2n}$. The existence of such a norm has now been established for general symplectic manifolds by Lalonde-McDuff [**406**] via pseudo-holomorphic curves and symplectic embedding techniques. The generalized Weinstein conjecture on the existence of a periodic orbits of Reeb flows for many 3-manifolds including the 3-sphere was also established in Hofer [**338**] by studying the finite energy pseudo-holomorphic plane in the symplectization of contact 3-manifolds.

Eliashberg-Givental-Hofer [**215**] recently introduced the concept of symplectic field theory, which is about invariants of punctured pseudo-holomorphic curves in a symplectic manifold with cylindrical ends. Though it has not been rigorously established, some applications in contact and symplectic topology have been found.

By analyzing the singularities of pseudo-holomorphic curves in a symplectic 4–manifold, D. McDuff [**494**] established rigorously the positivity of intersections of two distinct curves and the adjunction formula of an irreducible curve. Applying these basic properties to symplectic 4-manifolds containing embedded pseudo-holomorphic spheres with self-intersections at least -1, she was able to construct minimal models of general symplectic 4–manifolds, and classify those containing embedded symplectic spheres with non-negative self-intersections.

A fundamental question in symplectic geometry is to decide which topological manifold admits a symplectic structure and how, as was pointed out by Smith-Thomas-Yau [**631**], mirrors of certain non-Kähler complex manifolds should be symplectic manifolds. Based on this point of view, they construct a large class of symplectic manifolds with trivial first Chern class by reversing the procedure of Clemens-Friedman on non-Kähler Calabi-Yau manifolds [**164, 232**]. In dimension four, the Betti numbers of such manifolds are determined by T. J. Li [**444**]. In the last ten years, there has been extensive work on symplectic manifolds, initiated by Gromov [**284**], Taubes [**657, 658, 659, 660**], Donaldson [**197, 198, 200**] and Gompf [**259**]. These works are based on the understanding of pseudo-holomorphic curves and Lefschetz fibrations. They are most successful for four dimensional manifolds. The major tools are Seiberg-Witten theory [**603, 604, 717**] and analysis. The work of Taubes on the existence of pseudo-holomorphic curves and the topological meaning of its counting is one of the deepest works in geometry. Based on this work, Taubes [**657**] was able to prove the old conjecture that there is only one symplectic structure on the standard $\mathbb{C}P^2$. However, the following question of mine is still unanswered: If M is a symplectic 4-manifold homotopic to $\mathbb{C}P^2$, is M symplectomorphic to the standard $\mathbb{C}P^2$? (The corresponding question for complex geometry was solved by me in [**731**].) On the other hand, based on the work of Taubes [**656**], T. J. Li and A. K. Liu [**445**] did find a wall crossing formula for four dimensional manifolds that admit metrics with a positive scalar curvature. Subsequently A. Liu [**460**] gave the classification of such manifolds. (The surgery result by Stolz [**637**] based on Schoen-Yau-Gromov-Lawson for manifolds with positive scalar curvature is not effective for four dimensional manifolds.) As another application of the general wall crossing formula in [**445**], it was proved by T. J. Li and A. Liu in [**446**] that there is a unique symplectic structure on S^2-bundles over any Riemann surface. A main result of D. McDuff in [**493**] is used here.

McDuff [**492**] also used a refined bordism type Gromov-Witten invariant to distinguish two cohomologous and deformation equivalent symplectic forms on $S^2 \times S^2 \times T^2$, showing that they are not isotopic. Notice that there are also cohomologous but non-deformation equivalent symplectic forms on $K3 \times S^2$ as shown by Y. Ruan [**574**]. In contrast, it is not known whether examples of this kind exist in dimension 4 or not. This phenomenon might be related to the special features of pseudo-holomorphic curves in a 4-manifold.

Fukaya and Oh [**238**] have developed an elaborate theory for symplectic manifolds with Lagrangian cycles. Pseudo-holomorphic disks appeared as trace of motions of curves according to Floer theory. Due to boundary bubbles, the Lagrangian Floer homology is not always defined. Oh [**539**] developed some works on pseudo-holomorphic curves with Lagrangian boundary conditions and extended the Lagrangian Floer homology to all monotone symplectic manifolds. In order to understand open string theory, Katz-Liu [**376**] and Melissa Liu [**461**] developed the theory in analogue of the Gromov-Witten invariant for a holomorphic curve with boundaries on a given Lagrangian submanifold. Fukaya [**237**] discovered the underlying A^∞ structure of the Lagrangian Floer homology on the chain level, leading to the Fukaya category. By carefully analyzing this A^∞ structure, Fukaya, Oh, Ohta and Ono in [**239**] have constructed a sequence of obstruction classes which elucidate the rather difficult Lagrangian Floer homology theory to a great extent.

Seidel-Thomas [**605**] and W. D. Ruan [**573**] discussed Fukaya's category in relation to Kontsevich's homological mirror conjecture [**393**]. One wonders whether Fukaya's theory can help to construct canonical metrics for symplectic structures. For symplectic manifolds that admitan almost complex structure with zero first Chern class, it would be nice to construct Hermitian metrics with torsion that admit parallel spinor. Such structures may be considered as a mirror to the system constructed by Strominger on non-Kähler complex manifolds. Perhaps one can also gain some knowledge by reduction of G_2 or Spin(7) structures to six dimensions.

> **Comment**: Geometry from the symplectic point of view has seen powerful development in the past twenty years. Its relation to Seiberg-Witten theory and mirror geometry is fruitful. More interesting development is expected.

5.6. Kähler structure. The most interesting geometric structure is the Kähler structure. There are two interesting pre-Kähler structures. One is the complex structure and the other is the symplectic structure. The complex structure is rather rigid for complex two dimensional manifolds. However it is much more flexible in dimension greater than two. For example, the twistor space of anti-self-dual four manifolds admit complex structures. Taubes [**655**] constructed a large class of such manifolds and hence a large class of complex three manifolds. There is also the construction of Clemens-Friedman for non-Kähler Calabi-Yau manifolds which will be explained later.

For quite a long time, it was believed that every compact Kähler manifold can be deformed to a projective manifold until C. Voisin [**693, 694**] found many counterexamples. We still need to digest the distinction between these two categories.

Besides some obvious topological obstruction from Hodge theory and the rational homotopic type theory of Deligne-Griffiths-Morgan-Sullivan [**187**], it has been difficult to decide which complex manifolds admit Kähler structure. Harmonic map argument does give some information. But it requires the fundamental group to be large.

Many years ago, Sullivan [**641**] proposed to use the Hahn-Banach theorem to construct Kähler metrics. This involves the concept of duality and hence closed currents. P. Gauduchon [**245**] has proposed those Hermitian metrics ω which is $\partial\bar{\partial}\omega^{n-1} = 0$. Siu [**623**] was able to use these ideas to prove that every $K3$ surface is Kähler. Demailly [**190**] did some remarkable work on regularization of closed positive currents. Singular Kähler metrics have been studied and used by many researchers. In fact, in my paper on proving the Calabi conjecture, I proved the existence of the Kähler metrics singular along subvarieties with control on volume element. They can be used to handle problems in algebraic geometry, including Chern number inequalities.

> **Comment**: The Kähler structure is one of the richest structures in geometry. Deeper understanding may require some more generalized structure such as a singular Kähler metric or balanced metrics.

5.6.1. *Calabi-Yau manifolds.* The construction of Calabi-Yau manifolds was based on the existence of a complex structure which can support a Kähler structure and a pluriharmonic volume form.

A fundamental question is whether an almost complex manifold admits an integrable complex structure when complex dimension is greater than two. The condition that the first Chern class is zero is equivalent to the existence of pluriharmonic volume for Kähler manifolds. Such a condition is no more true for non-Kähler manifolds. It would be nice to find a topological method to construct an integrable complex structure with pluriharmonic volume form.

Once we have an integrable complex structure, we can start to search for Hermitian metrics with special properties. As was mentioned earlier, if we would like the geometry to have supersymmetries, then a Kähler metric is the only choice if the connection is torsion free. Further supersymmetry would then imply the manifold to be Calabi-Yau. However if we do not require the connection to be torsion free, Strominger [**638**] did derive a set of equations that exhibit supersymmetries without requiring the manifold to be Kähler. It is a coupled system of Hermitian Yang-Mills connections with Hermitian metrics. Twenty years ago, I tried to develop such a coupled system. The attempt was unsuccessful as I restricted myself to Kähler geometry. My student Bartnik with Mckinnon [**42**] did succeed in doing so in the Lorentzian case. They found non-singular solutions for such a coupled system. (The mathematically rigorous proof was provided by Smoller-Wasserman-Yau-Mcleod [**635**] and [**634**]).

The Strominger's system was shown to be solvable in a neighborhood of a Calabi-Yau structure by Jun Li and myself [**432**]. Fu and I [**234**] were also able to solve it for certain complex manifolds which admit no Kähler structure. These manifolds are balanced manifolds and were studied by M. Michelsohn [**513**]. These manifolds can be used to explain some questions of flux in string theory (see, e.g., [**46, 105**]). Since Strominger has shown such manifolds admits parallel spinors, I have directed my student C.C. Wu to decompose cohomology group of such manifolds correspondingly. It is expected that many theorems in Kähler geometry may have counterparts in such geometry.

Such a structure may help to understand a proposal of Reid [**564**] in connecting Calabi-Yau manifolds with a different topology. This was initiated by a construction of Clemens [**164**] who proposed to perform complex surgery by blowing down rational curves with negative normal bundles in a Calabi-Yau manifold to rational double points. Friedman [**232**] found the condition to smooth out such singularities. Based on this Clemens-Friedman procedure, one can construct a complex structure on connected sums of $S^3 \times S^3$. It would be nice to construct Strominger's system on these manifolds.

The Calabi-Yau structure was used by me and others to solve important problems in algebraic geometry before it appeared in string theory. For example, the proof of the Torelli theorem (by Piatetskii-Shapiro and Shafarevich [**555**]) for a $K3$ surface by Todorov [**679**]-Siu [**623**] and the surjectivity of the period map of a $K3$ surface (by Kulikov [**402**]) by Siu [**622**]-Todorov [**679**] are important works for algebraic surfaces. The proof of the Bogomolov [**59**]-Tian [**671**]-Todorov [**680**] theorem also requires the metric. The last theorem helps us to understand the moduli space of Calabi-Yau manifolds. It is important to understand the global behavior of the Weil-Petersson geometry for Calabi-Yau manifolds. C. L. Wang [**697**] was able to characterize these points which have finite distance in terms of the metric.

In my talk [**733**] in the Congress in 1978, I outlined the program and the results of classifying noncompact Calabi-Yau manifolds. Some of this work was written up in Tian-Yau [**677, 678**] and Bando-Kobayashi [**32, 33**]. During the period of 1984, there was an urgent request by string theorists to construct Calabi-Yau threefolds with a Euler number equal to ± 6. During the Argonne Lab conference, I [**736**] constructed such a manifold with a $\mathbb{Z}_3$ fundamental group by taking the quotient of a bi-degree $(1,1)$ hypersurface in the product of two cubics. Soon afterwards, more examples were constructed by Tian and myself [**676**]. However, it was pointed out by Brian Greene that all the manifolds constructed by Tian-Yau can be deformed to my original manifold. The idea of producing Calabi-Yau manifolds by the complete intersection of hypersurfaces in products of weighted projective space was soon picked up by Candelas et al [**96**]. By now, on the order of ten thousand examples of different homotopic types had been constructed. The idea of using toric geometry for construction was first performed by S. Roan and myself [**565**]. A few years later, the systematic study by Batyrev [**44**] on toric geometry allowed one to construct mirror pairs for a large class of Calabi-Yau manifolds, generalizing the construction of Greene-Plesser [**269**] and Candelas et al [**96**]. Tian and I [**676**] were also the first one to apply flop construction to change topology of Calabi-Yau manifolds. Greene-Morrison-Plesser [**270**] then made the remarkable discovery of isomorphic quantum field theory on two topological distinct Calabi-Yau manifolds. Most Calabi-Yau threefolds are a complete intersection of some toric varieties and they admit a large set of rational curves. It will be important to understand the reason behind it. Up to now all the Calabi-Yau manifolds that have a Euler number ± 6 and a nontrivial fundamental group can be deformed from the birational model of the manifold (or their mirrors) that I constructed. It would be important if one could give a proof of this statement.

The most spectacular advancement on Calabi-Yau manifolds come from the work of Greene-Plesser, Candelas et al on construction of pairs of mirror manifolds with isomorphic conformal field theories attached to them. It allows one to calculate Gromov-Witten invariants. Existence of such mirror pairs was conjectured by Lerche-Vafa-Warner [**416**] and rigorous proof of mirror conjecture was due to Givental [**256**] and Lian-Liu-Yau [**447**] independently. The deep meaning of the symmetry is still being pursued.

In [**639**], Strominger, Yau and Zaslow proposed a mathematical explanation for the mirror symmetry conjecture for Calabi-Yau manifolds. Roughly speaking, mirror Calabi-Yau manifolds should admit special Lagrangian tori fibrations and the mirror transformation is a nonlinear analog of the Fourier transformation along these tori.

This proposal has opened up several new directions in geometric analysis. The first direction is the geometry of special Lagrangian submanifolds in Calabi-Yau manifolds. This includes constructions of special Lagrangian submanifolds ([**415**] and others) and (special) Lagrangian fibrations by Gross [**291, 292**] and W.D. Ruan [**572**], mean curvature of Lagrangian submanifolds in Calabi-Yau manifolds by Thomas and Yau [**666**] [**667**], structures of singularities on such submanifolds by Joyce [**374**] and Fourier transformations along special Lagrangian fibration by Leung-Yau-Zaslow [**422**] and Leung [**420**].

The second direction is affine geometry with singularities. As explained in [**639**], the mirror transformation at the large structure limit corresponds to a Legendre transformation of the base of the special lagrangian fibration which carries a natural special affine structure with singularities. Solving these affine problem is not trivial in geometric analysis [**469**] [**470**] and much work is still needed to be done here.

The third direction is the geometry of special holonomy and duality and triality transformation in M-theory. In [**303**], Gukov, Yau and Zaslow proposed a similar picture to explain the duality in M-theory. The corresponding differential geometric structures are fibrations on G_2 manifolds by coassociative submanifolds. These structures are studied by Kovalev [**398**], Leung and others [**414**] [**421**].

> **Comment**: Although the first demonstration of the existence of Kähler Ricci flat metric was shown by me in 1976, it was not until the first revolution of string theory in 1984 that a large group of researchers did extensive calculations and changed the face of the subject. It is a subject that provides a good testing ground for analysis, geometry, physics, algebraic geometry, automorphic forms and number theory.

5.6.2. *Kähler metric with harmonic Ricci form and stability.* The existence of a Kähler Einstein metric with negative scalar curvature was proved by Aubin [**23**] and me [**732**] independently. I [**731**] did find its important applications to solve classical problems in algebraic geometry, e.g., the uniqueness of complex structure over $\mathbb{C}P^2$ [**731**], the Chern number inequality of Miyaoka [**516**]-Yau [**731**] and the rigidity of algebraic manifolds biholomorphic to Shimura varieties. The problem of existence of Kähler Einstein metrics with positive scalar curvature in the general case is not solved. However, my proof of the Calabi conjecture already provided all the necessary estimates except some integral estimate on the unknown. This of course can be turned into hypothesis. I conjectured that an integral estimate of this sort is related to the stability of manifolds. Tian [**674**] called it K-stability. Mabuchi's functional [**485**] made the integral estimate more intrinsic and it gave rise to a natural variational formulation of the problem. Siu has pointed out that the work of Tian [**673**] on two dimensional surfaces is not complete. The work of Nadel [**531**] on the multiplier ideal sheaf did give useful methods for the subject of the Kähler-Einstein metric.

For Kähler Einstein manifolds with positive scalar curvature, it is possible that they admit a continuous group of automorphisms. Matsushima [**490**] was the first one to observe that such a group must be reductive. Futaki [**241**] introduced a remarkable invariant for general Kähler manifolds and proved that it must vanish for such manifolds. In my seminars in the eighties, I proposed that Futaki's theorem should be generalized to understand the projective group acting on the embedding of the manifold by a high power of anti-canonical embedding and that Futaki's invariant should be relevant to my conjecture [**739**] relating the Kähler Einstein manifold to stability. Tian asked what happens when manifolds have no group actions. I explained that the shadow of the group action is there once it is inside the projective space and one should deform the manifold to a possibly singular variety to obtain more information. The connection of Futaki invariant to stability of manifolds has finally appeared in the recent work of Donaldson [**201, 202**]. Donaldson introduced a remarkable concept of stability based purely on concept of algebraic geometry. It is not clear that Donaldson's algebraic definition has anything to do with Tian's analytic definition of stability. Donaldson proved that the existence of Kähler-Einstein did imply his K-stability which in turn implies Hilbert stability and asymptotic Chow stability of the manifold. This theorem of

Donaldson already gives nontrivial information for manifolds with negative first Chern class and Calabi-Yau manifolds, where existence of Kähler-Einstein metrics was proved. Some part of the deep work of Gieseker [**252**] and Viehweg [**690**] can be recovered by these theorems. One should also mention the recent interesting work of Ross-Thomas [**568, 569**] on the stability of manifolds. Phong-Strum [**551**] also studied solutions of certain degenerate Monge-Ampére equations and [**552**] the convergence of the Kähler-Ricci flow.

A Kähler metric with constant scalar curvature is equivalent to the harmonicity of the first Chern form. The uniqueness theorem for harmonic Kähler metric was due to X. Chen [**134**], Donaldson [**201**] and Mabuchi for various cases. (Note that the most important case of the uniqueness of the Kähler Einstein metric with positive scalar curvature was due to the remarkable argument of Bando-Mabuchi [**34**].) My general conjecture for existence of harmonic Kähler manifolds based on stability of such manifolds is still largely unknown. In my seminar in the mid-eighties, this problem was discussed extensively. Several students of mine, including Tian [**672**], Luo [**477**] and Wang [**705**] had written thesis related to this topic. Prior to them, my former students Bando [**31**] and Cao [**99**] had made attempts to study the problem of constructing Kähler-Einstein metrics by Ricci flow. The fundamental curvature estimate was due to Cao [**100**]. The Kähler Ricci flow may either converge to Kähler Einstein metric or Kähler solitons. Hence in order for the approach, based on Ricci flow, to be successful, stability of the projective manifold should be related to such Kähler solitons. The study of harmonic Kähler metrics with constant scalar curvature on toric variety was initiated by S. Donaldson [**202**], who proposed to study the existence problem via the real Monge-Ampère equation. This problem of Donaldson in the Kähler-Einstein case was solved by Wang-Zhu [**704**]. LeBrun and his coauthors [**380**] also have found special constructions, based on twistor theory, for harmonic Kähler surfaces. Bando was also interesting in Kähler manifolds with harmonic i-th Chern form. (There should be an analogue of stability of algebraic manifolds associated to manifolds with harmonic i-th Chern form.)

In the early 90's, S.W. Zhang [**750**] studied heights of manifolds. By comparing metrics on Deligne pairings, he found that a projective variety is Chow semistable if and only if it can be mapped by an element of a special linear group to a balanced subvariety. (Note that a subvariety in $\mathbb{C}\mathrm{P}^N$ is called balanced if the integral of the moment map with respect to $SU(N+1)$ is zero, where the measure for the integral is induced from the Fubini-Study metric.) Zhang communicated his results to me. It is clearly related to Kähler-Einstein metric and I urged my students, including Tian, to study this connection.

Zhang's work has a nontrivial consequence on the previous mentioned development of Donaldson [**201, 202**]. Assume the projective manifold is embedded by an ample line bundle L into projective space. If the manifold has a finite automorphism group and admits a harmonic Kähler metric in $c_1(L)$, then Donaldson showed that for k large, L^k gives rise to an embedding which is balanced. Furthermore, the induced Fubini-Study form divided by k will converge to the harmonic Kähler form. Combining the work of Zhang and Luo, he then proved that the manifold is given by the embedding of L^k is stable in the sense of geometric invariant theory. Recently, Mabuchi generalized Donaldson's theorem to certain case which allow nontrivial projective automorphism.

Donaldson considered the problem from the point of view of symplectic geometry (Kähler form is a natural symplectic form). The Hamiltonian group then acts on the Hilbert space H of square integrable sections of the line bundle L where the first Chern class is the Kähler form. For each integrable complex structure on the manifold compatible with the symplectic form, the finite dimensional space of holomorphic sections gives a subspace of H. The Hamiltonian group acts on the Grassmannian of such subspaces. The moment map can be computed to be related to the Bergman kernel $\sum_\alpha s_\alpha(x) \otimes s^*_\alpha(y)$ where s_α form an orthonormal basis of the holomorphic sections. On the other hand, Fujiki [**235**] and Donaldson [**199**] computed the moment map for the Hamiltonian group action on the space of integrable complex structure, which turns out to be the scalar curvature of the Kähler metric. These two moment maps may not match, but for the line bundle L^k with large k, one can show that they converge to each other after normalization. Lu [**475**] has shown the first term of the expansion (in terms of $1/k$) of the Bergman kernel gives rise to scalar curvature. Hence we see the relevance of constant scalar curvature for a Kähler metric to these witha constant Bergman kernel function. S.W. Zhang's result says that the manifold is Chow semistable if and only if it is balanced. The balanced condition implies that there is a Kähler metric where the Bergman kernel is constant. With the work of Zhang and Donaldson, what remains to settle my conjecture is the convergence of the balanced metric when k is large. In general, we should not expect this to be true. However, for toric manifolds, this might be the case.

It may be noted that in my paper with Bourguignon and P. Li [**73**] on giving an upper estimate of the first eigenvalue of an algebraic manifold, this balanced condition also appeared. Perhaps first eigenfunction may play a role for questions of stability.

> **Comment**: Kähler metrics with constant scalar curvature is a beautiful subject as it is related to structure questions of algebraic varieties including the concept of stability of manifolds. The most effective application of such metrics to algebraic geometry are still restricted to the Kähler-Einstein metric. The singular Kähler-Einstein metric as was initiated by my paper on Calabi conjecture should be studied further in application to algebraic geometry.

5.7. Manifolds with special holonomy group. Besides Kähler manifolds, there are manifolds with special holonomy groups. Holonomy groups of Riemannian manifolds were classified by Berger [**48**]. The most important ones are $O(n)$, $U(n)$, $SU(n)$, G_2 and $Spin(7)$. The first two groups correspond to Riemannian and Kähler geometry respectively. $SU(n)$ corresponds to Calabi-Yau manifolds. A G_2 manifold is seven dimensional and a $Spin(7)$ is eight dimensional (assuming they are irreducible manifolds). These last three classes of manifolds have zero Ricci curvature. It may be noted that before I [**732**] proved the Calabi conjecture in 1976, there was no known nontrivial compact Ricci flat manifold. Manifolds with a special holonomy group admit nontrivial parallel spinors and they correspond to supersymmetries in the language of physics. The input of ideas from string theory did give a lot of help to understand these manifolds. However, the very basic question of constructing these structures on a given topological space is still not well understood. In the case of G_2 and $Spin(7)$, it was initiated by Bryant (see [**83, 85**]).

The first set of compact examples was given by Joyce [**371, 372, 373**]. Recently Dai-Wang-Wei [**182**] proved the stability of manifolds with parallel spinors.

The nice construction of Joyce was based on a singular perturbation which is similar to the construction of Taubes [**651**] on anti-self-dual connections. However, it is not global enough to give a good parametrization of G_2 or $Spin(7)$ structures. A great deal more work is needed. The beautiful theory of Hitchin [**335, 336**] on three forms and four forms may lead to a resolution of these important problems.

> **Comment**: Recent interest in M-theory has stimulated a lot of activities on manifolds with special holonomy group. We hope a complete structure theorem for such manifolds can be found.

5.8. Geometric structures by reduction. One can also obtain new geometric structures by imposing some singular structures on a manifold with a special holonomy group. For example, if we require a metric cone to admit a G_2, $Spin(7)$ or Calabi-Yau structure, the link of the cone will be a compact manifold with special structures. They give interesting Einstein metrics. When the cone is Calabi-Yau, the structure on the odd dimensional manifold is called Sasakian Einstein metric.

There is a natural Killing field called the Reeb vector field defined on a Sasakian Einstein manifold. If it generates a circle action, the orbit space gives rise to a Kähler Einstein manifold with positive scalar curvature. However, it need not generate a circle action and J. Sparks, Gauntlelt, Martelli and Waldram [**246**] gave many interesting explicit examples of non-regular Sasakian Einstein structures. They have interesting properties related to conformal field theory. For quasi-regular examples, there was work by Boyer, Galicki and Kollár [**75**]. The procedure gave many interesting examples of Einstein metrics on odd dimensional manifolds.

Sparks, Matelli and I have been pursuing general theory of Sasakian Einstein manifolds. I would like to consider them as a natural generalization of Kähler manifolds.

> **Comment**: The recent development of the Sasakian Einstein metric show that it gives a natural generalization of the Kähler-Einstein metric. Its relation with the recent activities on ADS/CFT theory is exciting.

5.9. Obstruction for existence of Einstein metrics on general manifolds. The existence of Einstein metrics on a fixed topological manifold is clearly one of the most important questions in geometry. Any metrics with a compact special holonomy group are Einstein. Besides Kähler geometry, we do not know much of their moduli space. For an Einstein metric with no special structures, we know only some topological constraints on four dimensional manifolds. There is work by Berger [**49**], Gray [**267**] and Hitchin [**330**] in terms of inequalities linking a Euler number and the signature of the manifold. (This is of course based on Chern's work [**148**] on the representation of characteristic classes by curvature forms.) Gromov [**283**] made use of his concept of Gromov volume to give further constraint. LeBrun [**413**] then introduced the ideas from Seiberg-Witten invariants to enlarge such classes and gave beautiful rigidity theorems on Einstein four manifolds. Unfortunately it is very difficult to understand moduli space of Einstein metrics when they admit no special structures. For example, it is still an open question of whether there is only one Einstein metric on the four dimensional sphere. M. Wang and Ziller [**703**] and C. Boehm [**57**] did use symmetric reductions to give many examples of Einstein metrics for higher dimensional manifolds. There may be much more examples of Einstein manifolds with negative Ricci curvature than we expected. This is certainly true for compact manifolds, with negative Ricci curvature. Gao-Yau [**242**] was the first one to demonstrate that such a metric exists on the three sphere. A few years later, Lokhamp [**471**] used the h-principle of Gromov to prove such a metric exists on any manifold with a dimension greater than three. It would be nice to prove that every manifold with a dimension greater than 4 admits an Einstein metric with negative Ricci curvature.

> **Comment**: The Einstein manifold without extra special structures is a difficult subject. Do we expect a general classification for such an important geometric structure?

5.10. Metric Cobordism. In the last five years, a great deal of attention was addressed by physicists on the holographic principle: boundary geometry should determine the geometry in the interior. The ADS/CFT correspondence studies the conformal boundary of the Einstein manifold which is asymptotically hyperbolic. Gauge theory on the boundary is supposed to be dual to the theory of gravity in the bulk. Much fascinating work was done in this direction. Manifolds with positive scalar curvature appeared as conformal boundary are important for physics. Graham-Lee [**265**] have studied a perturbation problem near the standard sphere which bounds the hyperbolic manifold. Witten-Yau [**718**] proved that for a manifold with positive scalar curvature to be a conformal boundary, it must be connected. It is not known whether there are further obstructions.

Cobordism theory had been a powerful tool to classify the topology of manifolds. The first fundamental work was done by Thom who determined the cobordism group. Characteristic numbers play important roles. When two manifolds are cobordant to each other, the theory of surgery helps us to deform one manifold to another. It is clear that any construction of surgery that may preserve geometric structures would play a fundamental role in the future of geometry.

There are many geometric structures that are preserved under a connected sum construction. This includes the category of conformally flat structures, metrics with positive isotropic curvature and metrics with positive scalar curvature. For the last category, there was work by Schoen-Yau-Gromov-Lawson where they perform surgery on spheres with a codimension greater than or equal to 3. A key part of the work of Hamilton-Perelman is to find a canonical neighborhood to perform surgery. If we can deform the spheres in the above SYGL construction to a more canonically defined position, one may be able able to create an extra geometric structure for the result of SYGL. In fact, the construction of Schoen-Yau did provide some information about the conformal structures of the manifold. In complex geometry, there are two important canonical neighborhoods given by the log transform of Kodaira and the operation of flop. There should be similar constructions for other geometric structures.

The theory of quasi-local mass mentioned in section 4.4 is another example of how boundary geometry can be controlled by the geometry in the bulk. The work of Choi-Wang [**154**] on the first eigenvalue is also based on the manifold that it bounds. There can be interesting theory of metric cobordism.

In the other direction, there are also beautiful rigidity of inverse problems for metric geometry by Gerver-Nadirashvili [**250**] and Pestov-Uhlmann [**550**] on recovering a Riemannian metric when one knows the distance functions between pair of points on the boundary, if the Riemannian manifold is reasonably convex.

> **Comment**: There should be a mathematical foundation of the holographic principle of physicists. Good understanding of metric cobordism may be useful.

References

[1] D. Abramovich, *A linear lower bound on the gonality of modular curves.* Internat. Math. Res. Notices **20**(1996), 1005–1011.

[2] I. Agol, *Finiteness of arithmetic Kleinian reflection groups.* math.GT/0512560.

[3] S. I. Al'ber, *Higher-dimensional problems in global variational calculus.* Dokl. Akad. Nauk SSSR **156**(1964), 727–730.

[4] W. K. Allard, *On the first variation of a varifold: boundary behavior.* Ann. of Math. (2) **101**(1975), 418–446.

[5] C. Allendoerfer and J. Eells, *On the cohomology of smooth manifolds.* Comment. Math. Helv. **32**(1958), 165–179.

[6] F. J. Almgren Jr., *Almgren's big regularity paper. Q-valued functions minimizing Dirichlet's integral and the regularity of area-minimizing rectifiable currents up to codimension 2.* With a preface by Jean E. Taylor and Vladimir Scheffer. World Scientific Monograph Series in Mathematics, **1**. World Scientific Publishing Co., Inc., River Edge, NJ, 2000.

[7] F. Almgren and L. Simon, *Existence of embedded solutions of Plateau's problem.* Ann. Scuola Norm. Sup. Pisa Cl. Sci. (4) **6** (1979), no. 3, 447–495.

[8] M. T. Anderson, *The Dirichlet problem at infinity for manifolds of negative curvature.* J. Differential Geom. **18**(1983), no. 4, 701–721 (1984).

[9] M. T. Anderson and R. Schoen, *Positive harmonic functions on complete manifolds of negative curvature.* Ann. of Math. (2) **121**(1985), no. 3, 429–461.

[10] M. Atiyah, *New invariants of* 3- *and* 4-*dimensional manifolds.* The mathematical heritage of Hermann Weyl, 285–299, Proc. Sympos. Pure Math., **48**, Amer. Math. Soc., Providence, RI, 1988.

[11] M. F. Atiyah and R. Bott, *The Yang-Mills equations over Riemann surfaces.* Philos. Trans. Roy. Soc. London Ser. A **308**(1983), no. 1505, 523–615.

[12] M. F. Atiyah and R. Bott, *The moment map and equivariant cohomology.* Topology **23**(1984), no. 1, 1–28.

[13] M. Atiyah, R. Bott and V. K. Patodi, *On the heat equation and the index theorem.* Invent. Math. **19**(1973), 279–330.

[14] M. F. Atiyah, H. Donnelly and I. M. Singer, *Eta invariants, signature defects of cusps, and values of L-functions.* Ann. of Math. (2) **118**(1983), no. 1, 131–177.

[15] M. F. Atiyah, N. J. Hitchin, V. G. Drinfel'd and Y. I. Manin, *Construction of instantons.* Phys. Lett. A **65**(1978), no. 3, 185–187.

[16] M. F. Atiyah, N. J. Hitchin and I. M. Singer, *Self-duality in four-dimensional Riemannian geometry.* Proc. Roy. Soc. London Ser. A **362**(1978), no. 1711, 425–461.

[17] M. F. Atiyah, V. K. Patodi and I. M. Singer, *Spectral asymmetry and Riemannian geometry.* I. Math. Proc. Cambridge Philos. Soc. **77**(1975), 43–69.

[18] M. F. Atiyah, V. K. Patodi and I. M. Singer, *Spectral asymmetry and Riemannian geometry.* II. Math. Proc. Cambridge Philos. Soc. **78**(1975), no. 3, 405–432.

[19] M. F. Atiyah, V. K. Patodi and I. M. Singer, *Spectral asymmetry and Riemannian geometry.* III. Math. Proc. Cambridge Philos. Soc. **79**(1976), no. 1, 71-99.

[20] M. F. Atiyah and I. M. Singer, *The index of elliptic operators on compact manifolds.* Bull. Amer. Math. Soc. **69**(1963), 422–433.

[21] T. Aubin, *Équations différentielles non linéaires et problème de Yamabe concernant la courbure scalaire.* J. Math. Pures Appl. (9) **55** (1976), no. 3, 269–296.

[22] T. Aubin, *Problèmes isopérimétriques et espaces de Sobolev.* J. Differential Geometry **11**(1976), no. 4, 573–598.

[23] T. Aubin, *Équations du type Monge-Ampère sur les variétés kähleriennes compactes.* C. R. Acad. Sci. Paris Sér. A-B **283**(1976), no. 3, Aiii, A119–A121.

[24] T. Aubin, *Espaces de Sobolev sur les variétés riemanniennes.* Bull. Sci. Math. (2) **100**(1976), no. 2, 149–173.

[25] V. M. Babič, *Eigenfunctions which are concentrated in the neighborhood of a closed geodesic.* Zap. Naučn. Sem. Leningrad. Otdel. Mat. Inst. Steklov. (LOMI) **9**(1968), 15–63.

[26] W.L. Baily Jr. and A. Borel, *Compactification of arithmetic quotients of bounded symmetric domains.* Ann. of Math. (2) **84**(1966) 442–528.

[27] W. Ballmann, *On the Dirichlet problem at infinity for manifolds of nonpositive curvature.* Forum Math. **1**(1989), no. 2, 201–213.

[28] W. Ballmann and F. Ledrappier, *The Poisson boundary for rank one manifolds and their cocompact lattices.* Forum Math. **6**(1994), no. 3, 301–313.

[29] W. Ballmann and J. Świątkowski, *On L^2-cohomology and property (T) for automorphism groups of polyhedral cell complexes.* Geom. Funct. Anal. **7**(1997), no. 4, 615–645.

[30] W. Ballmann, G. Thorbergsson and W. Ziller, *Closed geodesics on positively curved manifolds.* Ann. of Math. (2) **116**(1982), no. 2, 213–247.

[31] S. Bando, *On the classification of three-dimensional compact Kaehler manifolds of nonnegative bisectional curvature.* J. Differential Geom. **19**(1984), no. 2, 283–297.

[32] S. Bando and R. Kobayashi, *Ricci-flat Kähler metrics on affine algebraic manifolds.* Geometry and analysis on manifolds (Katata/Kyoto, 1987), 20–31, Lecture Notes in Math., **1339**, Springer, Berlin, 1988.

[33] S. Bando and R. Kobayashi, *Ricci-flat Kähler metrics on affine algebraic manifolds. II.* Math. Ann. **287**(1990), no. 1, 175–180.

[34] S. Bando and T. Mabuchi, *Uniqueness of Einstein Kähler metrics modulo connected group actions.* Algebraic geometry, Sendai, 1985, 11–40, Adv. Stud. Pure Math., **10**, North-Holland, Amsterdam, 1987.

[35] V. Bangert, *On the existence of closed geodesics on two-spheres.* Internat. J. Math. **4**(1993), no. 1, 1–10.

[36] V. Bangert and Y. M. Long, *The existense of two closed geodesics on every Finsler 2-sphere.* Preprint (2005). Submitted.

[37] D. Bao, S.-S. Chern and Z. Shen, *An introduction to Riemann-Finsler geometry.* Graduate Texts in Mathematics, **200**, Springer-Verlag, New York, 2000.

[38] M. Barlow, T. Coulhon and A. Grigor'yan, *Alexander Manifolds and graphs with slow heat kernel decay.* Invent. Math. **144**(2001), no. 3, 609–649.

[39] W. Barth and A. Van de Ven, *A decomposability criterion for algebraic 2-bundles on projective spaces.* Invent. Math. **25**(1974), 91–106.

[40] R. Bartnik, *Existence of maximal surfaces in asymptotically flat spacetimes.* Comm. Math. Phys. **94** (1984), no. 2, 155–175.

[41] R. Bartnik, *Quasi-spherical metrics and prescribed scalar curvature.* J. Differential Geom. **37**(1993), no. 1, 31–71.

[42] R. Bartnik and J. McKinnon, *Particlelike solutions of the Einstein-Yang-Mills equations.* Phys. Rev. Lett. **61**(1988), no. 2, 141–144.

[43] P. de Bartolomeis and G. Tian, *Stability of complex vector bundles.* J. Differential Geom. **43**(1996), no. 2, 231–275.

[44] V. V. Batyrev, *Dual polyhedra and mirror symmetry for Calabi-Yau hypersurfaces in toric varieties.* J. Algebraic Geom. **3**(1994), no. 3, 493–535.

[45] A. Beauville, *Variétés Kähleriennes dont la première classe de Chern est nulle.* J. Differential Geom. **18**(1983), no. 4, 755–782 (1984).

[46] K. Becker and K. Dasgupta, *Heterotic strings with torsion.* J. High Energy Phys. **11**(2002), 006, 30 pp.

[47] E. Berger, R. Bryant and P. Griffiths, *The Gauss equations and rigidity of isometric embeddings.* Duke Math. J. **50**(1983), no. 3, 803–892.

[48] M. Berger, *Sur les groupes d'holonomie homogène des variétés à connexion affine et des variétés riemanniennes.* Bull. Soc. Math. France 83 (1955), 279–330.

[49] M. Berger, *Sur quelques variétés d'Einstein compactes.* Ann. Mat. Pura Appl. (4) **53**(1961),89–95.

[50] M. Bershadsky, S. Cecotti, H. Ooguri and C. Vafa, *Kodaira-Spencer theory of gravity and exact results for quantum string amplitudes.* Comm. Math. Phys. **165**(1994), no. 2, 311–427.

[51] A. L. Besse, *Manifolds all of whose geodesics are closed.* With appendices by D. B. A. Epstein, J.-P. Bourguignon, L. Bérard-Bergery, M. Berger and J. L. Kazdan. Ergebnisse der Mathematik und ihrer Grenzgebiete, **93**. Springer-Verlag, Berlin-New York, 1978.
[52] G. Besson, G. Courtois and S. Gallot, *Entropies et rigidités des espaces localement symétriques de courbure strictement négative.* Geom. Funct. Anal. **5**(1995), no. 5, 731–799.
[53] J. M. Bismut, *The Atiyah-Singer index theorem for families of Dirac operators: two heat equation proofs.* Invent. Math. **83**(1986), no. 1, 91–151.
[54] J. M. Bismut and S. Goette, *Equivariant de Rham torsions.* Ann. of Math. (2) **159**(2004), no. 1, 53–216.
[55] J. M. Bismut and W. P. Zhang, *An extension of a theorem by Cheeger and Müller.* With an appendix by François Laudenbach. Astérisque No. **205**(1992), 235 pp.
[56] J. M. Bismut and W. P. Zhang, *Milnor and Ray-Singer metrics on the equivariant determinant of a flat vector bundle.* Geom. Funct. Anal. **4**(1994), no. 2, 136–212.
[57] C. Boehm, *Inhomogeneous Einstein metrics on low-dimensional spheres and other low-dimensional spaces.* Inv. Math. **134**(1998), 145-176.
[58] S. Bochner, *Curvature and Betti numbers.* Ann. of Math. (2) **49**(1948). 379–390.
[59] F. A. Bogomolov, *Hamiltonian Kälerian manifolds.* Dokl. Akad. Nauk SSSR **243**(1978), no. 5, 1101–1104.
[60] M. Boileau, B. Leeb and J. Porti, *Uniformization of small 3-orbifolds.* C. R. Acad. Sci. Paris Sr. I Math. **332**(2001), no. 1, 57–62.
[61] E. Bombieri and E. Giusti, *Harnack's inequality for elliptic differential equations on minimal surfaces.* Invent. Math. **15**(1972), 24–46.
[62] E. Bombieri, E. De Giorgi and E. Giusti, *Minimal cones and the Bernstein problem.* Invent. Math. **7**(1969), 243–268.
[63] R. E. Borcherds, *Automorphic forms on* $\mathrm{O}_{s+2,2}(R)$ *and infinite products.* Invent. Math. **120**(1995), no. 1, 161–213.
[64] R. E. Borcherds, *The moduli space of Enriques surfaces and the fake Monster Lie superalgebra.* Topology **35**(1996), no. 3, 699–710.
[65] A. Borel, *Some metric properties of arithmetic quotients of symmetric spaces and an extension theorem.* Collection of articles dedicated to S. S. Chern and D. C. Spencer on their sixtieth birthdays. J. Differential Geometry **6**(1972), 543–560.
[66] A. Borel and J.-P. Serre, *Corners and arithmetic groups. Avec un appendice: Arrondissement des variétés à coins, par A. Douady et L. Hérault.* Comment. Math. Helv. **48**(1973), 436–491.
[67] R. Bott, *Nondegenerate critical manifolds.* Ann. of Math. (2) **60**(1954), 248-261.
[68] R. Bott, *On manifolds all of whose geodesics are closed.* Ann. of Math. (2) **60**(1954), 375–382.
[69] R. Bott, *On the iteration of closed geodesics and the Sturm intersection theory.* Comm. Pure Appl. Math. **9**(1956), 171–206.
[70] R. Bott, *The stable homotopy of the classical groups.* Ann. of Math. (2) **70**(1959), 313–337.
[71] R. Bott, *A residue formula for holomorphic vector-fields.* J. Differential Geometry **1**(1967), 311–330.
[72] R. Bott and C. H. Taubes, *Clifford On the rigidity theorems of Witten.* J. Amer. Math. Soc. **2**(1989), no. 1, 137–186.
[73] J.-P. Bourguignon, P. Li and S.-T. Yau, *Upper bound for the first eigenvalue of algebraic submanifolds.* Comment. Math. Helv. **69**(1994), no. 2, 199–207.
[74] L. Boutet de Monvel and J. Sjöstrand, *Sur la singularité des noyaux de Bergman et de Szegö.* Journées: Équations aux Dérivées Partielles de Rennes (1975), Asterisque 34-35, Soc. Math. France, Paris, 1976, 123–164.
[75] C. P. Boyer, K. Galicki and J. Kollár, *Einstein Metrics on Spheresmath.* arXiv:math.DG/0309408.
[76] P. J. Braam and S. K. Donaldson, *Floer's work on instanton homology, knots and surgery.* The Floer memorial volume, 195–256, Progr. Math., **133**, Birkhäuser, Basel, 1995.
[77] K. A. Brakke, *The motion of a surface by its mean curvatur.* Mathematical Notes, **20**. Princeton University Press, Princeton, N.J., 1978.
[78] H. Brascamp and E. Lieb, *On extensions of the Brunn-Minkowski and Prkopa-Leindler theorems, including inequalities for log concave functions, and with an application to the diffusion equation.* J. Functional Analysis **22**(1976), no. 4, 366–389

[79] H. L. Bray, *Proof of the Riemannian Penrose inequality using the positive mass theorem.* J. Differential Geom. **59**(2001), no. 2, 177–267.

[80] R. Brooks, *The fundamental group and the spectrum of the Laplacian.* Comment. Math. Helv. **56**(1981), no.4, 581C598.

[81] J. D. Brown and J. W. York, *Quasilocal energy and conserved charges derived from the gravitational action.* Phys. Rev. D (3) **47**(1993), no. 4, 1407–1419.

[82] R. Bryant, *Conformal and minimal immersions of compact surfaces into the 4-sphere.* J. Differential Geom. **17**(1982), no. 3, 455–473.

[83] R. Bryant, *Metrics with exceptional holonomy.* Ann. of Math. (2) **126**(1987), no. 3, 525–576.

[84] R. Bryant, P. Griffiths and D. Yang, *Characteristics and existence of isometric embeddings.* Duke Math. J. **50**(1983), no. 4, 893–994.

[85] R. Bryant and S. Salamon, *On the construction of some complete metrics with exceptional holonomy,* Duke Math. J. **58**(1989), 829–850.

[86] N. P. Buchdahl, *Hermitian-Einstein connections and stable vector bundles over compact complex surfaces.* Math. Ann. **280**(1988), no. 4, 625–648.

[87] Y. Burago, M. Gromov and G. A. D. Perelman, *Aleksandrov spaces with curvatures bounded below.* Uspekhi Mat. Nauk **47**(1992), no. 2(284), 3–51, 222; translation in Russian Math. Surveys **47**(1992), no. 2, 1–58.

[88] F. E. Burstall, D. Ferus, F. Pedit and U. Pinkall, *Harmonic tori in symmetric spaces and commuting Hamiltonian systems on loop algebras.* Ann. of Math. (2) **138**(1993), no. 1, 173–212.

[89] A. Butscher, *Regularizing a singular special Lagrangian variety.* Comm. Anal. Geom. **12**(2004), no. 4, 733–791.

[90] L. Caffarelli, R. Hardt and L. Simon, *Minimal surfaces with isolated singularities.* Manuscripta Math. **48**(1984), no. 1-3, 1–18.

[91] L. Caffarelli, L. Nirenberg and J. Spruck, *The Dirichlet problem for nonlinear second order elliptic equations, III. Functions of the eigenvalues of the Hessian.* Acta Math. **155**(1985), 261–301.

[92] E. Calabi, *On Kähler manifolds with vanishing canonical class.* Algebraic geometry and topology. A symposium in honor of S. Lefschetz, pp. 78–89. Princeton University Press, Princeton, N. J., 1957.

[93] E. Calabi, *Improper affine hyperspheres of convex type and a generalization of a theorem by K. Jörgens.* Michigan Math. J. **5**(1958) 105–126.

[94] E. Calabi and E. Vesentini, *Sur les variétés complexes compactes localement symétriques.* Bull. Soc. Math. France **87**(1959) 311–317.

[95] C. Callias, *Axial anomalies and index theorems on open spaces.* Comm. Math. Phys. **62**(1978), no. 3, 213–234.

[96] P. Candelas, A. M. Dale, C. A. Lütken and R. C. Schimmrigk, *Complete intersection Calabi-Yau manifolds.* Frontiers of high energy physics (London, 1986), 88–134, Hilger, Bristol, 1987.

[97] P. Candelas, G. T. Horowitz, A. Strominger and E. Witten, *Vacuum configurations for superstrings.* Nuclear Phys. B **258**(1985), no. 1, 46–74.

[98] P. Candelas, de la Ossa, X. C., P. S. Green and L. Parkes, *A pair of Calabi-Yau manifolds as an exactly soluble superconformal theory.* Nuclear Phys. B **359**(1991), no. 1, 21–74.

[99] H. D. Cao, *Deformation of Kähler metrics to Kähler-Einstein metrics on compact Kähler manifolds.* Invent. Math. **81**(1985), no. 2, 359–372.

[100] H. D. Cao, *On Harnack inequalities for the Kähler-Ricci flow.* Invent. Math. **109**(1992), 247–263.

[101] H. D. Cao, *Limits of solutions to the Khler-Ricci flow.* J. Differential Geom. **45**(1997), no. 2, 257–272.

[102] H. D. Cao, X. P. Zhu, *Ricci flow and its application.*

[103] J. G. Cao and J. F. Escobar, *A new 3-dimensional curvature integral formula for PL-manifolds of non-positive curvature.* Comm. Anal. Geom. **11**(2003), no. 3, 489–551.

[104] J. G. Cao, H. J. Fan and F. Ledrappier, *Martin points on open manifolds of non-positive curvature.* To appear in Trans. AMS.

[105] G. L. Cardoso, G. Curio, G. Dall'Agata and D. Lüst, *Heterotic string theory on non-Kähler manifolds with H-flux and gaugino condensate.* Proceedings of the 36th International

Symposium Ahrenshoop on the Theory of Elementary Particles. Fortschr. Phys. **52**(2004), no. 6-7, 483–488.

[106] M. do Carmo and N. Wallach, *Minimal immersions of spheres into spheres.* Proc. Nat. Acad. Sci. U.S.A. **63**(1969) 640–642.

[107] E. Cartan, *Sur des familles remarquables d'hypersurfaces isoparamétriques dans les espaces sphériques.* Math. Z. **45**(1939), 335–367.

[108] B. Carter, *Global Structure of the Kerr Family of Gravitational Fields.* Phys. Rev. **174**(1968), 1559–1571.

[109] D. Catlin, *The Bergman kernel and a theorem of Tian.* Analysis and geometry in several complex variables (Katata, 1997), 1–23, Trends Math., Birkhäuser Boston, Boston, MA, 1999.

[110] S. C. Chang, *Recent developments on the Calabi flow.* Geometric evolution equations, 17–42, Contemp. Math., **367**, Amer. Math. Soc., Providence, RI, 2005.

[111] S. Y. A. Chang, M. J. Gursky and P. Yang, *An a priori estimate for a fully nonlinear equation on four-manifolds.* Dedicated to the memory of Thomas H. Wolff. J. Anal. Math. **87**(2002), 151–186.

[112] A. Chau and L.-F. Tam, *On the complex structure of Kähler manifolds with nonnegative curvature.* arXiv:math.DG/0504422.

[113] J. Cheeger, *A lower bound for the smallest eigenvalue of the Laplacian.* Problems in analysis (Papers dedicated to Salomon Bochner, 1969), pp. 195–199. Princeton Univ. Press, Princeton, N. J., 1970.

[114] J. Cheeger, *Analytic torsion and Reidemeister torsion.* Proc. Nat. Acad. Sci. U.S.A. **74**(1977), no. 7, 2651–2654.

[115] J. Cheeger and T. H. Colding, *On the structure of spaces with Ricci curvature bounded below. I.* J. Differential Geom. **46**(1997), no. 3, 406–480.

[116] J. Cheeger and T. H. Colding, *On the structure of spaces with Ricci curvature bounded below. II.* J. Differential Geom. **54**(2000), no. 1, 13–35.

[117] J. Cheeger and T. H. Colding, *On the structure of spaces with Ricci curvature bounded below. III.* J. Differential Geom. **54**(2000), no. 1, 37–74.

[118] J. Cheeger and D. Gromoll, *The splitting theorem for manifolds of nonnegative Ricci curvature.* J. Differential Geometry **6**(1971/72), 119–128.

[119] J. Cheeger and M. Gromov, *Collapsing Riemannian manifolds while keeping their curvature bounded. I.* J. Differential Geom. **23**(1986), no. 3, 309–346.

[120] J. Cheeger and M. Gromov, *Chopping Riemannian manifolds.* Differential geometry, 85–94, Pitman Monogr. Surveys Pure Appl. Math., **52**, Longman Sci. Tech., Harlow, 1991.

[121] J. Cheeger, M. Gromov and M. Taylor *Finite propagation speed, kernel estimates for functions of the Laplace operator, and the geometry of complete Riemannian manifolds.* J. Differential Geom. **17**(1982), no. 1, 15–53.

[122] J. Cheeger and J. Simons, *Differential characters and geometric invariants.* Geometry and topology (College Park, Md., 1983/84), 50–80, Lecture Notes in Math., **1167**, Springer, Berlin, 1985.

[123] J. Cheeger and S.-T. Yau, *A lower bound for the heat kernel.* Comm. Pure Appl. Math. **34**(1981), no. 4, 465–480.

[124] B. L. Chen, S. H. Tang and X. P. Zhu, *A uniformization theorem for complete non-compact Kähler surfaces with positive bisectional curvature.* J. Differential Geom. **67**(2004), no. 3, 519–570.

[125] B. L. Chen and X. P. Zhu, *Complete Riemannian manifolds with pointwise pinched curvature.* Invent. Math. **140**(2000), no. 2, 423–452

[126] B. L. Chen and Z. X. Ping, *On complete noncompact Kähler manifolds with positive bisectional curvature.* Math. Ann. **327**(2003), no. 1, 1–23.

[127] B. L. Chen and X. P. Zhu, *Volume growth and curvature decay of positive curved Kähler manifold,* arXiv:math.DG/0211374.

[128] B. L. Chen and X. P. Zhu, *Flow with Surgery on Four-manifolds with Positive Isotropic Curvature.* arXiv:math.DG/0504478.

[129] B. L. Chen and X. P. Zhu, *Uniqueness of the Ricci Flow on Complete Noncompact Manifolds.* arXiv:math.DG/0505447.

[130] C. C. Chen and C. S. Lin, *A spherical Harnack inequality for singular solutions of nonlinear elliptic equations.* Ann. Scuola Norm. Sup. Pisa Cl. Sci. (4) **30**(2001), no. 3-4, 713–738 (2002).

[131] K.-T. Chen, *Differential forms and homotopy groups.* J. Differential Geometry **6**(1971/72), 231–246.

[132] M. F. Chen and F. Y. Wang, *Application of coupling method to the first eigenvalue on manifold.* Progr. Natur. Sci. (English Ed.) **5**(1995), no. 2, 227–229.

[133] Q. Chen, J. Jost, G. F. Wang and J. Y. Li, *Regularity theorems and energy identities for Dirac-harmonic maps.* Math.DG/0411327.

[134] X. X. Chen, *The space of Kähler metrics.* J. Differential Geom. **56**(2000), no. 2, 189–234.

[135] Y. G. Chen, Y. Giga and S. Goto, *Uniqueness and existence of viscosity solutions of generalized mean curvature flow equations.* J. Differential Geom. **33**(1991), no. 3, 749–786.

[136] S. Y. Cheng, *Eigenvalue comparison theorems and its geometric applications.* Math. Z. **143**(1975), no. 3, 289–297.

[137] S. Y. Cheng, *Eigenfunctions and nodal sets.* Comment. Math. Helv. **51**(1976), no. 1, 43–55.

[138] S. Y. Cheng, P. Li and S.-T. Yau, *On the upper estimate of the heat kernel of a complete Riemannian manifold.* Amer. J. Math. **103**(1981), no. 5, 1021–1063.

[139] S. Y. Cheng, P. Li and S.-T. Yau, *Heat equations on minimal submanifolds and their applications.* Amer. J. Math. **106**(1984), no. 5, 1033–1065.

[140] S. Y. Cheng and S.-T. Yau, *Differential equations on Riemannian manifolds and their geometric applications*, Comm. Pure Appl. Math. **28**(1975), no. 3, 333–354.

[141] S. Y. Cheng and S.-T. Yau, *On the regularity of the Monge-Ampère equation* $\det(\partial^2 u/\partial x_i \partial s x_j) = F(x,u)$. Comm. Pure Appl. Math. **30**(1977), no. 1, 41–68.

[142] S. Y. Cheng and S.-T. Yau, *Maximal space-like hypersurfaces in the Lorentz-Minkowski spaces.* Ann. of Math. (2) **104**(1976), no. 3, 407–419.

[143] S. Y. Cheng and S. T. Yau, *On the regularity of the solution of the* n*-dimensional Minkowski problem.* Commun. Pure Appl. Math. **29**(1976), 495–516.

[144] S. Y. Cheng and S.-T. Yau, *On the existence of a complete Kähler metric on noncompact complex manifolds and the regularity of Fefferman's equation.* Comm. Pure Appl. Math. **33**(1980), no. 4, 507–544.

[145] S. Y. Cheng and S.-Y. Yau, *The real Monge-Ampère equation and affine flat structures*, in Proceedings of the 1980 Beijing Symposium on Differential Geometry and Differential Equations, Science Press, Beijing, China (1982), Gordon and Breach Science Publishing Company, New York , pp. 339–370.

[146] S. Y. Cheng and S. T. Yau, *Complete affine hypersurfaces. I. The completeness of affine metrics.* Comm. Pure Appl. Math. **39**(1986), no. 6, 839–866.

[147] *Proceedings of the 1980 Beijing Symposium on Differential Geometry and Differential Equations.* Vol. **1**, **2**, **3**. Held in Beijing, August 18–September 21, 1980. Edited by S. S. Chern and Wen Tsün Wu [Wen Jun Wu]. Science Press, Beijing; Gordon and Breach Science Publishers, New York, 1982.

[148] S.-S. Chern, *Characteristic classes of Hermitian manifolds.* Ann. of Math. (2) **47**(1946). 85–121.

[149] S.-S. Chern, *On minimal spheres in the four-sphere.* 1970 Studies and Essays (Presented to Yu-why chen on his 60th Birthday, April 1, 1970) pp. 137–150 Math. Res. Center, Nat. Taiwan Univ., Taipei.

[150] S. S. Chern, M. de Carmo and S. Kobayashi, *Minimal submanifolds of a sphere with second fundamental form of constant length*, Functional Analysis and Related Fields (Proc. Conf. for M. Stone, Univ. Chicago, Chicago, Ill., 1968), Springer, New York, 1970, pp. 59C75.

[151] S. S. Chern and J. Simons, *Characteristic forms and geometric invariants.* Ann. of Math. (2) **99**(1974), 48–69.

[152] H. I. Choi and R. Schoen, *The space of minimal embeddings of a surface into a three-dimensional manifold of positive Ricci curvature.* Invent. Math. **81**(1985), no. 3, 387–394.

[153] H. I. Choi and A. Treibergs, *Gauss maps of spacelike constant mean curvature hypersurfaces of Minkowski space.* J. Differential Geom. **32**(1990), no. 3, 775–817.

[154] H. I. Choi and A. N. Wang, *A first eigenvalue estimate for minimal hypersurfaces.* J. Differential Geom. **18**(1983), no. 3, 559–562.

[155] S. Y. Choi and W. M. Goldman, *The classification of real projective structures on compact surfaces.* Bull. Amer. Math. Soc. (N.S.) **34**(1997), no. 2, 161–171.

[156] Y. Choquet-Bruhat and V. Moncrief, *Future global in time Einsteinian spacetimes with* U(1) *isometry group.* Ann. Henri Poincare **2**(2001), no. 6, 1007–1064.

[157] Y. Choquet-Bruhat and J. W. York, *On H. Friedrich's formulation of the Einstein equations with fluid sources.* Topol. Methods Nonlinear Anal. **18**(2001), no. 2, 321–335.

[158] D. Christodoulou, *The instability of naked singularities in the gravitational collapse of a scalar field.* Ann. of Math. (2) **149**(1999), no. 1, 183–217.

[159] D. Christodoulou and S. Klainerman, *The global nonlinear stability of the Minkowski space.* Princeton Mathematical Series, **41**. Princeton University Press, Princeton, NJ, 1993.

[160] D. Christodoulou and A. Tahvildar-Zadeh, *On the regularity of spherically symmetric wave maps.* Comm. Pure Appl. Math. **46**(1993), no. 7, 1041–1091.

[161] D. Christodoulou and S.-T. Yau, *Some remarks on the quasi-local mass.* Mathematics and general relativity, 9–14, Contemp. Math., **71**, Amer. Math. Soc., Providence, RI, 1988.

[162] P. T. Chrúsciel, *Semi-global existence and convergence of solutions of the Robinson-Trautman (2-dimensional Calabi) equation.* Comm. Math. Phys. **137**(1991), no. 2, 289–313.

[163] F. Chung, *Discrete Isoperimetric Inequalities*, Surveys in Differential Geometry, IX, Inrernational Press, pp.53-82.

[164] C. H. Clemens, *Double solids.* Adv. in Math. **47**(1983), no. 2, 107–230.

[165] J. Cogdell, J. S. Li, I. Piatetski-Shapiro and P. Sarnak, Poincaré series for $SO(n, 1)$. Acta Math. **167**(1991), no. 3-4, 229–285.

[166] T. H. Colding, *Shape of manifolds with positive Ricci curvature.* Invent. Math. **124**(1996), no. 1-3, 175–191.

[167] T. H. Colding and W. P. Minicozzi II, *Harmonic functions on manifolds.* Ann. of Math. (2) **146**(1997), no. 3, 725–747.

[168] T. H. Colding and W. P. Minicozzi II *The space of embedded minimal surfaces of fixed genus in a 3-manifold. I. Estimates off the axis for disks.* Ann. of Math. (2) **160**(2004), no. 1, 27–68.

[169] T. H. Colding and W. P. Minicozzi II *The space of embedded minimal surfaces of fixed genus in a 3-manifold. II. Multi-valued graphs in disks.* Ann. of Math. (2) **160**(2004), no. 1, 69–92.

[170] T. H. Colding and W. P. Minicozzi II *The space of embedded minimal surfaces of fixed genus in a 3-manifold. III. Planar domains.* Ann. of Math. (2) **160**(2004), no. 2, 523–572.

[171] T. H. Colding and W. P. Minicozzi II *The space of embedded minimal surfaces of fixed genus in a 3-manifold. IV. Locally simply connected.* Ann. of Math. (2) **160**(2004), no. 2, 573–615.

[172] T. H. Colding and W. P. Minicozzi II, *Estimates for the extinction time for the Ricci flow on certain 3-manifolds and a question of Perelman.* J. Amer. Math. Soc. **18**(2005), no. 3, 561–569.

[173] T. H. Colding and W. P. Minicozzi II, *An excursion into geomtric analysis.* Surveys in Differential Geomtry **IX**, International Press, 83-146.

[174] Colin de Verdière, Y. *Sur la multiplicitë de la première valeur propre non nulle du laplacien.* Comment. Math. Helv. **61**(1986), no. 2, 254–270.

[175] R. Connelly, *The rigidity of suspensions.* J. Differential Geom. **13**(1978), no. 3, 399–408.

[176] K. Corlette, *Archimedean superrigidity and hyperbolic geometry.* Ann. of Math. (2) **135**(1992), no. 1, 165–182.

[177] C. B. Croke, *Some isoperimetric inequalities and eigenvalue estimates.* Ann. Sci. ecole Norm. Sup. (4) **13**(1980), no. 4, 419–435.

[178] C. B. Croke, *A sharp four-dimensional isoperimetric inequality.* Comment. Math. Helv. **59**(1984), no. 2, 187–192.

[179] M. Dafermos, *Exhaustions of complete manifolds of bounded curvature.* Tsing Hua lectures n geometry and analysis, 319–322, Internat. Press, Cambridge, MA, 1997.

[180] M. Dafermos, *Stability and instability of the Cauchy horizon for the spherically symmetric Einstein-Maxwell-scalar field equations.* Ann. of Math. (2) **158**(2003), no. 3, 875–928.

[181] X. Z. Dai, K. F. Liu and X. N. Ma, *On the asymptotic expansion of Bergman kernel.* C. R. Math. Acad. Sci. Paris **339**(2004), no. 3, 193-198. The full version: J. Differential Geom. to appear, math.DG/0404494.

[182] X. Z. Dai, X. D. Wang and G. F. Wei, *On the stability of Riemannian manifold with parallel spinors.* Invent. Math. **161**(2005), 151 - 176.

[183] G. D. Daskalopoulos and R. A. Wentworth, *Geometric quantization for the moduli space of vector bundles with parabolic structure.* Geometry, topology and physics, 119–155, de Gruyter, Berlin, 1997.

[184] E. B. Davies, *Gaussian upper bounds for the heat kernels of some second-order operators on Riemannian manifolds.* J. Funct. Anal. **80**(1988), no. 1, 16–32.

[185] E. B. Davies, *An eigenvalue bound for compact manifolds with varying curvature.* Ann. Global Anal. Geom. **7**(1989), no. 2, 107–114.

[186] P. Deligne, *Théorie de Hodge. II.* Inst. Hautes études Sci. Publ. Math. No. **40**(1971), 5–57.

[187] P. Deligne, P. Griffiths, J. Morgan and D. Sullivan, *Real homotopy theory of Kähler manifolds.* Invent. Math. **29**(1975), no. 3, 245–274.

[188] P. Deligne and D. Mumford, *The irreducibility of the space of curves of given genus.* Inst. Hautes études Sci. Publ. Math. No. **36**(1969), 75–109.

[189] J. Demailly, *Holomorphic Morse inequalities. Several complex variables and complex geometry.* 93–114, Proc. Sympos. Pure Math., **52**, Part 2, Amer. Math. Soc., Providence, RI, 1991.

[190] J. Demailly, *Regularization of closed positive currents and intersection theory.* J. Algebraic Geom. **1**(1992), no. 3, 361–409.

[191] J. Demailly, *A numerical criterion for very ample line bundles.* J. Differential Geom. **37**(1993), no. 2, 323–374.

[192] J. Demailly, *Effective bounds for very ample line bundles.* Invent. Math. **124**(1996), no. 1-3, 243–261.

[193] R. Dijkgraaf, H. Verlinde and E. Verlinde, *Topological strings in $d < 1$.* Nuclear Phys. B **352**(1991), no. 1, 59–8.

[194] S. K. Donaldson, *An application of gauge theory to four-dimensional topology.* J. Differential Geom. **18**(1983), no. 2, 279–315.

[195] S. K. Donaldson, *Anti self-dual Yang-Mills connections over complex algebraic surfaces and stable vector bundles.* Proc. London Math. Soc. (3) **50**(1985), no. 1, 1–26.

[196] S. K. Donaldson, *Infinite determinants, stable bundles and curvature.* Duke Math. J. **54**(1987), no. 1, 231–247.

[197] S. K. Donaldson, *The Seiberg-Witten equations and 4-manifold topology.* Bull. Amer. Math. Soc. (N.S.) **33**(1996), no. 1, 45–70.

[198] S. K. Donaldson, *Symplectic submanifolds and almost-complex geometry.* J. Differential Geom. **44**(1996), no. 4, 666–705.

[199] S. K. Donaldson, *Remarks on gauge theory, complex geometry and 4-manifold topology.* Fields Medallists' lectures, 384–403, World Sci. Ser. 20th century Math., **5**, World Sci. Publishing, River Edge, NJ, 1997.

[200] S. K. Donaldson, *Lefschetz pencils on symplectic manifolds.* J. Differential Geom. **53**(1999), no. 2, 205–236.

[201] S. K. Donaldson, *Scalar curvature and projective embeddings.* I. J. Differential Geom. **59**(2001), no. 3, 479–522.

[202] S. K. Donaldson, *Scalar curvature and stability of toric varieties.* J. Differential Geom. **62**(2002), no. 2, 289–349.

[203] S. K. Donaldson and P. B. Kronheimer, The geometry of four-manifolds. Oxford Mathematical Monographs. Oxford Science Publications. *The Clarendon Press, Oxford University Press, New York,* 1990.

[204] S. K. Donaldson and R. D. Friedman, *Connected Sums of Self-Dual Manifolds and Deformations of Singular Spaces.* Nonlinearity **2**(1989), 197–239.

[205] R. T. Dong, *Nodal sets of eigenfunctions on Riemann surfaces.* J. Differential Geom. **36**(1992), no. 2, 493–506.

[206] H. Donnelly and C. Fefferman, *Nodal sets of eigenfunctions on Riemannian manifolds.* Invent. Math. **93**(1988), no. 1, 161–183.

[207] J. J. Duistermaat and G. J. Heckman, *On the variation in the cohomology of the symplectic form of the reduced phase space.* Invent. Math. **69**(1982), no. 2, 259–268.

[208] J. J. Duistermaat and L. Hörmander, *Fourier integral operators. II.* Acta Math. **128**(1972), no. 3-4, 183–269.

[209] J. J. Duistermaat and V. W. Guillemin, *The spectrum of positive elliptic operators and periodic bicharacteristics.* Invent. Math. **29**(1975), no. 1, 39–79.

[210] K. Ecker, *Interior estimates and longtime solutions for mean curvature flow of noncompact spacelike hypersurfaces in Minkowski space.* J. Differential Geom. **46**(1997), no. 3, 481–498.

[211] J. Eells Jr. and J. H. Sampson, *Harmonic mappings of Riemannian manifolds.* Amer. J. Math. **86**(1964), 109–160.

[212] L. Ein and R. Lazarsfeld, *Global generation of pluricanonical and adjoint linear series on smooth projective threefolds.* J. Amer. Math. Soc. **6**(1993), no. 4, 875–903.

[213] I. Ekeland, H. Hofer, *Symplectic topology and Hamiltonian dynamics.* Math. Z. **200**(1989), 355–378.

[214] Y. Eliashberg, *The theorem on structure of wave fronts and its appilcations in Symplectic geometry.* Func. Ana. and its App. **21**(1987), 65-72.

[215] Y. Eliashberg, A. Givental and H. Hofer, *Introduction to symplectic field theory.* GAFA (2000), 560-673.

[216] L. C. Evans and J. Spruck, *Motion of level sets by mean curvature. I.* J. Differential Geom. **33**(1991), no. 3, 635–681.

[217] F. Farrell and L. Jones, *Negatively curved manifolds with exotic smooth structures.* J. Amer. Math. Soc. **2**(1989), no. 4, 899–908.

[218] H. Federer and W. H. Fleming, *Normal and integral currents.* Ann. of Math. (2) **72**(1960), 458–520.

[219] C. Fefferman, *The Bergman kernel and biholomorphic mappings of pseudoconvex domains.* Invent. Math. **26**(1974), 1–65.

[220] C. Fefferman and D. H. Phong, *The uncertainty principle and sharp Gårding inequalities.* Comm. Pure Appl. Math. **34**(1981), no. 3, 285–331.

[221] F. Finster, N. Kamran, J. Smoller and S.-T. Yau, *Decay rates and probability estimates for massive Dirac particles in the Kerr-Newman black hole geometry.* Comm. Math. Phys. **230**(2002), no. 2, 201–244.

[222] D. Fischer-Colbrie and R. Schoen, *The structure of complete stable minimal surfaces in 3-manifolds of nonnegative scalar curvature.* Comm. Pure Appl. Math. **33**(1980), no. 2, 199–211.

[223] A. Floer, *An instanton-invariant for 3-manifolds.* Comm. Math. Phys. **118**(1988), no. 2, 215–240.

[224] A. Floer, *Morse theory for Lagrangian intersections.* J. Differential Geom. **28**(1988), no. 3, 513–547.

[225] A. Floer, *Witten's complex and infinite-dimensional Morse theory.* J. Differential Geom. **30**(1989), no. 1, 207–221.

[226] S. Frankel, Complex geometry of convex domains that cover varieties. Acta Math. **163**(1989), no. 1-2, 109–149.

[227] J. Franks, *Geodesics on S^2 and periodic points of annulus homeomorphisms.* Invent. Math. **108**(1992), no. 2, 403–418.

[228] D.S. Freed, *Characteristic numbers and generalized path integrals.* Geometry, topology and physics, 126–138, Conf. Proc. Lecture Notes Geom. Topology, **IV**, Internat. Press, Cambridge, MA, 1995.

[229] M. Freedman, *The topology of four-dimensional manifolds.* J. Differential Geom. **17**(1982), no. 3, 357–453.

[230] M. Freedman, J. Hass and P. Scott, *Least area incompressible surfaces in 3-manifolds.* Invent. Math. **71**(1983), no. 3, 609–642.

[231] A. Freire, *On the Martin boundary of Riemannian products.* J. Differential Geom. **33**(1991), no. 1, 215–232.

[232] R. Friedman, *Simultaneous resolution of threefold double points.* Math. Ann. **274**(1986), no. 4, 671–689.

[233] H. Friedrich, *Hyperbolic reductions for Einstein's equations.* Classical Quantum Gravity. **13**(1996), no. 6, 1451–1469.

[234] J. X. Fu and S.-T. Yau, *Existence of Supersymmetric Hermitian Metrics with Torsion on Non-Kaehler Manifolds.* arXiv:hep-th/0509028.

[235] A. Fujiki, *The moduli spaces and Kähler metrics of polarized algebraic varieties.* Sugaku **42**(1990), no.3, 231–243.

[236] T. Fujita, *On polarized manifolds whose adjoint bundles are not semipositive.* Algebraic geometry, Sendai, 1985, 167–178, Adv. Stud. Pure Math., **10**, North-Holland, Amsterdam, 1987.

[237] K. Fukaya, *Morse homotopy, A^∞-category, and Floer homologies.* Proceedings of GARC Workshop on Geometry and Topology '93 (Seoul, 1993), 1–102, Lecture Notes Ser., **18**, Seoul Nat. Univ., Seoul, 1993.

[238] K. Fukaya and Y. G. Oh, *Zero-loop open strings in the cotangent bundle and Morse homotopy.* Asian J. Math. **1**(1997), no. 1, 96–180.

[239] K. Fukaya, Y.-G Oh, H. Ohta and K. Ono, *Lagrangian Intersection Theory–Anomaly and Obstruction.* book in preparation.

[240] K. Fukaya and K. Ono, *Arnold conjecture and Gromov-Witten invariant.* Topology **38**(1999), no. 5, 933–1048.

[241] A. Futaki, *An obstruction to the existence of Einstein-Kähler metrics.* Inv. Math., **73**(1983), 437–443.

[242] Z.Y. Gao and S.-T. Yau, *The existence of negatively Ricci curved metrics on three-manifolds.* Invent. Math. **85**(1986), no. 3, 637–652.

[243] C. S. Gardner, J. M. Greene, M. D. Kruskal and R. M. Miura, *Korteweg-deVries equation and generalization. VI. Methods for exact solution.* Comm. Pure Appl. Math. **27**(1974), 97–133.

[244] H. Garland, *p-adic curvature and the ohomology of discrete subgroups of p-adic groups.* Ann. of Math. (2) **97**(1973), 375–423.

[245] P. Gauduchon, *1-forme de torsion d'une variété hermitienne compacte.* Math. Ann. **267**(1984), no. 4, 495–518.

[246] J. P. Gauntlett, D. Martelli, J. Sparks and D. Waldram, *Sasaki-Einstein metrics on $S^2 \times S^3$.* Adv. Theor. Math. Phys. **8**(2004), no. 4, 711–734.

[247] C. Gerhardt, *H-surfaces in Lorentzian manifolds.* Comm. Math. Phys. **89**(1983), no. 4, 523–553.

[248] C. Gerhardt, *Hypersurfaces of prescribed scalar curvature in Lorentzian manifolds.* J. Reine Angew. Math. **554**(2003), 157–199.

[249] R. Geroch, *Energy extraction.* Ann. New York Acad. Sci., **224**, 108–117.

[250] M. L. Gerver and N. S. Nadirashvili, *A condition for isometry of Riemannian metrics in the disc.* Dokl. Akad. Nauk SSSR **275**(1984), no. 2, 289–293.

[251] M. Giaquinta and E. Giusti, *The singular set of the minima of certain quadratic functionals.* Ann. Scuola Norm. Sup. Pisa Cl. Sci. (4) **11**(1984), no. 1, 45–55.

[252] D. Gieseker, *On moduli of vector bundles on an algebraic surface,* Ann. of Math. **106**(1977) 45–60.

[253] D. Gieseker, *Global moduli for surfaces of general type.* Invent. Math. **43**(1977), no. 3, 233–282.

[254] D. Gieseker and J. Li, *Irreducibility of moduli of rank-2 vector bundles on algebraic surfaces.* J. Differential Geom. **40**(1994), no. 1, 23–104.

[255] A. B. Givental, *Equivariant Gromov-Witten invariants.* Internat. Math. Res. Notices 1996, no. **13**, 613–663.

[256] A. Givental, *A mirror theorem for toric complete intersections.* Topological field theory, primitive forms and related topics (Kyoto, 1996), 141–175, Progr. Math., **160**, Birkhüser Boston, Boston, MA, 1998.

[257] W. M. Goldman, *On the polynomial cohomology of affine manifolds.* Invent. Math. **65**(1981/82), no. 3, 453–457.

[258] R. Gompf, *A new construction of symplectic manifolds.* Ann. of Math. (2) **142**(1995), no. 3, 527–595.

[259] R. E. Gompf, *Toward a topological characterization of symplectic manifolds.* J. Symplectic Geom. **2**(2004), no. 2, 177–206.

[260] C. Gordon, D. Webb and S. Wolpert, *Isospectral plane domains and surfaces via Riemannian orbifolds.* Invent. Math. **110**(1992), no. 1, 1–22.

[261] C. Gordon and E. Wilson, *Continuous families of isospectral Riemannian metrics which are not locally isometric.* J. Differential Geom. **47**(1997), no. 3, 504–529.

[262] M. Goresky, G. Harder and R. MacPherson, *Weighted cohomology.* Invent. Math. **116**(1994), no. 1-3, 139–213.

[263] M. Goresky and R. MacPherson, *Intersection homology theory.* Topology **19**(1980), no. 2, 135–162.

[264] M. Goresky and R. MacPherson, *Intersection homology. II.* Invent. Math. **72**(1983), no. 1, 77–129.

[265] C. Graham and J. Lee, *Einstein metrics with prescribed conformal infinity on the ball.* Adv. Math. **87**(1991), no. 2, 186–225.

[266] H. Grauert and H. Reckziegel, *Hermitesche Metriken und normale Familien holomorpher Abbildungen.* (German) Math. Z. **89**(1965), 108–125.

[267] A. Gray, *Invariants of curvature operators of four-dimensional Riemannian manifolds.* Proceedings of the Thirteenth Biennial Seminar of the Canadian Mathematical Congress Vol. **2**, pp. 42–65. Canad. Math. Congr., Montreal, Que., 1972.

[268] L. W. Green, *Proof of Blaschke's sphere conjecture.* Bull. Amer. Math. Soc. **67**(1961), 156–158.

[269] B. Greene and M. Plesser, *Duality in Calabi-Yau moduli space.* Nuclear Phys. B **338**(1990), no. 1, 15–37.

[270] B. Greene, D. Morrison and M. Plesser, *Mirror manifolds in higher dimension.* Comm. Math. Phys. **173**(1995), no. 3, 559–597.

[271] R. E. Greene and H. Wu, *Analysis on noncompact Kähler manifolds.* Proc. Sympos. Pure Math., Vol. **XXX**, Part 2, 69–100. Amer. Math. Soc., Providence, R.I., 1977.

[272] P. A. Griffiths, *Complex-analytic properties of certain Zariski open sets on algebraic varieties.* Ann. of Math. (2) **94**(1971), 21–51.

[273] P. A. Griffiths and W. Schmid, *Locally homogeneous complex manifolds.* Acta Math. **123**(1969) 253–302.

[274] A. Grigor'yan, *The heat equation on noncompact Riemannian manifolds.* Mat. Sb. **182**(1991), no. 1, 55–87; translation in Math. USSR-Sb. **72**(1992), no. 1, 47–77

[275] A. Grigor'yan, *Heat kernel of a noncompact Riemannian manifold.* Proc. Sympos. Pure Math., **57**, 239–263. Amer. Math. Soc., Providence, RI, 1995.

[276] A. Grigor'yan, Y. Netrusov and S.-T. Yau, *Eigenvalues of elliptic operators and geometric applications*, in Servey in Differential Geometry, **IX**, 147-217. A. Grigor'yan and S.-T. Yau ed., International Press.

[277] D. Gromoll and W. Meyer, *On complete open manifolds of positive curvature.* Ann. of Math. (2) **90**(1969) 75–90.

[278] D. Gromoll and W. Meyer, *Periodic geodesics on compact riemannian manifolds.* J. Differential Geometry **3**(1969), 493–510.

[279] D. Gromoll and J. A. Wolf, *Some relations between the metric structure and the algebraic structure of the fundamental group in manifolds of nonpositive curvature.* Bull. Amer. Math. Soc. **77**(1971), 545–552.

[280] M. Gromov, *Manifolds of negative curvature.* J. Differential Geom. **13**(1978), no. 2, 223–230.

[281] M. Gromov, *Hyperbolic groups.* Essays in group theory, 75–263, Math. Sci. Res. Inst. Publ., **8**, Springer, New York, 1987.

[282] M. Gromov, *Curvature, diameter and Betti numbers.* Comment. Math. Helv. **56**(1981), no. 2, 179–195.

[283] M. Gromov, *Volume and bounded cohomology.* Inst. Hautes Études Sci. Publ. Math. **56**(1982), 5–99.

[284] M. Gromov, *Pseudoholomorphic curves in symplectic manifolds.* Invent. Math. **82**(1985), no. 2, 307–347.

[285] M. Gromov, *Partial differential relations.* Ergebnisse der Mathematik und ihrer Grenzgebiete (3), **9**. Springer-Verlag, Berlin, 1986.

[286] M. Gromov, *Kähler hyperbolicity and L_2-Hodge theory.* J. Differential Geom. **33**(1991), no. 1, 263–292.

[287] M. Gromov, *Metric structures for Riemannian and non-Riemannian spaces.* Based on the 1981 French original. With appendices by M. Katz, P. Pansu and S. Semmes. Translated from the French by Sean Michael Bates. Progress in Mathematics, **152**. Birkhäuser Boston, Inc., Boston, MA, 1999.

[288] M. Gromov and H. B. Lawson Jr., *Spin and scalar curvature in the presence of a fundamental group. I.* Ann. of Math. (2) **111**(1980), no. 2, 209–230.

[289] M. Gromov and H. B. Lawson Jr., *The classification of simply connected manifolds of positive scalar curvature.* Ann. of Math. (2) **111**(1980), no. 3, 423–434.

[290] M. Gromov and R. Schoen, *Harmonic maps into singular spaces and p-adic superrigidity for lattices in groups of rank one.* Inst. Hautes tudes Sci. Publ. Math. **76**(1992), 165–246.

[291] M. Gross, *Special Lagrangian fibrations. I. Topology.* Integrable systems and algebraic geometry (Kobe/Kyoto, 1997), 156–193, World Sci. Publishing, River Edge, NJ, 1998.

[292] M. Gross, *Special Lagrangian fibrations. II. Geometry.* A survey of techniques in the study of special Lagrangian fibrations. Surveys in differential geometry: differential geometry inspired by string theory, 341–403, Surv. Differ. Geom., **5**, Int. Press, Boston, MA, 1999.
[293] K. Grove and K. Shiohama, *A generalized sphere theorem.* Ann. Math. (2) **106**(1977), no. 2, 201–211.
[294] C. H. Gu, *On the initial-boundary value problem for harmonic maps from the 2-dimensional Minkowski space.* Manuscripta Math. **33**(1980/81), no. 1, 51–58.
[295] B. Guan and P. Guan, *Hypersurfaces with prescribed curvatures.* Ann. Math. **156**(2002), 655–674.
[296] P. F. Guan and X. N. Ma, *The Christoffel-Minkowski problem. I. Convexity of solutions of a Hessian equation.* Invent. Math. **151**(2003), no. 3, 553–577.
[297] V. Guillemin, *Wave-trace invariants.* Duke Math. J. **83**(1996), no. 2, 287–352.
[298] V. Guillemin and D. Kazhdan, *Some inverse spectral results for negatively curved 2-manifolds.* Topology **19**(1980), no. 3, 301–312.
[299] V. Guillemin and S. Sternberg, *Convexity properties of the moment mapping.* Invent. Math. **67**(1982), no. 3, 491–513.
[300] V. Guillemin and A. Weinstein, *Eigenvalues associated with a closed geodesic.* Bull. Amer. Math. Soc. **82**(1976), no. 1, 92–94 and Bull. Amer. Math. Soc. **82**(1976), no. 6, 966.
[301] Y. Guivarc'h, L. Z. Ji and J. C. Taylor, *Compactifications of symmetric spaces.* Progress in Mathematics, **156**. Birkhäuser Boston, Inc., Boston, MA, 1998.
[302] S. Gukov, *Three-dimensional quantum gravity, Chern-Simons theory, and the A-polynomial.* Comm. Math. Phys. **255**(2005), no. 3, 577–627.
[303] X. Gukov, S.-T. Yau and E. Zaslow, *Duality and fibrations on G_2 manifolds.* Turkish J. Math. **27**(2003), no. 1, 61–97.
[304] R. Gulliver, *Regularity of minimizing surfaces of prescribed mean curvature.* Ann. of Math. (2) **97**(1973), 275–305.
[305] M. Günther, *On the perturbation problem associated to isometric embeddings of Riemannian manifolds.* Ann. Global Anal. Geom. **7**(1989), no. 1, 69–77.
[306] L. Habermann and J. Jost, *Riemannian metrics on Teichmüller space.* Preprint.
[307] L. Habermann and J. Jost, *Metrics on Riemann surfaces and the geometry of moduli space.* Preprint.
[308] W. Haken, *Various aspects of the three-dimensional Poincarè problem.* 1970 Topology of Manifolds (Proc. Inst., Univ. of Georgia, Athens, Ga., 1969) pp. 140–152 Markham, Chicago, **Ill** .
[309] R. S. Hamilton, *Three-manifolds with positive Ricci curvature.* J. Differential Geom. **17**(1982), no. 2, 255–306.
[310] R. S. Hamilton, *The Ricci flow on surfaces.* Mathematics and general relativity, 237–262, Contemp. Math., **71**, Amer. Math. Soc., Providence, RI, 1988.
[311] R. S. Hamilton, *The Harnack estimate for the Ricci flow.* J. Differential Geom. **37**(1993), no. 1, 225–243.
[312] R. S. Hamilton, *The formation of singularities in the Ricci flow.* Surveys in differential geometry, Vol. II (Cambridge, MA, 1993), 7–136, Internat. Press, Cambridge, MA, 1995.
[313] R. S. Hamilton, *Four-manifolds with positive isotropic curvature.* Comm. Anal. Geom. **5**(1997), no. 1, 1–92.
[314] R. S. Hamilton, *Non-singular solutions of the Ricci flow on three-manifolds.* Comm. Anal. Geom. **7**(1999), no. 4, 695–729.
[315] Q. Han, R. Hardt and F. H. Lin, *Geometric measure of singular sets of elliptic equations.* Comm. Pure Appl. Math. **51**(1998), no. 11-12, 1425–1443.
[316] Q. Han, J. X. Hong and C. S. Lin, *Local isometric embedding of surfaces with nonpositive Gaussian curvature.* J. Differential Geom. **63**(2003), no. 3, 475–520.
[317] R. Hardt, M. Hoffmann-Ostenhof, T. Hoffmann-Ostenhof and N. Nadirashvili, *Critical sets of solutions to elliptic equations.* J. Differential Geom. **51**(1999), no. 2, 359–373.
[318] R. Hardt and L. Simon, *Boundary regularity and embedded solutions for the oriented Plateau problem.* Ann. of Math. (2) **110**(1979), no. 3, 439–486.
[319] F. R. Harvey and H. B. Lawson Jr., *On boundaries of complex analytic varieties. I.* Ann. of Math. (2) **102**(1975), no. 2, 223–290.
[320] F. R. Harvey and H. B. Lawson Jr., *On boundaries of complex analytic varieties. II.* Ann. Math. (2) **106**(1977), no. 2, 213–238.

[321] R. Harvey and H. B. Lawson, Jr. *Calibrated geometries.* Acta Math. **148**(1982), 47–157.
[322] R. Harvey and B. Shiffman, *A characterization of holomorphic chains.* Ann. of Math. (2) **99**(1974), 553–587.
[323] A. Hassell, R. Melrose and A. Vasy, *Spectral and scattering theory for symbolic potentials of order zero.* Adv. Math. **181**(2004), no. 1, 1–87.
[324] S. W. Hawking, *Black hole in general relativity.* Commun. Math. Phys. **25**(1972), 152–166.
[325] J. Hersch, *Quatre propriétés isopérimétriques de membranes sphériques homogènes.* C. R. Acad. Sci. Paris Sér. A-B **270**(1970), A1645–A1648.
[326] N. Hingston, *Equivariant Morse theory and closed geodesics.* J. Differential Geom. **19**(1984), no. 1, 85–116.
[327] N. Hingston, *On the growth of the number of closed geodesics on the two-sphere.* Internat. Math. Res. Notices 1993, no. **9**, 253–262.
[328] F. Hirzebruch and K. Kodaira, *On the complex projective spaces.* J. Math. Pures Appl. (9) **36**(1957), 201–216.
[329] N. Hitchin, *Harmonic spinors.* Advances in Math. **14**(1974), 1–55.
[330] N. Hitchin, *Compact four-dimensional Einstein manifolds.* J. Differential Geometry **9**(1974), 435–441.
[331] N. J. Hitchin, *The self-duality equations on a Riemann surface.* Proc. London Math. Soc. (3) **5**(1987), no. 1, 59–126.
[332] N. J. Hitchin, *Harmonic maps from a 2-torus to the 3-sphere.* J. Differential Geom. **31**(1990), no. 3, 627–710.
[333] N. J. Hitchin, *Lie groups and Teichmüller space.* Topology **31**(1992), no. 3, 449–473.
[334] N. J. Hitchin, *The moduli space of special Lagrangian submanifolds.* Ann. Scuola Norm. Sup. Pisa Cl. Sci. (4) **25**(1997), no. 3-4, 503–515.
[335] N. Hitchin, *The geometry of three-forms in six dimensions.* J. Differential Geom. **55**(2000), no. 3, 547–576.
[336] N. Hitchin, *Stable forms and special metrics.* Global differential geometry: the mathematical legacy of Alfred Gray (Bilbao, 2000), 70–89, Contemp. Math., **288** , Amer. Math. Soc., Providence, RI, 2001.
[337] H. Hofer, *On the topological properties of symplectic maps.* Proc. Royal Soc. Edinburgh, Sect. A **115**(1990), 25–38.
[338] H. Hofer, *Pseudo-holomorphic curves and Weinstein conjecture in dimension three.* Invent. math. **114**(1993), no. 3, 515–563.
[339] H. Hofer, K. Wysocki and E. Zehnder, *The dynamics on three-dimensional strictly convex energy surfaces.* Ann. of Math. (2) **148**(1998), no. 1, 197–289.
[340] H. Hofer, K. Wysocki and E. Zehnder, *Finite energy foliations of tight three-spheres and Hamiltonian dynamics.* Ann. of Math. (2) **157**(2003), 125–255.
[341] H. Hofer and E. Zehnder, *Symplectic capacities and Hamiltonian dynamics.* Birkhauser, Basel, 1994.
[342] M. Hoffmann-Ostenhof, T. Hoffmann-Ostenhof and N. Nadirashvili, *Critical sets of smooth solutions to elliptic equations in dimension 3*, Indiana Univ. Math. J. **45**(1996), 15–37.
[343] J. X. Hong, *Realization in R^3 of complete Riemannian manifolds with negative curvature.* Comm. Anal. Geom. **1**(1993), no. 3-4, 487–514.
[344] M. J. Hopkins, *Topological modular forms, the Witten genus, and the theorem of the cube.* Proceedings of the International Congress of Mathematicians, Vol. **1**, **2** (Zürich, 1994), 554–565, Birkhäuser, Basel, 1995.
[345] L. Hörmander, *An introduction to complex analysis in several variables.* Third edition. North-Holland Mathematical Library, **7**. North-Holland Publishing Co., Amsterdam, 1990.
[346] L.K. Hua, *Harmonic analysis of functions of several variables in the classical domains.* American Mathematical Society, Providence, RI, 1963.
[347] G. Huisken, *Flow by mean curvature of convex surfaces into spheres.* J. Differential Geom. **20**(1984), no. 1, 237–266.
[348] G. Huisken, *Contracting convex hypersurfaces in Riemannian manifolds by their mean curvature.* Invent. Math. **84**(1986), no. 3, 463–480.
[349] G. Huisken and T. Ilmanen, *The inverse mean curvature flow and the Riemannian Penrose inequality.* J. Differential Geom. **59**(2001), no. 3, 353–437.
[350] G. Huisken and C. Sinestrari, *Mean curvature flow singularities for mean convex surfaces.* Calc. Var. Partial Differential Equations **8**(1999), no. 1, 1–14.

[351] G. Huisken and C. Sinestrari, *Convexity estimates for mean curvature flow and singularities of mean convex surfaces.* Acta Math. **183**(1999), no. 1, 45–70.

[352] G. Huisken and S.-T. Yau, *Definition of center of mass for isolated physical systems and unique foliations by stable spheres with constant mean curvature.* Invent. Math. **124**(1996), no. 1-3, 281–311.

[353] W. Israël, *Event Horizons in Static Vacuum Space-Times.* Phys. Rev. **164**(1967), 1776–1779.

[354] T. Ivey, *Ricci solitons on compact 3-manifolds.* Differential Geom. Appl. **3**(1993) 301–307.

[355] H. Iwaniec and P. Sarnak, *L^∞ norms of eigenfunctions of arithmetic surfaces.* Ann. of Math. (2) **141**(1995), no. 2, 301–320.

[356] D. Jakobson, M. Levitin, N. Nadirashvili, N. Nigam and I. Polterovich, *How large can the first eigenvalue be on a surface of genus two?* Arxiv: math. SP/0509398.

[357] D. Jakobson and N. Nadirashvili, *Eigenfunctions with few critical points.* J. Differential Geom. **53**(1999), no. 1, 177–182.

[358] L. C. Jeffrey and F. C. Kirwan, *Localization and the quantization conjecture.* Topology **36**(1997), no. 3, 647-693.

[359] L. Z. Ji and R. MacPherson, *Geometry of compactifications of locally symmetric spaces.* Ann. Inst. Fourier (Grenoble) **52**(2002), no. 2, 457–559.

[360] M. Ji and G. Y. Wang, *Minimal surfaces in Riemannian manifolds.* Mem. Amer. Math. Soc. **104**(1993), no. 495.

[361] V. Jones, *Hecke algebra representations of braid groups and link polynomials.* Ann. of Math. (2) **126**(1987), no. 2, 335–388.

[362] K. Jörgens, *Über die Losungen der Differentialgleichung $rt-s^2=1$.* Math. Ann. **127**(1954), 130–134.

[363] J. Jorgenson and A. Todorov, *Enriques surfaces, analytic discriminants, and Borcherds's Φ function.* Comm. Math. Phys. **191**(1998), no. 2, 249–264.

[364] J. Jost, *Embedded minimal surfaces in manifolds diffeomorphic to the three-dimensional ball or sphere.* J. Differential Geom. **30**(1989), no. 2, 555–577.

[365] J. Jost, *Convex functionals and generalized harmonic maps into spaces of nonpositive curvature.* Comment. Math. Helv. **70**(1995), no. 4, 659–673.

[366] J. Jost and S.-T. Yau, *Harmonic mappings and Kähler manifolds.* Math. Ann. **262**(1983), no. 2, 145–166.

[367] J. Jost and S.-T. Yau, *Harmonic maps and group representations.* Differential geometry, 241–259, Pitman Monogr. Surveys Pure Appl. Math., **52**, Longman Sci. Tech., Harlow, 1991.

[368] J. Jost, and S.-T. Yau, *Applications of quasilinear PDE to algebraic geometry and arithmetic lattices.* Algebraic geometry and related topics, 169–193, Conf. Proc. Lecture Notes Algebraic Geom., I, Internat. Press, Cambridge, MA, 1993.

[369] J. Jost and S.-T. Yau, *Harmonic maps and superrigidity.* Differential geometry: partial differential equations on manifolds, 245–280, Proc. Sympos. Pure Math., **54**, Part 1, Amer. Math. Soc., Providence, RI, 1993.

[370] J. Jost and S.-T. Yau, *Harmonic maps and rigidity theorems for spaces of nonpositive curvature.* Comm. Anal. Geom. **7**(1999), no. 4, 681–694.

[371] D. D. Joyce, *Compact 8-manifolds with holonomy* Spin(7). Invent. Math. **123**(1996), no. 3, 507–552.

[372] D. D. Joyce, *Compact Riemannian 7-manifolds with holonomy G_2. I, II.* J. Differential Geom. **43**(1996), no. 2, 291–328, 329–375.

[373] D. Joyce, *A new construction of compact 8-manifolds with holonomy* Spin(7). J. Differential Geom. **53**(1999), no. 1, 89–130.

[374] D. Joyce, *Singularities of special Lagrangian fibrations and the SYZ conjecture.* Comm. Anal. Geom. **11**(2003), no. 5, 859–907.

[375] A. B. Katok, *Ergodic properties of degenerate integrable Hamiltonian systems.* Izv. Akad. Nauk SSSR. **37**(1973) (Russian). Math. USSR-Izv. **7**(1973) 535-571.

[376] X. Katz and C.-C. Liu, *Enumerative geometry of stable maps with Lagrangian boundary conditions and multiple covers of the disc.* Adv. Theor. Math. Phys. **5**(2001), no. 1, 1–49.

[377] J. L. Kazdan and F. W. Warner, *Existence and conformal deformation of metrics with prescribed Gaussian and scalar curvatures.* Ann. of Math. (2) **101**(1975), 317–331.

[378] D. A. Kazhdan, *On the connection of the dual space of a group with the structure of its closed subgroups* (Russian), Funkcional. Anal. i Prilozhen. **1**(1967), 71–74.

[379] L. Keen, *Collars on Riemann surfaces.* Discontinuous groups and Riemann surfaces (Proc. Conf., Univ. Maryland, College Park, Md., 1973), pp. 263–268. Ann. of Math. Studies, No. 79, Princeton Univ. Press, Princeton, N.J., 1974.

[380] J. Kim, C. LeBrun and M. Pontecorvo, *Scalar-flat Kähler surfaces of all genera.* J. Reine Angew. Math. **486**(1997), 69–95.

[381] J. R. King, *The currents defined by analytic varieties.* Acta Math. **127**(1971), no. 3-4, 185–220.

[382] S. Klainerman and F. Nicolò, *Peeling properties of asymptotically flat solutions to the Einstein vacuum equations".* Classical Quantum Gravity **20**(2003), no. 14, 3215–3257.

[383] S. Klainerman and S. Selberg, *Bilinear estimates and applications to nonlinear wave equations.* Commun. Contemp. Math. **4**(2002), no. 2, 223–295.

[384] W. Klingenberg, *Lectures on closed geodesics.* Third edition. Mathematisches Institut der Universität Bonn, Bonn, 1977.

[385] S. Kobayashi, *On compact Kähler manifolds with positive definite Ricci tensor.* Ann. of Math. (2) **74**(1961) 570–574.

[386] S. Kobayashi, *Intrinsic metrics on complex manifolds.* Bull. Amer. Math. Soc. **73**(1967) 347–349.

[387] S. Kobayashi and T. Ochiai, *Meromorphic mappings onto compact complex spaces of general type.* Invent. Math. **31**(1975), no. 1, 7–16.

[388] K. Kodaira, *On a differential-geometric method in the theory of analytic stacks.* Proc. Nat. Acad. Sci. U. S. A. **39**(1953). 1268–1273.

[389] K. Kodaira, *On Kähler varieties of restricted type (an intrinsic characterization of algebraic varieties).* Ann. of Math. (2) **60**(1954). 28–48.

[390] J. Kollár, *Effective base point freeness.* Math. Ann. **296**(1993), no. 4, 595–605.

[391] M. Kontsevich, *Enumeration of rational curves via torus actions.* The moduli space of curves (Texel Island, 1994), 335–368, Progr. Math., 129, Birkhäuser, Boston, 1995.

[392] M. Kontsevich, *Feynman diagrams and low-dimensional topology.* First European Congress of Mathematics, Vol. **II** (Paris, 1992), 97–121, Progr. Math., **120**, Birkhäuser, Basel, 1994.

[393] M. Kontsevich, *Homological algebra of mirror symmetry.* Proceedings of the International Congress of Mathematicians, Vol. **1**, **2** (Zürich, 1994), 120–139, Birkhäuser, Basel, 1995.

[394] N. Korevaar, *An easy proof of the interior gradient bound for solutions to the prescribed mean curvature equation.* Proc. Sympos. Pure Math., **45**, Part 2, 81–89, Amer. Math. Soc., Providence, RI, 1986.

[395] N. Korevaar, *Upper bounds for eigenvalues of conformal metrics.* J. Differential Geom. **37**(1993), no. 1, 73–93.

[396] N. Korevaar, R. Mazzeo, F. Pacard, and R. Schoen, *Refined asymptotics for constant scalar curvature metrics with isolated singularities.* Invent. Math. **135**(1999), no. 2, 233–272.

[397] N. Korevaar and R. Schoen, *Sobolev spaces and harmonic maps for metric space targets.* Comm. Anal. Geom. **1**(1993), no. 3-4, 561–659

[398] A. Kovalev, *Twisted connected sums and special Riemannian holonomy.* J. Reine Angew. Math. **565**(2003), 125–160.

[399] P. Kramer, *Group actions on compact hyperbolic manifolds and closed geodesics.* Symmetry in physics, 113–124, CRM Proc. Lecture Notes, **34**, Amer. Math. Soc., Providence, RI, 2004.

[400] P. B. Kronheimer, and T. S. Mrowka, *Embedded surfaces and the structure of Donaldson's polynomial invariants.* J. Differential Geom. **41**(1995), no. 3, 573–734.

[401] M. Kruskal, *The birth of the soliton.* Nonlinear evolution equations solvable by the spectral transform (Internat. Sympos., Accad. Lincei, Rome, 1977), pp. 1–8, Res. Notes in Math., **26**, Pitman, Boston, Mass.-London, 1978.

[402] V. Kulikov, *Epimorphicity of the period mapping for surfaces of type K3.* Uspehi Mat. Nauk **32**(1977), no. 4, 257–258.

[403] M. Kuranishi, *Strongly pseudoconvex CR structures over small balls. I. An a priori estimate.* Ann. of Math. (2) **115**(1982), no. 3, 451–500.

[404] M. Kuranishi, *Strongly pseudoconvex CR structures over small balls. II. A regularity theorem.* Ann. of Math. (2) **116**(1982), no. 1, 1–64.

[405] M. Kuranishi, *Strongly pseudoconvex CR structures over small balls. III. An embedding theorem.* Ann. of Math. (2) **116**(1982), no. 2, 249–330.

[406] F. Lalonde and D. McDuff, *The geometry of symplectic energy, Ann. of Math.* **141**(1995), 349-371.
[407] H. B. Lawson, Jr., *Local rigidity theorems for minimal hypersurfaces*, Ann. of Math. (2) **89** (1969), 187C197.
[408] H. B. Lawson, Jr., *Compact minimal surfaces in* S^3. 1970 Global Analysis (Proc. Sympos. Pure Math., Vol. XV, Berkeley, Calif., 1968) pp. 275–282 Amer. Math. Soc., Providence, R.I.
[409] H. B. Lawson and J. Simons, *On stable currents and their application to global problems in real and complex geometry.* Ann. of Math. (2) **98**(1973), 427–450.
[410] H. B. Lawson Jr. and S.-T. Yau, *Compact manifolds of nonpositive curvature.* J. Differential Geometry **7**(1972), 211–228.
[411] H. B. Lawson Jr. and S.-T. Yau, *Scalar curvature, non-abelian group actions, and the degree of symmetry of exotic spheres.* Comment. Math. Helv. **49**(1974), 232–244.
[412] C. R. LeBrun, *Explicit Self-Dual Metrics on* $\mathbb{CP}_2\#\cdots\#\mathbb{CP}_2$. J. Diff. Geom. **34**(1991), 223–253.
[413] C. LeBrun, *Ricci curvature, minimal volumes, and Seiberg-Witten theory.* Invent. Math. **145**(2001), no. 2, 279–316.
[414] J.-H. Lee and N. C. Leung, *Geometric structures on* G_2 *and* $Spin(7)$*-manifolds.* Arxiv: math.DG/0202045.
[415] Y.-I. Lee, *Embedded special Lagrangian submanifolds in Calabi-Yau manifolds.* Comm. Anal. Geom. **11**(2003), no. 3, 391–423.
[416] W. Lerche, C. Vafa and N. Warner, *Chiral rings in* $N = 2$ *superconformal theories.* Nuclear Phys. B **324**(1989), no. 2, 427–474.
[417] N. C. Leung, *Einstein type metrics and stability on vector bundles.* J. Differential Geom. **45**(1997), no. 3, 514–546.
[418] N. C. Leung, *Lagrangian submanifolds in hyperKähler manifolds, Legendre transformation.* J. Differential Geom. **61**(2002), no. 1, 107–145.
[419] N. C. Leung, *Topological quantum field theory for Calabi-Yau threefolds and* G_2*-manifolds.* Adv. Theor. Math. Phys. **6**(2002), no. 3, 575–591.
[420] N. C. Leung, *Mirror symmetry without corrections.* Comm. Anal. Geom. **13**(2005), no. 2, 287–331.
[421] N. C. Leung and X. W. Wang *Intersection theory of coassociative submanifolds in* G_2*-manifolds and Seiberg-Witten invariants.* Arxiv: math.DG/0401419.
[422] N. C. Leung, S.-T. Yau and E. Zaslow, *From special Lagrangian to Hermitian-Yang-Mills via Fourier-Mukai transform.* Winter School on Mirror Symmetry, Vector Bundles and Lagrangian Submanifolds (Cambridge, MA, 1999), 209–225, AMS/IP Stud. Adv. Math., **23**, Amer. Math. Soc., Providence, RI, 2001.
[423] J. S. Li, *Kloosterman-Selberg zeta functions on complex hyperbolic spaces.* Amer. J. Math. **113**(1991), no. 4, 653–731.
[424] J. Li, *Algebraic geometric interpretation of Donaldson's polynomial invariants.* J. Differential Geom. **37**(1993), no. 2, 417–466.
[425] J. Li, *The geometry of moduli spaces of vector bundles over algebraic surfaces.* Proceedings of the International Congress of Mathematicians, Vol. **1**, **2** (Zürich, 1994), 507–516, Birkhäuser, Basel, 1995.
[426] J. Li, *Stable morphisms to singular schemes and relative stable morphisms.* J. Differential Geom. **57**(2001), no. 3, 509–578.
[427] J. Li, *A degeneration formula of GW-invariants.* J. Differential Geom. **60**(2002), no. 2, 199–293.
[428] J. Li, C. C. M. Liu, K. F. Liu and J. Zhou, *A Mathematical Theory of the Topological Vertex Authors*, arXiv:math.AG/0408426.
[429] J. Li and G. Tian, *Virtual moduli cycles and Gromov-Witten invariants of algebraic varieties.* J. Amer. Math. Soc. **11**(1998), no. 1, 119–174.
[430] J. Li and G. Tian, *Virtual moduli cycles and Gromov-Witten invariants of general symplectic manifolds.* Topics in symplectic 4-manifolds, 47–83, First Int. Press Lect. Ser., **I**, Internat. Press, Cambridge, MA, 1998.
[431] J. Li and S. T. Yau, *Hermitian-Yang-Mills connection on non-Kähler manifolds.* Mathematical aspects of string theory (San Diego, Calif., 1986), 560–573, Adv. Ser. Math. Phys., **1**, World Sci. Publishing, Singapore, 1987.

[432] J. Li and S.-T. Yau, *The Existence of Supersymmetric String Theory with Torsion.* arXiv:hep-th/0411136.

[433] J. Li, S.-T. Yau and F. Y. Zheng, *A simple proof of Bogomolov's theorem on class* VII_0 *surfaces with* $b_2 = 0$. Illinois J. Math. **34**(1990), no. 2, 217–220.

[434] P. Li, *A lower bound for the first eigenvalue of the Laplacian on a compact manifold.* Indiana Univ. Math. J. **28**(1979), no. 6, 1013–1019.

[435] P. Li, *Harmonic sections of polynomial growth.* Math. Res. Lett. **4**(1997), no. 1, 35–44.

[436] P. Li, R. Schoen and S.-T. Yau, *On the isoperimetric inequality for minimal surfaces.* Ann. Scuola Norm. Sup. Pisa Cl. Sci. (4) **11**(1984), no. 2, 237–244.

[437] P. Li and G. Tian, *On the heat kernel of the Bergmann metric on algebraic varieties.* J. Amer. Math. Soc. **8**(1995), no. 4, 857–877.

[438] P. Li and J. P. Wang, *Counting dimensions of L-harmonic functions.* Ann. of Math. (2) **152** (2000), no. 2, 645–658.

[439] P. Li and J. P. Wang, *Complete manifolds with positive spectrum. II.* J. Differential Geom. **62**(2002), no. 1, 143–162.

[440] P. Li and S.-T. Yau, *Estimates of eigenvalues of a compact Riemannian manifold.* Geometry of the Laplace operator, 205–239, Proc. Sympos. Pure Math., **XXXVI**, Amer. Math. Soc., Providence, R.I., 1980.

[441] P. Li and S.-T. Yau, *A new conformal invariant and its applications to the Willmore conjecture and the first eigenvalue of compact surfaces.* Invent. Math. **69**(1982), no. 2, 269–291.

[442] P. Li and S.-T. Yau, *On the Schrödinger equation and the eigenvalue problem.* Comm. Math. Phys. **88**(1983), no. 3, 309–318.

[443] P. Li and S.-T. Yau, *On the parabolic kernel of the Schrödinger operator.* Acta Math. **156**(1986), no. 3-4, 153–201.

[444] T. J. Li, *Symplectic 4-manifolds with Kodaira dimension zero.* Preprint.

[445] T. J. Li and A. Liu, *General wall crossing formula.* Math. Res. Lett. **2**(1995), no. 6, 797–810

[446] T. J. Li and A. Liu, *Symplectic structure on ruled surfaces and a generalized adjunction formula.* Math. Res. Lett. **2**(1995), no. 4, 453–471.

[447] B. H. Lian, K. F. Liu and S.-T. Yau, *Mirror principle. I.* Asian J. Math. **1**(1997), no. 4, 729–763.

[448] B. H. Lian, K. F. Liu and S.-T. Yau, *Mirror principle. II.* Sir Michael Atiyah: a great mathematician of the twentieth century. Asian J. Math. **3**(1999), no. 1, 109–146.

[449] B. H. Lian, K. F. Liu and S.-T. Yau, *Mirror principle. III.* Asian J. Math. **3**(1999), no. 4, 771–800.

[450] B. H. Lian, K. F. Liu anf S.-T. Yau, *Mirror principle. IV.* Surveys in differential geometry, 475–496, Surv. Differ. Geom., **VII**, Int. Press, Somerville, MA, 2000.

[451] A. Lichnerowicz, *Spineurs harmoniques.* C. R. Acad. Sci. Paris, **257**(1963), 7–9.

[452] A. Lichnerowicz, *Magnetohydrodynamics: waves and shock waves in curved space-time.* Mathematical Physics Studies, **14**. Kluwer Academic Publishers Group, Dordrecht, 1994.

[453] C. S. Lin, *The local isometric embedding in* R^3 *of 2- dimensional Riemannian manifolds with nonnegative curvature.* J. Differential Geometry **21**(1985), 213–230.

[454] C. S. Lin, *The local isometric embedding in* R^3 *of two-dimensional Riemannian manifolds with Gaussian curvature changing sign cleanly.* Comm. Pure Appl. Math. **39**(1986) 867–887.

[455] F. H. Lin, *Asymptotic behavior of area-minimizing currents in hyperbolic space.* Comm. Pure Appl. Math. **42**(1989), no. 3, 229–242.

[456] F. H. Lin, *Nodal sets of solutions of elliptic and parabolic equations.* Comm. Pure Appl. Math. **44**(1991), no. 3, 287–308.

[457] F. H. Lin, *Gradient estimates and blow-up analysis for stationary harmonic maps.* Ann. of Math. (2) **149**(1999), no. 3, 785–829.

[458] F. H. Lin, *Mapping problems, fundamental groups and defect measures.* Acta Math. Sin. (Engl. Ser.) **15**(1999), no. 1, 25–52.

[459] H. Lindblad and I. Rodnianski, *Global existence for the Einstein vacuum equations in wave coordinates.* Comm. Math. Phys. **256**(2005), no. 1, 43–110.

[460] A. K. Liu, *Some new applications of general wall crossing formula, Gompf's conjecture and its applications.* Math. Res. Lett. **3**(1996), no. 5, 569–585.

[461] C.-C. Liu, *Moduli of J-Holomorphic Curves with Lagrangian Boundary Conditions and Open Gromov-Witten Invariants for an* S^1*-Equivariant Pair.* Arxiv: math.SG/0210257.

[462] C. C. M. Liu and S.-T. Yau, *Positivity of Quasilocal Mass*, Phys. Rev. Lett. **90**(2003), no. 23, 231102 4 pp.

[463] C. C. M. Liu and S.-T. Yau, *Positivity of quasi-local mass II*, J. Amer. Math. Soc. **19**(2006), no. 1, 181–204.

[464] G. Liu and G. Tian, *Floer homology and Arnold conjecture.* J. Differential Geom. **49**(1998), no. 1, 1–74.

[465] K. F. Liu, *On modular invariance and rigidity theorems.* J. Differential Geom. **41**(1995), no. 2, 343–396.

[466] K. F. Liu, *On elliptic genera and theta-functions.* Topology **35**(1996), no. 3, 617–640.

[467] K. F. Liu, X. F. Sun and S.-T. Yau, *Canonical metrics on the moduli space of Riemann surfaces. I.* J. Differential Geom. **68**(2004), no. 3, 571–637.

[468] K. F. Liu, X. F. Sun and S.-T. Yau, *Canonical Metrics on the Moduli Space of Riemann Surfaces. II.* J. Differential Geom. **69**(2005), no. 1, 163–216.

[469] J. C. Loftin, *Affine spheres and convex* $\mathbb{RP}^n$*-manifolds.* Amer. J. Math. **123**(2001), no. 2, 255–274.

[470] J. Loftin, S.-T. Yau and E. Zaslow, *Affine Manifolds, SYZ Geometry, and the "Y" Vertex Authors.* arXiv:math.DG/0405061.

[471] J. Lohkamp, *Curvature h-principles.* Ann. of Math. (2) **142**(1995), no. 3, 457–498.

[472] E. Looijenga, L^2*-cohomology of locally symmetric varieties.* Compositio Math. **67**(1988), no. 1, 3–20.

[473] S. Y. Lu and S.-T. Yau, *Holomorphic curves in surfaces of general type.* Proc. Nat. Acad. Sci. U.S.A. **87**(1990), no. 1, 80–82.

[474] P. Lu and G. Tian, *The complex structure on a connected sum of* $S^3 \times S^3$ *with trivial canonical bundle.* Math. Ann. **298**(1994), no. 4, 761–764.

[475] Z. Q. Lu, *On the lower order terms of the asymptotic expansion of Tian-Yau-Zelditch*, Amer. J. Math. **122**(2000), 235 – 273.

[476] A. Lubotzky, R. Phillips and P. Sarnak, *Ramanujan graphs.* Combinatorica **8**(1988), no. 3, 261–277.

[477] H. Luo, *Geometric criterion for Gieseker-Mumford stability of polarized manifolds*, J. Differential Geom. **49**(1998), no. 3, 577–599.

[478] W. Luo, Z. Rudnick and P. Sarnak, *On Selberg's eigenvalue conjecture.* Geom. Funct. Anal. **5**(1995), no. 2, 387–401.

[479] G. Lusztig, *Novikov's higher signature and families of elliptic operators.* J. Differential Geometry **7**(1972), 229–256.

[480] E. Lutwak, *Selected affine isoperimetric inequalities. Handbook of convex geometry*, Vol. A, B, 151–176, North-Holland, Amsterdam, 1993.

[481] E. Lutwak, D. Yang and G. Y. Zhang, L_p *affine isoperimetric inequalities.* J. Differential Geom. **56**(2000), no. 1, 111–132.

[482] M, Lübke, *Stability of Einstein-Hermitian vector bundles.* Manuscripta Math. **42**(1983), no. 2-3, 245–257.

[483] T. Lyons and D. Sullivan, *Function theory, random paths and covering spaces*, J. Differential Geom. **19**(1984), no. 2, 299–323.

[484] X. N. Ma and G. Marinescu, *Genegalized Bergman kernels on symplectic manifolds.* C. R. Math. Acad. Asi. Paris **339**(2004), no. 7, 493–498. The full version: math.DG/0411559.

[485] T. Mabuchi, *K-energy maps integrating Futaki invariants.* Tohoku Math. J. (2) **38**(1986), no. 4, 575–593.

[486] G. A. Margulis, *Discrete groups of motions of manifolds of nonpositive curvature.* Transl., I. Ser., Am. Math. Soc. **109**(1977), 33-45.

[487] G. A. Margulis, *Explicit constructions of graphs without short cycles and low density codes.* Combinatorica **2**(1982), no. 1, 71–78.

[488] J. Marsden and A. Weinstein, *Reduction of symplectic manifolds with symmetry.* Rep. Mathematical Phys. **5**(1974), no. 1, 121–130.

[489] T. Matsusaka, *Polarized varieties with a given Hilbert polynomial.* Amer. J. Math. **94**(1972), 1027–1077.

[490] Y. Matsushima, *Sur la structure du group d'homeomorphismes analytiques d'une certaine varietie Kaehleriennes.* Nagoya Math. J., **11**(1957), 145–150.

[491] Y. Matsushima, *On the first Betti number of compact quotient spaces of higher-dimensional symmetric spaces.* Ann. of Math. (2) **75**(1962), 312–330.

[492] D. McDuff, *Examples of symplectic structures.* Invent. Math. **89**(1987), 13-36.
[493] D. McDuff, *The structure of rational and ruled symplectic 4-manifolds.* J. Amer. Math. Soc. **3**(1990), no. 3, 679–712.
[494] D. McDuff. *Singularities and positivity of intersections of J-holomorphic curves.* Progr. Math., **117**, Holomorphic curves in symplectic geometry, 191–215, Birkhäuser, Basel, 1994.
[495] H. P. McKean, Jr and I. M. Singer, *Curvature and the eigenvalues of the Laplacian.* J. Differential Geometry **1**(1967), no. 1, 43–69.
[496] C. McMullen, *Iteration on Teichmüller space.* Invent. Math. **99**(1990), no. 2, 425–454.
[497] C. McMullen, *Riemann surfaces and the geometrization of 3-manifolds.* Bull. Amer. Math. Soc. (N.S.) **27**(1992), no. 2, 207–216.
[498] C. T. McMullen, *The moduli space of Riemann surfaces is Kähler hyperbolic.* Ann. of Math. (2) **151**(2000), no. 1, 327–357.
[499] W. H. Meeks, III and H. Rosenberg, *The uniqueness of the helicoid.* Ann. of Math. (2) **161**(2005), no. 2, 727–758.
[500] W. H. Meeks, L. Simon and S.-T. Yau, *Embedded minimal surfaces, exotic spheres, and manifolds with positive Ricci curvature.* Ann. of Math. (2) **116**(1982), no. 3, 621–659.
[501] W. H. Meeks and S.-T. Yau, *Topology of three-dimensional manifolds and the embedding problems in minimal surface theory.* Ann. of Math. (2) **112**(1980), no. 3, 441–484.
[502] W. H. Meeks and S. T. Yau, *The equivariant Dehn's lemma and loop theorem.* Comment. Math. Helv. **56**(1981), no. 2, 225–239.
[503] W. H. Meeks and S.-T. Yau, *The existence of embedded minimal surfaces and the problem of uniqueness.* Math. Z. **179**(1982), no. 2, 151–168.
[504] W. H. Meeks and S. T. Yau, *The classical Plateau problem and the topology of three-dimensional manifolds. The embedding of the solution given by Douglas-Morrey and an analytic proof of Dehn's lemma.* Topology **21**(1982), no. 4, 409–442.
[505] V. Mehta and A. Ramanathan, *Restriction of stable sheaves and representations of the fundamental group.* Invent. Math. **77**(1984), no. 1, 163–172.
[506] A. d. Melas, *On the nodal line of the second eigenfunction of the Laplacian in R^2.* J. Differential Geom. **35**(1992), no. 1, 255–263.
[507] R. Melrose, *The inverse spectral problem for planar domains.* Instructional Workshop on Analysis and Geometry, Part I (Canberra, 1995), 137–160, Proc. Centre Math. Appl. Austral. Nat. Univ., **34**, Austral. Nat. Univ., Canberra, 1996.
[508] M. J. Micallef, *Stable minimal surfaces in Euclidean space.* J. Differential Geom. **19**(1984), no. 1, 57–84.
[509] M. J. Micallef and J. D. Moore, *Minimal two-spheres and the topology of manifolds with positive curvature on totally isotropic two-planes.* Ann. of Math. (2) **127**(1988), no. 1, 199–227.
[510] M. J. Micallef and M. Y. Wang, *Metrics with nonnegative isotropic curvature.* Duke Math. J. **72**(1993), no. 3, 649–672.
[511] J. H. Michael and L. M. Simon, *Sobolev and mean-value inequalities on generalized submanifolds of R^n.* Comm. Pure Appl. Math. **26**(1973), 361–379.
[512] M. L. Michelsohn, *Kähler manifolds with vanishing first Chern class.* Seminar on Differential Geometry, pp. 359–361, Ann. of Math. Stud., **102**, Princeton Univ. Press, Princeton, N.J., 1982.
[513] M. L. Michelsohn, *On the existence of special metrics in complex geometry.* Acta Math. **149**(1982), no. 3-4, 261–295.
[514] J. Milnor, *Morse theory.* Based on lecture notes by M. Spivak and R. Wells. Annals of Mathematics Studies, No. **51**, Princeton University Press, Princeton, N.J. 1963.
[515] J. Milnor, *Eigenvalues of the Laplace operator on certain manifolds*, Proc. Nat. Acad. Sci. USA **51**(1964), 542.
[516] Y. Miyaoka, *On the Chern numbers of surfaces of general type.* Invent. Math. **42**(1977), 225–237.
[517] N. Mok, *Courbure bisectionnelle positive et variétés algébriques affines.* C. R. Acad. Sci. Paris Sér. I Math. **296**(1983), no. 11, 473–476.
[518] N. Mok, *An embedding theorem of complete Kähler manifolds of positive bisectional curvature onto affine algebraic varieties.* Bull. Soc. Math. France **112**(1984), 197–250.
[519] N. Mok, *The uniformization theorem for compact Kähler manifolds of nonnegative holomorphic bisectional curvature.* J. Differential Geom. **27**(1988), no. 2, 179–214.

[520] N. Mok, Y. T. Siu and S.-T. Yau, *The Poincar-Lelong equation on complete Kähler manifolds.* Compositio Math. **44**(1981), no. 1-3, 183–218.
[521] N. Mok, Y. T. Siu and S. K. Yeung, *Geometric superrigidity.* Invent. Math. **113**(1993), no. 1, 57–83.
[522] N. Mok and S.-T. Yau, *Completeness of the Kähler-Einstein metric on bounded domains and the characterization of domains of holomorphy by curvature conditions.* Proc. Sympos. Pure Math., **39**, pp.41-59, Amer. Math. Soc., Providence, RI, 1983.
[523] N. Mok and J. Q. Zhong, *Curvature characterization of compact Hermitian symmetric spaces.* J. Differential Geom. **23**(1986), no. 1, 15–67.
[524] N. Mok and J. Q. Zhong, *Compactifying complete Kähler-Einstein manifolds of finite topological type and bounded curvature.* Ann. of Math. (2) **129**(1989), no. 3, 427–470.
[525] J. Morgan, *On Thurston's uniformization theorem for three-dimensional manifolds.* The smith conjecture (New York, 1979), 37–115, Pure Appl. Math., **112**, Academic press, Orlando, FL, 1984.
[526] S. Mori, *Projective manifolds with ample tangent bundles.* Ann. of Math. (2) **110**(1979), no. 3, 593–606.
[527] G. D. Mostow, *Strong rigidity of locally symmetric spaces. Annals of Mathematics Studies,* No. **78**. Princeton University Press, Princeton, N.J.; University of Tokyo Press, Tokyo, 1973.
[528] M. Mulase, *KP equations, strings, and the Schottky problem.* Algebraic analysis, Vol. II, 473–492, Academic Press, Boston, MA, 1988.
[529] W. Müller, *Analytic torsion and R-torsion of Riemannian manifolds.* Adv. in Math. **28**(1978), no. 3, 233–305.
[530] D. Mumford, J. Fogarty and F. Kirwan, *Geometric invariant theory.* Third edition. Ergebnisse der Mathematik und ihrer Grenzgebiete (2) [Results in Mathematics and Related Areas (2)], **34**. Springer-Verlag, Berlin, 1994.
[531] A. M. Nadel, *Multiplier ideal sheaves and Kähler-Einstein metrics of positive scalar curvature.* Ann. of Math. (2) **132**(1990), no. 3, 549–596.
[532] J. Nash, *C^1 isometric imbeddings.* Ann. of Math. (2) **60**(1954). 383–396.
[533] J. Nash, *The imbedding problem for Riemannian manifolds.* Ann. of Math. (2) **63**(1956), 20–63.
[534] R. Neel and D. Stroock, *Analysis of the cutlocus via the heat kernel.* Surveys in Differential Geomtry **IX**, International Press, 337-350.
[535] L. Ni, *Ancient solution to Kahler-Ricci flow.* arXiv:math.DG/0502494.
[536] L. Nirenberg, *The Weyl and Minkowski problems in differential geometry in the large.* Commun. Pure Appl. Math. **6**(1953), 337–394.
[537] S. P. Novikov, *Multivalued functions and functionals. An analogue of the Morse theory.* Dokl. Akad. Nauk SSSR **260**(1981), no. 1, 31–35.
[538] K. O'Grady, *The irreducible components of moduli spaces of vector bundles on surfaces.* Invent. Math. **112**(1993), no. 3, 585–613.
[539] Y.-G Oh, *Floer cohomology of Lagrangian intersections discs and pseudo-holomorphic curves with Lagrangian boundary conditions, I, II.* Comm. Pure Appl. Math. **46**(1992), (1993), 949-1012.
[540] Y.-G. Oh, *Floer cohomology of Lagrangian intersections and pseudo-holomorphic disks. III.* Arnol'd-Givental conjecture. The Floer memorial volume, 555–573, Progr. Math., **133**, Birkhäuser, Basel, 1995.
[541] S. Osher, *A level set formulation for the solution of the Dirichlet problem for Hamilton-Jacobi equations.* SIAM J. Math. Anal. **24**(1993), no. 5, 1145–1152.
[542] R. Osserman, *A proof of the regularity everywhere of the classical solution to Plateau's problem.* Ann. of Math. (2) **91**(1970), 550–569.
[543] J. Otal, *Thurston's hyperbolization of Haken manifolds.* Surveys in differential geometry, Vol. III(Cambridge, MA, 1996), 77–194, Int. Press, Boston, MA, 1998
[544] P. Ozsváth and Z. Szabó, *The symplectic Thom conjecture.* Ann. of Math. (2) **151**(2000), no. 1, 93–124.
[545] T. H. Parker and J. G. Wolfson, *Pseudo-holomorphic maps and bubble trees.* (English. English summary) J. Geom. Anal. **3**(1993), no. 1, 63–98.
[546] C. K. Peng and C. L. Terng, *Minimal hypersurfaces of spheres with constant scalar curvature,* Seminar on minimal submanifolds, Ann. of Math. Stud., vol. **103**, Princeton Univ. Press, Princeton, NJ, 1983, pp. 177C198.

[547] G. Perelman, *The entropy formula for the Ricci flow and its geometric applications.* arXiv:math.DG/0211159.

[548] G. Perelman, *Ricci flow with surgery on three-manifolds*, arXiv:math.DG/0303109.

[549] G. Perelman, *Finite extinction time for the solutions to the Ricci flow on certain three-manifolds*, arXiv:math.DG/0307245.

[550] L. Pestov and G. Uhlmann, *The boundary distance function and the Dirichlet-to-Neumann map.* Math. Res. Lett. **11**(2004), no. 2-3, 285–297.

[551] D. H. Phong and J. Sturm, *On stability and the convergence of the Kähler-Ricci flow.* arXiv:math.DG/0412185

[552] D. H. Phong and J. Sturm, *The Monge-Ampère operator and geodesics in the space of Kähler potentials.* arXiv:math.DG/0504157.

[553] J. T. Pitts, *Existence and regularity of minimal surfaces on Riemannian manifolds.* Mathematical Notes, **27**. Princeton University Press, Princeton, N.J.; University of Tokyo Press, Tokyo, 1981.

[554] J. T. Pitts and J. H. Rubinstein, *Equivariant minimax and minimal surfaces in geometric three-manifolds.* Bull. Amer. Math. Soc. (N.S.) **19**(1988), no. 1, 303–309.

[555] I. I. Pjateckiĭ-Šapiro and I. R. Šafarevič, *Torelli's theorem for algebraic surfaces of type* K3. (Russian) Izv. Akad. Nauk SSSR Ser. Mat. **35**(1971) 530–572; (English Translation) Math. USSR Izvestija **5**(1971) 547–588.

[556] A. V. Pogorelov, *On the question of the existence of a convex surface with a given sum principal radii of curvature.* Usp. Mat. Nauk **8**(1953), 127–130.

[557] A. V. Pogorelov, *On the isometric immersion of a two-dimensional Riemannian manifold homeomorphic to a sphere into a three-dimensional Riemannian space.* (Russian) Dokl. Akad. Nauk SSSR **139**(1961), 818–820.

[558] A. V. Pogorelov, *Isometric immersion in the large of a two-dimensional Riemannian manifold into a three-dimensional one.* (Russian) Dokl. Akad. Nauk SSSR **137**(1961), 282–283.

[559] A. V. Pogorelov, *On the improper convex affine hyperspheres.* Geometriae Dedicata **1**(1972), no. 1, 33–46.

[560] A. V. Pogorelov, *Existence of a closed convex hypersurface with prescribed curvature.* Dokl. Akad. Nauk SSSR **274**(1984), no. 1, 28–31.

[561] G. Pólya and G. Szegö, *Isoperimetric Inequalities in Mathematical Physics.* Annals of Mathematics Studies, no. **27**, Princeton University Press, Princeton, N. J., 1951.

[562] J. Porti, *Geometric structures on three-dimensional manifolds.* (Catalan) First Conference on Mathematics(Bellaterra, 1998). Butl. Soc. Catalana Mat. **14**(1999), no. 1, 99–114.

[563] D. R. Ray and I. M. Singer, *Analytic torsion for complex manifolds.* Ann. of Math. (2) **98**(1973), 154–177.

[564] M. Reid, *The moduli space of* 3*-folds with* $K = 0$ *may nevertheless be irreducible.* Math. Ann. **278**(1987), no. 1-4, 329–334.

[565] S.-S. Roan, S.-T. Yau, *On Ricci flat* 3*-fold.* Acta Math. Sinica (N.S.) **3**(1987), no. 3, 256–288.

[566] D. C. Robinson, *Uniqueness of the Kerr Black hole.* Phys. Rev. Lett., **34**(1975), 905-906.

[567] J. Rosenberg, C^**-algebras, positive scalar curvature and the Novikov conjecture.* Publ. I.H.E.S. No. **58**(1983), 197–212.

[568] J. Ross and R. P. Thomas, *An obstruction to the existence of constant scalar curvature Kähler metricsmath.* arXiv:math.DG/0412518.

[569] J. Ross and R. P. Thomas, *A study of the Hilbert-Mumford criterion for the stability of projective varieties.* arXiv:math.AG/0412519.

[570] H. L. Royden, *Report on the Teichmüler metric.* Proc. Nat. Acad. Sci. U.S.A. **65**(1970) 497–499.

[571] W. D. Ruan, *Canonical coordinates and Bergmann metrics.* Comm. Anal. Geom. **6**(1998), no. 3, 589–631.

[572] W. D. Ruan, *Generalized special Lagrangian torus fibration for Calabi-Yau hypersurfaces in toric varieties I,II,III.* Arxiv: math.DG/0303114, math.DG/0303278, math.DG/0309450.

[573] W. D. Ruan, *The Fukaya category of symplectic neighborhood of a non-Hausdorff manifold.* math.SG/0602119

[574] Y. B. Ruan, *Symplectic topology on algebraic 3–folds.* JDG. **39**(1994), 215-227.

[575] Y. B. Ruan, *Topological sigma model and Donaldson-type invariants in Gromov theory.* Duke Math. J. **83**(1996), no. 2, 461–500.

[576] Y. B. Ruan and G. Tian, *A mathematical theory of quantum cohomology.* J. Differential Geom. **42**(1995), no. 2, 259–367.
[577] J. Sacks and K. Uhlenbeck, *The existence of minimal immersions of 2-spheres.* Ann. of Math. (2) **113**(1981), no. 1, 1–24.
[578] L. Saloff-Coste, *A note on Poincaré, Sobolev, and Harnack inequalities.* Internat. Math. Res. Notices 1992, no. 2, 27–38.
[579] L. Saloff-Coste, *Analysis on Riemannian co-compact covers.* Surveys in Differential Geometry, **IX**, 351-385. International Press, Somerville, MA, 02143.
[580] L. Saper, *On the cohomology of locally symmetric spaces and of their compactifications.* Current developments in mathematics, 2002, 219–289, Int. Press, Somerville, MA, 2003.
[581] L. Saper and M. Stern, *L_2-cohomology of arithmetic varieties.* Ann. of Math. (2) **132**(1990), no. 1, 1–69.
[582] P. Sarnak, *Additive number theory and Maass forms.* Lecture Notes in Math., **1052**, 286–309, Springer, Berlin, 1984.
[583] P. Sarnak, *Spectra of hyperbolic surfaces.* Bull. Amer. Math. Soc. (N.S.) **40**(2003), no. 4, 441–478.
[584] W. Schmid, *Variation of Hodge structure: the singularities of the period mapping.* Invent. Math. **22**(1973), 211–319.
[585] M. Schmidt, *A proof of the Willmore conjecture.* math.DG/0203224
[586] R. Schoen, *Conformal deformation of a Riemannian metric to constant scalar curvature.* J. Differential Geom. **20** (1984), no. 2, 479–495.
[587] R. Schoen, L. Simon and S.-T. Yau, *Curvature estimates for minimal hypersurfaces.* Acta Math. **134** (1975), no. 3-4, 275–288.
[588] R. Schoen and K. Uhlenbeck, *A regularity theory for harmonic maps.* J. Differential Geom. **17**(1982), no. 2, 307–335.
[589] R. Schoen and K. Uhlenbeck, *Boundary regularity and the Dirichlet problem for harmonic maps.* J. Differential Geom. **18**(1983), no. 2, 253–268.
[590] R. Schoen and J. Wolfson, *Minimizing area among Lagrangian surfaces: the mapping problem.* J. Differential Geom. **58**(2001), no. 1, 1–86.
[591] R. Schoen, S. Wolpert and S.-T. Yau, *Geometric bounds on the low eigenvalues of a compact surface.* Geometry of the Laplace operator (Proc. Sympos. Pure Math., Univ. Hawaii, Honolulu, Hawaii, 1979), pp. 279–285, Proc. Sympos. Pure Math., XXXVI, Amer. Math. Soc., Providence, R.I., 1980.
[592] R. Schoen and S.-T. Yau, *On univalent harmonic maps between surfaces.* Invent. Math. **44**(1978), no. 3, 265–278.
[593] R. Schoen and S. T. Yau, *Existence of incompressible minimal surfaces and the topology of three-dimensional manifolds with nonnegative scalar curvature.* Ann. of Math. (2) **110**(1979), no. 1, 127–142.
[594] R. Schoen and S.-T. Yau, *On the proof of the positive mass conjecture in general relativity.* Comm. Math. Phys. **65**(1979), no. 1, 45–76.
[595] R. Schoen and S.-T. Yau, *Positivity of the total mass of a general space-time.* Phys. Rev. Lett. **43**(1979), no. 20, 1457–1459.
[596] R. Schoen and S.-T. Yau, *On the structure of manifolds with positive scalar curvature.* Manuscripta Math. **28**(1979), no. 1-3, 159–183.
[597] R. Schoen and S.-T. Yau, *Compact group actions and the topology of manifolds with non-positive curvature.* Topology **18**(1979), no. 4, 361–380.
[598] R. Schoen and S.-T. Yau, *Proof of the positive mass theorem. II.* Comm. Math. Phys. **79**(1981), no. 2, 231–260.
[599] R. Schoen and S.-T. Yau, *Complete three-dimensional manifolds with positive Ricci curvature and scalar curvature.* Seminar on Differential Geometry, pp. 209–228, Ann. of Math. Stud., **102**, Princeton Univ. Press, Princeton, N.J., 1982.
[600] R. Schoen and S.-T. Yau, *The existence of a black hole due to condensation of matter.* Comm. Math. Phys. **90**(1983), no. 4, 575–579.
[601] R. Schoen and S.-T. Yau, *Conformally flat manifolds, Kleinian groups and scalar curvature.* Invent. Math. **92**(1988), no. 1, 47–71.
[602] R. Schoen and S.-T. Yau, *Lectures on differential geometry.* Lecture notes prepared by Wei Yue Ding, Kung Ching Chang [Gong Qing Zhang], Jia Qing Zhong and Yi Chao Xu. Translated from the Chinese by Ding and S. Y. Cheng. Preface translated from the Chinese

by Kaising Tso. Conference Proceedings and Lecture Notes in Geometry and Topology, **I**. International Press, Cambridge, MA, 1994.

[603] N. Seiberg and E. Witten, *Electric-magnetic duality, monopole condensation, and confinement in $N = 2$ supersymmetric Yang-Mills theory.* Nuclear Phys. B **426**(1994), no. 1, 19–52.

[604] N. Seiberg and E. Witten, *Monopoles, duality and chiral symmetry breaking in $N = 2$ supersymmetric QCD.* Nuclear Phys. B **431**(1994), no. 3, 484–550.

[605] P. Seidel and R. Thomas, *Braid group actions on derived categories of coherent sheaves.* Duke Math. J. **108**(2001), no. 1, 37–108.

[606] J. Shatah and M. Struwe, *Regularity results for nonlinear wave equations.* Ann. of Math. (2) **138**(1993), no. 3, 503–518.

[607] W. X. Shi, *Complete noncompact Kähler manifolds with positive holomorphic bisectional curvature.* Bull. Amer. Math. Soc. (N.S.) **23**(1990), no. 2, 437–440.

[608] W.-X. Shi, *Ricci flow and the uniformization on complete noncompact Kähler manifolds.* J. Differential Geom. **45**(1997), no. 1, 94–220.

[609] W. X. Shi, *A uniformization theorem for complete Kähler manifolds with positive holomorphic bisectional curvature.* J. Geom. Anal. **8**(1998), no. 1, 117–142.

[610] Y. G. Shi and L. F. Tam, *Positive mass theorem and the boundary behaviors of compact manifolds with nonnegative scalar curvature.* J. Differential Geom. **62**(2002), no. 1, 79–125.

[611] B. Shiffman, *Complete characterization of holomorphic chains of codimension one.* Math. Ann. **274**(1986), no. 2, 233–256.

[612] T. Shioya and T. Yamaguchi, *Collapsing three-manifolds under a lower curvature bound.* J. Differential Geom. **56**(2000), no. 1, 1–66.

[613] T. Shioya and T. Yamaguchi, *Volume collapsed three-manifolds with a lower curvature bound.* math.DG/0304472

[614] L. Simon, *Theorems on the regularity and singularity of minimal surfaces and harmonic maps.* Geometry and global analysis (Sendai, 1993), 111–145,

[615] L. Simon, *Existence of Willmore surfaces.* Miniconference on geometry and partial differential equations (Canberra, 1985), 187–216, Proc. Centre Math. Anal. Austral. Nat. Univ., **10**, Austral. Nat. Univ., Canberra, 1986.

[616] L. Simon, *Theorems on regularity and singularity of energy minimizing maps.* Lectures in Mathematics ETH Zürich. Birkhäuser Verlag, Basel, 1996.

[617] C. T. Simpson, *Constructing variations of Hodge structure using Yang-Mills theory and applications to uniformization.* J. Amer. Math. Soc. **1**(1988), no. 4, 867–918.

[618] C. T. Simpson, *Moduli of representations of the fundamental group of a smooth projective variet.* I. Inst. Hautes Études Sci. Publ. Math. No. **79**(1994), 47–129.

[619] C. T. Simpson, *Moduli of representations of the fundamental group of a smooth projective variety.* II. Inst. Hautes tudes Sci. Publ. Math. No. **80**(1994), 5–79 (1995).

[620] I. M. Singer, B. Wong, S.-T. Yau and S.-T. S. Yau, *An estimate of the gap of the first two eigenvalues in the Schrödinger operator.* Ann. Scuola Norm. Sup. Pisa Cl. Sci. (4) **12**(1985), no. 2, 319–333.

[621] Y.-T. Siu, *Complex-analyticity of harmonic maps and strong rigidity of compact Kähler manifolds.* Ann. of Math. **112**(1980), 73-111.

[622] Y. T. Siu, *A simple proof of the surjectivity of the period map of $K3$ surfaces.* Manuscripta Math. **35**(1981), no. 3, 311–321.

[623] Y. T. Siu, *Every $K3$ surface is Kähler.* Invent. Math. **73**(1983), no. 1, 139–150.

[624] Y. T. Siu, *Effective very ampleness.* Invent. Math. **124**(1996), no. 1-3, 563–571.

[625] Y. T. Siu, *Invariance of plurigenera.* Invent. Math. **134**(1998), no. 3, 661–673.

[626] Y. T. Siu, *Very ampleness part of Fujita's conjecture and multiplier ideal sheaves of Kohn and Nadel.* Complex analysis and geometry (Columbus, OH, 1999), 171–191, Ohio State Univ. Math. Res. Inst. Publ., **9**, de Gruyter, Berlin, 2001.

[627] Y. T. Siu, *Some recent transcendental techniques in algebraic and complex geometry.* Proceddings of the International Congress of Mathematicians, Vol. **I**(Beijing, 2002), 439–448.

[628] Y. T. Siu and S.-T. Yau, *Complete Kähler manifolds with nonpositive curvature of faster than quadratic decay.* Ann. of Math. (2) **105**(1977), no. 2, 225–264.

[629] Y. T. Siu and S.-T. Yau, *Compact Kähler manifolds of positive bisectional curvature.* Invent. Math. **59**(1980), no. 2, 189–204.

[630] Y. T. Siu and S.-T. Yau, *Compactification of negatively curved complete Kähler manifolds of finite volume.* Seminar on Differential Geometry, pp. 363–380, Ann. of Math. Stud., **102**, Princeton Univ. Press, Princeton, N.J., 1982.

[631] I. Smith, R. P. Thomas and S.-T. Yau, *Symplectic conifold transitions.* J. Differential Geom. **62**(2002), no. 2, 209–242.

[632] K. Smoczyk, *Longtime existence of the Lagrangian mean curvature flow.* Calc. Var. Partial Differential Equations **20**(2004), no. 1, 25–46.

[633] K. Smoczyk and M. T. Wang, *Mean curvature flows of Lagrangians submanifolds with convex potentials.* J. Differential Geom. **62**(2002), no. 2, 243–257.

[634] J. A. Smoller, A. G. Wasserman and S.-T. Yau, *Existence of black hole solutions for the Einstein-Yang/Mills equations.* Comm. Math. Phys. **154**(1993), no. 2, 377–401.

[635] J. A. Smoller, A. G. Wasserman, S.-T. Yau, J. B. McLeod, *Smooth static solutions of the Einstein/Yang-Mills equations.* Comm. Math. Phys. **143**(1991), no. 1, 115–147.

[636] M. Stern, *L^2-index theorems on locally symmetric spaces.* Invent. Math. **96**(1989), no. 2, 231–282.

[637] S. Stolz, *Simply connected manifolds of positive scalar curvature.* Ann. of Math. (2) **136**(1992), no. 3, 511–540.

[638] A. Strominger, *Superstrings with torsion.* Nuclear Phys. B **274**(1986), no. 2, 253–284.

[639] A. Strominger, S.-T. Yau and E. Zaslow, *Mirror symmetry is T-duality.* Nuclear Phys. B **479**(1996), no. 1-2, 243–259.

[640] D. Sullivan, *A generalization of Milnor's inequality concerning affine foliations and affine manifolds.* Comment. Math. Helv. **51**(1976), no. 2, 183–189.

[641] D. Sullivan, *Cycles for the dynamical study of foliated manifolds and complex manifolds.* Invent. Math. **36**(1976), 225–255.

[642] D. Sullivan, *Infinitesimal computations in topology.* Inst. Hautes études Sci. Publ. Math. No. **47**(1977), 269–331

[643] D. Sullivan, *The Dirichlet problem at infinity for a negatively curved manifold.* J. Differential Geom. **18**(1983), no. 4, 723–732.

[644] D. Sullivan, *Closed string operators and Feynman graphs.* To appear.

[645] T. Sunada, *Riemannian coverings and isospectral manifolds.* Ann. of Math. (2) **121**(1985), no. 1, 169–186.

[646] Z. Szabó, *Locally non-isometric yet super isospectral spaces.* Geom. Funct. Anal. **9**(1999), no. 1, 185–214.

[647] G. Szegö, G. *Inequalities for certain eigenvalues of a membrane of given area.* J. Rational Mech. Anal. **3**(1954), 343–356.

[648] G. Talenti, *Best constant in Sobolev inequality.* Ann. Mat. Pura Appl. (4) **110**(1976), 353–372.

[649] T. Tao, *Low-regularity global solutions to nonlinear dispersive equations.* Surveys in analysis and operator theory (Canberra, 2001), 19–48, Proc. Centre Math. Appl. Austral. Nat. Univ., **40**, Austral. Nat. Univ., Canberra, 2002.

[650] D. Tataru, *On global existence and scattering for the wave maps equation.* Amer. J. Math. **123**(2001), no. 1, 37–77.

[651] C. H. Taubes, *Self-dual Yang-Mills connections on non-self-dual* 4*-manifolds.* J. Differential Geom. **17**(1982), no. 1, 139–170.

[652] C. H. Taubes, *Self-dual connections on* 4*-manifolds with indefinite intersection matrix.* J. Differential Geom. **19**(1984), no. 2, 517–560.

[653] C. H. Taubes, *S^1 actions and elliptic genera.* Comm. Math. Phys. **122**(1989), no. 3, 455–526.

[654] C. H. Taubes, *Casson's invariant and gauge theory.* J. Differential Geom. **31**(1990), no. 2, 547–599.

[655] C. H. Taubes, *The existence of anti-self-dual conformal structures.* J. Differential Geom. **36**(1992), no. 1, 163–253.

[656] C. H. Taubes, *The Seiberg-Witten invariants and symplectic forms.* Math. Res. Lett. **1**(1994), no. 6, 809–822.

[657] H. C. Taubes, SW $\Rightarrow$ Gr*: from the Seiberg-Witten equations to pseudo-holomorphic curves.* J. Amer. Math. Soc. **9**(1996), no. 3, 845–918.

[658] C. H. Taubes, *Counting pseudo-holomorphic submanifolds in dimension* 4. J. Differential Geom. **44**(1996), no. 4, 818–893.

[659] C. H. Taubes, Gr $\Longrightarrow$ SW: *from pseudo-holomorphic curves to Seiberg-Witten solutions.* J. Differential Geom. **51**(1999), no. 2, 203–334.

[660] C. H. Taubes, GR = SW: *counting curves and connections.* J. Differential Geom. **52**(1999), no. 3, 453–609.

[661] C. H. Taubes, *Seiberg Witten and Gromov invariants for symplectic 4-manifolds.* Edited by Richard Wentworth. First International Press Lecture Series, **2**. International Press, Somerville, MA, 2000.

[662] C. L. Terng, *A higher dimension generalization of the sine-Gordon equation and its soliton theory.* Ann. of Math. (2)**111**(1980), no. 3, 491–510.

[663] C. L. Terng, *Recent progress in submanifold geometry. Differential geometry: partial differential equations on manifolds,* 439–484, Proc. Sympos. Pure Math., **54**, Part 1, Amer. Math. Soc., Providence, RI, 1993.

[664] C. L. Terng and K. Uhlenbeck, *Poisson actions and scattering theory for integrable systems.* Surv. Differ. Geom., **V**, 315–402. Int. Press, Boston, MA, 1998.

[665] C. L. Terng and K. Uhlenbeck, *Bäcklund transformations and loop group actions.* Comm. Pure Appl. Math. **53**(2000), no. 1, 1–75.

[666] R. P. Thomas, *Moment maps, monodromy and mirror manifolds.* Symplectic geometry and mirror symmetry (Seoul, 2000), 467–498, World Sci. Publishing, River Edge, NJ, 2001.

[667] R.P. Thomas and S.-T. Yau, *Special Lagrangians, stable bundles and mean curvature flow.* (English. English summary) Comm. Anal. Geom. **10**(2002), no. 5, 1075–1113.

[668] G. Thorbergsson, *A survey on isoparametric hypersurfaces and their generalizations.* Handbook of differential geometry, Vol. **I**, 963–995, North-Holland, Amsterdam, 2000.

[669] W. P. Thurston, *Three-dimensional manifolds, Kleinian groups and hyperbolic geometry.* Bull. Amer. Math. Soc. (N.S.) **6**(1982), no. 3, 357–381.

[670] W. Thurston, *Three-dimensional Geometry and Topology,* Vol. **1** (Silvio Levy, ed.) Princeton Math. Ser., vol. 35, Princeton Univ. Press, 1997.

[671] G. Tian, *Smoothness of the universal deformation space of compact Calabi-Yau manifolds and its Petersson-Weil metric.* Mathematical aspects of string theory (San Diego, Calif., 1986), 629–646, Adv. Ser. Math. Phys., **1**, World Sci. Publishing, Singapore, 1987.

[672] G. Tian, *On a set of polarized Kähler metrics on algebraic manifolds.* J. Differential Geom. **32**(1990), no. 1, 99–130.

[673] G. Tian, *Calabi's conjecture for complex surfaces with positive first Chern class.* Invent. Math. **101**(1990), no. 1, 101–172.

[674] G. Tian, *Kähler-Einstein metrics with positive scalar curvature.* Invent. Math. **130**(1997), no. 1, 1–37.

[675] G. Tian, *Gauge theory and calibrated geometry.* I. Ann. of Math. (2) **151**(2000), no. 1, 193–268.

[676] G. Tian and S.-T. Yau, *Three-dimensional algebraic manifolds with* $C_1 = 0$ *and* $\chi = -6$. Mathematical aspects of string theory (San Diego, Calif., 1986), 543–559, Adv. Ser. Math. Phys., **1**, World Sci. Publishing, Singapore, 1987.

[677] G. Tian and S.-T. Yau, *Complete Kähler manifolds with zero Ricci curvature. I.* J. Amer. Math. Soc. **3**(1990), no. 3, 579–609.

[678] G. Tian and S.-T. Yau, *Complete Kähler manifolds with zero Ricci curvature. II.* Invent. Math. **106**(1991), no. 1, 27–60.

[679] A. N. Todorov, *Applications of the Kähler-Einstein-Calabi-Yau metric to moduli of K3 surfaces.* Invent. Math. **61**(1980), no. 3, 251–265.

[680] A. N. Todorov, *The Weil-Petersson geometry of the moduli space of* $\mathrm{SU}(n \geq 3)$ *(Calabi-Yau) manifolds. I.* Comm. Math. Phys. **126**(1989), no. 2, 325–346.

[681] A. E. Treibergs, *Entire spacelike hypersurfaces of constant mean curvature in Minkowski space.* Invent. Math. **66**(1982), no. 1, 39–56.

[682] N. S. Trudinger, *Remarks concerning the conformal deformation of Riemannian structures on compact manifolds.* Ann. Scuola Norm. Sup. Pisa (3) **22**(1968), 265–274.

[683] N. S. Trudinger and X. J. Wang, *The Bernstein problem for affine maximal hypersurfaces.* Invent. Math. **140**(2000), no. 2, 399–422.

[684] K. K. Uhlenbeck, *Removable singularities in Yang-Mills fields.* Comm. Math. Phys. **83**(1982), no. 1, 11–29.

[685] K. K. Uhlenbeck, *Connections with* L^p *bounds on curvature.* Comm. Math. Phys. **83**(1982), no. 1, 31–42.

[686] K. Uhlenbeck, *Harmonic maps into Lie groups: classical solutions of the chiral model.* J. Differential Geom. **30**(1989), no. 1, 1–50.
[687] K. Uhlenbeck, and S.-T. Yau, *On the existence of Hermitian-Yang-Mills connections in stable vector bundles.* Comm. Pure Appl. Math. **39**(1986), no. S, suppl., S257–S293.
[688] C. Vafa, *Topological mirrors and quantum rings.* Essays on mirror manifolds, 96–119, Internat. Press, Hong Kong, 1992.
[689] N. Th. Varopoulos, L. Saloff-Coste and T. Coulhon, *Analysis and geometry on groups.* Cambridge Tracts in Mathematics, **100**. Cambridge University Press, Cambridge, 1992.
[690] E. Viehweg, *Quasi-projective moduli for polarized manifolds.* Ergebnisse der Mathematik und ihrer Grenzgebiete (3) [Results in Mathematics and Related Areas (3)], **30**. Springer-Verlag, Berlin, 1995.
[691] M. Vigué-Poirrier and D. Sullivan, *The homology theory of the closed geodesic problem.* J. Differential Geometry **11**(1976), no. 4, 633–644.
[692] C. Voisin, *A counterexample to the Hodge conjecture for Kähler varieties* Int. Math. Res. Not. 2002, no. 20, 1057–1075.
[693] C. Voisin, *On the homotopy types of compact Kähler and complex projective manifolds.* Invent. Math. **157**(2004), no. 2, 329–343.
[694] C. Voisin, *On the homotopy types of Kaehler manifolds and the birational Kodaira problem.* Arxiv: math.AG/0410040.
[695] T. Wan, *Constant mean curvature surface, harmonic maps, and universal Teichmüller space.* J. Differential Geom. **35**(1992), no. 3, 643–657.
[696] C. P. Wang, *Some examples of complete hyperbolic affine 2-spheres in R^3.* Global differential geometry and global analysis (Berlin, 1990), 271–280, Lecture Notes in Math., **1481**, Springer, Berlin, 1991.
[697] C. L. Wang, *On the incompleteness of the Weil-Petersson metric along degenerations of Calabi-Yau manifolds.* Math. Res. Lett. **4**(1997), no. 1, 157–171.
[698] H. C. Wang, *Complex parallisable manifolds* Proc. Amer. Math. Soc. **5**(1954), 771–776.
[699] M. T. Wang, *A fixed point theorem of discrete group actions on Riemannian manifolds.* J. Differential Geom. **50**(1998), no. 2, 249–267.
[700] M. T. Wang, *Generalized harmonic maps and representations of discrete groups.* Comm. Anal. Geom. **8**(2000), no. 3, 545–563.
[701] M. T. Wang, *Mean curvature flow of surfaces in Einstein four-manifolds.* J. Differential Geom. **57**(2001), no. 2, 301–338.
[702] M. T. Wang, *Deforming area preserving diffeomorphism of surfaces by mean curvature flow.* Math. Res. Lett. **8**(2001), no. 5-6, 651–661.
[703] M. Y. Wang and W. Ziller, *Existence and nonexistence of homogeneous Einstein metrics.* Invent. Math. **84**(1986), no. 1, 177–194.
[704] X. J. Wang and X. H. Zhu, *Kähler-Ricci solitons on toric manifolds with positive first Chern class.* Adv. Math. **188**(2004), no. 1, 87–103.
[705] X. W. Wang, *Balance point and stability of vector bundles over a projective manifold.* Math. Res. Lett. **9**(2002), no. 2-3, 393–411.
[706] A. Weil, *Sur les thormes de de Rham.* Comment. Math. Helv. **26**(1952), 119–145.
[707] H. Weyl, *Das asymptotische Verteilungsgesetz der Eigenwerte linearer partieller Differentialgleichungen* (mit einer Anwendung auf die Theorie der Hohlraumstrahlung). (German) Math. Ann. **71**(1912), no. 4, 441–479.
[708] S. Weinberger, *Aspects of the Novikov conjecture, in Geometric and Topological Invariants of Elliptic Operators.* J. Kaminker, editor, (1990), 281–297, A.M.S., Providence, R.I.
[709] A. Weinstein, *On the volume of manifolds all of whose geodesics are closed.* J. Differential Geometry **9**(1974), 513–517.
[710] N. Wickramasekera, *A rigidity theorem for stable minimal hypercones.* J. Differential Geom. **68**(2004), no. 3, 433–514.
[711] N. Wickramasekera, *On the singularities and a Bernstein property of immersed stable minimal hypersurfaces.* Calc. Var. Partial Differential Equations **22**(2005), no. 1, 1–20.
[712] E. Witten, *A new proof of the positive energy theorem.* Comm. Math. Phys. **80**(1981), no. 3, 381–402.
[713] E. Witten, *Supersymmetry and Morse theory* J. Differential Geom. **17**(1982), no. 4, 661–692 (1983).

[714] E. Witten, *Holomorphic Morse inequalities.* Algebraic and differential topology–global differential geometry, 318–333, Teubner-Texte Math., **70**, Teubner, Leipzig, 1984.
[715] E. Witten, *Elliptic genera and quantum field theory.* Comm. Math. Phys. **109**(1987), no. 4, 525–536.
[716] E. Witten, *On the structure of the topological phase of two-dimensional gravity.* Nuclear Phys. B **340**(1990), no. 2-3, 281–332.
[717] E. Witten, *Monopoles and four-manifolds.* Math. Res. Lett. **1**(1994), no. 6, 769–796.
[718] E. Witten and S.-T. Yau, *Connectedness of the boundary in the AdS/CFT correspondence.* Adv. Theor. Math. Phys. **3**(1999), no. 6, 1635–1655 (2000).
[719] M. Wolf, *The Teichmuller theory of harmonic maps.* J. Differential Geom. **29**(1989), no. 2, 449–479.
[720] S. A. Wolpert, *Geometry of the Weil-Petersson completion of Teichmüller space.* Surveys in differential geometry, Vol. **VIII**, 357–393. Int. Press, Somerville, MA, 2003.
[721] S. A. Wolpert, *Disappearance of cusp forms in special families.* Ann. of Math. (2) **139**(1994), 239–291.
[722] H. Wu, *Normal families of holomorphic mappings.* Acta Math. **119**(1967) 193–233.
[723] C. T. Yang, *Odd-dimensional wiedersehen manifolds are spheres.* J. Differential Geom. **15**(1980), no. 1, 91–96 (1981).
[724] C. T. Yang, *Smooth great circle fibrations and an application to the topological Blaschke conjecture.* Trans. Amer. Math. Soc. **320**(1990), no. 2, 507–524.
[725] C. T. Yang, *Any Blaschke manifold of the homotopy type of* CP^n *has the right volume.* Pacific J. Math. **151**(1991), no. 2, 379–394.
[726] P. C. Yang and S.-T. Yau, *Eigenvalues of the Laplacian of compact Riemann surfaces and minimal submanifolds.* Ann. Scuola Norm. Sup. Pisa Cl. Sci. (4) **7**(1980), no. 1, 55–63.
[727] S.-T. Yau, *On the fundamental group of compact manifolds of non-positive curvature.* Ann. of Math. (2) **93**(1971), 579–585.
[728] S.-T. Yau, *Harmonic functions on complete Riemannian manifolds.* Comm. Pure Appl. Math. **28**(1975), 201–228.
[729] S.-T. Yau, *Isoperimetric constants and the first eigenvalue of a compact Riemannian manifold.* Ann. Sci. cole Norm. Sup. (4) **8**(1975), no. 4, 487–507.
[730] S.-T. Yau, *Some function-theoretic properties of complete Riemannian manifold and their applications to geometry.* Indiana Univ. Math. J. **25**(1976), no. 7, 659–670.
[731] S.-T. Yau, *Calabi's conjecture and some new results in algebraic geometry.* Proc. Nat. Acad. Sci. U.S.A. **74**(1977), no. 5, 1798–1799.
[732] S.-T. Yau, *On the Ricci curvature of a compact Kähler manifold and the complex Monge-Ampère equation. I.* Comm. Pure Appl. Math. **31**(1978), no. 3, 339–411.
[733] S.-T. Yau, *The role of partial differential equations in differential geometry.* Proceedings of the International Congress of Mathematicians (Helsinki, 1978), pp. 237–250, Acad. Sci. Fennica, Helsinki, 1980.
[734] S.-T. Yau, *A general Schwarz lemma for Kähler manifolds.* Amer. J. Math. **100**(1978), no. 1, 197–203.
[735] S.-T. Yau, *Problem section.* Seminar on Differential Geometry, pp. 669–706, Ann. of Math. Stud., **102**, Princeton Univ. Press, Princeton, N.J., 1982.
[736] S.-T. Yau, *Compact three-dimensional K ähler manifolds with zero Ricci curvature.* Symposium on anomalies, geometry, topology (Chicago, Ill., 1985), 395–406, World Sci. Publishing, Singapore, 1985.
[737] S.-T. Yau, *Nonlinear analysis in geometry.* Monographies de L'Enseignement Mathmatique, **33**. Série des Conférences de l'Union Mathématique Internationale, 8. L'Enseignement Mathématique, Geneva, 1986. 54 pp.
[738] S.-T. Yau, *A review of complex differential geometry.* Proc. Symp. Pur Math. **52**(1991), 619-625.
[739] S.-T. Yau, *Open problems in Geometry.* Proc. Symp. Pure Math. **54**(1993) 1–28.
[740] S.-T. Yau, *A splitting theorem and an algebraic geometric characterization of locally Hermitian symmetric spaces.* Comm. Anal. Geom. **1**(1993), no. 3-4, 473–486.
[741] S.-T. Yau, *An application of eigenvalue estimate to algebraic curves defined by congruence subgroups.* Math. Res. Lett. **3**(1996), no. 2, 167–172.

[742] S.-T. Yau, *A note on the distribution of critical points of eigenfunctions.* Collection of papers on geometry, analysis and mathematical physics, 173–175, World Sci. Publishing, River Edge, NJ, 1997.

[743] S.-T. Yau, *An estimate of the gap of the first two eigenvalues in the Schrödinger operator. Lectures on partial differential equations*, 223–235, New Stud. Adv. Math., **2**, Int. Press, Somerville, MA, 2003.

[744] S.-T. Yau and W. H. Meeks, *The equivariant loop theorem for three-dimensional manifolds and a review of the existence theorems for minimal surfaces.* The Smith conjecture (New York, 1979), 153–163, Pure Appl. Math., **112**, Academic Press, Orlando, FL, 1984.

[745] R. G. Ye, *Gromov's compactness theorem for pseudo holomorphic curves.* Trans. Amer. Math. Soc. **342**(1994), no. 2, 671–694.

[746] R. G. Ye, *Foliation by constant mean curvature spheres on asymptotically flat manifolds.* Geometric analysis and the calculus of variations, 369–383, Internat. Press, Cambridge, MA, 1996.

[747] G. L. Yu, *Novikov conjecture for groups with finite asymptotic dimension.* Ann. of Math. (2) **147**(1998), no. 2, 325–355.

[748] S. Zelditch, *Szegö kernels and a theorem of Tian.* Int. Math. Res. Not. **6**(1998), 317-331.

[749] S. Zelditch, *Spectral determination of analytic bi-axisymmetric plane domains.* Geom. Funct. Anal. **10**(2000), no. 3, 628–677.

[750] S. W. Zhang, *Heights and reductions of semi-stable varieties.* Compositio Math. **104**(1996), no. 1, 77–105.

[751] J. Q. Zhong and H. C. Yang, *On the estimate of the first eigenvalue of a compact Riemannian manifold.* Sci. Sinica Ser. A **27**(1984), no. 12, 1265–1273.

[752] A. Zinger, *Enumeration of genus-two curves with a fixed complex structure in* $\mathbb{P}^2$ *and* $\mathbb{P}^3$, arXiv:math.SG/0201254.

[753] A. Zinger, *Enumerative vs. Invariants and obstruction bundles*, arXiv:math.SG/0201255.

[754] P. Zograf, *A spectral proof of Rademacher's conjecture for congruence subgroups of the modular group.* J. Reine Angew. Math. **414**(1991), 113–116.

[755] K. Zuo, *Representations of fundamental groups of algebraic varieties.* Lecture Notes in Mathematics, **1708**. Springer-Verlag, Berlin, 1999.

DEPARTMENT OF MATHEMATICS, HARVARD UNIVERSITY, CAMBRIDGE, MA 02138
E-mail address: yau@math.harvard.edu

AMS/IP Studies in Advanced Mathematics
Volume **39**, 2007

Automorphisms of K3 surfaces

De-Qi Zhang

ABSTRACT. In this note, we report some progress we made recently on the automorphisms groups of $K3$ surfaces. A short and straightforward proof of the impossibility of $\mathbb{Z}/(60)$ acting purely non-symplectically on a $K3$ surface, is also given, by using Lefschetz fixed point formula for vector bundles.

1. Introduction

We work over the complex numbers field $\mathbf{C}$. A smooth projective surface X is a $K3$ surface if the canonical line bundle (is trivial) $\omega_X \cong \mathcal{O}_X$ and if the irregularity $h^1(X, \mathcal{O}_X) = 0$. In this note we shall report some progress we made recently (mostly with K. Oguiso) on automorphisms of $K3$ surfaces.

By abuse of notation, we denote by the same ω_X a nowhere vanishing global holomorphic 2-form on X. So $H^0(X, \omega_X) = \mathbf{C}\omega_X$. Now the Hodge decomposition says that $H^2(X, \mathbf{C}) = \mathbf{C}\omega_X \oplus H^{1,1}(X) \oplus \mathbf{C}\overline{\omega_X}$.

Let $\mathrm{Aut}(X)$ be the group of all automorphisms of a $K3$ surface X. Take a subgroup $G \subseteq \mathrm{Aut}(X)$. For every g in G, we have $g^*\omega_X = \alpha(g)\omega_X$ for some $\alpha(g)$ in the multiplicative group $\mathbf{C}^* = \mathbf{C} - \{0\}$. One sees that $\alpha : G \to \mathbf{C}^*$ is a group homomorphism. Let $G_N = \mathrm{Ker}(\alpha)$ be the kernel. By Nikulin [**Ni80**], Theorem 0.1 or Sterk [**St**], Lemma 2.1, the image $\mathrm{Im}(\alpha)$ is a finite subgroup (of order $I = I(G) = I(X, G)$) in $\mathbf{C}^*$, so it equals the multiplicative group $\mu_I = \langle \zeta_I \rangle$, where $\zeta_I = \exp(2\pi\sqrt{-1}/I)$ is an I-th primitive root of 1. Then one has the **basic exact sequence** of groups:

$$1 \to G_N \to G \to \mu_I \to 1.$$

One may say that the study of G is reduced to that of G_N and μ_I, where G_N acts on X symplectically (see Notation and Terminology below) while $h^*\omega_X = \zeta_I\omega_X$ for some pre-image h in G of a generator in μ_I.

PROBLEM 1.1. [**Extension Problem**] Given symplectic subgroup H and a cyclic subgroup μ_I of $\mathrm{Aut}(X)$ on a $K3$ surface X, find the conditions on H and I such that there is a subgroup G of $\mathrm{Aut}(X)$ with $G_N = H$ and $I = I(X, G)$.

In the forthcoming paper Ivanov-Oguiso-Zhang [**IOZ**], we will tackle this problem.

2000 *Mathematics Subject Classification.* 14J28, 14J50, 14L30.
The author was partially supported by an Academic Research Fund of NUS.

Now a summary of the contents. We start with symplectic actions on $K3$ surfaces, then purely non-symplectic actions, and actions of mixed type. Finally we mention the connection between $K3$ surfaces with a non-symplectic action and log Enriques surfaces initially defined in Zhang [**Z91**] and [**Z93**]. Examples of actions are either included or mentioned about their whereabouts. In §3, we also provide a new proof of the impossibility of $\mathbb{Z}/(60)$ acting purely non-symplectically on a $K3$ surface, by using Lefschetz fixed point formula for vector bundles.

Notation and Terminology

For a $K3$ surface X, we use the same symbol ω_X ($\cong \mathcal{O}_X$) to denote the canonical line bundle and also a generator of $H^0(X, \omega_X) = \mathbf{C}\omega_X$. Let $S_X = \mathrm{Pic}(X) < H^2(X, \mathbf{Z})$ be the Neron Severi lattice and $T_X = (S_X)^{\perp} = \{t \in H^2(X, \mathbf{Z}) \,|\, (t, s) = 0, \forall s \in S_X\}$ the transcendental lattice.

A group G is called a $K3$ group if G is a subgroup of $\mathrm{Aut}(X)$ for some $K3$ surface X. An element g of G is simplectic if $g^*\omega_X = \omega_X$. A group G is symplectic if every element of G is symplectic. An element g (or the group $\langle g \rangle$) of order n is purely non-symplectic if $g^*\omega_X = \eta_n \omega_X$ for some primitive n-th root of 1.

For an integer $I \geq 2$, we denote the Euler function by $\varphi(I) = \#\{n \,|\, 1 \leq n \leq I - 1; \gcd(n, I) = 1\}$. Set $\zeta_I = \exp(2\pi\sqrt{-1}/I)$, a primitive I-th root of 1. Set $\mu_I = \langle \zeta_I \rangle$, a cyclic group of order I.

For groups H and K, the symbol $H.K$ denotes a group G such that H is normal in G and $G/H \cong K$, i.e., G is an extension of K by H. When K can be chosen to be a subgroup of G, then we denote $H : K$, i.e., G is a split extension of K by H.

2^n, 3^n, etc denote elementary abelian groups $(\mathbf{Z}/(2))^{\oplus n}$, $(\mathbf{Z}/(3))^{\oplus n}$, etc.

We shall use the notation in Mukai [**Mu**]. S_n is the symmetric group (of order $n!$) in n letters. A_n is the alternating group (of order $n!/2$) in n letters. M_{24}, M_{23} are Mathieu groups while M_{20}, M_{10} are some stabilizer subgroups (see [**Mu**] for details). Q_8 is the quaternion group of order 8, while D_n is the dihedral group of order n.

Acknowledgement.
The author would like to thank Professors S. Kondo and K. Oguiso for the valuable comments.

2. Symplectic $K3$ groups

The classification of abelian symplectic groups was done by Nikulin [**Ni80**]. Especially, he proved:

THEOREM 2.1. [**Ni80**] *Let g be a symplectic automorphism of a $K3$ surface X. Then $\mathrm{ord}(g) \leq 8$. Moreover, the fixed locus (point wise) X^g has cardinality $8, 6, 4, 4, 2, 3,$ or 2, if $\mathrm{ord}(g)$ equals $2, 3, 4, 5, 6, 7,$ or 8, respectively.*

Mukai [**Mu**] has determined all maximum finite symplectic $K3$ groups (11 of them); see also Kondo [**Ko98b**] for a lattice-theoretic proof. We need some preparation of terminology in [**Mu**]. For a nutural number n, denote by $\varepsilon(n) = 24(n\prod_{p|n}(1 + \frac{1}{p}))^{-1}$. Let μ be a $\mathbf{Q}$-valued function on a finite group G such that $\mu(g) = \varepsilon(\mathrm{ord}(g))$ for all g in G. A 24-dimensional representation of G is a *Mathieu representation* if its character is equal to μ. Take a symplectic automorphism h of a $K3$ surface X and take an element σ of the Mathieu group M_{23} acting naturally on $\Omega = \{1, 2, \ldots, 24\}$; suppose that $n = \mathrm{ord}(h) = \mathrm{ord}(\sigma)$. Then the cardinality of fixed point sets satisfies $|X^h| = |\Omega^\sigma| = \varepsilon(n)$ (depending only on the order of the element). This coincidence may have inspired Mukai's result below:

THEOREM 2.2. [**Mu**] *For a finite group G, the following are equivalent:*
(1) G has a (faithful) symplectic action on a $K3$ surface.
(2) G has an embedding $G \subset M_{23}$ into the Mathieu group and decomposes $\{1, 2, \ldots, 24\}$ into at least 5 orbits. (The bigger Mathieu group M_{24} acts naturally on the 24-element set with the smaller M_{23} as a one-point stabilizer subgroup.)
(3) G has a Mathieu representation V over $\mathbf{Q}$ with $\dim V^G \geq 5$ and the 2-Sylow subgroups of G can be embedded into M_{23}.
(4) G is isomorphic to one of the 11 (maximum symplectic) groups below where the order is indicated in the parenthesis:

$$\mathrm{PSL}_2(7)\ (168), \quad A_6\ (360), \quad S_5\ (120), \quad M_{20} = 2^4 : A_5\ (960),$$

$$F_{384} = 2^4 : S_4\ (384), \quad A_{4,4} = (S_4 \times S_4) \cap A_8\ (288),$$

$$T_{192} = (Q_8 * Q_8) : S_3\ (192), \quad H_{192} = 2^4 : D_{12}\ (192),$$

$$N_{72} = 3^2 : D_8\ (72), \quad M_9 = 3^2 : Q_8\ (72), \quad T_{48} = Q_8 : S_3\ (48).$$

The result above completely classifies all finite symplectic actions on $K3$ surfaces. The possible subgroups G (exactly 80 of them) of the 11 maximum symplectic $K3$ groups acting on some $K3$ surface X together with the data of $\mathrm{Sing}(X/G)$ has been given by Xiao [**Xi96**], the list.

3. Purely non-symplectic $K3$ groups

We now turn to the purely non-symplectic $K3$ groups. Suppose that a cyclic group $\mu_I = \langle g \rangle$ $(I \geq 2)$ acts on a $K3$ surface X such that $g^*\omega_X = \eta_I \omega_X$, where η_I is a primitive I-th root of 1. By Nikulin [**Ni80**] Theorem 0.1, we have $\varphi(I) \mid \mathrm{rank}(T_X)$ (see also Sterk [**St**], Lemma 2.1). Since $\mathrm{rank}(T_X) = 22 - \mathrm{rank}(S_X) \leq 21$, we get:

LEMMA 3.1. *Suppose g is an order I $(I \geq 2)$ automorphism of a $K3$ surface X such that $g^*\omega_X = \eta_I \omega_X$ with η_I a primitive I-th root of 1. Then $\varphi(I) \leq 21$ and hence I equals one of the following:*
Case $\varphi(I) = 20$. $I = 66, 50, 44, 33, 25$.
Case $\varphi(I) = 18$. $I = 54, 38, 27, 19$.
Case $\varphi(I) = 16$. $I = 60, 48, 40, 34, 32, 17$.
Case $\varphi(I) = 12$. $I = 42, 36, 28, 26, 21, 13$.
Case $\varphi(I) = 10$. $I = 22, 11$.
Case $\varphi(I) = 8$. $I = 30, 24, 20, 16, 15$.
Case $\varphi(I) = 6$. $I = 18, 14, 9, 7$.
Case $\varphi(I) = 4$. $I = 12, 10, 8, 5$.
Case $\varphi(I) = 2$. $I = 6, 4, 3$.
Case $\varphi(I) = 1$. $I = 2$.

Conversely, each I with $I \neq 60$ in the lemma above is geometrically realizable (see Kondo [**Ko92**], or Machida-Oguiso [**MO**], Proposition 4]).

PROPOSITION 3.2. *For each I with $I \neq 60$ in the lemma above, there is a $K3$ surface X such that* $\mathrm{Aut}(X) \geq \mu_I = \langle g \rangle$ *and* $g^*\omega_X = \zeta_I \omega_X$*, where* $\zeta_I = \exp(2\pi\sqrt{-1}/I)$ *is a primitive I-th root of 1.*

Indeed, μ_{60} can not act purely non-symplectically by the result below (the early proofs were first given by Xiao [**Xi95**], Theorem, page 1, and then by Machida-Oguiso [**MO**], Theorem 5.1]); our proof here comes from the very simple idea: the calculation of Lefschetz number by Atiyah, Segal and Singer, [**AS1**], [**AS2**], for the trivial line bundle, and hence more straightforward.

Therefore, the lemma and proposition above give the complete classification for the μ_I appearing in the basic exact sequence in the Introduction.

PROPOSITION 3.3. *(see* [**Xi95**] *and* [**MO**]*). There is no $K3$ surface with a purely non-symplectic action by the group μ_{60}. Namely, there is no pair $(X, \langle g \rangle)$ of a $K3$ surface X and an order 60 automorphism g with $g^*\omega_X = \zeta_{60}\omega_X$.*

PROOF. Suppose the contrary that there is a pair $(X, \langle g \rangle)$ as in the statement. Set $\omega = \omega_X$ and $\zeta = \zeta_{60}$, so $g^*\omega = \zeta\omega$. Let $\{C_i\}$ be the set of all g-fixed (point wise) irreducible curves. So at a point Q on C_i, one can diagonalize g^* as $(1, \zeta)$ with appropriate coordinates. Let $n = \sum(1 - g(C_i))$; we set $n = 0$ when there is no such g-fixed curve C_i. Let $M_j = \{P_{jk}\}$ be the set of points P at which g^* can be diagonalized as $(\zeta^{-j}, \zeta^{j+1})$ with $1 \leq j < 60/2 = 30$. Set $m_j = |M_j|$ and $m = \sum m_j$. Then the fixed locus X^g (point wise) is the disjoint union of smooth curves C_i and m isolated points P_{jk} (see, e.g., Oguiso - Zhang [**OZ96**], Lemma 2.1 (3) for the smoothness). By Atiyah - Segal [**AS1**] and Atiyah - Singer [**AS2**] at pages 542 and 567, we can calculate the Lefschetz number $L(g)$ in two different ways:

$$L(g) = \sum_{i=0}^{2}(-1)^i \operatorname{Tr}(g^*|H^i(X, \mathcal{O}_X)),$$

$$L(g) = \sum_{j,k} a(P_{jk}) + \sum_i b(C_i),$$

$$a(P) = \frac{1}{\det(1 - g^*|T_P)},$$

$$b(C_i) = \frac{1 - g(C_i)}{1 - \zeta} - \frac{\zeta C_i^2}{(1 - \zeta)^2} = \frac{(1 - g(C_i))(1 + \zeta)}{(1 - \zeta)^2}.$$

Here $T_P = T_{X,P}$ is the tangent space at P, and ζ^{-1} is the eigenvalue of the action g_* on the normal bundle of C_i. We see then that $\sum_i b(C_i) = n(1+\zeta)/(1-\zeta)^2$. Also $\sum a(P_{jk}) = \sum_j m_j/(1 - \zeta^{-j})(1 - \zeta^{j+1})$. On the other hand, by the Serre duality $H^2(X, \mathcal{O}_X) = H^0(X, \omega)^\vee$, we have $L(g) = 1 + \zeta^{-1}$. Identify the two $L(g)$, so $x := \zeta$ satisfies the equation below:

$$F_1(x) = -(1 + x^{-1}) + \sum_{j=1}^{29} \frac{m_j}{(1 - x^{-j})(1 - x^{j+1})} + \frac{n(1 + x)}{(1 - x)^2} = 0.$$

Note that the minimal polynomial $\Phi_{60}(x)$ of $x = \zeta$ over $\mathbf{Q}$ is given as follows; this is also obtained by factorizing $x^{30} + 1$:

$$\Phi_{60}(x) = x^{16} + x^{14} - x^{10} - x^8 - x^6 + x^2 + 1.$$

Our $x = \zeta$ also satisfies the following equations (factorizing $(x^{10})^3 + 1$):

$$x^{30} + 1 = 0 = x^{20} - x^{10} + 1.$$

Since $x^{\pm 30} = -1$, we have $(1-x^{-i})(1-x^{i+1}) = (1+x^{30-i})(1+x^{i+1-30})$. Substituting these equalities for $15 \le i \le 29$ into the equation $F_1(x) = 0$, we obtain a new equation $0 = F_1(x) = F_2(x)/G_2(x)$, where $F_2(x)$ is a polynomial of degree 146 in x with coefficients in m_j, and $G_2(x)$ is the product of polynomials each of which is of degree ≤ 12 in x. Since the minimal polynomial $\Phi_{60}(x)$ of $x = \zeta$ is of degree 16, our $x = \zeta$ satisfies $G_2(x) \ne 0$ and $F_2(x) = 0$. Substituting $x^{30k} = 1$ (resp. -1) when k is even (resp. odd) into the equation $F_2(x) = 0$, we obtain a polynomial equation $F_3(x) = 0$ of degree ≤ 29 in x. Substituting $x^{20} = x^{10} - 1$ into the equation $F_3(x) = 0$, we obtain an equation $F_4(x) = 0$ of degree ≤ 19 in x. Finally substituting $\Phi_{60}(x) = 0$ into the equation $F_4(x) = 0$, we obtain an equation $F_5(x) = \sum_{i=0}^{15} d_i x^i = 0$ of degree ≤ 15 in x, where each d_i is an integer linear combination of m_j's. Note that $x^i = \zeta^i$ $(0 \le i \le 15)$ are linearly independent over $\mathbf{Q}$. So we get linear equations in m_j: $d_0 = d_1 = \cdots = d_{15} = 0$. Solving them, we arrive at the following relations:

$$(*)\ 4m_1 = -1 + 2m_2 + 2m_{20} - 2m_{21} + 4m_{22} - 4m_{23} + 2m_{26}$$
$$-2m_{27} + 4m_{28} - 4m_{29} - 2m_3 + 4m_6 - 4m_7 + 2m_8 - 2m_9 + 8n,$$

$$9m_{10} = 60 - 12m_{17} + 18m_{18} - 17m_{19} - 141m_2 - 94m_{20} + 81m_{21} - 189m_{22}$$
$$+198m_{23} + 6m_{24} - 12m_{25} - 81m_{26} + 81m_{27} - 156m_{28} + 156m_{29}$$
$$+63m_3 - 6m_4 - 24m_5 - 198m_6 + 189m_7 - 105m_8 + 96m_9 + 312n,$$

$$36m_{11} = 267 - 48m_{17} + 108m_{18} - 104m_{19} - 582m_2 - 394m_{20} + 342m_{21} - 792m_{22}$$
$$+828m_{23} + 24m_{24} - 48m_{25} - 342m_{26} + 342m_{27} - 624m_{28} + 624m_{29}$$
$$+270m_3 - 24m_4 - 96m_5 - 828m_6 + 792m_7 - 438m_8 + 402m_9 + 1248n,$$

$$36m_{12} = 57 + 12m_{17} - 72m_{18} + 56m_{19} - 282m_2 - 134m_{20} + 162m_{21} - 252m_{22}$$
$$+216m_{23} - 24m_{24} + 12m_{25} - 162m_{26} + 162m_{27} - 384m_{28} + 384m_{29}$$
$$+90m_3 - 84m_4 + 24m_5 - 288m_6 + 252m_7 - 138m_8 + 102m_9 + 768n,$$

$$72m_{13} = 39 + 48m_{17} - 144m_{18} + 104m_{19} - 426m_2 - 182m_{20} + 234m_{21} - 252m_{22}$$
$$+180m_{23} - 60m_{24} + 84m_{25} - 270m_{26} + 234m_{27} - 600m_{28} + 600m_{29}$$
$$+126m_3 - 156m_4 + 60m_5 - 324m_6 + 252m_7 - 138m_8 + 102m_9 + 1200n,$$

$$18m_{14} = -15 + 12m_{17} - 36m_{18} + 32m_{19} - 48m_2 - 2m_{20} +$$
$$18m_{21} - 18m_{23} - 24m_{24} + 30m_{25} - 36m_{26} + 36m_{27} - 96m_{28}$$
$$+87m_{29} - 30m_4 + 24m_5 + 18m_6 - 36m_7 + 6m_8 - 6m_9 + 174n,$$

$$18m_{15} = 30 + 12m_{17} - 36m_{18} + 32m_{19} - 66m_2 - 56m_{20} + 72m_{21} - 144m_{22}$$
$$+126m_{23} + 12m_{24} - 6m_{25} - 54m_{26} + 54m_{27} - 96m_{28} + 87m_{29}$$
$$+18m_3 + 6m_4 - 12m_5 - 126m_6 + 108m_7 - 48m_8 + 48m_9 + 174n,$$

$$72m_{16} = -39 + 96m_{17} - 8m_{19} + 138m_2 + 86m_{20} - 90m_{21} + 180m_{22}$$
$$-180m_{23} - 12m_{24} + 60m_{25} + 54m_{26} - 90m_{27} + 168m_{28} - 168m_{29}$$
$$-54m_3 + 12m_4 + 12m_5 + 180m_6 - 180m_7 + 138m_8 - 102m_9 - 336n.$$

To double check that there is no any human error in the above substitutions, we substitute the above relations back into the original equation and get $F_1(x) = \Phi_{60}(x)F_6(x)/G_6(x)$, where $F_6(x)$ is a polynomial of degree 130 in x and $G_6(x)$ is a product of polynomials each of which is of degree ≤ 12 in x. So no error should have occurred during the substitution process, noting that $x = \zeta$ satisfies its minimal polynomial (over $\mathbf{Q}$) $\Phi_{60}(x) = 0$. The actual simple linear algebra calculation is done by the "Maple", though a calculation by hand is possible if the reader is patient and careful enough to avoid any error in doing the simple arithmetics and Gaussian elimination.

Now to conclude the proof, we note that the first equality (*) in the final relations above says that $4m_1 = -1+2s$, with s an integer (a integer linear combination of m_i and n). This is impossible because m_1 is an integer. We reach a contradiction. So there is no such pair (X, μ_{60}) and the proposition is proved. □

REMARK 3.4. As pointed out by Oguiso (and done in [**MO**]), indeed, one can say a bit stronger than the proposition above (consider X/G_N): If X is a $K3$ surface and $G \leq Aut(X)$ is finite then $\mu_{I(X,G)} \neq \mu_{60}$. This is because one always has $G_N = (1)$ and $G = \mu_{I(X,G)}$ whenever $\varphi(I(X,G)) \geq 10$ and $I(X,G) \neq 28$; see Ivanov-Oguiso-Zhang [**IOZ**], or Machida-Oguiso [**MO**] the proof of Lemma (4.1).

For a $K3$ surface X, the cohomology $H^2(X, \mathbf{Z})$ is an even lattice. It contains $S_X \oplus T_X$ as a sublattice of finite index. So the study of $\mathrm{Aut}(X)$ is reduced to that of $\mathrm{Aut}(S_X)$ to some extent. Indeed, by Pjateckii-Shapiro and Shafarevich [**PS**] (see also Burns - Rapoport [**BR**]) up to finite index and finite co-index, $\mathrm{Aut}(X)$ is isomrphic to $\mathrm{Aut}(S_X)/\Gamma(X)$ where $\Gamma(X)$ is generated by reflections associated to (-2)-vectors.

Let us consider the natural exact sequence

$$(1) \to H_X \to \mathrm{Aut}(X) \to \mathrm{Aut}(S_X).$$

Note that H_X acts faithfully on T_X and on the 1-dimensional space $\mathbf{C}\omega_X$ (see Nikulin [**Ni80**] Cor. 3.3, or Sterk [**St**] Lemma 2.1). Indeed, H_X is a cyclic finite group of order h_X with the Euler function $\varphi(h_X) \,|\, \mathrm{rank}(T_X)$. The results below show that these X with maximum $\varphi(h_X) = rk(T_X)$ are unique up to isomorphisms. For the unique examples of the X's in the results below, we refer to Kondo [**Ko92**]; see also Examples 3.7 and 3.10 below.

THEOREM 3.5. *(see Kondo* [**Ko92**]*, Main Th).*
Set $\Sigma := \{66, 44, 42, 36, 28, 12\}$*.*
(1) *Let* X *be a* $K3$ *surface with* $\varphi(h_X) = \mathrm{rank}(T_X)$ *and with unimodular transcendental lattice* T_X*. Then* $h_X \in \Sigma$*.*
(2) *Conversely, for each* $N \in \Sigma$*, there exists, modulo isomorphisms, a unique* $K3$ *surface* X *such that* $h_X = N$ *and* $\varphi(h_X) = \mathrm{rank}(T_X)$*. Moreover,* T_X *is unimodular for this* X*.*

THEOREM 3.6. *(see Oguiso - Zhang* [**OZ00a**]*, Theorem 2).*
Set $\Omega := \{3^k(1 \le k \le 3), 5^\ell(\ell = 1, 2), 7, 11, 13, 17, 19\}$*.*
(1) *Let* X *be a* $K3$ *surface with* $\varphi(h_X) = \mathrm{rank}(T_X)$ *and with non-unimodular transcendental lattice* T_X*. Then* $h_X \in \Omega$*.*
(2) *Conversely, for each* $N \in \Sigma$*, there exists, modulo isomorphisms, a unique* $K3$ *surface* X *such that* $h_X = N$ *and* $\varphi(h_X) = \mathrm{rank}(T_X)$*. Moreover,* T_X *is non-unimodular for this* X*. For* $I = 19, 17$*, or* 13*, the unique pair is given in Example 3.7 below.*

By Theorem 2.1 and Lemma 3.1, every prime order p automorphism g of a $K3$ surface X satisfies $p \le 19$. Let us give examples for the largest three p (see Kondo [**Ko92**]).

EXAMPLE 3.7. Here are examples of pairs (X_p, μ_p) ($p = 19, 17$ or 13) of the $K3$ surface X_p with a $\mu_p = \langle g_p \rangle$ action, given by its Weierstrass equation and the action on the coordinates.

$$X_{19} : \ y^2 = x^3 + t^7x + t, \quad g_{19}^*(x, y, t) = (\zeta_{19}^7 x, \zeta_{19} y, \zeta_{19}^2 t).$$
$$X_{17} : \ y^2 = x^3 + t^7x + t^2, \quad g_{17}^*(x, y, t) = (\zeta_{17}^7 x, \zeta_{17}^2 y, \zeta_{17}^2 t).$$
$$X_{13} : \ y^2 = x^3 + t^5x + t^4, \quad g_{13}^*(x, y, t) = (\zeta_{13}^5 x, \zeta_{13} y, \zeta_{13}^2 t).$$

Indeed, these pairs are unique with no conditions on h_X or so on; but they turn out to satisfy the conditions in the above theorem:

THEOREM 3.8. *(Oguiso-Zhang* [**OZ00a**] *Cor. 3). Let* X *be a* $K3$ *surface with an automorphism* g *of order* $p = 19$*,* 17 *or* 13 *(the three largest possible prime orders. Then the pair* $(X, \langle g \rangle)$ *is isomorphic to the pair* $(X_p, \langle g_p \rangle)$ *in the above example.*

REMARK 3.9. A similar result as above does not hold for $p = 11$. Indeed, there are three 1-dimensional families of pairs (X_t, μ_{11}) of $K3$ surfaces X_t with a μ_{11} action; see Oguiso-Zhang [**OZ99b**].

Now we give Kondo's examples of purely non-symplectic μ_I action on $K3$ surfaces with I a composite (see Kondo [**Ko92**], or Oguiso-Machida [**MO**], Prop. 4).

EXAMPLE 3.10. Here are examples of pairs (X_I, μ_I) (the first few bigger I in Lemma 3.1) of the $K3$ surface X_I with a $\mu_I = \langle g_I \rangle$ purely non-symplectic action, given by its Weierstrass equation or as weighted hypersurface and the action on the coordinates; see Kondo [**Ko92**].

$$X_{66} : \ y^2 = x^3 + t(t^{11} - 1), \quad g_{66}^*(x, y, t) = (\zeta_{66}^{40} x, \zeta_{66}^{27} y, \zeta_{66}^{54} t).$$

$$X_{50} = \{z^2 = x_0^6 + x_0 x_1^5 + x_1 x_2^5\} \subset \mathbf{P}(1, 1, 1, 3),$$
$$g_{50}^*[x_0 : x_1 : x_2 : z] = [x_0 : \zeta_{25}^{20} x_1 : \zeta_{25} x_2 : -z].$$

$X_{44}: \ y^2 = x^3 + x + t^{11}, \quad g_{44}^*(x,y,t) = (\zeta_{44}^{22}x, \zeta_{44}^{11}y, \zeta_{44}^{34}t).$

$X_{54}: \ y^2 = x^3 + t(t^9 - 1), \quad g_{54}^*(x,y,t) = (\zeta_{27}^{2}x, -\zeta_{27}^{3}y, \zeta_{27}^{6}t).$

$X_{38}: \ y^2 = x^3 + t^7x + t, \quad g_{38}^*(x,y,t) = (\zeta_{19}^{7}x, -\zeta_{19}y, \zeta_{19}^{2}t).$

$X_{48}: \ y^2 = x^3 + t(t^8 - 1), \quad g_{48}^*(x,y,t) = (\zeta_{48}^{2}x, \zeta_{48}^{3}y, \zeta_{48}^{6}t).$

Indeed, these surfaces (and the action to some extent) are unique:

THEOREM 3.11. *(Machida-Oguiso* [**MO**], *Main Th* 1*). Suppose that X is a $K3$ surface with a finite group $G \le \mathrm{Aut}(X)$ such that $I = I(X,G) \in \{66, 50, 44, 33, 25\}$ (i.e., $\varphi(I) = 20$).*
(1) The pair (X,G) is isomorphic to the pair $(X_I, \langle g_I\rangle)$ in the example above if I is even, and (X,G) is isomorphic to the pair $(X_{2I}, \langle g_{2I}^2\rangle)$ if I is odd.
(2) We have $\mathrm{Aut}(X) = G \cong \mu_I$ when I is even, and $\mathrm{Aut}(X) \cong \mu_{2I}$ when I is odd.

REMARK 3.12. A slight weak result holds for $I = 54, 38, 48$ (with the unique surface X_I); see Xiao [**Xi95**], Theorem, page 1.

So far, we have considered automorphisms of $K3$ surfaces with bigger order. We now consider the opposite cases. We start with some examples.

EXAMPLE 3.13. Let $\zeta = \zeta_4 = \sqrt{-1}$. Let $E = E_\zeta = \mathbf{C}/(\mathbf{Z} + \mathbf{Z}\zeta)$ be the elliptic curve of period ζ. Let $X_2 \to \overline{X}_2 := (E \times E)/\mathrm{diag}(\zeta, -\zeta)$ be the minimal resolution of the quotient surface $\overline{X}_2$. Then X_2 is a $K3$ surface. Let g_2 be the automorphism of X_2 induced by the action $\mathrm{diag}(-1, 1)$ on $E \times E$. Then $g_2^*\omega_{X_2} = -\omega_{X_2}$ and the fixed locus (point wise) X^{g_2} consists of 10 smooth rational curves (see Oguiso-Zhang [**OZ96**], Example 2 for details).

EXAMPLE 3.14. Let $\zeta = \zeta_3$. Let $E = E_\zeta = \mathbf{C}/(\mathbf{Z} + \mathbf{Z}\zeta)$ be the elliptic curve of period ζ. Let $X_3 \to \overline{X}_3 := (E \times E)/\mathrm{diag}(\zeta, \zeta^2)$ be the minimal resolution of the quotient surface $\overline{X}_3$. Then X_3 is a $K3$ surface. Let g_3 be the automorphism of X_3 induced by the action $\mathrm{diag}(\zeta, 1)$ on $E \times E$. Then $g_3^*\omega_{X_3} = \zeta\omega_{X_3}$ and the fixed locus (point wise) X^{g_3} consists of 6 smooth rational curves and 9 isolated points (see Oguiso-Zhang [**OZ96**], Example 1 for details).

From the result above or intuitively, we see that a $K3$ surface X with a non-symplectic action of bigger order is uniquely determined by the action. The result below shows that the other extreme case may happen too.

THEOREM 3.15. *(Oguiso-Zhang* [**OZ96**], *Theorem* 4*). Let (X, g) be a pair of a $K3$ surface X and an automorphism g of X satisfying (where the conditin* (3) *is removable by Nikulin* [**Ni83**], *or Zhang* [**Z98**] *Th* 3*):*
(1) $g^2 = id$,
(2) $g^*\omega_X = -\omega_X$,
(3) The fixed locus (point wise) X^g consists of only (smooth) rational curves, and
(4) X^g contains at least 10 *rational curves.*
Then (X, g) is isomorphic to the pair (X_2, g_2) up to isomorphisms.

THEOREM 3.16. *(Oguiso-Zhang* [**OZ96**], *Theorem* 3*). Let (X, g) be a pair of a $K3$ surface X and an automorphism g of X satisfying:*
(1) $g^3 = id$,

(2) $g^*\omega_X = \zeta_3\omega_X$,
(3) *The fixed locus (point wise)* X^g *consists of only (smooth) rational curves and possibly some isolated points, and*
(4) X^g *contains at least* 6 *rational curves.*
Then (X, g) *is isomorphic to the pair* (X_3, g_3) *up to isomorphisms.*

REMARK 3.17. (1) The $K3$ surfaces X_3 and X_2 in the theorems above have Picard number 20 and discriminants $|\det(\mathrm{Pic}(X_3))| = 3$ and $|\det\,(\mathrm{Pic}(X_2))| = 4$. Indeed, X_3 (resp. X_2) is the only $K3$ surface with Picard number 20 and discriminant 3 (resp. 4). They are called the *most algebraic* $K3$ *surfaces* in Vinberg [**Vi**]. For these two $X = X_i$, the infinite group $\mathrm{Aut}(X)$ has been calculated by Vinberg.
(2) A smooth $K3$ surface X is called *singular* if the Picard number $\rho(X) = 20$ (maximum). Such X is uniquely determined by the transendental lattice T_X; see Shioda-Inose [**SI**]. Moreover, it has been proved there that a singular $K3$ surface has infinite group $\mathrm{Aut}(X)$. See Oguiso's theorem below for a stronger result.
(3) Nikulin and Vinberg have determined all $K3$ surfaces X with finite $\mathrm{Aut}(X)$; see Nikulin [**Ni83**] and [**Ni85**]; see also Kondo [**Ko89**] for the detailed description of those finite $\mathrm{Aut}(X)$.
(4) In general, it is very difficult to calculate the full automorphism group $\mathrm{Aut}(X)$ for a $K3$ surface X. This has been done by Kondo [**Ko98a**] for a generic Jacobian Kummer surface (by using the results of Keum [**Ke97**]) and Keum-Kondo [**KK**] for Kummer surfaces associated with products of some elliptic curves.
(5) In their paper [**DK02**], Dolgachev and Keum have successfully calculated the full automorphism group $\mathrm{Aut}(X)$ for general quartic Hessian surfaces, by embedding the rank 16 Picard lattice into the rank 26 lattice L of signature $(1, 25)$ (the sum of the rank 24 Leech lattice and the standard hyperbolic plane). See also Kondo [**Ko99**] for the early employment of such L for Kummer surfaces.

We end this section with the following striking result in (Oguiso [**Og06**]); see the same paper for more results in the new directions.

THEOREM 3.18. *(Oguiso* [**Og06**]*). Let* X *be a singular* $K3$ *surface and let* $H := \mathrm{Hilb}^n(X)$*. Then each of* $\mathrm{Aut}(X)$*,* $\mathrm{Aut}(H)$ *and* $\mathrm{Bir}(H)$ *contains the non-commutative free group* $\mathbf{Z} * \mathbf{Z}$*.*

4. $K3$ groups of mixed type

So far, we have considered $K3$ groups of purely symplectic or purely non-symplectic type. In this section, we shall consider $K3$ groups G of the mixed type, i.e., the case where neither G_N nor μ_I (with $I = I(X, G)$) in the basic exact sequence of the Introduction becomes trivial. On the one hand, $|G_N| \leq |M_{20}| = 960$ by Theorem 2.2. On the other hand, $|\mu_I| \leq 66$ by Lemma 3.1. A finite $K3$ group G (and the pair (X, G)) of the largest order has been determined by Kondo:

THEOREM 4.1. *(Kondo* [**Ko99**]*). Let* X *be a* $K3$ *surface and* $G \leq \mathrm{Aut}(X)$ *a finite subgroup.*
(1) *We have* $|G| \leq 4 \cdot 960$.
(2) *Suppose that* $|G| = 4 \cdot 960$. *Then* X *equals the Kummer surface* $Km(E_{\sqrt{-1}} \times E_{\sqrt{-1}})$ *where* $E_{\sqrt{-1}} = \mathbf{C}/(\mathbf{Z} + \mathbf{Z}\sqrt{-1})$*, and* $G = M_{20}.\mu_4$*. Moreover, the group* G *and the action of* G *on* X *are unique up to isomorphisms.*

Besides the fact that M_{20} is the finite symplectic $K3$ group of the largest order, it is also the the finite perfect $K3$ group of the largest order. Here are all of the finite perfect $K3$ groups (cf. Xiao [**Xi96**], the list):

$$M_{20}, \quad A_6, \quad L_2(7) = \mathrm{PSL}_2(7), \quad A_5.$$

The first three groups are among the 11 maximum ones in Theorem [**Mu**]. The last three groups above are also the only non-abelian finite simple $K3$ groups. The last two groups happen to be the two smallest (in terms of order) non-abelian finite simple groups. These four groups (together with a bigger μ_I) determine the surface X uniquely (see the theorems below).

Let us start with an example of $K3$ group of mixed type.

EXAMPLE 4.2. Let $C_{168} = \{x_1x_2^3 + x_2x_3^3 + x_3x_1^3 = 0\} \subset \mathbf{P}^2$ be the Klein quartic curve of genus $g = 3$. A well known theorem says that $|\mathrm{Aut}(C)| \le 84(g-1)$ for any curve C of genus $g \ge 2$. This Klein quartic attains the maximum $|\mathrm{Aut}(C_{168})| = 84(3-1) = 168 = |L_2(7)|$ (see Oguiso-Zhang [**OZ00b**] for the action of $L_2(7)$ on C_{168}). Consider the quartic $K3$ surface $X_{168} = \{x_0^4 + x_1x_2^3 + x_2x_3^3 + x_3x_1^3 = 0\} \subset \mathbf{P}^3$. This is also the Galois $\mathbf{Z}/(4)$-cover of $\mathbf{P}^2$ branched along the Klein quartic C_{168}. Clearly, there is a faithful $L_2(7) \times \mu_4$ action on X_{168}. We shall call the pair $(X_{168}, L_2(7) \times \mu_4)$ the Klein-Mukai pair.

THEOREM 4.3. *(Oguiso-Zhang* [**OZ00b**]*, Main Th, Prop. 1). Let X be a $K3$ surface such that $G := L_2(7) \le \mathrm{Aut}(X)$. Let $\widetilde{G} \le \mathrm{Aut}(X)$ be a finite subgroup containing G.*
(1) G is normal in $\widetilde{G}$ with $|\widetilde{G}/G| \le 4$.
(2) Suppose that $|\widetilde{G}/G| = 4$. Then $(X, \widetilde{G})$ is isomorphic to the Klein-Mukai pair $(X_{168}, L_2(7) \times \mu_4)$.

Next we turn to the another simple group A_6, also called the Valentiner's group (see [**Bli**]). It is a very important group and also the root of troubles in the theory of simple groups. Indeed, the A_6 is at the junction of the three simple groups series: recall that $A_6 = [M_{10}, M_{10}] = \mathrm{PSL}(2, 9)$.

THEOREM 4.4. *(Keum-Oguiso-Zhang* [**KOZ1**]*, Th 3.1, Prop 4.1, Th 5.1). (1) Let F be the unique $K3$ surface of Picard number 20 whose transcendental lattice T_F has the intersection matrix $\mathrm{diag}[6, 6]$ (see Shioda-Inose* [**SI**]*). Then there is a group $\widetilde{A}_6 := A_6 : \mu_4$ (a split extension of A_6 by μ_4) and a faithful action of it on F (see Keum-Oguiso-Zhang* [**KOZ1**] *Def 2.7, Th 3.1 or* [**KOZ2**] *Th 3.2 for the action and the group structure).*
(2) Suppose that $\widetilde{G}$ is a finite $K3$ group containing A_6. Then A_6 is normal in $\widetilde{G}$, and the group $\widetilde{G}/A_6$ is cyclic of order ≤ 4.
(3) Suppose that $(X, \widetilde{G})$ is a pair of a $K3$ surface and a subgroup $\widetilde{G} \le \mathrm{Aut}(X)$ containing A_6 as a subgroup of maximum index 4. Then $(X, \widetilde{G})$ is isomorphic to the pair $(F, \widetilde{A}_6)$ in (1).

REMARK 4.5. The F in the theorem above is also the minimal resolution of the hypersurface in $\mathbf{P}^1 \times \mathbf{P}^2$ (with coordinates $([S : T], [X : Y : Z])$) given by the equation (see Keum-Oguiso-Zhang [**KOZ1**] Prop 3.5):

$$S^2(X^3 + Y^3 + Z^3) - 3(S^2 + T^2)XYZ = 0.$$

The action of $\widetilde{A}_6$ is still invisible from the equation (due to the trouble nature of A_6?) It is indeed constructed lattice-theoretically, where the Leech lattice and some deep results of S. Kondo [**Ko98b**] and R. E. Borcherds [**Bo**] are utilized.

In the case of the smallest non-abelian simple group A_5, we have the complete classification of finite $K3$ groups containing A_5 as a normal subgroup.

THEOREM 4.6. *(Zhang* [**Z06b**] *Th A, or* [**Z06a**] *Th E). Suppose that G is a finite $K3$ group containing A_5 as a normal subgroup. Then G equals one of the following (for their geometric realization, see Zhang* [**Z06b**] *Example* 1.10*):*

$$A_5, \quad S_5, \quad A_5 \times \mu_2, \quad S_5 \times \mu_2.$$

We now give an example of the largest (in terms of the order) finite solvable $K3$ group.

EXAMPLE 4.7. Let $X_4 := \{x_1^4 + x_2^4 + x_3^4 + x_4^4 = 0\} \subset \mathbf{P}^3$ be the Fermat quartic $K3$ surface. Then $(\mu_4)^4$ acts diagonally on the coordinates while S_4 permutes the coordinates. So there is a natural action of $\widetilde{F}_{384} = (\mu_4^4 : S_4)/\mu_4 = (\mu_4^4/\mu_4) : S_4$ on X_4 (see Mukai [**Mu**]). This $\widetilde{F}_{384}$ is a solvable group of order $4^3 \cdot 4! = 2^9 \cdot 3 = 384 \cdot 4$. Note that the 2-Sylow subgroup $\widetilde{F}_{128}$ of $\widetilde{F}_{384}$ is of course nilpotent and of order $2^9 = 128 \cdot 4$. Let $G = \widetilde{F}_{128}$, or $\widetilde{F}_{384}$, then in notation of the basic exact sequence in the Introduction, we have $\mu_{I(X,G)} = \mu_4$ and $G_N = F_{128}$, or F_{384} respectively (as defined in Mukai [**Mu**]; note also that F_{384} is also the second largest among the 11 maximum ones in Theorem 2.2). For both G, one can show that $\mathrm{Pic}(X_4)^G = \mathbf{Z}H$ and $H^2 = 4$ (see Oguiso [**Og05**], Th 1.2).

Now we state a pretty result due to Oguiso:

THEOREM 4.8. *(Oguiso* [**Og05**]*, Th* 1.2*).*
(1) *Let X be a $K3$ surface and $G \le \mathrm{Aut}(X)$ a finite solvable subgroup. Then $|G| \le 2^9 \cdot 3$. Moreover, if $|G| = 2^9 \cdot 3$ then (X, G) is isomorphic to the pair $(X_4, \widetilde{F}_{384})$ (with the standard action) in the example above.*
(2) *Let X be a $K3$ surface and $G \le \mathrm{Aut}(X)$ a finite nilpotent subgroup. Then $|G| \le 2^9$. Moreover, if $|G| = 2^9$ then (X, G) is isomorphic to the pair $(X_4, \widetilde{F}_{128})$ (with the standard action) in the example above.*

5. Log Enriques surfaces

The concept of Log Enriques surface was first introduced in Zhang [**Z91**], Def 1.1. In the smooth case, they are just abelian surface (index 1), $K3$ surface (index 2), Enriques surface (index 2) and hyperelliptic surfaces (index $2, 3, 4$, or 6). Log Enriques surface is closely related to the study of purely non-symplectic $K3$ groups.

DEFINITION 5.1. A normal projective surface Y with at worst quotient singularities is called a *log Enriques surface* if the irregularity $h^1(Y, \mathcal{O}_Y) = 0$ and if a positive multiple of the canonical divisor $mK_Y \sim 0$ (linearly equivalence). $I = I(Y) = \min\{m \in \mathbf{Z}_{>0} \mid mK_Y \sim 0\}$ is called the *index* of Y.

There is a Galois $\mathbf{Z}/(I)$-cover $\pi : X = Spec \oplus_{i=0}^{I-1} \mathcal{O}_Y(-iK_Y) \to Y$ which is unramified outside non-Du Val singular locus of Y, i.e., outside the set of singular points which is not rational double. This X is called the *canonical cover* of Y. One sees that $g^*\omega_X = \eta_I \omega_X$, where $\mathrm{Gal}(X/Y) = \langle g \rangle \cong \mu_I$ and η_I is a primitive I-th root of 1.

REMARK 5.2. (1) In general, the canonical cover X is a normal surface with at worst Du Val singularities and with $K_X \sim 0$. So either X is a (smooth) abelian surface, or a $K3$ surface with at worst Du Val singularities.
(2) Log Enriques surfaces are degenerate fibres of families of Kodaira dimension 0. Also the base space of elliptically fibred Calabi-Yau threefold $\Phi_D : Z \to Y$ with $D.c_2(Z) = 0$ are necessarily log Enriques surfaces (see Oguiso [**Og93**]).
(3) When the canonical cover X of Y is abelian, one has $I(Y) = 3$, or 5 and such Y is unique up to isomorphisms; for examples or proof, see Zhang [**Z91**] Example 4.2 and Theorem 4.1, or Blache [**Bl**], (1.2) and Theorem C.
(4) In general, one has $I(Y) \leq 21$; see Blache [**Bl**] Th C, and also Zhang [**Z91**] Lemma 2.3. If $I(Y)$ is prime, then $I(Y) \leq 19$ (Lemma 3.1). Examples of prime $I(Y)$ are given in Zhang [**Z91**] and Blache [**Bl**]. When $I(Y)$ is prime, the classification of Sing(X) for the canonical cover X of Y, is done in Zhang [**Z93**], Main Th.
(5) When the canonical cover X of Y is a normal $K3$, we let $\widetilde{X} \to X$ be the minimal resolution. Then the determination of Y is almost equivalent to that of $(\widetilde{X}, \mu_I)$, where $\mu_I = \mathrm{Gal}(X/Y)$ acts purely non-symplectically on the (smooth) $K3$ surface $\widetilde{X}$. Clearly, $20 \leq \rho(\widetilde{X}) = r(Y) + \rho(X) \geq r(Y) + 1$ with $r(Y)$ the number of components in the exceptional divisor of the resolution $\widetilde{X} \to X$. Here $\rho(Z)$ denotes the Picard number of Z. Hence $r \leq 19$.
(6) For rational log Enriques surfaces Y with maximum $r(Y) = 19$, the type Sing(X) has been determined. They are one of 7 Dynkin types:

$$D_{19},\ D_{16} + A_3,\ D_{13} + A_6,\ D_7 + A_{12},\ D_7 + D_{12},\ D_4 + A_{15},\ A_{19}.$$

For each of the 7 Dynkin types above, there is exactly one (up to isomorphisms) rational log Enriques surface Y such that the singular locus Sing(X) of the canonical cover X of Y has that Dynkin type; see Oguiso-Zhang [**OZ96**] Th 1, Th 2 and [**OZ99a**] Main Th.

Start with a log Enriques surface Y, by passing to a maximum crepant partial resolution (i.e., resolving Du Val singularities), we may assume the maximality condition for Y:

(*) Any birational morphism $Y' \to Y$ from another log Enriques surface Y' must be an isomorphism.

We now state a uniqueness result for the three largest prime indices:

THEOREM 5.3. *(Oguiso-Zhang* [**OZ00a**], *Cor. 4). Let Y be a log Enriques surface with $I = I(Y) = 19, 17$ or 13 and satisfying the maximality condition (*) above.*

(1) *We have $Y \cong X_I/\langle g_I \rangle$ when $I = 19$ or 17.*

(2) *We have $Y \cong \overline{X}_{13}/\langle g_{13} \rangle$ when $I = 13$.*

Here the pairs $(X, \langle g_I \rangle)$ are given in Example 3.7, and $\overline{X}_{13}$ is obtained from X_{13} by contracting the unique rational curve in the fixed locus (point wise) $X_{13}^{g_{13}}$. (see Zhang [**Z91**] *Ex 5.7 – 5.8,* [**Z93**] *Ex 7.3 for different constructions).*

We end the paper with the final remark.

REMARK 5.4. (1) In their paper [**DK04**], Dolgachev-Keum have extended Mukai's classification Theorem 2.2 to characteristic p case. For smaller p, some new groups of automorphisms (not in Mukai's list) appear, see ibid and Dolgachev-Kondo [**DKo**].
(2) In his paper [**Og03**], Theorem 1.5, Oguiso considers the subtle behavior of $\mathrm{Aut}(X_t)$ for $K3$ surfaces in a 1-dimensional family. On the one hand, for generic t the group $\mathrm{Aut}(X_t)$ contains a fixed subgroup with bounded index. On the other hand, he also produces an example where $\mathrm{Aut}(X_t)$ is infinite for generic t but is finite for some special t.
(3) Many interesting results could not be included in the paper due to the limitation of the knowledge of the author and the constraint of the space.

References

[AS1] M. F. Atiyah and G. B. Segal, *The index of elliptic operators. II.* Ann. of Math. (2) **87** (1968) 531–545.

[AS2] M. F. Atiyah and I. M. Singer, *The index of elliptic operators. III.* Ann. of Math. (2) **87** (1968) 546–604.

[Bl] R. Blache, The structure of l.c. surfaces of Kodaira dimension zero. I. J. Algebraic Geom. **4** (1995), no. 1, 137–179.

[Bli] H. F. Blichfeldt, *Finite collineation groups,* the Univ. Chicago Press 1917.

[Bo] R. E. Borcherds, *Automorphism groups of Lorentzian lattices.* J. Algebra **111** (1987), no. 1, 133–153.

[BR] D. Burns and M. Rapoport, *On the Torelli problem for kahlerian $K-3$ surfaces.* Ann. Sci. cole Norm. Sup. (4) **8** (1975), no. 2, 235–273.

[DK02] I. Dolgachev and J. Keum, *Birational automorphisms of quartic Hessian surfaces.* Trans. Amer. Math. Soc. **354** (2002), no. 8, 3031–3057.

[DK04] I. Dolgachev and J. Keum, *Finite groups of symplectic automorphisms of K3 surfaces in positive characteristic,* math.AG/0403478.

[DKo] I. Dolgachev and S. Kondo, *A supersingular $K3$ surface in characteristic 2 and the Leech lattice.* Int. Math. Res. Not. 2003, no. 1, 1–23.

[IOZ] A. A. Ivanov, K. Oguiso and D. -Q. Zhang, *The monster and $K3$ surfaces.* in preparation.

[Ke97] J. Keum, *Automorphisms of Jacobian Kummer surfaces.* Compositio Math. **107** (1997), no. 3, 269–288.

[KK] J. Keum and S. Kondo, *The automorphism groups of Kummer surfaces associated with the product of two elliptic curves.* Trans. Amer. Math. Soc. **353** (2001), no. 4, 1469–1487.

[KOZ1] J. Keum, K. Oguiso and D. -Q. Zhang, *The Alternating Group of degree 6 in Geometry of the Leech Lattice and $K3$ Surfaces,* Proc. London Math. Soc. **90** (2005) 371 - 394.

[KOZ2] J. Keum, K. Oguiso and D. -Q. Zhang, *Extensions of the alternating group of degree 6 in the geometry of $K3$ surfaces,* math.AG/0311462, European J. Combinatorics (special issue on Groups and Geometry), to appear.

[Ko89] S. Kondo, *Algebraic $K3$ surfaces with finite automorphism groups.* Nagoya Math. J. **116** (1989), 1–15.

[Ko92] S. Kondo, *Automorphisms of algebraic $K3$ surfaces which act trivially on Picard groups.* J. Math. Soc. Japan **44** (1992), no. 1, 75–98.

[Ko98a] S. Kondo, *The automorphism group of a generic Jacobian Kummer surface.* J. Algebraic Geom. **7** (1998), no. 3, 589–609.

[Ko98b] S. Kondo, *Niemeier lattices, Mathieu groups, and finite groups of symplectic automorphisms of $K3$ surfaces. With an appendix by Shigeru Mukai.* Duke Math. J. **92** (1998), no. 3, 593–603.

[Ko99] S. Kondo, *The maximum order of finite groups of automorphisms of $K3$ surfaces.* Amer. J. Math. **121** (1999), no. 6, 1245–1252.

[Mu] S. Mukai, *Finite groups of automorphisms of $K3$ surfaces and the Mathieu group.* Invent. Math. **94** (1988), no. 1, 183–221.

[MO] N. Machida and K. Oguiso, *On $K3$ surfaces admitting finite non-symplectic group actions.* J. Math. Sci. Univ. Tokyo **5** (1998), no. 2, 273–297.

[Ni80] V. V. Nikulin, *Finite automorphism groups of Kahler $K3$ surfaces.* Trans. Moscow Math. Soc. **38** (1980), 71 - 135.

[Ni83] V. V. Nikulin, *Factor groups of groups of automorphisms of hyperbolic forms with respect to subgroups generated by 2-reflections. Algebrogeometric applications.* J. Soviet Math. **22** (1983), 1401 - 1475.

[Ni85] V. V. Nikulin, *Surfaces of type $K3$ with a finite automorphism group and Picard group of rank three,* Proc. Steklov Inst. Math. **3** (1985), 131 - 155.

[Og93] K. Oguiso, *On algebraic fiber space structures on a Calabi-Yau 3-fold. With an appendix by Noboru Nakayama.* Internat. J. Math. **4** (1993), no. 3, 439–465.

[Og03] K. Oguiso, *Local families of $K3$ surfaces and applications.* J. Algebraic Geom. **12** (2003), no. 3, 405–433.

[Og05] K. Oguiso, *A characterization of the Fermat quartic $K3$ surface by means of finite symmetries,* Compositio Math. **141** (2005), 404 - 424.

[Og06] K. Oguiso *Tits alternative in hypekahler manifolds,* math.AG/0504098, to appear.

[OZ96] K. Oguiso and D. -Q. Zhang, *On the most algebraic $K3$ surfaces and the most extremal log Enriques surfaces.* Amer. J. Math. **118** (1996), no. 6, 1277–1297.

[OZ99a] K. Oguiso and D. -Q. Zhang, *On the complete classification of extremal log Enriques surfaces.* Math. Z. **231** (1999), no. 1, 23–50.

[OZ99b] K. Oguiso and D. -Q. Zhang, *$K3$ surfaces with order 11 automorphisms.* math.AG/9907020.

[OZ00a] K. Oguiso and D. -Q. Zhang, *On Vorontsov's theorem on $K3$ surfaces with non-symplectic group actions.* Proc. Amer. Math. Soc. **128** (2000), no. 6, 1571–1580.

[OZ00b] K. Oguiso and D. -Q. Zhang, *The simple group of order 168 and $K3$ surfaces. Complex geometry,* Collection of Papers Dedicated to Hans Grauert, I. Bauer et al (ed), Gottingen, 2000, 165–184, Springer, Berlin, 2002.

[PS] I. I. Pjateckii - Sapiro and I. R. Shafarevich, *A Torelli theorem for algebraic surfaces of type $K3$.* Math. USSR Izv. Vol **5** (1971), 547 -588.

[SI] T. Shioda and H. Inose, *On singular $K3$ surfaces. Complex analysis and algebraic geometry,* pp. 119–136. Iwanami Shoten, Tokyo, 1977.

[St] H. Sterk, *Finiteness results for algebraic $K3$ surfaces.* Math. Z. **189** (1985), no. 4, 507–513.

[Vi] E. B. Vinberg, *The two most algebraic $K3$ surfaces.* Math. Ann. **265** (1983), no. 1, 1–21.

[Xi95] G. Xiao, *Non-symplectic involutions of a $K3$ surface,* alg-geom/9512007.

[Xi96] G. Xiao, *Galois covers between $K3$ surfaces.* Ann. Inst. Fourier (Grenoble) **46** (1996), no. 1, 73–88.

[Z91] D. -Q. Zhang, *Logarithmic Enriques surfaces.* J. Math. Kyoto Univ. **31** (1991), no. 2, 419–466.

[Z93] D. -Q. Zhang, *Logarithmic Enriques surfaces. II.* J. Math. Kyoto Univ. **33** (1993), no. 2, 357–397.

[Z98] D. -Q. Zhang, *Quotients of $K3$ surfaces modulo involutions.* Japan. J. Math. (N.S.) **24** (1998), no. 2, 335–366.

[Z06a] D. -Q. Zhang, *Niemeier Lattices and $K3$ Groups,* math.AG/0408106, Proceedings of Conference on Algebraic Geometry – in honour of Professor I. Dolgachev, Contemporary Math. Amer. Math. Soc. to appear.

[Z06b] D. -Q. Zhang, *The alternating groups and $K3$ surfaces,* J. Pure and Appl. Alg. to appear; math.AG/0506610.

Department of Mathematics, National University of Singapore, 2 Science Drive 2, Singapore 117543, Singapore

E-mail address: `matzdq@math.nus.edu.sg`

AMS/IP Studies in Advanced Mathematics
Volume **39**, 2007

Vector Bundles on Certain Surfaces without Divisors

Ling ZHAO, Xiangyu ZHOU, and Qingzhong LI

ABSTRACT. On non-algebraic surfaces, the following problems are in general open, i.e. on a given topological complex vector bundle of rank 2 whether a holomorphic structure exists or not and in what circumstances a known holomorphic vector bundle is filtrable. In the present paper, we obtain some remarks on vector bundles on certain surfaces without divisors.

1. Introduction

It is well-known that every holomorphic vector bundle is filtrable on a projective algebraic manifold, and any complex vector bundle E of rank 2 on a projective algebraic surface admits a holomorphic structure iff $c_1(E) = c_1(L)$ for some holomorphic line bundle L. And this is not true for the non-algebraic case. Hence, as we speak of surfaces, we mean non-algebraic compact complex surfaces unless mentioned. In the non-algebraic case, the following problems are in general open, i.e. on a given topological vector bundle of rank 2 whether a holomorphic structure exists or not and in what circumstances a known holomorphic vector bundle is filtrable. Hence holomorphic vector bundles on non-algebraic surface are deserved to receive increasing attention. In the present paper, we obtain some results on holomorphic vector bundles on surfaces without divisors.

2. Preliminaries

2.1. Notation.

(1) $IS_2(X,0)$:= set of isomorphism classes of holomorphic, filtrable, indecomposable rank-2 vector bundles E over X with vanishing Chern classes.
(2) sub-$IS_2(X,0) := \{\ [E] \in IS_2(X,0) \mid \exists : 0 \longrightarrow \mathcal{O}_X \longrightarrow E \longrightarrow M \longrightarrow 0$ with $[M] \in \mathrm{Pic}(X)\}$.
(3) $\mathcal{I}_Z$ is the ideal sheaf of a locally complete intersection $Z \subset X$ of codimension 2.

1991 *Mathematics Subject Classification.* 32H10.
Key words and phrases. surface without divisors, holomorphic structure, filtrable vector bundle.
Supported by the National Natural Science Foundation of China (Grant No.10071051) and Outstanding Youth Foundation of NNSFC (Grant No.19825102.
Supported by the Natural Science Foundation of Beijing(Grant No.1002004).

2.2. Known results. In the classification theory of compact complex manifold, the Kodaira dimension and the algebraic dimension play important role. It is well-known that for a compact complex surface X, X is algebraic iff its algebraic dimension $a(X) = \dim X = 2$, and X is Kählerian iff the first betti number of X is even. Furthermore, for a non-algebraic surface X, $a(X) = 1$ iff X is elliptic, and $a(X) = 0$ iff X is non-elliptic.

PROPOSITION 2.1. *(See [K]) Every Hopf surface contains at least one non-singular elliptic curve.*

PROPOSITION 2.2. *(See [BH]) Let X be a non-algebraic surface with $K_X \cong \mathcal{O}_X$, then every curve on X is of canonical type. Hence X is elliptic iff it contains a curve.*

PROPOSITION 2.3. *(See [BF]) Let X be a non-algebraic compact complex surface and let $a \in NS(X)$ be fixed. Then, for every holomorphic 2-vector bundle E on X with $c_1(E) = a$, we have*

$$c_2(E) \geq min\{2\chi(X, \mathcal{O}_X) + \frac{1}{2}(c_1(X) \cdot a + a^2), \frac{1}{4}a^2\}.$$

PROPOSITION 2.4. *(See [EF]) Let X be a Kählerian surface with $NS(X) = 0$. Then for any vector bundle E of rank 2 on X, we have $c_2(E) \geq 0$.*

PROPOSITION 2.5. *(See [BP]) Let X be a non-algebraic surface, and E be a vector bundle of rank r, with $c_1(E) = c_1$ and $c_2(E) = c_2$, over X. Then*

$$\Delta = \frac{1}{r}(c_2 - \frac{r-1}{2r}c_1^2) \geq 0.$$

PROPOSITION 2.6. *(See [I]) Let X be a surface satisfying $b_1(X) = 1$ and $b_2(X) = 0$. Then $c_1^2(X) = 0 = c_2(X)$, $q(X) = 1$, $p_g(X) = 0$, $h^{1,0}(X) = \dim H^0(\Omega^1) = 0$ and $H^1(\mathbb{C}^*) \simeq H^1(\mathcal{O}^*)$ canonically. Moreover*

(1) *$\sum_{\nu=0}^{2}(-1)^\nu \dim H^\nu(\mathcal{O}(F)) = 0$ for any line bundle F, and*
(2) *$\sum_{\nu=0}^{2}(-1)^\nu \dim H^\nu(\Theta) = 0$ for the sheaf Θ of holomorphic tangent vector fields.*

PROPOSITION 2.7. *(See [EF]) Let X be a compact connected complex manifold without divisors. Then we have the following classification of vector bundles of rank 2 on X:*

(1) *Filtrable bundles.*
  (a) *Indecomposable bundles.*
  A filtrable 2-bundle is indecomposable iff its devissage is uniquely determined.
    (i) *Simple bundles. They have a devissage*

$$0 \longrightarrow L \longrightarrow E \longrightarrow M \otimes \mathcal{I}_Z \longrightarrow 0$$

with $L \ncong M$ and endomorphism ring $\mathrm{End}E \cong \mathbb{C}$.
    (ii) *Non-simple bundles. Their devissage is*

$$0 \longrightarrow L \longrightarrow E \longrightarrow L \otimes \mathcal{I}_Z \longrightarrow 0$$

and $\mathrm{End}E \cong \mathbb{C}[\varepsilon]$, $\varepsilon^2 = 0$, is the ring of dual numbers.
  (b) *Decomposable bundles.*

(i) *Bundles of the form* $E \cong L \oplus M$ *with* $L \ncong M$.
In this case, $\mathrm{End}E \cong \mathbb{C} \oplus \mathbb{C}$ *with componentwise multiplication.*
(ii) *Bundles of the form* $E \cong L \oplus L$.
In this case $\mathrm{End}E \cong \mathbf{M}_2(\mathbb{C})$ *is the full matrix ring.*

(2) *Non-filtrable bundles.*
These bundles are all simple, i.e. $\mathrm{End}E = \mathbb{C}$.

PROPOSITION 2.8. *(See [I]) Let X be a compact complex analytic surface satisfying the following two conditions:*

(A): $b_1(X) = 1$, $b_2(X) = 0$.
(B): *X contains no curves.*

Then

(1) *Any unramified covering surface X' of X satisfies the conditions (A) and (B).*
(2) *Any deformation X'' of X satisfies (A) and (B).*
(3) $mK_X \neq 0$ *for any* $m \neq 0$ *in* $\mathbb{Z}$, *where* K_X *denotes the canonical bundle of* X.
(4) $H^\nu(\mathcal{O}(F)) = 0, \nu = 0, 1, 2,$ *for* $F \neq 0, K_X$.

PROPOSITION 2.9. *(See [EF]) A vector bundle E of rank 2 on a connected complex manifold X is filtrable iff there exists a line bundle L on X such that* $\Gamma(X, L^* \otimes E) \neq 0$.

PROPOSITION 2.10. *(See [K]) Let X be a compact complex analytic surface free from singularities. If there exists on X a complex line F with $c(F)^2 > 0$, then X is an algebraic surface.*

LEMMA 2.11. *Let X be a compact complex surface without divisors. Then for an arbitrary line bundle* $L \in \mathrm{Pic}(X)$, *we have*

$$H^0(X, L) = \begin{cases} 0, & L \ncong \mathcal{O}_X, \\ \mathbb{C}, & L \cong \mathcal{O}_X. \end{cases}$$

PROOF. Assume $H^0(X, L)$ has non-constant sections, then take two non-vanishing sections of $H^0(X, L)$, s_1 and s_2, say, then $s_1/s_2 \in \mathcal{O}_X$. Since X is compact, s_1/s_2 must be a constant. Hence, $H^0(X, L) \cong \mathbb{C}$ which implies $L \cong \mathcal{O}_X$. If $L \ncong \mathcal{O}_X$, assume $H^0(X, L) \neq 0$. Take a non-vanishing section $s \in H^0(X, L)$, then (s) is the divisor associated to the section s, which contradicts to the assumption that X has no divisors. In particular,

$$H^0(X, mK_X) = \begin{cases} 0, & mK_X \ncong \mathcal{O}_X, \\ \mathbb{C}, & mK_X \cong \mathcal{O}_X. \end{cases}$$

where $m \in \mathbb{Z}$, which implies that the Kodaira dimension of X must be either $-\infty$ or 0. □

Enriques classified algebraic surfaces. Kodaira (see [K]) extended the Enriques classification to the Enriques-Kodaira classification for all compact complex surfaces.

Therefore, according to the Enriques-Kodaira classification of compact complex surfaces, we have the following table for the classification of non-algebraic surfaces X with $\mathrm{kod}(X) = -\infty$ or 0.

TABLE 1

surface X	kod(X)	$a(X)$	$\chi(X, \mathcal{O}_X)$	b_1	b_2	p_g	q
Elliptic Hopf surfaces	$-\infty$	1	0	1	0	0	1
Non-elliptic Hopf surfaces	$-\infty$	0	0	1	0	0	1
Inoue's surfaces	$-\infty$	0	0	1	0	0	1
Inoue-Hirzebruch surfaces	$-\infty$	0	0	1	>0	0	1
Elliptic $K3$ surfaces	0	1	2	0	22	1	0
Non-elliptic $K3$ surfaces	0	0	2	0	22	1	0
Elliptic 2-tori	0	1	0	4	6	1	2
Non-elliptic 2-tori	0	0	0	4	6	1	2
Primary Kodaira surfaces	0	1	0	3	4	1	2
Secondary Kodaira surfaces	0	1	0	1	0	0	1

On the other hand, X has no divisors implies that the only meromorphic function on X are constant, i.e. the algebraic dimension of X is zero, which implies that X is non-elliptic.

Up to now, we know that for a compact complex non-algebraic surface X to be without divisors, it is necessary that $a(X) = 0$ and $\text{kod}(X) = -\infty$ or 0. However, those restrictions on $a(X)$ and kod(X) are not sufficient to guarantee that X has no divisors. Since we have counterexamples here.

EXAMPLE 2.12. Although non-elliptic Hopf surfaces, which have $a(X) = 0$ and $\text{kod}(X) = -\infty$, we know that every Hopf surface (we mean primary and non-primary Hopf surface here) contains at least one non-singular elliptic curve.

EXAMPLE 2.13. It was M. Inoue who first constructed examples of surfaces of class VII_0 with $b_2 \geq 0$ which has $a(X) = 0$ and $\text{kod}(X) = -\infty$ and contains curves, where surfaces of class VII_0 means the class of minimal surfaces with $b_1 = q = 1$, $p_g = 0$.

However, according to Prop. 2.1, every curve on a compact complex surface X with canonical bundle is of canonical type, hence X is elliptic iff it contains a curve. We induced that if X is a compact complex surface with trivial canonical bundle, then X is non-elliptic if and only if X does not contain a curve.

There do exist non-elliptic $K3$ surfaces with Kodaira dimension zero. Summing up, we know that non-elliptic $K3$ surfaces are the candidates we are looking for.

A compact connected complex manifold without divisors has algebraic dimension zero, i.e. the only meromorphic functions are constant. The converse is not true, e.g. non-elliptic Hopf surfaces (i.e. the generic and the exceptional cases) do have algebraic dimension zero, but they also do have divisors. However, a torus has algebraic dimension zero if and only if it admits no divisors.

LEMMA 2.14. *Let X be a compact complex surface without divisors. Then for any holomorphic vector bundle E of rank 2 with $c_1(E) = 0$, we have $c_2(E) \geq 0$.*

PROOF. The Riemann-Roch theorem gives

$$\chi(X, E) = 2\chi(X, \mathcal{O}_X) + \frac{1}{2}(c_1(X) \cdot c_1(E) + c_1(E)^2) - c_2(E).$$

In our case,

$$\chi(X, E) = 2\chi(X, \mathcal{O}_X) - c_2(E).$$

Moreover, for a non-algebraic surface X, we have $\chi(X, \mathcal{O}_X) \geq 0$.

(1) If E is non-filtrable, then by Prop. 2.9, we have

$$H^0(X, E^* \otimes K_X) = 0,$$

otherwise E^* and therefore E would be filtrable. By Serre duality,

$$H^2(X, E)^* = H^0(X, E^* \otimes K_X) = 0.$$

And we have $H^0(X, E) = 0$, otherwise, $H^0(X, E \otimes L^*) \neq 0$, where L^* is the trivial line bundle, which is contradictory to the definition of non-filtrability.

It follows that

$$\chi(X, E) = -h^1(X, E) \leq 0.$$

Hence, $c_2(E) \geq 2\chi(X, \mathcal{O}_X) \geq 0$.

(2) If E is filtrable, then there exists an exact sequence

$$0 \longrightarrow L \longrightarrow E \longrightarrow M \otimes \mathcal{I}_Z \longrightarrow 0$$

where L, M are holomorphic line bundles and Z is a locally complete intersection of codimension 2 in X or empty.

$$0 = c_1(E) = c_1(L) + c_1(M),$$

$$c_2(E) = c_1(L) \cdot c_1(M) + \deg Z.$$

It follows that the discriminant

$$\begin{aligned} \Delta(E) &= -c_1(E)^2 + 4c_2(E) \\ &= 4 \deg Z - (c_1(L) - c_1(M))^2 \\ &= 4 \deg Z - c_1(L \otimes M^*)^2 \end{aligned}$$

Since X is non-algebraic, by Prop. 2.10, we have $c_1(L \otimes M^*)^2 \leq 0$, hence, $\Delta(E) \geq 0$.

On the other hand,

$$\Delta(E) = 4c_2(E) - c_1(E)^2 = 4c_2(E).$$

Therefore, $c_2(E) \geq 0$.

□

3. Main result

THEOREM 3.1. *Let X be a non-elliptic 2-torus or an Inoue's surface, and E be a $SL(2,\mathbb{C})$-bundle. Then E admits a holomorphic structure iff $c_2(E) \geq 0$. Consequently, every rank-2 bundle E on X with $c_1(E) = 0$ has a holomorphic structure. Furthermore, if it happens that $c_2(E) > 0$, then E has a filtrable holomorphic structure.*

PROOF. By Prop. 2.5, if E admits a holomorphic structure, then

$$\Delta(E) = 4c_2(E) - c_1(E)^2 \geq 0,$$

Hence $c_2(E) \geq 0$.

Conversely, assume E be a $SL(2,\mathbb{C})$-bundle, we shall deal with the problem case by case.

Case 1: X is a non-elliptic 2-torus. Let E be a topological 2-vector bundle with $\Lambda^2 E \cong \mathbb{C}_X$ (i.e. $\Lambda^2 E$ is topologically trivial) and $c_2 = c_2(E) \geq 0$. It follows that $c_1(E) = c_1(\Lambda^2 E) = 0$. The case $c_2 = 0$ is obvious: by the result of Wu (see [W]), E is (topologically) trivial. Hence it has a holomorphic structure (the trivial one).

Now suppose that $c_2 > 0$. Take $M \in \mathrm{Pic}_0(X)$ with $M \ncong \mathcal{O}_X$, the existence of such a line bundle M is guaranteed by the fact that $q = 2$ and $\mathrm{Pic}_0(X) \cong H^1(X, \mathcal{O}_X)/H^1(X, \mathbb{Z})$ is a 2-torus. And take a locally complete intersection $Z \subset X$ of co-dimension 2 with $\deg Z = c_2 > 0$. And consider the extension $0 \longrightarrow \mathcal{O}_X \longrightarrow E'' \longrightarrow M \otimes \mathcal{I}_Z \longrightarrow 0$, we want to convince ourselves that such a 2-bundle E'' with $c_1(E'') = 0$ and $c_2(E'') = c_2$ does exist.

The extensions of $M \otimes \mathcal{I}_Z$ by $\mathcal{O}_X$ are classified by the group $\mathrm{Ext}^1(M \otimes \mathcal{I}_Z, \mathcal{O}_X)$. There is an exact sequence

$$0 \longrightarrow H^1(X, \mathrm{Hom}(M \otimes \mathcal{I}_Z, \mathcal{O}_X)) \longrightarrow \mathrm{Ext}^1(M \otimes \mathcal{I}_Z, \mathcal{O}_X) \longrightarrow$$
$$\Gamma(X, \mathcal{E}xt^1(M \otimes \mathcal{I}_Z, \mathcal{O}_X)) \longrightarrow H^2(X, \mathrm{Hom}(M \otimes \mathcal{I}_Z, \mathcal{O}_X))$$

Since Z has co-dimension 2, and Z is a locally complete intersection consisting of discrete points, $\mathcal{I}_Z$ is finitely generated by constant functions. Hence, we have $\mathrm{Hom}(M \otimes \mathcal{I}_Z, \mathcal{O}_X) \cong M^*$. And we have $H^0(X, M^*) = 0$, since $M^* \ncong \mathcal{O}_X$ and X has no divisors. Similarly, $H^0(X, M) = 0$, by Serre duality, $H^2(X, M^*) = H^0(X, M) = 0$, since the canonical bundle of X is trivial. As Z is a locally complete intersection consisting of discrete points, we have $\mathcal{E}xt^1(M \otimes \mathcal{I}_Z, \mathcal{O}_X) \cong \mathcal{O}_Z$, which proves

$$\mathrm{Ext}^1(M \otimes \mathcal{I}_Z, \mathcal{O}_X) \longrightarrow \Gamma(X, \mathcal{I}_Z) \cong \overset{c_2}{\oplus} \mathbb{C} \longrightarrow 0$$

Now take an element in $\mathrm{Ext}^1(M \otimes \mathcal{I}_Z, \mathcal{O}_X)$ which has the element 1 of $\overset{c_2}{\oplus} \mathbb{C}$ as its image. By Serre, the corresponding extension $0 \longrightarrow \mathcal{O}_X \longrightarrow E'' \longrightarrow M \otimes \mathcal{I}_Z \longrightarrow 0$ gives a holomorphic bundle E'' with $c_1(E'') = c_1(\mathcal{O}_X) + c_1(M) = 0$, and $c_2(E'') = c_1(\mathcal{O}_X) \cdot c_1(M) + \deg Z = c_2$. Then E'' is topologically isomorphic with E.

Case 2: X is an Inoue's surface. Let E be a topological 2-vector bundle with $\Lambda^2 E \cong \mathbb{C}_X$ and $c_2 = c_2(E) \geq 0$. It follows that $c_1(E) = c_1(\Lambda^2 E) = 0$, The case $c_2 = 0$ is obvious.

Now suppose that $c_2 > 0$. Take $L, M \in \mathrm{Pic}_0(X) \cong \mathbb{C}^*$ with $L \ncong M$ and $M \ncong L \otimes K^*$, and take a locally complete intersection $Z \subset X$ of co-dimension 2 with $\deg Z = c_2 > 0$. And consider the extension $0 \longrightarrow$

$L \longrightarrow E'' \longrightarrow M \otimes \mathcal{I}_Z \longrightarrow 0$, we want to convince ourselves that such a 2-bundle E'' with $c_1(E'') = 0$ and $c_2(E'') = c_2$ does exist.

The extension of $M \otimes \mathcal{I}_Z$ by L are classified by the group $\mathrm{Ext}^1(M \otimes \mathcal{I}_Z, L)$. There is an exact sequence

$$0 \longrightarrow H^1(X, \mathrm{Hom}(M \otimes \mathcal{I}_Z, L)) \longrightarrow \mathrm{Ext}^1(M \otimes \mathcal{I}_Z, L) \longrightarrow \Gamma(X, \mathrm{Ext}^1(M \otimes \mathcal{I}_Z, L)) \longrightarrow H^2(X, \mathrm{Hom}(M \otimes \mathcal{I}_Z, L))$$

Since Z has co-dimension 2, we similarly have $\mathrm{Hom}(M \otimes \mathcal{I}_Z, L) \cong M^* \otimes L$. And we have $H^0(X, M^* \otimes L) = 0$ since $L \ncong M$ and X has no divisors. Similarly, $H^0(X, M \otimes L^* \otimes K) = 0$. By Serre duality, we have $H^2(X, M^* \otimes L) = 0$. By Riemann-Roch and Prop. 2.6, we get $H^1(X, M^* \otimes L) = 0$.

On the other hand, since Z is a locally complete intersection consisting of discrete points, we have $\mathrm{Ext}^1(M \otimes \mathcal{I}_Z, L) \cong \mathcal{O}_Z$, which proves

$$\mathrm{Ext}^1(M \otimes \mathcal{I}_Z, L) \cong \Gamma(X, \mathcal{O}_Z) \cong \overset{c_2}{\oplus}\, \mathbb{C}$$

Now take an element in $\mathrm{Ext}^1(M \otimes \mathcal{I}_Z, L)$ which has the element 1 of $\overset{c_2}{\oplus}\, \mathbb{C}$ as its image. By Serre, the corresponding extension $0 \longrightarrow L \longrightarrow E'' \longrightarrow M \otimes \mathcal{I}_Z \longrightarrow 0$ gives a holomorphic bundle E'' with $c_1(E'') = 0$ and $c_2(E'') = \deg Z = c_2$.

□

Elencwajg-Forster showed that simple vector bundle of rank $r \geq 2$ with vanishing Chern classes on a 2-torus is homogeneous.

Related to these facts, we have the following:

THEOREM 3.2. *Let X be a non-elliptic 2-torus or an Inoue's surface. If $E \in IS_2(X, 0)$, then E is quasi-homogeneous.*

PROOF. By Prop. 2.7, we have a devissage

$$0 \longrightarrow L \longrightarrow E \longrightarrow M \otimes \mathcal{I}_Z \longrightarrow 0$$

with $L, M \in \mathrm{Pic}(X)$ and $Z \subset X$ a subspace of co-dimension 2. Since $c_2(E) = 0$, Z must be empty. The extension of M by L are classified by the elements of

$$H^1(X, \mathrm{Hom}(M, L)) \cong H^1(X, M^* \otimes L).$$

We shall deal with the problem case by case.

Case 1: X is the non-elliptic 2-torus. we claim that

$$\dim H^1(X, M^* \otimes L) = \begin{cases} 2, & L \cong M, \\ 0, & L \ncong M. \end{cases}$$

In fact, if $L \cong M$, then $\dim H^1(X, M^* \otimes L) = q = 2$. And if $L \ncong M$, then $H^0(X, M^* \otimes L) = 0$. Otherwise, take $s \in H^0(X, M^* \otimes L)$ with $s \neq 0$, then $M^* \otimes L \cong \mathcal{O}_X(D)$, where D is the divisor determined by the not identically zero section s. Since X has no divisors, we have $M^* \otimes L \cong \mathcal{O}_X$, which contradicts to the assumption that $L \ncong M$. Similarly, $H^0(X, M \otimes L^*) = 0$, by Serre duality, we have $H^2(X, M^* \otimes L) = H^0(X, M \otimes L^*) = 0$, since the canonical bundle of X is trivial.

Since $\chi(X, M^* \otimes L) = \chi(X, \mathcal{O}_X) + \frac{1}{2}(c^2 + c_1 c) = \frac{1}{2}c^2$, where $c_1 = c_1(X)$, $c = c_1(M^* \otimes L)$, on the other hand,

$$\begin{aligned} 0 = \Delta(E) &= 4c_2(E) - c_1(E)^2 \\ &= 4c_1(L) \cdot c_1(M) - (c_1(L) + c_1(M))^2 \\ &= -c_1(M^* \otimes L)^2, \end{aligned}$$

hence $\chi(X, M^* \otimes L) = 0$. Therefore $H^1(X, M^* \otimes L) = 0$.

Now we have two possibilities, we will show that the second one is excluded. Actually, if $\dim H^1(X, M^* \otimes L) = 0$, then the extension of M by L are always trivial, i.e. $E \cong M \oplus L$, which is contradictory to the assumption that E is indecomposable. Hence, we have $0 \longrightarrow L \longrightarrow E \longrightarrow L \longrightarrow 0$. The extension of L by L are classified by the elements of $H^1(X, \mathrm{Hom}(L, L)) \cong H^1(X, \mathcal{O}_X)$.

For an arbitrary $\tau \in \mathrm{Aut}(X)$, let τ^* operate on the exact sequence

$$0 \longrightarrow L \longrightarrow E \longrightarrow L \longrightarrow 0$$

we get

$$0 \longrightarrow \tau^* L \longrightarrow \tau^* E \longrightarrow \tau^* L \longrightarrow 0$$

And the extension of $\tau^* L$ by $\tau^* L$ are classified by

$$H^1(X, \mathrm{Hom}(\tau^* L, \tau^* L)) \cong H^1(X, \mathcal{O}_X).$$

Hence, τ^* operates trivially on $H^1(X, \mathcal{O}_X)$, which shows that $\tau^* E \cong E$. Therefore, E is quasi-homogeneous.

Case 2: X is the Inoue's surface, similarly we obtain:

$$\dim H^1(X, M^* \otimes L) = \begin{cases} 1, & L \cong M, \\ 0, & L \ncong M \text{ and } M^* \otimes L \ncong K, \\ 1, & L \ncong M \text{ and } M^* \otimes L \cong K. \end{cases}$$

Now we have three possibilities. We shall show that the second and the third one are excluded. In fact, the exclusion of the second possibility is pretty similar to the case of non-elliptic 2-torus. We omit the details here.

As for the third one, if $\dim H^1(X, M^* \otimes L) = 1$, and $L \ncong M$, $M^* \otimes L \cong K$. Tensor the exact sequence $0 \longrightarrow L \longrightarrow E \longrightarrow M \longrightarrow 0$ by M^*, we obtain

$$0 \longrightarrow M^* \otimes L \longrightarrow M^* \otimes E \longrightarrow \mathcal{O}_X \longrightarrow 0$$

namely,

$$0 \longrightarrow K \longrightarrow M^* \otimes E \longrightarrow \mathcal{O}_X \longrightarrow 0$$

By Prop.2.8, we get $H^1(X, K^*) = 0$. The extension of $\mathcal{O}_X$ by K are classified by the elements of

$$H^1(X, \mathrm{Hom}(K, \mathcal{O}_X)) \cong H^1(X, K^*) = 0$$

which implies that $M^* \otimes E \cong K \oplus \mathcal{O}_X$, i.e. $E \cong M \otimes K \oplus M$, where K is nontrivial. This contradicts to the assumption that E is indecomposable. Hence, we have $0 \longrightarrow L \longrightarrow E \longrightarrow L \longrightarrow 0$. And the remain of our proof can be done by repeating verbatim as in the case of non-elliptic 2-torus.

□

THEOREM 3.3. *Let X be a $K3$ surface, not necessarily non-algebraic, then there does not exist simple vector bundle of rank $r \geq 2$ on X with vanishing Chern classes or simple bundle E of rank $r \geq 2$ on X with $c_1(E) = 0$ and $c_2(E) = 1$. Consequently, every rank-2 bundle E on a $K3$ surface X with $c_1(E) = 0 = c_2(E)$ is filtrable. $E \cong \mathcal{O}_X^2$, to be exact.*

PROOF. Firstly, assume E be a simple bundle of rank r with vanishing Chern classes, then $H^0(X, \mathrm{End}E) \cong \mathbb{C}$, by Serre duality, $H^2(X, \mathrm{End}E) \cong H^0(X, \mathrm{End}E)$

since the canonical bundle is trivial. By Riemann-Roch,

$$\begin{aligned}\chi(X,\mathrm{End}E) &= h^0(X,\mathrm{End}E) - h^1(X,\mathrm{End}E) + h^2(X,\mathrm{End}E)\\ &= r^2\chi(X,\mathcal{O}_X) + (r-1)c_1(E)^2 - r^2c_2(E)\\ &= 2r^2.\end{aligned}$$

Hence, $h^1(X,\mathrm{End}E) = -(2r^2-2) < 0$. This is impossible.

Secondly, assume E is a simple rank-r bundle with $c_1(E) = 0$ and $c_2(E) = 1$, then

$$h^1(X,\mathrm{End}E) = 2 - r^2 < 0,$$

This is also impossible.

Finally, assume E be a rank-2 bundle on a $K3$ surface X with $c_1(E) = 0 = c_2(E)$. Since every rank-2 bundle with vanishing Chern classes is non-simple, which implies filtrability. Since $NS(X)$ is trivial, by Prop.2.7, there exists $L \in \mathrm{Pic}(X)$ such that E fits into an exact sequence $0 \longrightarrow L \longrightarrow E \longrightarrow L \longrightarrow 0$. And the extension of L by L are classified by the elements of $H^1(X,\mathrm{Hom}(L,L)) \cong H^1(X,\mathcal{O}_X)$. But $h^1(X,\mathcal{O}_X) = q = 0$, hence $E \cong L \oplus L$. Furthermore, $NS(X)$ is trivial implies that every line bundle of $\mathrm{Pic}(X)$is trivial, since $q = 0$, and therefore we have an injective map

$$0 \longrightarrow H^1(X,\mathcal{O}^*) \xrightarrow{c_1} H^2(X,\mathbb{Z}).$$

where c_1 is the standard map of the Chern class. Hence, $E \cong \mathcal{O}_X^2$. □

COROLLARY 3.4. *Let X be a $K3$ surface and E be a rank-2 bundle on X with $c_2(E) = 1$, then E is a decomposable filtrable bundle.*

PROOF. Because of Theorem 3.3, E must be non-simple, hence E is filtrable. According to Prop.2.7, there exists $L \in \mathrm{Pic}(X)$ such that E fits into an exact sequence, in our case, $L \cong \mathcal{O}_X$, i.e. $0 \longrightarrow \mathcal{O}_X \longrightarrow E \longrightarrow \mathcal{O}_X \otimes \mathcal{I}_Z \longrightarrow 0$, where $Z \subset X$ is a locally complete intersection of co-dimension 2 and $\deg Z = 1$. Since Z has co-dimension 2, we have $\mathrm{Hom}(\mathcal{O}_X \otimes \mathcal{I}_Z, \mathcal{O}_X) \cong \mathcal{O}_X$. And the extension of $\mathcal{O}_X \otimes \mathcal{I}_Z$ by $\mathcal{O}_X$ are classified by the elements of $H^1(X,\mathrm{Hom}(\mathcal{O}_X \otimes \mathcal{I}_Z, \mathcal{O}_X)) \cong H^1(X,\mathcal{O}_X)$, moreover, $h^1(X,\mathcal{O}_X) = q = 0$. Hence, $E \cong \mathcal{O}_X \oplus \mathcal{O}_X \otimes \mathcal{I}_Z$. □

THEOREM 3.5. (1) *If X is an Inoue's surface. Then there is a one-to-one correspondence between sub-$IS_2(X,0)$ and $\{0,1\}$.*

(2) *If X is a non-elliptic 2-torus. Then sub-$IS_2(X,0)$ just has one element E which fits into the exact sequence*

$$0 \longrightarrow \mathcal{O}_X \longrightarrow E \longrightarrow \mathcal{O}_X \longrightarrow 0$$

PROOF. We shall deal with the problem case by case.

Case 1: Let X be an Inoue's surface, assume $[E] \in$ sub-$IS_2(X,0)$, then there exists $[L] \in \mathrm{Pic}(X)$, such that E fits into an exact sequence

$$0 \longrightarrow \mathcal{O}_X \longrightarrow E \longrightarrow L \longrightarrow 0$$

since E is indecomposable, it is necessary that $H^1(X,\mathrm{Hom}(L,\mathcal{O}_X)) \cong H^1(X,L^*) \neq 0$. We have either $L^* \cong \mathcal{O}_X$ or $L^* \cong K_X$. More precisely, we have $h^1(X,\mathcal{O}_X) = q = 1$, and

$$h^1(X,K_X) = h^0(X,K_X) + h^2(X,K_X) = 0 + h^0(X,\mathcal{O}_X) = 1.$$

For a manifold Y without divisors, the extension

$$0 \longrightarrow N \longrightarrow E \longrightarrow M \otimes \mathcal{I}_Z \longrightarrow 0$$

is unique for a given rank-2 vector bundle E, where N, $M \in \mathrm{Pic}(X)$, and $Z \subset Y$ is a 2-codimensional locally complete intersection or empty. Hence, the map

$$d\colon \text{sub-}IS_2(X,0) \longrightarrow \{0,1\},$$
defined by $d(E)=0$, if $E \in \text{sub-}IS_2(X,0)$ fits into the exact sequence
$$0 \longrightarrow \mathcal{O}_X \longrightarrow E \longrightarrow \mathcal{O}_X \longrightarrow 0,$$
and $d(E)=1$, if E fits into the exact sequence
$$0 \longrightarrow \mathcal{O}_X \longrightarrow E \longrightarrow K^* \longrightarrow 0,$$
is one-to-one.

Case 2: Let X be a non-elliptic 2-torus, assume $[E] \in \text{sub-}IS_2(X,0)$, then there exist $L \in \mathrm{Pic}(X)$, such that E fits into an exact sequence
$$0 \longrightarrow \mathcal{O}_X \longrightarrow E \longrightarrow L \longrightarrow 0,$$
we claim that
$$\dim H^1(X,L^*) = \begin{cases} 2, & L \cong \mathcal{O}_X, \\ 0, & L \ncong \mathcal{O}_X. \end{cases}$$
Actually, $h^1(X,\mathcal{O}_X) = q = 2$. If $L \ncong \mathcal{O}_X$, then $H^0(X,L^*) = 0 = H^0(X,L)$, since X has no divisors. By Serre duality, $H^2(X,L^*)=0$. By Riemann-Roch, $H^1(X,L^*) = 0$. But the second possibility is excluded, otherwise, we have $E \cong \mathcal{O}_X \oplus L$, which contradicts to the fact that E is indecomposable. Hence , E must fits into the exact sequence
$$0 \longrightarrow \mathcal{O}_X \longrightarrow E \longrightarrow \mathcal{O}_X \longrightarrow 0.$$
And E has a unique extension since non-elliptic 2-torus admits no divisors.

□

References

[BF] V. Brinzanescu, P. Flondor, *Holomorphic 2-vector bundles on non-algebraic 2-tori*, J. Reine Angew. Math., **363** (1985), 47–58.

[BH] E. Bombieri, D. Husemoller, *Classification and embeddings of surfaces*, AMS Proc. Symp. Pure Math., **29** (1975), 329–420.

[BP] C. Banica, J. Le Potiers, *Sur l'existence des fibres vectoriels holomorphes sur les surfaces non-algebriques*, J. Reine Angew. Math., **378** (1987), 1–31.

[EF] G. Elencwajg, O. Forster, *Vector bundles on manifolds without divisors and a theorem on deformations*, Ann. L'Inst. Fourier, **32** (1982), 25–51.

[I] M. Inoue, *On surfaces of class* VII_0, Inventions Math., **24** (1974), 269–310.

[K] K. Kodaira, *On the structure of compact complex analytic surfaces, I, II, III, IV*, Amer. J. Math., **86** (1964), 751–798; **88** (1966), 682–721; 1968, **90** (1968), 55–83; ibid, 1048–1066.

[LZ] Qingzhong Li, Jinhao Zhang, *On the moduli space of non-primary Hopf surfaces*, Advances in Math., **29**(3) (2000), 245–252.

[W] Wen-tsien Wu, *Sur les espaces fibres*, Publ. Inst. Univ. Strassbourg, **11** (1952), Paris.

[ZL] Jinhao Zhang, Qingzhong Li, *The moduli space of Hopf surfaces*, Chinese Science Bulletin, **38**(24) (1993), 2219–2222.

[Zh] Xiang-Yu Zhou, *A remark on the Douady sequence for non-primary Hopf manifolds*, Asian J. Math., **8** (2004), no. 1, 131—136.

Department of Mathematics and Physics, China University of Petroleum, Beijing 102249, China

E-mail address: zero007@126.com

Institute of Mathematics, Academy of Mathematics and System Sciences, Chinese Academy of Sciences, Beijing 100080, China

E-mail address: xyzhou@math.ac.cn

Department of Mathematics, Capital Normal University, Beijing 100037, China

E-mail address: liqzh@mail.cnu.edu.cn

Titles in This Series

39 **Zhijie Chen, Sheng-Li Tan, Jianpan Wang, and Stephen S.-T. Yau, Editors,** Proceedings of the International Conference on Complex Geometry and Related Fields, 2007

38 **Noriko Yui, Shing-Tung Yau, and James D. Lewis, Editors,** Mirror Symmetry V, 2006

37 **Lizhen Ji, Jian-Shu Li, H. W. Xu, and Shing-Tung Yau, Editors,** Lie Groups and Automorphic Forms, 2006

36 **Chuu-Lian Terng, Editor,** Integrable Systems, Geometry, and Topology, 2006

35 **Felix Finster,** The Principle of the Fermionic Projector, 2006

34 **Ren-Hong Wong, Editor,** Computational Geometry, 2003

33 **Eric D'Hoker, Duong Phong, and Shing-Tung Yau,** Mirror Symmetry IV, 2002

32 **Xi-Ping Zhu,** Lectures on Mean Curvature Flows, 2002

31 **Kiyoshi Igusa,** Higher Franz-Reidemeister Torsion, 2002

30 **Weiman Han and Mircea Sofonea,** Quasistatic Contact Problems in Viscoelasticity and Viscoplasticity, 2002

29 **S. T. Yau and Shuxing Chen, Editors,** Geometry and Nonlinear Partial Differential Equations, 2002

28 **Valentin Afraimovich and Sze-Bi Hsu,** Lectures on Chaotic dynamical Systems, 2002

27 **M. Ram Murty,** Introduction to p-adic Analytic Number Theory, 2002

26 **Raymond Chan, Yue-Kuen Kwok, David Yao, and Qiang Zhang, Editors,** Applied Probability, 2002

25 **Donggao Deng, Daren Huang, Rong-Qing Jia, Wei Lin, and Jian Zhong Wong, Editors,** Wavelet Analysis and Applications, 2002

24 **Jane Gilman, William W. Menasco, and Xiao-Song Lin, Editors,** Knots, Braids, and Mapping Class Groups—Papers Dedicated to Joan S. Birman, 2001

23 **Cumrun Vafa and S.-T. Yau, Editors,** Winter School on Mirror Symmetry, Vector Bundles and Lagrangian Submanifolds, 2001

22 **Carlos Berenstein, Der-Chen Chang, and Jingzhi Tie,** Laguerre Calculus and Its Applications on the Heisenberg Group, 2001

21 **Jürgen Jost,** Bosonic Strings: A Mathematical Treatment, 2001

20 **Lo Yang and S.-T. Yau, Editors,** First International Congress of Chinese Mathematicians, 2001

19 **So-Chin Chen and Mei-Chi Shaw,** Partial Differential Equations in Several Complex Variables, 2001

18 **Fangyang Zheng,** Complex Differential Geometry, 2000

17 **Lei Guo and Stephen S.-T. Yau, Editors,** Lectures on Systems, Control, and Information, 2000

16 **Rudi Weikard and Gilbert Weinstein, Editors,** Differential Equations and Mathematical Physics, 2000

15 **Ling Hsiao and Zhouping Xin, Editors,** Some Current Topics on Nonlinear Conservation Laws, 2000

14 **Jun-ichi Igusa,** An Introduction to the Theory of Local Zeta Functions, 2000

13 **Vasilios Alexiades and George Siopsis, Editors,** Trends in Mathematical Physics, 1999

12 **Sheng Gong,** The Bieberbach Conjecture, 1999

11 **Shinichi Mochizuki,** Foundations of p-adic Teichmüller Theory, 1999

10 **Duong H. Phong, Luc Vinet, and Shing-Tung Yau, Editors,** Mirror Symmetry III, 1999

9 **Shing-Tung Yau, Editor,** Mirror Symmetry I, 1998

[redacted]